数学定数事典

スティーヴン・R・フィンチ [著]

一松 信 [監訳]

朝倉書店

Mathematical Constants

STEVEN R. FINCH

CAMBRIDGE UNIVERSITY PRESS

© Steven R. Finch 2003

This book is in copyright. Subject to statutory exception
and to the provisions of relevant collective licensing agreements,
no reproduction of any part may take place without
the written permission of Cambridge University Press.

First published 2003

Japanese translation rights
arranged with Cambridge University Press

監訳者まえがき

　この本は

Steven R. Finch：*Mathematical Constants*, Cambridge University Press, 2003
の全訳です．同出版会はときおり「面白い」本を出版するといわれており，これもその一つと思います．

　数学的定数というと，まず e や π が思い浮かびますが，それらは序の口です．第1章だけでも多様な定数が続出しています．整数論・複素関数論・幾何学など数学の諸分野はもちろんのこと，物理学・化学・生物学・経済学・情報科学など幅広く関連諸分野に及んでいます．この種の「雑学」は貴重であり，私自身も拝見して随分多くの新しい結果を学習しました．

　訳は諸先生方に分担していただき，私が全編に眼を通しました．用語や文体などの統一に心がけましたが，各章ごとに訳者の持ち味が現れているかもしれません．訳文は「原文に忠実であること」が当然ですが，内容を正しく日本の読者にわかりやすく伝えることを優先しましたので，若干の意訳・補充もあります．諸先生方（一部は私も）が多くの有益な訳注を付加してくださったので，原著よりもよくなったのかと自負しております．

　この本では正しく証明された結果だけでなく，多くの予想にも言及しているので，数学研究者にも有益な題材が豊富と思います．一つの逸話をあげると，4.5節の「1/9定数」があります．当初1/9だと予想されたある定数が，実は別の（それに近い）値だった，しかし誤った名が残った．という次第です．

　文献は各節ごとにまとめられているので，重複はありますが参照には便利でしょう．個々の論文にはその後公表されたものもあり，単行本で日本語訳が出版されているものも少なくないようですが，すべて原著のままです．

　用語や記号で現在の日本での慣用と異なる例が散見されます．しかし記号表もあり，理解に苦しむことはないと判断して原著のままにしました．個々の結果に関する記号の説明や引用は，くどいくらいに記述されております．

　この本で使っている表現法で早速にも利用するとよいと思われる例がいくつか

あります．引用文献で共著者は&で結び，複数の著者が別々の論文で同じ題材を扱った場合と区別しているのが典型例です．また引用した文献番号と，本書の他の節番号の引用とをうまく区別しています．訳ではそれを一層明確にするため，後者の角カッコを【6.5】のように太い角カッコに改めました．

　この本の内容は高度に専門的ですが，しかし話題の中には高等学校までの学校教育に活用できそうな素材が含まれております．数学教育の面からも，いろいろな段階で有効に利用されることを期待します．

　最後にと申すと失礼かもしれませんが，朝倉書店編集部の諸氏に，終始お世話になりました．厚い感謝の詞をもって結びとします．

2010年1月

一松　信

序

　数はすべて平等に創造されたわけではない．定数というものがともかくも現れ，一見互いに無関係であるかのように数学の全分野にこだまを響かせているという事実こそ興味の源泉である．φ や e や π や γ を含むさまざまな公式が，おわかりのようにこの本の大きな部分を占めている．

　より特殊な目的をもった定数も多く存在する．こうした風変わりな数はたいてい，狭い分野の専門家が知っているだけで，一般の人々の目に触れることもなく文献の中に埋もれたままになっている．すぐに計算できる定数がある一方，正確にわかっているのはたった1桁（あるいは，1桁もない）で，存在するという厳密な証明さえおぼつかないというような定数もある．

　後者のような定数は，孤立しているように見えるかもしれないが，私はそうではないと信じている．このような定数に関連する分野では，(φ, e, π, γ のようなものとは違い）これらの数を自然なものとして表現するための言語，関数，対称性などを作り上げるのにまだ時間がかかるだけなのであろう．いいかえると，十分に研究し，じっくりと耳を傾ければ，「こだま」を聞くことができるようになるということである．

　数学における定数のしっかりした分類法はまだ確立されていない．本書の（私の学習してきた方法による）構成法も主観的なものにならざるをえない．巻末に示した小数近似の表も，（もし小さい数から順に並べることが役立つのならば）構成法の1つである．私の興味は小数点以下の桁数を伸ばすことではなく，定数がどのような数学の中から現れたのか，そして定数の間の関係は何かということである．すなわち，定数の単なる表ではなく，定数のもつ物語が本書の内容を互いに結びつけていることになる．

　数学的な知識をあまり多くおもちでない読者のために，有名な定数についての事項は早めに，かつ入念に提示するようにした．しかし，もっと深い内容になると，簡潔に書かざるをえなかった．読者としては，学部の上級生あるいはそれ以上の方を想定している（そのため読者が，微積分，行列，微分方程式，確率，抽象代数学の基礎，解析学をすでに学んでいることを仮定している）．明瞭で完全

であること，特定の定数や着想が重要である理由，どの文献に正確な証明や，より深い結果を見出しうるかを示すことなどを常に心がけた．

本書では，共同研究であることを示すのに Richard Guy（リチャード・ガイ）の「&」の使い方を踏襲した．たとえば，「Hardy & Ramanujan，そして Rademacher の研究による」と書いたとき，ハーディとラマヌジャンの共同研究と，ラーデマッヘルの研究は別である．[3,7] とは文献 3 と文献 7 という意味であり，【3.7】とはこの本の 3.7 節という意味である．ピリオドかカンマかの区別はとても重要である（訳注：日本版ではカッコの使い分け（[ ] と 【 】）によっても区別している）．

多くの人々が教育，研究におけるインターネットの役割に期待している．私もウェブの影響力は長く続くと思ってはいるが，本書でウェブサイトのアドレスを示したとして，それが 5 年先でさえ存在しているかどうかは疑問である．ただし，今日利用できるすべての数学に関するウェブサイトのうち，少なくとも次の 3 つは永く生き残っていくものと期待している．

・ロスアラモス国立研究所のプレプリントサーバー，ArXiv（アーカイブ）
「math.CA/9910045」とか「solv-int/9801008」を見よ，というのが何を意味するのか，ArXiv を訪れればわかるはずである．

・MathSciNet
アメリカ数学会のサイト．このサイトのサービスを受ければ，*Mathematical Reviews* や「MR 3,270 e」「MR 33 ♯ 3320」とか「MR 87 h : 51043」の意味について精通できる．

・On-Line Encyclopedia of Integer Sequence（正数列のオンライン百科事典）
Neil Sloan（ネイル・スローン）によって作られた（「A 000688」のような表記で，どのような整数列なのかを同定するのに十分になる）．

しかし，そんなに多くは生き残らないだろう．生き残ったとしても，色々な新しい場所に移動し，古いアドレスが時として消滅もするだろう．それゆえ，この本には URL を載せないことにした．ウェブサイトを引用するとき（たとえば "Numbers, Constants and Computation（数，定数，計算）" "Prime Pages（素数のページ）" "MathPages（数学のページ）" "Plouffe's Tables（プルーフェの表）" あるいは "Geometry Junkyard（幾何学のガラクタ）" など，参照箇所と

しては名前だけをあげることにした．

　この膨大な計画は私一人の仕事ではありえない．本書は，友人たちの多くの思いやりに満ちている．すべての人々に対する感謝を表明すると，この序がとても長くなってしまうので，それは止めることにした．この計画のはじめから貴重な激励をしてくださった，師である Philippe Flajolet（フィリッペ・フラジョレット）に特別な謝意を表したい．

　以下の諸氏に感謝したい．Victor Adamchik, Christian Bower, Anthony Guttmann, Joe Keane, Pieter Moree, Gerhard Niklasch, Simon Plouffe, Pascal Sebah, Craig Tracy, JohnWetzel, and Paul Zimmermann．また，私のオンラインでの研究ノートのためにウェブサイト（私の世界への窓！）を利用させてくれた MathSoft Inc., the Algorithms Group at INRIA, CECM at Simon Fraser University にも感謝したい．

　本書を出版するという冒険に付き合ってくれたケンブリッジ大学出版にも感謝したい．

　読者の方々からの意見，訂正，提案はいつでも歓迎する．Steven.Finch@inria.fr にeメールで送ってほしい．よろしくお願いします．

　本書のある部分は© 2000-2003 MathSoft Engineering & Education Inc. 社のものである．ここに許可を得て再掲した．

スティーヴン・R・フィンチ

■監訳者
一 松　　信　　　　　　　　　　　京都大学名誉教授

■訳者（五十音順，*幹事）
鹿 野　　健* （6章）　　　　　　　前山形大学教育学部・教授
神 谷 茂 保* （1.9～1.11節, 7章,　　岡山理科大学工学部・教授
　　　　　　　8章, 補遺, 定数一覧）
倉 坪 茂 彦　（3章, 4章）　　　　　弘前大学大学院理工学研究科・教授
後 藤 和 雄　（2章, 5章）　　　　　鳥取大学教育センター・准教授
山 野 剛 助　（1.1～1.8節）　　　　金沢工業大学基礎教育部・教授

目　　次

1. 有名な定数

1.1 ピタゴラスの定数, $\sqrt{2}$ ··1
    1.1.1 一般連分数　3
    1.1.2 二重根号の簡易化　4
1.2 黄金比, φ ··5
    1.2.1 根号による展開法　8
    1.2.2 黄金比の立方根版　8
    1.2.3 一般連分数　9
    1.2.4 ランダム・フィボナッチ数列　10
    1.2.5 フィボナッチ数の積　10
1.3 自然対数の底, e ··11
    1.3.1 極限値の考察　14
    1.3.2 連分数　15
    1.3.3 数 2 の自然対数　15
1.4 アルキメデスの定数, π ··17
    1.4.1 無限級数　20
    1.4.2 無限積　21
    1.4.3 定積分　22
    1.4.4 連分数　22
    1.4.5 無限根号　23
    1.4.6 楕円関数　23
    1.4.7 意外な所での π　24
1.5 オイラー-マスケロニの定数, γ ···28
    1.5.1 級数と無限積　30
    1.5.2 積　分　31
    1.5.3 一般オイラー級数　31
    1.5.4 ガンマ関数　32
1.6 アペリの定数, $\zeta(3)$ ···39
    1.6.1 ベルヌーイ数　40
    1.6.2 リーマン予想　40

1.6.3　級　数　41
   1.6.4　無限積　44
   1.6.5　積　分　45
   1.6.6　連分数　45
   1.6.7　スターリングのサイクル数　46
   1.6.8　多重対数関数　46
1.7　カタランの定数，G …………………………………………………… 52
   1.7.1　オイラー数　53
   1.7.2　級　数　53
   1.7.3　積　55
   1.7.4　積　分　55
   1.7.5　連分数　56
   1.7.6　逆正接積分　56
1.8　ヒンチン-レヴィ定数 …………………………………………………… 58
   1.8.1　他の表現法　60
   1.8.2　それから導かれる定数　61
   1.8.3　複素数への拡張　61
1.9　ファイゲンバウム-クーレ-トレサー定数 …………………………… 64
   1.9.1　一般化されたファイゲンバウム定数　66
   1.9.2　2次平面写像　67
   1.9.3　クピタノビッチ-ファイゲンバウム関数方程式　68
   1.9.4　黄金比と白銀比の円写像　69
1.10　マデラングの定数 ……………………………………………………… 74
   1.10.1　格子和とオイラー定数　76
1.11　チャイティンの定数 …………………………………………………… 79

2.　数論に関する定数

2.1　ハーディ-リトルウッド定数 …………………………………………… 83
   2.1.1　2次多項式で表される素数　86
   2.1.2　ゴールドバッハ予想　86
   2.1.3　3次式で表される素数　88
2.2　マイセル-メルテンス定数 ……………………………………………… 93
   2.2.1　平方剰余　95
2.3　ランダウ-ラマヌジャン定数 …………………………………………… 97
   2.3.1　類似の定数　98
2.4　アルティン定数 ………………………………………………………… 103
   2.4.1　関連する定数　105

- 2.4.2 補正因子　106
- 2.5 ハフナー-サルナック-マッカレー定数 …………………………… 109
  - 2.5.1 無関心な対　109
- 2.6 ニーヴェン定数 ……………………………………………………… 111
  - 2.6.1 充平方と充立方整数　112
- 2.7 オイラーの約数関数 ………………………………………………… 114
- 2.8 ペル-スティーヴンハーゲン定数 …………………………………… 118
- 2.9 アラディ-グリンステッド定数 ……………………………………… 120
- 2.10 シェルピンスキー定数 ……………………………………………… 122
  - 2.10.1 円の問題と約数問題　123
- 2.11 過剰数の密度定数 …………………………………………………… 125
- 2.12 リンニク定数 ………………………………………………………… 127
- 2.13 ミル定数 ……………………………………………………………… 130
- 2.14 ブルン定数 …………………………………………………………… 132
- 2.15 グレイシャー-キンケリン定数 ……………………………………… 135
  - 2.15.1 一般化グレイシャー定数　136
  - 2.15.2 多重バーンズ関数　137
  - 2.15.3 GUE 仮設　138
- 2.16 ストラルスキ-ハルボルス定数 ……………………………………… 145
  - 2.16.1 デジタル和　146
  - 2.16.2 ウラムの1加法数列　147
  - 2.16.3 交代ビット集合　148
- 2.17 ガウス-クズミン-ヴィルジング定数 ……………………………… 151
  - 2.17.1 リュール-マイヤー作用素　152
  - 2.17.2 漸近的正規性　154
  - 2.17.3 有界な部分分母　154
- 2.18 ポーター-ヘンズリー定数 …………………………………………… 156
  - 2.18.1 2進ユークリッドアルゴリズム　158
  - 2.18.2 最悪な場合の考察　159
- 2.19 ヴァレー定数 ………………………………………………………… 160
  - 2.19.1 継続多項式　162
- 2.20 エルデーシュの逆数和定数 ………………………………………… 163
  - 2.20.1 A 数列　163
  - 2.20.2 B_2 数列　164
  - 2.20.3 非平均数列　164
- 2.21 スティルチェス定数 ………………………………………………… 166
  - 2.21.1 一般ガンマ関数　169
- 2.22 リューヴィル-ロス定数 ……………………………………………… 171

- 2.23 ディオファンタス近似定数 ……………………………………… 173
- 2.24 自己数密度定数 ………………………………………………… 179
- 2.25 カメロンの無和集合定数 ……………………………………… 181
- 2.26 無三重集合定数 ………………………………………………… 183
- 2.27 エルデーシュ-レーベンソールド定数 ………………………… 186
  - 2.27.1 有限の場合　186
  - 2.27.2 無限の場合　186
  - 2.27.3 一般化　187
- 2.28 エルデーシュの和違い集合定数 ……………………………… 189
- 2.29 高速行列乗法の定数 …………………………………………… 191
- 2.30 ピゾー-ヴィジャヤラハヴァン-サーレム定数 ………………… 193
  - 2.30.1 $(3/2)^n$ の小数部分　195
- 2.31 フレイマンの定数 ……………………………………………… 201
  - 2.31.1 ラグランジュスペクトル，L　201
  - 2.31.2 マルコフスペクトル，M　201
  - 2.31.3 マルコフ-フルヴィッツの方程式　202
  - 2.31.4 ホール線　203
  - 2.31.5 L と M の比較　204
- 2.32 ド・ブルイン-ニューマン定数 ………………………………… 205
- 2.33 ホール-モンゴメリー定数 ……………………………………… 207

3. 解析学の不等式に関する定数

- 3.1 シャピロ-ドリンフェルドの定数 ……………………………… 210
  - 3.1.1 ジョコヴィッチの予想　212
- 3.2 カールソン-レヴィンの定数 …………………………………… 213
- 3.3 ランダウ-コルモゴロフの定数 ………………………………… 214
  - 3.3.1 $L_\infty(0, \infty)$ の場合　215
  - 3.3.2 $L_\infty(-\infty, \infty)$ の場合　215
  - 3.3.3 $L_2(-\infty, \infty)$ の場合　216
  - 3.3.4 $L_2(0, \infty)$ の場合　216
- 3.4 ヒルベルト定数 ………………………………………………… 218
- 3.5 コプソン-ド・ブルイン定数 …………………………………… 219
- 3.6 ソボレフの等周定数 …………………………………………… 221
  - 3.6.1 弦の不等式　222
  - 3.6.2 棒の不等式　222
  - 3.6.3 膜の不等式　223
  - 3.6.4 板の不等式　223

3.6.5　その他の変形　224
3.7　コーン定数 ……………………………………………………227
3.8　ウィットニー-ミクリンの拡張定数 …………………………229
3.9　ゾロタリョーフ-シュア定数 …………………………………231
　　3.9.1　楕円に関するシーウェル問題　231
3.10　クネーザー-マーラー多項式定数 ……………………………233
3.11　グロタンディークの定数 ………………………………………237
3.12　デュボア・レイモン定数 ……………………………………239
3.13　シュタイニッツ定数 …………………………………………241
　　3.13.1　動　機　241
　　3.13.2　定　義　242
　　3.13.3　結　果　243
3.14　ヤング-フェイエール-ジャクソン定数 ……………………244
　　3.14.1　余弦和の非負性　244
　　3.14.2　正弦和の正値性　245
　　3.14.3　一様有界性　245
3.15　ファン・デル・コルプト定数 ………………………………247
3.16　トゥラーンの指数和定数 ……………………………………248

4. 関数の近似に関する定数

4.1　ギッブス-ウィルブラハム定数 ………………………………250
4.2　ルベーグ定数 …………………………………………………252
　　4.2.1　三角フーリエ級数　252
　　4.2.2　ラグランジュ補間　254
4.3　アキェゼル-クレイン-ファヴァール定数 ……………………257
4.4　ベルンシュテイン定数 …………………………………………259
4.5　「1/9」予想の定数 ……………………………………………261
4.6　フランセン-ロビンソン定数 …………………………………264
4.7　ベリー-エッセン定数 …………………………………………266
4.8　ラプラス限界定数 ……………………………………………268
4.9　整係数チェビシェフ定数 ……………………………………271
　　4.9.1　超越直径　273

5. 離散構造数え上げに関する定数

5.1　アーベル群数え上げに関する定数 ……………………………275
　　5.1.1　半単純結合環　276

| 目　次

- 5.2　ピタゴラス3数定数 ……………………………………………278
- 5.3　レーニィの駐車定数 ……………………………………………280
  - 5.3.1　ランダム逐次吸着　282
- 5.4　ゴロム-ディックマン定数 ……………………………………286
  - 5.4.1　対称群　288
  - 5.4.2　ランダム写像の統計量　289
- 5.5　カルマールの合成定数 …………………………………………294
- 5.6　オッターの木の数え上げ定数 …………………………………297
  - 5.6.1　化学での異性体　300
  - 5.6.2　木のいろいろな種類　302
  - 5.6.3　属　性　305
  - 5.6.4　林　306
  - 5.6.5　カクタスと2-木　307
  - 5.6.6　写像パターン　308
  - 5.6.7　種々のグラフ　309
  - 5.6.8　データ構造　311
  - 5.6.9　ゴルトン-ワトソン分岐過程　312
  - 5.6.10　エルデーシュ-レーニィの進化過程　313
- 5.7　レンジェル定数 …………………………………………………317
  - 5.7.1　スターリングの分割数　317
  - 5.7.2　Sの部分集合束における鎖　318
  - 5.7.3　Sを分割した束における鎖　319
  - 5.7.4　ランダム鎖　320
- 5.8　竹内-プレルバーグ定数 ………………………………………322
- 5.9　ポリヤの酔歩定数 ………………………………………………323
  - 5.9.1　交差と捕獲　328
  - 5.9.2　ホロノミー性　329
- 5.10　自己回避歩行定数 ………………………………………………332
  - 5.10.1　多角形と小道　334
  - 5.10.2　チェス盤上のルークの通り道　335
  - 5.10.3　うねり図形とスタンプの折り曲げ　335
- 5.11　フェラーの硬貨投げ定数 ………………………………………340
- 5.12　強正方形エントロピー定数 ……………………………………344
  - 5.12.1　気体モデル格子における相転移　345
- 5.13　2分検索木定数 …………………………………………………351
- 5.14　デジタル検索木定数 ……………………………………………355
  - 5.14.1　他との関連　358
  - 5.14.2　近似数え上げ　359

- 5.15　最適停止定数 ……………………………………………………………362
- 5.16　極値定数 ………………………………………………………………365
- 5.17　パターンをもたない語の定数 …………………………………………369
- 5.18　浸透のクラスター密度定数 ……………………………………………372
  - 5.18.1　臨界確率　374
  - 5.18.2　級数展開　375
  - 5.18.3　変　種　376
- 5.19　クラナーのポリオミノ定数 ……………………………………………380
- 5.20　最長部分数列の定数 ……………………………………………………385
  - 5.20.1　増加部分数列　385
  - 5.20.2　共通部分数列　387
- 5.21　k-充足可能の定数 ………………………………………………………390
- 5.22　レンツ-イジング定数 …………………………………………………394
  - 5.22.1　低温級数展開　395
  - 5.22.2　高温級数展開　396
  - 5.22.3　強磁性体モデルにおける相転移　397
  - 5.22.4　臨界温度　399
  - 5.22.5　磁化率　399
  - 5.22.6　Q モーメントと P モーメント　401
  - 5.22.7　パンルヴェIII型方程式　403
- 5.23　モノマー・ダイマー定数 ………………………………………………408
  - 5.23.1　2次元ドミノによる充填　409
  - 5.23.2　菱形とバイボーン　410
  - 5.23.3　3次元ドミノによる充填　411
- 5.24　リーブの正方氷定数 ……………………………………………………414
  - 5.24.1　塗り分け　415
  - 5.24.2　折り曲げ　416
  - 5.24.3　氷の結晶内の原子配列　417
- 5.25　タット-ベラハ定数 ……………………………………………………419

6.　関数の反復に関する定数

- 6.1　ガウスのレムニスケート定数 ……………………………………………423
  - 6.1.1　ワイエルシュトラスのペー関数　425
- 6.2　オイラー-ゴンパーツ定数 ………………………………………………426
  - 6.2.1　指数積分　427
  - 6.2.2　対数積分　428
  - 6.2.3　発散級数　428

6.2.4　生存解析　428
6.3　ケプラー–ボウカンプ定数 ……………………………………………431
6.4　グロスマンの定数 ………………………………………………………433
6.5　プルーフェの定数 ………………………………………………………433
6.6　レーマーの定数 …………………………………………………………436
6.7　カーエンの定数 …………………………………………………………438
6.8　プルーエ–トゥエ–モース定数 …………………………………………440
　6.8.1　確率的数え上げ　441
　6.8.2　非整数の基　442
　6.8.3　外　角　442
　6.8.4　フィボナッチ語　443
　6.8.5　折り紙　443
6.9　ミンコフスキー–バウアーの定数 ……………………………………445
6.10　2次数列の定数 ………………………………………………………447
6.11　反復指数の定数 ………………………………………………………452
　6.11.1　指数的再帰数列　454
6.12　コンウェイの定数 ……………………………………………………456

7.　複素解析に関する定数

7.1　ブロック–ランダウ定数 ………………………………………………459
7.2　マサー–グラメイン定数 ………………………………………………462
7.3　ホイッタカー–ゴンチャロフ定数 ……………………………………465
　7.3.1　ゴンチャロフ多項式　467
　7.3.2　剰余多項式　467
7.4　ジョン定数 ………………………………………………………………469
7.5　ヘイマン定数 ……………………………………………………………472
　7.5.1　ヘイマン–キェルベリ　472
　7.5.2　ヘイマン–コレンブラム　472
　7.5.3　ヘイマン–スチュワート　473
　7.5.4　ヘイマン–ウー　474
7.6　リトルウッド–クルニー–ポメレンケ定数 ……………………………475
　7.6.1　アルファ　475
　7.6.2　ベータとガンマ　475
　7.6.3　予想される関係　476
7.7　リース–コルモゴロフ定数 ……………………………………………478
7.8　グレッチュ環状領域定数 ………………………………………………479
　7.8.1　$a(r)$ に対する公式　481

8. 幾何学に関する定数

- 8.1 幾何確率定数 …………………………………………483
- 8.2 円による被覆定数 ………………………………………488
- 8.3 普遍被覆定数 ……………………………………………493
  - 8.3.1 平行移動による被覆　494
- 8.4 モザーのミミズ定数 ……………………………………496
  - 8.4.1 単位長さの最大幅曲線　498
  - 8.4.2 閉じたミミズ　499
  - 8.4.3 平行移動による被覆　500
- 8.5 巡回セールスマン定数 …………………………………502
  - 8.5.1 ランダムリンク TSP　503
  - 8.5.2 最小全域木　504
  - 8.5.3 最小マッチング　505
- 8.6 シュタイナー木定数 ……………………………………508
- 8.7 エルミートの定数 ………………………………………511
- 8.8 タムスの定数 ……………………………………………513
- 8.9 双曲的体積定数 …………………………………………516
- 8.10 ルーロー三角形定数 ……………………………………519
- 8.11 光探知定数 ………………………………………………521
- 8.12 ソファー移動定数 ………………………………………525
- 8.13 カラビの三角形定数 ……………………………………528
- 8.14 デヴィッチの 4 次元立方体定数 ………………………529
- 8.15 グラハムの六角形定数 …………………………………531
- 8.16 ハイルブロンの三角形定数 ……………………………532
- 8.17 掛谷-ベシコヴィッチ定数 ………………………………535
- 8.18 直線的交点定数 …………………………………………537
- 8.19 外半径-内半径定数 ………………………………………540
- 8.20 アポロニウス充填定数 …………………………………542
- 8.21 出会い定数 ………………………………………………544

補　　遺 ……………………………………………………………547
付録　定数一覧 ……………………………………………………549
定 数 索 引 …………………………………………………………575
事 項 索 引 …………………………………………………………579

記 号 表

$\lfloor x \rfloor$	床関数：x 以下の最大整数
$\lceil x \rceil$	天井関数：x 以上の最小整数
$\{x\}$	小数部分：$x - \lfloor x \rfloor$
$\ln x$	自然対数：$\log_e x$
$\binom{n}{k}$	二項係数：$\dfrac{n!}{k!(n-k)!}$
$b_0 + \dfrac{a_1}{\lceil b_1 \rceil} + \dfrac{a_2}{\lceil b_2 \rceil} + \dfrac{a_3}{\lceil b_3 \rceil} + \cdots$	連分数：$b_0 + \dfrac{a_1}{b_1 + \dfrac{a_2}{b_2 + \dfrac{a_3}{b_3 + \cdots}}}$
$f(x) = O(g(x))$	大文字 O：$x \to x_0$ のとき $[f(x)/g(x)]$ が有界
$f(x) = o(g(x))$	小文字 o：$x \to x_0$ のとき $f(x)/g(x) \to 0$
$f(x) \sim g(x)$	漸近的に等値：$x \to x_0$ のとき $f(x)/g(x) \to 1$
$\sum_{p}$	素数 $p = 2, 3, 5, 7, 11 \cdots$ 全体に対する和（文字 p のみを使用する）
$\prod_{p}$	$\sum_{p}$ と同様に，素数に対する積（和の代わりに）
$f(x)^n$	累乗：n が整数のとき $(f(x))^n$
$f^n(x)$	反復：$\underbrace{f(f(\cdots\cdots (f(x) \cdots\cdots)))}_{n \text{ 個}}$，ここに $n \geq 0$ は整数

有名な定数

1.1 ピタゴラスの定数，$\sqrt{2}$

　一辺の長さが 1 の正方形の対角線の長さは $\sqrt{2}=1.4142135623\cdots$ である．ピタゴラス学派が提唱していた学説は，すべての幾何学的な量は有理数で表されるべきである，とするものであった．正方形の辺は，一辺のある整数倍が対角線の整数倍に等しいという意味で，対角線と同じ単位で測ることができると思われていた．この学説は「$\sqrt{2}$ は有理数でない」という発見で打ち砕かれた [1-4]．

　ここで $\sqrt{2}$ が有理数でないという証明を 2 つ紹介しよう．1 つは整数の分解法則を利用するものであり，もう 1 つは整列性を用いるものである．

・$\sqrt{2}$ が有理数とすると，方程式 $p^2=2q^2$ は整数解 p,q をもつ．ここで，p,q は最小の項[†]で表されているものとする．p^2 は偶数であるから，p 自身は偶数でなければならない．よって，$p=2r$ と書ける．これより，$2q^2=4r^2$ であるから，q は偶数でなければならない．これは p と q が最小の項であったという仮定に矛盾する．

・$\sqrt{2}$ が有理数とすると，$s\sqrt{2}$ が整数である最小の正の整数 s が存在する．$1<2$ であるから $1<\sqrt{2}$ であり，よって，$t=s(\sqrt{2}-1)$ は正の整数である．さらに，$t\sqrt{2}=s(\sqrt{2}-1)\sqrt{2}=2s-s\sqrt{2}$ は整数であり，明らかに $t<s$ である．これは s が上記の条件を満たす最小の正の整数であることに矛盾する．

　方程式を近似的に解くためのニュートン法を $x^2-2=0$ に適用して，次の 1 階の漸化式

[†]訳注：これ以上約分できない形．

1. 有名な定数

$$x_0=1, \quad x_k=\frac{x_{k-1}}{2}+\frac{1}{x_{k-1}} \quad k\geq 1 \text{ に対し}, \quad \lim_{k\to\infty}x_k=\sqrt{2}$$

が得られ，それは非常に収束が速く，$\sqrt{2}$ の近似値を求めるためによく使われる．もう 1 つの 1 階の漸化式

$$y_0=\frac{1}{2}, \quad y_k=y_{k-1}\left(\frac{3}{2}-y_{k-1}^2\right) \quad k\geq 1 \text{ に対し}, \quad \lim_{k\to\infty}y_k=\frac{1}{\sqrt{2}}$$

は $\sqrt{2}$ の逆数を与える[5]．また，ニュートンによる二項級数も，次の 2 つの興味深い和公式を与える[6]．

$$1+\sum_{n=1}^{\infty}\frac{(-1)^{n-1}}{2^{2n}(2n-1)}\binom{2n}{n}=1+\frac{1}{2}-\frac{1}{2\cdot 4}+\frac{1\cdot 3}{2\cdot 4\cdot 6}-+\cdots=\sqrt{2}$$

$$1+\sum_{n=1}^{\infty}\frac{(-1)^{n}}{2^{2n}}\binom{2n}{n}=1-\frac{1}{2}+\frac{1\cdot 3}{2\cdot 4}-\frac{1\cdot 3\cdot 5}{2\cdot 4\cdot 6}+-\cdots=\frac{1}{\sqrt{2}}$$

2 番目の式は【1.5.4】で拡張されている．次の 2 つの美しい無限積[5,7,8]

$$\prod_{n=1}^{\infty}\left(1+\frac{(-1)^{n+1}}{2n-1}\right)=\left(1+\frac{1}{1}\right)\left(1-\frac{1}{3}\right)\left(1+\frac{1}{5}\right)\left(1-\frac{1}{7}\right)\cdots=\sqrt{2}$$

$$\prod_{n=1}^{\infty}\left(1-\frac{1}{4(2n-1)^2}\right)=\frac{1\cdot 3}{2\cdot 2}\cdot\frac{5\cdot 7}{6\cdot 6}\cdot\frac{9\cdot 11}{10\cdot 10}\cdot\frac{13\cdot 15}{14\cdot 14}\cdots=\frac{1}{\sqrt{2}}$$

と正則連分数展開[9]

$$2+\cfrac{1}{2+\cfrac{1}{2+\cfrac{1}{2+\cdots}}}=2+\frac{1|}{|2}+\frac{1|}{|2}+\frac{1|}{|2}+\cdots=1+\sqrt{2}=(-1+\sqrt{2})^{-1}$$

にも注意しておこう．これは Pell（ペル）の数列

$$a_0=0, \quad a_1=1, \quad a_n=2a_{n-1}+a_{n-2}, \quad n\geq 2$$

と極限公式

$$\lim_{n\to\infty}\frac{a_{n+1}}{a_n}=1+\sqrt{2}$$

を介して関係している．この関係は黄金比 φ とフィボナッチ数列の間で知られている関係とよく似たものとなっている【1.2】．図 1.1 も参照せよ．

Viète（ヴィエト）によるアルキメデスの定数[†] π に対する注目すべき成果は，数 2 のみを用い，くり返し平方根を適用するというところにある【1.4.2】．π や累乗根に関する表現は【1.4.5】で取り上げる．

最後に無理数性に関する結果について述べよう．x,y が有理数でも x^y が無理数であるような x,y は明らかに存在する（たとえば，$x=2, y=1/2$）．それでは，x^y が有理数であるような無理数 x,y は存在するであろうか．これに対する答には唖然とさせられる．

$$z=\sqrt{2}^{\sqrt{2}}$$

とおく．もし z が有理数ならば，$x=y=\sqrt{2}$ とすればよい．もし z が無理数ならば，

[†] 訳注：円周率のこと．【1.4】参照．

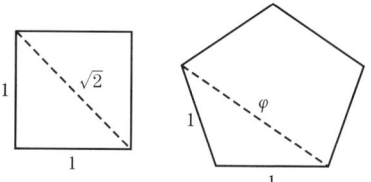

図 1.1 一辺の長さが 1 の正 5 角形の 2 つの隣接しない頂点を結ぶ対角線は，黄金比 φ で与えられた長さをもつ（ピタゴラスの定数 $\sqrt{2}$ と対照せよ）．

$x=z$, $y=\sqrt{2}$ とすれば，明らかに $x^y=2$ である．このように z の数論的な性質について調べることなしに，この問題に（正しいと）答えることができる．実際，1934 年に証明された Gel'fond-Schneider（ゲルフォンド-シュナイダー）の定理[10]によって z は超越数であることがわかっており，したがってこれは無理数である．この分野では未解決な問題がたくさんあり，たとえば，

$$\sqrt{2}^z = \sqrt{2}^{\sqrt{2}^{\sqrt{2}}}$$

が無理数（ましてや超越数）かどうかはわかっていない．

1.1.1 一般連分数

任意の 2 次の無理数は循環正則連分数展開をもち，逆も成り立つことはよく知られている．一方，少数の研究者がフラクタル的な構成法で表される一般連分数

$$w(p,q) = q + \cfrac{p + \cfrac{1}{q + \cfrac{p+\cdots}{q+\cdots}}}{q + \cfrac{p + \cfrac{1+\cdots}{q+\cdots}}{q + \cfrac{p+\cdots}{q+\cdots}}}$$

について考察してきた[11-17]．

1 つの生成項（部分近似分数）において，それぞれの新しい項は，規則

$$p \to p + \frac{1}{q}, \quad q \to q + \frac{p}{q}$$

によって次の生成項に置きかえられる．明らかに

$$w = q + \frac{p + \frac{1}{w}}{w}, \quad \text{すなわち} \quad w^3 - qw^2 - pw - 1 = 0$$

である．特に $p=q=3$ のとき，この高次の連分数は $(-1+\sqrt[3]{2})^{-1}$ に収束する．3 次の無理数の正則連分数は，ほとんどの実数に対してこのようにふるまうと予測されるが，どのような具体例も証拠として上がってきていない[18-21]．2 次の無理数の研究に対

しては，通常の変換規則

$$r \to r + \frac{1}{r}$$

だけで充分であるが，代数的数のより広範囲なクラスに対しては，この規則を拡張することが必要になる．

$\sqrt[3]{2}$ に対する交代表現には，次の2つのものがあり，それらは

$$\sqrt[3]{2} = 1 + \cfrac{1}{3 + \cfrac{3}{a} + \cfrac{1}{b}}, \quad \text{ここで} \quad a = 3 + \frac{3}{a} + \frac{1}{b}, \quad b = 12 + \frac{10}{a} + \frac{3}{b}$$

[22]と

$$\sqrt[3]{2} = 1 + \frac{1}{|3} + \frac{2}{|2} + \frac{4}{|9} + \frac{5}{|2} + \frac{7}{|15} + \frac{8}{|2} + \frac{10}{|21} + \frac{11}{|2} + \cdots$$

[23]である．一般連分数という言葉は，[24]では同時ディオファンタス近似への応用で取り扱われ，[25]では凸閉包の境界を含む幾何学的な解釈に使われている．

1.1.2 二重根号の簡易化

ここで，Ramanujan（ラマヌジャン）による注目すべき二重根号の簡易化の式を2つ取り上げておこう．

$$\sqrt[3]{\sqrt[3]{2}-1} = \sqrt[3]{\frac{1}{9}} - \sqrt[3]{\frac{2}{9}} + \sqrt[3]{\frac{4}{9}}, \quad \sqrt[2]{\sqrt[3]{5}-\sqrt[3]{4}} = \frac{1}{3}(\sqrt[3]{2} + \sqrt[3]{20} - \sqrt[3]{25})$$

このような簡易化は計算機代数システム[†]の重要な分野となっている[26]．

---

[1] G. H. Hardy and E. M. Wright, *An Introduction to the Theory of Numbers*, 5th ed., Oxford Univ. Press, 1985, pp. 38-45; MR 81i:10002.
[2] F. J. Papp, $\sqrt{2}$ is irrational, *Int. J. Math. Educ. Sci. Technol.* 25 (1994) 61-67; MR 94k:11081.
[3] O. Toeplitz, *The Calculus: A Genetic Approach*, Univ. of Chicago Press, 1981, pp. 1-6; MR 11, 584e.
[4] K. S. Brown, Gauss' lemma without explicit divisibility arguments (MathPages).
[5] X. Gourdon and P. Sebah, The square root of 2 (Numbers, Constants and Computation).
[6] K. Knopp, *Theory and Application of Infinite Series*, Hafner, 1951, pp. 208-211, 257-258; MR 18, 30c.
[7] I. S. Gradshteyn and I. M. Ryzhik, *Tables of Integrals, Series and Products*, Academic Press, 1980, p. 12; MR 97c:00014.
[8] F. L. Bauer, An infinite product for square-rooting with cubic convergence, *Math. Intellig.* 20 (1998) 12-13, 38.
[9] L. Lorentzen and H. Waadeland, *Continued Fractions with Applications*, North-Holland, 1992, pp. 10-16, 564-565; MR 93g:30007.
[10] C. L. Siegel, *Transcendental Numbers*, Princeton Univ. Press, 1949, pp. 75-84; MR 11,330c.
[11] D. Gómez Morin, *La Quinta Operación Aritmética: Revolución del Número*, 2000.
[12] A. K. Gupta and A. K. Mittal, Bifurcating continued fractions, math.GM/0002227.
[13] A. K. Mittal and A. K. Gupta, Bifurcating continued fractions II, math.GM/0008060.
[14] G. Berzsenyi, Nonstandardly continued fractions, *Quantum Mag.* (Jan./Feb. 1996) 39.
[15] E. O. Buchman, Problem 4/21, *Math. Informatics Quart.*, v. 7 (1997) n. 1, 53.

[†] 訳注：数式処理のソフトウェアを含むシステムと理論を統合した分野．

[16] A. Dorito and K. Ekblaw, Solution of problem 2261, *Crux Math.*, v. 24 (1998) n. 7, 430-431.
[17] W. Janous and N. Derigiades, Solution of problem 2363, *Crux Math.*, v. 25 (1999) n. 6, 376-377.
[18] J. von Neumann and B. Tuckerman, Continued fraction expansion of $2^{1/3}$, *Math. Tables Other Aids Comput.* 9 (1955) 23-24; MR 16,961d.
[19] R. D. Richtmyer, M. Devaney, and N. Metropolis, Continued fraction expansions of algebraic numbers, *Numer. Math.* 4 (1962) 68-84; MR 25 #44.
[20] A. D. Brjuno, The expansion of algebraic numbers into continued fractions (in Russian), *Zh. Vychisl. Mat. Mat. Fiz.* 4 (1964) 211-221; Engl. transl. in *USSR Comput. Math. Math. Phys.*, v. 4 (1964) n. 2, 1-15; MR 29 #1183.
[21] S. Lange and H. Trotter, Continued fractions for some algebraic numbers, *J. Reine Angew. Math.* 255 (1972) 112-134; addendum 267 (1974) 219-220; MR 46 #5258 and MR 50 #2086.
[22] F. O. Pasicnjak, Decomposition of a cubic algebraic irrationality into branching continued fractions (in Ukrainian), *Dopovidi Akad. Nauk Ukrain. RSR Ser. A* (1971) 511-514, 573; MR 45 #6765.
[23] G. S. Smith, Expression of irrationals of any degree as regular continued fractions with integral components, *Amer. Math. Monthly* 64 (1957) 86-88; MR 18,635d.
[24] W. F. Lunnon, Multi-dimensional continued fractions and their applications, *Computers in Mathematical Research*, Proc. 1986 Cardiff conf., ed. N. M. Stephens and M. P. Thorne, Clarendon Press, 1988, pp. 41-56; MR 89c:00032.
[25] V. I. Arnold, Higher-dimensional continued fractions, *Regular Chaotic Dynamics* 3 (1998) 10-17; MR 2000h:11012.
[26] S. Landau, Simplification of nested radicals, *SIAM J. Comput.* 21 (1992) 85-110; MR 92k:12008.

1.2 黄金比, φ

次の線分を見てみよう．

「この線分を2つの部分に分ける最も気持のよい分割は何か」と聞かれたとき，ある人は中点

---------------- • -----------------

と答えるかもしれない．また，ある人は1/4あるいは3/4の点と答えるかもしれない．しかしながら，「正しい答」はおそらく古代ギリシャ以後の西洋芸術において発見されたと思われる（美学者はそれを動的調和（dynamic symmetry）の原理と呼んでいる）．

--------------------- • ------------

右側の部分が長さ $v=1$ であれば，左側の部分は長さ $u=1.618\cdots$ である．このように分割された線分は，黄金の部分，あるいは，神の部分に分割されたといわれる．この特別な分割にこのような高い地位を付与するという理由は何であろうか？ 長さ u と全体の長さ $u+v$ の比は，長さ v と u の比と同等になる．すなわち，

$$\frac{u}{u+v}=\frac{v}{u}$$

である．$\varphi = u/v$ と置き，
$$1 + \frac{1}{\varphi} = 1 + \frac{v}{u} = \frac{u+v}{u} = \frac{u}{v} = \varphi$$
に注意して φ を求めてみよう．すると，2次方程式 $\varphi^2 - \varphi - 1 = 0$ の正の解が
$$\varphi = \frac{1+\sqrt{5}}{2} = 1.6180339887\cdots$$
である．この数は**黄金比** (Golden mean)，あるいは，**神の比率** (Divine proportion) とよばれている[1,2]．

定数 φ は**フィボナッチ数列**
$$f_0 = 0, \quad f_1 = 1, \quad f_n = f_{n-1} + f_{n-2}, \quad n \geq 2$$
と込み入った関係にある．この数列はウサギの個体数の増加を素朴な意味でモデル化している．ウサギは生まれてより2カ月経過してから，月に1回子ウサギを生むと仮定する．ウサギは常に1つがいの子ウサギ（1匹は雄，もう1匹は雌）を生み，決して死なず，繁殖を停止しないものとする．ウサギの n カ月後のつがいの数が f_n である．

もし，φ と $\{f_n\}$ に共通したものがあるとすれば，それは何なのか？ この疑問こそが全数学の中で見られる最も顕著な考え方の1つである．φ の正則連分数展開
$$\varphi = 1 + \cfrac{1}{1 + \cfrac{1}{1 + \cdots}} = 1 + \frac{1|}{|1} + \frac{1|}{|1} + \frac{1|}{|1} + \cdots$$
を導く部分近似分数は，すべて連続するフィボナッチ数の比である．ゆえに，
$$\lim_{n \to \infty} \frac{f_{n+1}}{f_n} = \varphi$$
である．この結果は，初期項 f_0, f_1 が異なるとしたときでも，同じ漸化式 $f_n = f_{n-1} + f_{n-2}$ をみたす数列であれば，もちろん成立する．

黄金比とフィボナッチ数列の間の幾何学的なおもしろい関係が，図1.2で見られる．まず，長さが φ で幅が1の1個の黄金長方形から出発し，この最初の長方形から，左側の目いっぱいの正方形を除き，次にこの2番目の長方形から，上側の目いっぱいの正

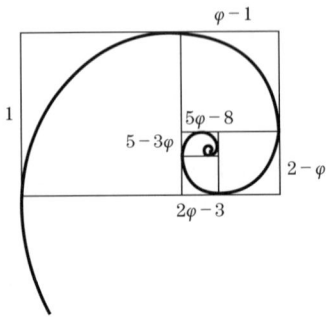

図1.2　黄金長方形の系列を外接させる黄金螺線

方形を除くということを続けて，縮小黄金長方形の自然な系列が得られる．n 番目の黄金長方形の長さと幅は，フィボナッチ数 a, b を係数とする 1 次式 $a + b\varphi$ で表すことができる．これらの黄金長方形は，図のように対数螺線に刻み込んでいくことができる．最初の黄金長方形の左下の点を xy 座標系の原点とする．螺線の集積点は $((1+3\varphi)/5, (3-\varphi)/5)$ であることが証明できる．集積点 (x_∞, y_∞) を通るすべての直線が螺線と定角 ξ で交わるという意味で，このような対数螺線は「等角」である．この点で，対数螺線は普通の円を一般化したものといえる（円に対しては $\xi = 90°$）．ここで描かれた対数螺線は，定角が $\xi = \mathrm{arccot}(2\ln(\varphi)/\pi) = 72.968\cdots°$ となる．対数螺線は自然界のいたる所に明瞭に見られる．たとえば，婉曲したオウム貝の貝殻，象の牙，ヒマワリやマツカサの模様などである[4-6]．

もう 1 つの黄金比の幾何学的な適用は，与えられた円に定規とコンパスで内接正 5 角形を作図するときである．これは

$$2\cos\left(\frac{\pi}{5}\right) = \varphi, \quad 2\sin\left(\frac{\pi}{5}\right) = \sqrt{3-\varphi}$$

という事実と関係している．黄金比そのものは，簡単な正則連分数展開をもっているが，さらに，次の多重根号展開

$$\varphi = \sqrt{1+\sqrt{1+\sqrt{1+\sqrt{1+\sqrt{1+\cdots}}}}}$$

をもつ[7]．この展開が φ に収束する様子は【1.2.1】で取り上げられる．ピタゴラスの定数と同様，黄金比も無理数であり，[8,9]に簡単な証明が見られる．

φ を含む級数[10]

$$\frac{2\sqrt{5}}{5}\ln(\varphi) = \left(1 - \frac{1}{2} - \frac{1}{3} + \frac{1}{4}\right) + \left(\frac{1}{6} - \frac{1}{7} - \frac{1}{8} + \frac{1}{9}\right) + \left(\frac{1}{11} - \frac{1}{12} - \frac{1}{13} + \frac{1}{14}\right) + \cdots$$

は，アルキメデスの定数 π と関係するある級数を思い起こさせる【1.4.1】．φ を和として表す直接的な表現は，φ のまわりで展開される平方根関数に対するテイラー級数から得られる．φ のもう 1 つの表現[11]

$$4 - \varphi = \sum_{n=0}^{\infty} \frac{1}{f_{2^n}} = \frac{1}{f_1} + \frac{1}{f_2} + \frac{1}{f_4} + \frac{1}{f_8} + \cdots$$

においては，やはり，フィボナッチ数が現れる．

φ を含む他の多くの公式の中から，次の 4 つの Rogers & Ramanujan (ロジャース＆ラマヌジャン) の連分数

$$\frac{1}{\alpha - \varphi}\exp\left(-\frac{2\pi}{5}\right) = 1 + \frac{e^{-2\pi}}{1} + \frac{e^{-4\pi}}{1} + \frac{e^{-6\pi}}{1} + \frac{e^{-8\pi}}{1} + \cdots,$$

$$\frac{1}{\beta - \varphi}\exp\left(-\frac{2\pi}{\sqrt{5}}\right) = 1 + \frac{e^{-2\pi\sqrt{5}}}{1} + \frac{e^{-4\pi\sqrt{5}}}{1} + \frac{e^{-6\pi\sqrt{5}}}{1} + \frac{e^{-8\pi\sqrt{5}}}{1} + \cdots,$$

$$\frac{1}{\kappa - (\varphi-1)}\exp\left(-\frac{\pi}{5}\right) = 1 - \frac{e^{-\pi}}{1} + \frac{e^{-2\pi}}{1} - \frac{e^{-3\pi}}{1} + \frac{e^{-4\pi}}{1} - \cdots,$$

$$\frac{1}{\lambda - (\varphi-1)}\exp\left(-\frac{\pi}{\sqrt{5}}\right) = 1 - \frac{e^{-\pi\sqrt{5}}}{1} + \frac{e^{-2\pi\sqrt{5}}}{1} - \frac{e^{-3\pi\sqrt{5}}}{1} + \frac{e^{-4\pi\sqrt{5}}}{1} - + \cdots$$

に注意しておこう．ここで，

$$\alpha = (\varphi\sqrt{5})^{1/2}, \quad \alpha' = \frac{1}{\sqrt{5}}((\varphi-1)\sqrt{5})^{5/2}, \quad \beta = \frac{\sqrt{5}}{1+\sqrt[5]{\alpha'-1}},$$

$$\kappa = ((\varphi-1)\sqrt{5})^{1/2}, \quad \kappa' = \frac{1}{\sqrt{5}}(\varphi\sqrt{5})^{5/2}, \quad \lambda = \frac{\sqrt{5}}{1+\sqrt[5]{\kappa'-1}}$$

である．4 番目の評価式は Ramanathan（ラマナタン）による[9, 12-16]．

1.2.1 根号による展開法

【1.2】における φ の根号による展開は，次の数列 $\{\varphi_n\}$ としても書き表すことができる．

$$\varphi_1 = 1, \quad \varphi_n = \sqrt{1+\varphi_{n-1}}, \quad n \geq 2$$

Paris（パリス）[17]は φ_n が φ に近づく割合が

$$\varphi - \varphi_n \sim \frac{2C}{(2\varphi)^n} \quad (n \to \infty)$$

で与えられることを証明した．ここで，$C = 1.0986419643\cdots$ であり，これは新しく出現した定数である．ここに，C の正確な特徴付けを示そう．次の関数等式

$$F(x) = 2\varphi F(\varphi - \sqrt{\varphi^2 - x}), \quad |x| < \varphi^2$$

の解析的な解で，初期条件 $F(0) = 0$，$F'(0) = 1$ をみたす関数を $F(x)$ とする．すると，$C = \varphi F(1/\varphi)$ である．これらの公式から，ベキ級数の手法を使って C を数値的に計算評価することができる．とはいえ，次の無限積

$$C = \prod_{n=2}^{\infty} \frac{2\varphi}{\varphi + \varphi_n}$$

を使う方が，もっと簡単である．これは着実で収束も速い[18]．

他に面白い定数が次の多重根号表現

$$\sqrt{1+\sqrt{2+\sqrt{3+\sqrt{4+\sqrt{5+\cdots}}}}} = 1.7579327566\cdots$$

によって定義される[7, 19]．しかし，この式が他の何らかの定数に関係するものかどうかは，何も知られていない．

1.2.2 黄金比の立方根版

Perrin（ペラン）数列は

$$g_0 = 3, \quad g_1 = 0, \quad g_2 = 2, \quad g_n = g_{n-2} + g_{n-3}, \quad n \geq 3$$

によって定義され，n が素数（$n > 1$）ならば，n が g_n を割り切るという性質をもっている[20, 21]．連続するペラン数の比の極限

$$\psi = \lim_{n \to \infty} \frac{g_{n+1}}{g_n}$$

は，$\psi^3 - \psi - 1 = 0$ を満たし

$$\psi = \left(\frac{1}{2} + \frac{\sqrt{69}}{18}\right)^{1/3} + \frac{1}{3}\left(\frac{1}{2} + \frac{\sqrt{69}}{18}\right)^{-1/3} = \frac{2\sqrt{3}}{3}\cos\left(\frac{1}{3}\arccos\left(\frac{3\sqrt{3}}{2}\right)\right)$$

$$= 1.3247179572\cdots$$

である．これは多重根号展開ももち

$$\psi = \sqrt[3]{1+\sqrt[3]{1+\sqrt[3]{1+\sqrt[3]{1+\sqrt[3]{1+\cdots}}}}}$$

である．ψ に関する面白い記事が[20]で見られる．そこでは，プラスチック（Plastic）定数という名で取り上げられている（黄金定数との対比として）．【2.30】も参照せよ．

トリボナッチ（三重フィボナッチ）**数列**といわれる数列

$$h_0=0, \quad h_1=0, \quad h_2=1, \quad h_n=h_{n-1}+h_{n-2}+h_{n-3}, \quad n \geq 3$$

は類似の極限比

$$\chi = \left(\frac{19}{27}+\frac{\sqrt{33}}{9}\right)^{1/3} + \frac{4}{9}\left(\frac{19}{27}+\frac{\sqrt{33}}{9}\right)^{-1/3} + \frac{1}{3} = \frac{4}{3}\cos\left(\frac{1}{3}\arccos\left(\frac{19}{8}\right)\right) + \frac{1}{3}$$

$$= 1.8392867552\cdots$$

をもっている．これは，$\chi^3-\chi^2-\chi-1=0$ の実数解である．【1.2.3】を参照せよ．また，**4数ゲーム**について考えてみよう．これは，まず非負の実数よりなる4次元のベクトル (a,b,c,d) から出発し，巡回的に差の絶対値のベクトル $(|b-a|,|c-b|,|d-c|,|a-d|)$ を作る．これを無限にくりかえす．ほとんどの場合（たとえば，a,b,c,d が正の整数の場合）有限段階の後に，4次元の零ベクトルに終結する．これは常に正しいだろうか？ いや，実は正しくない．$v=(1,\chi,\chi^2,\chi^3)$ や，v のスカラー倍，あるいは，v と4次元のベクトル $(1,1,1,1)$ の線形結合が反例として知られている[24]．また，$w=(\chi^3,\chi^2+\chi,\chi^2,0)$ や，w のスカラー倍，あるいは，w と4次元のベクトル $(1,1,1,1)$ の線形結合も反例である．これらが例外のすべてである．w から出発すると1段階後に v が得られることに注意しておこう．

1.2.3 一般連分数

一般連分数は，次の置きかえの規則

$$p \to p+\frac{1}{q}, \quad q \to q+\frac{p}{q}$$

を生成項の新しい項に適用することによって構成されるという【1.1.1】での話を思い起こそう．特に，$p=q=1$ のとき，部分近似分数はトリボナッチ数列の連続する2項の比に等しい．よって，それは χ に収束する．これとは対照的に，置きかえ規則

$$r \to r+\cfrac{1}{r+\cfrac{1}{r}}$$

は $x^3-rx^2-r=0$ の解と関係している[25,26]．もし $r=1$ ならば，この極限値は

$$\left(\frac{29}{54}+\frac{\sqrt{93}}{18}\right)^{1/3} + \frac{1}{9}\left(\frac{29}{54}+\frac{\sqrt{93}}{18}\right)^{-1/3} + \frac{1}{3} = \frac{2}{3}\cos\left(\frac{1}{3}\arccos\left(\frac{29}{2}\right)\right)^{\dagger} + \frac{1}{3} = 1.4655712318\cdots$$

---

†訳注：arccos(29/2) は複素数だが項全体は実数になる．

である.黄金比に関する高次の類比については[27-29]で取り上げられている.

1.2.4 ランダム・フィボナッチ数列

次の確率変数列

$$x_0=1, \quad x_1=1, \quad x_n=\pm x_{n-1}\pm x_{n-2}, \quad n\geq 2$$

について考えてみよう.ここで,±の符号は独立で等確率で取られるものとする.Viswanath(ヴィスワナト)[30-32]は

$$\lim_{n\to\infty}\sqrt[n]{|x_n|}=1.13198824\cdots$$

が確率1で成立するという驚くべき結果を証明した.Embree & Trefethen(エンブリー&トレフェセン)[33]は,次の一般化されたランダム漸化式

$$x_n=x_{n-1}\pm\beta x_{n-2}$$

について,$0<\beta<0.70258\cdots$ならば確率1で指数関数的に減少し,$\beta>0.70258\cdots$ならば確率1で指数関数的に増大することを証明した.

1.2.5 フィボナッチ数の積

$n\to\infty$のとき$\prod_{k=1}^{n}f_k\sim c\varphi^{n(n+1)/2}\cdot 5^{-n/2}$という漸近的な結果が成立していることに注意しておこう[34,35].ここで

$$c=\prod_{n=1}^{\infty}\left(1-\frac{(-1)^n}{\varphi^{2n}}\right)=1.2267420107\cdots$$

である.これと関係する【5.14】にある式を参照せよ.

[1] H. E. Huntley, *The Divine Proportion: A Study in Mathematical Beauty*, Dover, 1970.
[2] G. Markowsky, Misconceptions about the Golden ratio, *College Math. J.* 23 (1992) 2-19.
[3] S. Vajda, *Fibonacci and Lucas numbers, and the Golden Section: Theory and Applications*, Halsted Press, 1989; MR 90h:11014.
[4] C. S. Ogilvy, *Excursions in Geometry*, Dover, 1969, pp. 122-134.
[5] E. Maor, *e: The Story of a Number*, Princeton Univ. Press, 1994, pp. 121-125, 134-139, 205-207; MR 95a:01002.
[6] J. D. Lawrence, *A Catalog of Special Plane Curves*, Dover, 1972, pp. 184-186.
[7] R. Honsberger, *More Mathematical Morsels*, Math. Assoc. Amer., 1991, pp. 140-144.
[8] J. Shallit, A simple proof that phi is irrational, *Fibonacci Quart.* 13 (1975) 32, 198.
[9] G. H. Hardy and E. M. Wright, *An Introduction to the Theory of Numbers*, 5th ed., Oxford Univ. Press, 1985, pp. 44-45, 290.295; MR 81i:10002.
[10] G. E. Andrews, R. Askey, and R. Roy, *Special Functions*, Cambridge Univ. Press, 1999, p. 58, ex. 52; MR 2000g:33001.
[11] R. Honsberger, *Mathematical Gems III*, Math. Assoc. Amer., 1985, pp. 102-138.
[12] K. G. Ramanathan, On Ramanujan's continued fraction, *Acta Arith.* 43 (1984) 209-226; MR 85d:11012.
[13] J. M. Borwein and P. B. Borwein, *Pi and the AGM: A Study in Analytic Number Theory and Computational Complexity*, Wiley, 1987, pp. 78-81; MR 99h:11147.
[14] B. C. Berndt, H. H. Chan, S.-S. Huang, S.-Y. Kang, J. Sohn, and S. H. Son, The Rogers-Ramanujan continued fraction, *J. Comput. Appl. Math.* 105 (1999) 9-24;MR2000b:11009.
[15] H. H. Chan and V. Tan, On the explicit evaluations of the Rogers-Ramanujan continued

fraction, *Continued Fractions: From Analytic Number Theory to Constructive Approximation*, Proc. 1998 Univ. of Missouri conf., ed. B. C. Berndt and F. Gesztesy, Amer. Math. Soc., 1999, pp. 127-136; MR 2000e:11161.
[16] S.-Y. Kang, Ramanujan's formulas for the explicit evaluation of the Rogers-Ramanujan continued fraction and theta-functions, *Acta Arith.* 90 (1999) 49-68; MR 2000f:11047.
[17] R. B. Paris, An asymptotic approximation connected with the Golden number, *Amer. Math. Monthly* 94 (1987) 272-278; MR 88d:39014.
[18] S. Plouffe, The Paris constant (Plouffe's Tables).
[19] J. M. Borwein and G. de Barra, Nested radicals, *Amer. Math. Monthly* 98 (1991) 735-739; MR 92i:11011.
[20] I. Stewart, Tales of a neglected number, *Sci. Amer.*, v. 274 (1996) n. 7, 102-103; v. 275 (1996) n. 11, 118.
[21] K. S. Brown, Perrin's sequence (MathPages).
[22] S. Plouffe, The Tribonacci constant (Plouffe's Tables).
[23] N. J. A. Sloane, On-Line Encyclopedia of Integer Sequences, A000073.
[24] E. R. Berlekamp, The design of slowly shrinking labelled squares, *Math. Comp.* 29 (1975) 25-27; MR 51 #10133.
[25] G. A. Moore, A Fibonacci polynomial sequence defined by multidimensional continued fractions and higher-order golden ratios, *Fibonacci Quart.* 31 (1993) 354-364; MR 94g:11014.
[26] G. A. Moore, The limit of the golden numbers is 3/2, *Fibonacci Quart.* 32 (1994) 211-217; MR 95f:11008.
[27] P. G. Anderson, Multidimensional Golden means, *Applications of Fibonacci Numbers*, v. 5, Proc. 1992 St. Andrews conf., ed. G. E. Bergum, A. N. Philippou, and A. F. Horadam, Kluwer, 1993, pp. 1-9; MR 95a:11004.
[28] E. I. Korkina, The simplest 2-dimensional continued fraction (in Russian), *Itogi Nauki Tekh. Ser. Sovrem. Mat. Prilozh. Temat. Obz.*, v. 20, *Topologiya*-3, 1994; Engl. transl. in *J. Math. Sci.* 82 (1996) 3680-3685; MR 97j:11032.
[29] E. I. Korkina, Two-dimensional continued fractions: The simplest examples, *Singularities of Smooth Mappings with Additional Structures* (in Russian), ed. V. I. Arnold, *Trudy Mat. Inst. Steklov.* 209 (1995) 143-166; Engl. transl. in *Proc. Steklov Inst. Math.* 209 (1995) 124-144; MR 97k:11104.
[30] D.Viswanath, Random Fibonacci sequences and the number 1.13198824..., *Math. Comp.* 69 (2000) 1131-1155; MR 2000j:15040.
[31] B. Hayes, The Vibonacci numbers, *Amer. Scientist*, v. 87 (1999) n. 4, 296-301.
[32] I. Peterson, Fibonacci at random: Uncovering a new mathematical constant, *Science News* 155 (1999) 376.
[33] M. Embree and L. N. Trefethen, Growth and decay of random Fibonacci sequences, *Proc. Royal Soc. London A* 455 (1999) 2471-2485; MR 2001i:11098.
[34] R. L. Graham, D. E. Knuth, and O. Patashnik, *Concrete Mathematics*, Addison-Wesley, 1989, pp. 478, 571; MR 97d:68003.
[35] S. Plouffe, The Fibonacci factorial constant (Plouffe's Tables).
[36] *The Fibonacci Quarterly* (virtually any issue).

1.3 自然対数の底, e

誰が最初に次の等式

$$\lim_{x\to 0}(1+x)^{1/x}=e=2.7182818284\cdots$$

を規定したのかについてはよく知られていない．我々は激しいせめぎ合いの結果をこの極限の中に見ることができる．一方ではベキ指数は無限大に発散する．他方，$1+x$ は急速に乗法単位元1に近づく．この極限と加法的に同値である次の極限式

$$\lim_{x\to 0}x\cdot\frac{1}{x}=1$$

が自明であることは，興味深い[†]．e の幾何学的な特徴付けは次のとおりである．e は方程式

$$\int_1^x \frac{1}{u}du=1$$

のただ1つの正の解である．この方程式は e が自然対数の底として使われる根拠となっている．言いかえれば，e は1より大きいただ1つの正数で，曲線 $v=1/u$ と $v=0$，$u=1$，$u=e$ で囲まれた平面領域が面積1をもつような値である．

e の定義より

$$\frac{d}{dx}(c\cdot e^x)=c\cdot e^x$$

であることが示され，さらに，1階の微分方程式

$$\frac{dy}{dx}=y(x)$$

の任意の解は，この形でなければならないことを示している．応用例としては，人口増加の問題や放射性物質の崩壊の問題がある．2階の微分方程式

$$\frac{d^2y}{dx^2}=y(x)$$

の解は $y(x)=ae^x+be^{-x}$ の形であることが必要である．特別な場合として，$y(x)=\coth(x)$（すなわち，$a=b=1/2$）は懸垂線と呼ばれ，一様に弾力性をもったケーブルが，自分自身の重みでつり下がったと考えたときの形である．さらに，x 軸のまわりに懸垂線の一部を回転すると，回転体の表面積は，懸垂線の両端の点を結ぶ任意の曲線の回転体の表面積より小さくなる[1,2]．

級数

$$e=\sum_{k=0}^{\infty}\frac{1}{k!}=1+\frac{1}{1}+\frac{1}{1\cdot 2}+\frac{1}{1\cdot 2\cdot 3}+\cdots$$

は急速に収束する．——列挙された項の通常の総和を求める計算は，実用的な目的に照らし合わせても，非常に迅速にできる．——してみると，n 項までの部分和を計算するための，さらに効率的な手法があるということを知れば，これはまた驚くべきことかもしれない[3,4]．2つの関数 $p(a,b)$，$q(a,b)$ を次のように帰納的に定義しよう．

[†]訳注：$\lim_{x\to 0}(0+x)\cdot 1/x$ の意味．

$$\binom{p(a,b)}{q(a,b)} = \begin{cases} \binom{1}{b} & , \quad b=a+1 \\ \binom{p(a,m)q(m,b)+p(m,b)}{q(a,m)q(m,b)} & , \quad \text{その他，ここで，} m=\left\lfloor \dfrac{a+b}{2} \right\rfloor \end{cases}$$

そうすると，$1+p(0,n)/q(0,n)$ が求めたい部分和である．これを示すことはそんなに難しいことではない．e の計算に対するこの**二項分割**（binary splitting）方式[†]の方法は，通常の計算法より少ない算術演算回数で済む（すなわち，ビット複雑度を減少させる）．このような演算速度を速める方法は[5-7]の論文から始まった．FFT[††]をベースとする整数乗算と結びつけるとき，このアルゴリズムは知られているいろいろな方法の何よりも漸近的に速いアルゴリズムとなる．

階乗級数は，次のマッチング問題の解を与える[8]．関数 $f:\{1,2,3,\cdots,n\} \to \{1,2,3,\cdots,n\}$ を1対1の関数（全単射）とする．f をランダムに選ぶとき，少なくとも1つの固定点をもつ，すなわち，少なくとも1つの整数 k で $f(k)=k$，$1 \leq k \leq n$ となる k をもつ確率を $P(n)$ としよう．このとき，

$$\lim_{n\to\infty} P(n) = \sum_{k=1}^{\infty} \frac{(-1)^{k+1}}{k!} = 1 - \frac{1}{e} = 0.6321205588\cdots$$

が成立する．図1.3を参照せよ．一般化が【5.4】で見られる．

また，$X_1, X_2, X_3, \cdots$ をそれぞれ，区間 $[0,1]$ で一様分布に従う独立な確率変数としよう．整数 N を

$$N = \min\left\{ n : \sum_{k=1}^{n} X_k > 1 \right\}$$

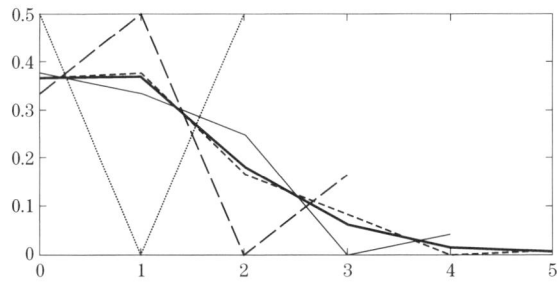

図1.3 n 個の文字の確率的な置換 f の固定点の数の分布．$n \to \infty$ とすると，固定点の数は平均値が1のポアソン分布に近づいていく．

[†]訳注：上記 $p(a,b)$，$q(a,b)$ を用いた計算のように，無限級数を高速に計算するための方法．級数の隣り合う項を計算して1つの項にする．これをくり返して，最終的に1つの項にして計算する方法．

[††]訳注：高速フーリエ変換．

で定義すれば，期待値は $E(N)=e$ となる．推測統計学の用語でいえば，一様な到着間隔時間 X_k をもつ更新過程においては，自然対数の底をその更新回数の期待値としてもつ[9]．

長さ r の棒を m 個に等分してみよう[10]．等分したものの長さの積が最大となる m は $[r/e]$ か $[r/e]+1$ である．秘書問題として知られている関連分野については【5.15】を参照せよ[4,11]．

Wallis（ウォリス）の公式に類似したいくつかの無限積[4,11]

$$e = \frac{2}{1} \cdot \left(\frac{4}{3}\right)^{1/2} \cdot \left(\frac{6 \cdot 8}{5 \cdot 7}\right)^{1/4} \cdot \left(\frac{10 \cdot 12 \cdot 14 \cdot 16}{9 \cdot 11 \cdot 13 \cdot 15}\right)^{1/8} \cdots,$$

$$\frac{e}{2} = \left(\frac{2}{1}\right)^{1/2} \cdot \left(\frac{2 \cdot 4}{3 \cdot 3}\right)^{1/4} \cdot \left(\frac{4 \cdot 6 \cdot 6 \cdot 8}{5 \cdot 5 \cdot 7 \cdot 7}\right)^{1/8} \cdots$$

や，【1.3.2】で取り上げる連分数展開，また，素数の理論との魅惑的な次の関係などがある[12]．もし，

$$n? = \prod_{\substack{p \leq n \\ p \text{ は素数}}} p \text{ と定義すれば，} \lim_{n \to \infty}(n?)^{1/n} = e$$

である．これは素数定理から導かれる．同様に魅惑的な結果が，スターリングの公式によって得られる

$$\lim_{n \to \infty} \frac{(n!)^{1/n}}{n} = \frac{1}{e}$$

である．このように，$n!$ の増加は，大きさとして $n?$ の増加を上回っている．また，

$$\lim_{n \to \infty}((n!)^{1/n} - ((n-1)!)^{1/(n-1)}) = \frac{1}{e}, \quad \lim_{n \to \infty} \prod_{k=1}^{n}(n^2+k)(n^2-k)^{-1} = e$$

という関係式も得られる[13-15]．

e が無理数であることは，Euler（オイラー）によって証明され，超越数であることは Hermite（エルミート）によって証明された．すなわち，自然対数の底 e は，整数係数の多項式を 0 にしない[4,16-18]．

スピゴットアルゴリズム（spigot algorithm）†という名で知られる，ちょっと変わった e に対する計算法が，[19]において最初に発表された．これが興味あるのは，アルゴリズムの速さにあるのではなくて（速さは遅い），別の特徴にある．たとえば，これは完全に整数演算のみで構成されている．

e をオイラー定数とよぶ人もいるが，同じ用語がしばしば Euler-Mascheroni（オイラー-マスケロニ）の定数 γ に対して使われるので，混乱は避けられないかもしれない．Napier（ネイピア）は 1614 年に e の発見直前までいった．それで e をネイピアの定数と呼ぶ人もある[1]．

1.3.1 極限値の考察

マクローリン級数

---

†訳注：前に計算した数を使ったりすることなしに一度に多量の数を生成するアルゴリズム．

$$\frac{1}{e}(1+x)^{1/x} = 1 - \frac{1}{2}x + \frac{11}{24}x^2 - \frac{7}{16}x^3 + \frac{2447}{5760}x^4 - \frac{959}{2304}x^5 + O(x^6)$$

は，e に対する極限での定義において何が起こっているかをよりくわしく明らかにしてくれる．たとえば，

$$\lim_{x \to 0} \frac{(1+x)^{1/x} - e}{x} = -\frac{1}{2}e, \quad \lim_{x \to 0} \frac{\frac{(1+x)^{1/x} - e}{x} + \frac{1}{2}e}{x} = \frac{11}{24}e$$

である．次の公式[20-24]

$$\lim_{x \to 0}\left(\frac{2+x}{2-x}\right)^{1/x} = e, \quad \lim_{n \to \infty}\left(\frac{(n+1)^{n+1}}{n^n} - \frac{n^n}{(n-1)^{n-1}}\right) = e$$

によって，より速い収束が達成できる．これは，それぞれの漸近展開の最初の項が $1 + x^2/12$ と $1 + 1/(24n^2)$ であるからである．さらなる改良も可能である．

1.3.2 連 分 数

e に対する正則連分数

$$e = 2 + \frac{1|}{|1} + \frac{1|}{|2} + \frac{1|}{|1} + \frac{1|}{|1} + \frac{1|}{|4} + \frac{1|}{|1} + \frac{1|}{|1} + \frac{1|}{|6} + \frac{1|}{|1} + \cdots$$

は，適当な変換をすれば，一連の連分数

$$\coth\left(\frac{1}{m}\right) = \frac{e^{2/m}+1}{e^{2/m}-1} = m + \frac{1|}{|3m} + \frac{1|}{|5m} + \frac{1|}{|7m} + \frac{1|}{|9m} + \cdots$$

に含まれるものの1つである[25-28]．ここで，m は任意の正の整数である．Davison（ダヴィソン）[29]は，たとえば，$\coth(3/2)$ と $\coth(2)$ の商を計算するアルゴリズムを得たが，何らかのパターンを見つけることはできなかった．その他の連分数として

$$e - 1 = 1 + \frac{2|}{|2} + \frac{3|}{|3} + \frac{4|}{|4} + \frac{5|}{|5} + \cdots, \quad \frac{1}{e-2} = 1 + \frac{1|}{|2} + \frac{2|}{|3} + \frac{3|}{|4} + \frac{4|}{|5} + \cdots$$

があり[1,26,30,31]，さらにいくつかを，[32,33]で見ることができる．

1.3.3 数 2 の自然対数

最後に，密接に関連する定数 $\ln(2)$

$$\ln(2) = \int_0^1 \frac{1}{1+t} dt = \lim_{n \to \infty} \sum_{k=1}^n \frac{1}{n+k} = 0.6931471805\cdots$$

に関して言及しておこう．これは e とよく似た極限公式

$$\lim_{x \to 0} \frac{2^x - 1}{x} = \ln(2) = \lim_{x \to 0} \frac{2^x - 2^{-x}}{2x}$$

をもつ．$\ln(1+x)$ のマクローリン級数の $x=1$ と $x=-1/2$ での結果より，よく知られた和

$$\ln(2) = \sum_{k=1}^\infty \frac{(-1)^{k-1}}{k} = \sum_{k=1}^\infty \frac{1}{k 2^k}$$

が導かれる．二進抽出アルゴリズム（binary digit extraction algorithm）は，級数

$$\ln(2) = \sum_{k=1}^{\infty} \left(\frac{1}{8k+8} + \frac{1}{4k+2} \right) \frac{1}{4^k}$$

にその基礎をおいている.この級数によって,それ以前の $(d-1)$ ビットをまったく計算することなしに,ln(2) の d 番目のビットを計算することが可能になる.【2.1】,【6.2】,【7.2】なども参照せよ.

[1] E. Maor, *e: The Story of a Number*, Princeton Univ. Press, 1994; MR 95a:01002.
[2] G. F. Simmons, *Differential Equations with Applications and Historical Notes*, McGraw-Hill, 1972, pp. 14-19, 52-54, 361-363; MR 58 #17258.
[3] X. Gourdon and P. Sebah, Binary splitting method (Numbers, Constants and Computation).
[4] J. M. Borwein and P. B. Borwein, *Pi and the AGM: A Study in Analytic Number Theory and Computational Complexity*, Wiley, 1987, pp. 329-300, 343, 347-362; MR 99h: 11147.
[5] R. P. Brent, The complexity of multiple-precision arithmetic, *Complexity of Computational Problem Solving*, Proc. 1974 Austral. Nat. Univ. conf., ed. R. S. Anderssen and R. P. Brent, Univ. of Queensland Press, 1976, pp. 126-165.
[6] R. P. Brent, Fast multiple-precision evaluation of elementary functions, *J. ACM* 23 (1976) 242-251; MR 52 #16111.
[7] E. A. Karatsuba, Fast evaluation of transcendental functions (in Russian), *Problemy Peredachi Informatsii* 27 (1991) 76-99; Engl. transl. in *Problems Information Transmission* 27 (1991) 339-360; MR 93c:65027.
[8] M. H. DeGroot, *Probability and Statistics*, Addison-Wesley, 1975, pp. 36-37.
[9] W. Feller, On the integral equation of renewal theory, *Annals of Math. Statist.* 12 (1941) 243-267; MR 3,151c.
[10] N. Shklov and C. E. Miller, Maximizing a certain product, *Amer. Math. Monthly* 61 (1954) 196-197.
[11] I. S. Gradshteyn and I. M. Ryzhik, *Table of Integrals, Series and Products*, Academic Press, 1980, p. 12; MR 97c:00014.
[12] T. Nagell, *Introduction to Number Theory*, Chelsea, 1981, pp. 38, 60-64; MR 30 #4714.
[13] J. Sandor, On the gamma function. II, *Publ. Centre Rech. Math. Pures* 28 (1997) 10-12.
[14] G. Pólya and G. Szegö, *Problems and Theorems in Analysis*, v. 1, Springer-Verlag, 1972, ex. 55; MR 81e:00002.
[15] Z. Sasvári and W. F. Trench, A Pólya-Szegö exercise revisited, *Amer. Math. Monthly* 106 (1999) 781-782.
[16] J. A. Nathan, The irrationality of exp(x) for nonzero rational x, *Amer. Math. Monthly* 105 (1998) 762-763; MR 99e:11096.
[17] G. H. Hardy and E. M. Wright, *An Introduction to the Theory of Numbers*, 5th ed., Oxford Univ. Press, 1985, pp. 46-47; MR 81i:10002.
[18] I. N. Herstein, *Topics in Algebra*, 2nd ed., Wiley, 1975, pp. 216-219; MR 50 #9456.
[19] A. H. J. Sale, The calculation of e to many significant digits, *Computer J.* 11 (1968) 229-230.
[20] L. F. Richardson and J. A. Gaunt, The deferred approach to the limit, *Philos. Trans. Royal Soc. London Ser. A* 226 (1927) 299-361.
[21] H. J. Brothers and J. A. Knox, New closed-for mapproximations to the logarithmic constant e, *Math. Intellig.*, v. 20 (1998) n. 4, 25-29; MR 2000c:11209.
[22] J. A. Knox and H. J. Brothers, Novel series-based approximations to e, *College Math. J.* 30 (1999) 269-275; MR 2000i:11198.
[23] J. Sandor, On certain limits related to the number e, *Libertas Math.* 20 (2000) 155-159; MR 2001k:26034.
[24] X. Gourdon and P. Sebah, The constant e (Numbers, Constants and Computation).
[25] L. Euler, De fractionibus continuis dissertatio, 1744, *Opera Omnia Ser. I*, v. 14, Lipsiae, 1911,

pp. 187-215; Engl. transl. in *Math. Systems Theory* 18 (1985) 295-328; MR 87d:01011b.
[26] L. Lorentzen and H. Waadeland, *Continued Fractions with Applications*, North-Holland, 1992, pp. 561-562; MR 93g:30007.
[27] P. Ribenboim, *My Numbers, My Friends*, Springer-Verlag, 2000, pp. 292-294; MR 2002d:11001.
[28] C. D. Olds, The simple continued fraction expansion of *e*, *Amer. Math. Monthly* 77 (1970) 968-974.
[29] J. L. Davison, An algorithm for the continued fraction of $e^{l/m}$, *Proc. 8th Manitoba Conf. on Numerical Mathematics and Computing*, Winnipeg, 1978, ed. D. McCarthy and H. C. Williams, Congr. Numer. 22, Utilitas Math., 1979, pp. 169-179; MR 80j:10012.
[30] L. Euler, *Introduction to Analysis of the Infinite. Book I*, 1748, transl. J. D. Blanton, Springer-Verlag, 1988, pp. 303-314; MR 89g:01067.
[31] H. Darmon and J. McKay, A continued fraction and permutations with fixed points, *Amer. Math. Monthly* 98 (1991) 25-27; MR 92a:05003.
[32] J. Minkus and J. Anglesio, A continued fraction, *Amer. Math. Monthly* 103 (1996) 605-606.
[33] M. J. Knight and W. O. Egerland, $F_n = (n+2)F_{n-1} - (n-1)F_{n-2}$, *Amer. Math. Monthly* 81 (1974) 675-676.
[34] X. Gourdon and P. Sebah, The logarithm constant log(2) (Numbers, Constants and Computation).

1.4 アルキメデスの定数，π

最も有名な超越的定数 π（円周率）について短く解説するということになれば，それは必然的に不充分なものとなろう[1-5]．数学の中で π が数えきれないほど出現することには，呆然とするばかりである．

半径 1 の円によって囲まれた面積は

$$A = \pi = 4\int_0^1 \sqrt{1-x^2}dx = \lim_{n\to\infty} \frac{4}{n^2}\sum_{k=0}^n \sqrt{n^2-k^2} = 3.1415926535\cdots$$

であり，円周は

$$C = 2\pi = 4\int_0^1 \frac{1}{\sqrt{1-x^2}}dx = 4\int_0^1 \sqrt{1+\left(\frac{d}{dx}\sqrt{1-x^2}\right)^2}dx$$

である．A の公式は，リーマン積分，すなわち，リーマン和の極限による面積の定義を基にしている．C の公式は，所与の連続で微分可能な曲線に対する弧長の定義を用いている．どうして，この 2 つの公式に，同じ神秘的な π が現れるのであろうか？簡単な部分積分を行えば三角法なしでその答を導くことができる[6]．

紀元前 3 世紀に，アルキメデスは円に内接する正 96 角形と外接する正 96 角形を考え，$3\frac{10}{71} < \pi < 3\frac{1}{7}$ であることを導いた．漸化式

$$a_0 = 2\sqrt{3}, \qquad b_0 = 3,$$
$$a_{n+1} = \frac{2a_n b_n}{a_n + b_n}, \qquad b_{n+1} = \sqrt{a_{n+1}b_n} \qquad n \geq 0$$

は，しばしば，Borchardo-Pfaff（ボルヒャルト-パフ）のアルゴリズムと呼ばれ，本質的にはアルキメデスの評価を4回目の反復で与える[7-11]．これはまさに線形的に収束するアルゴリズムである（線形的とは，反復回数が求めたい値の正しい桁数に概略比例するという意味）．これは，【6.1】でガウスのレムニスケート（連珠形）定数に関して取り上げる算術幾何平均（arithmetic-geometric-mean, AGM）と似ている．

πの効用は何も平面幾何に限ったことではない．n次元のユークリッド空間における半径1の球の体積は

$$V = \begin{cases} \dfrac{\pi^k}{k!} & (n=2k \text{ のとき}) \\ 2^{2k+1}\dfrac{k!}{(2k+1)!}\pi^k & (n=2k+1 \text{ のとき}) \end{cases}$$

であり，表面積は

$$S = \begin{cases} \dfrac{2\pi^k}{(k-1)!} & (n=2k \text{ のとき}) \\ 2^{2k+1}\dfrac{k!}{(2k)!}\pi^k & (n=2k+1 \text{ のとき}) \end{cases}$$

である．これらの式は【1.5.4】で取り上げるガンマ関数によって表現されることが多い．平面の場合（円）は$n=2$に対応する．

幾何学とπとのもう1つの関係がBuffon（ビュッフォン）の針の問題に見られる[1,12-15]．長さ1の1本の針が，距離1の平行線が画かれた平面上にランダムに投げられたとする．このとき，この針が平行線の1つと交わる確率はいくらか？答えは$2/\pi = 0.6366197723\cdots$である．

ここにπのまったく異なった1つの確率論的な解釈がある．2つの整数をランダムに選んでくるとする．それらが互いに素，すなわち，1より大なる共通因数をもたない確率はいくらか？答えは$6/\pi^2 = 0.6079271018\cdots$（大きな区間上で考え，その極限において）である．次のようにも表現できよう．$0 < a, b \leq N$を満たす整数a, bで，有理数a/bが相異なるものの個数を$R(N)$とする．$0 < a, b \leq N$を満たす整数a, bに対して，対(a, b)の総数はN^2個であるが，多くの分数が最小項の形ではないから，$R(N)$はこれよりは相当小さくなる．前の表現の形で明確にいうと$R(N) \sim 6N^2/\pi^2$である．

数学における最も有名な極限にStirling（スターリング）の公式

$$\lim_{n\to\infty} \frac{n!}{e^{-n}n^{n+1/2}} = \sqrt{2\pi} = 2.5066282746\cdots$$

がある[18]．アルキメデスの定数もまた，この他，いろいろな表現をもっている．その中のいくつかは後ほど紹介する．πが無理数であることは，Lambert（ランベルト）によって証明され，超越数であることは，Lindemann（リンデマン）によって証明された[2,16,19]．πの小数桁を計算するための最初の真に魅力的な公式は，Machin（マチン）によって発見された次の公式である[1,13]．

$$\frac{\pi}{4}=4\arctan\left(\frac{1}{5}\right)-\arctan\left(\frac{1}{239}\right)$$
$$=4\sum_{k=0}^{\infty}\frac{(-1)^k}{(2k+1)\cdot 5^{2k+1}}-\sum_{k=0}^{\infty}\frac{(-1)^k}{(2k+1)\cdot 239^{2k+1}}$$

この公式の長所は，2項目が急速に収束するということと，最初の項が十進法の計算に適しているということである．1706 年，Machin は π の 100 桁を正確に計算した最初の人となった．

計算に関する長い歴史をさておいて，Salamin & Brent（サラミン&ブレント）によるもう 1 つの重要なアルゴリズムについて取り上げてみよう．$n \geq 0$ として

$$a_0=1, \quad b_0=1/\sqrt{2}, \quad c_0=1/2, \quad s_0=1/2,$$
$$a_{n+1}=\frac{a_n+b_n}{2}, \quad b_{n+1}=\sqrt{a_n b_n}, \quad c_{n+1}=\left(\frac{c_n}{4a_{n+1}}\right)^2, \quad s_{n+1}=s_n-2^{n+1}c_{n+1}$$

によって漸化式を定義しよう．このとき，$2a_n^2/s_n$ は π に 2 次的に収束する（各段階での反復計算が漸近的に毎回前の 2 倍の正確な桁数を与えるという意味）．さらに速い 3 次，4 次のアルゴリズムが，Borwein（ボーウィン）& Borwein によって得られている[2,22,24,25]．これらは，モジュラ方程式についての Ramanujan の仕事を引き出してきて発展させたものである．これらはそれぞれ，計算に関してはアルキメデスの方法からはるかに進んでいる．このような方法を用いて，Kanada（金田康正）は π の 1 兆桁近くまでを計算した[†]．

e に対してと同様，π を計算するための spigot アルゴリズムがある[26]．しかしながら，さらに重要なのは，Bailey（ベイリー），Borwein & Plouffe（ボーウィン&プルーフェ）によって発見された桁ごとの抽出アルゴリズム（digit-extraction algorithm）である[27-29]．それは，次の公式（はじめは $r=0$ と考えて）

$$\pi=\sum_{k=0}^{\infty}\frac{1}{16^k}\times\left(\frac{4+8r}{8k+1}-\frac{8r}{8k+2}-\frac{4r}{8k+3}-\frac{2+8r}{8k+4}-\frac{1+2r}{8k+5}-\frac{1+2r}{8k+6}+\frac{r}{8k+7}\right)$$

に基礎をおいており，事実上計算機上の記憶を必要としない．（複素数 $r \neq 0$ への拡張は Adamchik & Wagon（アダムチック&ワゴン）によって行われている[30,31]．）この進展の結果，今では π の二進展開において 1000 兆目の桁を知ることができ Bellard（ベラルド）や Percival（パーシバル）に大いに感謝するところである．同じように 3 を基底にする公式が Broadhurst（ブロードハースト）によって見つけられた[32]．

数学者 Ludolph van Ceulen（ルドルフ・ファン・ケーレン）が，彼の人生のほとんどを費して π の小数 35 桁までを計算したので，π を**ルドルフの定数**と呼ぶ人もある．

この小節における公式は，自然対数の底 e の性質と質的に相異なる性質をもっている．Wimp（ウィンプ）はこのことを詳しく述べている[33]．彼が e-数学（e-mathematics）とよぶものは，線形で，明白で，容易に抽象化され得るものである．これに対して，π-数学（π-mathematics）は非線形で，神秘的で通常一般化は困難である．しかしながら，Cloitre（クロアトル）は e と π の間の，ある調和を暗示する公式を与

[†] 訳注：その後約 1 兆 2 千億桁まで計算され，さらに近年 4 兆桁まで計算された．

えた[34]. $u_1=v_1=0$, $u_2=v_2=1$ とし

$$u_{n+2}=u_{n+1}+\frac{u_n}{n}, \quad v_{n+2}=\frac{v_{n+1}}{n}+v_n, \quad n\geq 0$$

とする．このとき $\lim_{n\to\infty} n/u_n = e$ であり，一方，$\lim_{n\to\infty} 2n/v_n^2 = \pi$ である．

1.4.1 無限級数

500年以上前に，インドの数学者 Madhava（マダーバ）が次の公式

$$\frac{\pi}{4}=\sum_{n=0}^{\infty}\frac{(-1)^n}{2n+1}=1-\frac{1}{3}+\frac{1}{5}-\frac{1}{7}+\frac{1}{9}-\frac{1}{11}+-\cdots$$

を発見した[35-38]．これは Gregory（グレゴリ）[39]と Leibniz（ライプニッツ）[40]によっても独立に見つけられた．この無限級数は条件収束する．ゆえに，項を並べ換えることで任意の和をもつ級数を作ることができるし，$+\infty$ や $-\infty$ に発散させることもできる．同じことは交代調和級数に対しても正しい【1.3.3】．たとえば，

$$\frac{1}{4}\ln(2)+\frac{\pi}{4}=1+\frac{1}{5}-\frac{1}{3}+\frac{1}{9}+\frac{1}{13}-\frac{1}{7}+\frac{1}{17}+\frac{1}{21}-\frac{1}{11}++-\cdots$$

である（1つの負の項に対して2つの正の項）．一般化は可能である．

Gregory-Leibniz の級数において，プラス，マイナスの符号のパターンを変えて，たとえば，

$$\frac{\pi}{4}\sqrt{2}=1+\frac{1}{3}-\frac{1}{5}-\frac{1}{7}+\frac{1}{9}+\frac{1}{11}-\frac{1}{13}-\frac{1}{15}++--\cdots$$

が得られる[41,42]．また，部分級数を取り出して，

$$\frac{\pi}{8}(1+\sqrt{2})=1-\frac{1}{7}+\frac{1}{9}-\frac{1}{15}+\frac{1}{17}-\frac{1}{23}+\frac{1}{25}-\frac{1}{31}+-\cdots$$

が得られる[43]．Euler（オイラー）の有名な級数

$$\sum_{n=1}^{\infty}\frac{1}{n^2}=\frac{\pi^2}{6}, \quad \sum_{n=1}^{\infty}\frac{(-1)^{n+1}}{(2n-1)^3}=\frac{\pi^3}{32}$$

については，ここでは取り上げず，【1.6】と【1.7】に回すこととする．Euler が導いた多くの級数の1つに，

$$\frac{\pi}{2}=\sum_{n=0}^{\infty}\frac{2^n}{(2n+1)\binom{2n}{n}}=1+\frac{1}{3}+\frac{1\cdot 2}{3\cdot 5}+\frac{1\cdot 2\cdot 3}{3\cdot 5\cdot 7}+\cdots$$

がある[1,44]．次の公式

$$\sum_{n=1}^{\infty}\frac{1}{n^2\binom{2n}{n}}=\frac{\pi^2}{18}, \quad \sum_{n=1}^{\infty}\frac{(-1)^{n+1}}{n^2\binom{2n}{n}}=2\ln(\varphi)^2$$

に注意しておこう[2,45]．このように，π と黄金比 φ が複雑にからまっているものが他にあるだろうか．

Ramanujan[23,24,46]と Chudnovsky（チュドノフスキ）& Chudnovsky[47-50]は，π を計算するための数々の級数を発見した．それらは最速として知られるいくつかのアルゴリズムの基礎を構成している．

1.4.2 無 限 積

Viète（ヴィエト）は π に対する初めての解析的な表現

$$\frac{2}{\pi}=\frac{\sqrt{2}}{2}\cdot\frac{\sqrt{2+\sqrt{2}}}{2}\cdot\frac{\sqrt{2+\sqrt{2+\sqrt{2}}}}{2}\cdot\frac{\sqrt{2+\sqrt{2+\sqrt{2+\sqrt{2}}}}}{2}\cdots$$

を得た[51]．彼は Archimedes の多角形の面積の極限を考えることによって，この表現を導いたのである．また，Wallis は次の公式

$$\frac{\pi}{2}=\frac{2}{1}\cdot\frac{2}{3}\cdot\frac{4}{3}\cdot\frac{4}{5}\cdot\frac{6}{5}\cdot\frac{6}{7}\cdot\frac{8}{7}\cdot\frac{8}{9}\cdots=\lim_{n\to\infty}\frac{2^{4n}}{(2n+1)\binom{2n}{n}^2}$$

を導いた[52]．これらの積公式は，実際同じ所からの帰結である．これらが正しいことは，いろいろ異なった方法で証明できる [54]．論証の一部は，sin と cos の定義とみなす人もいる次式とからんでいる．次の無限積分解

$$\sin(x)=x\prod_{n=1}^{\infty}\left(1-\frac{x^2}{n^2\pi^2}\right),\quad \cos(x)=\prod_{n=1}^{\infty}\left(1-\frac{4x^2}{(2n-1)^2\pi^2}\right)$$

が成立する[55]．正弦と余弦関数は，三角法の基礎と数学における周期現象の研究に貢献している．応用として，機械や電気系における振幅が減少しない単振動や，太陽を巡る惑星の軌道の運動や，その他，いろいろなものがある[56]．よく知られているように，

$$\frac{d^2}{dx^2}(a\cdot\sin(x)+b\cdot\cos(x))+(a\cdot\sin(x)+b\cdot\cos(x))=0$$

である．さらに，次の2階の微分方程式

$$\frac{d^2y}{dx^2}+y(x)=0$$

の任意の解はこの sin, cos の形でなければならない．定数 π は，自然対数の底 e が指数関数を定めるときに担う役割と同じ役割を，sin と cos を定めるときに果たしている．この2つの過程が相互に関係付けられて，オイラーの公式 $e^{i\pi}+1=0$ として具現されるのである．ここで，i は虚数単位である．

π と素数に関する有名な積表示は，ゼータ関数の理論の帰結として，【1.6】と【1.7】で取りあつかう．このような積表示の1つとして，Euler による

$$\frac{\pi}{2}=\prod_{p\,\text{奇}}\frac{p}{p+(-1)^{(p-1)/2}}=\frac{3}{2}\cdot\frac{5}{6}\cdot\frac{7}{6}\cdot\frac{11}{10}\cdot\frac{13}{14}\cdot\frac{17}{18}\cdot\frac{19}{18}\cdots$$

がある[57]．ここで，分子は奇素数であり，分母はその素数に最も近い $4n+2$ 型の整数である．【2.1】も参照せよ．他に，数論における π の出現には，次の漸近表現

$$p(n)\sim\frac{1}{4\sqrt{3}n}\exp\left(\pi\sqrt{\frac{2n}{3}}\right)$$

がある[58]．これは Hardy & Ramanujan（ハーディ&ラマヌジャン）によるものであり，$p(n)$ は正の整数 n の制限を付けない分割の仕方の総数を表す（順序は問題にしない）．Hardy & Ramanujan[58]，そして，Rademacher（ラーデマッヘル）[59]が

$p(n)$ に対する精密な解析的公式を証明したが[60,61]，ここで取り上げることは，あまりにも本題を外れてしまうことになる．

1.4.3 定積分

π を表す有名な積分には

$$\int_0^\infty e^{-x^2}dx = \frac{\sqrt{\pi}}{2} \qquad \text{(Gaussの確率密度積分)}$$

$$\int_0^\infty \frac{1}{1+x^2}dx = \frac{\pi}{2} \qquad \text{(arctanの極限値)}$$

$$\int_0^\infty \sin(x^2)\,dx = \int_0^\infty \cos(x^2)\,dx = \frac{\pi\sqrt{2}}{4} \qquad \text{(フレネル積分)}$$

$$\int_0^{\pi/2} \ln(\sin(x))\,dx = \int_0^{\pi/2} \ln(\cos(x))\,dx = -\frac{\pi}{2}\ln(2)$$

$$\int_0^1 \sqrt{\ln\left(\frac{1}{x}\right)}\,dx = \frac{\sqrt{\pi}}{2}$$

などがある[62,63]．

次の積分

$$\int_0^\infty \frac{\cos(x)}{1+x^2}dx = \frac{\pi}{2e}, \quad \int_0^\infty \frac{x\sin(x)}{1+x^2}dx = \frac{\pi}{2e}$$

は単純な表現をしているが，$\cos(x)$ と $\sin(x)$ を入れ換えると複雑な結果となるのは好奇心をそそられる．詳細は【6.2】を参照せよ．

さらに，次の数列

$$s_n = \int_0^\infty \left(\frac{\sin(x)}{x}\right)^n dx, \quad n=1,2,3,\cdots$$

を考えてみよう．最初のいくつかの値は，$s_1 = s_2 = \pi/2$，$s_3 = 3\pi/8$，$s_4 = \pi/3$，$s_5 = 115\pi/384$ そして $s_6 = 11\pi/40$ である．すべての n に対する精密な公式が[64]に見られる．

1.4.4 連分数

Wallisの公式から出発して，Brounker（ブラウンカー）[1,2,52]は次の連分数展開

$$1 + \cfrac{4}{\pi} = 2 + \cfrac{1^2}{2} + \cfrac{3^2}{2} + \cfrac{5^2}{2} + \cfrac{7^2}{2} + \cfrac{9^2}{2} + \cdots$$

を発見した．これはその後，Eulerによって証明された[41]．この公式と，これと関連する展開式，たとえば，

$$\frac{4}{\pi} = 1 + \cfrac{1^2}{3} + \cfrac{2^2}{5} + \cfrac{3^2}{7} + \cfrac{4^2}{9} + \cfrac{5^2}{11} + \cdots$$

$$\frac{6}{\pi^2 - 6} = 1 + \cfrac{1^2}{1} + \cfrac{1\cdot 2}{1} + \cfrac{2^2}{1} + \cfrac{2\cdot 3}{1} + \cfrac{3^2}{1} + \cfrac{3\cdot 4}{1} + \cfrac{4^2}{1} + \cdots$$

$$\frac{2}{\pi - 2} = 1 + \cfrac{1\cdot 2}{1} + \cfrac{2\cdot 3}{1} + \cfrac{3\cdot 4}{1} + \cfrac{3\cdot 4}{1} + \cfrac{4\cdot 5}{1} + \cfrac{5\cdot 6}{1} + \cfrac{6\cdot 7}{1} + \cdots$$

$$\frac{12}{\pi^2} = 1 + \frac{1^4}{|3} + \frac{2^4}{|5} + \frac{3^4}{|7} + \frac{4^4}{|9} + \frac{5^4}{|11} + \cdots$$

$$\pi + 3 = 6 + \frac{1^2}{|6} + \frac{3^2}{|6} + \frac{5^2}{|6} + \frac{7^2}{|6} + \frac{9^2}{|6} + \cdots$$

などと比較すると，興味津々である[65-67]．

1.4.5 無限根号

単位円に内接する正 2^{n+1} 角形の 1 辺の長さを S_n とする．明らかに $S_1 = \sqrt{2}$ であり，一般に $S_n = 2\sin(\pi/2^{n+1})$ である．ゆえに半角の公式より

$$S_n = \sqrt{2 - \sqrt{4 - S_{n-1}^2}}$$

である．（この漸化式に対する純幾何学的な議論が，[68,69] で行われている．）正 2^{n+1} 角形の周の長さは $2^{n+1} S_n$ であり，$n \to \infty$ のとき 2π に収束する．よって，

$$\pi = \lim_{n \to \infty} 2^n S_n = \lim_{n \to \infty} 2^n \sqrt{2 - \sqrt{2 + \sqrt{2 + \sqrt{2 + \cdots \sqrt{2}}}}}$$

が成立する．ここで，右辺（の n 番目の項）は n 個の平方根をもつ．

この π に対する根号表現は，面白いものではあるが，数値計算のためには少数回の反復に対してのみ意味のある式である．これは，2 つのほぼ等しい量を引き算するときに起こる浮動小数点精度の損失（桁落ち）についての従来からある例である．π を近似するための方法はたくさんある．しかし，この方法はその中の 1 つとは言い難い．

1.4.6 楕円関数

長軸の半分の長さが 1 で，短軸の半分の長さが r $(0 < r \leq 1)$ である楕円を考えてみよう．この楕円で囲まれた面積は πr であるが，他方，その周は $4E(\sqrt{1-r^2})$ である．ここで

$$K(x) = \int_0^{\pi/2} \frac{1}{\sqrt{1 - x^2 \sin(\theta)^2}} d\theta = \int_0^1 \frac{1}{\sqrt{(1-t^2)(1-x^2 t^2)}} dt,$$

$$E(x) = \int_0^{\pi/2} \sqrt{1 - x^2 \sin(\theta)^2} d\theta = \int_0^1 \sqrt{\frac{1 - x^2 t^2}{1 - t^2}} dt$$

は，それぞれ第 1 種，第 2 種の**完全楕円積分**である（$K(x)$ と最初に出合うのは，しばしば，物理振子（大振幅）の周期を計算するときである[56]）．正弦関数の類似として，Jacobi（**ヤコビ**）の楕円関数 $\operatorname{sn}(x,y)$ がある．これは

$$x = \int_0^{\operatorname{sn}(x,y)} \frac{1}{\sqrt{(1-t^2)(1-y^2 t^2)}} dt \quad (0 \leq y \leq 1)$$

によって定義される．図 1.4 を参照せよ．

明らかに，$-\pi/2 \leq x \leq \pi/2$ のとき $\operatorname{sn}(x,0) = \sin(x)$ であり，$\operatorname{sn}(x,1) = \tanh(x)$ である．sn やその対になる関数 cn と dn を含む拡張された三角法のもろもろの性質が証明されている．固定された $0 < y < 1$ に対して，関数 $\operatorname{sn}(x,y)$ は全複素数平面に解析接続され，2 重周期をもつ有理型関数になる．すべての複素数 z に対して $\sin(z) =$

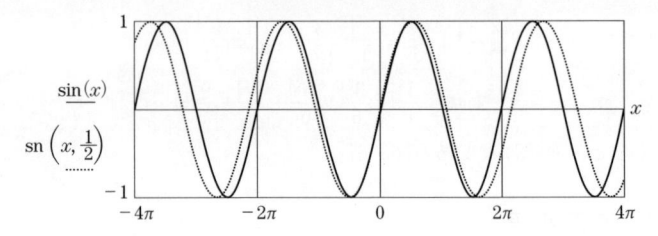

図1.4 円関数 $\sin(x)$ は周期 $2\pi \approx 6.28$ をもつが、楕円関数 $\text{sn}(x,1/2)$ は（実の）周期 $4K(1/2) \approx 6.74$ をもつ。

$\sin(z+2\pi)$ であるように、$\text{sn}(z) = \text{sn}(z+4K(y)+2iK(\sqrt{1-y^2}))$ である。ゆえに、定数 $K(y)$ と $K(\sqrt{1-y^2})$ は、円関数での π の働きと類似の役割を楕円関数に対して担っているといえる[2,70]。

1.4.7 意外な所での π

興味をそそる数論的関数 $f(n)$ が[71-77]で取り上げられている。正の整数 n に対して、$n-1$ の倍数でこの n に一番近い整数に増やす（$n-1$ の倍数で n より大きい最小のものを採る）。さらに、$n-2$ の倍数で、こうして得られた整数に一番近い整数に増やす。さらに一般に、k 回目で得られた整数を、$n-k-1$ の倍数で、この得られた整数に一番近い整数に増やす。こうして、$k=n-1$ で止めて、最後の値を $f(n)$ とする。たとえば、

$$10 \to 18 \to 24 \to 28 \to 30 \to 30 \to 32 \to 33 \to 34 \to 34$$

であるから、$f(10)=34$ である。比 $n^2/f(n)$ は、n が限りなく増大するとき π に近づく。同じ方向で、Matiyasevich & Guy（マチヤシェビッチ&ガイ）は

$$\pi = \lim_{m\to\infty} \sqrt{\frac{6 \cdot \ln(f_1 \cdot f_2 \cdot f_3 \cdots f_m)}{\ln(\text{lcm}(f_1, f_2, f_3, \cdots, f_m))}}$$

を得た[78]。ここで、$f_1, f_2, f_3, \cdots$ はフィボナッチ数列で【1.2】、lcm は最小公倍数 (least common multiple) を表す。フィボナッチ数列を、いろいろな2階の線形漸化式に置き換えても、極限値 π を変えないということがわかる。

【1.4.1】と【1.4.2】で $\binom{2n}{n}/(n+1)$ と似たような式を取り上げた。これらは**カタラン数**として知られているもので、たとえば、$2n+1$ 個の頂点をもつ狭義の2分木 (strictly binary trees) を数え上げるときのような組み合せ論において重要となる。このような木の平均の高さを h_n とすれば、Flajolet & Odlyzko（フラジョレット&オドリズコ）の定理によって、

$$\lim_{n\to\infty} \frac{h_n}{\sqrt{n}} = 2\sqrt{\pi}$$

が成立する[79,80]（木に関する用語の説明は【5.6】で行う）。これが π が意外な所に現れる例の1つである。

1.4 アルキメデスの定数, π

[1] P. Beckmann, *A History of* π, St. Martin's Press, 1971; MR 56 #8261.
[2] J. M. Borwein and P. B. Borwein, *Pi and the AGM: A Study in Analytic Number Theory and Computational Complexity*, Wiley, 1987, pp. 46-52, 169-177, 337-362, 385-386; MR 99h:11147.
[3] D. Blatner, *The Joy of Pi*, Walker, 1997.
[4] P. Eymard and J.-P. Lafon, *Autour du nombre* π, Hermann, 1999; MR 2001a:11001.
[5] X. Gourdon and P. Sebah, The constant pi (Numbers, Constants and Computation).
[6] E. F. Assmus, Pi, *Amer. Math. Monthly* 92 (1985) 213-214.
[7] Archimedes, Measurement of a circle, 250 BC, in *Pi: A Source Book*, 2nd ed., ed. L. Berggren, J. M. Borwein, and P. B. Borwein, Springer-Verlag, 2000, pp. 7-14; MR 98f:01001.
[8] O. Toeplitz, *The Calculus: A Genetic Approach*, Univ. of Chicago Press, 1981, pp. 18-22; MR 11, 584e.
[9] G. M. Phillips, Archimedes the numerical analyst, *Amer. Math. Monthly* 88 (1981) 165-169; also in *Pi: A Source Book*, pp. 15-19; MR 83e:01005.
[10] G. Miel, Of calculations past and present: The Archimedean algorithm, *Amer. Math. Monthly* 90 (1983) 17-35; MR 85a:01006.
[11] G. M. Phillips, Archimedes and the complex plane, *Amer. Math. Monthly* 91 (1984) 108-114; MR 85h:40003.
[12] S. D. Dubey, Statistical determination of certain mathematical constants and functions using computers, *J. ACM* 13 (1966) 511-525; MR 34 #2149.
[13] H. Dörrie, *100 Great Problems of Elementary Mathematics: Their History and Solution*, Dover, 1965, pp. 73-77; MR 84b:00001.
[14] E. Waymire, Buffon noodles, *Amer. Math. Monthly* 101 (1994) 550-559; addendum 101 (1994) 791; MR 95g:60021a and MR 95g:60021b.
[15] E. Wegert and L. N. Trefethen, From the Buffon needle problem to the Kreiss matrix theorem, *Amer. Math. Monthly* 101 (1994) 132-139; MR 95b:30036.
[16] G. H. Hardy and E. M. Wright, *An Introduction to the Theory of Numbers*, 5th ed., Oxford Univ. Press, 1985, pp. 47, 268-269; MR 81i:10002.
[17] A. M. Yaglom and I. M. Yaglom, *Challenging Mathematical Problems with Elementary Solutions*, v. I, Holden-Day, 1964, ex. 92-93; MR 88m:00012a.
[18] A. E. Taylor and R. Mann, *Advanced Calculus*, 2nd ed., Wiley, 1972, pp. 740-745; MR 83m:26001.
[19] T. Nagell, *Introduction to Number Theory*, Chelsea, 1981, pp. 38, 60-64; MR 30 #4714.
[20] R. P. Brent, Fast multiple-precision evaluation of elementary functions, *J. ACM* 23 (1976) 242-251; MR 52 #16111.
[21] E. Salamin, Computation of π using arithmetic-geometric mean, *Math. Comp.* 30 (1976) 565-570; MR 53 #7928.
[22] D. C. van Leijenhorst, Algorithms for the approximation of π, *Nieuw Arch. Wisk.* 14 (1996) 255-274; MR 98b:11130.
[23] G. Almkvist and B. Berndt, Gauss, Landen, Ramanujan, the arithmetic-geometric mean, ellipses, π, and the Ladies Diary, *Amer. Math. Monthly* 95 (1988) 585-608; MR89j:01028.
[24] J. M. Borwein, P. B. Borwein, and D. H. Bailey, Ramanujan, modular equations, and approximations to pi, or how to compute one billion digits of pi, *Amer. Math. Monthly* 96 (1989) 201-219; also in *Organic Mathematics*, Proc. 1995 Burnaby workshop, ed. J. Borwein, P. Borwein, L. Jörgenson, and R. Corless, Amer. Math. Soc., 1997, pp. 35-71; MR 90d:11143.
[25] J. M. Borwein and F. G. Garvan, Approximations to pi via the Dedekind eta function, *Organic Mathematics*, Proc. 1995 Burnaby workshop, Amer. Math. Soc., 1997, pp. 89-115; MR 98j:11030.
[26] S. Rabinowitz and S. Wagon, A spigot algorithm for the digits of π, *Amer. Math. Monthly* 102 (1995) 195-203; MR 96a:11152.
[27] D. Bailey, P. Borwein, and S. Plouffe, On the rapid computation of various polylogarithmic constants, *Math. Comp.* 66 (1997) 903-913; MR 98d:11165.
[28] D. H. Bailey, J. M. Borwein, P. B. Borwein, and S. Plouffe, The quest for pi, *Math. Intellig.* 19 (1997) 50-57; CECM preprint 96:070; MR 98b:01045.

[29] M. D. Hirschhorn, A new formula for π, *Austral. Math. Soc. Gazette* 25 (1998) 82-83; MR 99d: 01046.
[30] V. S. Adamchik and S. Wagon, π: A 2000-year-old search changes direction, *Mathematica Educ. Res.* 5 (1996) 11-19.
[31] V. S. Adamchik and S. Wagon, A simple formula for π, *Amer. Math. Monthly* 104 (1997) 852-854; MR 98h:11166.
[32] D. J. Broadhurst, Massive 3-loop Feynman diagrams reducible to SC* primitives of algebras of the sixth root of unity, *Europ. Phys. J. C Part. Fields* 8 (1999) 313-333; hep-th/9803091; MR 2002a:81180.
[33] J. Wimp, Book review of "Pi and the AGM," *SIAM Rev.* 30 (1988) 530-533.
[34] B. Cloitre, e and π in a mirror, unpublished note (2002).
[35] Madhava, The power series for arctan and π, 1400, in *Pi: A Source Book*, pp. 45-50.
[36] R. Roy, The discovery of the series formula for π by Leibniz, Gregory and Nilakantha, *Math. Mag.* 63 (1990) 291-306; also in *Pi: A Source Book*, pp. 92-107; MR 92a:01029.
[37] J. M. Borwein, P. B. Borwein, and K. Dilcher, Pi, Euler numbers, and asymptotic expansions, *Amer. Math. Monthly* 96 (1989) 681-687; MR 91c:40002.
[38] G. Almkvist, Many correct digits of π, revisited, *Amer. Math. Monthly* 104 (1997) 351-353; MR 98a:11189.
[39] J. Gregory, correspondence with J. Collins, 1671, in *Pi: A Source Book*, pp. 87-91.
[40] G. W. Leibniz, Schediasma de serierum summis, et seriebus quadraticibus, 1674, in J. M. Child, *The Early Mathematical Manuscripts of Leibniz*, transl. from texts published by C. I. Gerhardt, Open Court Publishing, 1920, pp. 60-61.
[41] L. Euler, *Introduction to Analysis of the Infinite. Book I*, 1748, transl. J. D. Blanton, Springer-Verlag, 1988, pp. 137-153, 311-312; MR 89g:01067.
[42] G. E. Andrews, R. Askey, and R. Roy, *Special Functions*, Cambridge Univ. Press, 1999, p. 58, ex. 52; MR 2000g:33001.
[43] L. B. W. Jolley, *Summation of Series*, 2nd rev. ed., Dover, 1961, pp. 14-17; MR 24 # B511.
[44] M. Beeler, R. W. Gosper, and R. Schroeppel, Series acceleration technique, HAKMEM, MIT AI Memo 239, 1972, item 120.
[45] G. Almkvist and A. Granville, Borwein and Bradley's Apéry-like formulae for $\zeta(4n+3)$, *Experim. Math.* 8 (1999) 197-203; MR 2000h:11126.
[46] S. Ramanujan, Modular equations and approximations to π, *Quart. J. Math.* 45 (1914) 350-72.
[47] D. V. Chudnovsky and G. V. Chudnovsky, The computation of classical constants, *Proc. Nat. Acad. Sci. USA* 86 (1989) 8178-8182; MR 90m:11206.
[48] D. V. Chudnovsky and G. V. Chudnovsky, Classical constants and functions: Computations and continued fraction expansions, *Number Theory: New York Seminar 1989-1990*, ed. D. V. Chudnovsky, G. V. Chudnovsky, H. Cohn, and M. B. Nathanson, Springer-Verlag, 1991, pp. 13-74; MR 93c:11118.
[49] J. M. Borwein and P. B. Borwein, More Ramanujan-type series for $1/\pi$, *Ramanujan Revisited*, Proc. 1987 Urbana conf., ed. G. E. Andrews, R. A. Askey, B. C. Berndt, K. G. Ramanathan, and R. A. Rankin, Academic Press, 1988, pp. 375-472; MR 89d: 11118.
[50] D. V. Chudnovsky and G. V. Chudnovsky, Approximations and complex multiplication according to Ramanujan, *Ramanujan Revisited*, Proc. 1987 Urbana conf., Academic Press, 1988, pp. 375-472; MR 89f:11099.
[51] F. Viète, Variorum de Rebus Mathematicis Reponsorum Liber VIII, 1593, in *Pi: A Source Book*, pp. 53-56, 690-706.
[52] J. Wallis, Computation of π by successive interpolations, Arithmetica Infinitorum, 1655, in *Pi: A Source Book*, pp. 68-80.
[53] T. J. Osler, The union of Vieta's and Wallis's products for pi, *Amer. Math. Monthly* 106 (1999) 774-776.
[54] A. M. Yaglom and I. M. Yaglom, *Challenging Mathematical Problems with Elementary Solu-*

tions, v. II, Holden-Day, 1967, ex. 139-147; MR 88m:00012b.
[55] J. B. Conway, *Functions of One Complex Variable*, 2nd ed. Springer-Verlag, 1978; MR 80c:30003.
[56] G. F. Simmons, *Differential Equations with Applications and Historical Notes*, McGraw-Hill, 1972, pp. 21-25, 83-86, 93-107; MR 58 #17258.
[57] A. Weil, *Number Theory: An Approach Through History from Hammurapi to Legendre*, Birkhäuser, 1984, p. 266; MR 85c:01004.
[58] G. H. Hardy and S. Ramanujan, Asymptotic formulae in combinatory analysis, *Proc. London Math. Soc.* 17 (1918) 75-115; also in *Collected Papers of G. H. Hardy*, v. 1, Oxford Univ. Press, 1966, pp. 306-339.
[59] H. Rademacher, On the partition function $p(n)$, *Proc. London Math. Soc.* 43 (1937) 241-254.
[60] G. E. Andrews, *The Theory of Partitions*, Addison-Wesley, 1976; MR 99c:11126.
[61] G. Almkvist and H. S. Wilf, On the coefficients in the Hardy-Ramanujan-Rademacher formula for $p(n)$, *J. Number Theory* 50 (1995) 329-334; MR 96e:11129.
[62] I. S. Gradshteyn and I. M. Ryzhik, *Table of Integrals, Series and Products*, 5th ed., Academic Press, 1980, pp. 342, 956; MR 97c:00014.
[63] M. Abramowitz and I. A. Stegun, *Handbook of Mathematical Functions*, Dover, 1972, pp. 78, 230, 302; MR 94b:00012.
[64] R. Butler, On the evaluation of $\int_0^\infty (\sin^m t)/t^m dt$ by the trapezoidal rule, *Amer. Math. Monthly* 67 (1960) 566-569; MR 22 #4841.
[65] L. Lorentzen and H. Waadeland, *Continued Fractions with Applications*, North-Holland, 1992, pp. 561-562; MR 93g:30007.
[66] M. A. Stern, Theorie der Kettenbrüche und ihre Anwendung. III, *J. Reine Angew. Math.* 10 (1833) 241-274.
[67] L. J. Lange, An elegant continued fraction for π, *Amer. Math. Monthly* 106 (1999) 456-458.
[68] R. Courant and H. Robbins, *What is Mathematics?*, Oxford Univ. Press, 1941, pp. 123-125, 299; MR 93k:00002.
[69] G. L. Cohen and A. G. Shannon, John Ward's method for the calculation of pi, *Historia Math.* 8 (1981) 133-144; MR 83d:01021.
[70] G. D. Anderson, M. K. Vamanamurthy, and M. Vuorinen, *Conformal Invariants, Inequalities, and Quasiconformal Maps*, Wiley, 1997, pp. 108-117; MR 98h:30033.
[71] K. Brown, Rounding up to pi (MathPages).
[72] P. Erdős and E. Jabotinsky, On sequences of integers generated by a sieving process, *Proc. Konink. Nederl. Akad. Wetensch. Ser. A* 61 (1958) 115-128; *Indag. Math.* 20 (1958) 115-128; MR 21 #2628.
[73] D. Betten, Kalahari and the sequence "Sloane No. 377," *Combinatorics '86*, Proc. 1986 Trento conf., ed. A. Barlotti, M. Marchi, and G. Tallini, Annals of Discrete Math. 37, North-Holland, 1988, pp. 51-58; MR 89f:05010.
[74] N. J. A. Sloane, On-Line Encyclopedia of Integer Sequences, A002491.
[75] Y. David, On a sequence generated by a sieving process, *Riveon Lematematika* 11 (1957) 26-31; MR 21 #2627.
[76] D. M. Broline and D. E. Loeb, The combinatorics of Mancala-type games: Ayo, Tchoukaitlon, and $1/\pi$, *UMAP J.* 16 (1995) 21-36.
[77] N. J. A. Sloane, My favorite integer sequences, *Sequences and Their Applications (SETA)*, Proc. 1998 Singapore conf., ed. C. Ding, T. Helleseth, and H. Niederreiter, Springer-Verlag, 1999, pp. 103-130; math.CO/0207175.
[78] Y. V. Matiyasevich and R. K. Guy, A new formula for π, *Amer. Math. Monthly* 93 (1986) 631-635; MR 2000i:11199.
[79] P. Flajolet and A. Odlyzko, The average height of binary trees and other simple trees, *J. Comput. Sys. Sci.* 25 (1982) 171-213; MR 84a:68056.
[80] A. M. Odlyzko, Asymptotic enumeration methods, *Handbook of Combinatorics*, v. II, ed. R. L. Graham, M. Grötschel, and L. Lovász, MIT Press, 1995, pp. 1063-1229; MR 97b:05012.

1.5 オイラー-マスケローニの定数, γ

Euler-Mascheroni（オイラー-マスケローニ）の定数 γ[†] は，次の極限値

$$\gamma=\lim_{n\to\infty}\Big(\sum_{k=1}^{n}\frac{1}{k}-\ln(n)\Big)=0.5772156649\cdots$$

で定義される [1-8]．言いかえれば，γ は，調和級数（最も単純な発散級数）の部分和が，対数関数（その近似積分）とどの程度差があるかを測っているものといえる．γ は π や e の陰にかくれた重要な定数である．γ は $\sum_{k=1}^{n}1/k$ の評価が必要であるときはいつでも，自然な形で現れる．たとえば，$X_1, X_2, \cdots, X_n$ を，連続な分布関数をもつ独立で同一の分布に従う確率変数の系列とする．そして，R_n をこの系列における**上位記録** (upper records) の数と定義する．すなわち，$X_k > \max\{X_1, X_2, \cdots, X_{k-1}\}$ となる場合の総数である [9-12]．X_1 もこれに含まれるものと約束する．確率変数 $R(n)$ は，期待値 $E(R_n)$ をもち，$\lim_{n\to\infty}(E(R_n)-\ln(n))=\gamma$ を満たす．もう1つの例として，**クーポン集めの問題**がある．$C=\{1, 2, \cdots, n\}$ をクーポンの集合としよう．クーポンは同じものがあれば交換をくり返して集めるものとし，C のすべてを収集するために必要とする試みの回数を S_n とすれば，$\lim_{n\to\infty}((E(S_n)-n\log(n))/n)=\gamma$ である [13-15]．

いくつかの応用があるが，そこでは，γ はまったく神秘的な形で現れる．n 個の記号のランダムな置換を，相異なる巡回置換に分解することを考えよう．たとえば，$\{0, 1, 2, \cdots, 8\}$ の置換 π を $\pi(x)=2x \bmod 9$ で定義すれば，$\pi=(0)(124875)(36)$ という巡回構造をもつ．$n\to\infty$ のとき，π のどの巡回置換も同じ長さをもたないような確率は何か？ この問に対する答は，$e^{-\gamma}=0.5614594835\cdots$ である．ランダムな置換に関するさらなる記述は，【5.4】に見られる．次に，素数 p を法とする n 次のランダムな整数係数多項式 $F(x)$ を分解することを考えよう．$p\to\infty$, $n\to\infty$ のとき，$F(x)$ のどの2つの既約因子も同じ次数でない確率はいくらか？ この問にも同じ答 $e^{-\gamma}$ が適合する [16-21]．しかし，これを証明することは，2重極限をあつかうので複雑になる．

オイラーの定数は，数論でしばしば現れる．たとえば，オイラーの関数との関係においてである【2.7】．ここで，いくつかの応用を見てみよう．$d(n)$ を n の相異なる約数の個数とする．このとき，分割関数の平均値は

$$\lim_{n\to\infty}\Big(\frac{1}{n}\sum_{k=1}^{n}d(k)-\ln(n)\Big)=2\gamma-1=0.1544313298\cdots$$

を満たす [22-24]．これは，【2.10】で再び議論される．de la Vallée Poussin（ド・ラ・ヴァレー・プサン）による驚くべき結果は，

[†] 訳注：オイラーの定数と略称することもある．

1.5 オイラー-マスケロニの定数，γ

$$\lim_{n\to\infty}\sum_{k=1}^{n}\left\{\frac{n}{k}\right\}=1-\gamma=0.4227843351\cdots$$

である[25-28]．ここで，$\{x\}$ は x の小数部分を表す．言い換えれば次のようである．大きい整数 n を $1\leqq k\leqq n$ なる k で割ったとする．商 n/k から，これより大きい次の整数までの小区間を考え，それぞれの k に対するこの小区間の長さを平均すると，その値は 1/2 ではなくて，γ である！ 同様に，n を等差数列の項を動くものとしても，また，素数列の項を動くものとしても，同じ平均値が得られる．また，n を超えない素数 p で，2^p-1 が素数となる p の個数を $M(n)$ とすれば，ほぼ $\ln(n)$ と同じ割合で $M(n)\to\infty$ となることが示唆されており，さらに，$\lim_{n\to\infty}M(n)/\ln(n)=e^\gamma/\ln(2)=2.5695443449\cdots$ も言及されている[29-32]．この主張を裏付ける実験データは極めて少ない．知られているメルセンヌ素数は，わずか 39 個である[33][†]．その他の数論的な応用は[34-37]に見られる．

オイラーの定数の計算は，π の計算のように多くの人の好奇心をそそるものとはなっていない．しかし，いまだに，いくらかの人達のあこがれの的となっている．γ の値の評価は難しく，数億桁だけが知られているにすぎない．π に対しては，Borwein の 4 次収束アルゴリズムがある．それは，逐次反復の各段階で正しく求められる桁数を 1 回の反復計算でほぼ 4 倍にする．これとは対照的に，γ に対しては，2 次の収束アルゴリズムすら知られていない[38-40]．

数値的に有用な結果を得るには，γ の定義自体からではあまりにも収束が遅い．このことは，両側からの評価を与える次の不等式

$$\frac{1}{2(n+1)}<\sum_{k=1}^{n}\frac{1}{k}-\ln(n)-\gamma<\frac{1}{2n}$$

によって説明される[41,42]．一方では，n 回で計算を打ち切ったとき，k 桁の精度で γ を求めたいならば，$n\geqq 10^{k+1}$ で十分である．他方，$n<10^k$ では十分大きいとは言えない．いくつかの交代評価や不等式が[43,44]で報告されている．最もよく知られている技法が Euler-Maclaurin (オイラー-マクローリン) の和公式で，それは次の式

$$\gamma=\sum_{k=1}^{n}\frac{1}{k}-\ln(n)-\frac{1}{2n}+\frac{1}{12n^2}-\frac{1}{120n^4}+\frac{1}{252n^6}-\frac{1}{240n^8}+\frac{1}{132n^{10}}-\frac{691}{32760n^{12}}+O\left(\frac{1}{n^{14}}\right)$$

を含むいろいろな評価の改良に貢献している．Euler はこの公式を用いて，$n=10$ として γ を 15 桁まで正しく求めた[45-48]．Karatsuba（カラツバ）の FEE (Fast E-function Evaluation) の方法[49,50]や Brent（ブレント）の二進分割方式[††][51]のような高速アルゴリズムは，最新の計算において絶対必要なものであった．Papanikolaou（パパニコラウ）は[56]における結果を使い，γ に対する正則連分数展開において，最初の 475006 項の部分商を計算した．そして，もし γ が有理数ならば，その分母は 10^{244663} を超えなければならないことを導いた．これはオイラー定数が有理数でな

[†] 訳注：原著の初版の時点である．2008 年には 46 個に増えたが，依然として少数である．
[††] 訳注：【1.3】参照．

いということに反論できない証拠であろう．無理数性を証明することは（超越性は別にしても）いまだに手のとどかない所にある[57]．[58,59]にある2つの徒労に終った企ても参照せよ．

その他，2つの未解決の問題がある．1つは調和級数に関するものであり，もう1つはクーポン収集家の問題とよく似た問題である．kを与えられた正の整数とし，$\sum_{j=1}^{n-1} 1/j < k < \sum_{j=1}^{n} 1/j$を満たす一意的に定まる整数を$n_k$とする．このとき，$n_k$は常に$e^{k-\gamma}$に一番近い整数であるか[60-65]？ そして次に，独立に正しいコインを投げることによって生成される二項数列Bと，正の整数nを考えることにする．すると，（Bの部分系列として）長さnの相異なる2^n個のパターンすべてが起こるまでの待ち時間[†] T_nは何か？ 平均待ち時間は（[66,67]の事実から）$\lim_{n\to\infty}((E(T_n)-2^n n\ln(2))/2^n)=\gamma$であることが予想される．しかし，これは未解決である．しかしながら，最小の待ち時間はde Bruijn（ド・ブルイン）数列と呼ばれるものと関係する事実の帰結として，たったの2^n+n-1となる[68]．

1.5.1 級数と無限積

次の級数はγの定義の明らかな言いかえになっている．

$$\gamma = \sum_{k=1}^{\infty}\left(\frac{1}{k} - \ln\left(1+\frac{1}{k}\right)\right)$$

その他のγを含む公式では，次の2つはEuler[1]によって得られたものである．

$$\gamma = \frac{1}{2}\cdot\left(1+\frac{1}{2^2}+\frac{1}{3^2}+\cdots\right) - \frac{1}{3}\cdot\left(1+\frac{1}{2^3}+\frac{1}{3^3}+\cdots\right) + \frac{1}{4}\cdot\left(1+\frac{1}{2^4}+\frac{1}{3^4}+\cdots\right) - +\cdots$$

$$\gamma = \frac{1}{2}\cdot\left(\frac{1}{2^2}+\frac{1}{3^2}+\frac{1}{4^2}+\cdots\right) + \frac{2}{3}\cdot\left(\frac{1}{2^3}+\frac{1}{3^3}+\frac{1}{4^3}+\cdots\right) + \frac{3}{4}\cdot\left(\frac{1}{2^4}+\frac{1}{3^4}+\frac{1}{4^4}+\cdots\right) + \cdots$$

また，

$$\gamma = \frac{1}{2} - \frac{1}{3} + 2\cdot\left(\frac{1}{4} - \frac{1}{5} + \frac{1}{6} - \frac{1}{7}\right) + 3\cdot\left(\frac{1}{8} - \frac{1}{9} + \cdots - \frac{1}{15}\right) + 4\cdot\left(\frac{1}{16} - \frac{1}{17} + \cdots - \frac{1}{31}\right) + \cdots$$

はVacca（ヴァッカ）[69-75]によるものであり，

$$\gamma = 1 - \left(\frac{1}{2}+\frac{1}{3}\right) + \frac{3}{4} - \left(\frac{1}{5}+\frac{1}{6}+\frac{1}{7}+\frac{1}{8}\right) + \frac{5}{9} - \left(\frac{1}{10}+\frac{1}{11}+\cdots+\frac{1}{15}\right) + \frac{7}{16} - + \cdots$$

はPólya（ポリヤ）[26,76]によって得られたものである．また，次の2つの公式

$$e^{\gamma} = \lim_{n\to\infty}\frac{1}{\ln(n)}\cdot\prod_{p\leq n}\frac{p}{p-1}, \quad \frac{6e^{\gamma}}{\pi^2} = \lim_{n\to\infty}\frac{1}{\ln(n)}\prod_{p\leq n}\frac{p+1}{p}$$

はMertens（メルテンス）[22,77]によって得られたものである．ここで，積はnを超えない素数pについて取られるものとする．Mertensの最初の公式は，

$$\gamma = \lim_{n\to\infty}\left(\sum_{p\leq n}\ln\left(\frac{p}{p-1}\right) - \ln(\ln(n))\right)$$

と書き直すことができる[55]．もし，この級数において，$\ln(p/(p-1))$を漸近的には

[†] 訳注：長さnのBの系列を順次シフトして調べていく回数．

等しい $1/p$ で置き換えると，この極限値は γ とは異なる定数となる【2.2】．その他の級数，無限積については[78,95]に紹介がある．

1.5.2 積　　分
多少注目される次の積分

$$\int_0^\infty e^{-x}\ln(x)\,dx=-\gamma,\quad \int_0^\infty e^{-x^2}\ln(x)\,dx=-\frac{\sqrt{\pi}}{4}(\gamma+2\ln(2))$$

$$\int_0^\infty e^{-x}\ln(x)^2\,dx=\frac{\pi^2}{6}+\gamma^2,\quad \int_0^1\ln\left(\ln\left(\frac{1}{x}\right)\right)dx=-\gamma$$

$$\int_0^\infty \frac{e^{-x^a}-e^{-x^b}}{x}dx=\frac{a-b}{ab}\gamma,\quad \int_0^\infty \frac{x}{1+x^2}\cdot\frac{1}{e^{2\pi x}-1}dx=\frac{1}{4}(2\gamma-1)$$

$$\int_0^1\left(\frac{1}{\ln(x)}+\frac{1}{1-x}\right)dx=\gamma,\quad \int_0^1\frac{1}{1+x}\left(\sum_{k=1}^\infty x^{2^k}\right)dx=1-\gamma$$

を初めとして，オイラーの定数を含む積分はたくさんある[55,75,96,97]．ここで，パラメータ a,b は $a>0$, $b>0$ を満たすものとする．$\{x\}$ を x の小数部分とすれば，

$$\int_1^\infty \frac{\{x\}}{x^2}dx=\int_0^1\left\{\frac{1}{y}\right\}dy=1-\gamma$$

が成り立つ[22,24]．また，【1.6.5】，【1.8】，【2.21】にも同じような積分が見られる．[98-101]も参照せよ．

1.5.3 一般オイラー定数
オイラー定数が関心を引き付けてきた反面，次の形の定数

$$\gamma(m,f)=\lim_{n\to\infty}\left(\sum_{k=m}^n f(k)-\int_m^n f(x)\,dx\right)$$

が比較的おろそかにされてきたのはなぜかと Boas（ボアス）[102-104]は考えた．$f(x)=x^{-q}$ ($0<q<1$) の場合は，ゼータ関数【1,6】の値を含む定数 $\zeta(q)+1/(1-q)$ となり，$f(x)=\ln(x)^r/x$ ($r\geq 0$) の場合は，スティルチェス定数 γ_r【2.21】となる．表 1.1 にいくつかの数値例を挙げる．Brigss（ブリッグス）[105]と Lehmer（レーマー）[106]は，等差数列 $a,a+b,a+2b,a+3b,\cdots$ に関して γ との類似で

$$\gamma_{a,b}=\lim_{n\to\infty}\left(\sum_{\substack{0<k\leq n\\ k\equiv a\bmod b}}\frac{1}{k}-\frac{1}{b}\ln(n)\right)$$

表 1.1　一般オイラー定数

m	$f(x)$	$\gamma(m,f)$
1	$1/x$	$0.5772156649\cdots=\gamma_0$
2	$1/\ln(x)$	$0.8019254372\cdots$
2	$1/(x\cdot\ln(x))$	$0.4281657248\cdots$
1	$1/\sqrt{x}$	$0.5396454911\cdots=\zeta(1/2)+2$
1	$\ln(x)/x$	$-0.0728158454\cdots=\gamma_1$

を研究した．たとえば，$\gamma_{0,b}=(\gamma-\ln(b))/b$，$\sum_{a=0}^{b-1}\gamma_{a,b}=\gamma$ であり，

$$\gamma_{1,3}=\frac{1}{3}\gamma+\frac{\sqrt{3}}{18}\pi+\frac{1}{6}\ln(3), \quad \gamma_{1,4}=\frac{1}{4}\gamma+\frac{1}{8}\pi+\frac{1}{4}\ln(2)$$

である．[107,108]も参照せよ．オイラー定数の2次元版は【7.2】で取り上げ，n次元の格子和版は【1.10】で議論される．

1.5.4 ガンマ関数

複素数 z に対して，オイラーのガンマ関数 $\Gamma(z)$ は

$$\Gamma(z)=\lim_{n\to\infty}\frac{n!\cdot n^z}{\prod_{k=0}^{n}(z+k)}$$

によって定義され，正でない整数で1位の極となる点を除いて全複素数平面で解析的となる．$x>0$ である実数に対しては，

$$\Gamma(x)=\int_0^{\infty}s^{x-1}e^{-s}ds=\int_0^1\left(\ln\left(\frac{1}{t}\right)\right)^{x-1}dt$$

と積分表現され，n が正の整数のとき，$\Gamma(n)=(n-1)!$ である．これが，我々がよく

$$\left(-\frac{1}{2}\right)!=\sqrt{\pi}=1.7724538509\cdots$$

という表現を見かける理由である．というのも，$\Gamma(1/2)$ は変数変換によって，よく知られているガウスの確率密度関数の積分に変形されるからである．

Bohr-Mollerup（ボア-モレラップ）の定理[109,110]は，$\Gamma(z)$ が階乗関数の複素数平面への（無限に多くある拡張の中で）最も自然な拡張であることを示している．

どのような変数値に対してガンマ関数が超越的になるであろうか？ Chudnovsky（チュドノフスキ）[111-114]は1975年に，$\Gamma(1/6)$，$\Gamma(1/4)$，$\Gamma(1/3)$，$\Gamma(2/3)$，$\Gamma(3/4)$，$\Gamma(5/6)$ が超越数であり，それらは π と代数的に独立であることを示した．奇妙なことに，何年か前に $\Gamma(1/4)^4/\pi$ と $\Gamma(1/3)^2/\pi$ が超越的であることが知られてはいた[115,116]．Nesterenko（ネステレンコ）[117-121]は1996年に，π, e^{π}，$\Gamma(1/4)=3.6256099082\cdots$ が代数的に独立であることを証明した．定数 $\Gamma(1/4)$ は【3.2】，【6.1】そして【7.2】でも取り上げられる．Nesterenkoはまた，$\pi, e^{\pi\sqrt{3}}$，$\Gamma(1/3)=2.6789385347\cdots$ が代数的に独立であることも証明した．同じような強い結果は $\Gamma(1/6)=5.5663160017\cdots$ や $\Gamma(1/5)=4.5908437119\cdots$ に対してはいまだに証明されておらず，無理数であるという論証さえもない．反転公式により

$$\Gamma\left(\frac{1}{4}\right)\Gamma\left(\frac{3}{4}\right)=\pi\sqrt{2}, \quad \Gamma\left(\frac{1}{3}\right)\Gamma\left(\frac{2}{3}\right)=\frac{2}{3}\pi\sqrt{3},$$

$$\Gamma\left(\frac{1}{6}\right)\Gamma\left(\frac{5}{6}\right)=2\pi, \quad \Gamma\left(\frac{1}{5}\right)\Gamma\left(\frac{4}{5}\right)=\frac{2}{5}\pi\sqrt{5}\sqrt{2+\varphi}$$

が得られる．ここで φ は黄金比【1.2】である．さらに，

$$\Gamma\left(\frac{1}{4}\right)=2^{1/2}\pi^{3/4}h_1^{1/2}, \quad \Gamma\left(\frac{1}{3}\right)=2^{4/9}3^{-1/12}\pi^{2/3}h_3^{1/3}, \quad \Gamma\left(\frac{1}{6}\right)=2^{5/9}3^{1/3}\pi^{5/6}h_3^{2/3}$$

である[122,123]．ここで，

$$h_1 = \frac{2}{\pi} K\left(\frac{\sqrt{2}}{2}\right) = \left(\sum_{n=-\infty}^{\infty} e^{-n^2\pi}\right)^2 = 1.1803405990\cdots$$

$$h_3 = \frac{2}{\pi} K\left(\frac{\sqrt{2}}{4}(\sqrt{3}-1)\right) = \left(\sum_{n=-\infty}^{\infty} e^{-n^2\sqrt{3}\pi}\right)^2 = 1.0174087975\cdots$$

であり，$K(x)$ は第1種の完全楕円積分【1.4.6】である．

ガンマ関数 $y = \Gamma(x)$ のグラフを描くとき，第1象限にある最小値となる点は x, y 座標で $(x_{\min}, \Gamma(x_{\min})) = (1.4616321449\cdots, 0.8856031944\cdots)$ である．θ を次の方程式

$$\frac{d}{dx}\ln(\Gamma(x))\Big|_{x=\theta} = \ln(\pi)$$

の一意的に定まる正の解とすれば，$d_S = 2\theta = 7.2569464048\cdots$ と $d_S = 2(\theta-1) = 5.2569464048\cdots$ は，d 次元の単位球の表面積と体積がそれぞれ最大化される分数次元である[124]．

関連した級数が[125-129]に見られる．たとえば，Ramanujan による次の2つの級数

$$\sum_{n=0}^{\infty} \frac{(-1)^n}{2^{4n}}\binom{2n}{n}^2 = (2\pi)^{-3/2}\Gamma\left(\frac{1}{4}\right)^2, \quad \sum_{n=0}^{\infty} \frac{(-1)^n}{2^{6n}}\binom{2n}{n}^3 = \left(\frac{\Gamma(9/8)}{\Gamma(5/4)\Gamma(7/8)}\right)^2$$

は，【1.1】でとり上げた級数を拡張したものである．また，2つの無限積

$$\prod_{n=1}^{\infty}\left(1 - \frac{1}{(4n+1)^2}\right) = \frac{4\cdot 6}{5\cdot 5}\cdot\frac{8\cdot 10}{9\cdot 9}\cdot\frac{12\cdot 14}{13\cdot 13}\cdot\frac{16\cdot 18}{17\cdot 17}\cdots = \frac{1}{8\sqrt{\pi}}\Gamma\left(\frac{1}{4}\right)^2$$

$$\prod_{n=1}^{\infty}\left(1 - \frac{1}{(2n+1)^2}\right)^{(-1)^n} = \frac{3^2}{3^2-1}\cdot\frac{5^2-1}{5^2}\cdot\frac{7^2}{7^2-1}\cdot\frac{9^2-1}{9^2}\cdots = \frac{1}{16\pi^2}\Gamma\left(\frac{1}{4}\right)^4$$

がある[96,133]．さらに，$u>0$，$v>0$ をパラメータとして，標本積分

$$\int_0^{\pi/2} \sin(x)^{u-1}\cos(x)^{v-1} dx = \int_0^1 y^{u-1}(1-y^2)^{v/2-1} dy = \frac{1}{2}\frac{\Gamma(u/2)\Gamma(v/2)}{\Gamma((u+v)/2)}$$

がある[96,134,135]．

Euler のガンマ関数に対するオイラー定数の重要性は，公式 $\psi(1) = -\gamma$ によって最もよく要約される．ここで[90]，

$$\psi(x) = \frac{d}{dx}\ln(\Gamma(x)) = -\gamma - \sum_{n=0}^{\infty}\left(\frac{1}{x+n} - \frac{1}{n+1}\right)$$

はディガンマ関数である．$x=1$ における高階微分係数には，ゼータ関数の値が含まれている【1.6】．このような微分係数（ポリガンマ関数）に関する情報は[134,136]に見られる．

[1] J. W. L. Glaisher, On the history of Euler's constant, *Messenger of Math.* 1 (1872) 25-30.
[2] R. Johnsonbaugh, The trapezoid rule, Stirling's formula, and Euler's constant, *Amer. Math. Monthly* 88 (1981) 696-698; MR 83a:26006.
[3] C. W. Barnes, Euler's constant and e, *Amer. Math. Monthly* 91 (1984) 428-430.
[4] D. Bushaw and S. C. Saunders, The third constant, *Northwest Sci.* 59 (1985) 147-158.
[5] J. Nunemacher, On computing Euler's constant, *Math. Mag.* 65 (1992) 313-322; MR 93j:65042.

[6] R. Barshinger, Calculus II and Euler also (with a nod to series integral remainder bounds), *Amer. Math. Monthly* 101 (1994) 244-249; MR 94k:26003.
[7] J. Sondow, An antisymmetric formula for Euler's constant, *Math. Mag.* 71 (1998) 219-220.
[8] T. M. Apostol, An elementary view of Euler's summation formula, *Amer. Math. Monthly* 106 (1999) 409-418.
[9] F. G. Foster and A. Stuart, Distribution-free tests in time-series based on the breaking of records, *J. Royal Stat. Soc. Ser. B* 16 (1954) 1-22; MR 16,385i.
[10] D. E. Knuth, *The Art of Computer Programming: Fundamental Algorithms*, v. 1, 2nd ed., Addison-Wesley, 1973, pp. 73-77, 94-99; MR 51 # 14624.
[11] W. Katzenbeisser, On the joint distribution of the number of upper and lower records and the number of inversions in a random sequence, *Adv. Appl. Probab.* 22 (1990) 957-960; MR 92a: 60046.
[12] B. C. Arnold, N. Balakrishnan, and H. N. Nagaraja, *Records*, Wiley, 1998, pp. 22-25; MR 2000b: 60127.
[13] B. Dawkins, Siobhan's problem: The coupon collector revisited, *Amer. Statist.* 45 (1991) 76-82.
[14] B. Levin, Regarding "Siobhan's problem: The coupon collector revisited," *Amer. Statist.* 46 (1992) 76.
[15] R. Sedgewick and P. Flajolet, *Introduction to the Analysis of Algorithms*, Addison-Wesley, 1996, pp. 425-427.
[16] D. H. Lehmer, On reciprocally weighted partitions, *Acta Arith.* 21 (1972) 379-388; MR 46 # 3437.
[17] D. E. Knuth, *The Art of Computer Programming: Seminumerical Algorithms*, v. 2, 2nd ed., Addison-Wesley, 1981, pp. 439, 629; MR 44 # 3531.
[18] D. H. Greene and D. E. Knuth, *Mathematics for the Analysis of Algorithms*, 3rd ed., Birkhäuser, 1990, pp. 48-54, 95-98; MR 92c:68067.
[19] A. Knopfmacher and R. Warlimont, Distinct degree factorizations for polynomials over a finite field, *Trans. Amer. Math. Soc.* 347 (1995) 2235-2243; MR 95i:11144.
[20] P. Flajolet, X. Gourdon, and D. Panario, Random polynomials and polynomial factorization, *Proc. 1996 Int. Colloq. on Automata, Languages and Programming (ICALP)*, Paderborn, ed. F. Meyer auf der Heide and B. Monien, Lect. Notes in Comp. Sci. 1099, Springer-Verlag, 1996, pp. 232-243; MR 98e:68123.
[21] P. Flajolet, X. Gourdon, and D. Panario, The complete analysis of a polynomial factorization algorithm over finite fields, *J. Algorithms* 40 (2001) 37-81; INRIA preprint RR3370; MR 2002f: 68193.
[22] G. H. Hardy and E. M. Wright, *An Introduction to the Theory of Numbers*, 5th ed., Oxford Univ. Press, 1985, pp. 264-265, 347, 351; MR 81i:10002.
[23] M. R. Schroeder, *Number Theory in Science and Communication: With Applications in Cryptography, Physics, Digital Information, Computing and Self-Similarity*, 2nd ed., Springer-Verlag, 1986, pp. 127-131; MR 99c:11165.
[24] T. M. Apostol, *Introduction to Analytic Number Theory*, Springer-Verlag, 1998, pp. 52-59; MR 55 # 7892.
[25] Ch. de la Vallée Poussin, Sur les valeurs moyennes de certaines fonctions arithmétiques, *Annales de la Societe Scientifique de Bruxelles* 22 (1898) 84-90.
[26] G. Pólya and G. Szegö, *Problems and Theorems in Analysis*, v. 1, Springer-Verlag 1998, problems 18, 19, 32, 42; MR 81e:00002.
[27] L. E. Dickson, *History of the Theory of Numbers*, v. 1, *Divisibility and Primality*, Chelsea, 1971; pp. 134, 136, 294, 317, 320, 328, and 330; MR 39 # 6807.
[28] J. H. Conway and R. K. Guy, *The Book of Numbers*, Springer-Verlag, 1996, p. 260; MR 98g: 00004.
[29] S. S. Wagstaff, Divisors of Mersenne numbers, *Math. Comp.* 40 (1983) 385-397; MR 84j:10052.
[30] R. K. Guy, *Unsolved Problems in Number Theory*, 2nd ed., Springer-Verlag, 1994, sect. A3; MR 96e:11002.

1.5 オイラー-マスケロニの定数, γ

[31] P. Ribenboim, *The New Book of Prime Number Records*, 3rd ed., Springer-Verlag, 1996, pp. 411-413; MR 96k:11112.
[32] R. Crandall and C. Pomerance, *Prime Numbers: A Computational Perspective*, Springer-Verlag, 2001, pp. 20-24; MR 2002a:11007.
[33] C. Caldwell, Mersenne Primes: History, Theorems and Lists (Prime Pages).
[34] P. Erdös and A. Ivić, Estimates for sums involving the largest prime factor of an integer and certain related additive functions, *Studia Sci. Math. Hungar.* 15 (1980) 183-199; MR 84a:10046.
[35] V. Sita Ramaiah and M. V. Subbarao, The maximal order of certain arithmetic functions, *Indian J. Pure Appl. Math.* 24 (1993) 347-355; MR 94i:11075.
[36] A. Knopfmacher and J. N. Ridley, Reciprocal sums over partitions and compositions, *SIAM J. Discrete Math.* 6 (1993) 388-399; MR 94g:11111.
[37] R. Warlimont, Permutations with roots, *Arch. Math. (Basel)* 67 (1996) 23-34; MR 97h:11109.
[38] D. H. Bailey, Numerical results on the transcendence of constants involving π, e and Euler's constant, *Math. Comp.* 50 (1988) 275-281; MR 88m:11056.
[39] D. H. Bailey and H. R. P. Ferguson, Numerical results on relations between fundamental constants using a new algorithm, *Math. Comp.* 53 (1989) 649-656; MR 90e:11191.
[40] J. M. Borwein and P. B. Borwein, On the complexity of familiar functions and numbers, *SIAM Rev.* 30 (1988) 589-601; MR 89k:68061.
[41] S. R. Tims and J. A. Tyrrell, Approximate evaluation of Euler's constant, *Math. Gazette* 55 (1971) 65-67.
[42] R. M. Young, Euler's constant, *Math. Gazette* 75 (1991) 187-190.
[43] D. W. DeTemple, A quicker convergence to Euler's constant, *Amer. Math. Monthly* 100 (1993) 468-470; MR 94e:11146.
[44] T. Negoi, A faster convergence to Euler's constant (in Romanian), *Gazeta Mat.* 15 (1997) 111-113; Engl. transl. in *Math. Gazette* 83 (1999) 487-489.
[45] D. E. Knuth, Euler's constant to 1271 places, *Math. Comp.* 16 (1962) 275-281; MR 26 #5763.
[46] D. W. Sweeney, On the computation of Euler's constant, *Math. Comp.* 17 (1963) 170-178; corrigenda 17 (1963) 488; MR 28 #3522.
[47] W. A. Beyer and M. S. Waterman, Error analysis of a computation of Euler's constant, *Math. Comp.* 28 (1974) 599-604; MR 49 #6555.
[48] C. Elsner, On a sequence transformation with integral coefficients for Euler's constant, *Proc. Amer. Math. Soc.* 123 (1995) 1537-1541; MR 95f:11111.
[49] E. A. Karatsuba, Fast evaluation of transcendental functions (in Russian), *Problemy Peredachi Informatsii* 27 (1991) 76-99; Engl. transl. in *Problems Information Transmission* 27 (1991) 339-360; MR 93c:65027.
[50] E. A. Karatsuba, On the computation of the Euler constant γ, *Numer. Algorithms* 24 (2000) 83-97; Univ. of Helsinki Math. Report 226 (1999); MR 2002f:33004.
[51] R. P. Brent, Fast multiple-precision evaluation of elementary functions, *J. ACM* 23 (1976) 242-251; MR 52 #16111.
[52] R. P. Brent, Computation of the regular continued fraction for Euler's constant, *Math. Comp.* 31 (1977) 771-777; MR 55 #9490.
[53] R. P. Brent and E. M. McMillan, Some new algorithms for high-precision computation of Euler's constant, *Math. Comp.* 34 (1980) 305-312; MR 82g:10002.
[54] B. Haible and T. Papanikolaou, Fast multiprecision evaluation of series of rational numbers, *Proc. 1998 Algorithmic Number Theory Sympos. (ANTS-III)*, Portland, ed. J. P. Buhler, Lect. Notes in Comp. Sci. 1423, Springer-Verlag, 1998, pp. 338-350; MR 2000i:11197.
[55] X. Gourdon and P. Sebah, The Euler constant gamma (Numbers, Constants and Computation).
[56] R. P. Brent, A. J. van der Poorten, and H. J. J. te Riele, A comparative study of algorithms for computing continued fractions of algebraic numbers, *Proc. 1996 Algorithmic Number Theory Sympos. (ANTS-II)*, Talence, ed. H. Cohen, Lect. Notes in Comp. Sci. 1122, Springer-Verlag, 1996, pp. 37-49; MR 98c:11144.

[57] J. Sondow, Criteria for irrationality of Euler's constant, math.NT/0209070.
[58] A. Froda, La constante d'Euler est irrationnelle, *Atti Accad. Naz. Lincei Rend. Cl. Sci. Fis. Mat. Natur.* 38 (1965) 338-344; MR 32 #5599.
[59] R. G. Ayoub, Partial triumph or total failure?, *Math. Intellig.*, v. 7 (1985) n. 2, 55-58; MR 86i: 01001.
[60] S. M. Zemyan, On two conjectures concerning the partial sums of the harmonic series, *Proc. Amer. Math. Soc.* 95 (1985) 83-86; MR 86m:40002.
[61] L. Comtet, About $\Sigma 1/n$, *Amer. Math. Monthly* 74 (1967) 209.
[62] R. P. Boas and J. W. Wrench, Partial sums of the harmonic series, *Amer. Math. Monthly* 78 (1971) 864-870; MR 44 #7179.
[63] R. P. Boas, An integer sequence from the harmonic series, *Amer. Math. Monthly* 83 (1976) 748-749.
[64] G. Pólya and G. Szegö, *Problems and Theorems in Analysis*, v. 2, Springer-Verlag 1976, problems 249-251, 260; MR 57 #5529.
[65] K. T. Atanassov, Remark on the harmonic series, *C. R. Acad. Bulgare Sci.*, v. 40 (1987) n. 5, 25-28; MR 88m:40004.
[66] A. Benczur, On the expected time of the first occurrence of every k bit long patterns in the symmetric Bernoulli process, *Acta Math. Hungar.* 47 (1986) 233-286; MR 87m:60029.
[67] T. F.Móri, On the expectation of the maximum waiting time, *Annales Univ. Sci. Budapest Sect. Comput.* 7 (1987) 111-115; MR 90e:60048.
[68] N. G. de Bruijn, A combinatorial problem, *Proc. Konink. Nederl. Akad. Wetensch. Sci. Sect.* 49 (1946) 758-764; *Indag. Math.* 8 (1946) 461-467; MR 8,247d.
[69] G. Vacca, A new series for the Eulerian constant, *Quart. J. Pure. Appl. Math.* 41 (1910) 363-368.
[70] H. F. Sandham and D. F. Barrow, Problem/Solution 4353, *Amer. Math. Monthly* 58 (1951) 116-117.
[71] A. W. Addison, A series representation for Euler's constant, *Amer. Math. Monthly* 74 (1967) 823-824; MR 36 #1397.
[72] I. Gerst, Some series for Euler's constant, *Amer. Math. Monthly* 76 (1969) 273-275; MR 39 #3181.
[73] M. Beeler, R. W. Gosper, and R. Schroeppel, Series acceleration technique, HAKMEM, MIT AI Memo 239, 1972, item 120.
[74] J. Sandor, On the irrationality of some alternating series, *Studia Univ. Babes-Bolyai Math.* 33 (1988) 8-12; MR 91c:40003.
[75] B. C. Berndt and D. C. Bowman, Ramanujan's short unpublished manuscript on integrals and series related to Euler's constant, *Constructive, Experimental, and Nonlinear Analysis*, Proc. 1999 Limoges conf., ed. M. Théra, Amer. Math. Soc., 2000, pp. 19-27; MR 2002d:33001.
[76] G. Pólya, A series for Euler's constant, *Research Papers in Statistics (J. Neyman Festschift)*, Wiley, 1966, pp. 259-261; also in *Collected Papers*, v. 3, ed. J. Hersch and G.-C. Rota, MIT Press, 1984, pp. 475-477; MR 34 #8026.
[77] G. Tenenbaum, *Introduction to Analytic and Probabilistic Number Theory*, Cambridge Univ. Press, 1995, pp. 14-18; MR 97e:11005b.
[78] A. M. Glicksman, Euler's constant, *Amer. Math. Monthly* 50 (1943) 575.
[79] O. Dunkel, Euler's constant, *Amer. Math. Monthly* 51 (1944) 99-102.
[80] F. Supnick, A geometric facet of the Eulerian constant, *Amer. Math. Monthly* 69 (1962) 208-209.
[81] V. P. Burlachenko, Representation of some constants by double series (in Ukrainian), *Dopovidi Akad. Nauk Ukrain. RSR A* 4 (1970) 303-305, 380; MR 43 #5204.
[82] D. P. Verma and A. Kaur, Summation of some series involving Riemann zeta function, *Indian J. Math.* 25 (1983) 181-184; MR 87a:11077.
[83] J. Choi and H. M. Srivastava, Sums associated with the zeta function, *J. Math. Anal. Appl.* 206 (1997) 103-120; MR 97i:11092.

[84] M. Hata, Farey fractions and sums over coprime pairs, *Acta Arith.* 70 (1995) 149-159; MR 96c: 11022.
[85] G. Xiong, On a kind of the best estimates for the Euler constant γ, *Acta Math. Scientia: English Ed.* 16 (1996) 458-468; MR 97m:11162.
[86] S. Ramanujan, A series for Euler's constant γ, *Messenger of Math.* 46 (1917) 73-80; also in *Collected Papers*, ed. G. H. Hardy, P. V. Seshu Aiyar, and B. M. Wilson, Cambridge Univ. Press, 1927, pp. 163-168, 325.
[87] B. C. Berndt, *Ramanujan's Notebooks: Part I*, Springer-Verlag, 1985, pp. 98-99, 196; MR 86c: 01062.
[88] R. P. Brent, An asymptotic expansion inspired by Ramanujan, *Austral. Math. Soc. Gazette* 20 (1993) 149-155; MR 95b:33006.
[89] R. P. Brent, Ramanujan and Euler's constant, *Mathematics of Computation 1943-1993: A Half-Century of Computational Mathematics*, Proc. 1993 Vancouver conf., ed. W. Gautschi, Amer. Math. Soc. 1994, pp. 541-545; MR 95k:01022.
[90] D. Bradley, Ramanujan's formula for the logarithmic derivative of the gamma function, *Math. Proc. Cambridge Philos. Soc.* 120 (1996) 391-401; MR 97a:11132.
[91] H. S. Wilf and D. A. Darling, An infinite product, *Amer. Math. Monthly* 105 (1998) 376.
[92] F. K. Kenter, A matrix representation for Euler's constant, γ, *Amer. Math. Monthly* 106 (1999) 452-454.
[93] F. Pittnauer, Eine Darstellung der Eulerschen Konstanten, *Math. Semesterber* 31 (1984) 26-27; MR 85j:11017.
[94] H. G. Killingbergto and C. P. Kirkebo, A new(?) formula for computing γ (in Norwegian), *Normat* 41 (1993) 120-124, 136; MR 94g:11117.
[95] P. Flajolet and I. Vardi, Zeta function expansions of classical constants, unpublished note (1996).
[96] I. S. Gradshteyn and I. M Ryzhik, *Tables of Integrals, Series and Products*, Academic Press, 1980; MR 97c:00014.
[97] J.-M. Arnaudiès, *Problèmes de préparation à l'Agrégation de Mathématiques*, v. 4, *Analyse, Intégrale, séries de Fourier, équations différentielles*, Edition Ellipses, 1998, pp. 63-80.
[98] S. K. Lakshmana Rao, On the sequence for Euler's constant, *Amer. Math. Monthly* 63 (1956) 572-573; also in *A Century of Calculus*, v. 1, ed. T. M. Apostol, H. E. Chrestenson, C. S. Ogilvy, D. E. Richmond, and N. J. Schoonmaker, Math. Assoc. Amer., 1992, pp. 389-390.
[99] J. Anglesio and D. A. Darling, An integral giving Euler's constant, *Amer. Math. Monthly* 104 (1997) 881.
[100] J. Anglesio, The integrals are Euler's constant, *Amer. Math. Monthly* 105 (1998) 278-279.
[101] J. Choi and T. Y. Seo, Integral formulas for Euler's constant, *Commun. Korean Math. Soc.* 13 (1998) 683-689; MR 2000i:11196.
[102] R. P. Boas, Growth of partial sums of divergent series, *Math. Comp.* 31 (1977) 257-264; MR 55 #13730.
[103] R. P. Boas, Partial sums of infinite series, and how they grow, *Amer. Math. Monthly* 84 (1977) 237-258; MR 55 #13118.
[104] J. V. Baxley, Euler's constant, Taylor's formula, and slowly converging series, *Math. Mag.* 65 (1992) 302-313; MR 93j:40001.
[105] W. E. Briggs, The irrationality of γ or of sets of similar constants, *Norske Vid. Selsk. Forh. (Trondheim)* 34 (1961) 25-28; MR 25 #3011.
[106] D. H. Lehmer, Euler constants for arithmetical progressions, *Acta Arith.* 27 (1975) 125-142; MR 51 #5468.
[107] W. Leighton, Remarks on certain Eulerian constants, *Amer. Math. Monthly* 75 (1968) 283-285; MR 37 #3235.
[108] T. Tasaka, Note on the generalized Euler constants, *Math. J. Okayama Univ.* 36 (1994) 29-34; MR 96k:11157.

[109] J. B. Conway, *Functions of One Complex Variable*, 2nd ed., Springer-Verlag, 1978, pp. 176-187; MR 80c:30003.
[110] W. Rudin, *Principles of Mathematical Analysis*, 3rd ed., McGraw-Hill, 1976, pp. 192-195; MR 52 #5893.
[111] G. V. Chudnovsky, Algebraic independence of constants connected with exponential and elliptic functions (in Russian), *Dokl. Akad. Nauk Ukrain. SSR Ser. A* 8 (1976) 698-701, 767; MR 54 #12670.
[112] M. Waldschmidt, Les travaux de G. V. Cudnovskii sur les nombres transcendants, *Séminaire Bourbaki: 1975/76*, Lect. Notes in Math. 567, Springer-Verlag, pp. 274-292; MR 55 #12650.
[113] G. V. Chudnovsky, *Contributions to the Theory of Transcendental Numbers*, Amer. Math. Soc., 1984; MR 87a:11004.
[114] M. Waldschmidt, Algebraic independence of transcendental numbers: Gel'fond's method and its developments, *Perspectives in Mathematics: Anniversary of Oberwolfach*, ed. W. Jäger, J. Moser, and R. Remmert, Birkhäuser, 1984, pp. 551-571; MR 86f:11054.
[115] P. J. Davis, Leonhard Euler's integral: A historical profile of the Gamma function, *Amer. Math. Monthly* 66 (1959) 849-869; MR 21 #5540.
[116] C. L. Siegel, *Transcendental Numbers*, Princeton Univ. Press, 1949, pp. 95-100; MR 11,330c.
[117] Yu. V. Nesterenko, Algebraic independence of π and e^π, *Number Theory and Its Applications*, Proc. 1996 Ankara conf., ed. C. Y. Yildirim and S. A. Stepanov, Dekker, 1999, pp. 121-149; MR 99k:11113.
[118] Yu. V. Nesterenko, Modular functions and transcendence problems, *C. R. Acad. Sci. Paris Sér. I Math.* 322 (1996) 909-914; MR 97g:11080.
[119] M. Waldschmidt, Algebraic independence of transcendental numbers: A survey, *Number Theory*, ed. R. P. Bambah,V. C. Dumir, and R. J. Hans Gill, Birkhäuser, 2000, pp. 497-527.
[120] M. Waldschmidt, Sur la nature arithmétique des valeurs des fonctions modulaires, *Séminaire Bourbaki*, 49ème année 1996-97, #824; MR 99g:11089.
[121] F. Gramain, Quelques résultats d'indépendance algébrique, *International Congress of Mathematicians*, v. 2, Proc. 1998 Berlin conf., *Documenta Mathematica* Extra Volume II (1998) 173-182; MR 99h:11079.
[122] J. M. Borwein and I. J. Zucker, Fast evaluation of the gamma function for small rational fractions using complete elliptic integrals of the first kind, *IMA J. Numer. Anal.* 12 (1992) 519-529; MR 93g:65028.
[123] A. Eagle, *The Elliptic Functions as They Should Be*, Galloway and Porter, 1958; MR 20 #123 and MR 55 #6767.
[124] D. Wells, *The Penguin Dictionary of Curious and Interesting Numbers*, Penguin, 1986, p. 6.
[125] G. H. Hardy, Srinivasa Ramanujan obituary notice, *Proc. London Math. Soc.* 19 (1921) xl-lviii.
[126] S. Ramanujan, On question 330 of Professor Sanjana, *J. Indian Math. Soc.* 4 (1912) 59-61; also in *Collected Papers*, ed. G. H. Hardy, P. V. Seshu Aiyar, and B. M. Wilson, Cambridge Univ. Press, 1927, pp. 15-17.
[127] W. D. Fryer and M. S. Klamkin, Comment on problem 612, *Math. Mag.* 40 (1967) 52-53.
[128] G. N.Watson, Theorems stated by Ramanujan. XI, *J. London Math. Soc.* 6 (1931) 59-65.
[129] E. R. Hansen, *A Table of Series and Products*, Prentice-Hall, 1975.
[130] B. C. Berndt, *Ramanujan's Notebooks: Part II*, Springer-Verlag, 1989, pp. 11, 24, 41; MR 90b:01039.
[131] G. H. Hardy, Some formulae of Ramanujan, *Proc. London Math. Soc.* 22 (1924) xii-xiii; also in *Collected Papers*, v. 4, Oxford Univ. Press, 1966, pp. 517-518.
[132] J. Todd, The lemniscate constants, *Commun. ACM* 18 (1975) 14-19, 462; MR 51 #11935.
[133] W. Magnus and F. Oberhettinger, *Formulas and Theorems for the Special Functions of Mathematical Physics*, Chelsea, 1949.
[134] A. Erdélyi, W. Magnus, F. Oberhettinger, and F. G. Tricomi, *Higher Transcendental Functions*, v. 1, McGraw-Hill, 1953, pp. 1-55; MR 15,419i.

[135] J. Choi and H. M. Srivastava, Gamma function representation for some definite integrals, *Kyungpook Math. J.* 37 (1997) 205-209; MR 98g:33002.
[136] X. Gourdon and P. Sebah, The Gamma function (Numbers, Constants and Computation).

1.6　アペリの定数，$\zeta(3)$

　Apéry（アペリ）の定数 $\zeta(3)$ は，Riemann（リーマン）のゼータ関数

$$\zeta(x) = \sum_{n=1}^{\infty} \frac{1}{n^x}, \quad x>1$$

の $x=3$ における値として定義される．$\zeta(3)$ がアペリの定数と命名されたのは近年のことであるが，当然，その名を受けるに値するものである．1979 年に，Apéry は $\zeta(3)$ $=1.2020569031\cdots$ が無理数であるという驚くべき証明で，数学界に衝撃を与えた[1-10]．Riemann のゼータ関数について簡単に述べた後で，このことに再度言及することにしよう．

　x が正の偶数のとき，ゼータ関数の値は正確に定まり，

$$\zeta(2k) = \frac{(-1)^{k-1}(2\pi)^{2k} B_{2k}}{2(2k)!}$$

である[11-14]．ここで，$\{B_n\}$ は**ベルヌーイ数**【1.6.1】である．たとえば，

$$\zeta(2)=\frac{\pi^2}{6}, \quad \zeta(4)=\frac{\pi^4}{90}, \quad \zeta(6)=\frac{\pi^6}{945}$$

である．調和級数は発散するから，$\zeta(x)$ のここでの定義から，$\zeta(1)$ は定義できないことは明らかである．ゼータ関数は関数等式

$$\zeta(1-z) = \frac{2}{(2\pi)^z} \cos\left(\frac{\pi z}{2}\right) \Gamma(z) \zeta(z)$$

によって全複素数平面に解析接続され，$z=1$ で 1 位の極をもつ[15-19]．ここで，$\Gamma(z)=(z-1)!$ はガンマ関数【1.5.4】である．$\zeta(x)$ と素数定理の間の関係は，次の 2 つの公式

$$\zeta(x) = \prod_{p \, \text{素}} \left(1-\frac{1}{p^x}\right)^{-1}, \quad \frac{\zeta(2x)}{\zeta(x)} = \prod_{p \, \text{素}} \left(1+\frac{1}{p^x}\right)^{-1}$$

によって最も手短に示されている．有名なリーマン予想が，いつの日にか証明されたならば，素数の分布に関するさらに進んだ情報が得られるであろう．

　密接に関係する関数が

$$\eta(x) = \sum_{n=1}^{\infty} \frac{(-1)^{n-1}}{n^x}, \quad x>0$$

であり[20-22]，これは $x \neq 1$ のとき $(1-2^{1-x})\zeta(x)$ に等しい．たとえば，

$$\eta(1)=\ln(2), \quad \eta(2)=\frac{\pi^2}{12}, \quad \eta(4)=\frac{7\pi^4}{720}$$

である.

定数 $\zeta(3)$ は確率論的な解釈ができる側面ももっている. 3個のランダムな整数が与えられたとき, そのすべての整数を割る1より大きい因数が存在しない確率は $1/\zeta(3)$ =0.8319073725… である (大きな区間で考えて, その極限として). これとは対照的に, 3個の整数が2個ずつ互いに素である確率は, わずか 0.2867474284… である.【2.5】の公式を参照せよ. n を 2 の累乗とし, $C(n)$ を p を素数として方程式 $n=p+ij$ を満たす整数解 (i,j,p) の個数とする[25,26]. このとき, $\lim_{n\to\infty}C(n)/n=105\,\zeta(3)/(2\pi^4)$ である. その他, 整数論における $\zeta(3)$ の登場は[27-30]に見られ,【2.7】でも取り上げられる. またそれは, 最小全域木 (spanning tree) の長さに関するランダムグラフの理論でも出現する【8.5】.

Apéry の成果を, 任意の $k>1$ に対して $\zeta(2k+1)$ に拡張することは, van der Poorten (ファン・デァ・ポールテン) が書いたように「謎につつまれた神秘」[2]として残されている. $\zeta(3)$ が超越数であるかどうかや $\zeta(3)/\pi^3$ が無理数であるかどうかさえも未解決の問題として残されている. 最近, Rivoal (リヴォール)[31,32]は $\zeta(2k+1)$ が無理数であるような k が無限にあることを証明し, さらに, Zudilin (ズディリン)[33,34]は $\zeta(5)$, $\zeta(7)$, $\zeta(9)$, $\zeta(11)$ の少なくとも 1 つが無理数であることを示した. Apéry による $\zeta(3)$ の無理数性の証明以後, 関連する話題の中で, これが最も劇的なものである.

1.6.1 ベルヌーイ数

Bernoulli (ベルヌーイ) 数 $\{B_n\}$ を生成関数

$$\frac{x}{e^x-1}=\sum_{k=0}^{\infty}B_k\frac{x^k}{k!}$$

によって定義しよう[7,19-22]. すると, $B_0=1$, $B_1=-1/2$, $B_2=1/6$, $B_4=-1/30$, $B_6=1/42$, そして, $n>0$ に対して $B_{2n+1}=0$ である.

(運の悪いことに, ベルヌーイ数についてはもう1つの定義があり, 我々を困惑させる. このもう1つの定義では, 添え字の使い方が多少異なっており, すべての数が正である. 論文を読むときは, どの定義が使われているかに注意しなくてはならない.)

ベルヌーイ数は

$$\tan(x)=\sum_{k=1}^{\infty}\frac{(-1)^{k+1}2^{2k}(2^{2k}-1)B_{2k}}{(2k)!}x^{2k-1}$$

のような級数展開にも現れる.

1.6.2 リーマン予想

Wiles (ワイルズ) による Fermat (フェルマー) の最終定理の証明が承認されたいま, 数学における最も名高い未解決問題は, リーマン予想になった. この予想は, 帯状

領域 $0 \leq \text{Re}(z) \leq 1$ (臨界帯) にある $\zeta(z)$ の零点はすべて $\text{Re}(z) = 1/2$ という直線上にあるというものである．

リーマン予想をまったく初等的に言い換えたものがある[35]．平方因子をもたない正の整数が，偶数個の相異なる素数の積であるとき，その整数は**赤**であると定義する．また，奇数個の積であるとき，**青**であると定義する．$R(n)$ を n を超えない赤の整数の個数とし，$B(n)$ を n を超えない青の整数の個数とすれば，リーマン予想は次の命題と同値である．任意の $\varepsilon > 0$ に対して，整数 N が存在して，すべての $n > N$ に対して
$$|R(n) - B(n)| < n^{1/2+\varepsilon}$$
が成立する．これは通常 Möbius（メビウス）のミュー関数の言葉で述べられる【2.2】．$\varepsilon = 0$ とすることは不可能であり，それは Mertens（メルテンス）予想として知られていた結果が誤りであったことに抵触する．

いくつかある同値な命題[36, 37]の中の1つに，次のものがある．リーマン予想が真であるための必要十分条件は，
$$\int_0^\infty \int_{1/2}^\infty \frac{1 - 12y^2}{(1 + 4y^2)^3} \ln|\zeta(x + iy)| dx dy = \frac{3 - \gamma}{32}\pi$$
が成立することである[38]．ここで γ はオイラー–マスケローニの定数である【1.5】．このような条件つきで導かれる等式と，条件なしで導かれる公式を比較してみるとおもしろい．たとえば，臨界帯における零点 ρ のすべての集合を Z とするとき，
$$\sum_{\rho \in Z} \frac{1}{\rho} = \frac{1}{2}\gamma + 1 - \ln(2) - \frac{1}{2}\ln(\pi) = 0.0230957089\cdots$$
である[39–41]．すなわち，零点の場所が謎として残ってはいるが，零点の逆数の和を正確に計算できる程度にはわかっている．注意すると，$\sum_\rho |\rho|^{-1}$ は発散するが，$\sum_\rho \rho^{-1}$ は複素共役な項を一緒に組み合わせて和をとれば収束する．

リーマン予想からの帰結（多くある中で特に[17]）の1つが，【2.13】で取り上げられる．いつの日か運よく証明が見つけられれば我々の素数分布に関する理解はより深いものになるであろう．de Bruijn-Newman（ド・ブルイン-ニューマン）定数についての小論は，計算という立場からのアプローチについて詳しく述べたものである【2.32】．GUE 予想（Gaussian unitary ensemble hypothesis）といわれるより深い予想は，零点と零点の間の垂直方向の間隔の分布を制約するものである【2.15.3】．

1.6.3 級　　数

等差数列上で和を考えると，少し異なった形のものが得られる[42, 43]．
$$\lambda(3) = \sum_{k=0}^\infty \frac{1}{(2k+1)^3} = \frac{7}{8}\zeta(3), \quad \sum_{k=0}^\infty \frac{1}{(3k+1)^3} = \frac{2\pi^3}{81\sqrt{3}} + \frac{13}{27}\zeta(3)$$
$$\sum_{k=0}^\infty \frac{1}{(4k+1)^3} = \frac{\pi^3}{64} + \frac{7}{16}\zeta(3), \quad \sum_{k=0}^\infty \frac{1}{(6k+1)^3} = \frac{\pi^3}{36\sqrt{3}} + \frac{91}{216}\zeta(3)$$
$\lambda(x)$ については後ほど【1.7】で取り上げる．二項係数の中央項を含む次の2つの級数

$$\sum_{k=1}^{\infty}\frac{(-1)^{k+1}}{k^3\binom{2k}{k}}=\frac{2}{5}\zeta(3), \quad \sum_{k=1}^{\infty}\frac{30k-11}{(2k-1)k^3\binom{2k}{k}^2}=4\zeta(3)$$

がある[42,44-47]．前の式は Apéry が彼の研究に活用したことで有名になった公式である．

$\zeta(2n)$ に対するきっちりとした公式に対応するような $\zeta(2n+1)$ の類似表現は何であろうか？　これは誰にも知られていないが，Grosswald（グロスワルド）[48-51]によって得られた級数

$$\zeta(3)=\frac{7}{180}\pi^3-2\sum_{k=1}^{\infty}\frac{1}{k^3(e^{2\pi k}-1)}, \quad \zeta(7)=\frac{19}{56700}\pi^7-2\sum_{k=1}^{\infty}\frac{1}{k^7(e^{2\pi k}-1)}$$

や Plouffe（プルーフェ）[52] と Borwein（ボーウィン）[26,53] によって得られた

$$\zeta(5)=\frac{1}{294}\pi^5-\frac{72}{35}\sum_{k=1}^{\infty}\frac{1}{k^5(e^{2\pi k}-1)}-\frac{2}{35}\sum_{k=1}^{\infty}\frac{1}{k^5(e^{2\pi k}+1)}$$

があり，これらは，先駆的な名誉を担うものとなろう．これらの公式は Ramanujan のノートブックにあるいくつかの記事によって示唆されたものであった[54]．

いくつかの多重級数

$$\sum_{i=1}^{\infty}\sum_{j=1}^{\infty}\frac{1}{ij(i+j)}=2\zeta(3), \quad \sum_{i=1}^{\infty}\sum_{j=1}^{\infty}\frac{(-1)^{i-1}}{ij(i+j)}=\frac{5}{8}\zeta(3)$$

$$\sum_{i=1}^{\infty}\sum_{j=1}^{\infty}\frac{(-1)^{i+j}}{ij(i+j)}=\frac{1}{4}\zeta(3), \quad \sum_{i=2}^{\infty}\sum_{j=1}^{i-1}\frac{1}{i^2j}=\zeta(3)$$

$$\sum_{i=3}^{\infty}\sum_{j=2}^{i-1}\sum_{k=1}^{j-1}\frac{1}{i^3j^2k}=-\frac{29}{6480}\pi^6+3\zeta(3)^2$$

が[55-62]で取り上げられており，また，これらに類した（いろいろな深さをもつ）評価式が知られている[63-75]．

$0<x<1$ であるとき，次の式

$$\lim_{n\to\infty}\Bigl(\sum_{k=1}^{n}\frac{1}{k^x}-\frac{n^{1-x}}{1-x}\Bigr)=\zeta(x)=(1-2^{1-x})^{-1}\eta(x)=\frac{-1}{2^{1-x}-1}\sum_{k=1}^{\infty}\frac{(-1)^{k-1}}{k^x}$$

は確かに成立する[19]．たとえば，$x=1/2$ のとき，この極限値はオイラーの定数【1.5.3】の項で注意したように

$$\lim_{n\to\infty}\Bigl(1+\frac{1}{\sqrt{2}}+\cdots+\frac{1}{\sqrt{n}}-2\sqrt{n}\Bigr)=(-\sqrt{2}+1)\Bigl(1-\frac{1}{\sqrt{2}}+\frac{1}{\sqrt{3}}-+\cdots\Bigr)$$
$$=-1.4603545088\cdots$$

である[76]．【1.5.1】から再考すれば，

$$\gamma=\sum_{k=2}^{\infty}(-1)^k\frac{\zeta(k)}{k}, \quad 1-\gamma=\sum_{k=2}^{\infty}\frac{\zeta(k)-1}{k}$$

である．ゼータ関数の値を含む級数の注目すべきものに

$$S(n)=\sum_{k=1}^{\infty}\frac{\zeta(2k)}{(2k+n)2^{2k-1}}, \quad n=0,1,2,\cdots$$

がある[77,78]．たとえば，

1.6 アペリの定数，$\zeta(3)$

$$S(0)=\ln(\pi)-\ln(2), \quad S(1)=-\ln(2)+1, \quad S(2)=\frac{7}{2\pi^2}\zeta(3)-\ln(2)+\frac{1}{2}$$

$$S(3)=\frac{9}{2\pi^2}\zeta(3)-\ln(2)+\frac{1}{3}, \quad S(4)=-\frac{93}{2\pi^4}\zeta(5)+\frac{9}{\pi^2}\zeta(3)-\ln(2)+\frac{1}{4}$$

である[79-83]．これらのものは，さらに収束の早い級数を作るために，（部分分数展開による）さまざまな方法で，組み合わせて使うことができる．たとえば，Euler による公式は，

$$\sum_{k=1}^{\infty}\frac{\zeta(2k)}{(2k+1)(2k+2)2^{2k}}=-\frac{7}{4\pi^2}\zeta(3)+\frac{1}{4}$$

であり[84-89]，Wilton（ウィルトン）による公式は，

$$\sum_{k=1}^{\infty}\frac{\zeta(2k)}{k(k+1)(2k+1)(2k+3)2^{2k}}=\frac{2}{\pi^2}\zeta(3)-\frac{11}{18}+\frac{1}{3}\ln(\pi)$$

である[90-92]．その他，多くの級数がある[93-102]．

Broodhurst（ブラッドハースト）[103]は，π に対する Bailey-Borwein-Plouffe（ベイリー–ボーウィン–プルーフェ）のアルゴリズム【1.4】に似た $\zeta(3)$ と $\zeta(5)$ に対する桁ごとの抽出アルゴリズムを提案した．$\zeta(3)$ に対する級数は

$$\zeta(3)=\frac{48}{7}\sum_{k=0}^{\infty}\frac{1}{2\cdot 16^k}\Big(\frac{1}{(8k+1)^3}-\frac{7}{(8k+2)^3}-\frac{1}{2(8k+3)^3}+\frac{10}{2(8k+4)^3}-\frac{1}{2^2(8k+5)^3}$$
$$-\frac{7}{2^2(8k+6)^3}+\frac{1}{2^3(8k+7)^3}\Big)+\frac{32}{7}\sum_{k=0}^{\infty}\frac{1}{8\cdot 16^{3k}}\Big(\frac{1}{(8k+1)^3}+\frac{1}{2(8k+2)^3}-\frac{1}{2^3(8k+3)^3}$$
$$-\frac{2}{2^4(8k+4)^3}-\frac{1}{2^6(8k+5)^3}+\frac{1}{2^7(8k+6)^3}+\frac{1}{2^9(8k+7)^3}\Big)$$

である．

Amdeberhan（アムデバーハン），Zeilberger（ザイルバーガー），Wilf（ウィルフ）[104-106]は，$\zeta(3)$ を計算するための非常に早く収束する級数を発見した．これにより，現在 $\zeta(3)$ の値が数億桁まで知られている．[107-110]も参照せよ．さらに，次の事実

$$\sum_{k=1}^{\infty}\frac{(-1)^k}{k^3(k+1)^3}=10-\frac{3}{2}\zeta(3)-12\ln(2),$$

$$\text{Li}_3\Big(\frac{1}{2}\Big)=\frac{7}{8}\zeta(3)+\frac{\pi^2}{12}\ln\Big(\frac{1}{2}\Big)-\frac{1}{6}\ln\Big(\frac{1}{2}\Big)^3,$$

$$\text{Li}_3(2-\varphi)=\frac{4}{5}\zeta(3)+\frac{\pi^2}{15}\ln(2-\varphi)-\frac{1}{12}\ln(2-\varphi)^3$$

に注意しよう[111-114]．ここで，Li_3 は【1.6.8】で定義する三重対数関数であり，φ は黄金比である【1.2】．

最後に，$\zeta(4n+3)$ の生成関数は

$$\sum_{n=0}^{\infty}\zeta(4n+3)x^n=\frac{5}{2}\sum_{i=1}^{\infty}\frac{(-1)^{i+1}}{i^3\binom{2i}{i}}\frac{1}{1-\frac{x}{i^4}}\prod_{j=1}^{i-1}\frac{j^4+4x}{j^4-x}, \quad |x|<1$$

である[115,116]．Apéryの級数は$x=0$の場合である．両辺をxで微分して$x=0$とおくと，$\zeta(7)$に対する級数

$$\zeta(7) = \frac{5}{2}\sum_{k=1}^{\infty}\frac{(-1)^{k+1}}{k^7\binom{2k}{k}} + \frac{25}{2}\sum_{k=1}^{\infty}\frac{(-1)^{k+1}}{k^3\binom{2k}{k}}\sum_{m=1}^{k-1}\frac{1}{m^4}$$

が得られ，その他のnに対しても同様の式が得られる．$\zeta(4n+1)$に対する類似の生成関数は知られていない．級数[117]

$$\zeta(5) = 2\sum_{k=1}^{\infty}\frac{(-1)^{k+1}}{k^5\binom{2k}{k}} - \frac{5}{2}\sum_{k=1}^{\infty}\frac{(-1)^{k+1}}{k^3\binom{2k}{k}}\sum_{m=1}^{k-1}\frac{1}{m^2}$$

はどのように拡張できるのであろうか？

1.6.4 無限積

Gosper（ゴスパー）[118]による行列の無限積の一族があり，ひときわ目立っている．最も単純な結果は

$$\prod_{k=1}^{\infty}\begin{pmatrix} -\frac{k}{2(2k+1)} & \frac{5}{4k^2} \\ 0 & 1 \end{pmatrix} = \begin{pmatrix} 0 & \zeta(3) \\ 0 & 1 \end{pmatrix}$$

であり，前出の二項係数の中央項の和と同値なものである．一般的な場合は，$n \geq 2$として$(n+1) \times (n+1)$の上三角行列

$$\prod_{k=1}^{\infty}\begin{bmatrix} -\frac{k}{2(2k+1)} & \frac{1}{2k(2k+1)} & 0 & \cdots & 0 & \frac{1}{k^{2n}} \\ 0 & -\frac{k}{2(2k+1)} & \frac{1}{2k(2k+1)} & \cdots & 0 & \frac{1}{k^{2n-2}} \\ \vdots & \vdots & \vdots & & \vdots & \vdots \\ 0 & 0 & 0 & \cdots & \frac{1}{2k(2k+1)} & \frac{1}{k^4} \\ 0 & 0 & 0 & \cdots & -\frac{k}{2(2k+1)} & \frac{5}{4k^2} \\ 0 & 0 & 0 & \cdots & 0 & 1 \end{bmatrix}$$

$$= \begin{bmatrix} 0 & \cdots & 0 & \zeta(2n+1) \\ 0 & \cdots & 0 & \zeta(2n-1) \\ \vdots & & \vdots & \vdots \\ 0 & \cdots & 0 & \zeta(5) \\ 0 & \cdots & 0 & \zeta(3) \\ 0 & \cdots & 0 & 1 \end{bmatrix}$$

である．対角要素と上対角要素が上記のように重複して並べられ，最終列がk^{2m}の逆数であり，その他の成分はすべて0となっている．

1.6.5 積 分

リーマンのゼータ関数は，$x>1$ として

$$\zeta(x) = \frac{1}{\Gamma(x)} \int_0^\infty \frac{t^{x-1}}{e^t - 1} dt$$

という別の表現をもつ[17]．$\{t\}$ を t の小数部分とすれば，

$$\int_1^\infty \frac{\{t\}}{t^{x+1}} dt = \begin{cases} \dfrac{1}{x-1} - \dfrac{\zeta(x)}{x} & (0<x<1 \text{ または } x>1 \text{ のとき}) \\ 1 - \gamma & (x=1 \text{ のとき}) \end{cases}$$

である[18,19]．その他の x に対しては，積分は発散する．しかしながら，次の区間での敏速な調整が可能で

$$\int_1^\infty \frac{\{t\} - \frac{1}{2}}{t^{x+1}} dt = \begin{cases} \dfrac{1}{x-1} - \dfrac{1}{2x} - \dfrac{\zeta(x)}{x} & (-1<x<0 \text{ のとき}) \\ \dfrac{1}{2}\ln(2\pi) - 1 & (x=0 \text{ のとき}) \end{cases}$$

である．

Munthe Hjortnaes（ムンテ・ヒョルトネス）[119]は

$$\zeta(3) = 10 \int_0^{\ln(\varphi)} x^2 \coth(x) \, dx = 10 \int_0^{1/2} \frac{\operatorname{arcsinh}(y)^2}{y} dy$$

を証明した．部分積分法を用いれば

$$\zeta(3) = -5 \int_0^{2\ln(\varphi)} \theta \ln\left(2 \sinh\left(\frac{\theta}{2}\right)\right) d\theta$$

が得られる[120]．Euler の積分表現

$$4 \int_0^\pi \theta \ln\left(\sin\left(\frac{\theta}{2}\right)\right) d\theta = 7\zeta(3) - 2\pi^2 \ln(2)$$

から出発すると，前と同様に（ただし逆向きに）部分積分法を用いれば

$$-8 \int_0^1 \frac{\arcsin(y)^2}{y} dy = -8 \int_0^{\pi/2} x^2 \cot(x) \, dx = 7\zeta(3) - 2\pi^2 \ln(2)$$

が得られる[80,81]．

1.6.6 連 分 数

Stieltjes[122] と Ramanujan[54]は連分数展開

$$\zeta(3) = 1 + \cfrac{1|}{|2 \cdot 2} + \cfrac{1^3|}{|1} + \cfrac{1^3|}{|6 \cdot 2} + \cfrac{2^3|}{|1} + \cfrac{2^3|}{|10 \cdot 2} + \cfrac{3^3|}{|1} + \cfrac{3^3|}{|14 \cdot 2} + \cdots$$

を発見した．2つずつの項をまとめることで

$$\zeta(3) = 1 + \cfrac{1|}{|5} - \cfrac{1^6|}{|21} - \cfrac{2^6|}{|55} - \cfrac{3^6|}{|119} - \cfrac{4^6|}{|225} - \cfrac{5^6|}{|385} - \cdots$$

を得る．ここで，部分分母は多項式 $2n^3 + 3n^2 + 11n + 5$ によって生成される．この連分数展開の収束する速さは，$\zeta(3)$ の無理数性を証明するのに充分なほど早くはない．Apéry は

$$\zeta(3) = \frac{6}{5} \bigg| \frac{1^6}{117} \bigg| \frac{2^6}{535} \bigg| \frac{3^6}{1463} \bigg| \frac{4^6}{3105} \bigg| \frac{5^6}{5665} \bigg| \cdots$$

として収束を速めることに成功した．ここで，部分分母は多項式 $34n^3+51n^2+27n+5$ によって生成される．

1.6.7 スターリングのサイクル数

n 個の記号の置換で，ちょうど m 個のサイクル（巡回置換）をもつものの数を $s_{n,m}$ とする[123]．この数 $s_{n,m}$ は**第1種のスターリング**（Stirling）**数**とよばれ，漸化式

$$s_{n,0} = \begin{cases} 1 & (n=0 \text{ のとき}) \\ 0 & (n \geq 1 \text{ のとき}) \end{cases}$$

$$s_{n,m} = (n-1)s_{n-1,m} + s_{n-1,m-1} \quad (n \geq m \geq 1 \text{ のとき})$$

を満たす．たとえば，(123)，(321) は相異なる置換であるから $s_{3,1}=2$ である．さらに一般に，$s_{n,1}=(n-1)!$ であり，$s_{n,2}=(n-1)!\sum_{k=1}^{n-1}1/k$ である．$m \geq 3$ である m に対して，高次の調和級数和を含む複雑な公式が適用される．この結果，$m \geq 1$ に対して

$$\sum_{n=1}^{\infty} \frac{s_{n,m}}{n!\,n} = \zeta(m+1)$$

である[124]．$m=2$ の場合は，先の（Euler[67]による）多重級数の1つから従う．$n \to \infty$ としたときの $s_{n,m}$ の漸近的挙動が[125]に見られる．

1.6.8 多重対数関数

多重対数関数 Li_n を定義する前に，次の問題について考えてみよう．次の式

$$(-1)^k k!\,\zeta(k+1) = \int_0^1 \frac{\ln(x)^k}{1-x}\,dx, \quad k=1,2,3,\cdots$$

が知られている．ここで，積分区間を $[0,1]$ から $[1,2]$ に変更したらどうなるであろうか？　Ramanujan[42]は，

$$a_k = \int_1^2 \frac{\ln(x)^k}{1-x}\,dx$$

とするとき，$a_1 = \zeta(2)/2 = \pi^2/12$，$a_2 = \zeta(3)/4$ を示した．このパターンが続くことが期待され，また，すべての $k \geq 1$ に対して，a_k が $\zeta(k+1)$ の有理数倍になることが期待された．しかしながら，このことは $k=3$ に対してさえも正しいということにはならない．$\text{Li}_1(x) = -\ln(1-x)$ とし，$n \geq 2$ なる整数に対して

$$\text{Li}_n(x) = \sum_{k=1}^{\infty} \frac{x^k}{k^n} = \int_0^x \frac{\text{Li}_{n-1}(t)}{t}\,dt \quad (|x| \leq 1)$$

と定義する[113,114]．明らかに $\text{Li}_n(1) = \zeta(n)$ である．我々は先に，Landen（ランデン）による三重対数関数 Li_3 のいくつかの値について注意した．四重対数関数についてはそれほど知られてはいないが，Levin（レビン）[126]は

$$a_3 = \frac{\pi^4}{15} + \frac{\pi^2 \ln(2)^2}{4} - \frac{\ln(2)^4}{4} - \frac{21\ln(2)}{4}\zeta(3) - 6\text{Li}_4\!\left(\frac{1}{2}\right)$$

などを証明した．したがって，我々の問題に対して完全に答えるには，$\text{Li}_n(1/2)$ の算術的な性質への理解が必要となる．多重対数関数についてのさらなる詳細は [127-131] に見られる．

[1] R. Apéry, Irrationalité de $\zeta(2)$ et $\zeta(3)$, *Astérisque* 61 (1979) 11-13.
[2] A. van der Poorten, A proof that Euler missed... Apéry's proof of the irrationality of $\zeta(3)$, *Math. Intellig.* 1 (1979) 196-203; MR 80i:10054.
[3] A. van der Poorten, Some wonderful formulae... footnotes to Apéry's proof of the irrationality of $\zeta(3)$, *Séminaire Delange-Pisot-Poitu (Théorie des nombres)*, 20e année, 1978/79, n. 29, pp. 1-7; MR 82a:10037.
[4] A. van der Poorten, Some wonderful formulas... an introduction to polylogarithms, *Queen's Papers in Pure and Applied Mathematics*, n. 54, Proc. 1979 Queen's Number Theory Conf., ed. P. Ribenboim, 1980, pp. 269-286; MR 83b:10043.
[5] F. Beukers, A note on the irrationality of $\zeta(2)$ and $\zeta(3)$, *Bull. London Math. Soc.* 11 (1979) 268-272; MR 81j:10045.
[6] M. Prévost, A new proof of the irrationality of $\zeta(3)$ using Padé approximants, *J. Comput. Appl. Math.* 67 (1996) 219-235; MR 97f:11056.
[7] J. M. Borwein and P. B. Borwein, *Pi and the AGM: A Study in Analytic Number Theory and Computational Complexity*, Wiley, 1987, pp. 362-386; MR 99h:11147.
[8] P. Borwein and T. Erdélyi, *Polynomials and Polynomial Inequalities*, Springer-Verlag, 1995, pp. 372-381; MR 97e:41001.
[9] G. E. Andrews, R. Askey, and R. Roy, *Special Functions*, Cambridge Univ. Press, 1999, pp. 391-394; MR 2000g:33001.
[10] D. Huylebrouck, Similarities in irrationality proofs for π, ln(2), $\zeta(2)$, and $\zeta(3)$, *Amer. Math. Monthly* 108 (2001) 222-231; MR 2002b:11095.
[11] B. C. Berndt, Elementary evaluation of $\zeta(2n)$, *Math. Mag.* 48 (1975) 148-153; MR 51 #3078.
[12] F. Beukers, J. A. C. Kolk, and E. Calabi, Sums of generalized harmonic series and volumes, *Nieuw Arch. Wisk.* 4 (1993) 217-224; MR 94j:11022.
[13] R. Chapman, Evaluating $\zeta(2)$, unpublished note (1998).
[14] N. D. Elkies, On the sums $\sum_{k=-\infty}^{\infty}(4k+1)^{-n}$, math.CA/0101168.
[15] J. B. Conway, *Functions of One Complex Variable*, 2nd ed., Springer-Verlag, 1978, pp. 187-194; MR 80c:30003.
[16] W. Ellison and F. Ellison, *Prime Numbers*, Wiley, 1985, pp. 147-152; MR 87a:11082.
[17] E. C. Titchmarsh, *The Theory of the Riemann Zeta Function*, 2nd ed., rev. by D. R. Heath-Brown, Oxford Univ. Press, 1986, pp. 18-19, 282-328; MR 88c:11049.
[18] A. Ivić, *The Riemann Zeta-Function*, Wiley, 1985, pp. 1-12; MR 87d:11062.
[19] T. M. Apostol, *Introduction to Analytic Number Theory*, Springer-Verlag, 1976, pp. 55-56, 249-267; MR 55 #7892.
[20] M. Abramowitz and I. A. Stegun, *Handbook of Mathematical Functions*, Dover, 1972, pp. 807-808; MR 94b:00012.
[21] J. Spanier and K. B. Oldham, *An Atlas of Functions*, Hemisphere, 1987, pp. 25-33.
[22] I. S. Gradshteyn and I. M Ryzhik, *Tables of Integrals, Series and Products*, Academic Press, 1980; MR 97c:00014.
[23] J. E. Nymann, On the probability that k positive integers are relatively prime, *J. Number Theory* 4 (1972) 469-473; MR 46 #3478.
[24] P. Moree, Counting carefree couples, unpublished note (1999).
[25] B. M. Bredihin, Applications of the dispersion method in binary additive problems (in Russian), *Dokl. Akad. Nauk SSSR* 149 (1963) 9-11; MR 26 #2419.
[26] J. M. Borwein, D. M. Bradley, and R. E. Crandall, Computational strategies for the Riemann

zeta function, *J. Comput. Appl. Math.* 121 (2000) 247-296; CECM preprint 98:118; MR 2001h: 11110.
[27] S. Akiyama, A new type of inclusion exclusion principle for sequences and asymptotic formulas for $\zeta(k)$, *J. Number Theory* 45 (1993) 200-214; MR 94k:11027.
[28] S. Akiyama, A criterion to estimate the least common multiple of sequences and asymptotic formulas for $\zeta(3)$ arising from recurrence relation of an elliptic function, *Japan. J. Math.* 22 (1996) 129-146; MR 97f:11021.
[29] V. Strehl, Recurrences and Legendre transform, *Sémin. Lothar. Combin.* 29 (1992) B29b; MR 96m:11017.
[30] Y. Lan, A limit formula for $\zeta(2k+1)$, *J. Number Theory* 78 (1999) 271-286; MR 2000f:11102.
[31] T. Rivoal, La fonction zêta de Riemann prend une infinité de valeurs irrationnelles aux entiers impairs, *C. R. Acad. Sci. Paris* 331 (2000) 267-270; math.NT/0008051; MR 2001k:11138.
[32] T. Rivoal, Irrationalité d'au moins un des neuf nombres $\zeta(5)$, $\zeta(7)$, ..., $\zeta(21)$, *Acta Arith.* 103 (2002) 157-167; math.NT/0104221.
[33] W. Zudilin, On the irrationality of the values of the zeta function at odd points (in Russian), *Uspekhi Mat. Nauk* 56 (2001) 215-216; Engl. transl. in *Russian Math. Surveys* 56 (2001) 423-424.
[34] W. Zudilin, One of the numbers $\zeta(5)$, $\zeta(7)$, $\zeta(9)$, $\zeta(11)$ is irrational (in Russian), *Uspekhi Mat. Nauk* 56 (2001) 149-150; Engl. transl. in *Russian Math. Surveys* 56 (2001) 774-776; MR 2002g: 11098.
[35] H. S. Wilf, A greeting; and a view of Riemann's hypothesis, *Amer. Math. Monthly* 94 (1987) 3-6; MR 88a:11082.
[36] F. T. Wang, A note on the Riemann zeta-function, *Bull. Amer. Math. Soc.* 52 (1946) 319-321; MR 7,417a.
[37] M. Balazard, E. Saias, and M. Yor, Notes sur la fonction ζ de Riemann. II, *Adv. Math.* 143 (1999) 284-287; MR 2000c:11140.
[38] V. V. Volchkov, On an equality equivalent to the Riemann hypothesis (in Ukrainian), *Ukrain. Math. Zh.* 47 (1995) 422-423; Engl. transl. in *Ukrainian Math. J.* 47 (1995) 491-493; MR 96g:11111.
[39] H. M. Edwards, *Riemann's Zeta Function*, Academic Press, 1974, pp. 67, 159-160; MR 57 # 5922.
[40] H. Davenport, *Multiplicative Number Theory*, 2nd ed., rev. by H. L. Montgomery, Springer-Verlag, 1980, pp. 79-83; MR 82m:10001.
[41] S. J. Patterson, *An Introduction to the Theory of the Riemann Zeta Function*, Cambridge Univ. Press, 1988, pp. 33-34; MR 89d:11072.
[42] B. C. Berndt, *Ramanujan's Notebooks: Part I*, Springer-Verlag, 1985, pp. 163-164, 232, 290-293; MR 86c:01062.
[43] E. R. Hansen, *A Table of Series and Products*, Prentice-Hall, 1975.
[44] J. M. Borwein, D. J. Broadhurst, and J. Kamnitzer, Central binomial sums, multiple Clausen values and zeta values, hep-th/0004153.
[45] S. Plouffe, The art of inspired guessing, unpublished note (1998).
[46] R. W. Gosper, A calculus of series rearrangements, *Algorithms and Complexity: New Directions and Recent Results*, Proc. 1976 Carnegie-Mellon conf., ed. J. F. Traub, Academic Press, 1976, pp. 121-151; MR 56 # 9899.
[47] R. W. Gosper, Strip mining in the abandoned orefields of nineteenth century mathematics, *Computers in Mathematics*, Proc. 1986 Stanford Univ. conf., ed. D. V. Chudnovsky and R. D. Jenks, Dekker, 1990, pp. 261-284; MR 91h:11154.
[48] E. Grosswald, DieWerte der Riemannschen Zetafunktion an ungeraden Argumentstellen, *Nachr. Akad. Wiss. Göttingen, Math.-Phys. Klasse* 2 (1970) 9-13; MR 42 # 7606.
[49] D. Shanks, Calculation and applications of Epstein zeta functions, *Math. Comp.* 29 (1975) 271-287; corrigenda 29 (1975) 1167 and 30 (1976) 900; MR 53 # 13114a-c.
[50] D. H. Bailey, J. M. Borwein, and R. E. Crandall, On the Khintchine constant, *Math. Comp.* 66 (1997) 417-431; CECM preprint 95:036; MR 97c:11119.
[51] H. Cohen, High precision computation of Hardy-Littlewood constants, unpublished note

(1999).
[52] S. Plouffe, Identities inspired from Ramanujan Notebooks. II, unpublished note (1998).
[53] J. M. Borwein, Experimental mathematics: Insight from computation, presentation at *AMS/ MAA Joint Meetings*, San Antonio, 1999.
[54] B. C. Berndt, *Ramanujan's Notebooks: Part II*, Springer-Verlag, 1989, pp. 153-155, 275-276, 293; MR 90b:01039.
[55] W. E. Briggs, S. Chowla, A. J. Kempner, and W. E. Mientka, On some infinite series, *Scripta Math.* 21 (1955) 28-30; MR 16,1014d.
[56] L. J. Mordell, On the evaluation of some multiple series, *J. London Math. Soc.* 33 (1958) 368-371; MR 20 # 6615.
[57] R. Sitaramachandrarao and A. Sivaramasarma, Some identities involving the Riemann zeta function, *Indian J. Pure Appl. Math.* 10 (1979) 602-607; MR 80h:10047.
[58] M. V. Subbarao and R. Sitaramachandrarao, On some infinite series of L. J. Mordell and their analogues, *Pacific J. Math.* 119 (1985) 245-255; MR 87c:11091.
[59] R. Sitaramachandrarao, A formula of S. Ramanujan, *J. Number Theory* 25 (1987) 1-19; MR 88c:11048.
[60] M. E. Hoffman, Multiple harmonic series, *Pacific J. Math.* 152 (1992) 275-290; MR 92i:11089.
[61] C. Markett, Triple sums and the Riemann zeta function, *J. Number Theory* 48 (1994) 113-132; MR 95f:11067.
[62] M. E. Hoffman and C. Moen, Sums of triple harmonic series, *J. Number Theory* 60 (1996) 329-331; MR 98a:11113.
[63] D. Zagier, Values of zeta functions and their applications, *First European Congress of Mathematics*, v. 2, Paris, 1992, ed. A. Joseph, F. Mignot, F. Murat, B. Prum, and R. Rentschler, Birkhäuser, 1994, pp. 497-512; MR 96k:11110.
[64] D. Borwein and J. Borwein, On an intriguing integral and some series related to $\zeta(4)$, *Proc. Amer. Math. Soc.* 123 (1995) 1191-1198; MR 95e:11137.
[65] D. H. Bailey, J. M. Borwein, and R. Girgensohn, Experimental evaluation of Euler sums, *Experim. Math.* 3 (1994) 17-30; MR 96e:11168.
[66] D. Borwein, J. M. Borwein, and R. Girgensohn, Explicit evaluation of Euler sums, *Proc. Edinburgh Math. Soc.* 38 (1995) 277-294; MR 96f:11106.
[67] P. Flajolet and B. Salvy, Euler sums and contour integral representations, *Experim. Math.* 7 (1998) 15-35; MR 99c:11110.
[68] J. M. Borwein and R. Girgensohn, Evaluation of triple Euler sums, *Elec. J. Combin.* 3 (1996) R23; MR 97d:11137.
[69] J. M. Borwein, D. M. Bradley, and D. J. Broadhurst, Evaluations of k-fold Euler/Zagier sums: A compendium of results for arbitrary k, *Elec. J. Combin.* 4 (1997) R5; MR 98b:11091.
[70] J. M. Borwein, D. M. Bradley, D. J. Broadhurst, and P. Lisonek, Combinatorial aspects of multiple zeta values, *Elec. J. Combin.* 5 (1998) R38; math.NT/9812020; MR 99g:11100.
[71] J. M. Borwein, D. M. Bradley, D. J. Broadhurst, and P. Lisonek, Special values of multidimensional polylogarithms, *Trans. Amer. Math. Soc.* 353 (2001) 907-941; math.CA/9910045; CECM preprint 98:106.
[72] N. R. Farnum and A. Tissier, Apéry's constant, *Amer. Math. Monthly* 106 (1999) 965-966.
[73] A. Granville, A decomposition of Riemann's zeta-function, *Analytic Number Theory*, Proc. 1996 Kyoto conf., ed. Y. Motohashi, Cambridge Univ. Press, 1997, pp. 95-101; MR 2000c:11134.
[74] O. M. Ogreid and P. Osland, Summing one- and two-dimensional series related to the Euler series, *J. Comput. Appl. Math.* 98 (1998) 245-271; hep-th/9801168; MR 99m:40003.
[75] O. M. Ogreid and P. Osland, More series related to the Euler series, hep-th/9904206.
[76] S. Ramanujan, On the sum of the square roots of the first n natural numbers, *J. Indian Math. Soc.* 7 (1915) 173-175; also in *Collected Papers*, ed. G. H. Hardy, P. V. Seshu Aiyar, and B. M. Wilson, Cambridge Univ. Press, 1927, pp. 47-49, 337.
[77] N.-Y. Zhang and K. S. Williams, Some series representations for $\zeta(2n+1)$, *Rocky Mount. J.*

Math. 23 (1993) 1581-1592; MR 94m:11099.
[78] N.-Y. Zhang and K. S.Williams, Some infinite series involving the Riemann zeta function, *Analysis, Geometry and Groups: A Riemann Legacy Volume*, ed. H. M. Srivastava and T. M. Rassias, Hadronic Press, 1993, pp. 691-712; MR 96a:11084.
[79] M. L. Glasser, Some integrals of the arctangent function, *Math. Comp.* 22 (1968) 445-447.
[80] C. Nash and D. J. O'Connor, Determinants of Laplacians, the Ray-Singer torsion on lens spaces and the Riemann zeta function, *J. Math. Phys.* 36 (1995) 1462-1505; erratum 36 (1995) 4549; MR 95k:58173.
[81] A. Dabrowski, A note on the values of the Riemann zeta function at odd positive integers, *Nieuw Arch. Wisk.* 14 (1996) 199-207; MR 97g:11142.
[82] M.-P. Chen and H. M. Srivastava, Some familes of series representations for the Riemann $\zeta(3)$, *Resultate Math.* 33 (1998) 179-197; MR 99b:11095.
[83] H. M. Srivastava, M. L. Glasser, and V. S. Adamchik, Some definite integrals associated with the Riemann zeta function, *Z. Anal. Anwendungen* 19 (2000) 831-846; MR 2001g:11136.
[84] L. Euler, Exercitationes Analyticae, 1772, *Opera Omnia Ser. I*, v. 15, Lipsiae, 1911, pp. 131-167.
[85] V. Ramaswami, Notes on Riemann's ζ-function, *J. London Math. Soc.* 9 (1934) 165-169.
[86] A. Terras, Some formulas for the Riemann zeta function at odd integer argument resulting from Fourier expansions of the Epstein zeta function, *Acta Arith.* 29 (1976) 181-189; MR 53 # 299.
[87] J. A. Ewell, A new series representation for $\zeta(3)$, *Amer. Math. Monthly* 97 (1990) 219-220; MR 91d:11103.
[88] J. A. Ewell, On values of the Riemann zeta function at integral arguments, *Canad. Math. Bull.* 34 (1991) 60-66; MR 92c:11087.
[89] J. A. Ewell, On the zeta function values $\zeta(2k+1)$, $k=1, 2, ...$, *Rocky Mount. J. Math.* 25 (1995) 1003-1012; MR 97i:11093.
[90] J. R. Wilton, A proof of Burnside's formula for $\log(\Gamma(x+1))$ and certain allied properties of Riemann's ξ-function, *Messenger of Math.* 52 (1922) 90-93.
[91] D. Cvijovic and J. Klinowski, New rapidly convergent series representations for $\zeta(2n+1)$, *Proc. Amer. Math. Soc.* 125 (1997) 1263-1271; MR 97g:11090.
[92] M. Katsurada, Rapidly convergent series representations for $\zeta(2n+1)$ and their χ-analogue, *Acta Arith.* 90 (1999) 79-89; MR 2000f:11101.
[93] U. Balakrishnan, A series for $\zeta(s)$, *Proc. Edinburgh Math. Soc.* 31 (1988) 205-210; MR 90g:11123.
[94] P. L. Butzer, C. Markett, and M. Schmidt, Stirling numbers, central factorial numbers, and representations of the Riemann zeta function, *Resultate Math.* 19 (1991) 257-274; MR 92a:11095.
[95] P. L. Butzer and M. Hauss, Riemann zeta function: rapidly converging series and integral representations, *Appl. Math. Lett.*, v. 5 (1992) n. 2, 83-88; MR 93b:11106.
[96] J. Choi, H. M. Srivastava, and J. R. Quine, Some series involving the zeta function, *Bull. Austral. Math. Soc.* 51 (1995) 383-393; MR 96d:11090.
[97] J. Choi and H. M. Srivastava, Sums associated with the zeta function, *J. Math. Anal. Appl.* 206 (1997) 103-120; MR 97i:11092.
[98] V. S. Adamchik and H. M. Srivastava, Some series of the zeta and related functions, *Analysis* 18 (1998) 131-144; MR 99d:11096.
[99] H. M. Srivastava, Further series representations for $\zeta(2n+1)$, *Appl. Math. Comput.* 97 (1998) 1-15; MR 99h:11148.
[100] H. M. Srivastava, Some rapidly converging series for $\zeta(2n+1)$, *Proc. Amer. Math. Soc.* 127 (1999) 385-396; MR 99c:11164.
[101] J. Choi, A. K. Rathie, and H. M. Srivastava, Some hypergeometric and other evaluations of $\zeta(2)$ and allied series, *Appl. Math. Comput.* 104 (1999) 101-108; MR 2000e:11104.
[102] J. Choi and H. M. Srivastava, Certain classes of series involving the zeta function, *J. Math. Anal. Appl.* 231 (1999) 91-117; MR 2000c:11143.
[103] D. J. Broadhurst, Polylogarithmic ladders, hypergeometric series and the ten millionth digits

of $\zeta(3)$ and $\zeta(5)$, math.CA/9803067.
[104] T. Amdeberhan, Faster and faster convergent series for $\zeta(3)$, *Elec. J. Combin.* 3 (1996) R13; MR 97b:11154.
[105] T. Amdeberhan and D. Zeilberger, Hypergeometric series acceleration via the WZ method, *Elec. J. Combin.* 4 (1997) R3; MR 99e:33018.
[106] H. Wilf, Accelerated series for universal constants, by the WZ method, *Discrete Math. Theoret. Comput. Sci.* 3 (1999) 155-158.
[107] E. A. Karatsuba, Fast evaluation of $\zeta(3)$ (in Russian), *Problemy Peredachi Informatsii* 29 (1993) 68-73; Engl. transl. in *Problems Information Transmission* 29 (1993) 58-62; MR 94e:11145.
[108] E. A. Karatsuba, Fast evaluation of the Riemann zeta function $\zeta(s)$ for integer values of the argument s (in Russian), *Problemy Peredachi Informatsii* 31 (1995) 69-80; Engl. transl. in *Problems Information Transmission* 31 (1995) 353-362; MR 96k:11155.
[109] X. Gourdon and P. Sebah, The Apéry constant (Numbers, Constants and Computation).
[110] B. Gourevitch, Une formule BBP pour $\zeta(3)$, unpublished note (2000).
[111] J. W. L. Glaisher, Summations of certain numerical series, *Messenger of Math.* 42 (1912) 19-34.
[112] P. Kesava Menon, Summation of certain series, *J. Indian Math. Soc.* 25 (1961) 121-128; MR 26 #2761.
[113] L. Lewin, *Polylogarithms and Associated Functions*, North-Holland, 1981, pp. 153-156; MR 83b:33019.
[114] L. Lewin, ed., *Structural Properties of Polylogarithms*, Amer. Math. Soc., 1991; MR 93b:11158.
[115] J. Borwein and D. Bradley, Empirically determined Apéry-like formulae for $\zeta(4n+3)$, *Experim. Math.* 6 (1997) 181-194; MR 98m:11142.
[116] G. Almkvist and A. Granville, Borwein and Bradley's Apéry-like formulae for $\zeta(4n+3)$, *Experim. Math.* 8 (1999) 197-203; MR 2000h:11126.
[117] M. Koecher, Letter, *Math. Intellig.* 2 (1980) 62-64.
[118] R. W. Gosper, Analytic identities from path invariant matrix multiplication, unpublished manuscript (1976).
[119] M. Munthe Hjortnaes, Transformation of the series $\sum_{k=1}^{\infty} 1/k^3$ to a definite integral (in Norwegian), *Tolfte Skandinaviska Matematikerkongressen*, Proc. 1953 Lunds conf., Lunds Univ. Mat. Instit., 1954, pp. 211-213; MR 16,343a.
[120] N.-Y. Zhang and K. S. Williams, Values of the Riemann zeta function and integrals involving $\log(2 \sinh(\theta/2))$ and $\log(2 \sin(\theta/2))$, *Pacific J. Math.* 168 (1995) 271-289; MR 96f:11170.
[121] R. Ayoub, Euler and the zeta function, *Amer. Math. Monthly* 71 (1974) 1067-1086; MR 50 # 12566.
[122] P. Flajolet, B. Vallée, and I. Vardi, Continued fractions from Euclid to the present day, École Polytechnique preprint (2000).
[123] L. Lovász, *Combinatorial Problems and Exercises*, 2nd ed., North-Holland, 1993; MR 94m:05001.
[124] V. S. Adamchik, On Stirling numbers and Euler sums, *J. Comput. Appl. Math.* 79 (1997) 119-130; MR 97m:11025.
[125] H. S. Wilf, The asymptotic behavior of the Stirling numbers of the first kind, *J. Combin. Theory Ser. A* 64 (1993) 344-349; MR 94m:11025.
[126] V. I. Levin, About a problem of S. Ramanujan (in Russian), *Uspekhi Mat. Nauk*, v. 5 (1950) n. 3, 161-166.
[127] N. Nielsen, Der Eulersche Dilogarithmus und seine Verallgemeinerungen, *Nova Acta Leopoldina, Abhandlungen der Kaiserlich Leopoldinisch-Carolinischen Deutschen Akademie der Naturforscher* 90 (1909) 121-212.
[128] K. S. Kölbig, Nielsen's generalized polylogarithms, *SIAM J. Math. Anal.* 17 (1986) 1232-1258; MR 88a:33028.
[129] M. J. Levine, E. Remiddi, and R. Roskies, Analytic contributions to the g factor of the electron in sixth order, *Phys. Rev. D* 20 (1979) 2068-2076.

[130] R. Gastmans andW. Troost, On the evaluation of polylogarithmic integrals, *Simon Stevin* 55 (1981) 205-219; MR 83c:65028.
[131] P. J. de Doelder, On some series containing $\psi(x)-\psi(y)$ and $(\psi(x)-\psi(y))^2$ for certain values of x and y, *J. Comput. Appl. Math.* 37 (1991) 125-141; MR 92m:40002.

1.7 カタランの定数, G

Catalan (カタラン) の定数 G は

$$G=\sum_{n=0}^{\infty}\frac{(-1)^n}{(2n+1)^2}=0.9159655941\cdots$$

で定義される．我々はアペリの定数【1.6】とともに議論を進めていくことにする．この2つを比較していくことは，充分意味のあることである．まず，Dirichlet (**ディリクレ**) のベータ関数

$$\beta(x)=\sum_{n=0}^{\infty}\frac{(-1)^n}{(2n+1)^x}, \quad x>0$$

(mod 4 の非単位指標についての Dirichlet の L 関数としても引用される) について調べてみる．$G=\beta(2)$ であることに注意する．

ベータ関数は x の値が正の奇数のところでは正確に数値を求めることができ，

$$\beta(2k+1)=\frac{(-1)^k E_{2k}}{2(2k)!}\left(\frac{\pi}{2}\right)^{2k+1}$$

である[1-3]．ここで，$\{E_n\}$ は**オイラー** (Euler) **数**【1.7.1】である．たとえば，

$$\beta(1)=\frac{\pi}{4}, \quad \beta(3)=\frac{\pi^3}{32}, \quad \beta(5)=\frac{5\pi^5}{1536}$$

である．ゼータ関数【1.6】と同様に，$\beta(x)$ は関数等式

$$\beta(1-z)=\left(\frac{2}{\pi}\right)^z \sin\left(\frac{\pi z}{2}\right)\Gamma(z)\beta(z)$$

によって全複素数平面に解析接続される[4-6]．ここで，$\Gamma(z)=(z-1)!$ はガンマ関数【1.5.4】である．ディリクレの関数はリーマンの関数とは異なり，すべての点で定義され，特異点はもたない．素数定理との関係は式

$$\beta(x)=\prod_{\substack{p\,\text{素}\\p\equiv 1\,\text{mod}\,4}}\left(1-\frac{1}{p^x}\right)^{-1}\cdot\prod_{\substack{p\,\text{素}\\p\equiv 3\,\text{mod}\,4}}\left(1+\frac{1}{p^x}\right)^{-1}=\prod_{p\,\text{奇,素}}\left(1-\frac{(-1)^{(p-1)/2}}{p^x}\right)^{-1}$$

によって最もよく要約されている[7]．項の整理は絶対収束であることが保証してくれる．これと密接に関係する関数に

$$\lambda(x)=\sum_{n=0}^{\infty}\frac{1}{(2n+1)^x}=\left(1-\frac{1}{2^x}\right)\zeta(x), \quad x>1$$

がある[8-10]．たとえば，

$$\lambda(2) = \frac{\pi^2}{8}, \quad \lambda(4) = \frac{\pi^4}{96}, \quad \lambda(6) = \frac{\pi^6}{960}$$

である．アペリの定数とは異なり，G が無理数かどうかはわかっていない[11,12]．また，G/π^2 の算術的な性質についても何も知られていない．統計工学では G/π はダイマー問題【5.23】の正確な解の一部分として現れる．Schmidt（シュミット）[13]は次の奇妙なよく似た事実

$$\frac{\pi^2}{12\ln(2)} = \left(1 - \frac{1}{2^2} + \frac{1}{3^2} - \frac{1}{4^2} + \cdots\right)\left(1 - \frac{1}{2} + \frac{1}{3} - \frac{1}{4} + \cdots\right)^{-1}$$

$$\frac{4G}{\pi} = \left(1 - \frac{1}{3^2} + \frac{1}{5^2} - \frac{1}{7^2} + \cdots\right)\left(1 - \frac{1}{3} + \frac{1}{5} - \frac{1}{7} + \cdots\right)^{-1}$$

について指摘した．ここで，最初の式（Lévy（レヴィ）の定数）は連分数近似において重要となる【1.8】．これと少し変わった形として

$$\frac{8G}{\pi^2} = \left(1 - \frac{1}{3^2} + \frac{1}{5^2} - \frac{1}{7^2} + \cdots\right)\left(1 + \frac{1}{3^2} + \frac{1}{5^2} + \frac{1}{7^2} + \cdots\right)^{-1}$$

があり，ある共役関数に関する不等式が成立するような最良の係数として現れる【7.7】．定数 $2G/(\pi\ln(2))$ は，2分木の平均ルート分岐率としても現れる【5.6】．

1.7.1 オイラー数

次の生成関数

$$\operatorname{sech}(x) = \frac{2e^x}{e^{2x}+1} = \sum_{k=0}^{\infty} E_k \frac{x^k}{k!}$$

によって，オイラー（Euler）数 $\{E_n\}$ を定義しよう．すると，すべてのオイラー数は整数で，$E_0 = 1$, $E_2 = -1$, $E_4 = 5$, $E_6 = -61$, … であり，$n > 0$ のとき $E_{2n-1} = 0$ であることがわかる．

（都合の悪いことに，混乱を引き起こすことになるオイラー数の別定義がある．この定義によると，添え字の付け方がいくらか異なるので，すべての数が正となる．論文を読むときは，どちらの定義が使われているかに注意しなければならない．）

オイラー数は

$$\sec(x) = \sum_{k=0}^{\infty} \frac{(-1)^k E_{2k}}{(2k)!} x^{2k}$$

のような級数展開にも出てくる．

1.7.2 級　　数

等差数列の上での和を考えると，少し異なった形のもの

$$\sum_{k=0}^{\infty} \frac{1}{(4k+1)^2} = \frac{1}{16}\pi^2 + \frac{1}{2}G, \quad \sum_{k=0}^{\infty} \frac{1}{(4k+3)^2} = \frac{1}{16}\pi^2 - \frac{1}{2}G$$

が得られる[14-16]．次の4つの公式

$$\sum_{k=0}^{\infty}\frac{2^{2k}}{(2k+1)^2\binom{2k}{k}}=2G, \quad \sum_{k=0}^{\infty}\frac{1}{2^{3k}(2k+1)^2}\binom{2k}{k}=\frac{\pi}{4\sqrt{2}}\ln(2)+\frac{1}{\sqrt{2}}G$$

$$\sum_{k=0}^{\infty}\frac{1}{(2k+1)^2\binom{2k}{k}}=\frac{8}{3}G-\frac{\pi}{3}\ln(2+\sqrt{3})$$

$$\sum_{k=0}^{\infty}\frac{2^{4k}}{(k+1)(2k+1)^2\binom{2k}{k}^2}=2\pi G-\frac{7}{2}\zeta(3)$$

は，二項係数の中央項を含むものである[1,17-19]．面白いことに，Berndt（ブレント）[17]も指摘したが，最初の式は有名なアペリの級数【1.6.3】を思い起こさせる．しかしながら，それはかなり以前に発見されていたのである．関係する級数族に

$$R(n)=\sum_{k=0}^{\infty}\frac{1}{2^{4k}(2k+n)}\binom{2k}{k}^2, \quad n=0,1,2,\cdots$$

がある[20,21,23]．これらは次の漸化式

$$R(0)=2\ln(2)-\frac{4G}{\pi}, \quad R(1)=\frac{4G}{\pi}$$

$$(n-1)^2R(n)=(n-2)^2R(n-2)+\frac{2}{\pi} \quad (n\geq 2 \text{ のとき})$$

を満たすことが証明されている[1,22,24]．

$\beta(2n+1)$ に対する精密な公式の，$\beta(2n)$ についての類似な公式は何であろうか？これについては誰も答えていないが，Ramanujan によって得られた級数[16,25]

$$G=\frac{5}{48}\pi^2-2\sum_{k=0}^{\infty}\frac{(-1)^k}{(2k+1)^2(e^{\pi(2k+1)}-1)}-\frac{1}{4}\sum_{k=1}^{\infty}\frac{\text{sech}(\pi k)}{k^2}$$

が研究の出発点となろう．

いくつかの多重級数

$$\sum_{n=1}^{\infty}\frac{(-1)^{n+1}}{n}\sum_{k=0}^{n-1}\frac{(-1)^k}{2k+1}=G, \quad \sum_{n=0}^{\infty}\frac{(-1)^n}{2n+1}\sum_{k=1}^{n}\frac{1}{k}=G-\frac{\pi}{2}\ln(2)$$

$$\sum_{n=0}^{\infty}\frac{(-1)^n}{2n+1}\sum_{k=0}^{n-1}\frac{1}{2k+1}=\frac{\pi}{8}\ln(2)-\frac{1}{2}G, \quad \sum_{n=1}^{\infty}\frac{(-1)^{n+1}}{n^2}\sum_{k=0}^{n-1}\frac{1}{2k+1}=\pi G-\frac{7}{4}\zeta(3)$$

$$\sum_{n=1}^{\infty}\frac{(-1)^{n+1}}{n^2}\sum_{k=1}^{n}\frac{1}{k+n}=\pi G-\frac{33}{16}\zeta(3), \quad \sum_{n=0}^{\infty}\frac{2^n}{(2n+1)\binom{2n}{n}}\sum_{k=0}^{n}\frac{1}{2k+1}=2G$$

がある[16,17,26-28]．次の2つの級数

$$\sum_{n=1}^{\infty}\frac{n\zeta(2n+1)}{2^{4n}}=1-G, \quad \sum_{n=1}^{\infty}\frac{\zeta(2n)}{2^{4n}(2n+1)}=\frac{1}{2}-\frac{1}{4}\ln(2)-\frac{1}{\pi}G$$

は，ゼータ関数の値を含むものである[29-31]．Broadhurst（ブロードハースト）は次の級数

$$G=3\sum_{k=0}^{\infty}\frac{1}{2\cdot 16^k}\left(\frac{1}{(8k+1)^2}-\frac{1}{(8k+2)^2}+\frac{1}{2(8k+3)^2}-\frac{1}{2^2(8k+5)^2}+\frac{1}{2^2(8k+6)^2}-\frac{1}{2^3(8k+7)^2}\right)$$
$$-2\sum_{k=0}^{\infty}\frac{1}{8\cdot 16^{3k}}\left(\frac{1}{(8k+1)^2}+\frac{1}{2(8k+2)^2}+\frac{1}{2^3(8k+3)^2}-\frac{1}{2^6(8k+5)^2}-\frac{1}{2^7(8k+6)^2}-\frac{1}{2^9(8k+7)^2}\right)$$

によって，G に対する桁ごとの抽出アルゴリズムを決定した[32-34]．

1.7.3 積

ゼータ関数の奇数における値に関して【1.6.4】と同様に，Gosper（ゴスパー）[35]はベータ関数の偶数の値を与える無限行列積を発見した．4×4 行列の場合のみについて示すと，

$$\prod_{k=1}^{\infty}\begin{bmatrix}\frac{4k^2}{(4k-1)(4k+1)} & \frac{-1}{(4k-1)(4k+1)} & 0 & \frac{1}{(2k-1)^5} \\ 0 & \frac{4k^2}{(4k-1)(4k+1)} & \frac{-1}{(4k-1)(4k+1)} & \frac{1}{(2k-1)^3} \\ 0 & 0 & \frac{4k^2}{(4k-1)(4k+1)} & \frac{6k-1}{2(2k-1)(4k-1)} \\ 0 & 0 & 0 & 1\end{bmatrix}$$
$$=\begin{bmatrix}0 & 0 & 0 & \beta(6) \\ 0 & 0 & 0 & \beta(4) \\ 0 & 0 & 0 & \beta(2) \\ 0 & 0 & 0 & 1\end{bmatrix}$$

である．$(n+1)\times(n+1)$ 行列，$\beta(2n)$ の場合に拡張したものは前と同じパターンになる．

1.7.4 積分

ベータ関数は $x>0$ として
$$\beta(x)=\frac{1}{2\Gamma(x)}\int_0^{\infty}\frac{t^{x-1}}{\cosh(t)}dt$$
という別表現をもつ[4]．カタランの定数を含む多くの積分がある[10,15,16,36,37]．
$$2\int_0^1\frac{\arctan(x)}{x}dx=\int_0^{\pi/2}\frac{x}{\sin(x)}dx=2G,\quad \frac{1}{2}\int_0^1 K(x)\,dx=\int_0^1 E(x)\,dx-\frac{1}{2}=G$$
$$\int_0^1\frac{\ln(x)}{1+x^2}dx=-\int_1^{\infty}\frac{\ln(x)}{1+x^2}dx=-G$$
$$\int_0^{\pi/4}\ln(2\cos(x))\,dx=-\int_0^{\pi/4}\ln(2\sin(x))\,dx=\frac{1}{2}G$$
$$4\int_0^1\frac{\arctan(x)^2}{x}dx=\int_0^{\pi/2}\frac{x^2}{\sin(x)}dx=2\pi G-\frac{7}{2}\zeta(3)$$
$$\int_0^{\pi/2}\operatorname{arcsinh}(\sin(x))\,dx=\int_0^{\pi/2}\operatorname{arcsinh}(\cos(x))\,dx=G$$

ここで，$K(x)$ と $E(x)$ は完全楕円積分である【1.4.6】．【1.7.6】も参照せよ．

1.7.5 連 分 数

次の連分数展開は,Stieltjes[38],Rogers[39],Ramanujan[40]によるものである.

$$2G = 2 - \cfrac{1|}{|3} + \cfrac{2^2|}{|1} + \cfrac{2^2|}{|3} + \cfrac{4^2|}{|1} + \cfrac{4^2|}{|3} + \cfrac{6^2|}{|1} + \cfrac{6^2|}{|3} + \cdots$$

$$2G = 1 + \cfrac{1|}{|\frac{1}{2}} + \cfrac{1^2|}{|\frac{1}{2}} + \cfrac{1\cdot 2|}{|\frac{1}{2}} + \cfrac{2^2|}{|\frac{1}{2}} + \cfrac{2\cdot 3|}{|\frac{1}{2}} + \cfrac{3^2|}{|\frac{1}{2}} + \cfrac{3\cdot 4|}{|\frac{1}{2}} + \cfrac{4^2|}{|\frac{1}{2}} + \cdots$$

1.7.6 逆正接積分

$\text{Ti}_1(x) = \arctan(x)$ とし,$n \geq 2$ である整数 n に対して,

$$\text{Ti}_n(x) = \sum_{k=0}^{\infty} \frac{(-1)^k}{(2k+1)^n} x^{2k+1} = \int_0^x \frac{\text{Ti}_{n-1}(s)}{s} ds$$

とする[41].ここで,$|x| \leq 1$ とする.明らかに $\text{Ti}_n(1) = \beta(n)$ である.$n=2$ の場合は,**逆正接積分**とよばれる.これは別表現をもち,$0 < \theta < \pi/2$ なる θ に対して

$$\text{Ti}_2(\tan(\theta)) = \frac{1}{2} \int_0^{2\theta} \frac{t}{\sin(t)} dt = \theta \ln(\tan(\theta)) - \int_0^{\theta} \ln(2\sin(t)) dt + \int_0^{\theta} \ln(2\cos(t)) dt$$

である.代表的な値としては,

$$\text{Ti}_2(2-\sqrt{3}) = \frac{2}{3} G + \frac{\pi}{12} \ln(2-\sqrt{3}), \quad \text{Ti}_2(2+\sqrt{3}) = \frac{2}{3} G + \frac{5\pi}{12} \ln(2+\sqrt{3})$$

がある[21,41].後の式では積分表現を用いる(なぜならば,級数は $x>1$ のとき発散するが,積分は収束するから).$n>2$ のとき,$\text{Ti}_n(x)$ については,ほとんど何も知られていない.

[1] J. M. Borwein and P. B. Borwein, *Pi and the AGM: A Study in Analytic Number Theory and Computational Complexity*, Wiley, 1987, pp. 198-199, 383-386; MR 99h:11147.
[2] F. Beukers, J. A. C. Kolk, and E. Calabi, Sums of generalized harmonic series and volumes, *Nieuw Arch. Wisk.* 4 (1993) 217-224; MR 94j:11022.
[3] N. D. Elkies, On the sums $\sum_{k=-\infty}^{\infty}(4k+1)^{-n}$, math.CA/0101168.
[4] A. Erdélyi, W. Magnus, F. Oberhettinger, and F. G. Tricomi, *Higher Transcendental Functions*, v. 1, McGraw-Hill, 1953, pp. 27-35; MR 15,419i.
[5] W. Ellison and F. Ellison, *Prime Numbers*, Wiley, 1985, p. 180; MR 87a:11082.
[6] T. M. Apostol, *Introduction to Analytic Number Theory*, Springer-Verlag, 1976, pp. 249-263; MR 55 #7892.
[7] P. Moree and J. Cazaran, On a claim of Ramanujan in his first letter to Hardy, *Expos. Math* 17 (1999) 289-312; MR 2001c:11103.
[8] M. Abramowitz and I. A. Stegun, *Handbook of Mathematical Functions*, Dover, 1972, pp. 807-808; MR 94b:00012.
[9] J. Spanier and K. B. Oldham, *An Atlas of Functions*, Hemisphere, 1987, pp. 25-33.
[10] I. S. Gradshteyn and I. M Ryzhik, *Tables of Integrals, Series and Products*, Academic Press, 1980; MR 97c:00014.
[11] W. Zudilin, Apéry-like difference equation for Catalan's constant, math.NT/0201024.
[12] T. Rivoal and W. Zudilin, Diophantine properties of numbers related to Catalan's constant, UMR preprint 317 (2002), l'Institut de Mathématiques de Jussieu.

[13] A. L. Schmidt, Ergodic theory of complex continued fractions, *Number Theory with an Emphasis on the Markoff Spectrum*, Proc. 1991 Provo conf., ed. A. D. Pollington and W. Moran, Dekker, 1993, pp. 215-226; MR 95f:11055.

[14] K. S. Kölbig, The polygamma function $\psi^{(k)}(x)$ for $x=1/4$ and $x=3/4$, *J. Comput. Appl. Math.* 75 (1996) 43-46; MR 98d:33001.

[15] V. S. Adamchik, 33 representations for the Catalan constant, unpublished note (1997).

[16] D. M. Bradley, Representations of Catalan's constant, unpublished note (2001).

[17] B. C. Berndt, *Ramanujan's Notebooks: Part I*, Springer-Verlag, 1985, pp. 264-267, 289-290, 293-294; MR 86c:01062.

[18] D. Bradley, A class of series acceleration formulae for Catalan's constant, *Ramanujan J.* 3 (1999) 159-173; MR 2000f:11163.

[19] R. W. Gosper, A calculus of series rearrangements, *Algorithms and Complexity: New Directions and Recent Results*, Proc. 1976 Carnegie-Mellon conf., ed. J. F. Traub, Academic Press, 1976, pp. 121-151; MR 56 #9899.

[20] E. P. Adams and R. L. Hippisley, *Smithsonian Mathematical Formulae and Tables of Elliptic Functions*, Smithsonian Institute, 1922, p. 142.

[21] E. R. Hansen, *A Table of Series and Products*, Prentice-Hall, 1975.

[22] J. Dutka, Two results of Ramanujan, *SIAM J. Math. Anal.* 12 (1981) 471; MR 83a: 05010.

[23] S. Yang, Some properties of Catalan's constant G, *Int. J. Math. Educ. Sci. Technol.* 23 (1992) 549-556; MR 93j:11058.

[24] V. S. Adamchik, A certain series associated with Catalan's constant, unpublished note (2000).

[25] S. Ramanujan, On the integral $\int_0^x \tan^{-1}(t)dt/t$, *J. Indian Math. Soc.* 7 (1915) 93-96; also in *Collected Papers*, ed. G. H. Hardy, P. V. Seshu Aiyar, and B. M. Wilson, Cambridge Univ. Press, 1927, pp. 40-43, 336-337.

[26] O. Espinosa and V. H. Moll, On some definite integrals involving the Hurwitz zeta function, *Ramanujan J.* 6 (2002) 159-188; math.CA/0012078.

[27] R. Sitaramachandrarao, A formula of S. Ramanujan, *J. Number Theory* 25 (1987) 1-19; MR 88c: 11048.

[28] G. J. Fee, Computation of Catalan's constant using Ramanujan's formula, *Proc. 1990 Int. Symp. Symbolic and Algebraic Computation (ISSAC)*, ed. S. Watanabe and M. Nagata, Tokyo, ACM, 1990, pp. 157-160.

[29] J. W. L. Glaisher, Numerical values of the series $1-1/3^n+1/5^n-1/7^n+1/9^n-\cdots$, *Messenger of Math.* 42 (1913) 35-58.

[30] M.-P. Chen and H. M. Srivastava, Some familes of series representations for the Riemann $\zeta(3)$, *Resultate Math.* 33 (1998) 179-197; MR 99b:11095.

[31] J. Choi, The Catalan's constant and series involving the zeta function, *Commun. Korean Math. Soc.* 13 (1998) 435-443; MR 2000h:11091.

[32] D. J. Broadhurst, Polylogarithmic ladders, hypergeometric series and the ten millionth digits of $\zeta(3)$ and $\zeta(5)$, math.CA/9803067.

[33] J. M. Borwein, Experimental mathematics: Insight from computation, presentation at *AMS/MAA Joint Meetings*, San Antonio, 1999.

[34] J. M. Borwein and R. M. Corless, Emerging tools for experimental mathematics, *Amer. Math. Monthly* 106 (1999) 889-909; MR 2000m:68186.

[35] R. W. Gosper, Analytic identities from path invariant matrix multiplication, unpublished manuscript (1976).

[36] H. M. Srivastava and E. A. Miller, A simple reducible case of double hypergeometric series involving Catalan's constant and Riemann's zeta function, *Int. J. Math. Educ. Sci. Technol.* 21 (1990) 375-377; MR 91d:33032.

[37] I. J. Zucker, G. S. Joyce, and R. T. Delves, On the evaluation of the integral $\int_0^{\pi/4} \ln(\cos^{m/n}\theta \pm \sin^{m/n}\theta)d\theta$, *Ramanujan J.* 2 (1998) 317-326; MR 99g:26019.

[38] T. J. Stieltjes, Recherches sur les fractions continues, *Annales Faculté Sciences Toulouse* 8

(1894) J1-J122; 9 (1895) A1-A47; also in *Oeuvres Complētes*, t. 2, ed. W. Kapteyn and J. C. Kluyver, Noordhoff, 1918, pp. 402-566; Engl. transl. in *Collected Papers*, v. 2, ed. G. van Dijk, Springer-Verlag, 1993, pp. 406-570, 609-745; MR 95g:01033.

[39] L. J. Rogers, Supplementary note on the representation of certain asymptotic series as convergent continued fractions, *Proc. London Math. Soc.* 4 (1907) 393-395.

[40] B. C. Berndt, *Ramanujan's Notebooks: Part II*, Springer-Verlag, 1989, pp. 150-153; MR 90b:01039.

[41] L. Lewin, *Polylogarithms and Associated Functions*, North-Holland, 1981, pp. 38-45, 106, 166, 190; MR 83b:33019.

1.8 ヒンチン-レヴィ定数

x を実数とする.x を正則連分数として（一意に）展開したものを

$$x = q_0 + \frac{1}{|q_1|} + \frac{1}{|q_2|} + \frac{1}{|q_3|} + \cdots$$

とする.ここで,q_0 は整数とし,$q_1, q_2, q_3, \cdots$ は正の整数である.小数展開とは異なり,正則連分数の諸性質は基底のとり方によらない.よって,数論研究者にとっては十進小数よりも連分数で扱う方がより自然である.

$k > 0$ を任意とするとき,q_k の平均的な挙動についてはどのようなことがいえるであろうか? たとえば,相乗平均

$$M(n, x) = (q_1 q_2 q_3 \cdots q_n)^{1/n}$$

について,$n \to \infty$ としたときの極限を考えてみよう.この極限値はかなり込み入った方法で x に依存するものと予期されるであろう.q の数列は1つの x を決定するから,想像できるある条件に従う数列 q に対していくつかの x が存在する.それゆえ,$\lim_{n \to \infty} M(n, x)$ を求める企ては,不可能と思われるほどの困難を極めるものと思われるだろう.

ここに,数学の世界で最も驚くべき事実の1つが出現する.Khintchine（ヒンチン）[1-4]はほとんどすべての実数 x に対して

$$\lim_{n \to \infty} M(n, x) = \prod_{k=1}^{\infty} \left(1 + \frac{1}{k(k+2)}\right)^{\ln(k)/\ln(2)} = K = e^{0.9878490568\cdots} = 2.6854520010\cdots$$

と「定数」であることを証明した.このことは Khintchine の結果が成立しない x の集合（たとえば,すべての有理数,2次の無理数,その他［後述］）のルベーグ測度が0であるということである.真にランダムに選ばれてきた x は Khintchine の法則に従うであろうということを,我々は確率的であるが確信することができる.これは実数の性質に関する深遠な命題である.エルゴート理論を利用する Ryll-Nardzewski（ルイルナルジェウスキ）[5]による別の証明が Kac（カッツ）[6]で見られる.

K を表現する無限積は収束が非常に遅い.K を計算するための速い数値計算法が［7-

13]に出ている．K のいくつかの相異なる表現の中に

$$\ln(2)\,\ln(K) = -\sum_{i=2}^{\infty}\ln\left(1-\frac{1}{i}\right)\ln\left(1+\frac{1}{i}\right) = \sum_{j=2}^{\infty}\frac{(-1)^j(2-2^j)}{j}\zeta'(j),$$

$$\ln(2)\,\ln(K) = \sum_{k=1}^{\infty}\frac{\zeta(2k)-1}{k}\left(1-\frac{1}{2}+\frac{1}{3}-+\cdots+\frac{1}{2k-1}\right),$$

$$\ln(2)\,\ln(K) = -\int_0^1\frac{1}{x(1+x)}\ln\left(\frac{\sin(\pi x)}{\pi x}\right)dx = \frac{\pi^2}{12}+\frac{\ln(2)^2}{2}+\int_0^{\pi}\frac{\ln|\theta\cot(\theta)|}{\theta}d\theta$$

がある [8,11,13,14]．ここで，$\zeta(x)$ は Riemann のゼータ関数であり，$\zeta'(x)$ はその導関数である．

多くの疑問が湧きあがる．K は無理数であろうか？ Khintchine の結果に対するわずかな［測度 0 の］例外の中で，よく知られている無理数は何か？ Lehmer (レーマー) [7,15] は e が例外数であることを明らかにしたが，$\sqrt[3]{2}$ や π そして K それ自身 (!) がそうであるかは未解決の問題として残されている．

関連する考え方の中に，互いに素な正の整数 P_n と Q_n の漸近的なふるまいが含まれる．ここで，P_n/Q_n は x の n 次の部分近似分数である．すなわち，P_n/Q_n は x の q_n までの有限正則連分数展開の値である．Lévy (レヴィ) [16,17] はほとんどすべての実数 x に対して

$$\lim_{n\to\infty}Q_n^{1/n} = e^{\pi^2/(12\ln(2))} = e^{1.1865691104\cdots} = 3.2758229187\cdots = \lim_{n\to\infty}\left(\frac{P_n}{x}\right)^{1/n}$$

となることを示した．Philipp (フィリップ) [18,19] は Khintchine と Lévy の両方の極限に関連する誤差限界を改良した．別の見解が [20-22] で与えられ

$$-\lim_{n\to\infty}\frac{1}{n}\log_{10}\left|x-\frac{P_n}{Q_n}\right| = \frac{\pi^2}{6\ln(2)\ln(10)} = 1.0306408341\cdots$$

である．これは連分数の典型的な項［連分数展開の n 番目までの項］が（ほとんどすべての実数に対して）ほぼ 1.03 桁の分まで近似されていることを示している．同じことであるが，$\{x\}$ を x の小数部分を表すものとすれば，連分数変換 $x\to\{1/x\}$ の計量エントロピー (metric entropy) が

$$\lim_{n\to\infty}\frac{Q_{n+1}^2}{Q_n^2} = e^{\pi^2/(6\ln(2))} = 10.7310157948\cdots = (0.0931878229\cdots)^{-1}$$

であるということである [23,24]．すなわち，連分数展開のそれ以後に引き続く項は 10.73 の因子で x の不確実さを減少させる．これに対して，桁移動の変換 $x\mapsto\{10x\}$ のエントロピーは 10 である．

Corless (コアレス) [13,25] は次の面白い対照を見せる式

$$\ln(K) = \int_0^1\frac{\ln\left\lfloor\frac{1}{x}\right\rfloor}{\ln(2)\,(1+x)}dx,\quad \frac{\pi^2}{12\ln(2)} = \int_0^1\frac{\ln\left(\frac{1}{x}\right)}{\ln(2)\,(1+x)}dx$$

に言及した．ここで $\lfloor x\rfloor$ は x を超えない最大整数を表す．

ここで，本来の問に立ち返ってみよう．$k>0$ とし，連分数の k 番目の部分分母 q_k の平均的な挙動について，何が言えるであろうか？ これまで，平均値の 1 つである相乗

平均についてのみ吟味してきた．平均値の一般化は[26]

$$M(s,n,x)=\Big(\frac{1}{n}\sum_{k=1}^{n}q_k^s\Big)^{1/s}$$

である．これは $s=-1, 0^{\dagger}, 1$，そして 2 に対して，それぞれ，調和平均，相乗平均，相加平均，そして二乗平均平方根となる．よく知られた平均は，このようにして連続的な階層の平均値中にうまく割り込ませることができる．$s \geq 1$ ならば，ほとんどすべての実数 x に対して $\lim_{n\to\infty} M(s,n,x) = \infty$ となることが知られている[3,27]．$s<1$，$s \neq 0$ である s に対する $M(s,n,x)$ の値について何が言えるであろうか？ Khintchin の公式の類似なものとして，ほとんどすべての実数 x に対して

$$\lim_{n\to\infty} M(s,n,x)=\Big[\frac{1}{\ln(2)}\sum_{k=1}^{\infty}k^s\ln\Big(1+\frac{1}{k(k+2)}\Big)\Big]^{1/s}=K_s$$

が成立する．$K_{-1}=1.7454056624\cdots$，$K_{-2}=1.4503403284\cdots$，$K_{-3}=1.3135070786\cdots$，そして $s \to -\infty$ のとき $K_s=1+O(1/s)$ であることが知られている[13,28]．

密接に関連する事柄が【2.17】，【2.18】，【2.19】で取り上げられる．

1.8.1 他の表現法

ヒンチン-レヴィ定数に結びつける，実数に対する正則連分数によく似た表現法がある．たとえば，$0<x<1$ であるすべての実数 x は，一意的に次の形

$$x=\frac{1}{a_1+1}+\sum_{n=2}^{\infty}\Big(\prod_{k=1}^{n-1}\frac{1}{a_k(a_k+1)}\Big)\frac{1}{a_n+1}=\frac{1}{b_1}+\sum_{n=2}^{\infty}\Big(\prod_{k=1}^{n-1}\frac{1}{b_k(b_k+1)}\Big)\frac{(-1)^{n-1}}{b_n}$$

に表される．ここで，$a_1, a_2, a_3, \cdots$ と $b_1, b_2, b_3, \cdots$ は正の整数である．これらは，それぞれ，x の Lüroth (**ルロス**) **表現**，**交代ルロス表現**と呼ばれている．どちらの場合も極限となる定数は同じで

$$\lim_{n\to\infty}(a_1a_2a_3\cdots a_n)^{1/n}=\prod_{k=1}^{\infty}k^{1/(k(k+1))}=e^{0.7885305659\cdots}=2.2001610580\cdots=U,$$

$$\lim_{n\to\infty}\Big|x-\frac{P_n}{Q_n}\Big|^{1/n}=\prod_{k=1}^{\infty}[k(k+1)]^{-1/(k(k+1))}=e^{-2.0462774528\cdots}=V$$

である[29-31]．ここで，P_n/Q_n は n 次の近似分数である．これと少し変わったものとして[32]，

$$\lim_{n\to\infty}((a_1+1)(a_2+1)\cdots(a_n+1))^{1/n}=\prod_{k=1}^{\infty}(k+1)^{1/(k(k+1))}=e^{1.2577468869\cdots}=W$$

があり，【2.9】でも取り上げられる．もちろん，$UVW=1$ であり，

$$\ln(U)=-\sum_{i=2}^{\infty}(-1)^i\zeta'(i), \quad \ln(V)=2\sum_{j=1}^{\infty}\zeta'(2j), \quad \ln(W)=-\sum_{k=2}^{\infty}\zeta'(k)$$

である．

2 番目の例[22]は $0<x<1$ である実数に対する Bolyai-Rényi (**ボリヤイ-レーニィ**) **の表現**

---

†訳注：$\lim_{s\to 0} M(s,n,x)=\sqrt[n]{q_1q_2\cdots q_n}$ の意味．

$$x = -1 + \sqrt{a_1 + \sqrt{a_2 + \sqrt{a_3 + \cdots}}}$$

である.ここで,$a_k \in \{0, 1, 2\}$ である.連分数のエントロピーに対しては,正確な表現 $\pi^2/(6\ln(2)) = 2.373138\cdots$ があるのに対して,根号展開のエントロピーに対しては,数値計算による $1.056313\cdots$ という結果だけが存在している.

3番目の例[34-41]は $-1/2 < x < 1/2$ である実数に対する**最近整数による連分数** (nearest integer continued fraction)

$$x = \frac{1|}{|c_1} + \frac{1|}{|c_2} + \frac{1|}{|c_3} + \cdots$$

である.ここで,$c_1, c_2, c_3, \cdots$ は,次の規則

$$c_1 = \left\lfloor \frac{1}{x} + \frac{1}{2} \right\rfloor, \quad x_1 = \frac{1}{x} - c_1, \quad c_2 = \left\lfloor \frac{1}{x_1} + \frac{1}{2} \right\rfloor, \quad x_2 = \frac{1}{x_1} - c_2, \quad \cdots$$

によって生成される.c_i のいくつかは負の値になることもある.この場合,ヒンチン-レヴィ定数に対応する公式は

$$\lim_{n \to \infty} |c_1 c_2 \cdots c_n|^{1/n} = \left(\frac{5\varphi+3}{5\varphi+2}\right)^{\ln(2)/\ln(\varphi)} \prod_{k=3}^{\infty} \left(\frac{8(k-1)\varphi + (2k-3)^2 + 4}{8(k-1)\varphi + (2k-3)^2}\right)^{\ln(k)/\ln(\varphi)}$$
$$= e^{1.6964441175\cdots} = 5.4545172445\cdots,$$
$$\lim_{n \to \infty} Q_n^{1/n} = e^{\pi^2/(12\ln(\varphi))} = e^{1.7091579853} = 5.5243079702\cdots$$

である.ここで,P_n/Q_n は n 次の部分近似分数であり,φ は黄金比【1.2】である.このような展開は**中心連分数** (centered continued fraction) とも呼ばれる[42].

1.8.2 それから導かれる定数

我々は(測度が0の集合に属している)Khintchine の法則に従わない例外の実数 x をいくつか知っているが,その法則を満たすと立証できる明白な定数 y を1つも知らない.これは,y が x より充分豊富にあるから,それを見つけることは容易にできると期待されることから考えると,注目すべきことである.y が「明白」であるという要請そのものが,難しい部分である.これは特に,y の正則連分数における部分分母 q_n が,任意の精度まで K を知ることに依存すべきでない [K を知って q_n を定めるのでない] ということを意味する.Robinson(ロビンソン)[43]は4つの明白でない定数を明らかにした.これらは単純な方法で K から帰納的に導かれる(表1.2を参照せよ).Bailey, Borwein & Crandall(ベイリー,ボーウィン&クランドール)[13]は少なくとも列挙する $q_0, q_1, \cdots$ が明白であるような精巧な構成法で,他の例を与えた(それでも定数 y は明白ではないのだが).

1.8.3 複素数への拡張

Schmidt(シュミット)[44-46]は連分数理論を複素数体上に一般化するための最も自然な接近の仕方と思われる方法を提唱した.たとえば[47-50]によって,Lévy の定数の複素数への拡張は $\exp(G/\pi)$ である.ここで,G はカタランの定数である【1.7】.

表1.2 K から帰納的に導かれる明白でない定数

$y=2.3038421962\cdots$	q_n は $\prod_{k=0}^{n} q_k < K^{n+1}$ を満たす正の最大整数
$y=3.3038421963\cdots$	q_n は $\prod_{k=0}^{n} q_k > K^{n+1}$ を満たす正の最小整数
$y=2.2247514809\cdots$	n が偶数のとき, $\prod_{k=0}^{n} q_k$ は K^{n+1} より小さくて,
	n が奇数のとき, $\prod_{k=0}^{n} q_k$ は K^{n+1} より大きい.
$y=3.4493588902\cdots$	n が偶数のとき, $\prod_{k=0}^{n} q_k$ は K^{n+1} より大きくて,
	n が奇数のとき, $\prod_{k=0}^{n} q_k$ は K^{n+1} より小さい.

Khintchine の定数は複素数への拡張をもっているのであろうか？

[1] A. Khintchine, Metrische Kettenbruchprobleme, *Compositio Math.* 1 (1935) 361-382.
[2] A. Khintchine, Zur metrischen Kettenbruchprobleme, *Compositio Math.* 3 (1936) 276-285.
[3] A. Khintchine, *Continued Fractions*, Univ. of Chicago Press, 3rd ed., 1961; MR 28 # 5038.
[4] A. M. Rockett and P. Szüsz, *Continued Fractions*, World Scientific, 1992; MR 93m:11060.
[5] C. Ryll-Nardzewski, On the ergodic theorems. II: Ergodic theory of continued fractions, *Studia Math.* 12 (1951) 74-79; MR 13,757b.
[6] M. Kac, *Statistical Independence in Probability, Analysis and Number Theory*, Math. Assoc. Amer., 1959; MR 22 # 996.
[7] D. H. Lehmer, Note on an absolute constant of Khintchine, *Amer. Math. Monthly* 46 (1939) 148-152.
[8] D. Shanks and J. W. Wrench, Khintchine's constant, *Amer. Math. Monthly* 66 (1959) 276-279; MR 21 # 1950.
[9] J. W. Wrench, Further evaluation of Khintchine's constant, *Math. Comp.* 14 (1960) 370-371; MR 30 # 693.
[10] J. W. Wrench and D. Shanks, Questions concerning Khintchine's constant and the efficient computation of regular continued fractions, *Math. Comp.* 20 (1966) 444-448.
[11] I. Vardi, *Computational Recreations in Mathematica*, Addison-Wesley, 1991; MR 93e:00002.
[12] P. Flajolet and I. Vardi, Zeta function expansions of classical constants, unpublished note (1996).
[13] D. H. Bailey, J. M. Borwein, and R. E. Crandall, On the Khintchine constant, *Math. Comp.* 66 (1997) 417-431; CECM preprint 95:036; MR 97c:11119.
[14] V. S. Adamchik and H. M. Srivastava, Some series of the zeta and related functions, *Analysis* 18 (1998) 131-144; MR 99d:11096.
[15] P. Shiu, Computations of continued fractions without input values, *Math. Comp.* 64 (1995) 1307-1317; MR 87h:11127.
[16] P. Lévy, Sur les lois probabilité dont dépendent les quotients complets et incomplets d'une fraction continue, *Bull. Soc. Math. France* 57 (1929) 178-194; also in *Oeuvres*, v. 6, ed. D. Dugué, Gauthier-Villars, 1980, pp. 266-282.
[17] P. Lévy, Sur le développement en fraction continue d'un nombre choisi au hasard, *Compositio Math.* 3 (1936) 286-303; also in *Oeuvres*, v. 6, ed. D. Dugué, Gauthier-Villars, 1980, pp. 285-302.
[18] W. Philipp, Some metrical theorems in number theory, *Pacific J. Math.* 20 (1967) 109-127; MR 34 # 5755.
[19] W. Philipp, Some metrical theorems in number theory. II, *Duke Math. J.* 37 (1970) 447-458; MR 42 # 7620.
[20] G. Lochs, Vergleich der Genauigkeit von Dezimalbruch und Kettenbruch, *Abh. Math. Sem.*

Univ. Hamburg 27 (1964) 142–144; MR 29 #57.
[21] C. Kraaikamp, A new class of continued fraction expansions, *Acta Arith.* 57 (1991) 1-39; MR 92a:11090.
[22] W. Bosma, K. Dajani, and C. Kraaikamp, Entropy and counting correct digits, Univ. of Nijmegen Math. Report 9925 (1999).
[23] V. A. Rohlin, Exact endomorphisms of a Lebesgue space (in Russian), *Izv. Akad. Nauk SSSR Ser. Mat.* 25 (1961) 499-530; MR 26 #1423.
[24] P. Flajolet, B. Vallée, and I. Vardi, Continued fractions from Euclid to the present day, École Polytechnique preprint (2000).
[25] R. M. Corless, Continued fractions and chaos, *Amer. Math. Monthly* 99 (1992) 203-215; also in *Organic Mathematics*, Proc. 1995 Burnaby workshop, ed. J. Borwein, P. Borwein, L. Jörgenson, and R. Corless, Amer. Math. Soc., 1997; MR 94g:58135.
[26] E. F. Beckenbach and R. Bellman, *Inequalities*, Springer-Verlag, 1965; MR 33 #236.
[27] D. E. Knuth, *The Art of Computer Programming*, v. 2, *Seminumerical Algorithms*, 2nd ed., Addison-Wesley, 1981, pp. 361-362, 604-606; MR 44 #3531.
[28] H. Riesel, On the metric theory of nearest integer continued fractions, *BIT* 27 (1987) 248-263; errata 28 (1988) 188; MR 88k:11048 and MR 89f:11114; Zbl. 617.10036 and Zbl. 644.10036.
[29] A. David and S. Dvorak, The estimations of the limit of the geometric means of Lüroth's digits, *Acta Fac. Rerum Natur. Univ. Comenian. Math.* 35 (1979) 95-107; MR 82c:10062.
[30] S. Kalpazidou, Khintchine's constant for Lüroth representation, *J. Number Theory* 29 (1988) 196-205; MR 89f:11105.
[31] S. Kalpazidou, A. Knopfmacher, and J. Knopfmacher, Metric properties of alternating Lüroth series, *Portugal. Math.* 48 (1991) 319-325; MR 92h:11068.
[32] H. Jager and C. de Vroedt, Lüroth series and their ergodic properties, *Proc. Konink. Nederl. Akad. Wetensch. Ser. A* 72 (1969) 31-42; *Indag. Math.* 31 (1969) 31-42; MR 39 #157.
[33] O. Jenkinson and M. Pollicott, Ergodic properties of the Bolyai-Rényi expansion, *Indag. Math.* 11 (2000) 399-418.
[34] G. J. Rieger, Über die mittlere Schrittanzahl bei Divisionsalgorithmen, *Math. Nachr.* 82 (1978) 157-180; MR 58 #533.
[35] G. J. Rieger, Ein Gauss-Kusmin-Levy-Satz für Kettenbrüche nach nächsten Ganzen, *Manuscripta Math.* 24 (1978) 437-448; MR 58 #27875.
[36] G. J. Rieger, Mischung und Ergodizität bei Kettenbrüchen nach nächsten Ganzen, *J. Reine Angew. Math.* 310 (1979) 171-181; MR 81c:10066.
[37] A. M. Rockett, The metrical theory of continued fractions to the nearer integer, *Acta Arith.* 38 (1980/81) 97-103; MR 82d:10074.
[38] S. Kalpazidou, Some asymptotic results on digits of the nearest integer continued fraction, *J. Number Theory* 22 (1986) 271-279; MR 87i:11101.
[39] L. Degeratu, Some metrical results for the nearest integer continued fraction, *Politehn. Univ. Bucharest Sci. Bull. Ser. A* 57/58 (1995/96) 61-67; MR 99e:11107.
[40] V. S. Adamchik, Evaluation of the nearest integer analog of Khintchine's constant, unpublished note (2001).
[41] J. Bourdon, On the Khintchine constant for centred continued fraction expansions, unpublished note (2001).
[42] P. Flajolet and B. Vallée, Continued fractions, comparison algorithms, and fine structure constants, *Constructive, Experimental, and Nonlinear Analysis*, Proc. 1999 Limoges conf., ed. M. Théra, Amer. Math. Soc., 2000, pp. 53-82; INRIA preprint RR4072; MR 2001h:11161.
[43] H. P. Robinson and E. Potter, *Mathematical Constants*, UCRL-20418 (1971), Univ. of Calif. at Berkeley; available through the National Technical Information Service, Springfield VA 22151.
[44] A. L. Schmidt, Diophantine approximation of complex numbers, *Acta Math.* 134 (1975) 1-85; MR 54 #10160.

[45] A. L. Schmidt, Ergodic theory for complex continued fractions, *Monatsh. Math.* 93 (1982) 39-62; MR 83g:10036.
[46] A. L. Schmidt, Ergodic theory of complex continued fractions, *Number Theory with an Emphasis on the Markoff Spectrum*, Proc. 1991 Provo conf., ed. A. D. Pollington and W. Moran, Dekker, 1993, pp. 215-226; MR 95f:11055.
[47] H. Nakada, On ergodic theory of A. Schmidt's complex continued fractions over Gaussian field, *Monatsh. Math.* 105 (1988) 131-150; MR 89f:11113.
[48] H. Nakada, On metrical theory of diophantine approximation over imaginary quadratic field, *Acta Arith.* 51 (1988) 393-403; MR 89m:11070.
[49] H. Nakada, The metrical theory of complex continued fractions, *Acta Arith.* 56 (1990) 279-289; MR 92e:11081.
[50] H. Nakada, Dynamics of complex continued fractions and geodesics over H^3, *Dynamical Systems and Chaos*, v. 1, Proc. 1994 Tokyo conf., ed. N. Aoki, K. Shiraiwa, and Y. Takahashi, World Scientific, 1995, pp. 192-199; MR 99c:11103.

1.9 ファイゲンバウム-クーレ-トレサー定数

$f(x)=ax(1-x)$ とする．ただし a は定数とする．各 $a\in[0,4]$ に対して区間 $[0,1]$ は，f によって $[0,1]$ それ自身の中へ写像される．a というパラメータをもつこれらの関数の族は，**ロジスティック写像**として知られている [1-8]．

f の 1 周期点（すなわち f の固定点）は何か．$x=f(x)$ を解くことにより

$$x=0 \quad (a<1 \text{ のとき吸引的}, a>1 \text{ のとき反発的}) \quad \text{と}$$

$$x=\frac{a-1}{a} \quad (1<a<3 \text{ のとき吸引的}, a>3 \text{ のとき反発的})$$

を得る．

f の 2 周期点は何か．すなわち，f の固定点ではなく反復合成写像 f^2 の固定点は何か．$x=f^2(x)$，$x\neq f(x)$ を解くことにより，2 周期点

$$x=\frac{a+1\pm\sqrt{a^2-2a-3}}{2a} \quad (3<a<1+\sqrt{6} \text{ のとき吸引的}, a>1+\sqrt{6} \text{ のとき反発的})$$

を得る．

$a>1+\sqrt{6}=3.4495\cdots$ に対して吸引的 4 周期点が現れる．$x=f^4(x)$，$x\neq f^2(x)$ を数値的に解くことにより 4 周期点が得られる．この 4 周期点は，$3.4495\cdots<a<3.5441\cdots$ のとき吸引的であり，$a>3.5441\cdots$ のときは反発的とわかる．

$a>3.5441\cdots$ に対して吸引的 8 周期点が現れる．$x=f^8(x)$，$x\neq f^4(x)$ を数値的に解くことにより 8 周期点が得られる．この 8 周期点は，$3.5441\cdots<a<3.5644\cdots$ のとき吸引的で，$a>3.5644\cdots$ のとき反発的であることがわかる．

周期倍化分岐の列は，どのくらい長く続くのか．これが 4 のはるか手前でとどまることは興味深い．

1.9 ファイゲンバウム-クーレ-トレサー定数

$$a_0=1, \quad a_1=3, \quad a_2=3.4495\cdots, \quad a_3=3.5441\cdots, \quad a_4=3.5644\cdots$$

が f の**分岐点**を表しているとすると，

$$a_\infty = \lim_{n\to\infty} a_n = 3.5699\cdots < 4$$

を証明することができる．この極限点は，「周期の生じるところ」と「カオスが生じるところ」の分かれ目を示している．カオス理論の発展や，物理的，化学的，生物学的な体系への応用をめざして多くの研究がなされてきた．この理論の狭い一部分，すなわち前述した指数関数的な集積点に関連した 2 つの「普遍定数」のみに焦点をあてよう．図 1.5 の分岐の図は，これから用いる記号を定義する際に助けになる．f の**超安定周期点**の列とは，

$$\tilde{a}_1=1+\sqrt{5}=3.2360\cdots, \quad \tilde{a}_2=3.4985\cdots, \quad \tilde{a}_3=3.5546\cdots, \quad \tilde{a}_4=3.5666\cdots$$

のことである．ここで $\tilde{a}_n$ は，2^n 周期点が臨界要素 1/2 を含むような最小のパラメータの値である．この周期点を $\tilde{c}(n)$ ということにする．f の**超安定幅**の列とは，

$$\tilde{w}_1=(\sqrt{5}-1)/4=0.3090\cdots, \quad \tilde{w}_2=0.1164\cdots, \quad \tilde{w}_3=0.0459\cdots$$

のことである．ここで $\tilde{w}_n$ は，1/2 と 1/2 に最も近い要素 $f^{2^{n-1}}(1/2) \in \tilde{c}(n)$ との距離である．また f の**分岐幅**とは，

$$w_1=\sqrt{2(\sqrt{6}-1)}/5=0.4099\cdots, \quad w_2=0.1603\cdots, \quad w_3=0.0636\cdots$$

のことである．ここで w_n は，a_{n+1} での対応する周期点距離である．超安定に関する $\tilde{a}_n$，$\tilde{w}_n$ は，a_n，w_n よりも数値的な計算が容易である．2 つの Feigenbaum-Coullet-Tresser（**ファイゲンバウム-クーレ-トレサー**）**定数**を

$$\delta = \lim_{n\to\infty} \frac{a_n - a_{n-1}}{a_{n+1} - a_n} = \lim_{n\to\infty} \frac{\tilde{a}_n - \tilde{a}_{n-1}}{\tilde{a}_{n+1} - \tilde{a}_n} = 4.6692016091\cdots$$

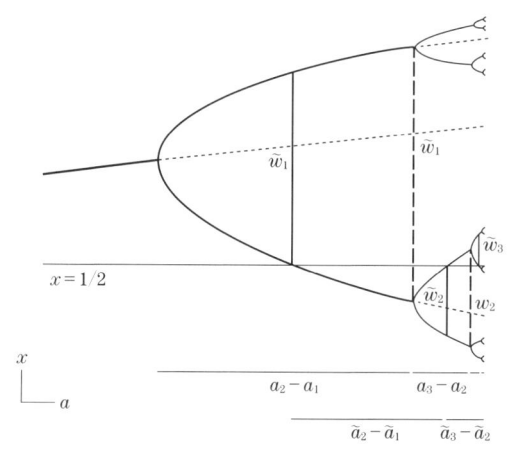

図 1.5 分岐の水平・垂直方向の特徴が a_n，w_n によってはかられる．

$$\alpha = \lim_{n \to \infty} \frac{w_n}{w_{n+1}} = \lim_{n \to \infty} \frac{\widetilde{w}_n}{\widetilde{w}_{n+1}} = 2.5029078750\cdots = (0.3995352805\cdots)^{-1}$$

と定義する[9-17]．ここで示したように，~ があってもなくても極限値 δ, α は変わらない．

どういう意味でこれらの定数が「普遍的」といわれるのか．ロジスティック写像 f を，たとえば $g(x) = b\sin(\pi x)$, $0 \leq b \leq 1$ でおきかえても，おもしろいことに同じ δ と α が出る． f と g の両方の関数が 2 次の最大点をもつ．この条件を拡張して一般化されたファイゲンバウム定数【1.9.1】が得られる．2 次元の例にも言及する【1.9.2】．

1 次元の 2 次最大値をもつ場合の普遍性のきちんとした証明が Lanford（ランフォード）[18-22] や Campanino & Epstein（カンパニーノ&エプステイン）[23-28]によって与えられた．明らかに前者が数学におけるこの種のコンピュータ支援による証明の最初のものである．

ファイゲンバウム定数のもっと単純な定義は存在するのか．それほど多くの説明を必要としないような，極限あるいは積分といった言葉での，より古典的な特徴づけがあるとよいのだが．これに最も近いのは，ある種の関数方程式を含む【1.9.3】．これは，実際に定数の計算を高精度で行う最も実用的なアルゴリズムを与えるようである[29-37]．円上の写像や【1.9.4】や異なる形のカオスにも言及する．

既に述べた数 3.5441… や 3.5644… は，[38,39]で議論されているように，それぞれ次数 12, 240 の代数的数であることが知られている．

Salamin（サラミン）[40]は，量子電磁力学からの（単位なし）微細構造定数 $(137.0359\cdots)^{-1}$ は，現在知られているよりも高度な理論において，ファイゲンバウム定数のような定数に関連するであろうと見当をつけている．

1.9.1 一般化されたファイゲンバウム定数

f, g を前に定義した関数とする．区間 $[-1,1]$ 上で定義された関数 $h(x) = 1 - c|x|^r$ を考える．ここで c と r は，$1 < c < 2$, $r > 1$ を満たす定数とする．各関数は，単峰，凹，対称で，h について $x = 0$ で生じうる例外を除いていたる所解析的である．さらに $r = 2$ のとき最大点での各 2 階導関数の値が厳密に負である．すなわち f, g, h は 2 次の最大点をもつ．これとは対照的に，$r = 3$ のとき h の最大値での次数は 3 次であり，$r = 4$ のときは 4 次である，など．これはファイゲンバウム定数の値に関して重大な違いである．

多数の著者たちが，δ と α を記述するのに「普遍な」という言葉を使ってきた．そしてこれは考察すべき対象が 2 次の最大値であるのならば適切である．しかし r を変えると δ と α の異なる値が生じる．数値的に考えると，δ は r について増加し，α は極限値 1 に減少していく[36,41]（表 1.3 を参照のこと）．実際[42-48]で

$$\lim_{r \to \infty} \delta(r) = 29.576303\cdots, \quad \lim_{r \to \infty} \alpha(r)^{-r} = 0.0333810598\cdots$$

とわかる．他方の限界では[15,31] $\lim_{r \to 1^+} \delta(r) = 2$, 一方 $\lim_{r \to 1^+} \alpha(r) = \infty$ となる．

いくぶん異なる一般化には，周期倍化ではなく周期 3 倍化が関係する[1,16,29,30,

表1.3 位数 r の関数としてのファイゲンバウム定数

r	3	4	5	6
$\delta(r)$	5.9679687038…	7.2846862171…	8.3494991320…	9.2962468327…
$\alpha(r)$	1.9276909638…	1.6903029714…	1.5557712501…	1.4677424503…

49-51]. ロジスティック写像に対しては，$3.8284\dots \leqq a \leqq 3.8540\dots$ のときパラメータの値が $\hat{a}_n$ での 3^n 周期点への三分岐の連鎖が，次のファイゲンバウム定数とともに生じる．すなわち

$$\hat{\delta}=\lim_{n\to\infty}\frac{\hat{a}_n-\hat{a}_{n-1}}{\hat{a}_{n+1}-\hat{a}_n}=55.247\cdots, \quad \hat{\alpha}=\lim_{n\to\infty}\frac{\widetilde{w}_n}{\widetilde{w}_{n+1}}=9.27738\cdots$$

3周期点は特別興味深い．なぜならこれらは，カオスの存在を保証するからである[2]．f が漸近周期的ではないような点 a の最小値は正確にはわからない．最初の6周期点は $3.6265\cdots$ で生じ[2]，最初の奇数周期点は，$3.6786\cdots$ で生じる[1]．

定数 $55.247\cdots$ と $9.27738\cdots$ は，「もともとの」ファイゲンバウム定数と同じ正確さでは計算されてはいない．存在に関する理論[27,28]が，周期倍化に対してのみ応用されたようである．周期3倍化に関する（われわれの）知識は，目下のところ明らかに数学的厳密さよりも数値からの発見の方に基礎づけがされている．

ところで $r=2$ のときの h の分岐点は

$$c_2=\frac{5}{4}=1.25, \quad c_3=1.3680\cdots, \quad c_4=1.3940\cdots, \quad \cdots, \quad c_\infty=1.4011\cdots$$

であり，変換 $c_n=a_n(a_n-2)/4$ により a_n と関係づけられる．極限点 $c_\infty=1.401$ は Myrberg（ミルベリ）[52]によるが，これは決して普遍ではない．同様にして，$r=2$ のとき h の連続する超安定幅比がわかる．それらは，

$$\alpha_1=3.2185\cdots, \quad \alpha_2=2.6265\cdots, \quad \alpha_3=2.5281\cdots, \quad \cdots, \quad \alpha_\infty=\alpha=2.5029\cdots$$

である．前に定義した記号を用いると $a_n=\widetilde{w}_n(\tilde{a}_{n+1}-2)\widetilde{w}_{n+1}^{-1}(\tilde{a}_n-2)^{-1}$ となる．数列 $\{c_n\}$，$\{a_n\}$ の両方とも【1.9.3】で必要となる．

1.9.2　2次平面写像

面積を保存する2次の（保存系）Hénon（エノン）写像[53,54]

$$\begin{pmatrix} x_{n+1} \\ y_{n+1} \end{pmatrix} = \begin{pmatrix} 1-ax_n^2+y_n \\ x_n \end{pmatrix}$$

も周期倍化の連鎖を導く．しかしファイゲンバウム定数は，$\alpha=4.0180767046\cdots$，$\beta=16.3638968792\cdots$（2方向に対して測る），$\delta=8.7210972\cdots$ となり1次元の場合よりも大きくなる．これらは2次の最大値をもつ2次元写像のある種の部分族の特徴である [50,55,56]．しかし，「もともとの」ファイゲンバウム定数 $\delta=4.6692016091\cdots$ が現れる異なる族が存在する．それは，面積を縮小する（散逸系）エノン写像[49,57,58]

$$\begin{pmatrix} x_{n+1} \\ y_{n+1} \end{pmatrix} = \begin{pmatrix} 1-ax_n^2+y_n \\ bx_n \end{pmatrix}$$

である．（ここで，付け加えられたパラメータ b は $|b|<1$ をみたす．）より高次元においても現れる．それゆえ δ の普遍性の範囲は，われわれの想像より広い！

【1.9.1】で議論した周期3倍化定数のように，4.01808…，16.36389… や 8.72109… は，「もともとの」ファイゲンバウム定数と同程度の精密さでは計算されていない．2次元保存系写像に対して Eckmann（エックマン）と Koch & Wittwer（コッホ&ウィットヴァ）[59,60]が，これらは実際普遍的であることを証明した．N 次元散逸系写像に対しては，Collet（コレ），Eckmann & Koch[61,62]が，定数 4.66920… が，普遍性をもつようだという証明の概略を示した．

1.9.3　クビタノビッチ-ファイゲンバウム関数方程式

D を複素数平面での連結開集合で区間 [0,1] を含むものとする．X を，D 上解析的，D の閉包で連続であり，[0,1] 上で実数値をとり，$F(0)=0$ を満たす関数に上限ノルムを考えたときの実バナッハ空間とする．

実数 $r>1$ を固定する．Ω_r を $f(x)=1+F(|x|^r)$ という形の関数 $f:[-1,1]\to(-1,1]$ の集合とする．ただし $F\in X$ であり，すべての $y\in[0,1]$ に対して $F'(y)<0$ を満たすものとする．いいかえると，Ω_r は，$|x|^r$ のベキ級数として書くことができ，$-1<f^2(0)<f(0)=1$ を満たす $[-1,1]$ 上の偶関数の折りたたみ自己写像 f の集合である．また $\Omega_{r,0}$ を Ω_r の部分集合で $f^2(0)<0<f^4(0)<-f^2(0)<f^3(0)<1$ という条件をつけ加えたものとする．

f と F との対応を使うと，集合 $\Omega_{r,0}$ と Ω_r が，X の入れ子になった開部分集合と自然に同一視できる．故に $\Omega_{r,0}$，Ω_r の両方とも X を基礎とするバナッハ多様体である．よっていわゆる**周期倍化作用素** $T_r:\Omega_{r,0}\to\Omega_r$ に関する微分法を行うことができる．これは，ある関数 φ の近傍で T_r に最も密着した線形作用素 $L_r:X\to X$ を得るものである．このことは，すぐにできるであろうが，これには，ファイゲンバウム定数の厳密な定式化が必要である[15,27,63]．

前に定義した関数 h を考えよう．パラメータ c によることを明確にしておこう．そして今後これを h_c と書くことにしよう．明らかに $h_c\in\Omega_r$ である．$r=2$ に対して【1.9.1】の最後に定義した数列 $\{c_n\}$，$\{a_n\}$ を思い出そう．任意の $r>1$ に対して似たような数列が定義できる．c が c_∞ に増加し，グラフの真中の部分が限りなく大きくなっていくときの h_c の反復写像の普遍性に興味がある．驚くべき極限

$$\lim_{n\to\infty}(-a_n)^n\cdot h_{c_n}^{2n}\left(\frac{x}{a_n^n}\right)=\varphi(x)$$

が存在する[6-7]．そして $\varphi\in\Omega_{r,0}$ であり，Cvitanovic-Feigenbaum（**クビタノビッチ-ファイゲンバウム**）**関数方程式**

$$\varphi(x)=\varphi(1)^{-1}\cdot\varphi(\varphi(\varphi(1)\cdot x))=T_r[\varphi](x)$$

を満たす．図1.6を見るとよい幾何学的な解釈が得られる．さらに解 φ は，r が偶数のとき一意的であることが証明されている[68-71]．

この一意性を任意の $r>1$ まで拡張することは未解決問題への一つの挑戦である

$$\varphi(x)$$
$$\varphi(\varphi(x))$$
$$\varphi(\varphi(\varphi(\varphi(x))))$$

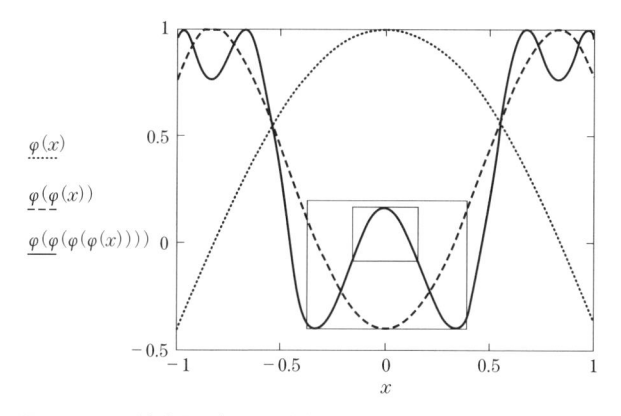

図 1.6 φ の反復合成写像の自己相似性が，だんだん小さくなる長方形の窓の中に示されている．条件 $\varphi(1)<0$ により，向きが裏返しになる．

[72]．結果として各 r に対して $a(r) = -\varphi(1)^{-1}$ を得る．

さて固定された φ での T_r の局所的な線形化（フレシェ微分）を考えよう．それは

$$L_r[\psi](x) = \varphi(1)^{-1} \cdot \{\varphi'(\varphi(\varphi(1) \cdot x)) \cdot \psi(\varphi(1) \cdot x) + \psi(\varphi(\varphi(1) \cdot x)) \\ + \psi(1) \cdot [\varphi'(x) \cdot x - \varphi(x)]\}$$

である．このとき，各 r に対して $\delta(r)$ は L_r に付随した最大の固有値であり，実際，単位円板の外にあるただ 1 つの固有値である．これが，$\delta(r)$ の正確な評価の基礎である．幸運なことに，この計算には $L_r[\psi](x)$ の 3 つの項の初めの 2 項のみが必要である．[27,36] あるいは，$\delta(r) = \lim_{n \to \infty} \sigma_{n+1}/\sigma_n$，ここで [45-47]

$$\sigma_n = \frac{1}{\xi(1)^n} \sum_{k=1}^{2^n-1} \xi^k(0) \cdot \left(\prod_{j=0}^{k-1} \xi'(\xi^j(0))\right)^{-1}$$

であり，$0 < r < 1$ に対して $\xi(x) = |\varphi(x^{1/r})|^r$ である．この公式は魅力的であるが，残念なことに高次の精密な結果を求めるためには数値的にふさわしくない．δ のさらなる公式が [73-75] にある．

周期 3 倍化【1.9.1】に対するクビタノビッチ-ファイゲンバウム方程式の類似

$$\varphi(x) = \varphi(1)^{-1} \cdot \varphi(\varphi(\varphi(1) \cdot x)))$$

は，$\hat{a}$ の評価を与え，右辺の線形化が $\hat{\delta}$ を導く．平面写像に対しては，行列による類似ができる．他の関数方程式もまもなく出現するであろう．

1.9.4 黄金比と白銀比の円写像

少し異なる例 [76-79] にふれておく．それは

$$\theta_{n+1} = k_a(\theta_n) = \theta_n + a - \frac{1}{2\pi} \sin(2\pi\theta_n)$$

であり，これは，円周 l からそれ自身の上への同相写像と考えられる．このような任意の円写像 l に対して極限

$$\rho(l) = \lim_{n\to\infty} \frac{l^n(\theta) - \theta}{n}$$

が存在し，これは，θ によらない．$\rho(l)$ は，l の**巻き数**あるいは**回転数**といわれる．ここでの関心事は，周期倍化ではなく擬周期の方にある．この話題は，カオスへのあらたな展開を提供し，ρ が無理数のとき，いろいろな条件の下で生じた異なる状況に起源をもつ．

$f_1 = f_2 = 1$，$f_3 = 2$ はフィボナッチ数を表すものとする【1.2】．そして[80,81]

$$k_{a_n}^{f_n}(0) = f_{n-1}, \quad w_n = k_{a_n}^{f_{n-1}}(0) = f_{n-2}$$

により数列 $\{a_n\}$，$\{w_n\}$ を定義する．$\rho(k_{a_\infty}) = (1-\sqrt{5})/2$ が証明される．よって円写像 k_{a_n} の族は**黄金比的**であり，対応するファイゲンバウム定数は $\alpha = 1.2885745539\cdots$ であり，$\delta = 2.8336106558\cdots$ である．さらに，ただ 1 つの 3 次変曲点をもつすべての黄金比円写像に対して定数 α, δ は普遍性をもつ．フィボナッチ数をペル数【1.1】に置きかえると，$\rho(k_{a_\infty}) = \sqrt{2}-1$ になる．よってこの円写像 k_{a_n} の族は $\alpha = 1.5868266790\cdots$，$\delta = 6.7992251609\cdots$ となる**白銀比円写像**である．同様の普遍性が 3 次白銀比円写像に対しても成立する．他の無理数の巻き数についても研究されている[30]．かわりに r 次の変曲点をもつ黄金比円写像を考えると，関数 $\alpha(r)$, $\delta(r)$ が出現し，これらは

$$\lim_{r\to\infty} \alpha(r) = 1, \quad \lim_{r\to\infty} \alpha(r)^r = 3.63600703\cdots,$$

すべての $r > 0$ に対して $\alpha\left(\dfrac{1}{r}\right) = \alpha(r)^r$, $\quad \lim_{r\to\infty} \delta(r) = 4.121326\cdots$

を満たす[47,80,82-86]．

すべての r に対して $\delta(1/r) = \delta(r)$ と予想されたが，まだ証明されてはいない．

区間写像のときと同様にある関数方程式が，円写像に対応する $\alpha(r)$, $\delta(r)$ を精密に計算するための数値的な鍵を与える[8]．すなわち黄金比写像の場合は

$$\varphi(\theta) = \varphi(1)^{-1} \cdot \varphi(\varphi(\varphi(1)^2 \cdot \theta))$$

であり，白銀比写像の場合は

$$\varphi(\theta) = \varphi(1)^{-1} \cdot \varphi(\varphi(1) \cdot \varphi(\varphi(\varphi(1)^2 \cdot \theta)))$$

である．

McCarthy（マッカーシー）[87]は，2 つの有名な関数方程式

$$\varphi(x) \cdot \varphi(y) = \varphi(x+y), \quad \varphi(\varphi(y)) = s^{-1} \varphi(sy)$$

を比較した．前者は，乗法が単に変数の加法の形であり，後者では，自己合成がちょうど計量の変換になっている．彼は，後者の解に対して「20 世紀指数関数」というふさわしい名前を提唱した．しかしながらこの分野の研究は，もっとずっと長く続いていくことになるだろう．

[1] R. M. May, Simple mathematical models with very complicated dynamics, *Nature* 261 (1976) 459-467.

[2] T. Y. Li and J. A. Yorke, Period three implies chaos, *Amer. Math. Monthly* 82 (1975) 985-992; MR 52 #5898.

[3] D. Singer, Stable orbits and bifurcations of maps on the interval, *SIAM J. Appl. Math.* 35 (1978)

260-271; MR 58 #13206.
[4] R. L. Devaney, *An Introduction to Chaotic Dynamical Systems*, 2nd ed., Addison-Wesley, 1989; MR 91a:58114.
[5] M. Martelli, *Discrete Dynamical Systems and Chaos*, Longman, 1992; MR 97i:58104.
[6] H. G. Schuster, *Deterministic Chaos: An Introduction*, 3rd ed., VCH Verlagsgesellschaft mbH, 1995; MR 97c:58103.
[7] R. A. Holmgren, *A First Course in Discrete Dynamical Systems*, 2nd ed., Springer-Verlag, 1996; MR 97f:58052.
[8] R. L. Kraft, Chaos, Cantor sets, and hyperbolicity for the logistic maps, *Amer. Math. Monthly* 106 (1999) 400-408; MR 2000f:37042.
[9] M. J. Feigenbaum, Quantitative universality for a class of nonlinear transformations, *J. Stat. Phys.* 19 (1978) 25-52; MR 58 #18601.
[10] M. J. Feigenbaum, The universal metric properties of nonlinear transformations, *J. Stat. Phys.* 21 (1979) 669-706; MR 82e:58072.
[11] M. J. Feigenbaum, Universal behavior in nonlinear systems, *Los Alamos Sci.* 1 (1980) 4-27; MR 82h:58031.
[12] C. Tresser and P. Coullet, Itérations d'endomorphismes et groupe de renormalisation, *C. R. Acad. Sci. Paris Sér. A-B* 287 (1978) A577-A580; MR 80b:58043.
[13] C. Tresser and P. Coullet, Itérations d'endomorphismes et groupe de renormalisation, *J. de Physique Colloque* C5, n. 8, t. 39 (1978) C5-25-C5-27.
[14] P. Collet and J.-P. Eckmann, Properties of continuous maps of the interval to itself, *Mathematical Problems in Theoretical Physics*, ed. K. Osterwalder, Lect. Notes in Physics 116, Springer-Verlag, 1979, pp. 331-339; MR 82b:58051.
[15] P. Collet, J.-P. Eckmann, and O. E. Lanford, Universal properties of maps on an interval, *Commun. Math. Phys.* 76 (1980) 211-254; MR 83d:58036.
[16] P. Collet and J.-P. Eckmann, *Iterated Maps on the Interval as Dynamical Systems*, Birkhaüser, 1980; MR 82j:58078.
[17] P. Cvitanovic, *Universality in Chaos*, Adam Hilger, 1984; MR 91e:58124.
[18] O. E. Lanford, Remarks on the accumulation of period-doubling bifurcations, *Mathematical Problems in Theoretical Physics*, ed. K. Osterwalder, Lect. Notes in Physics 116, Springer-Verlag, 1979, pp. 340-342; MR 82b:58052.
[19] O. E. Lanford, Smooth transformations of intervals, *Séminaire Bourbaki 1980/81*, n. 563, Lect. Notes in Math. 901, Springer-Verlag, 1981, pp. 36-54; MR 83k:58066.
[20] O. E. Lanford, A computer-assisted proof of the Feigenbaum conjectures, *Bull. Amer. Math. Soc.* 6 (1982) 427-434; MR 83g:58051.
[21] O. E. Lanford, Computer-assisted proofs in analysis, *Physica A* 124 (1984) 465-470; MR 86a:00013.
[22] O. E. Lanford, A shorter proof of the existence of the Feigenbaum fixed point, *Commun. Math. Phys.* 96 (1984) 521-538; MR 86c:58121.
[23] M. Campanino and H. Epstein, On the existence of Feigenbaum's fixed point, *Commun. Math. Phys.* 79 (1981) 261-302; MR 82j:58099.
[24] M. Campanino, H. Epstein, and D. Ruelle, On Feigenbaum's functional equation $g \circ g(\lambda x) + \lambda g(x) = 0$, *Topology* 21 (1982) 125-129; MR 83g:58039.
[25] J.-P. Eckmann and P. Wittwer, A complete proof of the Feigenbaum conjectures, *J. Stat. Phys.* 46 (1987) 455-475; MR 89b:58131.
[26] H. Koch, A. Schenkel, and P. Wittwer, Computer-assisted proofs in analysis and programming in logic: A case study, *SIAM Rev.* 38 (1996) 565-604; Université de Genève physics preprint 94-394; MR 97k:39001.
[27] E. B. Vul, Ya. G. Sinai, and K. M. Khanin, Feigenbaum universality and thermodynamic formalism (in Russian), *Uspekhi Mat. Nauk*, v. 39 (1984) n. 3, 3-37; Engl. transl. in *Russian Math. Surveys*, v. 39 (1984) n. 3, 1-40; MR 86g:58106.

[28] D. Rand, Universality and renormalisation, *New Directions in Dynamical Systems*, ed. T. Bedford and J. Swift, Lect. Notes 127, London Math. Soc., 1988, pp. 1-56; MR 89j:58082.
[29] R. Delbourgo and B. G. Kenny, Universality relations, *Phys. Rev. A* 33 (1986) 3292-3302.
[30] R. Delbourgo, Relations between universal scaling constants in dissipative maps, *Nonlinear Dynamics and Chaos*, Proc. Fourth Physics Summer School, ed. R. L. Dewar and B. I. Henry, World Scientific, 1992, pp. 231-256; MR 93c:58121.
[31] J. P. van derWeele, H. W. Capel, and R. Kluiving, Period doubling in maps with a maximum of order z, *Physica A* 145 (1987) 425-460; MR 89f:58109.
[32] J. W. Stephenson and Y. Wang, Numerical solution of Feigenbaum's equation, *Appl. Math. Notes* 15 (1990) 68-78; MR 92c:58085.
[33] J. W. Stephenson and Y. Wang, Relationships between the solutions of Feigenbaum's equations, *Appl. Math. Lett.* 4 (1991) 37-39; MR 92g:39003.
[34] J. W. Stephenson and Y. Wang, Relationships between eigenfunctions associated with solutions of Feigenbaum's equation, *Appl. Math. Lett.* 4 (1991) 53-56; MR 92g:39004.
[35] K. M. Briggs, How to calculate the Feigenbaum constants on your PC, *Austral. Math. Soc. Gazette* 16 (1989) 89-92; also in *Feigenbaum Scaling in Discrete Dynamical Systems*, Ph.D. thesis, Univ. of Melbourne, 1997.
[36] K. M. Briggs, A precise calculation of the Feigenbaum constants, *Math. Comp.* 57 (1991) 435-439; MR 91j:11117.
[37] D. J. Broadhurst, Feigenbaum constants to 1018 decimal places, unpublished note (1999).
[38] D. H. Bailey, Multiprecision translation and execution of Fortran programs, *ACM Trans. Math. Software* 19 (1993) 288-319.
[39] D. H. Bailey and D. J. Broadhurst, Parallel integer relation detection: Techniques and applications, *Math. Comp.* 70 (2001) 1719-1736; math.NA/9905048; MR 2002b:11174.
[40] E. Salamin, Fine structure constant, unpublished note (2000).
[41] A. P. Kuznetsov, S. P. Kuznetsov, and I. R. Sataev, A variety of period-doubling universality classes in multi-parameter analysis of transition to chaos, *Physica D* 109 (1997) 91-112; MR 98k:58159.
[42] J.-P. Eckmann and P. Wittwer, *Computer Methods and Borel Summability Applied to Feigenbaum's Equation*, Lect. Notes in Physics 227, Springer-Verlag, 1985; MR 86m:58129.
[43] J.-P. Eckmann and H. Epstein, Bounds on the unstable eigenvalue for period doubling, *Commun. Math. Phys.* 128 (1990) 427-435; MR 91a:582134.
[44] J. Groeneveld, On constructing complete solution classes of the Cvitanovic-Feigenbaum equation, *Physica A* 138 (1986) 137-166; MR 88k:58079.
[45] J. B. McGuire and C. J. Thompson, Asymptotic properties of sequences of iterates of nonlinear transformations, *J. Stat. Phys.* 27 (1982) 183-200; MR 84h:58123.
[46] C. J. Thompson and J. B. McGuire, Asymptotic and essentially singular solutions of the Feigenbaum equation, *J. Stat. Phys.* 51 (1988) 991-1007; MR 90c:58134.
[47] K. M. Briggs, T. W. Dixon, and G. Szekeres, Analytic solutions of the Cvitanovic-Feigenbaum and Feigenbaum-Kadanoff-Shenker equations, *Int. J. Bifurcat. Chaos Appl. Sci. Eng.* 8 (1998) 347-357; MR 99h:58054.
[48] H. Thunberg, Feigenbaum numbers for certain flat-top families, *Experim. Math.* 3 (1994) 51-57; MR 96f:58085.
[49] B. Derrida, A. Gervois, and Y. Pomeau, Universal metric properties of bifurcations of endomorphisms, *J. Phys. A* 12 (1979) 269-296; MR 80k:58078.
[50] B. Derrida and Y. Pomeau, Feigenbaum's ratios of two-dimensional area preserving maps, *Phys. Lett. A* 80 (1980) 217-219; MR 81m:58051.
[51] A. I. Golberg, Ya. G. Sinai, and K. M. Khanin, Universal properties of sequences of periodtripling bifurcations (in Russian), *Uspekhi Mat. Nauk*, v. 38 (1983) n. 1, 159-160; Engl. transl. in *Russian Math. Surveys*, v. 38 (1983) n. 1, 187-188; MR 84i:58085.
[52] P. J. Myrberg, Iteration der reellen Polynome zweiten Grades. III, *Annales Acad. Sci. Fenn. A*,

v. 336 (1963) n. 3, 1-18; MR 27 #1552.
[53] M. Hénon, A two-dimensional mapping with a strange attractor, *Commun. Math. Phys.* 50 (1976) 69-77; MR 54 #10917.
[54] K. M. Briggs, G. R. W. Quispel, and C. J. Thompson, Feigenvalues for Mandelsets, *J. Phys. A* 24 (1991) 3363-3368; MR 92h:58167.
[55] P. Collet, J.-P. Eckmann, and H. Koch, On universality for area-preserving maps of the plane, *Physica D* 3 (1981) 457-467; MR 83b:58055.
[56] J. M. Greene, R. S. MacKay, F. Vivaldi, and M. J. Feigenbaum, Universal behaviour in families of area-preserving maps, *Physica D* 3 (1981) 468-486; MR 82m:58041.
[57] J.-P. Eckmann, Roads to turbulence in dissipative dynamical systems, *Rev. Mod. Phys.* 53 (1981) 643-654; MR 82j:58079.
[58] J. M. T. Thompson and H. B. Stewart, *Nonlinear Dynamics and Chaos*, Wiley, 1986; MR 87m:58098.
[59] J.-P. Eckmann, H. Koch, and P. Wittwer, Existence of a fixed point of the doubling transformation for area-preserving maps of the plane, *Phys. Rev. A* 26 (1982) 720-722.
[60] J.-P. Eckmann, H. Koch, and P. Wittwer, *A Computer-Assisted Proof of Universality for Area-Preserving Maps*, Amer. Math. Soc., 1984; MR 85e:58119.
[61] P. Collet, J.-P. Eckmann, and H. Koch, Period doubling bifurcations for families of maps on $\mathbb{R}^n$, *J. Stat. Phys.* 25 (1981) 1-14; MR 82i:58052.
[62] O. E. Lanford, Period doubling in one and several dimensions, *Physica D* 7 (1983) 124-125.
[63] Y. Jiang, T. Morita, and D. Sullivan, Expanding direction of the period doubling operator, *Commun. Math. Phys.* 144 (1992) 509-520; MR 93c:58169.
[64] H. Epstein, New proofs of the existence of the Feigenbaum functions, *Commun. Math. Phys.* 106 (1986) 395-426; MR 88i:58124.
[65] J.-P. Eckmann and H. Epstein, Fixed points of composition operators, *Proc. 1986 International Congress on Mathematical Physics*, ed. M. Mebkhout and R. Seneor, World Scientific, 1987, pp. 517-530; MR 89d:58080.
[66] H. Epstein, Fixed points of composition operators, *Nonlinear Evolution and Chaotic Phenomena*, ed. G. Gallavotti and P. F. Zweifel, Plenum Press, 1988, pp. 71-100; MR 92f:58003.
[67] H. Epstein, Fixed points of composition operators. II, *Nonlinearity* 2 (1989) 305-310; MR 90j:58086.
[68] D. Sullivan, Bounds, quadratic differentials, and renormalization conjectures, *Mathematics into the Twenty-first Century*, ed. F. Browder, Amer. Math. Soc., 1992, pp. 417-466; MR 93k:58194.
[69] W. de Melo and S. van Strien, *One-Dimensional Dynamics*, Springer-Verlag, 1993; MR 95a:58035.
[70] C. T. McMullen, *Complex Dynamics and Renormalization*, Princeton Univ. Press, 1994; MR 96b:58097.
[71] M. Lyubich, The quadratic family as a qualitatively solvable model of chaos, *Notices Amer. Math. Soc.* 47 (2000) 1042-1052; MR 2001g:37063.
[72] Y. Jiang, Renormalization on one-dimensional folding maps, *Dynamical Systems and Chaos*, v. 1, Proc. 1994 Hachioji conf., ed. N. Aoki, K. Shiraiwa, and Y. Takahashi, World Scientific, 1995, pp. 116-125; MR 98g:58045.
[73] F. Christiansen, P. Cvitanovic, and H. H. Rugh, The spectrum of the period-doubling operator in terms of cycles, *J. Phys. A* 23 (1990) L713-L717; MR 91d:58136.
[74] R. Artuso, E. Aurell, and P. Cvitanovic, Recycling of strange sets. II: Applications. *Nonlinearity* 3 (1990) 361-386; MR 92c:58105.
[75] M. Pollicott, A note on the Artuso-Aurell-Cvitanovic approach to the Feigenbaum tangent operator, *J. Stat. Phys.* 62 (1991) 257-267; MR 92d:58166.
[76] S. J. Shenker, Scaling behavior in a map of a circle onto itself: Empirical results, *Physica D* 5 (1982) 405-411; MR 84f:58066.
[77] M. J. Feigenbaum, L. P. Kadanoff, and S. J. Shenker, Quasiperiodicity in dissipative systems:

A renormalization group analysis, *Physica D* 5 (1982) 370-386; MR 84f:58101.
[78] S. Östlund, D. Rand, J. Sethna, and E. Siggia, Universal properties of the transition from quasiperiodicity to chaos in dissipative systems, *Physica D* 8 (1983) 303-342; MR 85d:58066.
[79] O. E. Lanford, Functional equations for circle homeomorphisms with golden ratio rotation number, *J. Stat. Phys.* 34 (1984) 57-73; MR 86a:58064.
[80] T. W. Dixon, B. G. Kenny, and K. M. Briggs, On the universality of singular circle maps, *Phys. Lett. A* 231 (1997) 359-366.
[81] T. W. Dixon, T. Gherghetta, and B. G. Kenny, Universality in the quasiperiodic route to chaos, *Chaos* 6 (1996) 32-42; also in *Chaos and Nonlinear Dynamics*, ed. R. C. Hilborn and N. B. Tufillaro, Amer. Assoc. Physics Teachers, 1999, pp 90-100; MR 97a:58162.
[82] L. Jonker and D. A. Rand, Universal properties of maps of the circle with ε-singularities, *Commun. Math. Phys.* 90 (1983) 273-292; MR 85j:58103.
[83] A. Arneodo and M. Holschneider, Fractal dimensions and homeomorphic conjugacies, *J. Stat. Phys.* 50 (1988) 995-1020; MR 90a:58092.
[84] B. Hu, A. Valinia, and O. Piro, Universality and asymptotic limits of the scaling exponents in circle maps, *Phys. Lett. A* 144 (1990) 7-10; MR 90k:58131.
[85] R. Delbourgo and B. G. Kenny, Relations between universal scaling constants for the circle map near the golden mean, *J. Math. Phys.* 32 (1991) 1045-1051; MR 92e:58109.
[86] T. W. Dixon and B. G. Kenny, Transition to criticality in circle maps at the golden mean, *J. Math. Phys.* 39 (1998) 5952-5963; MR 99h:58169.
[87] P. J. McCarthy, Ultrafunctions, projective function geometry, and polynomial functional equations, *Proc. London Math. Soc.* 53 (1986) 321-339; MR 88c:39009.

1.10 マデラングの定数

整数格子点 $(i,j) \neq (0,0)$ に符号 $(-1)^{i+j}$ つきの単位負荷をもつ平面における正方格子を考える．(i,j) での負荷によって生じる原点での静電気ポテンシャルは $(-1)^{i+j}/\sqrt{i^2+j^2}$ である．よってすべての負荷から生じる原点での全静電気ポテンシャルは[1]

$$M_2 = \sum_{i,j=-\infty}^{\infty}{}' \frac{(-1)^{i+j}}{\sqrt{i^2+j^2}}$$

である．ここで'（プライム）は，$(0,0)$ を和から除くことを示している．
この無限格子和は，どのように解釈されるべきだろうか．これは微妙な問題である．なぜなら $i=j$ とした部分級数は発散するからである．よって全体の級数の別の性質を注意深く研究する必要がある[2-7]．それにもかかわらず拡大していく円あるいは拡大していく正方形を，使うことができ，いずれの場合でも同じ収束和

$$M_2 = 4(\sqrt{2}-1)\zeta\left(\frac{1}{2}\right)\beta\left(\frac{1}{2}\right) = -1.6155426267\cdots$$

を得る．ここで $\zeta(x)$ は，Riemann のゼータ関数【1.6】であり，$\beta(x)$ は Dirichlet のベータ関数【1.7】である．この和 M_2 を2次元塩化ナトリウム結晶に対する Madelung

(マデラング) の定数という．このようによく知られた関数を用いて格子和を書き直したことが本質的なことである．なぜならもしそうしなければ，収束の比率が極めて遅いからである．

3次元の類似

$$M_3 = \sum_{i,j,k=-\infty}^{\infty}{}' \frac{(-1)^{i+j+k}}{\sqrt{i^2+j^2+k^2}}$$

は，驚くべきことに，拡大していく球を使った計算法では，「発散」を導いてしまうのでより扱いにくい．この驚くべき事実は，Emersleben（エメルスレーベン）[16]が初めて気づいた．かわりに拡大立方体を用いると，Benson-Mackenzie（ベンソン-マッケンジー）公式[17,18]

$$M_3 = -12\pi \sum_{m,n=1}^{\infty} \mathrm{sech}\left(\frac{\pi}{2}\sqrt{(2m-1)^2+(2n-1)^2}\right)^2 = -1.7475645946\cdots$$

を得る．これは，とても速く収束する．別のたくさんの言い方があるが，その中にHautot（ハウトート）による公式[19]

$$M_3 = \frac{\pi}{2} - \frac{9}{2}\ln(2) + 12\sum_{m,n=1}^{\infty}(-1)^m \frac{\mathrm{csch}(\pi\sqrt{m^2+n^2})}{\sqrt{m^2+n^2}}$$

がある．これは，後で議論する別の格子和と形式的には一致するが，収束はそれほど速くない．M_3 は，3次元塩化ナトリウム結晶に対するマデラングの定数，あるいは，より単純に**マデラングの定数**といわれる．Glasser & Zucker（グラサー&ツッカー）[20]は，彼らのすばらしい概説の中で$\pm 2M_3$を同じ名でよんでいるので，文献を調べる際は注意すべきである．M_3 の別の表現が[21-23]にある．

4次元，6次元，8次元の類似も[24]にあり

$$M_4 = \sum_{i,j,k,l=-\infty}^{\infty}{}' \frac{(-1)^{i+j+k+l}}{\sqrt{i^2+j^2+k^2+l^2}} = -8(5-3\sqrt{2})\zeta\left(\frac{1}{2}\right)\zeta\left(-\frac{1}{2}\right) = -1.8393990840\cdots,$$

$$M_6 = \frac{3}{\pi^2}\left[4(\sqrt{2}-1)\zeta\left(\frac{1}{2}\right)\beta\left(\frac{5}{2}\right) - (4\sqrt{2}-1)\zeta\left(\frac{5}{2}\right)\beta\left(\frac{1}{2}\right)\right] = -1.9655570390\cdots,$$

$$M_8 = \frac{15}{4\pi^3}(8\sqrt{2}-1)\zeta\left(\frac{1}{2}\right)\zeta\left(\frac{7}{2}\right) = -2.0524668272\cdots$$

である．Borwein & Borwein（ボーウィン&ボーウィン）[4]による一般的な結果によるとマデラング定数の n 次元類似は，任意の $n\geq 1$ に対して収束することがわかる．もちろん $M_1 = -2\ln(2)$ である．$M_5 = -1.9093378156\cdots$ あるいは $M_7 = -2.0124059897\cdots$ に対する速く収束する級数表現は，つかみどころのない形をしているようである[25]．しかしすべての n に対して

$$M_n = \frac{1}{\sqrt{\pi}}\int_0^{\infty}\left\{\left(\sum_{k=-\infty}^{\infty}(-1)^k e^{-k^2 t}\right)^n - 1\right\}\frac{dt}{\sqrt{t}}$$

が知られている．これから高い精度をもつ数値計算が可能である[26,27]．この積分を用いると，$n\to\infty$ のとき $M_n \sim -\sqrt{4\ln(n)/\pi}$ であることが証明できる[28]．

これらの格子和にはたくさんの変種がありうる．たとえば，分母の $\sqrt{}$ を除くことが

でき，次の式が得られる[15, 20]．

$$N_1 = \sum_{i=-\infty}^{\infty}{}' \frac{(-1)^i}{i^2} = -\frac{\pi^2}{6}, \quad N_2 = \sum_{i,j=-\infty}^{\infty}{}' \frac{(-1)^{i+j}}{i^2+j^2} = -\pi \ln(2),$$

$$N_3 = \sum_{i,j,k=-\infty}^{\infty}{}' \frac{(-1)^{i+j+k}}{i^2+j^2+k^2}$$

$$= \frac{\pi^2}{3} - \pi \ln(2) - \frac{\pi}{\sqrt{2}} \ln(2(\sqrt{2}+1)) + 8\pi \sum_{m,n=1}^{\infty} (-1)^n \frac{\operatorname{csch}(\pi\sqrt{m^2+2n^2})}{\sqrt{m^2+2n^2}}$$

$$= -2.5193561520\cdots,$$

$$N_4 = \sum_{i,j,k,l=-\infty}^{\infty}{}' \frac{(-1)^{i+j+k+l}}{i^2+j^2+k^2+l^2} = -4\ln(2)$$

漸近的に $N_n \sim -\ln(n)$ が同様に求められる．異なる格子上の和，たとえば正方形格子ではなく平面上の基本ベクトル $(1,0)(1/2, \sqrt{3}/2)$ をもつ正六角形格子に変えることができる．これは次の表現

$$H_2 = \frac{4}{3} \sum_{i,j=-\infty}^{\infty}{}' \frac{\sin((i+1)\theta)\sin((j+1)\theta) - \sin(i\theta)\sin((j-1)\theta)}{\sqrt{i^2+ij+j^2}},$$

を導く．ここで $\theta = 2\pi/3$ であり，H_2 は

$$H_2 = -3(\sqrt{3}-1)\zeta\left(\frac{1}{2}\right)\left(1 - \frac{1}{\sqrt{2}} + \frac{1}{\sqrt{4}} - \frac{1}{\sqrt{5}} + \frac{1}{\sqrt{7}} - \frac{1}{\sqrt{8}} + \frac{1}{\sqrt{10}} - \frac{1}{\sqrt{11}} + - \cdots\right)$$

$$= 1.5422197217\cdots.$$

と書きかえることができる．これは平面六角形格子に対するマデラングの定数であり，これの3次元の類似 H_3 はたぶん M_3 によく似た化学的な重要性をもつ．同じように分母の $\sqrt{}$ を外すと

$$K_2 = \frac{4}{3} \sum_{i,j=-\infty}^{\infty}{}' \frac{\sin((i+1)\theta)\sin((j+1)\theta) - \sin(i\theta)\sin((j-1)\theta)}{i^2+ij+j^2} = \sqrt{3}\pi \ln(3)$$

となる．

オイラー-マスケローニの定数【1.5】の格子和の一般化が【1.10.1】に現れる．ところで，これは Stieltjes（スティルチェス）[22] あるいは Masser（マサー）や Gramain（グラメイン）【7.2】による異なる一般化とは関連がない．

Forrester & Glasser（フォレスター&グラサー）[29] は，

$$\sum_{i,j,k=-\infty}^{\infty} \frac{(-1)^{i+j+k}}{\sqrt{\left(i-\frac{1}{6}\right)^2 + \left(j-\frac{1}{6}\right)^2 + \left(k-\frac{1}{6}\right)^2}} = \sqrt{3}$$

を見つけた．これは（このような公式が原点の近くの任意の点では知られていないという意味で）可能なかぎり M_3 の正確な評価に近いといえる．三角関数を含むある変種が [30, 31] で研究された．ここで挙げることができた [20, 32] 以外にも，もっとたくさんの関連した和が存在する．

1.10.1 格子和とオイラー定数

任意の整数 $p \geq 2$ に対して

$$\varDelta(n,p) = \sum_{i_1,i_2,\cdots,i_p=-n}^{n}{}' \frac{1}{\sqrt{i_1^2+i_2^2+\cdots+i_p^2}} - \int_{x_1,x_2,\cdots,x_p=-n-1/2}^{n+1/2} \frac{dx_1 dx_2 \cdots dx_p}{\sqrt{x_1^2+x_2^2+\cdots+x_p^2}}$$

と定義する．この積分は原点に特異点をもつが収束する．2次元では $n \to \infty$ のとき

$$\varDelta(n,2) = \sum_{i,j=-n}^{n}{}' \frac{1}{\sqrt{i^2+j^2}} - 4\ln\left(\frac{\sqrt{2}+1}{\sqrt{2}-1}\right)\left(n+\frac{1}{2}\right)$$

$$\to 4\zeta\left(\frac{1}{2}\right)\beta\left(\frac{1}{2}\right) = (\sqrt{2}+1)M_2 = -3.9002649200\cdots = \delta_2$$

となる[33]．関数 $f(z)$ を

$$f(z) = \sum_{i,j=-\infty}^{\infty}{}' \frac{1}{(i^2+j^2)^z}, \quad \mathrm{Re}(z) > 1$$

と定義すると，f は $z=1$ で1位の極という特異点を1つだけもち $F(z)=4\zeta(z)\beta(z)$ という公式を用いて全複素数平面上の関数 F に解析接続できるということは興味深い．よって格子和は $f(1/2)=\infty$ であるが，$\delta_2 = F(1/2) = -3.90026\cdots$ を得る．すなわち積分は最後の答には「何の役割もはたさない」．

同様に，関数

$$g(z) = \sum_{i,j,k=-\infty}^{\infty}{}' \frac{1}{(i^2+j^2+k^2)^z}, \quad \mathrm{Re}(z) > \frac{3}{2}$$

から始めることにより，g は $z=3/2$ で1位の極をもつことを除いて，いたる所解析的な関数 G に解析接続しうる．しかし2次元の場合と違って $n \to \infty$ のとき

$$\varDelta(n,3) = \sum_{i,j,k=-n}^{n}{}' \frac{1}{\sqrt{i^2+j^2+k^2}} - 12\left(-\frac{\pi}{6}+\ln\left(\frac{\sqrt{3}+1}{\sqrt{3}-1}\right)\right)\left(n+\frac{1}{2}\right)^2$$

$$\to G\left(\frac{1}{2}\right) + \frac{\pi}{6} = -2.3136987039\cdots = \delta_3$$

を得る．すなわちここでの積分は役目があり，「補整項」$\pi/6$ が必要である．$G(1/2)$ を評価する早い表現は[20, 34]

$$G\left(\frac{1}{2}\right) = \frac{7\pi}{6} - \frac{19}{2}\ln(2) + 4\sum_{m,n=1}^{\infty}[3+3(-1)^m+(-1)^{m+n}]\frac{\mathrm{csch}(\pi\sqrt{m^2+n^2})}{\sqrt{m^2+n^2}}$$

$$= -2.8372974794\cdots$$

で，これは M_3 に対する Hautot 公式との類似性をもつものである．

さて任意の整数 $p \geq 1$ に対して

$$\gamma_p = \lim_{n\to\infty}\left(\sum_{i_1,i_2,\cdots,i_p=1}^{n}\frac{1}{\sqrt{i_1^2+i_2^2+\cdots+i_p^2}} - \int_{x_1,x_2,\cdots,x_p=1}^{n}\frac{dx_1 dx_2 \cdots dx_p}{\sqrt{x_1^2+x_2^2+\cdots+x_p^2}}\right)$$

と定義する．$\gamma_1 = \gamma$ がオイラー–マスケロニの定数【1.5】であることは，誰もが知っているが[35-37]，それに対して

$$\gamma_2 = \frac{1}{4}\left\{\delta_2 + 2\ln\left(\frac{\sqrt{2}+1}{\sqrt{2}-1}\right) - 4\gamma_1\right\} = -0.6709083078\cdots,$$

$$\gamma_3 = \frac{1}{8}\left\{\delta_3 + 3\left[-\frac{\pi}{6}+\ln\left(\frac{\sqrt{3}+1}{\sqrt{3}-1}\right)\right] + 12\gamma_2 - 6\gamma_1\right\} = 0.5817480456\cdots$$

であることは，あまり知られていない．任意の $p \geq 4$ に対する γ_p の値は誰も計算していない．

[1] E. Madelung, Das elektrische Feld in Systemen von regelmässig angeordneten Punktladungen, *Phys. Z.* 19 (1918) 524-533.
[2] D. Borwein, J. M. Borwein, and K. F. Taylor, Convergence of lattice sums and Madelung's constant, *J. Math. Phys.* 26 (1985) 2999-3009; MR 86m:82047.
[3] K. F. Taylor, On Madelung's constant, *J. Comput. Chem.* 8 (1987) 291-295; MR88h:82066.
[4] D. Borwein and J. M. Borwein, A note on alternating series in several dimensions, *Amer. Math. Monthly* 93 (1986) 531-539; MR 87j:40008.
[5] J. P. Buhler and R. E. Crandall, On the convergence problem for lattice sums, *J. Phys. A* 23 (1990) 2523-2528; MR 91h:82008.
[6] J. Buhler and S. Wagon, Secrets of the Madelung constant, *Mathematica Educ. Res.* 5 (1996) 49-55.
[7] D. Borwein, J. M. Borwein, and C. Pinner, Convergence of Madelung-like lattice sums, *Trans. Amer. Math. Soc.* 350 (1998) 3131-3167; CECM preprint 95:040; MR 98k:11135.
[8] G. H. Hardy and E. M. Wright, *An Introduction to the Theory of Numbers*, 5th ed., Oxford Univ. Press, 1985, pp. 256-257; MR 81i:10002.
[9] M. L. Glasser, The evaluation of lattice sums. I: Analytic procedures, *J. Math. Phys.* 14 (1973) 409-413; comments by A. Hautot, 15 (1974) 268; MR 47 #5328 and MR 55 #741b.
[10] M. L. Glasser, The evaluation of lattice sums. II: Number-theoretic approach, *J. Math. Phys.* 14 (1973) 701-703; erratum 15 (1974) 520; MR 48 #8409 and MR 49 #6849.
[11] I. J. Zucker, A note on lattice sums in two dimensions, *J. Math. Phys.* 15 (1974) 187; MR 55 #741a.
[12] I. J. Zucker and M. M. Robertson, Exact values for some two-dimensional lattice sums, *J. Phys. A* 8 (1975) 874-881; MR 54 #9515.
[13] I. J. Zucker and M. M. Robertson, Some properties of Dirichlet L-series, *J. Phys. A* 9 (1976) 1207-1214; MR 54 #253.
[14] I. J. Zucker and M. M. Robertson, A systematic approach to the evaluation of $\sum_{(m,n \neq 0,0)} (am^2 + bmn + cn^2)^{-s}$, *J. Phys. A* 9 (1976) 1215-1225; MR 54 #244.
[15] J. M. Borwein and P. B. Borwein, *Pi and the AGM: A Study in Analytic Number Theory and Computational Complexity*, Wiley, 1987, pp. 288-305; MR 99h:11147.
[16] O. Emersleben, Über die Konvergenz der Reihen Epsteinscher Zetafunktionen, *Math. Nachr.* 4 (1951) 468-480.
[17] G. C. Benson, A simple formula for evaluating the Madelung constant of an NaCl-type crystal, *Canad. J. Phys.* 34 (1956) 888-890.
[18] J. K. Mackenzie, A simple formula for evaluating the Madelung constant of an NaCl-type crystal, *Canad. J. Phys.* 35 (1957) 500-501.
[19] A. Hautot, A new method for the evaluation of slowly convergent series, *J. Math. Phys* 15 (1974) 1722-1727; MR 53 #9575.
[20] M. L. Glasser and I. J. Zucker, Lattice sums, *Theoretical Chemistry: Advances and Perspectives*, v. 5, ed. H. Eyring and D. Henderson, Academic Press, 1980, pp. 67-139.
[21] R. E. Crandall and J. P. Buhler, Elementary function expansions for Madelung constants, *J. Phys. A* 20 (1987) 5497-5510; MR 88m:82034.
[22] R. E. Crandall, New representations for the Madelung constant, *Experim. Math.* 8 (1999) 367-379; MR 2000m:11125.
[23] R. E. Crandall, Fast evaluation for a certain class of lattice sums, Perfectly Scientific Inc. preprint (2000).
[24] I. J. Zucker, Exact results for some lattice sums in 2, 4, 6 and 8 dimensions, *J. Phys. A* 7 (1974) 1568-1575; MR 53 #11258.
[25] M. L. Glasser, Evaluation of lattice sums. IV: A five-dimensional sum, *J. Math. Phys.* 16 (1975)

1237-1238; MR 55 #7228.
[26] H. Essén and A. Nordmark, Some results on the electrostatic energy of ionic crystals, *Canad. J. Chem.* 74 (1996) 885-891.
[27] J. Keane, Computations of M_n and N_n for $n \leq 64$, unpublished note (1997).
[28] P. Flajolet, A problem in asymptotics, unpublished note (2001).
[29] P. J. Forrester and M. L. Glasser, Some new lattice sums including an exact result for the electrostatic potential within the NaCl lattice, *J. Phys. A* 15 (1982) 911-914; MR 83d:82047.
[30] J. Boersma and P. J. De Doelder, On alternating multiple series, *SIAM Rev.* 35 (1993) 497-500.
[31] D. Borwein and J. M. Borwein, On some trigonometric and exponential lattice sums, *J. Math. Anal. Appl.* 188 (1994) 209-218; MR 95i:11089.
[32] Q. C. Johnson and D. H. Templeton, Madelung constants for several structures, *J. Chem. Phys.* 34 (1961) 2004-2007.
[33] D. Borwein, J. M. Borwein, and R. Shail, Analysis of certain lattice sums, *J. Math. Anal. Appl.* 143 (1989) 126-137; MR 90j:82038.
[34] I. J. Zucker, Functional equations for poly-dimensional zeta functions and the evaluation of Madelung constants, *J. Phys. A* 9 (1976) 499-505; MR 54 #6800.
[35] V. V. Kukhtin and O. V. Shramko, Lattice sums within the Euler-Maclaurin approach, *Phys. Lett. A* 156 (1991) 257-259; MR 92d:82101.
[36] V. V. Kukhtin and O. V. Shramko, A new evaluation of the energy of Wigner BCC and FCC crystals, *J. Phys. A* 26 (1993) L963-L965.
[37] V. V. Kukhtin and O. V. Shramko, Analogues of the Euler constant for two- and three-dimensional spaces (in Ukrainian), *Dopovidi Akad. Nauk Ukrain.* (1993) n. 8, 42-43; MR 94m:11152.

1.11 チャイティンの定数

ここでアルゴリズム的情報理論[1-4]を少し述べる．ここでの見方は数論的であり，取り扱いは形式ばらないものである．たとえば，ここでコンピュータ（チューリングマシン）の定義を与えようとは思わない．

ディオファンタス方程式は整数係数の多項式 $p(x_1, x_2, \cdots, x_n)$ にかかわっている．ヒルベルトの第10問題は，任意の p が与えられたとき $p(x_1, x_2, \cdots, x_n) = 0$ が正の整数解 $x_1, x_2, \cdots, x_n$ をもつかどうかを確かめることを可能にする一般的なアルゴリズムについて問うている．Matiyasevic（マティヤセビッツ），Davis（デービス），Putnam（プトナム）および Robinson（ロビンソン）の研究[5]は，このようなアルゴリズムは存在しないということをついに証明した．実際，パラメータ N を変化させることにより解 x の対応する集合 D_N を正の整数の帰納的可算（再帰的に枚挙可能な）集合にできるような**普遍ディオファンタス方程式** $P(N, x_1, x_2, \cdots, x_n) = 0$ を見出すことができる．これと同値のことだが，決定系のコンピュータプログラムの出力となりうる正の整数の任意の集合は，ある N に対する D_N に違いない．P の存在は数理論理学でのゲーデルの不完全性定理や計算可能性理論における停止問題のチューリングによる否定的な解と関連

がある．
　さて，実数 A を

$$A_N = \begin{cases} 1 & (D_N \neq \emptyset \text{ のとき}) \\ 0 & (D_N = \emptyset \text{ のとき}) \end{cases}$$

となるような二項展開 $0.A_1A_2A_3\cdots$ により定義する．任意に N を与えたとき，$A_N=1$ か $A_N=0$ かを決定するアルゴリズムは存在しない．よって A は計算不可能な実数である．A についてさらに何か述べることができるか．

　計算可能性とランダム性の間には興味深い関連がある．実数 Z のはじめの N ビットが N ビットよりも短いプログラムに圧縮しえないとき**ランダム**であるということにする．Z の連続するビットが，ごまかしのない公正な硬貨の独立した（硬貨）投げの結果と区別しえないことがわかる．数論においてランダム性が生じるという考えは，理解をふらつかせる．計算可能な実数 Z でランダム性をもつものはない[6,7]．よって A もランダム性をもたないことになる！　算術における予測不可能性を見出すことは，少しばかり，より難しいと思わなくてはならない．

　指数的ディオファンタス方程式は，前と同じように整数係数の多項式 $q(x_1, x_2, \cdots, x_n)$ にかかわっている．そしてある正の整数 c と $x_j = c^{x_i}$ となる $1 \leq i < j \leq n$ があってもよいこと，また $x_k = x_i^{x_j}$ となる $1 \leq i \leq j < k \leq n$ があってもよいという自由度が付け加わっている．すなわち，指数は変数であることも許容される．Jones（ジョーンズ）とマティヤセビッツの研究で始まり，チャイティン[6,7]が，次の驚くべき性質をもった指数的ディオファンタス方程式 $Q(N, x_1, x_2, \cdots, x_n) = 0$ を発見した．E_N は，各 N に対して $Q=0$ の正の整数の解の集合を表す．実数 Ω を次のように決めた Ω_N を用いた $0.\Omega_1\Omega_2\Omega_3\cdots$ で定義する．ここで

$$\Omega_N = \begin{cases} 1 & (E_N \text{ が無限集合のとき}) \\ 0 & (E_N \text{ が有限集合のとき}) \end{cases}$$

とする．

　すると Ω は単に計算不可能なだけではなく，ランダムでもある．よって方程式 $P=0$ は計算不可能な A を与えたが，方程式 $Q=0$ はランダムな Ω を与える．これは，数学における正真正銘の不確かさのかけらを示す．チャイティンは $Q=0$ を明示的に書き下している．これは 17000 変数をもち印字するのに 200 ページを要するものである．対応する定数 Ω は**チャイティンの定数**とよばれる．表現 Q の別の選択も可能であるので，別のランダムな Ω が存在する．チャイティンの Q の選び方に対する基本は，ゲーデルの番号づけに似ている（チャイティンの修正した LISP（リスプ）による実装がこのことを非常に具体的にした）．しかし，とても手のこんだものなので詳細をここで説明することはできない．

　チャイティンの定数は，ある自己限定的万能コンピュータの停止確率である．前のように，異なるマシンは，多くの場合異なる定数を与える．よってチューリングの基本的な結果は停止**問題**が非可解であるというのに対して，チャイティンの結果は，停止**確率**がランダムであるということである．驚くべき公式[2-4]

1.11 チャイティンの定数

$$\Omega = \sum_\pi 2^{-|\pi|}$$

がある．ここでの無限和は，自己限定的プログラム π すべてにわたる．これがチャイティンの万能コンピュータが時として停止する原因である．ここで $|\pi|$ は π の（プログラムをビットの連鎖と考えたときの）長さを表している．

チャイティンのもともとの Ω の初めの数ビットはわかっている．そしてこれまでのところ，すべては 1 であることがわかる．この観察から興味深い哲学的な進展が生じる．ZFC（選択公理つきの Zermelo-Frenkel（ツェルメロ-フレンケル）集合論）が算術的に正しいと仮定する．すなわち ZFC によって証明された任意の算術の定理は正しいものと仮定する．この条件のもとで，ZFC が決定できるような Ω のビットの数に関する明確な有限な限界が存在する．Solovay（ソロヴェイ）[11,12] は，劇的に最悪なマシン U を構成した．そのマシン U に対しては，ZFC が $\Omega(U)$ のどのビットもまったく計算することができない．さらに ZFC は，任意のチャイティン定数 Ω に対して最初の 0 のブロックより先を予測できない．k 番目のビットは実は 0 かもしれないが，この事実を ZFC では証明不可能である．Calude（カルデ）[13] が書いているように「1 つの 0 を得るや，すべては終了する．」Solovey の Ω は，0 から出発する．よって不可知なものになる．ごく最近 $\Omega(V) = \Omega(U)$ を満たすが U より取り扱いやすい非ソロヴェイマシン V を構成することにより，このような Ω のはじめの 64 ビットを計算するための手順が実行された[14]．

計算可能な可算ランダム実数の集合は，チャイティン万能コンピュータ[15-17] の Ω の停止確率すべての集合と一致する．それを記述することが確実性を弱めてしまうほどに複雑ではないような「より単純な」ランダム Ω を定義することができるだろうか．この定理は，すべてのこのような数がディオファンタス表現 $Q = 0$ をもつといっている．方程式の大きさをかなり減らすことができるかどうかは，未解決の問題である．

[1] G. J. Chaitin, *Algorithmic Information Theory*, Cambridge Univ. Press, 1987; MR 89g:68022.

[2] C. S. Calude, *Information and Randomness*, Springer-Verlag, 1994; MR 96d:68103.

[3] G. J. Chaitin, *The Limits of Mathematics*, Springer-Verlag, 1997; MR 98m:68056.

[4] G. J. Chaitin, *The Unknowable*, Springer-Verlag, 1999; MR 2000h:68071.

[5] M. Davis, *Computability and Unsolvability*, Dover, 1973; MR 23 # A1525 (Appendix 2, "Hilbert's tenth problem is unsolvable," appeared originally in *Amer. Math. Monthly* 80 (1973) 233-269; MR 47 #6465).

[6] G. J. Chaitin, Randomness and Gödel's theorem, *Mondes en Développement*, Proc. 1985 Brussels Symp. on Laws of Nature and Human Conduct, n. 54-55 (1986) 125-128; also *Information, Randomness and Incompleteness: Papers on Algorithmic Information Theory*, World Scientific, 1987, pp. 66-69; MR 89f:01089.

[7] G. J. Chaitin, Incompleteness theorems for random reals, *Adv. Appl. Math.* 8 (1987) 119-146; MR 88h:68038.

[8] G. J. Chaitin, Randomness and mathematical proof, *Sci. Amer.*, v. 232 (1975) n. 5, 47-52.

[9] C. H. Bennett, On random and hard-to-describe numbers: Chaitin's Ω, quoted by M. Gardner, *Fractal Music, Hypercards and More...*, W. H. Freeman, 1992; MR 92m:00005 (appeared originally in *Sci. Amer.*, v. 241 (1979) n. 5, 20-34).

[10] G. J. Chaitin, Randomness in arithmetic, *Sci. Amer.*, v. 259 (1988) n. 1, 80-85.

[11] R. M. Solovay, A version of Ω for which ZFC cannot predict a single bit, *Finite Versus Infinite: Contributions to an Eternal Dilemma*, ed. C. S. Calude and G. Paun, Springer-Verlag, 2000, pp. 323-334; CDMTCS report 104.
[12] C. S. Calude and G. J. Chaitin, Randomness everywhere, *Nature* 400 (1999) 319-320.
[13] C. S. Calude, Chaitin Ω numbers, Solovay machines and Gödel incompleteness, *Theoret. Comput. Sci.* 284 (2002) 269-277; CDMTCS report 114.
[14] C. S. Calude, M. J. Dinneen, and C.-K. Shu, Computing a glimpse of randomness, *Experim. Math.*, to appear; CDMTCS report 167; nlin.CD/0112022.
[15] C. S. Calude, P. H. Hertling, B. Khoussainov, and Y. Wang, Recursively enumerable reals and Chaitin Ω numbers, *Theoret. Comput. Sci.* 255 (2002) 125-149; also in *Proc. 1998 Symp. on Theoretical Aspects of Computer Science (STACS)*, Paris, ed. M. Morvan, C. Meinel, and D. Krob, Lect. Notes in Comp. Sci. 1373, Springer-Verlag, 1998, pp. 596-606; CDMTCS report 59; MR 99h:68089 and MR 2002f:68065.
[16] A. Kucera and T. A. Slaman, Randomness and recursive enumerability, *SIAM J. Comput.* 31 (2001) 199-211.
[17] C. S. Calude, A characterization of c.e. random reals, *Theoret. Comput. Sci.* 271 (2002) 3-14; CDMTCS report 95.

2 数論に関する定数

2.1 ハーディ-リトルウッド定数

素数列 2, 3, 5, 7, 11, 13, 17, …は何世紀にもわたり，数学者を魅了している．たとえば，素数の個数の関数

$$P_n = \sum_{p \leq n} 1 = \text{素数の個数} \leq n$$

を考える．ただし，和はすべての素数にわたる．$P_n(p) = P_n$ と書く．見慣れない記号に隠された動機は，すぐに明らかになる．**素数定理** (prime number theorem) として知られている

$$P_n(p) \sim \frac{n}{\ln(n)} \quad (n \to \infty)$$

は（多くの研究に基づき）Hadamard（アダマール）と de la Vallée Poussin（ド・ラ・ヴァレー・プサン）により証明される1896年まで証明されなかった．しかしながら，素数理論で証明された多くの問題に対して，未解決問題もいくつかある．最も有名な問題は，次の2つである．

Goldbach（**ゴールドバッハ**）**予想**：2より大きいすべての偶数は，2つの素数の和で表すことができる．

双子素数予想：p と $p+2$ がともに素数であるような素数 p が無限に存在する．

これは次のように書きかえることができる．

$P_n(p, p+2)$ を，n 以下の双子素数の組の個数とする．このとき，$\lim_{n \to \infty} P_n(p, p+2) = \infty$ が成り立つ．

際立った理論的な進歩は，これらの予想を解くことでもたらされている．しかし，困

難が克服できない空隙が依然として残されている．Hardy & Littlewood（ハーディ＆リトルウッド）[1]によって発見された，発見的な式に焦点をあてる．その式は，次の問題に答えようとしている．存在の問題は脇において，さまざまな追加的な制限をつけた素数の分布はどうなるのだろうか．本当は，素数定理の漸近的分布公式と類似した公式を望みたい．

拡張された双子素数予想[2-6]は，

$$P_n(p,p+2) \sim 2C_{\text{twin}}\frac{n}{\ln(n)^2}$$

である．ただし，$C_{\text{twin}} = \prod_{p>2}\frac{p(p-2)}{(p-1)^2} = 0.6601618158\cdots = \frac{1}{2}\times(1.3203236316\cdots)$ である．

3つの素数の組に関する2つの予想[2]は，

$$P_n(p,p+2,p+6) \sim P_n(p,p+4,p+6) \sim D\frac{n}{\ln(n)^3}$$

である．ただし，$D = \frac{9}{2}\prod_{p>3}\frac{p^2(p-3)}{(p-1)^3} = 2.8582485957\cdots$ である．

4つの素数の組に関する2つの予想[2]は，

$$P_n(p,p+2,p+6,p+8) \sim \frac{1}{2}P_n(p,p+4,p+6,p+10) \sim E\frac{n}{\ln(n)^4}$$

である．ただし，$E = \frac{27}{2}\prod_{p>3}\frac{p^3(p-4)}{(p-1)^4} = 4.1511808632\cdots$ である．

m^2+1 で表される素数を含む予想[3,4,7-9]は，Q_n を，ある整数 m について $p=m^2+1$ を満たす n 以下の素数 $p(\leq n)$ の個数とする．このとき，

$$Q_n \sim 2C_{\text{quad}}\frac{\sqrt{n}}{\ln(n)}$$

である．ただし，$C_{\text{quad}} = \frac{1}{2}\prod_{p>2}\left(1-\frac{(-1)^{(p-1)/2}}{p-1}\right) = 0.6864067314\cdots = \frac{1}{2}\times 1.3728134628\cdots$ である．

拡張されたゴールドバッハ予想[3,4,10,11]：（順序も考慮した）2つの素数の和として，ある偶数 n の表現の個数を R_n と定義する．このとき，

$$R_n \sim 2C_{\text{twin}} \cdot \prod_{\substack{p>2 \\ p|n}}\frac{p-1}{p-2} \cdot \frac{n}{\ln(n)^2}$$

である．ただし，n を割るすべての素数 p にわたって積をとる．

拡張された双子素数予想と拡張されたゴールドバッハ予想が同じ定数 C_{twin} を含むことに興味がわく．ゴールドバッハ予想は，双子素数予想[12]と「共役（conjugate）」である，とよくいわれる．Q_n【2.1.1】の評価と R_n【2.1.2】の評価における最近の成果について述べる．Shar & Wilson（シャー＆ウィルソン）[13]は，R_n に対する漸近公式を大規模に調べた．C_{twin} をシャー-ウィルソン定数[14]とよぶこともある．C_{twin} を計算する式は【2.4】で得られる．

ここで議論されるハーディ-リトルウッド型の定数はすべて，素数の上をわたる無限

積を含む．本書でこのような積が起こる他のものは，Landau-Ramanujan（ランダウ-ラマヌジャン）定数【2.3】，Artin（アルティン）定数【2.4】，Hafner-Sarnak-McCurley（ハフナー-サルナック-マッカレー）定数【2.5】，Bateman-Grosswald（ベイトマン-グロスワルド）定数【2.6.1】，Euler（オイラー）のφ関数とPell-Stevenhagen（ペル-スティーヴンハーゲン）定数【2.8】である．

Riesel（リーゼル）[2]は，3つの素数や4つの素数の組の一般化である素数の集まりを議論し，それに対応するハーディ-リトルウッド定数の計算の方法を示した．素数の列の全体はわからないけれども，Riemannのゼータ関数$\zeta(x)$【1.6】のある変換を用いて，ハーディ-リトルウッド定数を任意の精度で計算できるというすばらしい結果を，彼は強調した．

上記の2次多項式で得られた素数値に対する予想の，3次多項式に対する類似の予想【2.1.3】がある．偶然にも，$2C_{\text{quad}}$をほんの少し動かすと，閉じた表現式

$$\prod_{p>2}\left(1-\frac{(-1)^{(p-1)/2}}{p}\right)=\frac{4}{\pi}=\frac{1}{\beta(1)}$$

を得る．ただし，$\beta(x)$はDirichlet（ディリクレ）のベータ関数【1.7】である．

よく知られたMertens（メルテンス）の式【2.2】は

$$\lim_{n\to\infty}\frac{1}{\ln(n)}\prod_{2<p<n}\frac{p}{p-1}=\frac{1}{2}e^\gamma=0.8905362089\cdots$$

である．ただし，γはEuler-Mascheroni（オイラー-マスケローニ）の定数【1.5】である．あまり有名でない結果[15-17]は

$$\lim_{n\to\infty}\frac{1}{\ln(n)^2}\prod_{2<p<n}\frac{p}{p-2}=\frac{1}{4C_{\text{twin}}}e^{2\gamma}=1.2013035599\cdots=\frac{1}{0.8324290656}\cdots$$

である．Hardy & Litttlewoodにより導入された$C_{\text{twin}}=C_2$の拡張は，

$$C_n=\prod_{p>n}\left(\frac{p}{p-1}\right)^{n-1}\frac{p-n}{p-1}=\prod_{p>n}\left(1-\frac{1}{p}\right)^{-n}\left(1-\frac{n}{p}\right)$$

であり，$C_3=0.635166346\cdots=2D/9$，$C_4=0.3074948787\cdots=2E/27$，$C_5=0.4098748850\cdots$，$C_6=0.1866142973\cdots$，$C_7=0.3694375103\cdots$である．

Waring（ウェアリング）問題の研究で，Bateman & Stemmler（ベイトマン&ステムラー）[21-24]は，予想

$$P_n(p,p^2+p+1)\sim H\frac{n}{\ln(n)^2}$$

を調べた．ただし，

$$H=\frac{1}{2}\prod_p\left(1-\frac{1}{p}\right)^{-2}\left(1-\frac{2+\chi(p)}{p}\right)=1.5217315350\cdots=2\cdot 0.7608657675\cdots$$

であり，$\chi(p)$は$p\equiv -1,0,1 \bmod 3$に対応して，それぞれ$\chi(p)=-1,0,1$である．なお[25-28]をも参照せよ．

ゴールドバッハ予想に漠然と関係する2つの問題を取り上げる．1つかそれ以上の「連続した」素数の和としてnを表現する個数を，関数$f(n)$と定義する．たとえば，$f(41)=3$である．$41=11+13+17=2+3+5+7+11+13$であるからである．Moser

(モザー)[29]は,
$$\lim_{N\to\infty}\frac{1}{N}\sum_{n=1}^{N}f(n)=\ln(2)=0.6931471805\cdots$$
を証明した．1つの素数と2のベキの和として表される，nを超えない整数の個数を$g(n)$とする．Romani(ロマニ)[30]は数値的に比 $g(n)/n$ を調べ，そのような整数の漸近的密度は，0.434…であることを得た．

2.1.1 2次多項式で表される素数

前に Q_n を定義した．$\tilde{Q}_n$ を，素因子が2個以下である n 以下の正の整数 k で，ある整数 m を用いて，$k=m^2+1$ となる整数 k の個数とする．Q_n の極限のふるまいに関するハーディ-リトルウッド予想は，証明されていない．この予想の正しさを支持する数値実験が，ずいぶん前[31,32]に発表されている．しかしながら，Iwaniec(イワニエツ)は，最近，漸近的不等式[4,33]

$$\tilde{Q}_n > \frac{1}{77}\cdot 2C_{\text{quad}}\cdot\frac{\sqrt{n}}{\ln(n)}=0.0178\cdots\frac{n}{\ln(n)}$$

を示した．これは，無限に多くの所要の形の**概素数**(almost primes)が存在することを示している．彼の結果は，任意の2次の既約多項式 am^2+bm+c ($a>0$, c は奇数)に拡張されている．Q_n の上からのよい評価式は，知られているとは思えない．

Shanks(シャンクス)[32]は式

$$C_{\text{quad}}=\frac{3}{4G}\frac{\zeta(6)}{\zeta(3)}\prod_{p\equiv 1\bmod 4}\left(1+\frac{2}{p^3-1}\right)\left(1-\frac{2}{p(p-1)^2}\right)$$

を述べた．ただし，$G=\beta(2)$ は Catalan (カタラン) 定数【1.7】である．彼は，等式

$$1=\frac{17}{16}\frac{\zeta(8)}{\zeta(4)\beta(4)}\prod_{p\equiv 1\bmod 4}\left(1+\frac{2}{p^4-1}\right)$$

を掛けることにより，より速く収束することを付け加えている．

2.1.2 ゴールドバッハ予想

ゴールドバッハ予想の証明で，すなわちある人の推測を定理にする過程で，いくつかの進展が最近あった．素数2つの組と3つの組の結果がある．

予想 G. すべての偶数 >2 は，2つの素数の和として表される．

予想 G′. すべての奇数 >5 は，3つの素数の和として表される．

もし，G が真であるならば G′ も真である，ことに注意する．これらに対応する漸近的な結果がある．

予想 AG. すべての偶数 $>N$ が2つの素数の和として表される，充分大きな N が存在する．

予想 AG′. すべての奇数 $>N'$ が3つの素数の和として表される，充分大きな N' が存在する．

Hardy & Littlewood[1]の円の方法(circle method)を用いて，Vinogradov(ヴィ

ノグラドフ）[34]は，AG′ が真であることを証明した．さらに，

$$S_n \sim \prod_p \left(1 + \frac{1}{(p-1)^3}\right) \cdot \prod_{\substack{p>2 \\ p|n}} \left(1 - \frac{1}{p^2 - 3p + 3}\right) \cdot \frac{n^2}{2\ln(n)^3}$$

を証明した．ただし，S_n は，3つの素数の和として表される充分大きな整数 n の表現の個数である．これは，予想ではなく，定理であることに注意する．さらに，Borodzkin（ボロズキン）[35]は，**ヴィノグラドフ数** $N′$ として $3^{3^{15}} \sim 10^{7000000}$ と取れることを示し，Chen & Wang（チャン＆ワン）[36,37]は 10^{7194} に改良した．現在のテクノロジーを用いても，この敷居値まですべての奇数を検証することは不可能である．よって，$G′$ についても検証できない．しかし，一般化 Riemann（リーマン）予想が正しいという仮定のもとで，$N′$ が 10^{20} にできることが，Zinoviev（ジノビエフ）[38]と Saouter（ソーター）[39]により証明され，Deshouillers（デショウィラー）ら[40]は $N′=5$ に減らすことに成功した．したがって，一般化リーマン予想が正しいという条件のもとで，命題 G′ は真である．

AG や AG′ に対する，類似の条件付きの証明はない．これらを弱めた2つの定理がある．

定理（Ramaré（ラマレ）[41,42]）すべての偶数は，6つかそれ以下の素数の和として表される．（いいかえれば Schnirelmann（**シュニレルマン**）数は ≤ 6 である．）

定理（Chen（チャン；陳景潤）[11,12,43,44]）充分大きいすべての偶数は，1つの素数と素因子の個数が2以下の1つの整数との和で表すことができる．

実際，Chen は漸近的不等式

$$\tilde{R}_n > 0.67 \cdot \prod_{p>2} \left(1 - \frac{1}{(p-1)^2}\right) \cdot \prod_{\substack{p>2 \\ p|n}} \frac{p-1}{p-2} \cdot \frac{n}{\ln(n)^2}$$

を証明した．ただし，$\tilde{R}_n$ は，対応する表現の個数である．Chen はまた，$p+2$ が概素数であるような素数 p が無限に多く存在することを証明した．これは双子素数の予想を弱めたものである．同じ係数 0.67 が現れる．

これらの結果に関して，付け加える詳しい結果がある．Kaniecki（カニエツキ）[45]は，リーマン予想が正しいと仮定して，すべての奇数は多くとも5つの素数の和で表現できることを証明した．大量な計算により，最終的に，多くとも4つの素数の和で表現できる，と改良できるであろう．対比すると，すべての偶数は多くとも6つの素数の和で表現できる，という Ramaré の結果には条件がついていない（リーマン予想に依存していない）．

Vinogradov の結果は次のように書きかえられる．

$$\liminf_{n \to \infty} \frac{\ln(n)^3}{n^2} S_n = \frac{1}{2} \prod_p \left(1 + \frac{1}{(p-1)^3}\right) \cdot \prod_{p>2} \left(1 - \frac{1}{p^2 - 3p + 3}\right) = C_{\text{twin}}$$
$$= 0.6601618158\cdots$$
$$\limsup_{n \to \infty} \frac{\ln(n)^3}{n^2} S_n = \frac{1}{2} \prod_p \left(1 + \frac{1}{(p-1)^3}\right) = 1.1504807723\cdots$$

すなわち, S_n は† 漸近的には変動するけれども, その増大度は同じオーダーである. Chen の結果:

$$\liminf_{n\to\infty} \frac{\ln(n)^2}{n}\tilde{R}_n > 0.67 \cdot C_{\text{twin}} = 0.44,$$

$$\limsup_{n\to\infty} \frac{\ln(n)}{n}\tilde{R}_n > 0.67 \cdot \frac{1}{2}e^\gamma = 0.59$$

に対しては, 同じことはいえない. 上極限は, 対数項 $\ln(n)$ 1個分だけ下極限よりも速く大きくなる. これらは Mertens (メルテンス) の公式を用いて得られる.

ゴールドバッハ予想[43]に対する Chen の係数 0.67 は, [46]で 0.81 に, [11]で 2 に置きかわった. 双子素数における Chen の不等式は, 同様に改良することができる. この場合を含んだ精密化は, [47]で 1.42, [48]で 1.94, [49]で 2.03, [50]で 2.1 となる. [52-54]を基礎として Chen[51]は, 上界

$$R_n \le 7.8342 \cdot \prod_{p>2}\left(1 - \frac{1}{(p-1)^2}\right) \cdot \prod_{\substack{p>2 \\ p|n}} \frac{p-1}{p-2} \cdot \frac{n}{\ln(n)^2}$$

を証明した.

Pan (パン)[55]は, 係数が 7.9880 である弱い結果であるが, 簡単な証明を与えた. 双子素数問題を含む, 対応する係数 7.8342 に関する改良は, [56]で 7.8156, [57]で 7.5555, [58]で 7.5294, [59]で 7, [47]で 6.9075, [50]で 6.8354, [60]で 6.8325 となる. ([3]で主張され[50]でレビューされている上界 6.26 は, 誤りである.)

双子素数問題の精密化の多くは, [59]に基づいている. それは複雑な理由でゴールドバッハ予想には適用「できない」.

ゴールドバッハ予想に対する可能な反例の集合は, 小さいに違いない[61-66], という見解がある. 正の偶数 $\le n$ で 2 つの素数の和「でない」数の個数 $\varepsilon(n)$ は, $\varepsilon(n) = o(n^{0.914})$ $(n \to \infty)$ を満たすことが証明できる. もちろん, すべての $n \ge 2$ について $\varepsilon(n) = 1$ を期待する. [67-69]を参照せよ.

2.1.3 3次式で表される素数

Hardy & Littlewood[1]は, k を固定された立方数でない数とするとき, $m^3 + k$ の形の素数が無限に多く存在する, ことを予想した. さらに, T_m を, ある整数 m で $p = m^3 + 2$ と表される素数 $p \le n$ の個数と定義する. このとき,

$$\lim_{n\to\infty} \frac{\ln(n)}{\sqrt[3]{n}} T_n = A = \prod_{p \equiv 1 \bmod 6} \frac{p - a(p)}{p - 1} = 1.2985395575\cdots$$

である. ただし,

$$a(p) = \begin{cases} 3, & 2 \text{ が } \bmod p \text{ で立方剰余であるとき } (x^3 \equiv 2 \bmod p \text{ に解があるとき}) \\ 0, & \text{その他の場合} \end{cases}$$

同様に, U_m を, ある整数 m により $p = m^3 + 3$ と表される素数 $p \le n$ の個数と定義する. このとき,

---

†訳注:原文では $S(n)$ だが, S_n とした.

$$\lim_{n\to\infty}\frac{\ln(n)}{\sqrt[3]{n}}U_n=B=\prod_{p\equiv 1\bmod 6}\frac{p-\beta(p)}{p-1}=1.3905439387\cdots$$

である. ただし,

$$\beta(p)=\begin{cases}3, & 3\text{ が mod }p\text{ で立方剰余であるとき }(x^3\equiv 3\bmod p\text{ に解があるとき})\\ 0, & \text{その他の場合}\end{cases}$$

である. 定数 A, B は, Bateman (**ベイトマン**) **定数**として知られ, Shanks & Lal (シャンクス&ラル)[3, 22, 70, 71]によりはじめて, 高い精度で計算された.

4乗数を含む例がある[72]. V_n を, ある整数 m により $p=m^4+1$ と表される素数 $p\leq n$ の個数と定義する. このとき,

$$\lim_{n\to\infty}\frac{\ln(n)}{\sqrt[4]{n}}V_n=4I=2.6789638796\cdots$$

である. ただし,

$$I=\frac{\pi^2}{16\ln(1+\sqrt{2})}\prod_{p\equiv 1\bmod 8}\left(1-\frac{4}{p}\right)\left(\frac{p+1}{p-1}\right)^2=0.6697409699\cdots$$

である. この定数は, Shanks (**シャンクス**) **定数**とよぶことが適当であろう. m^5+2 または m^5+3 の形の素数に対する同じような評価は, 文献には明らかに現れていない.

Batemann-Horn (ベイトマン-ホーン) 予想[3,21,73]は, この定理を任意の次数の多項式に拡張したものである. また, いくつかのそのような多項式が同時に素数である状況の場合に, 適用できる. たとえば[74-77], F_n を, $(m-1)^2+1<n$ を満たす2つの素数 $(m-1)^2+1$ と $(m+1)^2+1$ の組の個数とする. このとき,

$$\lim_{n\to\infty}\frac{\ln(n)^2}{\sqrt{n}}F_n=4J=1.9504911124\cdots$$

である. ただし,

$$J=\frac{\pi^2}{8}=\prod_{p\equiv 1\bmod 4}\left(1-\frac{4}{p}\right)\left(\frac{p+1}{p-1}\right)^2=0.4876227781\cdots$$

である. F_n はまた, 複素数平面において直線 $x+i$ の上にある, ガウス双子素数 $(m-1+i, m+1+i)$ の個数であることに注意せよ. したがって, J は**ガウス双子素数定数**とよぶ. (これらは平面上にあるガウス双子素数のすべてではない. 直線 $x+2i$ 上で, $m=179984$ を考えよ.)

他の例として, $(m-1)^4+1\leq n$ を満たす, 2つの素数 $(m-1)^4+1$ と $(m+1)^4+1$ の組の個数を G_n とする. このとき,

$$\lim_{n\to\infty}\frac{\ln(n)^2}{\sqrt[4]{n}}G_n=16K=12.6753318106\cdots$$

である. ただし,

$$K=2I^2\prod_{p\equiv 1\bmod 8}\frac{p(p-8)}{(p-4)^2}=0.7922082381\cdots$$

である. これは Lal (**ラル**) **定数**として知られている. Sebah (セバ) [77]はこの値と多くの定数の値を計算した.

[1] G. H. Hardy and J. E. Littlewood, Some problems of 'Partitio Numerorum.' III: On the expression of a number as a sum of primes, Acta Math. 44 (1923) 1-70; also in *Collected Papers of G. H. Hardy*, v. 1, Oxford Univ. Press, 1966, pp. 561-630.
[2] H. Riesel, *Prime Numbers and Computer Methods for Factorization*, Birkhäuser, 1985; MR 95h: 11142.
[3] P. Ribenboim, *The New Book of Prime Number Records*, Springer-Verlag, 1996, pp. 259-265, 291-299, 403-411; MR 96k:11112.
[4] R. K. Guy, *Unsolved Problems in Number Theory*, 2nd ed., Springer-Verlag, 1994, sect. A1, A8, C1; MR 96e:11002.
[5] G. H. Hardy and E. M. Wright, *An Introduction to the Theory of Numbers*, 5th ed., Oxford Univ. Press, 1985, pp. 371-373; MR 81i:10002.
[6] S. W. Golomb, The twin prime constant, *Amer. Math. Monthly* 67 (1960) 767-769.
[7] D. Shanks, On the conjecture of Hardy & Littlewood concerning the number of primes of the form $n^2 + a$, *Math. Comp.* 14 (1960) 321-332; MR 22 #10960.
[8] D. Shanks, Supplementary data and remarks concerning a Hardy-Littlewood conjecture, *Math. Comp.* 17 (1963) 188-193; MR 28 #3013.
[9] D. Shanks, Polylogarithms, Dirichlet series, and certain constants, *Math. Comp.* 18 (1964) 322-324; MR 30 #5460.
[10] W. Ellison and F. Ellison, *Prime Numbers*, Wiley, 1985, p. 333; MR 87a:11082.
[11] M. B. Nathanson, *Additive Number Theory: The Classical Bases*, Springer-Verlag, 1996, pp. 151-298; MR 97e:11004.
[12] H. Halberstam and H.-E. Richert, *Sieve Methods*, Academic Press, 1974, pp. 116-117; MR 54 # 12689.
[13] N. M. Shah and B. M. Wilson, On an empirical formula connected with Goldbach's theorem, *Proc. Cambridge Philos. Soc.* 19 (1919) 238-244.
[14] F. Le Lionnais, *Les Nombres Remarquables*, Hermann, 1983.
[15] B. Rosser, The nth prime is greater than $n \log(n)$, *Proc. London Math. Soc.* 45 (1938) 21-44.
[16] J. W. Wrench, Evaluation of Artin's constant and the twin prime constant, *Math. Comp.* 15 (1961) 396-398; MR 23 #A1619.
[17] A. Fletcher, J. C. P. Miller, L. Rosenhead, and L. J. Comrie, *An Index of Mathematical Tables*, 2nd ed., v. 1, Addison-Wesley, 1962; MR 26 #365a-b.
[18] H. P. Robinson and E. Potter, *Mathematical Constants*, UCRL-20418 (1971), Univ. of Calif. at Berkeley; available through the National Technical Information Service, Springfield VA 22151.
[19] R. Harley, Some estimates due to Richard Brent applied to the "high jumpers" problem, unpublished note (1994).
[20] G. Niklasch and P. Moree, Some number-theoretical constants: Products of rational functions over primes, unpublished note (2000).
[21] P. T. Bateman and R. A. Horn, A heuristic asymptotic formula concerning the distribution of prime numbers, *Math. Comp.* 16 (1962) 363-367; MR 26 #6139.
[22] P. T. Bateman and R. A. Horn, Primes represented by irreducible polynomials in one variable, *Theory of Numbers*, ed. A. L. Whiteman, Proc. Symp. Pure Math. 8, Amer. Math. Soc., 1965, pp. 119-132; MR 31 #1234.
[23] D. Shanks and J. W. Wrench, The calculation of certain Dirichlet series, *Math. Comp.* 17 (1963) 136-154; corrigenda 17 (1963) 488 and 22 (1968) 699; MR 28 #3012 and MR 37 #2414.
[24] P. T. Bateman and R. M. Stemmler, Waring's problem for algebraic number fields and primes of the form $(p^r-1)/(p^d-1)$, *Illinois J. Math.* 6 (1962) 142-156; MR 25 #2059.
[25] E. Grosswald, Arithmetic progressions that consist only of primes, *J. Number Theory* 14 (1982) 9-31; MR 83k:10081.
[26] N. Kurokawa, Special values of Euler products and Hardy-Littlewood constants, *Proc. Japan Acad. Ser. A. Math. Sci.* 62 (1986) 25-28; MR 87j:11127.

[27] E. Bogomolny and P. Leboeuf, Statistical properties of the zeros of zeta functions - beyond the Riemann case, *Nonlinearity* 7 (1994) 1155-1167; MR 95k:11108.
[28] R. Gross and J. H. Smith, A generalization of a conjecture of Hardy and Littlewood to algebraic number fields, *Rocky Mount. J. Math.* 30 (2000) 195-215; MR 2001g:11175.
[29] L. Moser, Notes on number theory. III: On the sum of consecutive primes, *Canad. Math. Bull.* 6 (1963) 159-161; MR 28 #75.
[30] F. Romani, Computations concerning primes and powers of two, *Calcolo* 20 (1983) 319-336: MR 86c:11082.
[31] A. E. Western, Note on the number of primes of the form n^2+1, *Cambridge Philos. Soc.* 21 (1922) 108-109.
[32] D. Shanks, A sieve method for factoring numbers of the form n^2+1, *Math. Comp.* 13 (1959) 78-86.
[33] H. Iwaniec, Almost-primes represented by quadratic polynomials, *Invent. Math.* 47 (1978) 171-188; MR 58 #5553.
[34] I. M. Vinogradov, Representation of an odd number as a sum of three primes (in Russian), *Dokl. Akad. Nauk SSSR* 15 (1937) 169-172; Engl. trans. in *Goldbach Conjecture*, ed. Y. Wang, World Scientific, 1984.
[35] K. G. Borozdkin, On the problem of I. M. Vinogradov's constant (in Russian), *Proc. Third All-Union Math. Conf.*, v. 1, Moscow, Izdat. Akad. Nauk SSSR, 1956, p. 3; MR 34 #5784.
[36] J. R. Chen and T. Z. Wang, On the odd Goldbach problem (in Chinese), *Acta Math. Sinica* 32 (1989) 702-718; addendum 34 (1991) 143-144; MR 91e:11108 and MR 92g:11101.
[37] J. R. Chen and T. Z. Wang, The Goldbach problem for odd numbers (in Chinese), *Acta Math. Sinica* 39 (1996) 169-174; MR 97k:11138.
[38] D. Zinoviev, On Vinogradov's constant in Goldbach's ternary problem, *J. Number Theory* 65 (1997) 334-358; MR 98f:11107.
[39] Y. Saouter, Checking the odd Goldbach conjecture up to 10^{20}, *Math. Comp.* 67 (1998) 863-866; MR 98g:11115.
[40] J.-M. Deshouillers, G. Effinger, H. te Riele, and D. Zinoviev, A complete Vinogradov 3-primes theorem under the Riemann hypothesis, *Elec. Res. Announce. Amer. Math. Soc.* 3 (1997) 99-104; MR 98g:11112.
[41] O. Ramaré and R. Rumely, Primes in arithmetic progressions, *Math. Comp.* 65 (1996) 397-425; MR 97a:11144.
[42] O. Ramaré, On Schnirelmann's constant, *Annali Scuola Norm. Sup. Pisa Cl. Sci.* 22 (1995) 645-706; MR 97a:11167.
[43] J. R. Chen, On the representation of a large even integer as the sum of a prime and the product of at most two primes, *Sci. Sinica* 16 (1973) 157-176; MR 55 #7959.
[44] P. M. Ross, On Chen's theorem that each large even number has the form $p_1 + p_2$ or $p_1 + p_2 p_3$, *J. London Math. Soc.* 10 (1975) 500-506; MR 52 #10646.
[45] L. Kaniecki, On Snirelman's constant under the Riemann hypothesis, *Acta Arith.* 72 (1995) 361-374; MR 96i:11112.
[46] J. R. Chen, On the representation of a large even integer as the sum of a prime and the product of at most two primes. II, *Sci. Sinica* 21 (1978) 421-430; MR 80e:10037.
[47] É. Fouvry and F. Grupp, On the switching principle in sieve theory, *J. Reine Angew. Math.* 370 (1986) 101-126; MR 87j:11092.
[48] J. H. Kan, On the sequence $p+h$, *Arch. Math. (Basel)* 56 (1991) 454-464; MR92d:11110.
[49] H. Q. Liu, On the prime twins problem, *Sci. China Ser. A*33 (1990) 281-298; MR91i:11125.
[50] J. Wu, Sur la suite des nombres premiers jumeaux, *Acta Arith.* 55 (1990) 365-394; MR 91j:11074.
[51] J. R. Chen, On the Goldbach's problem and the sieve methods, *Sci. Sinica* 21 (1978) 701-739; MR 80b:10069.
[52] A. Selberg, On elementary methods in primenumber-theory and their limitations, *Den 11*[te] *Skandinaviske Matematikerkongress*, Proc. 1949 Trondheim conf., Johan Grundt Tanums Forlag,

1952, pp. 13-22; MR 14,726k.
[53] C. D. Pan, A new application of the Ju. V. Linnik large sieve method (in Chinese), *Acta Math. Sinica* 14 (1964) 597-606; Engl. transl. in *Chinese Math.* - *Acta* 5 (1964) 642-652; MR 30 #3871.
[54] E. Bombieri and H. Davenport, Small differences between prime numbers, *Proc. Royal Soc. Ser. A* 293 (1966) 1-18; MR 33 #7314.
[55] C. B. Pan, On the upper bound of the number of ways to represent an even integer as a sum of two primes, *Sci. Sinica* 23 (1980) 1368-1377; MR 82j:10078.
[56] D. H. Wu, An improvement of J. R. Chen's theorem (in Chinese), *Shanghai Keji Daxue Xuebao* (1987) n. 1, 94-99; MR 88g:11072.
[57] É. Fouvry and H. Iwaniec, Primes in arithmetic progressions, *Acta Arith.* 42 (1983) 197-218; MR 84k:10035.
[58] É. Fouvry, Autour du théorème de Bombieri-Vinogradov, *Acta Math.* 152 (1984) 219-244; MR 85m:11052.
[59] E. Bombieri, J. B. Friedlander, and H. Iwaniec, Primes in arithmetic progressions to large moduli, *Acta Math.* 156 (1986) 203-251; MR 88b:11058.
[60] J. K. Haugland, *Application of Sieve Methods to Prime Numbers*, Ph.D. thesis, Oxford Univ., 1999.
[61] H. L. Montgomery and R. C. Vaughan, The exceptional set in Goldbach's problem, *Acta Arith.* 27 (1975) 353-370; MR 51 #10263.
[62] J. R. Chen and C. D. Pan, The exceptional set of Goldbach numbers. I, *Sci. Sinica* 23 (1980) 416-430; MR 82f:10060.
[63] J. R. Chen, The exceptional set of Goldbach numbers. II, *Sci. Sinica Ser. A* 26 (1983) 714-731; MR 85m:11059.
[64] J. R. Chen and J. M. Liu, The exceptional set of Goldbach numbers. III, *Chinese Quart. J. Math.* 4 (1989) 1-15; MR 90k:11129.
[65] H. Li, The exceptional set of Goldbach numbers, *Quart. J. Math.* 50 (1999) 471-482; MR 2001a:11172.
[66] H. Li, The exceptional set of Goldbach numbers. II, *Acta Arith.* 92 (2000) 71-88; MR 2001c:11110.
[67] J.-M. Deshouillers, A. Granville, W. Narkiewicz, and C. Pomerance, An upper bound in Goldbach's problem, *Math. Comp.* 61 (1993) 209-213; MR 94b:11101.
[68] T. Oliveira e Silva, Goldbach conjecture verification, unpublished note (2001).
[69] J. Richstein, Verifying Goldbach's conjecture up to $4 \cdot 10^{14}$, *Math. Comp.* 70 (2001) 1745-1749; MR 2002c:11131.
[70] D. Shanks and M. Lal, Bateman's constants reconsidered and the distribution of cubic residues, *Math. Comp.* 26 (1972) 265-285; MR 46 #1734.
[71] D. Shanks, Calculation and applications of Epstein zeta functions, *Math. Comp.* 29 (1975) 271-287; corrigenda 29 (1975) 1167 and 30 (1976) 900; MR 53 #13114a-c.
[72] D. Shanks, On numbers of the form n^4+1, *Math. Comp.* 15 (1961) 186-189; corrigenda 16 (1962) 513; MR 22 #10941.
[73] D. Shanks, *Solved and Unsolved Problems in Number Theory*, 2nd ed., Chelsea, 1978; MR 86j:11001.
[74] D. Shanks, A note on Gaussian twin primes, *Math. Comp.* 14 (1960) 201-203; MR 22 #2586.
[75] M. Lal, Primes of the form n^4+1, *Math. Comp.* 21 (1967) 245-247; MR 36 #5059.
[76] D. Shanks, Lal's constant and generalizations, *Math. Comp.* 21 (1967) 705-707; MR 36 #6363.
[77] X. Gourdon and P. Sebah, Some constants from number theory (Numbers, Constants and Computation).

2.2 マイセル-メルテンス定数

この節と【2.14】で取り上げる無限級数のすべては，素数 2, 3, 5, 7, 11, 13, 17, …の逆数を含む．すべての素数の逆数の和は発散する．実際[1-6]

$$\lim_{n \to \infty}\left(\sum_{p \leq n}\frac{1}{p} - \ln(\ln(n))\right) = M = \gamma + \sum_{p}\left[\ln\left(1 - \frac{1}{p}\right) + \frac{1}{p}\right] = 0.2614972128\cdots$$

である．ただし，2つの和は素数すべてについて和をとり，γ はオイラー定数【1.5】である．[7,8]によれば，M の定義は，1866 年 Meissel（マイセル）と独立に 1874 年 Mertens（メルテンス）によって，妥当なものになった．M の値は，時には Kronecker（クロネッカー）定数[9]または素数逆数定数（prime reciprocal constant）[10]とよばれる．M への急速な収束数列は[11-13]

$$M = \gamma + \sum_{k=2}^{\infty}\frac{\mu(k)}{k}\ln(\zeta(k))$$

である．ただし，$\zeta(k)$ は Riemann のゼータ関数【1.6】であり，$\mu(k)$ は Möbius（メビウス）のミュー関数

$$\mu(k) = \begin{cases} 1 & (k=1 \text{ のとき}) \\ (-1)^r & (k \text{ が } r \text{ 個の異なる素数の積のとき}) \\ 0 & (k \text{ が平方数} > 1 \text{ で割り切れるとき}) \end{cases}$$

である．

$\omega(n)$ を，任意の整数 n の「異なる」素因子の個数とする．このとき，おもしろいことに $\omega(1), \omega(2), \cdots, \omega(n)$ の平均値

$$\mathrm{E}_n(\omega) = \frac{1}{n}\sum_{k=1}^{n}\omega(k)$$

は，式[2,9,14-16]

$$\lim_{n \to \infty}(\mathrm{E}_n(\omega) - \ln(\ln(n))) = M$$

によって漸近的に表すことができる．（くりかえし因数の個数を数えた）n の素因子の「全部の」個数 $\Omega(n)$ に対するいくぶん大きな平均値は，次のものである．

$$M' = \lim_{n \to \infty}(\mathrm{E}_n(\omega) - \ln(\ln(n))) = M + \sum_{p}\frac{1}{p(p-1)}$$

$$= \gamma + \sum_{p}\left[\ln\left(1 - \frac{1}{p}\right) + \frac{1}{p-1}\right] = \gamma + \sum_{k=2}^{\infty}\frac{\varphi(k)}{k}\ln(\xi(k))$$

$$= 1.0346538818\cdots$$

である．ただし，$\varphi(k)$ はオイラーの約数関数【2.7】である．関連する極限[1,17]は，

$$\lim_{n \to \infty}\left(\sum_{p \leq n}\frac{\ln(p)}{p} - \ln(n)\right) = -M'' = -\gamma - \sum_{p}\frac{\ln(p)}{p(p-1)} = -1.3325822757\cdots$$

である．M'' を計算する速い方法は，級数[18]

$$M''=\gamma+\sum_{k=2}^{\infty}\mu(k)\frac{\zeta'(k)}{\zeta(k)}$$

を用いることである．

Dirichlet の有名な定理は，「a, b が互いに素な正の整数とするとき，$a+bl$ という形からなる素数は無限個存在する」である．このような素数の逆数のすべての和について何がいえるだろうか．極限

$$m_{a,b}=\lim_{n\to\infty}\left(\sum_{\substack{p\leq n\\p\equiv a\bmod b}}\frac{1}{p}-\frac{1}{\varphi(b)}\ln(\ln(n))\right)$$

は，各 a, b に対して存在し，有限の値であることが証明できる．たとえば，[19-23]

$$m_{1,4}=\ln\left(\frac{\sqrt{\pi}}{4K}\right)+\frac{\gamma}{2}+\sum_{p\equiv 1\bmod 4}\left[\ln\left(1-\frac{1}{p}\right)+\frac{1}{p}\right]=-0.2867420562\cdots$$

$$m_{3,4}=\ln\left(\frac{2K}{\sqrt{\pi}}\right)+\frac{\gamma}{2}+\sum_{p\equiv 3\bmod 4}\left[\ln\left(1-\frac{1}{p}\right)+\frac{1}{p}\right]=0.0482392690\cdots$$

である．ただし，K は Landau-Ramanujan（ランダウ-ラマヌジャン）定数【2.3】である．もちろん，$m_{1,4}+m_{3,4}+1/2=M$ である．

素数を平方した逆数の和は

$$N=\sum_{p}\frac{1}{p^2}=\sum_{k=1}^{\infty}\frac{\mu(k)}{k}\ln(\zeta(2k))=0.4522474200\cdots$$

である．これは，公式[9,14]

$$\lim_{n\to\infty}(\mathrm{Var}_n(\omega)-\ln(\ln(n)))=M-N-\frac{\pi^2}{6}=-1.8356842740\cdots$$

により，$\omega(1),\omega(2),\cdots,\omega(n)$ の分散

$$\mathrm{Var}_n(\omega)=\mathrm{E}_n(\omega^2)-\mathrm{E}_n(\omega)^2$$

に関係している．同様に

$$N'=\sum_{p}\frac{1}{(p-1)^2}=1.3750649947\cdots$$

は，

$$\lim_{n\to\infty}(\mathrm{Var}_n(\Omega)-\ln(\ln(n)))=M'+N'-\frac{\pi^2}{6}=0.7647848097\cdots$$

に現れる．N と N' の詳しい評価については[15,24]を，ω と Ω の漸近的確率分布については[25-27]を見よ．

与えられた正の整数 n に対して，$D_n=\max\{d:d^2|n\}$ とおく．S を，D_n が素数となる n の集合と定義し，$\tilde{S}$ を，$D_n^3\nmid n$ となる $n\in S$ の集合と定義する．S と $\tilde{S}$ の漸近密度は，それぞれ[28-30]

$$\frac{6}{\pi^2}\sum_{p}\frac{1}{p^2}=0.2749334633\cdots,\quad\frac{6}{\pi^2}\sum_{p}\frac{1}{p(p+1)}=0.2007557220\cdots$$

である．言いかえると，S は，素因子の各々がちょうど1つの例外を除いて単純（simple) であるような整数の集合である．$\tilde{S}$ において，例外は素数の平方である．関連した論文で，無平方集合†については【2.5】を，充平方集合については【2.6】を参照せよ．

Bach（バッハ）[12]は，アルティン定数 C_{Artin}【2.4】と双子素数定数 C_{twin}【2.1】のみならず，M を計算する場合の計算複雑度（computational complexity）を見積もった．

$p_1=2$，$p_2=3$，$p_3=5$，…とするとき，交代級数

$$\sum_{k=1}^{\infty}(-1)^k\frac{1}{p_k}=-0.2696063519\cdots$$

は，明らかに収束する[31]．これは，交代級数でない 2 つの級数[32-35]

$$\sum_{k=2}^{\infty}\varepsilon_k\frac{1}{p_k}=0.3349813253\cdots, \quad \sum_{k=1}^{\infty}\varepsilon'_k\frac{1}{p_k}=0.6419448385\cdots$$

に比べて，たいして興味をひくものではない．ただし，

$$\varepsilon_k=\begin{cases}-1 & (p_k\equiv 1 \bmod 4 \text{ のとき}) \\ 1 & (p_k\equiv 3 \bmod 4 \text{ のとき})\end{cases} \quad \varepsilon'_k=\begin{cases}-1 & (p_k\equiv 1 \bmod 3 \text{ のとき}) \\ 1 & (p_k\equiv 2 \bmod 3 \text{ のとき}) \\ 0 & (p_k\equiv 0 \bmod 3 \text{ のとき})\end{cases}$$

である．もちろん，次の級数も収束する[36]．

$$\sum_{k=2}^{\infty}\varepsilon_k\frac{1}{p_k^2}=0.0946198928\cdots$$

Erdös（エルデーシュ）[37,38]は，級数 $\sum_{k=1}^{\infty}(-1)^k k/p_k$ が収束することは真である，と考えた．

Merrified（メリフィールド）[39]と Lienard（リナード）[40]は，M と $\gamma-M=0.3157184521\cdots$ のみならず，$2\leq n\leq 167$ に対して，級数 $\sum_p p^{-n}$ の値を表にした．

2.2.1 平方剰余

p を素数とする．法 p における正の平方非剰余のうち，最小な数を $f(p)$ とする．$f(p)$ の平均値の値は[41,42]

$$\lim_{n\to\infty}\frac{\sum_{p\leq n}f(p)}{\sum_{p\leq n}1}=\lim_{n\to\infty}\frac{\ln(n)}{n}\sum_{p\leq n}f(p)=\sum_{k=1}^{\infty}\frac{p_k}{2^k}=3.6746439660\cdots$$

である．さらに一般的に，m を奇数とし，m が非平方数のとき，Jacobi（ヤコビ）記号 $(k/m)<1$ を満たす最小の正の整数 k を $f(m)$ とし，m が平方数のとき $f(m)=0$ とする．$(k/m)=-1$ であるとき，k は法 m で平方非剰余であるという．$f(m)$ の平均値は[41,43,44]

$$\lim_{n\to\infty}\frac{2}{n}\sum_{\substack{m\leq n \\ m \text{ は奇数}}}f(m)=1+\sum_{j=2}^{\infty}\frac{p_j+1}{2^{j-1}}\prod_{i=1}^{j-1}\left(1-\frac{1}{p_i}\right)=3.1477551485\cdots$$

---

†訳注：無平方（square-free）とは 1 より大きい平方数で割り切れない数，充平方（square-full）とはすべての素因子の 2 乗で割り切れる数を意味する．無立方（cube-free），充立方（cube-full）なども同様である．

である.

[1] J. B. Rosser and L. Schoenfeld, Approximate formulas for some functions of prime numbers, *Illinois J. Math.* 6 (1962) 64-94; MR 25 #1139.
[2] G. H. Hardy and E. M. Wright, *An Introduction to the Theory of Numbers*, 5th ed., Oxford Univ. Press, 1985, pp. 351-358; MR 81i:10002.
[3] A. E. Ingham, *The Distribution of Prime Numbers*, Cambridge Univ. Press, 1932; MR 91f:11064.
[4] H. M. Terrill and L. Sweeny, Two constants connected with the theory of prime numbers, *J. Franklin Inst.* 239 (1945) 242-243; MR 6,169a.
[5] D. H. Greene and D. E. Knuth, *Mathematics for the Analysis of Algorithms*, 3rd ed., Birkhäuser, 1990, pp. 60-63; MR 92c:68067.
[6] G. Tenenbaum, *Introduction to Analytic and Probabilistic Number Theory*, Cambridge Univ. Press, 1995, pp. 14-18; MR 97e:11005b.
[7] P. Lindqvist and J. Peetre, On a number theoretic sum considered by Meissel-A historical observation, *Nieuw Arch. Wisk.* 15 (1997) 175-179; MR 99b:11141.
[8] P. Lindqvist and J. Peetre, On the remainder in a series of Mertens, *Expos. Math.* 15 (1997) 467-478; MR 98i:11110.
[9] M. R. Schroeder, *Number Theory in Science and Communication: With Applications in Cryptography, Physics, Digital Information, Computing and Self-Similarity*, 2nd ed., Springer-Verlag, 1986; MR 99c:11165.
[10] E. Bach and J. Shallit, *Algorithmic Number Theory, v. 1. Efficient Algorithms*, MIT Press, 1996, pp. 233-237, 263; MR 97e:11157.
[11] P. Flajolet and I. Vardi, Zeta function expansions of classical constants, unpublished note (1996).
[12] E. Bach, The complexity of number-theoretic constants, *Inform. Process. Lett.* 62 (1997) 145-152; MR 98g:11148.
[13] X. Gourdon and P. Sebah, Some constants from number theory (Numbers, Constants and Computation).
[14] P. Diaconis, F. Mosteller and H. Onishi, Second-order terms for the variances and covariances of the number of prime factors-including the square free case, *J. Number Theory* 9 (1977) 187-202; MR 55 #7953.
[15] H. Cohen, High precision computation of Hardy-Littlewood constants, unpublished note (1999).
[16] R. L. Duncan, A class of arithmetical functions, *Amer. Math. Monthly* 69 (1962) 34-36.
[17] H. L. Montgomery, *Topics in Multiplicative Number Theory*, Lect. Notes in Math. 227, Springer-Verlag, 1971, p. 43; MR 49 #2616.
[18] T. Jameson, Asymptotics of $\sum_{n<x}\mu(n)^2/\varphi(n)$, unpublished note (1999).
[19] S. Uchiyama, On some products involving primes, *Proc. Amer. Math. Soc.* 28 (1971) 629-630; MR 43 #3227.
[20] K. S.Williams, Mertens' theorem for arithmetic progressions, *J. Number Theory* 6 (1974) 353-359; MR 51 #392.
[21] E. Grosswald, Some number theoretical products, *Rev. Colombiana Mat.* 21 (1987) 231-242; MR 90e:11129.
[22] E. A. Vasil'kovskaja, Mertens' formula for an arithmetic progression (in Russian), *Voprosy Mat. Sbornik Naucn. Trudy Taskent. Gos. Univ.* (1977) n. 548, 14-17, 139-140; MR 58 #27848.
[23] G. Niklasch and P. Moree, Generalized Meissel-Mertens constants, unpublished note (2000).
[24] J. W. L. Glaisher, On the sums of the inverse powers of the prime numbers, *Quart. J. Pure Appl. Math.* 25 (1891) 347-362.
[25] P. Erdös and M. Kac, The Gaussian law of errors in the theory of additive number theoretic functions, *Amer. J. Math.* 62 (1940) 738-742; MR 2,42c.
[26] D. E. Knuth and L. Trabb Pardo, Analysis of a simple factorization algorithm, *Theoret.*

Comput. Sci. 3 (1976) 321-348; also in *Selected Papers on Analysis of Algorithms*, CSLI, 2000, pp. 303-339; MR 58 #16485.
[27] S. Guiasu, On the distribution of the number of distinct prime factors, *J. Inst. Math. Comp. Sci. (Math. Ser.)* 4 (1991) 171-179; MR 92j:11106.
[28] A. Rényi, On the density of certain sequences of integers, *Acad. Serbe Sci. Publ. Inst. Math.* 8 (1955) 157-162; also in Selected Papers, v. 1, Akadémiai Kiadó, 1976, pp. 506-512; MR 17,944f.
[29] E. Cohen, Arithmetical notes. VIII: An asymptotic formula of Rényi, *Proc. Amer. Math. Soc.* 13 (1962) 536-539; MR 25 #2049.
[30] E. Cohen, Some asymptotic formulas in the theory of numbers, *Trans. Amer. Math. Soc.* 112 (1964) 214-227; MR 29 #3458.
[31] H. P. Robinson and E. Potter, *Mathematical Constants*, UCRL-20418 (1971), Univ. of Calif. at Berkeley; available through the National Technical Information Service, Springfield VA 22151.
[32] L. Euler, De summa seriei ex numeris primis formatae $1/3-1/5+1/7+1/11-1/13-\cdots$ ubi numeri primi formae $4n-1$ habent signum positivum formae autem $4n+1$ signum negativum, 1775, *Opera Omnia Ser.* I, v. 4, Lipsiae, 1911, pp. 146-162.
[33] J. W. L. Glaisher, On the series $1/3+1/5+1/7+1/11-1/13+\cdots$, *Quart. J. Pure Appl. Math.* 25 (1891) 375-383.
[34] A. Weil, *Number Theory: An Approach Through History from Hammurapi to Legendre*, Birkhäuser, 1984, pp. 266-267; MR 85c:01004.
[35] J. W. L. Glaisher, On the series $1/2+1/5-1/7+1/11 . 1/13+\cdots$, *Quart. J. Pure Appl. Math.* 25 (1891) 48-65.
[36] J.W. L. Glaisher,On the series $1/3^2-1/5^2+1/7^2+1/11^2-1/13^2-\cdots$, *Quart. J. Pure Appl. Math.* 26 (1893) 33-47.
[37] P. Erdös, Some of my new and almost new problems and results in combinatorial number theory, *Number Theory: Diophantine, Computational and Algebraic Aspects*, Proc. 1996 Elger conf., ed. K. Györy, A. Pethö, and V. T. Sós, Gruyter, 1998, pp. 169-180; MR 2000a:11001.
[38] R. K. Guy, *Unsolved Problems in Number Theory*, 2nd ed., Springer-Verlag, 1994, sect. E7; MR 96e:11002.
[39] C. W. Merrifield, The sums of the series of the reciprocals of the prime numbers and of their powers, *Proc. Royal Soc. London* 33 (1881) 4-10.
[40] R. Lienard, *Tables fondamentales à 50 décimales des sommes S_n, u_n, Σ_n*, Paris, 1948; MR 10, 149i.
[41] P. Ribenboim, *The New Book of Prime Number Records*, 3rd ed., Springer-Verlag, 1996, p. 142; MR 96k:11112.
[42] P. Erdös, Remarks on number theory. I, *Mat. Lapok* 12 (1961) 10-17; MR 26 #2410.
[43] R. Baillie and S. S. Wagstaff, Lucas pseudoprimes, *Math. Comp.* 35 (1980) 1391-1417; MR 81j:10005.
[44] N. J. A. Sloane, On-Line Encyclopedia of Integer Sequences, A053760 and A053761.

2.3 ランダウ-ラマヌジャン定数

$B(x)$ を，2つの平方数の和として表される x を超えない正の整数の個数とする．明らかに $B(x) \to \infty$ $(x \to \infty)$ であるが，その発散の速度はかなりおもしろい！
Landau (ランダウ) [1-3]と Ramanujan (ラマヌジャン) [4,5]は，次の極限が存

在することを独立に証明した．
$$\lim_{n\to\infty}\frac{\ln(x)}{x}B(x)=K$$
が成り立つ．ただし，K は注目すべき定数
$$K=\frac{1}{\sqrt{2}}\prod_{p=3\bmod 4}\left(1-\frac{1}{p^2}\right)^{-1/2}=\frac{\pi}{4}\prod_{p=3\bmod 4}\left(1-\frac{1}{p^2}\right)^{1/2}$$
である．2つの積は素数 p にかぎる．この極限の数値実験による検証は[6]で行われている．Shanks[7,8]による K に速く収束する式は，
$$K=\frac{1}{\sqrt{2}}\prod_{k=1}^{\infty}\left[\left(1-\frac{1}{2^{2k}}\right)\frac{\zeta(2^k)}{\beta(2^k)}\right]^{1/2^{k+1}}=0.7642236535\cdots$$
である．ただし，$\zeta(z)$ は Riemann のゼータ関数【1.6】で，$\beta(x)$ は Dirichlet のベータ関数【1.7】である．Landau による，より強い結果は
$$\lim_{x\to\infty}\frac{\ln(x)^{3/2}}{Kx}\left(B(x)-\frac{Kx}{\sqrt{\ln(x)}}\right)=C$$
である．ただし，C は[7,9-12]で
$$C=\frac{1}{2}+\frac{\ln(2)}{4}-\frac{\gamma}{4}-\frac{\beta'(1)}{4\beta(1)}+\frac{1}{4}\frac{d}{ds}\ln\left(\prod_{p=3\bmod 4}\left(1-\frac{1}{p^{2s}}\right)\right)\Big|_{s=1}$$
$$=\frac{1}{2}\left(1-\ln\left(\frac{\pi e^{\gamma}}{2L}\right)\right)-\frac{1}{4}\sum_{k=1}^{\infty}\left(\frac{\zeta'(2^k)}{\zeta(2^k)}-\frac{\beta'(2^k)}{\beta(2^k)}+\frac{\ln(2)}{2^{2k}-1}\right)$$
$$=0.5819486593\cdots$$
であり，γ はオイラー定数【1.5】，$L=2.6220575542\cdots$ はガウスのレムニスケート(lemniscate)定数【6.1】である．これらの公式は，Flajolet & Vardi（フラジョレット＆ヴァルディ），Zimmermann（ツィンメルマン），Adamchik（アダムチック），Golden & Gosper（ゴールデン＆ゴスパー），MacLeod（マックレード）と Hare（ハレ）による，最近いくつか行われている高い精度の計算の基礎となった．

2.3.1 類似の定数

いくつかの変形（類似の定数）がある．K_n を，a^2+nb^2 の形をした正の整数の個数のときの，K と類似なものと定義する．C_n も同様に定義する．[10,13-16]で次のことが証明された．

$$K_2=\frac{1}{\sqrt[4]{2}}\prod_{p=5\text{または}7\bmod 8}\left(1-\frac{1}{p^2}\right)^{-1/2}=0.8728875581\cdots$$

$$K_3=\frac{1}{\sqrt{2}\sqrt[3]{3}}\prod_{p=2\bmod 3}\left(1-\frac{1}{p^2}\right)^{-1/2}=0.6389094054\cdots$$

$$K_4=\frac{3}{4}K=0.5731677401\cdots,\quad C_4=C=0.5819486593\cdots$$

Moree & te Riele（モレ＆テリーレ）[17]は最近 $C_3=0.5767761224\cdots$ を計算した．しかし，$n=2$ と $n>4$ に対する C_n の値は，未だに誰も見つけていない．

$n=3$ の場合，a^2+3b^2 の形の正の整数の個数は，a^2+ab+b^2 の形をした正の整数の個数と同値である．

K の類似としてその代わりに，l と m が互いに素であり，a^2+b^2 と $lc+m$ の形を同時に満たす正の整数の個数 $K_{l,m}$ を定義する．ただし，$K_{l,m}$ は単純に，l だけに依存する K の有理数倍である[18,19]．

さらに，変わった数がある．2つの平方数の和として表すことのできる，x を超えない無平方な正の整数の個数を $B_{\mathrm{sqfr}}(x)$ とする．また，$B_{\mathrm{copr}}(x)$ を，互いに素な2つの平方数の和として表すことのできる，x を超えない正の整数の個数とする．[20-22]で，次のことが証明された．

$$\lim_{x\to\infty}\frac{\sqrt{\ln(x)}}{x}B_{\mathrm{sqfr}}(x)=\frac{6K}{\pi^2}=0.4645922709\cdots$$

$$\lim_{x\to\infty}\frac{\sqrt{\ln(x)}}{x}B_{\mathrm{copr}}(x)=\frac{3}{8K}=0.4906940504\cdots$$

最初の極限からの結論は，無平方であることと2つの平方数の和であることとは，漸近的に独立な性質であるということである．もちろん，2つの平方数は互いに素でなければならない．そうでなければ，和は無平方にはならない．

2番目の式で1番目の式を割ることにより，2番目の集合の部分集合としての1番目の集合の漸近的な相対密度として[22]，

$$\lim_{x\to\infty}\frac{B_{\mathrm{sqfr}}(x)}{B_{\mathrm{copr}}(x)}=\frac{16K^2}{\pi^2}=\prod_{p\equiv 1\bmod 4}\left(1-\frac{1}{p^2}\right)=0.9468064072\cdots$$

を得る．これは大きな密度である．一方，ランダムに互いに素な2つの整数を選び，平方し，それらを加えるとき，和はきわめて無平方になりやすい．他方，これには無限に多くの反例がある．たとえば，原始ピタゴラス数の組（primitive Pythagorean triples）【5.2】を考えよ．

$j=1, 3$ とする．素因子のすべてが法4で j と合同になる，x までの正の整数の個数を $B_j(x)$ とする．[20,21,23,24]で，次のことが示されている．

$$\lim_{x\to\infty}\frac{\sqrt{\ln(x)}}{x}B_1(x)=\frac{1}{4K}=0.3271293669\cdots$$

$$\lim_{x\to\infty}\frac{\sqrt{\ln(x)}}{x}B_3(x)=\frac{2K}{\pi}=0.4865198884\cdots$$

これらが等しくないことは，おもしろい！ これは Rubenstein & Sarnak（ルーベンスタイン&サルナック）[25]により説明された Chebyshev（**チェビシェフ**）効果の現れである．関連する議論は【2.8】を参照せよ．

Uchiyama（内山三郎）[26]により発見された2つの極限は

$$\lim_{x\to\infty}\sqrt{\ln(x)}\prod_{\substack{p\le x\\p\equiv 1\bmod 4}}\left(1-\frac{1}{p}\right)=\frac{4}{\sqrt{\pi}}\exp\left(-\frac{\gamma}{2}\right)K=1.2923041571\cdots$$

$$\lim_{x\to\infty}\sqrt{\ln(x)}\prod_{\substack{p\le x\\p\equiv 3\bmod 4}}\left(1-\frac{1}{p}\right)=\frac{\sqrt{\pi}}{2}\exp\left(-\frac{\gamma}{2}\right)\frac{1}{K}=0.8689277682\cdots$$

である.2つを掛けると,有名なメルテンスの定理【2.2】が得られる.これらの結果の拡張は[27-29]にある.系として,

$$\lim_{x\to\infty}\frac{1}{\sqrt{\ln(x)}}\prod_{\substack{p\leq x \\ p\equiv 1\bmod 4}}\left(1+\frac{1}{p}\right)=\frac{4}{\pi^{3/2}}\exp\left(\frac{\gamma}{2}\right)K=0.7326498193\cdots$$

$$\lim_{x\to\infty}\frac{1}{\sqrt{\ln(x)}}\prod_{\substack{p\leq x \\ p\equiv 3\bmod 4}}\left(1+\frac{1}{p}\right)=\frac{1}{\sqrt{\pi}}\exp\left(\frac{\gamma}{2}\right)\frac{1}{K}=0.9852475810\cdots$$

を得る.以前に,$16K/\pi^2$ の表現を補足する公式があった.

$$\prod_{p\equiv 3\bmod 4}\left(1-\frac{1}{p^2}\right)=\frac{1}{2K^2}=0.8561089817\cdots,$$

$$\prod_{p\equiv 1\bmod 4}\left(1+\frac{1}{p^2}\right)=\frac{192K^2G}{\pi^4}=1.0544399448\cdots,$$

$$\prod_{p\equiv 3\bmod 4}\left(1+\frac{1}{p^2}\right)=\frac{\pi^2}{16K^2G}=1.1530805616\cdots$$

である.$G=\beta(2)$ は Catalan(カタラン)の定数【1.7】である.同じ表現が次の状況を扱うときに現れる.数列 n^2+1, $1\leq n\leq x$ に属する正の無平方な数の個数を,$\widehat{B}(x)$ とする.このとき,[30,31]

$$\lim_{n\to\infty}\frac{\widehat{B}(x)}{x}\prod_{p\equiv 1\bmod 4}\left(1-\frac{1}{p^2}\right)=0.8948412245\cdots$$

である.この結果の幅広い一般化は,[32-34]で述べられている.

0 でない異なる2つの平方数の和として表すことができない n^2 で,x を超えない正の整数 n の個数を,$\widetilde{B}(x)$ とする.Shanks[35,36]はこれらを**非斜辺**(non-hypotenuse)**数**とよび,

$$\widetilde{K}=\lim_{x\to\infty}\frac{\sqrt{\ln(x)}}{x}\widetilde{B}(x)=\frac{4K}{\pi}=0.9730397768\cdots$$

$$\lim_{x\to\infty}\frac{\ln(x)^{3/2}}{\widetilde{K}x}\left(\widetilde{B}(x)-\frac{\widetilde{K}x}{\sqrt{\ln(x)}}\right)=C+\frac{1}{2}\ln\left(\frac{\pi e^\gamma}{2L^2}\right)=0.7047534517\cdots$$

を証明した.さらに,3次の項は正であることが知られている,と言及した(しかし,この値を計算していない).

2つの平方数の和として表すことのできる,x を超えない「素数」の個数を $A(x)$ とする.a^2+b^2 の形の奇素数は,法 4 でちょうど 1 に等しいから,

$$\lim_{x\to\infty}\frac{\ln(x)}{x}A(x)=\frac{1}{2}$$

を得る.

a^2+b^4 の形で表される x を超えない素数の個数を,$U(x)$ と定義する.Friedlander & Iwaniec(フリードランダー&イワニエツ)[37,38]は,

$$\lim_{x\to\infty}\frac{\ln(x)}{x^{3/4}}U(x)=\frac{4L}{3\pi}=1.1128357889\cdots$$

を証明した.偶然の一致であるが,同じ定数 L が $B(x)$ の第 2 次近似に現れた.このことに示唆されて,Heath-Brown(ヒース-ブラウン)[39]は最近,a^3+2b^2 の形の素

数に対する類似の結果を証明した.
a^2+b^4 の形で表される x を超えない正の整数の個数を, $V(x)$ と定義する. ほとんどすべての整数に対して, 条件を満たす表し方はただ1通りであることがわかる. したがって, [38]での公式が適用可能であり,

$$\lim_{x\to\infty} x^{-3/4} V(x) = \frac{L}{3} = 0.8740191847\cdots$$

である. a^3+2b^3 の形の正の整数に対する, 対応した漸近式を調べるのはよいことだろう. 関連する資料は[40, 41]にある.

3つの平方数の和として表される x を超えない正の整数の個数を, $Q(x)$ とする. Landau[1]は $Q(x)/x \to 5/6$ $(x\to\infty)$ を証明した. 誤差項 $\Delta(x) = Q(x) - 5x/6$ は,

$$0 = \liminf_{x\to\infty} \Delta(x) < \limsup_{x\to\infty} \Delta(x) = \frac{1}{3\ln(2)}$$

であるという意味で, 漸近的に, よいふるまいではない[42-44]. $\Delta(x)$ の平均値は, 周期的でいたるところ微分不可能な連続な関数で, 正確に測ることができる. このような定式化のさらなる結果は, 【2.16】にある. $a^3+b^3+c^3$ または $a^4+b^4+c^4+d^4$ の形をした x の個数に対する漸近式は, 未解決のままである[45].

[1] E. Landau, Über die Einteilung der positiven ganzen Zahlen in vier Klassen nach der Mindestzahl der zu ihrer additiven Zusammensetzung erforderlichen Quadrate, *Archiv Math. Phys.* 13 (1908) 305-312; also in *Collected Works*, v. 4, ed. L. Mirsky, I. J. Schoenberg, W. Schwarz, and H. Wefelscheid, Thales Verlag, 1985, pp. 59-66.
[2] G. H. Hardy, *Ramanujan: Twelve Lectures on Subjects Suggested by His Life and Work*, Chelsea, 1940, pp. 60-63; MR 21 #4881.
[3] W. J. LeVeque, *Topics in Number Theory, II*, Addison-Wesley, 1956, pp. 257-263; MR 18, 283d.
[4] B. C. Berndt, *Ramanujan's Notebooks: Part IV*, Springer-Verlag, 1994, pp. 52, 60-66; MR 95e: 11028.
[5] P. Moree and J. Cazaran, On a claim of Ramanujan in his first letter to Hardy, *Expos. Math* 17 (1999) 289-312; MR 2001c:11103.
[6] P. Shiu, Counting sums of two squares: The Meissel-Lehmer method, *Math. Comp.* 47 (1986) 351-360; MR 87h:11127.
[7] D. Shanks, The second-order term in the asymptotic expansion of $B(x)$, *Math. Comp.* 18 (1964) 75-86; MR 28 #2391.
[8] P. Flajolet and I. Vardi, Zeta function expansions of classical constants, unpublished note (1996).
[9] G. K. Stanley, Two assertions made by Ramanujan, *J. London Math. Soc.* 3 (1928) 232-237; corrigenda 4 (1929) 32.
[10] W. Heupel, Die Verteilung der ganzen Zahlen, die durch quadratische Formen dargestellt werden, *Arch. Math. (Basel)* 19 (1968) 162-166; MR 37 #2686.
[11] A. MacLeod, The Landau-Ramanujan second-order constant, unpublished note (1996).
[12] P. Moree, Chebyshev's bias for composite numbers with restricted prime divisors, *Math. Comp.*, to appear; math.NT/0112100.
[13] R. D. James, The distribution of integers represented by quadratic forms, *Amer. J. Math.* 60 (1938) 737-744.
[14] G. Pall, The distribution of integers represented by binary quadratic forms, *Bull. Amer. Math. Soc.* 49 (1943) 447-449; MR 4, 240g.

[15] D. Shanks and L. P. Schmid, Variations on a theorem of Landau. Part I, *Math. Comp.* 20 (1966) 551-569; MR 35 #1564.
[16] K. S. Williams, Note on integers representable by binary quadratic forms, *Canad. Math. Bull.* 18 (1975) 123-125; MR 52 #269.
[17] P. Moree and H. J. J. te Riele, The hexagonal versus the square lattice, *Math. Comp.*, to appear; math.NT/0204332.
[18] K. Prachar, Über Zahlen der Form $a^2 + b^2$ in einer arithmetischen Progression, *Math. Nachr.* 10 (1953) 51-54; MR 15,289b.
[19] H. Iwaniec, The half dimensional sieve, *Acta Arith.* 29 (1976) 69-95; MR 54 #261.
[20] P. Moree, Variations on the Landau-Ramanujan problem, unpublished note (2000).
[21] G. J. Rieger, Über die Anzahl der als Summe von zwei Quadraten darstellbaren und in einer primen Restklasse gelegenen Zahlen unterhalb einer positiven Schranke. II, *J. Reine Angew. Math.* 217 (1965) 200-216; MR 30 #4734.
[22] W. Bosma and P. Stevenhagen, Density computations for real quadratic units, *Math. Comp.* 65 (1996) 1327-1337; MR 96j:11171.
[23] P. Stevenhagen, A density conjecture for the negative Pell equation, *Computational Algebra and Number Theory*, Proc. 1992 Sydney conf., ed. W. Bosma and A. van der Poorten, Kluwer, 1995, pp. 187-200; MR 96g:11137.
[24] G. Tenenbaum, *Introduction to Analytic and Probabilistic Number Theory*, Cambridge Univ. Press, 1995, p. 265; MR 97e:11005b.
[25] M. Rubinstein and P. Sarnak, Chebyshev's bias, *Experim. Math.* 3 (1994) 173-197; MR 96d:11099.
[26] S. Uchiyama, On some products involving primes, *Proc. Amer. Math. Soc.* 28 (1971) 629-630; MR 43 #3227.
[27] K. S.Williams, Mertens' theorem for arithmetic progressions, *J. Number Theory* 6 (1974) 353-359; MR 51 #392.
[28] E. Grosswald, Some number theoretical products, *Rev. Colombiana Mat.* 21 (1987) 231- 242; MR 90e:11129.
[29] E. A. Vasil'kovskaja, Mertens' formula for an arithmetic progression (in Russian), *Voprosy Mat. Sbornik Naucn. Trudy Taskent. Gos. Univ.* (1977) n. 548, 14-17, 139-140; MR 58 #27848.
[30] P. Moree, Square-free polynomial values, unpublished note (2000).
[31] P. Sebah, Approximation of an infinite product, unpublished note (2000).
[32] P. Erdös, Arithmetical properties of polynomials, *J. London Math. Soc.* 28 (1953) 416-425; MR 15,104f.
[33] C. Hooley, On the power free values of polynomials, *Mathematika* 14 (1967) 21-26; MR 35 #5405.
[34] A. Granville, ABC allows us to count squarefrees, *Int. Math. Res. Notices* (1998) 991-1009; MR 99j:11104.
[35] D. Shanks, Non-hypotenuse numbers, *Fibonacci Quart.* 13 (1975) 319-321;MR52 #8062.
[36] N. J. A. Sloane, On-Line Encyclopedia of Integer Sequences, A001481, A004144, A009003, and A022544.
[37] J. Friedlander and H. Iwaniec, Using a parity-sensitive sieve to count prime values of a polynomial, *Proc. Nat. Acad. Sci. USA* 94 (1997) 1054-1058; MR 98b:11097.
[38] J. Friedlander and H. Iwaniec, The polynomial $X^2 + Y^4$ captures its primes, *Annals of Math.* 148 (1998) 945-1040; MR 2000c:11150a.
[39] D. R. Heath-Brown, Primes represented by $x^3 + 2y^3$, *Acta Math.* 186 (2001) 1-84; MR 2002b:11122.
[40] P. Erdös and K. Mahler, On the number of integers which can be represented by a binary form, *J. London Math. Soc.* 13 (1938) 134-139.
[41] R. K. Guy, *Unsolved Problems in Number Theory*, 2nd ed., Springer-Verlag, 1994, sect. D4; MR 96e:11002.
[42] M. C. Chakrabarti, On the limit points of a function connected with the three-square problem, *Bull. Calcutta Math. Soc.* 32 (1940) 1-6; MR 3,162i.
[43] P. Shiu, Counting sums of three squares, *Bull. London Math. Soc.* 20 (1988) 203-208; MR 89c:

11054.
[44] A. H. Osbaldestin and P. Shiu, A correlated digital sum problem associated with sums of three squares, *Bull. London Math. Soc.* 21 (1989) 369-374; MR 90f:11023.
[45] J.-M. Deshouillers, F. Hennecart, and B. Landreau, Do sums of 4 biquadrates have a positive density?, *Proc. 1998 Algorithmic Number Theory Sympos. (ANTS-III)*, Portland, ed. J. P. Buhler, Lect. Notes in Comp. Sci. 1423, Springer-Verlag, 1998, pp. 196-203; MR 2000k:11051.

2.4 アルティン定数

Fermat（フェルマー）の小定理は，「p が素数であり，n は p で割れないとする．このとき，$n^{p-1}-1$ は p で割り切れる」というものである．

素数 p で n^e-1 が割り切れる，正の整数 e のすべてからなる集合を考える．もし，最小の $e=p-1$ であるとき，n は**法 p の原始根**とよばれる．

たとえば，11 で 6^1, 6^2, 6^3, $\cdots$, 6^9 を割った剰余のすべては，1 ではないから，6 は法 11 の原始根であり，このとき $e=10=11-1$ である．しかし，6^9-1 は 19 で割り切れ，かつ $e=9<19-1$ であるから，6 は法 19 の原始根ではない．

もっと代数的用語を使った解釈がある．集合 $Z_p=\{0,1,2,\cdots,p-1\}$ は，法 p で和と積に関して体になる．さらに，部分集合 $U_p=\{1,2,\cdots,p-1\}$ は，法 p で積に関して巡回群となる．したがって，整数 n（より正確には，法 p の剰余類）が法 p での原始根である必要十分条件は，n が群 U_p の生成元となることである．

別の解釈がある．$p>5$ を素数とする．このとき，分数 $1/p$ の 10 進小数展開が最大周期（$=p-1$）をもつための必要十分条件は，10 が p の原始根であることである．この条件を満たす素数は，**長い素数**（long primes）[1-4] としても知られている．

Artin（アルティン）[5] は 1927 年に，「$n \neq -1, 0, 1$ が整数の平方でないとき，n が原始根であるような素数全体の集合 $S(n)$ は，無限集合であるに違いない」，と予想した．この予想を証明する方向でいくつかの注目すべき進歩は，[6-9] で示されている．たとえば，集合 $S(2)$, $S(3)$, $S(5)$ の少なくとも 1 つは無限集合であることが知られている．

さらに付け加えて，n は任意の $r>1$ に対して整数の r 乗ではないと仮定する．n' を n の無平方な部分とする，すなわち，d^2 という形のすべての因数を取り除いた後に残る，n の約数である．Artin はさらに，次のことを予想した．その素数と互いに素である集合 $S(n)$ の密度が存在し，もし $n' \equiv 1 \bmod 4$ であるとき n の選び方に「無関係に」

$$C_{\text{Artin}}=\prod_p\left(1-\frac{1}{p(p-1)}\right)=0.3739558136\cdots$$

に等しい．この信じられない予想の証明は，未だに知られていない．他の場合では，有理数の補正因子が必要である（【2.4.2】を参照せよ）．しかし，アルティン定数はこのよ

うな公式の中心的特徴をとどめている．Hooley（ホーレイ）[10,11]は，一般化リーマン予想が正しいという条件の下で，このような公式が正しいことを証明した．

アルティン定数に速く収束する式は，次のものである[12-18]．Lucas（リュカ）**数列**を，
$$l_0=2, \quad l_1=1, \quad l_n=l_{n-1}+l_{n-2}, \quad n \geq 2$$
とする．φ を黄金比（Golden mean）【1.2】とするとき，$l_n = \varphi^n + (1-\varphi)^n$ であることに注意する．このとき，
$$C_{\text{Artin}} = \prod_{n \geq 2} \zeta(n)^{-1/n \sum_{k|n} l_k \cdot \mu(n/k)}$$
$$= \zeta(2)^{-1} \zeta(3)^{-1} \zeta(4)^{-1} \zeta(5)^{-2} \zeta(6)^{-2} \zeta(7)^{-4} \zeta(8)^{-5} \zeta(9)^{-8} \cdots$$
である．ただし，$\zeta(n)$ は Riemann のゼータ関数【1.6】で，$\mu(n)$ は Möbius のミュー関数【2.2】である．比較として，双子素数定数【2.1】の類似な表現式は
$$C_{\text{twin}} = \prod_{n \geq 2} \left[\left(1 - \frac{1}{2^n}\right)\zeta(n)\right]^{-1/n \sum_{k|n} 2^k \cdot \mu(n/k)}$$
$$= \left(\frac{3\zeta(2)}{4}\right)^{-1} \left(\frac{7\zeta(3)}{8}\right)^{-2} \left(\frac{15\zeta(4)}{16}\right)^{-3} \left(\frac{31\zeta(5)}{32}\right)^{-6} \left(\frac{63\zeta(6)}{64}\right)^{-9} \left(\frac{127\zeta(7)}{128}\right)^{-18} \cdots$$
である．

技術的な詳細は省略して，アルティン定数の k 次元への2つの拡張を簡単に調べる．まずはじめに，$S(n_1, n_2, \cdots, n_k)$ を，整数 $n_1, n_2, \cdots, n_k$ が同時に法 p の原始根となるすべての素数の集合とする．Matthews（マシューズ）[19,20]は，素数と互いに素な $S(n_1, n_2, \cdots, n_k)$ の密度に対応する C_{Artin} に類似なものを求めた[21]．

$$C_{\text{Matthews},k} = \prod_p \left(1 - \frac{p^k - (p-1)^k}{p^k(p-1)}\right) = \begin{cases} 0.1473494003\cdots & (k=2 \text{ のとき}) \\ 0.0608216553\cdots & (k=3 \text{ のとき}) \\ 0.0261074464\cdots & (k=4 \text{ のとき}) \end{cases}$$

これらは，有理数の補正因子を除いて正しい．第2に，集合 $\{n_1, n_2, \cdots, n_k\} \subseteq U_p$ により生成される巡回群 U_p の部分群を N とし，$\tilde{S}(n_1, n_2, \cdots n_k)$ を $N = U_p$ であるようなすべての素数の集合とする．Pappalardi（パパラルディ）[22,23]は，素数と互いに素な $\tilde{S}(n_1, n_2, \cdots, n_k)$ の密度に対応する C_{Artin} に類似なものを得た[17]．

$$C_{\text{Pappalardi},k} = \prod_p \left(1 - \frac{1}{p^k(p-1)}\right) = \begin{cases} 0.6975013584\cdots & (k=2 \text{ のとき}) \\ 0.8565404448\cdots & (k=3 \text{ のとき}) \\ 0.9312651841\cdots & (k=4 \text{ のとき}) \end{cases}$$

これもまた有理数の補正因子を除いて正しい．Niklasch & Moree（ニクラシュ&モレ）[17]は，論文で $C_{\text{Pappalardi},k}$ と多くの定数を計算した．

二次体では[24,25]，うまく拡張されたアルティン予想は，定数
$$\frac{8C_{\text{twin}}}{\pi^2} = \prod_{p>2}\left(1 - \frac{1}{p(p-1)}\right) = 0.5351070126\cdots$$
のみならず $C_{\text{Pappalardi},k}$ も含む．

任意の代数数体への拡張は，未解決であるように思われる．フィボナッチ原始根を含んだ C_{Artin} の好奇心をそそる変形は[26-28]を，同様に，擬素数と Carmichael（カーマ

イケル）数については[29]を参照せよ．

未解決問題を述べる．奇素数 p に対して，法 p の原始根で最小の正の整数を $g(p)$ と定義し，法 p の原始根で最小の正の素数を $G(p)$ と定義する．$g(p)$ と $G(p)$ の期待値はいくらであるのか？ Murata（村田玲音）[21,30]は発見的に，ほとんどすべての p に対して，$g(p)$ は

$$1+C_{\text{Murata}} = 1+\prod_p \left(1+\frac{1}{(p-1)^2}\right) = 3.8264199970\cdots$$

からあまり離れていないことを論じた．この見積もりは，低すぎることが明らかになった．実験データ[21,31,32]から，$E(g(p))=4.9264\cdots$，$E(G(p))=5.9087\cdots$ である，と示されている．

Matthews（マシュー）定数[21]を含む $E(g(p))$ に対する複雑な無限級数があるが，おそらくコンピュータでは実行不可能であろう．別のところで C_{Murata} が現れることについては，【2.7】を参照せよ．

2.4.1 関連する定数

数論のさまざまな分野からの関連した定数がある．$m=n=0$ を除いて $a^m b^n \neq 1$ という意味において，乗法的に独立な整数を a, b とする．ある非負整数 k に対して，$p|(a^k-b)$ であるようなすべての素数の集合を，$T(a,b)$ とする．一般化リーマン予想を仮定して，Stephens（ステファン）[33]は，素数に対する $T(a,b)$ の相対密度は，有理数の補正因子を除いて

$$\prod_p \left(1-\frac{p}{p^3-1}\right) = 0.5759599688\cdots$$

であることを証明した．Moree & Stevenhagen（モレ&スティーヴンハーゲン）[34]は Stephens の仕事を拡張し，補正因子の調整をした．さらに，彼らは条件をつけないで，$T(a,b)$ の密度が正であることを証明した．ステファン定数へ速く収束する式は[16,17]で与えられている．

Feller-Tornier（フェラー-トーニー）定数[35-37]

$$\frac{1}{2}+\frac{1}{2}\prod_p \left(1-\frac{2}{p^2}\right) = \frac{1}{2}+\frac{3}{\pi^2}\prod_p \left(1-\frac{1}{p^2-1}\right) = 0.6613170494\cdots$$

は，整数の因数分解において素数のベキが偶数であるような整数の密度である．「ベキ乗」というときは，1次より高い累乗を意味する．したがって，$2\cdot 3^2 \cdot 5^3$ は素数のベキ乗を2つもち，密度に寄与する．しかしながら，$3\cdot 7\cdot 19\cdot 31^2$ は素数のベキ乗が1つであるので，密度には寄与しない．

方程式 $x_0^3 = x_1 x_2 x_3$，と条件 $0 < x_j \leq X$ ($1\leq j \leq 3$) および $\gcd(x_1,x_2,x_3)=1$ を満たす整数値ベクトル (x_0,x_1,x_2,x_3) の集合を考える．$X\to\infty$ のとき，この集合の濃度 $N(X)$ の漸近値はどのようになるのか？ Heath-Brown & Moroz（ヒース・ブラウン&モロズ）[38]は

$$\lim_{X\to\infty}\frac{2880N(X)}{X\ln(X)^6}=\prod_p\left(1-\frac{1}{p^7}\right)\left(1+\frac{7}{p}+\frac{1}{p^2}\right)=0.0013176411\cdots$$

を証明した．任意の3次曲面に対するこのような数え上げの問題は，とても難しい．

与えられた正の整数 n に対して，n の最も大きな平方因子を $D_n^2=n/n'$ とする．D_n と n' が互いに素であるような n の集合を Σ と定義する．このとき，Σ は漸近密度[37]

$$\chi=\prod_p\left(1-\frac{1}{p^2(p+1)}\right)=0.8815138397\cdots$$

をもつ．おもしろいことに，定数 χ は次のところにも現れる．

d は虚二次体 $(d<0)$ の基本判別式であり，$h(d)$ をその類数とする．このとき，比 $2h(d)/\sqrt{-d}$ の平均値は χ に等しい[39,40]．この定数は実二次体 $(d>0)$ に対しても，役割がある．不定2進二次形式（indefinite binary quadratic forms）に関連して，c^2-4 の形の判別式 $0<d<D$ のまばらな部分集合（thin subset）の上を動く，$h(d)$ の平均値が，$D>1$ のとき漸近的に

$$\frac{5\pi^2}{48}\prod_p\left(1-\frac{1}{p^2}-\frac{1}{p^3}\right)\cdot\frac{\sqrt{D}}{\ln(D)}=0.7439711933\cdots\cdot\frac{\sqrt{D}}{\ln(D)}$$

であることを，Sarnak（サルナック）[41]は得た．$c^{2\nu}-4$, $\nu\geq 2$ の形の $0<d<D$ に対する類似の定数に関して，同様な公式をもつとは思えない．

Riemann のゼータ関数の（臨界直線上の）$2k$ 次モーメント

$$m_{2k}(T)=\frac{1}{T}\int_0^T\left|\zeta\left(\frac{1}{2}+it\right)\right|^{2k}dt$$

は，$m_2(T)\sim\ln(T)$, $m_4\sim 1/(2\pi^2)\ln(T)^4$, $(T\to\infty)$ を満たすことが知られている．$m_{2k}(T)\sim\gamma_k\ln(T)^{k^2}$ であると予想され，さらに[42-44]

$$\frac{9!}{42}\gamma_6=\prod_p\left(1-\frac{1}{p}\right)^4\left(1+\frac{4}{p}+\frac{1}{p^2}\right)$$

$$\frac{16!}{24024}\gamma_8=\prod_p\left(1-\frac{1}{p}\right)^9\left(1+\frac{9}{p}+\frac{9}{p^2}+\frac{1}{p^3}\right)$$

と予想されている．この解析は Dirichlet の L-関数へ拡張できる．これらのようなモーメントを求めることは，数論へ多くの恩恵がある．

2.4.2 補正因子

任意の $r>1$ に対して，$n\neq -1,0,1$ は整数の r 乗ではないと仮定し，n' を n の無平方な部分と仮定する．$n'\not\equiv 1\bmod 4$ であるとき，素数に対する集合 $S(n)$ の相対密度は，[8,10,14,45,46]で

$$\left(1-\mu(|n'|)\prod_{q|n'}\frac{1}{q^2-q-1}\right)\cdot C_{\text{Artin}}$$

である，と予想されている．ただし，積は素数 q にかぎる．たとえば，$n'=u$ が素数であるとき，この公式は簡単に

2.4 アルティン定数

$$\left(1+\frac{1}{u^2-u-1}\right) \cdot C_{\text{Artin}}$$

となる. $u \equiv 1 \bmod 4$, $v \equiv 1 \bmod 4$ がともに素数であるとき, $n'=uv$ とすると, 公式は

$$\left(1-\frac{1}{u^2-u-1}-\frac{1}{v^2-v-1}\right) \cdot C_{\text{Artin}}$$

となる. n が整数の r 乗であるときには, もう少し複雑な公式になる.

[1] D. H. Lehmer, A note on primitive roots, Scripta Math. 26 (1961) 117-119; MR 26 #7128.
[2] D. Shanks, *Solved and Unsolved Problems in Number Theory*, 2nd ed., Chelsea, 1978, pp. 80-83, 222-225; MR 86j:11001.
[3] J. H. Conway and R. K. Guy, *The Book of Numbers*, Springer-Verlag, 1996, pp. 157-163, 166-171; MR 98g:00004.
[4] N. J. A. Sloane, On-Line Encyclopedia of Integer Sequences, A001122, A001913, A006883, A019334-A019421.
[5] E. Artin, *Collected Papers*, ed. S. Lang and J. T. Tate, Springer-Verlag, 1965, pp. viii.ix.
[6] M. Ram Murty, Artin's conjecture for primitive roots, *Math. Intellig.* 10 (1988) 59-67; MR 89k:11085.
[7] K. Ireland and M. Rosen, *A Classical Introduction to Modern Number Theory*, 2nd ed., Springer-Verlag, 1990, pp. v, 39-47; MR 92e:11001.
[8] P. Ribenboim, *The New Book of Prime Number Records*, Springer-Verlag, 1996, pp. 22-25, 379-386; MR 96k:11112.
[9] R. K. Guy, *Unsolved Problems in Number Theory*, 2nd ed., Springer-Verlag, 1994, sect. F9; MR 96e:11002.
[10] C. Hooley, On Artin's conjecture, *J. Reine Angew. Math.* 225 (1967) 209-220; MR 34 #7445.
[11] C. Hooley, *Applications of Sieve Methods to the Theory of Numbers*, Cambridge Univ. Press, 1976; MR 53 #7976.
[12] J.W. Wrench, Evaluation of Artin's constant and the twin prime constant, *Math. Comp.* 15 (1961) 396-398; MR 23 #A1619.
[13] P. Flajolet and I. Vardi, Zeta function expansions of classical constants, unpublished note (1996).
[14] E. Bach, The complexity of number-theoretic constants, *Inform. Process. Lett.* 62 (1997) 145-152; MR 98g:11148.
[15] H. Cohen, High precision computation of Hardy-Littlewood constants, unpublished note (1999).
[16] P. Moree, Approximation of singular series and automata, *Manuscripta Math.* 101 (2000) 385-399; MR 2001f:11204.
[17] G. Niklasch and P. Moree, Some number-theoretical constants: Products of rational functions over primes, unpublished note (2000).
[18] X. Gourdon and P. Sebah, Some constants from number theory (Numbers, Constants and Computation).
[19] K. R. Matthews, A generalisation of Artin's conjecture for primitive roots, *Acta Arith.* 29 (1976) 113-146; MR 53 #313.
[20] R. N. Buttsworth, An inclusion-exclusion transform, *Ars Combin.* 15 (1983) 279-300; MR 85k:05007.
[21] P. D. T. A. Elliott and L. Murata, On the average of the least primitive root modulo p, *J. London Math. Soc.* 56 (1997) 435-454 (computations by A. Paszkiewicz);MR98m:11094.
[22] F. Pappalardi, On the r-rank Artin conjecture, *Math. Comp.* 66 (1997) 853-868; MR 97f:11082.
[23] L. Cangelmi and F. Pappalardi, On the r-rank Artin conjecture. II, *J. Number Theory* 75 (1999) 120-132; MR 2000i:11149.

[24] H. Roskam, A quadratic analogue of Artin's conjecture on primitive roots, *J. Number Theory* 81 (2000) 93-109; errata 85 (2000) 108; MR 2000k:11128.
[25] H. Roskam, Artin's primitive root conjecture for quadratic fields, *J. Théorie Nombres Bordeaux*, to appear; report MI-2000-22, Univ. of Leiden.
[26] D. Shanks, Fibonacci primitive roots, *Fibonacci Quart.* 10 (1972) 163-168, 181; MR 45 #6747.
[27] H. W. Lenstra, Jr., On Artin's conjecture and Euclid's algorithm in global fields, *Invent. Math.* 42 (1977) 201-224; MR 58 #576.
[28] J. W. Sander, On Fibonacci primitive roots, *Fibonacci Quart.* 28 (1990) 79-80; MR 91b:11094.
[29] S. S. Wagstaff, Pseudoprimes and a generalization of Artin's conjecture, *Acta Arith.* 41 (1982) 141-150; MR 83m:10004.
[30] L. Murata, On the magnitude of the least prime primitive root, *J. Number Theory* 37 (1991) 47-66; MR 91j:11082.
[31] T. Oliveira e Silva, Least primitive root of prime numbers, unpublished note (2001).
[32] A. Paszkiewicz and A. Schinzel, On the least prime primitive root modulo a prime, *Math. Comp.* 71 (2002) 1307-1321.
[33] P. J. Stephens, Prime divisors of second-order linear recurrences. I, *J. Number Theory* 8 (1976) 313-332; MR 54 #5142.
[34] P. Moree and P. Stevenhagen, A two-variable Artin conjecture, *J. Number Theory* 85 (2000) 291-304; MR 2001k:11188.
[35] W. Feller and E. Tornier, Mengentheoretische Untersuchungen von Eigenschaften der Zahlenreihe, *Math. Annalen* 107 (1933) 188-232.
[36] I. J. Schoenberg, On asymptotic distributions of arithmetical functions, *Trans. Amer. Math. Soc.* 39 (1936) 315-330.
[37] E. Cohen, Some asymptotic formulas in the theory of numbers, *Trans. Amer. Math. Soc.* 112 (1964) 214-227; MR 29 #3458.
[38] D. R. Heath-Brown and B. Z. Moroz, The density of rational points on the cubic surface $X_0^3 = X_1 X_2 X_3$, *Math. Proc. Cambridge Philos. Soc.* 125 (1999) 385-395; MR 2000f:11080.
[39] R. Ayoub, *An Introduction to the Analytic Theory of Numbers*, Amer. Math. Soc., 1963, pp. 320-322; MR 28 #3954.
[40] H. Cohen, *A Course in Computational Algebraic Number Theory*, Springer-Verlag, 1993, pp. 290-293; MR 94i:11105.
[41] P. C. Sarnak, Class numbers of indefinite binary quadratic forms. II, *J. Number Theory* 21 (1985) 333-346; MR 87h:11027.
[42] J. B. Conrey and A. Ghosh, A conjecture for the sixth power moment of the Riemann zeta-function, *Int. Math. Res. Notices* (1998) 775-780; MR 99h:11096.
[43] J. B. Conrey and S. M. Gonek, High moments of the Riemann zeta-function, *Duke Math. J.* 107 (2001) 577-604; MR 2002b:11112.
[44] J. B. Conrey, L-functions and random matrices, *Mathematics Unlimited-2001 and Beyond*, ed. B. Engquist and W. Schmid, Springer-Verlag, 2001, pp. 331-352; MR 2002g: 11134.
[45] D. H. Lehmer and E. Lehmer, Heuristics, anyone?, *Studies in Mathematical Analysis and Related Topics*, ed. G. Szegö, C. Loewner, S. Bergman, M. M. Schiffer, J. Neyman, D. Gilbarg, and H. Solomon, Stanford Univ. Press, 1962, pp. 202-210; MR 26 #2409.
[46] A. E. Western and J. C. P. Miller, *Tables of Indices and Primitive Roots*, Royal Soc. Math. Tables, v. 9, Cambridge Univ. Press, 1968, pp. xxxvii.xlii; MR 39 #7792.

2.5 ハフナー-サルナック-マッカレー定数

よく知られた定理[1]からはじめる．ランダムに選んだ2つの整数が互いに素となる確率は，（長い区間上で極限をとることにより）$6/\pi^2 = 0.6079271018\cdots$である．整数を，成分が整数である正方行列に置きかえると，何が起こるだろうか？ ランダムに選んだ成分が整数の2つの正方行列に対して，対応する2つの行列の行列式が互いに素である確率$\Delta(n)$は，いくらであるのか？

Hafner, Sarnak & McCurley（ハフナー，サルナック&マッカレー）[2]は，各nに対して

$$\Delta(n) = \prod_p \left[1 - \left(1 - \prod_{k=1}^n (1 - p^{-k})\right)^2\right]$$

であることを示した．ただし，外側の積は素数pにかぎる．さらに，

$$\Delta(1) = \frac{6}{\pi^2} > \Delta(2) > \Delta(3) > \cdots > \Delta(n-1) > \Delta(n) > \cdots$$

であることが証明されている．Vardi（ヴァルディ）[3,4]は極限値

$$\lim_{n\to\infty} \Delta(n) = \prod_p \left[1 - \left(1 - \prod_{k=1}^\infty (1 - p^{-k})\right)^2\right] = 0.3532363719\cdots$$

を計算した．

2.5.1 無関心な対

ランダムに選んだ整数xが無平方である[1]，すなわちxが1より大きいどのような平方数でも割り切れない，確率も$6/\pi^2$であることが知られている．Schroeder（シュレーダー）[5]は次のような問題を出した．無平方な事象と2数が互いに素である事象は，統計的に互いに独立か？ 答えは，否定的である．2つの性質には正の相関関係がある．さらに詳しくいおう．ランダムに選ばれた2つの整数xとyとが**無関心**（carefree）[5,6]とは，x,yが互いに素であり，xが無平方である，と定義する．このとき，xとyが**無関心**である確率は，$36/\pi^4 = 0.3695\cdots$よりもいくぶん大きい．正確には

$$P = \frac{6}{\pi^2} \prod_p \left(1 - \frac{1}{p(p+1)}\right) = 0.4282495056\cdots$$

である．Moore[7]はSchroederの公式が正しいことを証明した．さらに，彼は，xとyが**強無関心**（strongly carefree）であるとは，x,yがともに無平方であり，かつx,yが互いに素であるとき，と定義した．この場合の確率は[8]

$$Q = \frac{6}{\pi^2} \prod_p \left(1 - \frac{2}{p(p+1)}\right) = \frac{36}{\pi^4} \prod_p \left(1 - \frac{1}{(p+1)^2}\right) = 0.2867474284\cdots$$

である．x,yが**弱無関心**（weakly carefree）とは，x,yが互いに素であり，xまたは

y が無平方であるとき,と定義する.系として,この確率は,$P(A \cup B) = P(A) + P(B) - P(A \cap B)$ を用いて,$2P - Q = 0.5697515829\cdots$ を得る.これらの結合確率と類似なものが,行列に存在するだろうか?

定数 P と Q は,数論[7]の別なところにも現れる.$D_n = \max\{d : d^2 | n\}$ とする.

$$\kappa(n) = \frac{n}{D_n^2}, \quad n \text{ の無平方の部分}$$

$$K(n) = \prod_{p|n} p, \quad n \text{ の無平方の核 (kernel)}$$

と定義する.このとき,[9-11]

$$\lim_{N \to \infty} \frac{1}{N^2} \sum_{n=1}^{N} \kappa(n) = \frac{\pi^2}{30} = 0.3289\cdots, \quad \lim_{N \to \infty} \frac{1}{N^2} \sum_{n=1}^{N} K(n) = \frac{\pi^2 P}{12} = 0.3522\cdots$$

である.(D_n の平均については【2.10】を参照せよ.) $\omega(n)$ を,【2.2】にあるように n の異なる素因数の数とする.このとき,[11-13]

$$\lim_{N \to \infty} \frac{1}{N \ln(N)} \sum_{n=1}^{N} 2^{\omega(n)} = \frac{6}{\pi^2} = 0.6079\cdots$$

$$\lim_{N \to \infty} \frac{1}{N \ln(N)^2} \sum_{n=1}^{N} 3^{\omega(n)} = \frac{Q}{2} = 0.1433\cdots$$

である.$\omega(n)$ を,n の素因数の全部の個数である $\Omega(n)$ に置きかえたとき,[11,14,15]

$$\lim_{N \to \infty} \frac{1}{N \ln(N)^2} \sum_{n=1}^{N} 2^{\Omega(n)} = \frac{1}{8 \ln(2) C_{\text{twin}}} = 0.2731707223\cdots$$

である.ただし,C_{twin} は双子素数定数【2.1】であり,これは P と Q には関係がないように思われる.

一般化で締めくくる.【1.6】で示されているように,ランダムに k 個を選んだ整数が全体として互いに素である確率は $1/\zeta(k)$ である.それらが2つずつ互いに素である確率は[5,7],$2 \leq k \leq 3$ に対して,

$$\prod_p \left(1 - \frac{1}{p}\right)^{k-1} \left(1 + \frac{k-1}{p}\right)$$

であることが知られている.しかし,$k > 3$ の証明は,いまだに知られていない.$k = 2$ のときは,この式は自然に $6/\pi^2$ になる.もっと驚くことに,$k = 3$ のときは Q に等しい.

[1] G. H. Hardy and E. M. Wright, *An Introduction to the Theory of Numbers*, 5th ed., Oxford Univ. Press, 1985, p. 269; MR 81i:10002.
[2] J. L. Hafner, P. Sarnak, and K. McCurley, Relatively prime values of polynomials, *A Tribute to Emil Grosswald: Number Theory and Related Analysis*, ed. M. Knopp and M. Sheingorn, Contemp. Math. 143, Amer. Math. Soc., 1993, pp. 437-443; MR 93m:11094.
[3] I. Vardi, *Computational Recreations in Mathematica*, Addison-Wesley, 1991, p. 174; MR 93e:00002.
[4] P. Flajolet and I. Vardi, Zeta function expansions of classical constants, unpublished note (1996).

[5] M. R. Schroeder, *Number Theory in Science and Communication: With Applications in Cryptography, Physics, Digital Information, Computing and Self-Similarity*, 2nd ed., Springer-Verlag, 1986, pp. 25, 48-51, 54; MR 99c:11165.

[6] M. R. Schroeder, Square-free and coprime, unpublished note (1998).

[7] P. Moree, Counting carefree couples, unpublished note (1999).

[8] G. Niklasch and P. Moree, Some number-theoretical constants: Products of rational functions over primes, unpublished note (2000).

[9] E. Cohen, Arithmetical functions associated with the unitary divisors of an integer, *Math. Z.* 74 (1960) 66-80; MR 22 #3707.

[10] E. Cohen, Some asymptotic formulas in the theory of numbers, *Trans. Amer. Math. Soc.* 112 (1964) 214-227; MR 29 #3458.

[11] G. Tenenbaum, *Introduction to Analytic and Probabilistic Number Theory*, Cambridge Univ. Press, 1995, pp. 53-54; MR 97e:11005b.

[12] A. Selberg, Note on a paper by L. G. Sathe, *J. Indian Math. Soc.* 18 (1954) 83-87; MR 16,676a.

[13] H. Delange, Sur des formules de Atle Selberg, *Acta Arith.* 19 (1971) 105-146 (errata insert); MR 44 #6623.

[14] E. Grosswald, The average order of an arithmetic function, *Duke Math. J.* 23 (1956) 41-44; MR 17,588f.

[15] P. Bateman, Proof of a conjecture of Grosswald, *Duke Math. J.* 25 (1957) 67-72; MR 19,1040a.

2.6 ニーヴェン定数

素因数分解 $p_1^{a_1}p_2^{a_2}p_3^{a_3}\cdots p_k^{a_k}$ をもつ正の整数を m とする．指数 $a_i \geq 1$ で，かつ，$p_i \neq p_j (i \neq j)$ を仮定する．2つの関数を

$$h(m)=\begin{cases} 1 & (m=1) \\ \min\{a_1,\cdots,a_k\} & (m>1) \end{cases}, \quad H(m)=\begin{cases} 1 & (m=1) \\ \max\{a_1,\cdots,a_k\} & (m>1) \end{cases}$$

と定義する．すなわち，m の最小と最大の指数である．Niven（ニーヴェン）[1,2]は

$$\lim_{n\to\infty}\frac{1}{n}\sum_{m=1}^{n}h(m)=1$$

であること，さらに

$$\lim_{n\to\infty}\frac{\left(\sum_{m=1}^{n}h(m)\right)-n}{\sqrt{n}}=\frac{\zeta\left(\frac{3}{2}\right)}{\zeta(3)}=2.1732543125\cdots$$

を証明した．ただし，$\zeta(x)$ は Riemann のゼータ関数【1.6】である．彼はまた

$$\lim_{n\to\infty}\frac{1}{n}\sum_{m=1}^{n}H(m)=C$$

であることを証明した．定数 C を Niven（**ニーヴェン**）**定数**とよび，値は

$$C=1+\sum_{n=2}^{\infty}\left(1-\frac{1}{\zeta(k)}\right)=1.7052111401\cdots$$

である．彼に続く研究者は，拡張した次の結果[3,4]を得た．

$$\sum_{m=1}^{n} h(m) = n + c_{02}n^{1/2} + (c_{12}+c_{03})n^{1/3} + (c_{13}+c_{04})n^{1/4} + (c_{23}+c_{14}+c_{05})n^{1/5} + O(n^{1/6})$$

$$\sum_{m=1}^{n} \frac{1}{h(m)} = n - \frac{c_{02}}{2}n^{1/2} - \frac{3c_{12}+c_{03}}{6}n^{1/3} - \frac{2c_{13}+c_{04}}{12}n^{1/4}$$
$$- \frac{10c_{23}+5c_{14}+3c_{05}}{60}n^{1/5} + O(n^{1/6})$$

ただし，係数 c_{ij} は【2.6.1】で与えられる．さらに，

$$\lim_{n\to\infty} \frac{1}{n}\sum_{m=1}^{n} \frac{1}{H(m)} = \sum_{k=2}^{\infty} \frac{1}{k(k-1)\zeta(k)} = 0.7669444905\cdots$$

である．H の平均の，漸近的な式は，h の平均ほどにはわかっていない．

定数 $c_{02}=\zeta(3/2)/\zeta(3)$ はまた，$c_{12}=\zeta(2/3)/\zeta(2)=-1.4879506635\cdots$，が示すように，充平方（square-full）整数【2.6.1】の個数の漸近的な増大を評価する場合に現れる．それに対して，定数 $6/\pi^2$ は無平方な整数【2.5】に関係して現れる．

ノルムつき自由アーベル半群（free aberian normed semigroup）への Niven の定理の一般化は，[5]にある．

C と同じような表現を与える問題がある．最初に，[6,7]

$$\sum_{l=2}^{\infty}\sum_{n=2}^{\infty} \frac{1}{n^l} = \sum_{l=2}^{\infty}(\zeta(l)-1) = 1, \quad \sum_{p}\sum_{n=2}^{\infty}\frac{1}{n^p} = \sum_{p}(\zeta(p)-1) = 0.8928945714\cdots$$

である．和は，素数 p すべてにわたる和である．2つの級数は，重複（たとえば，$2^4=4^2$，$4^3=8^2$）がある自明でない整数ベキの逆数を含む．$S=\{4, 8, 9, 16, 25, 27, 32, 36, 49, 64, 81, \cdots\}$ を，重複がない整数の累乗の集合とする．[8]から，

$$\sum_{s\in S} \frac{1}{s} = -\sum_{k=2}^{\infty} \mu(k)(\zeta(k)-1) = 0.8744643684\cdots$$

である．$\mu(k)$ は Möbius（メビウス）関数である【2.2】．また，[8,9]から

$$\sum_{s\in S} \frac{1}{s-1} = 1, \quad \sum_{s\in S} \frac{1}{s+1} = \frac{\pi^2}{3} - \frac{5}{2}$$

である．任意の整数 $c\notin S$ に対して，$\sum_{s\in S}(s-c)^{-1}$ についてどんなことがいえるだろうか？（Mihailescu（ミハイルスク）によるカタラン予想の最近の証明から，S の中で1だけ異なる整数の組は，8と9のみである．）他の表現については【5.1】を参照せよ．

2.6.1 充平方と充立方整数

$k\geq 2$ を整数とする．正の整数 m が k-**充**（または**型 k の充ベキ**（powerfull）とは，$m=1$ または任意の素数 p に対して，$p|m$ ならば $p^k|m$ であるときをいう[†]．

$N_k(x)$ を，x を超えない k-充整数の個数とする．$k=2$ の場合，Erdös & Szekeres（エルデーシュ & セケレス）[10]は

[†]訳注：特に，2-充を充平方（square-full），3-充を充立方（cube-full）という．

2.6 ニーヴェン定数

$$N_2(x) = \frac{\zeta\left(\frac{3}{2}\right)}{\zeta(3)} x^{1/2} + O(x^{1/3})$$

を証明し，Bateman & Grosswald（ベイトマン&グロスワルド）[11-13]はより正確な結果

$$N_2(x) = \frac{\zeta\left(\frac{3}{2}\right)}{\zeta(3)} x^{1/2} + \frac{\zeta\left(\frac{2}{3}\right)}{\zeta(2)} x^{1/3} + O(x^{1/6})$$

を証明した．これは，未解決問題であるリーマン予想に関する追加的な知識なしで，本質的に可能なかぎり正確な誤差評価である．多くの研究者はこの問題を研究している．誤差で，一番よい評価として知られている最近のものは[14,15]であり，リーマン予想を仮定して，任意の $\varepsilon>0$ に対して $O(x^{1/7+\varepsilon})$ である．幾人かの研究者が，1/7 が 1/10 にできると予想している．

$k=3$ の場合，Bateman & Grosswald[12]と Krätzel（クレーツェル）[16,17]は条件を付けないで，

$$N_3(x) = c_{03} x^{1/3} + c_{13} x^{1/4} + c_{23} x^{1/5} + o(x^{1/8})$$

を示した．

リーマン予想を仮定して，誤差項[15]は $O(x^{97/804+\varepsilon})$ に改良できる．係数 c_{ij} に対する公式は，[3,12,18-20]にある．

$$c_{0j} = \prod_p \left(1 + \sum_{m=j+1}^{2j-1} p^{-m/j}\right) = \begin{cases} 4.6592661225\cdots & (j=3) \\ 9.6694754843\cdots & (j=4) \\ 19.4455760839\cdots & (j=5) \end{cases}$$

$$c_{1j} = \zeta\left(\frac{j}{j+1}\right) \prod_p \left(1 + \sum_{m=j+2}^{2j-1} p^{-m/(j+1)} - \sum_{m=2j+2}^{3j} p^{-m/(j+1)}\right)$$

$$= \begin{cases} -5.8726188208\cdots & (j=3) \\ -16.9787814834\cdots & (j=4) \end{cases}$$

$$c_{23} = \zeta\left(\frac{3}{5}\right) \zeta\left(\frac{4}{5}\right) \prod_p (1 - p^{-8/5} - p^{-9/5} - p^{-10/5} - p^{-13/5} - p^{-14/5})$$

$$= 1.6824415102\cdots$$

ただし，すべての積は，素数 p にかぎってとる．ここでの Bateman-Grosswald（**ベイトマン-グロスワルド**）**定数**の近似値は，Niklasch & Moree（ニクラシュ&モレ）[21]と Sebah（セバ）[22]による．高い次数の係数は，$N_k(x)$ ($k\geq 4$) の展開式に現れる．

Erdös & Szekeres の論文[10]はまた，アーベル群の数え上げ【5.1】の漸近式に決定的な役割をするものと認められている．Ivić（イビッチ）[23]と Krätzel[24]の本は，詳しい解析と背景を説明したものである．m の最小および最大の素因子の議論については，【5.4】も参照せよ．

[1] I. Niven, Averages of exponents in factoring integers, *Proc. Amer. Math. Soc.* 22 (1969) 356-360; MR 39 #2713.

[2] F. Le Lionnais, *Les Nombres Remarquables*, Hermann, 1983.
[3] D. Suryanarayana and R. Sitaramachandrarao, On the maximum and minimum exponents in factoring integers, *Arch. Math. (Basel)* 28 (1977) 261-269; MR 55 #10368.
[4] H. Z. Cao, The asymptotic formulas related to exponents in factoring integers, *Math. Balkanica* 5 (1991) 105-108; MR 93e:11107.
[5] S. Porubský, On exponents in arithmetical semigroups, *Monatsh. Math.* 84 (1977) 49-53; MR 56 #15591.
[6] G. Pólya and G. Szegö, *Problems and Theorems in Analysis*, v. 2, Springer-Verlag, 1976, ex. 262-1; MR 81e:00002.
[7] G. Salamin and D. Wilson, Sum of reciprocal powers, unpublished note (1999).
[8] V. F. Lev and P. Pleasants, Sum of reciprocal powers, unpublished note (2002).
[9] J. D. Shallit and K. Zikan, A theorem of Goldbach, *Amer. Math. Monthly* 93 (1986) 402-403.
[10] P. Erdös and G. Szekeres, Über die Anzahl der Abelschen Gruppen gegebener Ordnung und über ein verwandtes zahlentheoretisches Problem, *Acta Sci. Math. (Szeged)* 7 (1934-35) 95-102.
[11] P. T. Bateman, Squarefull integers, *Amer. Math. Monthly* 61 (1954) 477-479.
[12] P. T. Bateman and E. Grosswald, On a theorem of Erdös and Szekeres, *Illinois J. Math.* 2 (1958) 88-98; MR 20 #2305.
[13] E. Cohen, On the distribution of certain sequences of integers, *Amer. Math. Monthly* 70 (1963) 516-521; MR 26 #6105.
[14] X. Cao, On the distribution of square-full integers, *Period. Math. Hungar.* 34 (1998) 169-175; MR 99a:11104.
[15] J.Wu, On the distribution of square-full and cube-full integers, *Monatsh. Math.* 126 (1998) 353-367; MR 2000a:11125.
[16] E. Krätzel, Zahlen k^{ter} art, *Amer. Math. J.* 94 (1972) 209-328; MR 45 #5093.
[17] E. Krätzel, Zweifache Exponentialsummen und dreidimensionale Gitterpunktprobleme, *Elementary and Analytic Theory of Numbers*, Proc. 1982 Warsaw conf., ed. H. Iwaniec, Banach Center Publ., 1985, pp. 337-369; MR 87m:11091.
[18] A. Ivić and P. Shiu, The distribution of powerful numbers, *Illinois J. Math.* 26 (1982) 576-690; MR 84a:10047.
[19] P. Shiu, The distribution of cube-full numbers, *Glasgow Math. J.* 33 (1991) 287-295; MR 92g:11091.
[20] P. Shiu, Cube-full numbers in short intervals, *Math. Proc. Cambridge Philos. Soc.* 112 (1992) 1-5; MR 93d:11097.
[21] G. Niklasch and P. Moree, Some number-theoretical constants: Products of rational functions over primes, unpublished note (2000).
[22] P. Sebah, Evaluating certain prime products, unpublished note (2001).
[23] A. Ivić, *The Riemann Zeta-Function*, Wiley, 1985, pp. 33-34, 407-413, 438-439; MR 87d:11062.
[24] E. Krätzel, *Lattice Points*, Kluwer, 1988, pp. 276-293; MR 90e:11144.

2.7 オイラーの約数関数

n を正の整数とする．オイラーのファイ関数（約数関数）$\varphi(n)$ は，n と互いに素である n を超えない正の整数の個数と定義される．たとえば，p と q が相異なる素数で，r, s を正の整数とする．このとき，

2.7 オイラーの約数関数

$$\varphi(p^r) = p^{r-1}(p-1)$$
$$\varphi(p^r q^s) = p^{r-1} q^{s-1}(p-1)(q-1)$$

である．群論の言葉では，$\varphi(n)$ は，次数 n の巡回群の生成元の個数である．Landau [1-4] は

$$\limsup_{n\to\infty} \frac{\varphi(n)}{n} = 1$$

であるが，

$$\liminf_{n\to\infty} \frac{\varphi(n)\ln(\ln(n))}{n} = e^{-\gamma} = 0.5614594835\cdots$$

であることを証明した．γ はオイラー-マスケロニの定数【1.5】である．

すべての正の整数での $\varphi(n)$ の平均的なふるまいには，多くの研究者が興味をもっている．Walfisz（ワルフィス）[5,6] は，Direchlet & Mertens[2] の仕事を基礎にして，$N\to\infty$ のとき，

$$\sum_{n=1}^{N} \varphi(n) = \frac{3N^2}{\pi^2} + O(N\ln(N)^{2/3}\ln(\ln(N))^{4/3})$$

を証明した．このような既知の漸近公式の中では，もっとも正確なものである．（指数 4/3 は任意の $\varepsilon>0$ に対して，$1+\varepsilon$ に置きかえることができるという[7]の主張は，間違いである[8]．さらに，誤差項は $o(N\ln(\ln(\ln(N))))$ ではない，と知られている [9,10]．

$\varphi(n)$ の逆数の級数を考えたとき，興味ある定数が現れる．Landau[11-13] は，

$$\sum_{n=1}^{N} \frac{1}{\varphi(n)} = A\cdot(\ln(N) + B) + O\left(\frac{\ln(N)}{N}\right)$$

を証明した．ただし，

$$A = \frac{\zeta(2)\zeta(3)}{\zeta(6)} = \frac{315}{2\pi^4}\zeta(3) = 1.9435964368\cdots$$

$$B = \gamma - \sum_{p} \frac{\ln(p)}{p^2 - p + 1} = \gamma - 0.6083817178\cdots = \frac{-0.0605742294\cdots}{A}$$

であり，$\zeta(x)$ は Riemann のゼータ関数【1.6】である．p に関する和と積は，素数 p にかぎる．B にある和は，Jameson（ジェイムソン）[14]，Moree[15]，Sebah[16] により，正確に計算された．Landau の誤差項 $O(\ln(N)/N)$ は，Sitaramachandrarao（シタラマチャンドララオ）[17,18] により $O(\ln(N)^{2/3}/N)$ に改良された．

$\varphi(n)\leq x$ を満たすすべての正の整数 n の個数を，$K(x)$ と定義する．次の分布関数的な結果が真であることが知られている[19-22]．任意の $0<c<1/\sqrt{2}$ に対して，

$$K(x) = Ax + O(x\exp(-c\sqrt{\ln(x)\ln(\ln(x))}))$$

が成り立つ．他の関連する式は[18,23,24]，

$$\sum_{n=1}^{N} \frac{\varphi(n)}{n} = \frac{6N}{\pi^2} + O(\ln(N)^{2/3}\ln(\ln(N))^{4/3}),$$

$$\sum_{n=1}^{N}\frac{n}{\varphi(n)}=AN-\frac{1}{2}\ln(N)-\frac{1}{2}C+O(\ln(N)^{2/3}),$$

$$\sum_{n=1}^{N}\frac{1}{n\varphi(n)}=D-\frac{A}{N}+O\Big(\frac{\ln(N)}{N^2}\Big)$$

である．ただし，

$$C=\ln(2\pi)+\gamma+\sum_{p}\frac{\ln(p)}{p(p-1)}=\ln(2\pi)+1.3325822757\cdots=3.1704593421\cdots$$

である．これは【2.2】で現れた．また

$$D=\frac{\pi^2}{6}\prod_{p}\Big(1+\frac{1}{p^2(p-1)}\Big)=2.2038565964\cdots$$

は，[25]にある評価の Moree[24]による精密化による．このような素数の積の数値の評価は[26]を参照せよ．定数 A は，ある種の素数約数関数の漸近的な平均として[27,28]に現れ，その他[29]にも現れる．定数 D はさらに，Chowla（チョウラ）[30]により証明された，ある種のハーディ-リトルウッド予想にも現れる．

A の別の表現を次にあげる．

$$A=\prod_{p}\frac{1-p^{-6}}{(1-p^{-2})(1-p^{-3})}=\prod_{p}\Big(1+\frac{1}{p(p-1)}\Big)$$

であり，アルティン定数【2.4】ときわだった類似がある．ただ1つの違いは，加法を減法に置きかえていることである．不思議なことに，アルティン定数と村田定数【2.4】は，次の漸近結果[31,32]にはっきり現れる．

$$\lim_{N\to\infty}\frac{\ln(n)}{N}\sum_{p\leq N}\frac{\varphi(p-1)}{p-1}=C_{\text{Artin}}=0.3739558136\cdots$$

$$\lim_{N\to\infty}\frac{\ln(n)}{N}\sum_{p\leq N}\frac{p-1}{\varphi(p-1)}=C_{\text{Murata}}=2.8264199970\cdots$$

n と $\varphi(n)$ が互いに素であるような，x を超えないすべての整数 n の個数を，$L(x)$ とする．Erdös[33,34]は

$$\lim_{n\to\infty}\frac{L(n)\ln(\ln(\ln(n)))}{n}=e^{-\gamma}$$

を証明した．オイラー-マスケローニの定数 γ のおもしろい別の現れである．

[1] E. Landau, Über den Verlauf der zahlentheoretischen Funktion $\varphi(x)$, *Archiv Math. Phys.* 5 (1903) 86-91; also in *Collected Works*, v. 1, ed. L. Mirsky, I. J. Schoenberg,W. Schwarz, and H. Wefelscheid, Thales Verlag, 1983, pp. 378-383.
[2] G. H. Hardy and E. M. Wright, *An Introduction to the Theory of Numbers*, 5th ed., Oxford Univ Press, 1985, pp. 267-268, 272; MR 81i:10002.
[3] N. J. A. Sloane, On-Line Encyclopedia of Integer Sequences, A000010 and A002088.
[4] M. Hausman, Generalization of a theorem of Landau, *Pacific J. Math.* 84 (1979) 91-95; MR 81d:10005.
[5] A.Walfisz, Über die Wirksamkeit einiger Abschätzungen trigonometrischer Summen, *Acta Arith.* 4 (1958) 108-180; MR 21 # 2623.
[6] A. Walfisz, *Weylsche Exponentialsummen in der neueren Zahlentheorie*, Mathematische For-

schungsberichte, XV, VEB Deutscher Verlag der Wissenschaften, 1963; MR 36 #3737.
[7] A. I. Saltykov, On Euler's function (in Russian), *Vestnik Moskov. Univ. Ser. I Mat. Mekh.* (1960) n. 6, 34-50; MR 23 #A2395.
[8] U. Balakrishnan and Y.-F. S. Pétermann, Errata to: "The Dirichlet series of $\zeta(s)\zeta^a(s+1)$ $f(s+1)$: On an error term associated with its coefficients," *Acta Arith.* 87 (1999) 287-289; MR 99m:11105.
[9] S. S. Pillai and S. D. Chowla, On the error terms in some asymptotic formulae in the theory of numbers, *J. London Math. Soc.* 5 (1930) 95-101.
[10] S. Chowla, Contributions to the analytic theory of numbers, *Math. Z.* 35 (1932) 279-299.
[11] E. Landau, Über die zahlentheoretische Function $\varphi(n)$ und ihre Beziehung zum Goldbachschen Satz, *Nachr. Königlichen Ges. Wiss. Göttingen, Math.-Phys. Klasse* (1900) 177-186; also in *Collected Works*, v. 1, ed. L. Mirsky, I. J. Schoenberg, W. Schwarz, and H. Wefelscheid, Thales Verlag, 1983, pp. 106-115.
[12] J.-M. DeKoninck and A. Ivić, *Topics in Arithmetical Functions: Asymptotic Formulae for Sums of Reciprocals of Arithmetical Functions and Related Fields*, North-Holland, 1980, pp. 1-3; MR 82a:10047.
[13] H. Halberstam and H.-E. Richert, *Sieve Methods*, Academic Press, 1974, pp. 110-111; MR 54 #12689.
[14] T. Jameson, Asymptotics of $\sum_{n<x} 1/\varphi(n)$, unpublished note (1999).
[15] P. Moree, Expressing B as $\gamma - \sum_{k\geq 2} e_k \zeta'(k)/\zeta(k)$, unpublished note (2000).
[16] X. Gourdon and P. Sebah, Some constants from number theory (Numbers, Constants and Computation).
[17] R. Sitaramachandrarao, On an error term of Landau, *Indian J. Pure Appl. Math.* 13 (1982) 882-885; MR 84b:10069.
[18] R. Sitaramachandrarao, On an error term of Landau. II, *Rocky Mount. J. Math.* 15 (1985) 579-588; MR 87g:11116.
[19] P. Erdös, Some remarks on Euler's φ-function and some related problems, *Bull. Amer. Math. Soc.* 51 (1945) 540-544; MR 7,49f.
[20] R. E. Dressler, A density which counts multiplicity, *Pacific J. Math.* 34 (1970) 371-378; MR 42 #5940.
[21] P. T. Bateman, The distribution of values of the Euler function, *Acta Arith.* 21 (1972) 329-345; MR 46 #1730.
[22] M. Balazard and A. Smati, Elementary proof of a theorem of Bateman, *Analytic Number Theory*, Proc. 1989 Allerton Park conf., ed. B. C. Berndt, H. G. Diamond, H. Halberstam, and A. Hildebrand, Birkhäuser, 1990, pp. 41-46; MR 92a:11104.
[23] W. G. Nowak, On an error term involving the totient function, *Indian J. Pure Appl. Math.* 20 (1989) 537-542; MR 90g:11135.
[24] P. Moree, On the error terms in Stephens' work, unpublished note (2000).
[25] P. J. Stephens, Prime divisors of second-order linear recurrences. I, *J. Number Theory* 8 (1976) 313-332; MR 54 #5142.
[26] G. Niklasch and P. Moree, Some number-theoretical constants: Products of rational functions over primes, unpublished note (2000).
[27] J. Knopfmacher, A prime-divisor function, *Proc. Amer. Math. Soc.* 40 (1973) 373-377; MR 48 #6036.
[28] J. Knopfmacher, Arithmetical properties of finite rings and algebras, and analytic number theory. VI, *J. Reine Angew. Math.* 277 (1975) 45-62; MR 52 #3088.
[29] D. S. Mitrinovic, J. Sándor, and B. Crstici, *Handbook of Number Theory*, Kluwer, 1996, pp. 9-37, 49-51, 332; MR 97f:11001.
[30] S. Chowla, The representation of a number as a sum of four squares and a prime, *Acta Arith.* 1 (1935) 115-122.
[31] P. Moree, On some sums connected with primitive roots, preprint MPI 98-42, Max-Planck

[32] P. Moree, On primes in arithmetic progression having a prescribed primitive root, *J. Number Theory* 78 (1999) 85-98; MR 2001i:11118.
[33] P. Erdös, Some asymptotic formulas in number theory, *J. Indian Math. Soc.* 12 (1948) 75-78; MR 10,594d.
[34] I. Z. Ruzsa, Erdös and the integers, *J. Number Theory* 79 (1999) 115-163; MR2002e:11002.

2.8 ペル-スティーヴンハーゲン定数

$d>1$ を平方数でないとするとき，Pell（ペル）方程式
$$x^2 - dy^2 = 1$$
は整数の解をもつ（実際は，無限個の解をもつ）．この事実は遠い昔から知られている [1-5]．ここではより難しい質問を問題とする．**負のペル方程式**
$$x^2 - dy^2 = -1$$
が整数の解をもつような，整数 $d>1$ の集合 D について，何がいえるのだろうか？ごく最近，これに次のように答えることで進展した．

最初に，以下で必要な**ペル定数**を
$$P = 1 - \prod_{\substack{j \geq 1 \\ j \text{ 奇数}}} \left(1 - \frac{1}{2^j}\right) = 0.5805775582\cdots$$
と定義する．定数 P は無理数である[6]と証明できるが，超越数であることは予想である．さらに関数
$$\psi(p) = \frac{2 + (1 + 2^{1-v_p})p}{2(p+1)}$$
を定義する．ただし，v_p は $p-1$ に含まれる素因数 2 の個数である．

正の整数からなる任意の集合 S に対して，$f_S(n)$ を n を超えない S の要素の個数と定義する．Stevenhagen（スティーヴンハーゲン）[6-8]は，D の分布関数に関するいくつかの予想を発表した．彼は，数え上げ関数（counting function）$f_D(n)$ が次の式 [7]
$$\lim_{n \to \infty} \frac{\sqrt{\ln(n)}}{n} f_D(n) = \frac{3P}{2\pi} \prod_{p \equiv 1 \bmod 4} \left(1 + \frac{\psi(p)}{p^2-1}\right)\left(1 - \frac{1}{p^2}\right)^{1/2} = 0.28136\cdots$$
を満たすという予想を立てた．ただし，積は素数 p にかぎる．

U を 4 で割り切れない正の整数の集合とし，V を法 4 で 3 と合同などのような素数でも割り切れない正の整数の集合とする．明らかに，D は $U \cap V$ の部分集合であり，$U \cap V$ は互いに素である 2 つの平方数の和で表すことのできる正の整数の集合である．Rieger（リーガー）[9]による，ここで述べた予想される極限および【2.3.1】で得られた互いに素であることに関する結果から，$U \cap V$ での D の密度は[7]

$$\lim_{n\to\infty}\frac{f_D(n)}{f_{U\cap V}(n)}=P\prod_{p\equiv 1\bmod 4}\left(1+\frac{\psi(p)}{p^2-1}\right)\left(1-\frac{1}{p^2}\right)=0.57339\cdots$$

である.

別の予想がある.Wを無平方な整数の集合とする.すなわち,1より大きい平方数で割り切れない整数である.Stevenhagen[6]は

$$\lim_{n\to\infty}\frac{\sqrt{\ln(n)}}{n}f_{U\cap W}(n)=\frac{6}{\pi^2}PK=0.2697318462\cdots$$

と仮説を立てた.ただし,Kはランダウ-ラマヌジャン定数【2.3】である.明らかに,$V\cap W$は,互いに素である2つの平方数の和で表される正の無平方な整数の集合である.2番目の予想された極限およびMoore[10]による【2.3.1】で得られた無平方の結果から,$V\cap W$での$D\cap W$の密度は[8]

$$\lim_{n\to\infty}\frac{f_{D\cap W}(n)}{f_{V\cap W}(n)}=P=0.5805775582\cdots$$

である.

連分数とのすばらしい関連は,次のようなものである[7].整数$d\geq 1$がDに属するための必要十分条件は$\sqrt{d}$が無理数で,かつ正則連分数に展開したときに周期の長さが「奇数」であることである.

Pに類似な定数Qは【5.14】に現れる.しかし,Qの指数は奇数にかぎらない.

[1] T. Nagell, *Introduction to Number Theory*, Chelsea, 1981, pp. 195-204; MR 30 #4714.
[2] K. Ireland and M. Rosen, *A Classical Introduction to Modern Number Theory*, 2nd ed., Springer-Verlag, 1990, pp. 276-278; MR 92e:11001.
[3] K. H. Rosen, *Elementary Number Theory and Its Applications*, Addison-Wesley, 1985, pp. 401-409; MR 93i:11002.
[4] D. A. Buell, *Binary Quadratic Forms: Classical Theory and Modern Computations*, Springer-Verlag, 1989, pp. 31-34; MR 92b:11021.
[5] M. R. Schroeder, *Number Theory in Science and Communication: With Applications in Cryptography, Physics, Digital Information, Computing and Self-Similarity*, 2nd ed., Springer-Verlag, 1986, pp. 98-99; MR 99c:11165.
[6] P. Stevenhagen, The number of real quadratic fields having units of negative norm, *Experim. Math.* 2 (1993) 121-136; MR 94k:11120.
[7] P. Stevenhagen, A density conjecture for the negative Pell equation, *Computational Algebra and Number Theory*, Proc. 1992 Sydney conf., ed. W. Bosma and A. van der Poorten, Kluwer, 1995, pp. 187-200; MR 96g:11137.
[8] W. Bosma and P. Stevenhagen, Density computations for real quadratic units, *Math. Comp.* 65 (1996) 1327-1337; MR 96j:11171.
[9] G. J. Rieger, Über die Anzahl der als Summe von zwei Quadraten darstellbaren und in einer primen Restklasse gelegenen Zahlen unterhalb einer positiven Schranke. II, *J. Reine Angew. Math.* 217 (1965) 200-216; MR 30 #4734.
[10] P. Moree, Variations on the Landau-Ramanujan problem, unpublished note (2000).

2.9 アラディ-グリンステッド定数

n を正の整数とする．よく知られた式
$$n! = 1 \cdot 2 \cdot 3 \cdot 4 \cdots (n-1) \cdot n$$
は，$n!$ を n 個の整数の積として分解するときに，たくさんの利用できる方法のうちの1つの形式である．1は因数として認めない，さらに n 個の因数の各々は素数の累乗
$$p_k^{b_k}, \quad \text{各 } p_k \text{ は素数で，} b_k \geq 1, \; k = 1, 2, \cdots, n$$
であるとする．（したがって，$n!$ のはじめに述べた自然な分解は認められない．）さらに，左から右に因数を等しいか大きくなる順に書くとする．たとえば，$n=9$ のとき，すべての認められる分解は

$$9! = 2 \cdot 2 \cdot 2 \cdot 2 \cdot 2 \cdot 2^2 \cdot 5 \cdot 7 \cdot 3^4$$
$$= 2 \cdot 2 \cdot 2 \cdot 2 \cdot 3 \cdot 5 \cdot 7 \cdot 2^3 \cdot 3^3$$
$$= 2 \cdot 2 \cdot 2 \cdot 2 \cdot 5 \cdot 7 \cdot 2^3 \cdot 3^2 \cdot 3^2$$
$$= 2 \cdot 2 \cdot 2 \cdot 3 \cdot 2^2 \cdot 2^2 \cdot 5 \cdot 7 \cdot 3^3$$
$$= 2 \cdot 2 \cdot 2 \cdot 2^2 \cdot 2^2 \cdot 5 \cdot 7 \cdot 3^2 \cdot 3^2$$
$$= 2 \cdot 2 \cdot 2 \cdot 3 \cdot 3 \cdot 5 \cdot 7 \cdot 3^2 \cdot 2^4$$
$$= 2 \cdot 2 \cdot 3 \cdot 3 \cdot 2^2 \cdot 5 \cdot 7 \cdot 2^3 \cdot 3^2$$
$$= 2 \cdot 2 \cdot 3 \cdot 3 \cdot 3 \cdot 3 \cdot 5 \cdot 7 \cdot 2^5$$
$$= 2 \cdot 3 \cdot 3 \cdot 2^2 \cdot 2^2 \cdot 2^2 \cdot 5 \cdot 7 \cdot 3^2$$
$$= 2 \cdot 3 \cdot 3 \cdot 3 \cdot 3 \cdot 2^2 \cdot 5 \cdot 7 \cdot 2^4$$
$$= 2 \cdot 3 \cdot 3 \cdot 3 \cdot 3 \cdot 5 \cdot 7 \cdot 2^3 \cdot 2^3$$
$$= 3 \cdot 3 \cdot 3 \cdot 3 \cdot 2^2 \cdot 2^2 \cdot 5 \cdot 7 \cdot 2^3$$

である．最も左側の因数の11個は2であり，1つは3であることに注意せよ．したがって，$9!$ を9つの素数ベキの積に分解する，すべての許容される分解（admissible decompositions）を考慮したとき，最左端の最大因子は3である．
$$a(9) = \frac{\ln(3)}{\ln(9)}$$
と定義する．

同じように，任意の n に対して，$n!$ を n 個の素数ベキにすべての許容される分解をするときの，最も左側にある因数の最大値を p^b とし，
$$a(n) = \frac{\ln(p^b)}{\ln(n)}$$
と定義する．明らかに，各 n で $a(n) < 1$ である．大きな n について，$a(n)$ について何がいえるだろうか？

Alladi & Grinstead（アラディ&グリンステッド）[1,2] は，$n > 1$ のときの $a(n)$ の

2.9 アラディ-グリンステッド定数

極限の存在を示し，値を

$$\lim_{n\to\infty} \alpha(n) = e^{c-1} = 0.8093940205\cdots$$

と決定した．ただし，

$$c = -\sum_{k=2}^{\infty} \frac{1}{k}\ln\left(1-\frac{1}{k}\right) = \sum_{j=2}^{\infty} \frac{\zeta(j)-1}{j-1} = 0.7885305659\cdots$$
$$= -\ln(0.4545121805\cdots).$$

であり，$\zeta(x)$ は Riemann のゼータ関数である【1.6】．

Alladi & Grinstead の結果は，$n!$ の分解にどのように強く依存しているのだろうか，他の関数 $f(n)$ に依存していないのだろうか？ f は，各 n に対して，十分にたくさんの小さな，かつ，さまざまな素因数が生じる関数であると仮定する．関連する未解決問題は[3]を参照せよ．

$d(m)$ を，m の正の整数の約数の個数とする．$d(n!)$ については何がいえるだろうか？ Erdös ら[4]は

$$\lim_{n\to\infty} \frac{\ln(\ln(n!))^2}{\ln(n!)}\ln(d(n!)) = C$$

を証明した．ただし，【1.8】で述べられているように

$$C = \sum_{k=2}^{\infty} \frac{1}{k(k-1)\ln(k)} = -\sum_{j=2}^{\infty} \zeta'(j) = 1.2577468869\cdots$$

である．c と C の類似性はかなりおもしろい．

関連する4つの無限積がある[5,6]．

$$\prod_{n\geq 2}\left(1+\frac{1}{n}\right)^{1/n} = 1.7587436279\cdots, \quad \prod_{n\geq 2}\left(1-\frac{1}{n}\right)^{1/n} = 0.4545121805\cdots$$

$$\prod_{p}\left(1+\frac{1}{p}\right)^{1/p} = 1.4681911223\cdots, \quad \prod_{p}\left(1-\frac{1}{p}\right)^{1/p} = 0.5598656169\cdots$$

最後の2つの積では，p は素数にかぎる．2番目の積は e^{-c} である．4番目の積は[7,8]に現れる．n の最小および最大の素因数の漸近的な性質に関する，関連する問題は【5.4】で議論される．

[1] K. Alladi and C. Grinstead, On the decomposition of $n!$ into prime powers, *J. Number Theory* 9 (1977) 452-458; MR 56 #11934.
[2] R. K. Guy, *Unsolved Problems in Number Theory*, 2nd ed., Springer-Verlag, 1994, sect. B22; MR 96e:11002.
[3] R. K. Guy and J. L. Selfridge, Factoring factorial n, *Amer. Math. Monthly* 105 (1998) 766-767.
[4] P. Erdös, S.W. Graham, A. Ivić, and C. Pomerance, On the number of divisors of $n!$, *Analytic Number Theory*, Proc. 1995 Allerton Park conf., v. 1, ed. B. C. Berndt, H. G. Diamond, and A. J. Hildebrand, Birkhäuser, 1996, pp. 337-355; MR 97d:11142.
[5] P. Sebah, Evaluating $\prod_{n\geq 2}(1\pm 1/p)^{1/n}$, unpublished note (2000).
[6] M. Deleglise, Computing $\prod_{p}(1\pm 1/p)^{1/p}$, unpublished note (1999).
[7] P. Erdös, Remarks on two problems of the Matematikai Lapok (in Hungarian), *Mat. Lapok* 7 (1956) 10-17; MR 20 #4534.
[8] I. Z. Ruzsa, Erdös and the integers, *J. Number Theory* 79 (1999) 115-163; MR2002e:11002.

2.10 シェルピンスキー定数

1908年の学位論文で，Sierpinski（シェルピンスキー）[1]は，順序と符号を考慮した2つの平方数の和として表される正の整数 n の表現の個数として定義される，関数 $r(n)$ を含むある級数を研究した．たとえば，$r(1)=4$ で，素数 $p\equiv 3 \bmod 4$ に対して $r(p)=0$，素数 $q\equiv 1 \bmod 4$ に対して $r(q)=8$ である．

$r(n)$ についてのある結果を見るのは難しくない．たとえば，[2-4]，$n\to\infty$ のとき，

$$\sum_{k=1}^{n} r(k) = \pi n + O(n^{1/2})$$

である．この評価のより詳しいものは，【2.10.1】にある．[1,5,6]にある Sierpinski の級数は

$$\sum_{k=1}^{n} \frac{r(k)}{k} = \pi(\ln(n)+S) + O(n^{-1/2})$$

$$\sum_{k=1}^{n} r(k^2) = \frac{4}{\pi}(\ln(n)+\widehat{S})n + O(n^{2/3})$$

$$\sum_{k=1}^{n} r(k)^2 = 4(\ln(n)+\widetilde{S}) + O(n^{3/4}\ln(n))$$

である．ただし，定数 $\widehat{S}$ と $\widetilde{S}$ は

$$\widehat{S} = \gamma + S - \frac{12}{\pi^2}\zeta'(2) + \frac{\ln(2)}{3} - 1, \quad \widetilde{S} = 2S - \frac{12}{\pi^2}\zeta'(2) + \frac{\ln(2)}{3} - 1$$

と S を用いて定義される．γ はオイラー-マスケローニの定数【1.5】であり，$\zeta(x)$ は Riemann のゼータ関数【1.6】である．$\zeta'(2)$ の他の出現については，【2.15】と【2.18】を参照せよ．

したがって，**シェルピンスキー定数**とよばれる定数 S は，3つすべての級数の和で役割を果たす．それは

$$S = \gamma + \frac{\beta'(1)}{\beta(1)} = \ln\left(\frac{\pi^2 e^{2\gamma}}{2L^2}\right) = \ln\left(\frac{4\pi^3 e^{2\gamma}}{\Gamma\left(\frac{1}{4}\right)^4}\right) = \frac{2.5849817595\cdots}{\pi}$$

としても定義される．$\beta(x)$ は Dirichlet のベータ関数【1.7】であり，$L=2.6220575542\cdots$ は Gauss のレムニスケート（lemniscate）定数【6.1】であり，$\Gamma(x)$ は Euler のガンマ関数【1.5.4】である．それはまたこの本で，Landau-Ramanujan 定数【2.3】と Masser-Gramain（マサー-グラメイン）定数【7.2】でも現れる．実際，Sierpinski は極限

$$S = \frac{1}{\pi}\lim_{z\to 1}\left(F(z) - \frac{\pi}{z-1}\right)$$

として S を定義した．関数 $F(z)=4\zeta(z)\beta(z)$ は，格子点での和【1.10.1】の議論の中

心である．定積分表示を含む S の別の式は，

$$S = 2\gamma + \frac{4}{\pi}\int_0^\infty \frac{e^{-x}\ln(x)}{1+e^{-2x}}dx$$

である．明らかに，これは多くのアイデアの出会いの場所であり，すべてはすぐに一緒になる．

2.10.1 円の問題と約数問題

より精密に[7-12]，1 から n までの r [前述]の和は

$$\sum_{k=1}^{n} r(k) = \pi n + O(n^{23/73}\ln(n)^{315/146})$$

を満たすことが証明される．任意の $\varepsilon > 0$ に対して

$$\sum_{k=1}^{n} r(k) = \pi n + O(n^{1/4+\varepsilon})$$

が予想されている．誤差項の評価の問題は，原点を中心とした半径 $\sqrt{n}$ の円盤の中に落ちる，順序をつけた整数の組の個数と同じなので，**円の問題** (circle problem) として知られている．

【1.5】で簡単にふれた，**約数問題** (divisor problem) として知られている関連する問題がある．$d(n)$ を n の相異なる約数の個数とする．このとき，

$$\sum_{k=1}^{n} d(k) = n\ln(n) + (2\gamma-1)n + O(n^{23/73}\ln(n)^{461/146})$$

は，1 から n までの d の和の知られているうちで最良の評価である．再び，誤差項の指数は，$1/4+\varepsilon$ と予想されているが，未解決のままである．Sierpinski の 3 番目の級数と類似なものは，たとえば[13-15]，

$$\sum_{k=1}^{n} d(k)^2 = (A\ln(n)^3 + B\ln(n)^2 + C\ln(n) + D)n + O(n^{1/2+\varepsilon})$$

である．ただし，

$$A = \frac{1}{\pi^2}, \quad B = \frac{12\gamma-3}{\pi^2} - \frac{36}{\pi^4}\zeta'(2)$$

であり，定数 C, D はもっと複雑な式である．Sierpinski の 1 番目の級数と類似なものは[16]，

$$\sum_{k=1}^{n} \frac{d(k)}{k} = \frac{1}{2}\ln(n)^2 + 2\gamma\ln(n) + (\gamma^2 - 2\gamma_1) + O(n^{-1/2})$$

である．$\gamma_1 = -0.0728158454\cdots$ は，第 1 Stieltjes（スティルチェス）定数【2.21】である．

$d(n)$ の変種に関して，無平方である n の約数に関心を限定する[17]．同様に，$r(n)$ に対して，u, v が互いに素であるような $n = u^2 + v^2$ の表現だけを数える，または和よりもむしろ差を調べる別の種類のものがある．u, v を任意の整数とする，$n = |u|^m + |v|^m$ の表現の個数を，$r_m(n)$ と定義する．$m \geq 3$ のとき，[12,18,19]

$$\sum_{k=1}^{n} r_m(k) = \frac{2\Gamma\left(\dfrac{1}{m}\right)^2}{m\Gamma\left(\dfrac{2}{m}\right)} n^{2/m} + O(n^{1/m(1-1/m)})$$

が知られている.さらに,誤差項は

$$2^{3-1/m}\pi^{-1-1/m}m^{1/m}\Gamma\left(1+\frac{1}{m}\right)\cdot \sum_{k=1}^{\infty} k^{-1-1/m}\sin\left(2\pi k n^{1/m}-\frac{\pi}{2m}\right)\cdot n^{1/m(1-1/m)}$$

$$+ O\left(n^{46/(73m)}\ln(n)^{315/146}\right)$$

に置きかえることができる.このような円の問題や約数問題の漸近的な性質の完全な解析は,ひじょうに難しく,すぐに解決できることは期待できない.

1908年の論文に関連して,Sierpinski[20-22]は次の事実を発見した.$D_n = \max\{d : d^2 | n\}$,すなわち,$D_n^2$ は n の最大平方因数とする.このとき,

$$\frac{1}{n}\sum_{k=1}^{n} D_k = \frac{3}{\pi^2}\ln(n) + \frac{9\gamma}{\pi^2} - \frac{36}{\pi^4}\zeta'(2) + o(1) \quad (n\to\infty)$$

である.比較として,n の無平方な項の平均は,【2.5】にある.

[1] W. Sierpinski, On the summation of the series $\sum_{n>a}^{n<b}\tau(n)f(n)$, where $\tau(n)$ denotes the number of decompositions of n into a sum of two integer squares (in Polish), *Prace Matematyczno-Fizyczne* 18 (1908) 1-59; French transl. in Oeuvres Choisies, t. 1, Editions Scientifiques de Pologne, 1974, pp. 109-154; MR 54 # 2405.

[2] G. H. Hardy and E. M. Wright, *An Introduction to the Theory of Numbers*, 5th ed., Oxford Univ. Press, 1985, pp. 241-243, 256-258, 270-271; MR 81i:10002.

[3] D. Shanks, *Solved and Unsolved Problems in Number Theory*, 2nd ed., Chelsea, 1978, pp. 162-165; MR 86j:11001.

[4] E. C. Titchmarsh, *The Theory of the Riemann Zeta Function*, Oxford Univ. Press, 1951, pp. 262-275; MR 88c:11049.

[5] A. Schinzel, Waclaw Sierpinski's papers on the theory of numbers, *Acta Arith.* 21 (1972) 7-13; MR 46 # 9b.

[6] S. Plouffe, Sierpinski constant (Plouffe's Tables).

[7] E. Grosswald, *Representations of Integers as Sums of Squares*, Springer-Verlag, 1985, pp. 20-21; MR 87g:11002.

[8] H. Iwaniec and C. J. Mozzochi, On the divisor and circle problems, *J. Number Theory* 29 (1988) 60-93; MR 89g:11091.

[9] M. N. Huxley, Exponential sums and lattice points. II, *Proc. London Math. Soc.* 66 (1993) 279-301; corrigenda 68 (1994) 264; MR 94b:11100 and MR 95d:11134.

[10] R. K. Guy, *Unsolved Problems in Number Theory*, 2nd ed., Springer-Verlag, 1994, sect. F1; MR 96e:11002.

[11] A. Ivić, *The Riemann Zeta-Function*, Wiley, 1985, pp. 35-36, 93-94, 351-384; MR 87d:11062.

[12] E. Krätzel, *Lattice Points*, Kluwer, 1988, pp. 140-142, 228-230; MR 90e:11144.

[13] S. Ramanujan, Some formulae in the analytic theory of numbers, *Messenger of Math.* 45 (1916) 81-84; also in *Collected Papers*, ed. G. H. Hardy, P. V. Seshu Aiyar, and B. M. Wilson, Cambridge Univ. Press, 1927, pp. 133-135, 339-340.

[14] B. M. Wilson, Proofs of some formulae enunciated by Ramanujan, *Proc. London Math.* Soc. 21 (1922) 235-255.

[15] D. Suryanarayana and R. Sitaramachandra Rao, On an asymptotic formula of Ramanujan,

Math. Scand. 32 (1973) 258-264; MR 49 # 2611.
[16] S. A. Amitsur, Some results on arithmetic functions, *J. Math. Soc. Japan* 11 (1959) 275-290; MR 26 # 67.
[17] R. C. Baker, The square-free divisor problem, *Quart. J. Math* 45 (1994) 269-277; part II, *Quart. J. Math* 47 (1996) 133-146; MR 95h:11098 and MR 97f:11080.
[18] W. G. Nowak, On sums and differences of two relative prime cubes, Analysis 15 (1995) 325-341; part II, *Tatra Mount. Math. Publ.* 11 (1997) 23-34; MR 96m:11085 and MR 98j:11073.
[19] M. Kühleitner, On sums of two k^{th} powers: An asymptotic formula for the mean square of the error term, *Acta Arith.* 92 (2000) 263-276; MR 2001a:11164.
[20] W. Sierpinski, On the average values of several numerical functions (in Polish), *Sprawozdania Towarzystwo Naukowe Warszawskie* 1 (1908) 215-226.
[21] S. M. Lee, On the sum of the largest k^{th} divisors, *Kyungpook Math. J.* 15 (1975) 105-108; MR 51 # 5528.
[22] K. Greger, Square divisors and square-free numbers, *Math. Mag.* 51 (1978) 211-219; MR 58 # 21916.

2.11 過剰数の密度定数

n を正の整数とする．$\sigma(n)$ を n の正の約数のすべての和とする．このとき，n は，$\sigma(n)=2n$ のとき**完全数** (perfect) といい，$\sigma(n)<2n$ のとき**不足数** (deficient) といい，$\sigma(n)>2n$ のとき**過剰数** (abundant) という．

完全数の最小の例は，6 と 28 である．もし，Mersenne (メルセンヌ) 数 $2^{m+1}-1$ が素数であるとき，$2^m(2^{m+1}-1)$ は完全数である．2 つの有名な未解決問題がある[1]．偶数の完全数は無限個存在するか？ 奇数の完全数は存在するか？ ([2]によれば，後者の反例は 10^{300} より小さい範囲にはない．)

正の実数 x に対して，密度関数
$$A(x)=\lim_{n\to\infty}\frac{|\{n : \sigma(n)\geq xn\}|}{n}$$
を定義する．Behrend (ベーレンド) [3,4]，Davenport (ダベンポート) [5]，Chowla (チョウラ) [6]らは独立に，$A(x)$ は存在し，すべての x で連続であることを証明した．Erdös[7,8]は，基本的な知識だけを必要とする証明を与えた．明らかに，$x \leq 1$ のとき $A(x)=1$ であり，$x \to \infty$ のとき $A(x) \to 0$ である．Behrend の技法を改良することにより，Wall (ウォール) [9,10]は，**過剰数密度定数** (abundant numbers density constant) が，
$$0.2441 < A(2) < 0.2909$$
であることを得た．Deléglise (デレグリース) [11]はこれを
$$|A(2)-0.2477|<0.0003$$
に改良した．さらに，[12]で，$A(x)$ はルベーグ測度 0 の集合を除いて，いたるところ微分可能であり，$\mathrm{Re}(s)>1$ を満たす複素数 s に対して

$$\int_0^\infty x^{s-1} A(x)\, dx = \frac{1}{s} \prod_p \left[\left(1-\frac{1}{p}\right)^{-s+1} \sum_{k=0}^n \frac{1}{p^k} \left(1-\frac{1}{p^{k+1}}\right)^s \right]$$

であることを証明した．積は，すべての素数 p をわたる．この等式 (Mellin (メリン) 変換) の逆変換は，理論的には可能であるが，いまだに数値的に実行可能でない[11].

余談として，$n = p_1^{a_1} \cdots p_r^{a_r}$ の**指数的約数** (exponential divisor) d を定義する．$d = p_1^{b_1} \cdots p_r^{b_r}$ で，各 j について $b_j | a_j$ であるとき，d を指数的約数と定義する．n の指数的約数のすべての和を，$\sigma^{(e)}(n)$ とする．簡単のために $\sigma^{(e)}(1) = 1$ とする．このとき，[13-16]

$$\lim_{N\to\infty} \frac{1}{N^2} \sum_{n=1}^N \sigma(n) = \frac{\pi^2}{12}, \quad \lim_{N\to\infty} \frac{1}{N^2} \sum_{n=1}^N \sigma^{(e)}(n) = B$$

である．ただし，

$$B = \frac{1}{2} \prod_p \left[1 + \frac{1}{p(p^2-1)} - \frac{1}{p^2-1} + \left(1-\frac{1}{p}\right) \sum_{k=2}^\infty \frac{p^k}{p^{2k}-1} \right]$$

$= 0.5682854937\cdots$

である．これに対応した密度関数 $A^{(e)}(x)$ の研究は，[17]で始まった．

[1] K. Ireland and M. Rosen, *A Classical Introduction to Modern Number Theory*, 2nd ed., Springer-Verlag, 1990; MR 92e：11001.
[2] R. K. Guy, *Unsolved Problems in Number Theory*, 2nd ed., Springer-Verlag, 1994, sect. B1, B2; MR 96e：11002.
[3] F. Behrend, Über numeri abundantes, *Sitzungsber. Preuss. Akad. Wiss.* (1932) 322-328.
[4] F. Behrend, Über numeri abundantes. II, *Sitzungsber. Preuss. Akad. Wiss.* (1933) 280-293.
[5] H. Davenport, Über numeri abundantes, *Sitzungsber. Preuss. Akad. Wiss.* (1933) 830-837.
[6] S. Chowla, On abundant numbers, *J. Indian Math. Soc.* 1 (1934) 41-44.
[7] P. Erdös, On the density of the abundant numbers, *J. London Math. Soc.* 9 (1934) 278-282.
[8] P. D. T. A. Elliott, *Probabilistic Number Theory. I: Mean-Value Theorems*, Springer-Verlag, 1979, pp. 3-4, 187-189, 203-213; MR 82h：10002a.
[9] C. R. Wall, Density bounds for the sum of divisors function, *The Theory of Arithmetic Functions*, Proc. 1971 Kalamazoo conf., ed. A. A. Gioia and D. L. Goldsmith, Lect. Notes in Math. 251, Springer-Verlag, 1971, pp. 283-287; MR 49 # 10650.
[10] C. R.Wall, P. L. Crews, andD. B. Johnson, Density bounds for the sum of divisors function, *Math. Comp.* 26 (1972) 773-777; errata 31 (1977) 616; MR 48 # 6042 and MR 55 # 286.
[11] M. Deléglise, Bounds for the density of abundant integers, *Experim. Math.* 7 (1998) 137-143; MR 2000a：11137.
[12] J. Martinet, J. M. Deshouillers, and H. Cohen, La fonction somme des diviseurs, *Séminaire de Théorie des Nombres*, 1972-1973, exp. 11, Centre Nat. Recherche Sci., Talence, 1973; MR 52 # 13607.
[13] M. V. Subbarao, On some arithmetic convolutions, *The Theory of Arithmetic Functions*, Proc. 1971 Kalamazoo conf., ed. A. A. Gioia and D. L. Goldsmith, Lect. Notes in Math. 251, Springer-Verlag, 1972, pp. 247-271; MR 49 # 2510.
[14] J. Fabrykowski and M. V. Subbarao, The maximal order and the average order of multiplicative function $\sigma^{(e)}(n)$, *Théorie des nombres*, Proc. 1987 Québec conf., ed. J.-M. De Koninck and C. Levesque, Gruyter, 1989, pp. 201-206; MR 90m：11012.
[15] J. Wu, Problème de diviseurs exponentiels et entiers exponentiellement sans facteur carré, *J. Théorie Nombres Bordeaux* 7 (1995) 133-141; MR 98e：11108.

[16] A. Smati and J. Wu, On the exponential divisor function, *Publ. Inst. Math. (Beograd)*, v. 61 (1997) n. 75, 21-32; MR 98k：11127.

[17] P. Hagis, Some results concerning exponential divisors, *Int. J. Math. Math. Sci.* 11 (1988) 343-349; MR 90d：11011.

2.12 リンニク定数

特別な数列が素数の値をとる問題を最初に議論する．ディリクレの定理は，$a \geq 1$，$b \geq 1$ を互いに素な整数とするとき，任意の等差数列 $\{an+b : n \geq 0\}$ は無限に多くの素数を含む，という定理である．自然な疑問が生じる．そのとき最初の素数 $p(a,b)$ はどのくらい大きいのか？

$1 \leq b < a$，$\gcd(a,b) = 1$ を満たすすべての b での $p(a,b)$ の最大値を $p(a)$ と定義し，

$$K = \sup_{a \geq 2} \frac{\ln(p(a))}{\ln(a)}, \quad L = \lim_{a \to \infty} \frac{\ln(p(a))}{\ln(a)}$$

とする．すなわち，K はすべての $a=2$ に対して $p(a) < a^\kappa$ を満たす κ の下限であり，L は充分大きなすべての a に対して $p(a) < a^\lambda$ を満たす λ の下限である．$p(a,b)$ の上からの評価と下からの評価の，別の形を決定することに加えて，K と L を評価するために，多くの研究[1,2]が行われた．

明らかに，$K > 1.82$（$p(5) = 19$ を証明して）である．Schinzel & Sierpinski（シンツェル&シェルピンスキー）[3]と Kanold（カノルド）[4,5]は，$K \leq 2$ であると予想した．もしこれが真であるとき，次の列

$$b, \ a+b, \ 2a+b, \ \cdots, \ (a-1)a+b$$

のどこかに素数が存在する，ことを意味する．ただし，$\gcd(a,b) = 1$ である．

このような命題は，現在の数学者の手の届かないものである．Shinzel & Sierpinski は，（いくつかある中で）彼らの仮説の運命がどうなるのか知らないと認めた．Ribenboim（リーベンボイム）[2]は，ペアノの公理の枠組み内でこのような仮説は決定不能ではないかと疑った．

Linnik（リンニク）[6,7]は，L が存在し有限である，ことを証明した．明らかに，$L \leq K$ である．一般リーマン予想を仮定すると，[8-10]から

$$p(a) = O(\varphi(a)^2 \ln(a))$$

であり，これから $L \leq 2$ が導かれる．ここに $\varphi(x)$ はオイラーの約数関数【2.7】である．**リンニク定数** L の条件を付けない上界の研究は，多くの研究者[11-13]が行っている．この仕事の頂点は，Heath-Brown（ヒース-ブラウン）の証明[14] $L \leq 5.5$ である．

$L \leq 2$ に対する部分的な証拠は，次に含まれる．任意の固定した正の整数 b,k に対して，Bombieri, Friedlander & Iwaniec（ボンビエリ，フリードランダー&イワニエ

ツ）[15]は，Granville（グランヴィル）[16,17]の仕事から気づいたように，密度0の集合に属さないすべての a に対して，

$$p(a,b) < \frac{a^2}{\ln(a)^k}$$

であることを証明した．したがって，「ほとんどすべて」の整数 a に対して $L \leq 2$ と結論できる．

Chowla[18]は $L=1$ であると信じていた．彼に続く研究者[19-23]は

$$p(a) = O(\varphi(a)\ln(a)^2)$$

であると予想した．これから $L=1$ が導かれる．Elliott & Halberstam（エリオット＆ハルバースタム）[24]の初期の研究は，この新しい評価に部分的な支持を与えている．

さて，特別な方程式の素数解に関心を変えよう．Liu & Tsang（リュー＆ツァン）[25-28]は，とりわけ，線型方程式 $ap+bq+cr=d$ の素数の解 p, q, r の存在問題を研究した．ただし，a,b,c は非負整数で，さらに $a+b+c-d$ は偶数で，$\gcd(a,b,c)=1$, $\gcd(d,a,b)=1$, $\gcd(d,b,c)=1$, $\gcd(d,a,c)=1$ を仮定する．（$c=0$ を許すとき，$a=b=1$ はゴールドバッハ予想と同値になり，$a=1$, $b=-1$, $d=2$ のときは，双子素数問題と同値になる[†]．）

a,b,c がすべて正であるか，そうでないかに依存して，2つの場合がある．ここでは1つの場合のみを議論する．a,b,c のすべてが同じ符号ではないと仮定する．このとき，方程式 $ap+bq+cr=d$ は

$$\max(p,q,r) \leq 3|d| + (\max(3,|a|,|b|,|c|))^\mu$$

を満たす素数 p,q,r の解をもたなければならない，という性質をもつ定数 μ が存在する．この結果は，Linnik のもとの定理の一般化の1つである．

このような μ のうちの下限 M は，Baker（ベイカー）定数[29]として知られている．$L \leq M$ が証明されている．知られている一番よい M の上界[30,31]は，（無条件で）45であり，（一般リーマン予想を仮定して）4である．Chowla と同様に，Liu & Tsang は $M=1$ であることを予想した．

[1] R. K. Guy, *Unsolved Problems in Number Theory*, 2nd ed., Springer-Verlag, 1994, sect. A4; MR 96e:11002.
[2] P. Ribenboim, *The New Book of Prime Number Records*, Springer-Verlag, 1996, pp. 277-284, 397-400; MR 96k:11112.
[3] A. Schinzel and W. Sierpinski, Sur certaines hypothèses concernant les nombres premiers, *Acta Arith.* 4 (1958) 185-208; erratum 5 (1959) 259; MR 21 #4936.
[4] H.-J. Kanold, Elementare Betrachtungen zur Primzahltheorie, *Arch. Math.* 14 (1963) 147-151; MR 27 #89.
[5] H.-J. Kanold, Über Primzahlen in arithmetischen Folgen, *Math. Annalen* 156 (1964) 393-395; 157 (1965) 358-362; MR 30 #70 and MR 37 #5168.
[6] U. V. Linnik, On the least prime in an arithmetic progression. I: The basic theorem, *Mat. Sbornik* 15 (1944) 139-178; MR 6,260b.
[7] U. V. Linnik, On the least prime in an arithmetic progression. II: The Deuring-Heilbronn

---

[†]訳注：ただし前者は解の存在が，後者は無限性が課題である．

phenomenon, *Mat. Sbornik* 15 (1944) 347-368; MR 6,260c.
[8] E. C. Titchmarsh, A divisor problem, *Rend. Circ. Mat. Palmero* 54 (1930) 414-429.
[9] Y. Wang, S.-K. Hsieh, and K.-J. Yu, Two results on the distribution of prime numbers (in Chinese), *J. China Univ. Sci. Technol.* 1 (1965) 32-38; MR 34 #7482.
[10] C. Hooley, The distribution of sequences in arithmetic progressions, *International Congress of Mathematicians*, v. 1, Proc. 1974 Vancouver conf., ed. R. D. James, Canad. Math. Congress, 1975, pp. 357-364; MR 58 #16560.
[11] M. Jutila, On Linnik's constant, *Math. Scand.* 41 (1977) 45-62; MR 57 #16230.
[12] S. Graham, On Linnik's constant, *Acta Arith.* 39 (1981) 163-179; MR 83d:10050.
[13] J. R. Chen and J. M. Liu, On the least prime in an arithmetical progression and theorems concerning the zeros of Dirichlet's L-functions. V, *International Symposium in Memory of Hua Loo Keng*, v. 1, Proc. 1988 Beijing conf., ed. S. Gong, Q. K. Lu, Y. Wang, and L. Yang, Springer-Verlag, 1991, pp. 19-42; MR 92m:11093.
[14] D. R. Heath-Brown, Zero-free regions for Dirichlet L-functons and the least prime in an arithmetic progression, *Proc. London Math. Soc.* 64 (1992) 265-338; MR 93a:11075.
[15] E. Bombieri, J. B. Friedlander, and H. Iwaniec, Primes in arithmetic progressions to large moduli. III, *J. Amer. Math. Soc.* 2 (1989) 215-224; MR 89m:11087.
[16] A. Granville, Least primes in arithmetic progressions, *Théorie des nombres*, Proc. 1987 Québec conf., ed. J. M. De Koninck and C. Levesque, Gruyter, 1989, pp. 306-321; MR 91c:11052.
[17] A. Granville, Some conjectures in analytic number theory and their connection with Fermat's last theorem, *Analytic Number Theory*, Proc. 1989 Allerton Park conf., ed. B. C. Berndt, H. G. Diamond, H. Halberstam, and A. Hildebrand, Birkhäuser, 1990, pp. 311-326; MR 92a:11031.
[18] S. Chowla, On the least prime in an arithmetical progression, *J. Indian Math. Soc.* 1 (1934) 1-3.
[19] D. R. Heath-Brown, Almost-primes in arithmetic progressions and short intervals, *Math. Proc. Cambridge Philos. Soc.* 83 (1978) 357-375; MR 58 #10789.
[20] S. S. Wagstaff, Greatest of the least primes in arithmetic progressions having a given modulus, *Math. Comp.* 33 (1979) 1073-1080; MR 81e:10038.
[21] C. Pomerance, A note on the least prime in an arithmetic progression, *J. Number Theory* 12 (1980) 218-223; MR 81m:10081.
[22] K. S. McCurley, The least r -free number in an arithmetic progression, *Trans. Amer. Math. Soc.* 293 (1986) 467-475; MR 87b:11016.
[23] A. Granville and C. Pomerance, On the least prime in certain arithmetic progressions, *J. London Math. Soc.* 41 (1990) 193-200; MR 91i:11119.
[24] P. D. T. A. Elliott and H. Halberstam, The least prime in an arithmetic progression, *Studies in Pure Mathematics*, ed. L. Mirsky, Academic Press, 1971, pp. 59-61; MR 42 #7609.
[25] M. C. Liu and K. M. Tsang, Small prime solutions of linear equations, *Théorie des nombres*, Proc. 1987 Québec conf., ed. J. M. De Koninck and C. Levesque, Gruyter, 1989, pp. 595-624; MR 90i:11112.
[26] K. K. Choi, M. C. Liu, and K. M. Tsang, Small prime solutions of linear equations. II, *Proc. Amalfi Conf. on Analytic Number Theory*, Maiori, 1989, ed. E. Bombieri. A. Perelli, S. Salerno, and U. Zannier, Univ. Salerno, 1992, pp. 1-16; MR 94i:11080.
[27] K. K. Choi, M. C. Liu, and K. M. Tsang, Conditional bounds for small prime solutions of linear equations, *Manuscripta Math.* 74 (1992) 321-340; MR 93a:11084.
[28] M. C. Liu and K. M. Tsang, Recent progress on a problem of A. Baker, *Séminaire de Théorie des Nombres, Paris*, 1991-92, ed. S. David, Birkhäuser, 1993, pp. 121-133; MR 95h:11107.
[29] A. Baker, On some diophantine inequalities involving primes, *J. Reine Angew. Math.* 228 (1967) 166-181; MR 36 #111.
[30] K. K. Choi, A numerical bound for Baker's constant-Some explicit estimates for small prime solutions of linear equations, *Bull. Hong Kong Math. Soc.* 1 (1997) 1-19; MR 98k:11137.
[31] M. C. Liu and T. Wang, A numerical bound for small prime solutions of some ternary linear equations, *Acta Arith.* 86 (1998) 343-383; MR 99m:11115.

2.13 ミル定数

Mill（ミル）[1]は，すべての正の整数 N に対して，表現 $\lfloor C^{3^N} \rfloor$ が素数のみを表すような正の定数 C が存在するという驚くべき結果を示した．（$\lfloor x \rfloor$ は x を超えない最大の整数を表す．）証明は，Hoheisel（ホーハイゼル）[2]により Ingham（インガム）[3]が精密化した，素数論の難しい定理による．$p<p'$ を引き続いた素数とするとき，任意の $\varepsilon>0$ に対して，

$$p'-p<p^{5/8+\varepsilon}$$

が充分大きい p に対して成り立つ．この不等式は次の漸化数列を定義するために用いる．$q_0=2$，q_{n+1} は各 $n\geqq 0$ に対して q_n^3 を超える素数のうち最小の数とする．例として[4,5]，$q_1=11$，$q_2=1361$，$q_3=2521008887$ である．ホーハイゼル-インガムの定理から，大きな n で

$$q_n^3<q_{n+1}<q_{n+1}+1<q_n^3+q_n^{15/8+3\varepsilon}+1<(q_n+1)^3$$

が成り立つ．だから，

$$q_n^{3^{-n}}<q_{n+1}^{3^{-(n+1)}}<(q_{n+1}+1)^{3^{-(n+1)}}<(q_n+1)^{3^{-n}}$$

を得る．これから，$C=\lim_{n\to\infty} q_n^{3^{-n}}$ の存在が導かれる．望んだ素数の表現結果である．ここで選んだ特別な数列に対して[4,6,7]，$C=1.3063778838\cdots$ であることの計算はやさしい．

初期値 q_0 または指数 3 の選び方を変えると，異なる C の値になる．このような量 C は無限にある．すなわち，ミル定数 $1.3063778838\cdots$ は，素数だけを与える C の唯一の値ではない．Mill の定理の一般化（増大の制限に従う任意の正の整数列）は，[8]での練習問題である．

別の定数 $c=1.9287800\cdots$ は，Wright（ライト）[9]で別の素数を表す関数

$$\lfloor 2^{2^{\cdot^{\cdot^{\cdot^{2^c}}}}} \rfloor$$

として現れる．指数 2 の繰り返しは N 回であり，一番上は c である．Mill とは違ったこの例は，研究するためには，深い定理を必要としない．必要なすべては，$p'<2p$ という，Bertrand の仮設として知られている定理だけである[†].

数人の研究者[6,7,10]は，Mill の式のような式はあまり役に立たないと，賢明にも指摘した．少しだけの素数を計算するために，正確な C の多くの桁の値を知る必要がある．もっと悪いことに，素数 q_1，q_2，q_3，$\cdots$（すなわち，論法は循環的である）を用いて計算する方法を「除いて」，C を計算するどのような方法もないと思われることである．Mill の式が有用となる唯一の方法は，C の「正確な」値が何とかして利用可能

[†] 訳注：これは Chebyshev によって証明されている．初等的な証明もある．

となることである．そうなると誰も予想していないことが起こるかもしれない．

それにもかかわらず，C が存在することは衝撃的である．C が無理数であることは必要であるか，については知られていない．Odlyzko & Wilf（オドリズコ&ウィルフ）による類似の定数 1.6222705028⋯ は，【2.30】にある．$\lfloor C^N \rfloor$ の形をした表現の関連する問題については[11]を参照せよ．

とりわけ Huxley（ハクスレー）[12]は，指数 5/8 を 7/12 に置きかえることに成功した．ホーハイゼル-インガムの定理の精密化に関する最近の結果は，[13-16]にある．現在知られている最良の結果は

$$p' - p = O(p^{0.525})$$

である．リーマン予想が真であると仮定すると，Cramér（クラメール）[17,18]は

$$p' - p = O(\sqrt{p} \ln(p))$$

を証明した．証明されていない主張が，いつか解析学の定理になるとき，劇的な改良となる．Cramér は続いて[19]，

$$p' - p = O(\ln(p)^2)$$

さらに，

$$\limsup_{p \to \infty} \frac{p' - p}{\ln(p)^2} = 1$$

であると予想した．Maiser（メイサー）[22]の仕事を基礎にして，Granville（グランヴィル）[20,21]は次のように予想を修正した．

$$\limsup_{p \to \infty} \frac{p' - p}{\ln(p)^2} \geq 2e^{-\gamma} = 1.122\cdots$$

である．ただし，γ はオイラー定数【1.5】である．ずいぶん前から[23]

$$\limsup_{p \to \infty} \frac{p' - p}{\ln(p)} = \infty$$

であると知られている．したがって，Cramér の上極限で $\ln(p)^2$ を $\ln(p)$ で置きかえることはできない．しかしながら，[24-26]

$$\liminf_{p \to \infty} \frac{p' - p}{\ln(p)} \leq 0.248$$

である．さらなる改良は可能であるのか？ もし双子素数予想が真であるならば，この下極限は明らかに 0 である．

[1] W. H. Mills, A prime-representing function, *Bull. Amer. Math. Soc.* 53 (1947) 604; MR 8,567d.
[2] G. Hoheisel, Primzahlprobleme in der Analysis, *Sitzungsber. Preuss. Akad. Wiss.* (1930) 580-588.
[3] A. E. Ingham, On the difference between consecutive primes, *Quart. J. Math.* 8 (1937) 255-266.
[4] C. K. Caldwell, Mills' theorem-A generalization (Prime Pages).
[5] N. J. A. Sloane, On-Line Encyclopedia of Integer Sequences, A051254 and A016104.
[6] P. Ribenboim, *The New Book of Prime Number Records*, Springer-Verlag, 1996, pp. 186- 187, 252-257; MR 96k:11112.
[7] R. L. Graham, D. E. Knuth, and O. Patashnik, *Concrete Mathematics*, 2nd ed., Addison- Wesley, 1994, pp. 109, 150, and 523; MR 97d:68003.
[8] W. Ellison and F. Ellison, *Prime Numbers*, Wiley, 1985, pp. 22, 31-32; MR 87a:11082.

[9] E. M. Wright, A prime-representing function, *Amer. Math. Monthly* 58 (1951) 616-618; MR 13, 321e.
[10] G. H. Hardy and E. M. Wright, *An Introduction to the Theory of Numbers*, 5th ed., Oxford Univ. Press, 1985, pp. 344-345, 414; MR 81i:10002.
[11] R. C. Baker and G. Harman, Primes of the form $\lfloor c^p \rfloor$, *Math. Z.* 221 (1996) 73-81; MR 96k:11115.
[12] M. N. Huxley, On the difference between consecutive primes, *Invent. Math.* 15 (1972) 164-170; MR 45 #1856.
[13] C. J. Mozzochi, On the difference between consecutive primes, *J. Number Theory* 24 (1986) 181-187; MR 88b:11057.
[14] S. T. Lou and Q. Yao, The number of primes in a short interval, *Hardy-Ramanujan J.* 16 (1993) 21-43; MR 94b:11089.
[15] R. C. Baker and G. Harman, The difference between consecutive primes, *Proc. London Math. Soc.* 72 (1996) 261-280; MR 96k:11111.
[16] R. C. Baker, G. Harman, and J. Pintz, The difference between consecutive primes. II, *Proc. London Math. Soc.* 83 (2001) 532-562; MR 2002f:11125.
[17] H. Cramér, Some theorems concerning prime numbers, *Ark. Mat. Astron. Fysik*, v. 15 (1920) n. 5, 1-32; also in *Collected Works*, v. 1, ed. A. Martin-Löf, Springer-Verlag, 1994, pp. 138-170.
[18] A. Ivić, *The Riemann Zeta-Function*, Wiley, 1985, pp. 321-330, 349-350; MR 87d:11062.
[19] H. Cramér, On the order of magnitude of the differences between consecutive prime numbers, *Acta Arith.* 2 (1936) 23-46; also in *CollectedWorks*, v. 2, ed. A. Martin-Löf, Springer-Verlag, 1994, pp. 871-894.
[20] A. Granville, Unexpected irregularities in the distribution of prime numbers, *International Congress of Mathematicians*, v. 1, Proc. 1994 Zürich conf., ed. S. D. Chatterji, Birkhäuser, 1995, pp. 388-399; MR 97d:11139.
[21] A. Granville, Harald Cramér and the distribution of prime numbers, *Scand. Actuar. J.* (1995) n. 1, 12-28; MR 96g:01002.
[22] H. Maier, Primes in short intervals, *Michigan Math. J.* 32 (1985) 221-225; MR 86i:11049.
[23] E.Westzynthius, Über die Verteilung der Zahlen die zu den n ersten Primzahlen teilerfremd sind, *Comment. Phys. Math. Soc. Sci. Fenn.*, v. 5 (1931) n. 25, 1-37.
[24] E. Bombieri and H. Davenport, Small differences between prime numbers, *Proc. Royal Soc. Ser. A* 293 (1966) 1-18; MR 33 #7314.
[25] M. N. Huxley, Small differences between consecutive primes. II, *Mathematika* 24 (1977) 142-152; MR 57 #5925.
[26] H. Maier, Small differences between prime numbers, *Michigan Math. J.* 35 (1988) 323-344; MR 90e:11126.

2.14 ブルン定数

Brun（ブルン）定数は双子素数のすべての逆数の和として定義される[1,2].

$$B_2 = \left(\frac{1}{3}+\frac{1}{5}\right)+\left(\frac{1}{5}+\frac{1}{7}\right)+\left(\frac{1}{11}+\frac{1}{13}\right)+\left(\frac{1}{17}+\frac{1}{19}\right)+\left(\frac{1}{29}+\frac{1}{31}\right)+\cdots$$

素数5は2度加えられていることに注意する（別な定義ではそうではない）．もしこの級数が発散するならば，双子素数予想【2.1】の証明はただちに成り立つ．しかしながら，Burnは，この級数は収束することを証明した．よって，B_2は有限の値である[3-8].

彼の結果は，すべての素数（逆数の和は発散する【2.2】）に対して双子素数が稀であることを示した．しかし，双子素数の個数が有限であろうと無限であろうとも，彼の結果は輝きを失わない．

 Selmer（セルマー）[9]，Fröberg（フレーベルグ）[10]，Bohman（ボーマン）[11]，Shanks & Wrench（シャンクス&レンチ）[12]，Brent（ブレント）[13,14]，Nicely（ナイスリー）[15-18]，Sebah（セバ）[19]その他は，B_2 の数値評価の改良に成功した．最も最近の計算では，双子素数の大きなデータベースおよび拡張された双子素数予想【2.1】が正しいと仮定して，

$$B_2 = 1.9021605831\cdots$$

である．後の議論でさらに詳しく述べる．Hardy & Littlewood(ハーディ&リトルウッド) の仮説からわかるように，双子素数の逆数の定義どおりの和は，ひじょうにゆっくりと収束する．

$$\sum_{\substack{\text{twin}\\p\leq n}}\frac{1}{p} - B_2 = O\Big(\frac{1}{\ln(n)}\Big)$$

しかし，次の補正は収束を加速するために役に立つ[10,12,15]．

$$\Big(\sum_{\substack{\text{twin}\\p\leq n}}\frac{1}{p} + \frac{4C_{\text{twin}}}{\ln(n)}\Big) - B_2 = O\Big(\frac{1}{\sqrt{n}\ln(n)}\Big)$$

ただし，$C_{\text{twin}} = 0.6601618158\cdots$ は双子素数定数である．高い次数の補外は存在するが，実用的な有用性はいまだにない．Nicely（ナイスリー）[15]の計算の中心部分において，彼はあまり知られていないインテルペンティアム CPU のバグ（設計ミス）を発見した．

 関連する3つの変形を議論する．A_3 を $(p, p+2, p+6)$ の形をした素数の3つの組の逆数の和とし，A_3' を $(p, p+4, p+6)$ の形をした素数の3つの組の逆数の和とし，A_4 を $(p, p+2, p+6, p+8)$ の形をした素数の4つの組の逆数の和とする．Nicely[2,20]は

$$A_3 = 1.0978510391\cdots, \quad A_3' = 0.8371132125\cdots, \quad A_4 = 0.8705883800\cdots$$

を計算した．B_h を，2つの素数の差が h である素数の逆数の和と定義する．ただし，$h \geq 2$ は偶数とする．$\tilde{B}_h$ を，「引き続く」2つの素数の差が h である素数の逆数の和と定義する．Segal（セガル）は，B_h はすべての h に対して有限であることを証明した[5,21,22]．したがって，$\tilde{B}_h$ も同様に有限である．明らかに，$B_2 = \tilde{B}_2$ であり，

$$B_4 = \Big(\frac{1}{3}+\frac{1}{7}\Big)+\Big(\frac{1}{7}+\frac{1}{11}\Big)+\Big(\frac{1}{13}+\frac{1}{17}\Big)+\Big(\frac{1}{19}+\frac{1}{23}\Big)+\cdots = \tilde{B}_4 + \frac{10}{21}$$

である．しかし，高い精度での B_h および $\tilde{B}_h$（$h \geq 4$）の計算は，いまだに行われていない．Wolf（ウルフ）[23]は，$h \geq 6$ に対して，少量のデータに基づいて

$$\tilde{B}_h = \frac{4C_{\text{twin}}}{h}\prod_{\substack{p|h\\p>2}}\frac{p-1}{p-2}$$

であると推測した．たとえ彼の予想がいつか誤りであると示されたとしても，この予想

は，数論で見られる他の定数と，このような一般化ブルン定数との関係を，さらに調べてみようという気にさせる．

[1] H. Riesel, *Prime Numbers and Computer Methods for Factorization*, Birkhäuser, 1985, pp. 64-65; MR 95h：11142.
[2] P. Ribenboim, *The New Book of Prime Number Records*, 3rd ed., Springer-Verlag, 1996, pp. 261, 509-510; MR 96k:11112.
[3] V. Brun, La série $1/5+1/7+1/11+1/13+1/17+1/19+1/29+1/31+1/41+1/43+1/59+1/61+\cdots$ où les dénominateurs sont "nombres premiers jumeaux" est convergente ou finie, *Bull Sci. Math.* 43 (1919) 100-104, 124-128.
[4] H. Rademacher, *Lectures on Elementary Number Theory*, Blaisdell, 1964, pp. 137-144; MR 58 # 10677.
[5] H. Halberstam and H.-E. Richert, *Sieve Methods*, Academic Press, 1974, pp. 50-52, 91-92, 116-117; MR 54 # 12689.
[6] W. J. LeVeque, *Fundamentals of Number Theory*, Addison-Wesley, 1977, pp. 173-177; MR 58 # 465.
[7] M. B. Nathanson, *Additive Number Theory: The Classical Bases*, Springer-Verlag, 1996, pp. 167-174; MR 97e:11004.
[8] K. D. Boklan, Fugitive pieces: Elementary results for twin primes, unpublished note (2000).
[9] E. S. Selmer, A special summation method in the theory of prime numbers and its application to "Brun's sum" (in Norwegian), *Norsk. Mat. Tidsskr.* 24 (1942) 74-81; MR 8,316g.
[10] C.-E. Fröberg, On the sum of inverses of primes and twin primes, *Nordisk Tidskr. Informat. (BIT)* 1 (1961) 15-20.
[11] J. Bohman, Some computational results regarding the prime numbers below 2,000,000,000, *Nordisk Tidskr. Informat.* (BIT) 13 (1973) 242-244; errata 14 (1974) 127; MR 48 # 217.
[12] D. Shanks and J.W. Wrench, Brun's constant, *Math. Comp.* 28 (1974) 293-299; corrigenda 28 (1974) 1183; MR 50 # 4510.
[13] R. P. Brent, Irregularities in the distribution of primes and twin primes, *Math. Comp.* 29 (1975) 43-56, correction 30 (1976) 198; MR 51 # 5522 and 53 # 302.
[14] R. P. Brent, Tables concerning irregularities in the distribution of primes and twin primes up to 10^{11}, *Math. Comp.* 29 (1975) 331; 30 (1976) 379.
[15] T. R. Nicely, Enumeration to 10^{14} of the twin primes and Brun's constant, *Virginia J. Sci.* 46 (1995) 195-204; MR 97e:11014.
[16] R. P. Brent, Review of T. R. Nicely's paper, *Math. Comp.* 66 (1997) 924-925.
[17] T. R. Nicely, Enumeration to $1-6 \cdot 10^{15}$ of the twin primes and Brun's constant, unpublished note (1999).
[18] T. R. Nicely, A new error analysis for Brun's constant, *Virginia J. Sci.* 52 (2001) 45-55.
[19] P. Sebah, Counting twin primes and estimating Brun's constant up to 10^{16} (Numbers, Constants and Computation).
[20] T. R. Nicely, Enumeration to $1.6 \cdot 10^{15}$ of the prime quadruplets, unpublished note (1999).
[21] B. Segal, Generalization of a theorem of Brun's (in Russian), *C. R. Acad. Sci. URSS A* (1930) 501-507.
[22] W. Narkiewicz, *Number Theory*, World Scientific, 1983, pp. 144-153; MR 85j:11002.
[23] M. Wolf, Generalized Brun's constants, unpublished note (1997).

2.15 グレイシャー-キンケリン定数

Stirling (スターリング) の公式[1]

$$\lim_{n\to\infty}\frac{n!}{e^{-n}n^{n+1/2}}=\sqrt{2\pi}$$

は，大きな数の階乗を評価するよく知られた式である．$n!=\Gamma(n+1)$ を違う式で置きかえる．たとえば

$$K(n+1)=\prod_{m=1}^{n}m^m \quad \text{または} \quad G(n+1)=\frac{(n!)^n}{K(n+1)}=\prod_{m=1}^{n-1}m!$$

このとき，近似式は異なる形となる．Kinkelin (キンケリン) [2], Jeffery (ジェフリー) [3], Glaisher (グレイシャー) [4-6]は，

$$\lim_{n\to\infty}\frac{K(n+1)}{e^{-(1/4)n^2}n^{(1/2)n^2+(1/2)n+1/12}}=A, \quad \lim_{n\to\infty}\frac{G(n+1)}{e^{-(3/4)n^2}(2\pi)^{(1/2)n}n^{(1/2)n^2-1/12}}=\frac{e^{1/12}}{A}$$

を示した．定数 A は，これらの近似において，Stirling の公式での $\sqrt{2\pi}$ と同じような役割をする．A は次の閉じた式をもつ．

$$A=\exp\left(\frac{1}{12}-\zeta'(-1)\right)=\exp\left(\frac{-\zeta'(2)}{2\pi^2}+\frac{\ln(2\pi)+\gamma}{12}\right)=1.2824271291\cdots$$

ただし，$\zeta'(x)$ は Riemann のゼータ関数【1.6】の導関数であり，γ はオイラー-マスケローニの定数【1.5】である．$\zeta'(2)$ の別のところでの出現については，【2.10】および【2.18】を参照せよ．

A を含む多くの美しい式が存在する．無限積を含む 2 つのものは，[6]

$$1^{1/1}\cdot 2^{1/4}\cdot 3^{1/9}\cdot 4^{1/16}\cdot 5^{1/25}\cdots=\left(\frac{A^{12}}{2\pi e^\gamma}\right)^{\pi^2/6}$$

$$1^{1/1}\cdot 3^{1/9}\cdot 5^{1/25}\cdot 7^{1/49}\cdot 9^{1/81}\cdots=\left(\frac{A^{36}}{2^4\pi^3 e^{3\gamma}}\right)^{\pi^2/24}$$

である．2 つの定積分は[4,7]

$$\int_0^\infty \frac{x\ln(x)}{e^{2\pi x}-1}dx=\frac{1}{24}-\frac{1}{2}\ln(A)$$

$$\int_0^{1/2}\ln(\Gamma(x+1))\,dx=-\frac{1}{2}-\frac{7}{24}\ln(2)+\frac{1}{4}\ln(\pi)+\frac{3}{2}\ln(A)$$

である．さらに多くの公式が[8-12]にある．最後の式の一般化

$$\int_0^x \ln(\Gamma(t+1))\,dt=\frac{1}{2}\ln(2\pi)x-\frac{1}{2}x(x+1)+x\ln(\Gamma(x+1))-\ln(G(x+1))$$

は，$G(x+1)$ の解析接続を用いて，Alexeiewsky (アレクセイエフスキー) [13], Hölder (ヘルダー) [14], Barnes (バーンズ) [15-17]により得られた．ガンマ関数が階乗関数 $\Gamma(n+1)$ を複素数平面に拡張したものであるように，**バーンズ G-関数**

$$G(z+1) = (2\pi)^{(1/2)z} e^{-(1/2)z(z+1)-(\gamma/2)z^2} \prod_{n=1}^{\infty} \left(1+\frac{z}{n}\right)^n e^{-z+(1/2n)z^2}$$

は $G(n+1)$ を拡張したものである。ガンマ関数が $z=1/2$ で特別な値

$$\Gamma\left(\frac{1}{2}\right) = \left(-\frac{1}{2}\right)! = \sqrt{\pi}$$

をとるように，バーンズ関数は

$$G\left(\frac{1}{2}\right) = 2^{1/24} e^{1/8} \pi^{-1/4} A^{-3/2}$$

を満たす。

$K(z+1) = \Gamma(z+1)^z / G(z+1)$ よるキンケリン関数の類似した自然な拡張は G の方が好まれていたため，$K(z)$ は研究者から比較的無視され続けていた。応用例がある。

$$D(x) = \lim_{n\to\infty} \prod_{k=1}^{2n+1} \left(1+\frac{x}{k}\right)^{(-1)^{k+1}k} = \exp(x) \cdot \lim_{n\to\infty} \prod_{k=1}^{2n} \left(1+\frac{x}{k}\right)^{(-1)^{k+1}k}$$

と定義する。Melzak (メルザック) [18]は $D(2) = (\pi e)/2$ を証明した。Borwein & Dykshoorn (ボーウィン&ダイクスホーン) [19]はこの結果を

$$D(x) = \left(\frac{\Gamma\left(\frac{x}{2}+\frac{1}{2}\right)}{\Gamma\left(\frac{x}{2}\right)}\right)^x \left(\frac{K\left(\frac{x}{2}\right)K\left(\frac{1}{2}\right)}{K\left(\frac{x}{2}+\frac{1}{2}\right)}\right)^2 \exp\left(-\frac{x}{2}\right)$$

と拡張した。ただし，$x>0$ である。特別な場合として，$D(1) = A^6/(2^{1/6}\pi^{1/2})$ である。

たまのうわさ話[20-27]は別として，グレイシャー-キンケリン定数の大部分は最近まで忘れられていた。Vignéras (ヴィニェラス) [28]，Voros (ヴォロス) [29]，Sarnak (サルナック) [30]，Vardi (ヴァルディ) [31]らは，G 関数が数理物理および微分幾何学におけるあるスペクトル関数に関連するという理由で，G 関数の興味が再びよみがえった。また，ランダム行列理論およびゼータ関数の零点の間隔[32-34]の問題とも関連している。同様に【2.15.3】および【5.22】を参照せよ。したがって，ここでの $\Gamma(1/2)$ および $G(1/2)$ の一般化は，最初の発見者が予想しなかった意味をもっている。

2.15.1 一般化グレイシャー定数

Bendersky (ベンダースキー) [35,36]は，積 $1^{1^k} \cdot 2^{2^k} \cdot 3^{3^k} \cdot 4^{4^k} \cdots n^{n^k}$ を研究した。$k=0$ のとき $n!$ で，$k=1$ のとき $K(n+1)$ である。さらに正確には，彼は積の対数を調べ，極限値

$$\ln(A_k) = \lim_{n\to\infty}\left(\sum_{m=1}^{n} m^k \ln(m) - p_k(n)\right)$$

を決定した。ただし，

$$p_k(n) = \left(\frac{n^{k+1}}{k+1} + \frac{n^k}{2} + \frac{B_{k+1}}{k+1}\right)\ln(n) - \frac{n^{k+1}}{(k+1)^2}$$

$$+ k! \sum_{j=1}^{k-1} \frac{B_{j+1}}{(j+1)!} \frac{n^{k-j}}{(k-j)!} \left(\ln(n) + \sum_{i=1}^{j} \frac{1}{k-i+1}\right)$$

で，B_n は n 次のベルヌーイ数【1.6.1】である．明らかに，$A_0=\sqrt{2\pi}$, $A_1=A$ である．Choudhury（チョウドゥリ）[37]と Adamchik（アダムチック）[38]は，すべての $k \geqq 0$ に対して次の正確な A_k の表現式を得た．

$$A_k = \exp\left(\frac{B_{k+1}}{k+1}\sum_{j=1}^{k}\frac{1}{j} - \zeta'(-k)\right) = \begin{cases} 1.0309167521\cdots & (k=2\text{ のとき}) \\ 0.9795555269\cdots & (k=3\text{ のとき}) \\ 0.9920479745\cdots & (k=4\text{ のとき}) \\ 1.0096803872\cdots & (k=5\text{ のとき}) \end{cases}$$

負の整数におけるゼータ関数の導関数は，次のように置きかえられる．$n>0$ とするとき[12, 39]，

$$\zeta'(-2n) = (-1)^n \frac{(2n)!}{2(2\pi)^{2n}} \zeta(2n+1),$$

$$\zeta'(-2n+1) = \frac{1}{2n}\left[(-1)^{n+1}\frac{2(2n)!}{(2\pi)^{(2n)}}\zeta'(2n) + \left(\sum_{j=1}^{2n-1}\frac{1}{j} - \ln(2\pi) - \gamma\right)B_{2n}\right]$$

である．これから，$\ln(A_2) = \zeta(3)/4\pi^2$ および $\ln(A_3) = 3\zeta'(4)/4\pi^4 - (\ln(2\pi)+\gamma)/120$ を得る．

2.15.2 多重バーンズ関数

Barnes[40]は複素数平面で

$$G_0(z) = z, \quad G_n(1) = 1, \quad G_{n+1}(z+1) = \frac{G_{n+1}(z)}{G_n(z)} \quad (n \geqq 0)$$

を満たす関数 $\{G_n(z)\}$ の列を定義した．Bohr-Mollerup（ボーア-モレラップ）の定理[41]と類似の議論により，さらに

$$(-1)^n \frac{d^{n+1}}{dx^{n+1}} \ln(G_n(x)) \geqq 0 \quad (x>0 \text{ について})$$

を仮定すると，この関数列 $\{G_n(z)\}$ はただ1つに決まる．明らかに，$G_1(z) = 1/\Gamma(z)$, $G_2(z) = G(z)$ である．$\{G_n(z)\}$ の性質は[31, 42, 43]で与えられている．特別な興味は $G_n(1/2)$ の値にある．Adamchik[42]は，最も簡単なものであると知られている式

$$\ln\left(G_n\left(\frac{1}{2}\right)\right) = \frac{1}{(n-1)!}\left[-\frac{\ln(\pi)}{2^n}\prod_{k=2}^{n}(2k-3) \right.$$
$$\left. + \sum_{m=1}^{n}\left(\ln(2)\frac{B_{m+1}}{m+1} + (2^{m+1}-1)\zeta'(-m)\right)\frac{q_{m,n}}{2^m}\right]$$

を決定した．ただし，$q_{m,n}$ は，多項式 $2^{1-n}\prod_{j=1}^{n-1}(2x+2j-1)$ の展開式における x^m の係数である．したがって，一般化グレイシャー定数 A_k を用いて，

$$\ln\left(G_3\left(\frac{1}{2}\right)\right) = \frac{1}{8} + \frac{1}{24}\ln(2) - \frac{3}{16}\ln(\pi) - \frac{3}{2}\ln(A_1) - \frac{7}{8}\ln(A_2),$$

$$\ln\left(G_4\left(\frac{1}{2}\right)\right) = \frac{265}{2304} + \frac{229}{5760}\ln(2) - \frac{5}{32}\ln(\pi) - \frac{23}{16}\ln(A_1) - \frac{21}{16}\ln(A_2) - \frac{5}{16}\ln(A_3)$$

と表される．

2.15.3 GUE 仮設

リーマン予想【1.6.2】が真と仮定する．

$\gamma_1 = 14.1347251417\cdots \leq \gamma_2 = 21.0220396387\cdots \leq \gamma_3 = 25.0108575801\cdots \leq \gamma_4 \leq \gamma_5 \leq \cdots$
を上半平面にある $\zeta(z)$ の自明でない零点の虚数部分であるとする．$N(T)$ を，虚数部分が $<T$ であるような零点の個数とする．このとき，Riemann-von Mangoldt（リーマン-フォン·マンゴルト）公式[44]は，$T \to \infty$ のとき

$$N(T) = \frac{T}{2\pi}\ln\left(\frac{T}{2e}\right) + O(\ln(T))$$

である．したがって，$n \to \infty$ のとき

$$\gamma_n \sim \frac{2\pi n}{\ln(n)}$$

が成り立つ．γ_n と γ_{n+1} の平均の間隔は，$n \to \infty$ のとき 0 に収束する．だから，相続く 2 数の差を

$$\delta_n = \frac{\gamma_{n+1} - \gamma_n}{2\pi}\ln\left(\frac{\gamma_n}{2\pi}\right)$$

と再正規化する（「間隔を広げる」）ことは役に立つ．したがって，δ_n は平均値 1 をもつ．

δ_n の確率分布については，何がいえるのだろうか？ すなわち，どのような密度関数 $p(s)$ が，すべての $0 < \alpha < \beta$ に対して，

$$\lim_{N \to \infty} \frac{1}{N}|\{n : 1 \leq n \leq N, \quad \alpha \leq \delta_n \leq \beta\}| = \int_\alpha^\beta p(s)\,ds$$

を満たすのだろうか？

予想のすばらしい答えがある．ランダムエルミート（Hermitian）$N \times N$ 行列 X が**ガウス・ユニタリー集合**（Gaussian unitary ensemble）（GUE）に属するとは，（実）対角要素 x_{jj} および（複素）上三角成分 $x_{jk} = u_{jk} + iv_{jk}$ が，$\mathrm{Var}(x_{jj}) = 2$ $(1 \leq j \leq N)$, $\mathrm{Var}(u_{jk}) = \mathrm{Var}(v_{jk}) = 1$ $(1 \leq j < k \leq N)$ をもつ平均値 0 のガウス分布から独立に選ばれたものである，ときをいう．$\lambda_1 \leq \lambda_2 \leq \lambda_3 \leq \cdots \leq \lambda_N$ を，X の（実数の）固有値とし，正規化された間隔（normalized spacings）

$$\tilde{\delta}_n = \frac{\lambda_{n+1} - \lambda_n}{4\pi}\sqrt{8N - \lambda_n^2} \quad \left(n \approx \frac{N}{2}\right)$$

を考える．

この尺度の選び方により，$\tilde{\delta}_n$ の平均値は 1 である．$N \to \infty$ のとき，$\tilde{\delta}_n$ の確率密度は Gaudin（**ガウディン**）**密度** $p(s)$ とよばれるものに収束する．Montgomery[45]によるいくつかの理論的な仕事により示唆されて，Odlyzko[46-50]は実験的に，δ_n と $\tilde{\delta}_n$ の分布はひじょうに近いことを示した．**GUE 仮設**（または Montgomery-Odlyzko（**モンゴメリー−オドリズコ**）**法則**）は，この 2 つの分布が同じであるという驚くべき予想である．図 2.1 を見よ．

さらに，関数 $p(s)$ に関する拡張した結果がある．

2.15 グレイシャー-キンケリン定数

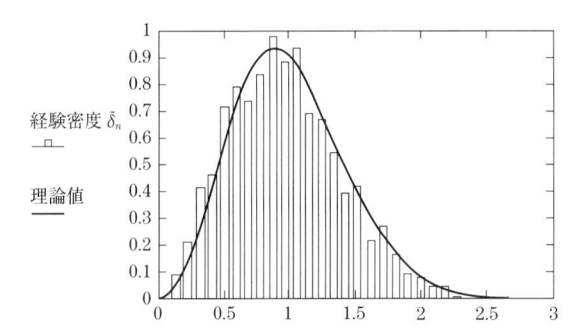

図 2.1 少数のシミュレーションで 120×120 ランダム GUE 行列を 50 個発生させたときの固有値の分布. $\tilde{\delta}_n$ と $p(s)$ のヒストグラムはよく一致.

$$E(s) = \exp\left(\int_0^{\pi s} \frac{\sigma(t)}{t}\, dt\right)$$

と定義する. ただし, $\sigma(t)$ は Painlevé (パンルヴェ) の第 V 微分方程式 (「シグマ形式」で)

$$(t\cdot\sigma'')^2 + 4(t\cdot\sigma' - \sigma)[t\cdot\sigma' - \sigma + (\sigma')^2] = 0$$

で, 境界条件

$$\sigma(t) \sim -\frac{t}{\pi} - \left(\frac{t}{\pi}\right)^2 \quad (t\to 0^+), \quad \sigma(t) \sim -\left(\frac{t}{2}\right)^2 - \frac{1}{4} \quad (t\to\infty)$$

を満たす. このとき, $p(s) = d^2 E/ds^2$ となる.

$E(s)$ に対する Fredholm (フレドホルム) 行列式表現[49,57,58]が, 歴史的にはより早いが, 上にある Painlevé 表現[51-56]で, $p(s)$ の直接的な数値計算ができる. (ちなみに, Painlevé の第 II 方程式は, 最長増加数列の問題 (the longest increasing subsequence probrem)【5.20】の議論に現れる. Painlevé の第 III 方程式は, イジングモデル (Ising model)【5.22】に関連して現れる.)

$p(s)$ を用いて, $\tilde{\delta}_n$ の中央値, モード, 分散を計算できる. もちろん高い次数のモーメントも計算できる.

ガウディン分布[59,60]の裾 (端) を扱うおもしろい問題がある. 関数 $E(s)$ は, 区間 $[0,s]$ が (基準化された) 固有値をまったく含まない, という確率として解釈できる. 特定の区間 $[0,s]$ を任意の長さ s の区間に置きかえても, 確率は同じままである. [49,61]より

$$E(s) \sim 1 - s + \frac{\pi^2 s^4}{360} \quad (s\to 0^+), \quad E(s) \sim C\cdot(\pi s)^{-1/4}\exp\left(-\frac{1}{8}(\pi s)^2\right) \quad (s\to\infty)$$

である. ただし, C はある定数である. Dyson (ダイソン) [49,62]は, Widom (ウィドン) [63]の結果を用いて, 厳密ではないが,

$$C = 2^{1/3} e^{3\zeta'(-1)} = 2^{1/4} e^{2B}$$

と特定した. ただし,

$$B = \frac{1}{24}\ln(2) + \frac{3}{2}\zeta'(-1) = -0.2192505830\cdots$$

である．今度は，これは公式[22]

$$e^{2B} = 2^{1/12} e^{1/4} A^{-3}$$

により，グレイシャー定数 A と関連する．$E(s)$ の完全漸近展開が現在知られて[60,64-66]，因子 C を除いてすべて正確に得られていることは不思議なことである．関連するある問題に関して[67-70]で，同じ現象が報告されている．

GUE 仮設を見る別の方法がある．正規化された連続するゼータ関数の零点の差 δ_n に戻る．

$$\Delta_{nk} = \sum_{j=0}^{k} \delta_{n+j}$$

と定義する．以前は，k は0にかぎられていた．$k \geq 0$ を変化させると，Δ_{nk} の「分布」はどうなるのだろうか？ Montgomery[45]は次の簡単な式

$$\lim_{N\to\infty} |\{(n,k): 1 \leq n \leq N, \ k \geq 0, \ \alpha \leq \Delta_{nk} \leq \beta\}| = \int_\alpha^\beta \left[1 - \left(\frac{\sin(\pi r)}{\pi r}\right)^2\right] dr$$

が真である，と予想した．いいかえると，$1-(\sin(\pi r)/(\pi r))^2$ は，Montgomery の部分的な結果により予言されるように，ゼータ関数の零点の**対相関関数**（pair correlation function）である．信じ難い話だが，GUE 固有値は同じ対相関関数をもつことが証明されている．Odlyzko[46-49]は再び，この予想を支持する広範囲にわたる数値的な証拠を集めている．素数論に対する対相関関数がもつ意味は，[71]で調べられている．Hejhal（ヘジャル）[72]は，三重相関（triple correlation）予想として知られている，3次元の類似な相関について研究した．3より高次の相関については，[73]で調べられている．

注意深い読者は，以前の δ_n の定義で登場した，制限 $n \approx N/2$ に気付くだろう．わずかなシミュレーションにおいて，スペクトルの「バルク (bulk)」として知られているサンプリングである，固有値の中央集団（the middle third，3つに分けた中央の集団）のみを取った．もし，その代わりにスペクトルの「端」をサンプルとすると，違った密度が現れる[69,70]．この「バルク」に対するフレドホルム行列式の正弦核は，「端」に対しては Airy（エアリ）核に置きかえられる．

Rudnick & Sarnak（ルドニック&サルナック）[73,74]と Katz & Sarnak（カッツ&サルナック）[75,76]は，より広く，より抽象的な条件でGUE仮設を一般化した．彼らは，ある重要な特別ないくつかの場合を証明をした．しかし，ここで議論した元の場合ではない．

δ_n の上極限および下極限に，興味がある．それらはそれぞれ∞および0である[77-80]と予想されている．

ひじょうに多くの研究は，ランダム行列（対称性は仮定しない）とランダム多項式の関連する分野に導かれている．一例だけを述べる．$q(x)$ を，標準ガウス分布から独立に選ばれた実数を係数とする，次数 n のランダム多項式とする．z_n を $q(x)$ の実数の

2.15 グレイシャー-キンケリン定数

零点の期待個数とする．Kac（カッツ）[81,82]は

$$\lim_{n\to\infty} \frac{z_n}{\ln(n)} = \frac{2}{\pi}$$

を証明した．[82-88]から

$$\lim_{n\to\infty} \left(z_n - \frac{2}{\pi}\ln(n) \right) = c$$

ただし，

$$c = \frac{2}{\pi}\left[\ln(2) + \int_0^\infty (\sqrt{x^{-2} - 4e^{-2x}(1-e^{-2x})^{-2}} - (x+1)^{-1})\,dx \right]$$
$$= 0.6257358072\cdots$$

であると知られている．漸近展開のもっと多くの項が知られている．展望としては[82,87]を参照せよ．

[1] A. E. Taylor and R. Mann, *Advanced Calculus*, 2nd ed., Wiley, 1972, pp. 740-745; MR 83m:26001.
[2] J. Kinkelin, Über eine mit der Gammafunction verwandte Transcendente und deren Anwendung auf die Integralrechnung, *J. Reine Angew. Math.* 57 (1860) 122-158.
[3] H. M. Jeffery, On the expansion of powers of the trigonometrical ratios in terms of series of ascending powers of the variable, *Quart. J. Pure Appl. Math.* 5 (1862) 91-108.
[4] J. W. L. Glaisher, On the product $1^1 \cdot 2^2 \cdot 3^3 \cdots n^n$, *Messenger of Math.* 7 (1878) 43-47.
[5] J. W. L. Glaisher, On certain numerical products in which the exponents depend upon the numbers, *Messenger of Math.* 23 (1893) 145-175.
[6] J. W. L. Glaisher, On the constant which occurs in the formula for $1^1 \cdot 2^2 \cdot 3^3 \cdots n^n$, *Messenger of Math.* 24 (1894) 1-16.
[7] G. Almkvist, Asymptotic formulas and generalized Dedekind sums, *Experim. Math.* 7 (1998) 343-359; MR 2000d:11126.
[8] R. W. Gosper, $\int_{n/4}^{m/6} \ln(\Gamma(z))\,dz$, *Special Functions, q-Series and Related Topics*, ed. M. Ismail, D. Masson, and M. Rahman, Fields Inst. Commun., v. 14, Amer. Math. Soc., 1997, pp. 71-76; MR 98m:33004.
[9] J. Choi and H. M. Srivastava, Sums associated with the zeta function, *J. Math. Anal. Appl.* 206 (1997) 103-120; MR 97i:11092.
[10] J. Choi and C. Nash, Integral representations of the Kinkelin's constant A, *Math. Japon.* 45 (1997) 223-230; MR 98j:33002.
[11] J. Choi, H. M. Srivastava, and N.-Y. Zhang, Integrals involving a function associated with the Euler-Maclaurin summation formula, *Appl. Math. Comput.* 93 (1998) 101-116; MR 99h:33004.
[12] J. Choi and H. M. Srivastava, Certain classes of series associated with the zeta function and multiple gamma functions, *J. Comput. Appl. Math.* 118 (2000) 87-109; MR 2001f:11147.
[13] W. Alexeiewsky, Über eine Classe von Functionen, die der Gammafunction analog sind, *Berichte über die Verhandlungen der Königlich Sächsischen Gesellschaft der Wissenschaften zu Leipzig*, Math.-Phys. Klasse 46 (1894) 268-275.
[14] O. Hölder, Über eine transcendente Function, *Nachrichten von der Königlichen Gesellschaft der Wissenschaften und der Georg Augusts Universität zu Göttingen* (1886) 514-522.
[15] E.W. Barnes, The theory of the G-function, *Quart. J. Pure Appl. Math.* 31 (1900) 264-314.
[16] E. W. Barnes, The genesis of the double gamma functions, *Proc. London Math. Soc.* 31 (1899) 358-381.
[17] E.W. Barnes, The theory of the double gamma function, *Philos. Trans. Royal Soc. London Ser. A* 196 (1901) 265-388.
[18] Z. A. Melzak, Infinite products for πe and π/e, *Amer. Math. Monthly* 68 (1961) 39-41; MR 23

A252.
[19] P. Borwein and W. Dykshoorn, An interesting infinite product, *J. Math. Anal. Appl.* 179 (1993) 203-207; MR 94j:11131.
[20] E. T. Whittaker and G. N.Watson, *A Course of Modern Analysis*, 4th ed., Cambridge Univ. Press, 1963, p. 264; MR 97k:01072.
[21] F. W. J. Olver, *Asymptotics and Special Functions*, Academic Press, 1974, pp. 285-292; MR 97i:41001.
[22] B. M. McCoy and T. T.Wu, *The Two-Dimensional Ising Model*, Harvard Univ. Press, 1973, Appendix B.
[23] D. H. Greene and D. E. Knuth, *Mathematics for the Analysis of Algorithms*, Birkhäuser, 1981, pp. 99-100; MR 92c:68067.
[24] G. H. Norton, On the asymptotic analysis of the Euclidean algorithm, *J. Symbolic Comput.* 10 (1990) 53-58; MR 91k:11088.
[25] T. Shintani, A proof of the classical Kronecker limit formula, *Tokyo J. Math.* 3 (1980) 191-199; MR 82f:10038.
[26] R. A MacLeod, Fractional part sums and divisor functions, *J. Number Theory* 14 (1982) 185-227; MR 83m:10080.
[27] B. C. Berndt, *Ramanujan's Notebooks: Part I*, Springer-Verlag, 1985, pp. 273-288; MR 86c:01062.
[28] M.-F.Vignéras,L'équation fonctionnelle de la fonction zêta de Selberg du groupe modulaire PSL(2, Z), *Astérisque* 61 (1979) 235-249; MR 81f:10040.
[29] A. Voros, Spectral functions, special functions and the Selberg zeta function, *Commun. Math. Phys.* 110 (1987) 439-465; MR 89b:58173.
[30] P. Sarnak, Determinants of Laplacians, *Commun. Math. Phys.* 110 (1987) 113-120; MR 89e:58116.
[31] I. Vardi, Determinants of Laplacians and multiple gamma functions, *SIAM J. Math. Anal.* 19 (1988) 493-507; MR 89g:33004.
[32] J. P. Keating and N. C. Snaith, Random matrix theory and $\zeta(1/2+it)$, *Commun. Math. Phys.* 214 (2000) 57-89; MR 2002c:11107.
[33] J. P.Keating andN. C. Snaith, Random matrix theory and L-functions at $s=1/2$, *Commun. Math. Phys.* 214 (2000) 91-110; MR 2002c:11108.
[34] P. J. Forrester, *Log-Gases and Random Matrices*, unpublished manuscript (2000).
[35] L. Bendersky, Sur la fonction gamma généralisée, *Acta Math.* 61 (1933) 263-322.
[36] H. T. Davis, *The Summation of Series*, Principia Press of Trinity Univ., 1962, pp. 85-91; MR 25 #5305.
[37] B. K. Choudhury, The Riemann zeta-function and its derivatives, *Proc. Royal Soc. London A* 450 (1995) 477-499; MR 97e:11095.
[38] V. S. Adamchik, Polygamma functions of negative order, *J. Comput. Appl. Math.* 100 (1999) 191-199; MR 99j:33001.
[39] J. Miller and V. S. Adamchik, Derivatives of the Hurwitz zeta function for rational arguments, *J. Comput. Appl. Math.* 100 (1998) 201-206; MR 2000g:11085.
[40] E. W. Barnes, On the theory of the multiple gamma function, *Trans. Cambridge Philos. Soc.* 19 (1904) 374-425.
[41] J. B. Conway, *Functions of One Complex Variable*, 2nd ed., Springer-Verlag, 1978; MR 80c:30003.
[42] V. S. Adamchik, On the Barnes function, *Proc. 2001 Int. Symp. Symbolic and Algebraic Computation (ISSAC)*, ed. B. Mourrain, Univ. of Western Ontario, ACM, 2001, pp. 15-20.
[43] J. Choi, H. M. Srivastava, and V. S. Adamchik, Multiple gamma and related functions, *Appl. Math. Comput.*, to appear.
[44] A. Ivić, *The Riemann Zeta-Function*, Wiley, 1985, pp. 17-21, 251-252; MR 87d:11062.
[45] H. L. Montgomery, The pair correlation of zeros of the zeta function, *Analytic Number Theory*, ed. H. G. Diamond, Proc. Symp. Pure Math. 24, Amer. Math. Soc., 1973, pp. 181-193; MR 49 #2590.
[46] A. M. Odlyzko, On the distribution of spacings between zeros of the zeta function, *Math. Comp.*

2.15 グレイシャー-キンケリン定数

48 (1987) 273-308; MR 88d:11082.
[47] A. M. Odlyzko, The $10^{20\text{th}}$ zero of the Riemann zeta function and 175 million of its neighbors, unpublished manuscript (1992).
[48] A. M. Odlyzko, The $10^{22\text{nd}}$ zero of the Riemann zeta function, *Dynamical, Spectral, and Arithmetic Zeta Functions*, Proc. 1999 San Antonio conf., ed. M. L. Lapidus and M. van Frankenhuysen, Amer. Math. Soc., 2001, pp. 139-144.
[49] M. L. Mehta, *Random Matrices*, Academic Press, 1991; MR 92f:82002.
[50] B. Cipra, A prime case of chaos, *What's Happening in the Mathematical Sciences*, v. 4, Amer. Math. Soc., 1998-1999.
[51] M. Jimbo, T. Miwa, Y. Mori, and M. Sato, Density matrix of an impenetrable Bose gas and the fifth Painlevé transcendent, *Physica D* 1 (1980) 80-158; MR 84k:82037.
[52] A. R. Its, A. G. Izergin, V. E. Korepin, and N. A. Slavnov, Differential equations for quantum correlation functions, *Int. J. Mod. Phys.* B 4 (1990) 1003-1037; MR 91k:82009.
[53] M. L. Mehta, A nonlinear differential equation and a Fredholm determinant, *J. Physique I* 2 (1992) 1721-1729; MR 93i:58163.
[54] M. L. Mehta, Painlevé transcendents in the theory of random matrices, *An Introduction to Methods of Complex Analysis and Geometry for Classical Mechanics and Non-Linear Waves*, Proc. Third Workshop on Astronomy and Astrophysics, ed. D. Benest and C. Froeschlé, Frontières, 1994, pp. 197-208; MR 96k:82007.
[55] C. A. Tracy and H. Widom, Introduction to random matrices, *Geometric and Quantum Aspects of Integrable Systems*, Proc. 1992 Scheveningen conf., ed. G. F. Helminck, Lect. Notes in Physics 424, Springer-Verlag, 1993, pp. 103-130; hep-th/9210073; MR 95a:82050.
[56] C. A. Tracy and H. Widom, Universality of the distribution functions of random matrix theory, *Integrable Systems: From Classical to Quantum*, Proc. 1999 Montréal conf., ed. J. Harnad, G. Sabidussi, and P. Winternitz, Amer. Math. Soc, 2000, pp. 251-264; mathph/9909001; MR 2002f:15036.
[57] M. Gaudin, Sur la loi limite de l'éspacement des valeurs propres d'une matrice aléatorie, *Nucl. Phys.* 25 (1961) 447-458; also in C. E. Porter, *Statistical Theories of Spectra: Fluctuations*, Academic Press, 1965.
[58] M. L. Mehta and M. Gaudin, On the density of eigenvalues of a random matrix, *Nucl. Phys.* 18 (1960) 420-427; MR 22 # 3741.
[59] E. L. Basor, C. A. Tracy, and H. Widom, Asymptotics of level-spacing distributions for random matrices, *Phys. Rev. Lett.* 69 (1992) 5-8, 2880; MR 93g:82004a-b.
[60] C. A. Tracy and H. Widom, Asymptotics of a class of Fredholm determinants, *Spectral Problems in Geometry and Arithmetic*, Proc. 1997 Iowa City conf., Amer. Math. Soc., 1999, pp. 167-174; solv-int/9801008; MR 2000e:47077.
[61] J. des Cloizeaux and M. L. Mehta, Asymptotic behavior of spacing distributions for the eigenvalues of random matrices, *J. Math. Phys.* 14 (1973) 1648-1650; MR 48 # 6500.
[62] F. J. Dyson, Fredholm determinants and inverse scattering problems, *Commun. Math. Phys.* 47 (1976) 171-183; MR 53 # 9993.
[63] H. Widom, The strong Szegö limit theorem for circular arcs, *Indiana Univ. Math. J.* 21 (1971) 277-283; MR 44 # 5693.
[64] H. Widom, The asymptotics of a continuous analogue of orthogonal polynomials, *J. Approx. Theory* 77 (1994) 51-64; MR 95f:42041.
[65] H. Widom, Asymptotics for the Fredholm determinant of the sine kernel on a union of intervals, *Commun. Math. Phys.* 171 (1995) 159-180; MR 96i:47050.
[66] P. A. Deift, A. R. Its, and X. Zhou, A Riemann-Hilbert approach to asymptotic problems arising in the theory of random matrix models, and also in the theory of integrable statistical mechanics, *Annals of Math.* 146 (1997) 149-235; MR 98k:47097.
[67] P. J. Forrester and A. M. Odlyzko, A nonlinear equation and its application to nearest neighbor spacings for zeros of the zeta function and eigenvalues of random matrices, *Organic Mathe-*

matics, Proc. 1995 Burnaby workshop, ed. J. Borwein, P. Borwein, L. Jörgenson, and R. Corless, Amer. Math. Soc., 1995, pp. 239-251; MR 99c:11107.

[68] P. J. Forrester and A. M. Odlyzko, Gaussian unitary ensemble eigenvalues and Riemann zeta function zeros: A non-linear equation for a new statistic, *Phys. Rev. E* 54 (1996) R4493-R4495; MR 98a:82054.

[69] C. A. Tracy and H. Widom, Level-spacing distributions and the Airy kernel, *Phys. Lett. B* 305 (1993) 115-118; hep-th/9210074; MR 94f:82046.

[70] C. A. Tracy and H. Widom, Level-spacing distributions and the Airy kernel, *Commun. Math. Phys.* 159 (1994) 151-174; hep-th/9211141; MR 95e:82003.

[71] D. A. Goldston and H. L. Montgomery, Pair correlation of zeros and primes in short intervals, *Analytic Number Theory and Diophantine Problems*, Proc. 1984 Stillwater conf., ed. A. C. Adolphson, J. B. Conrey, A. Ghosh, and R. I. Yager, Birkhäuser, 1987, pp. 183-203; MR 90h: 11084.

[72] D. A. Hejhal, On the triple correlation of zeros of the zeta function, *Int. Math. Res. Notices* (1994) n. 7, 293-302; MR 96d:11093.

[73] Z. Rudnick and P. Sarnak, The n-level correlations of zeros of the zeta function, *C. R. Acad. Sci. Paris Ser. I Math.* 319 (1994) 1027-1032; MR 96b:11124.

[74] Z. Rudnick and P. Sarnak, Zeros of principal L-functions and random matrix theory, *Duke Math. J.* 81 (1996) 269-322; MR 97f:11074.

[75] N. M. Katz and P. Sarnak, *Random Matrices, Frobenius Eigenvalues, and Monodromy*, Amer. Math. Soc., 1999; MR 2000b:11070.

[76] N. M. Katz and P. Sarnak, Zeroes of zeta functions and symmetry, *Bull. Amer. Math. Soc.* 36 (1999) 1-26; MR 2000f:11114.

[77] H. L. Montgomery and A. M. Odlyzko, Gaps between zeros of the zeta function, *Topics in Classical Number Theory*, v. II, Proc. 1981 Budapest conf., Colloq. Math. Soc. János Bolyai 34, North-Holland, 1984, pp. 1079-1106; MR 86e:11072.

[78] J. B. Conrey, A. Ghosh, and S. M. Gonek, A note on gaps between zeros of the zeta function, *Bull. London Math. Soc.* 16 (1984) 421-424; MR 86i:11048.

[79] J. B. Conrey, A. Ghosh, D. Goldston, S. M. Gonek, and D. R. Heath-Brown, On the distribution of gaps between zeros of the zeta function, *Quart. J. Math.* 36 (1985) 43-51; MR 86j:11083.

[80] J. B. Conrey, A. Ghosh, and S. M. Gonek, Large gaps between zeros of the zeta function, *Mathematika* 33 (1986) 212-238; MR 88g:11057.

[81] A. T. Bharucha-Reid and M. Sambandham, *Random Polynomials*, Academic Press, 1986, pp. 11 -14, 90-91; MR 87m:60118.

[82] K. Farahmand, *Topics in Random Polynomials*, Longman, 1998, pp. 34-35, 59-71; MR 2000d: 60092.

[83] B. R. Jamrom, The average number of real roots of a random algebraic polynomial (in Russian), *Vestnik Leningrad. Univ.* (1971) n. 19, 152-156; MR 45 #7791.

[84] B. R. Jamrom, The average number of real zeros of random polynomials (in Russian), *Dokl. Akad. Nauk SSSR* 206 (1972) 1059-1060; Engl. transl. in *Soviet Math. Dokl.* 13 (1972) 1381-1383; MR 47 #2666.

[85] Y. J. Wang, Bounds on the average number of real roots of a random algebraic equation, *Chinese Annals Math. Ser. A* 4 (1983) 601-605; Engl. summary in *Chinese Annals Math. Ser. B* 4 (1983) 527; MR 85c:60081.

[86] Z. M. Yu, Bounds on the average number of real roots for a class of random algebraic equations, *J. Math. Res. Expos.* 2 (1982) 81-85; MR 84h:60081.

[87] J. E.Wilkins, An asymptotic expansion for the expected number of real zeros of a random polynomial, *Proc. Amer. Math. Soc.* 103 (1988) 1249-1258; MR 90f:60105.

[88] A. Edelman and E. Kostlan, How many zeros of a random polynomial are real?, *Bull. Amer. Math. Soc.* 32 (1995) 1-37; erratum 33 (1996) 325; MR 95m:60082.

2.16 ストラルスキ-ハルボルス定数

 与えられた正の整数 k に対して，k を 2 進表示したとき現れる 1 の個数を $b(k)$ とする．Glaisher（グレイシャー）[1-6] は，$\binom{k}{j}$，$0 \leq j \leq k$ の形の奇数の二項係数の個数は，$2^{b(k)}$ であることを示した．その結果として，パスカル三角形の最初の n 列の中の奇数要素の個数は，

$$f(n) = \sum_{k=0}^{n-1} 2^{b(k)}$$

であり，漸化式

$$f(0) = 0, \quad f(1) = 1,$$
$$f(n) = \begin{cases} 3f(m), & n = 2m \text{ のとき} \\ 2f(m) + f(m+1), & n = 2m+1 \text{ のとき} \end{cases} \quad (n \geq 2)$$

を満たす．ここでの疑問は，「$f(n)$ に対する簡単な近似式を見つけることができるか？」である．答えは肯定的である．$\theta = \ln(3)/\ln(2) = 1.5849625007\cdots$ とおく．これは，法 2 でのパスカル三角形のフラクタル次元である [7,8]．n^θ が，$f(n)$ に対する妥当な近似であることがわかる．また，$f(n)$ は漸近的によい挙動をしないことがわかる．Stolarski（ストラルスキ）[9] と Harborth（ハルボルス）[10] は

$$0.812556 < \lambda = \liminf_{n \to \infty} \frac{f(n)}{n^\theta} < 0.812557 < \limsup_{n \to \infty} \frac{f(n)}{n^\theta} = 1$$

であると決定した．$\lambda = 0.8125565590\cdots$ を Stolarsky-Harborth（**ストラルスキ-ハルボルス**）定数とよぶ．

 一般化がある．p を素数とし，パスカル三角形の最初の n 列の中にある，p で割り切れない要素の個数を $f_p(n)$ とする．

$$\theta_p = \frac{\ln\left(\dfrac{p(p+1)}{2}\right)}{\ln(p)}$$

と定義する．$\lim_{n \to \infty} \theta_p = 2$ に注意する．もちろん，$f_2(n) = f(n)$，$\theta_2 = \theta$ である．[11-14] より

$$\lambda_p = \liminf_{n \to \infty} \frac{f_p(n)}{n^{\theta_p}} < \limsup_{n \to \infty} \frac{f_p(n)}{n^{\theta_p}} = 1$$

$$\lambda_3 = \left(\frac{3}{2}\right)^{1-\theta_3} = 0.7742\cdots, \quad \lim_{p \to \infty} \lambda_p = \frac{1}{2}$$

が知られている．さらに

$$\lambda_5 = \left(\frac{3}{2}\right)^{1-\theta_5} = 0.7582\cdots, \quad \lambda_7 = \left(\frac{3}{2}\right)^{1-\theta_7} = 0.7491\cdots,$$

$$\lambda_{11} = \frac{59}{44}\left(\frac{22}{31}\right)^{\theta_{11}} = 0.7364\cdots$$

と予想されている．不思議なことに，$\lambda_2 = \lambda$ に対する厳密な式は見つかっていない．より広い一般化は，多項係数[15-17]を含む．

2.16.1 デジタル和

前出の関数 $f_2(n)$ は**デジタル和の指数和**（exponential sum of digital sums）である．別の例は

$$m_p(n) = \sum_{k=0}^{n-1}(-1)^{b(pk)}$$

である．これは $p=3$ の場合には，3の倍数は数字1を偶数個もつことが多いという実験的観察結果を測ることになる．しかしながら，最初に**デジタル和のベキ和**（power sum of digital sums）

$$s_q(n) = \sum_{k=0}^{n-1} b(k)^q$$

について議論する．具体的なために $q=1$ とおく．

Trollope（トロロペ）[18]と Delange（ドゥランジュ）[19]は，[20-26]に基づいて，「正確に」

$$s_1(n) = \frac{1}{2\ln(2)} n \ln(2) + nS\left(\frac{\ln(n)}{\ln(2)}\right)$$

であることを証明した．ただし，$S(x)$ は，周期1をもついたるところ微分不可能なある連続関数で

$$-0.2075\cdots = \frac{\ln(3)}{2\ln(2)} - 1 = \inf_x S(x) < \sup_x S(x) = 0$$

を満たす．$S(x)$ のフーリエ係数はすべて知られている．図2.2を見よ．

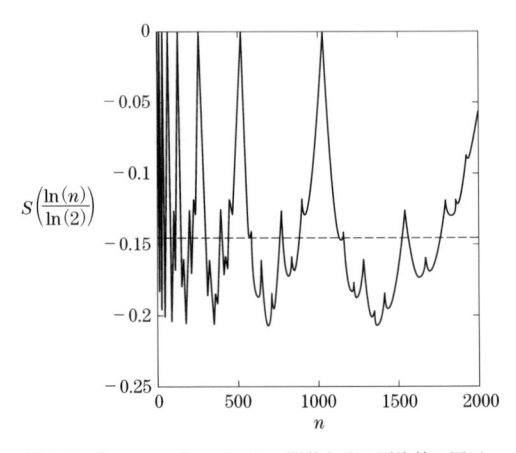

図2.2 トロロペ-ドゥランジュ関数とその平均値の図示．

$S(x)$ の平均は，[19,27]
$$\int_0^1 S(x)\,dx = \frac{1}{2\ln(2)}(\ln(2\pi)-1) - \frac{3}{4} = -0.1455\cdots$$
である．この顕著な結果の任意の q への拡張は，[28-36]にある．

$\omega = \theta/2$ とし，n が奇数のとき $\varepsilon(n) = (-1)^{b(3n-1)}$，その他のとき 0 とする．Newman [37-39]は，つねに $m_3(n) > 0$ で，$O(n^\omega)$ であることを証明した．Coquet（コケ）[40]は，これを
$$m_3(n) = n^\omega M\left(\frac{\ln(n)}{2\ln(2)}\right) + \frac{1}{3}\varepsilon(n)$$
と強めた．ただし，$M(x)$ はいたるところ微分不可能な周期 1 の連続関数で，
$$1.1547\cdots = \frac{2\sqrt{3}}{3} = \inf_x M(x) < \sup_x M(x) = \frac{55}{3}\left(\frac{3}{65}\right)^\omega = 1.6019\cdots$$
である．再び，$M(x)$ のフーリエ係数はすべて知られている．$M(x)$ の平均値[27]は，1.4092203477…であるが，複雑な積分表示をもつ．この結果の $p=5$ および 17 への拡張は，[41-43]にある．$\{(-1)^{b(k)}\}$ のパターンは，よく知られている Prouhet-Thue-Morse（プルーエ-トゥエ-モース）数列{6.8}に従う．$\{(-1)^{b(pk+r)}\}$ の形の部分列に付随した和（associated sums）は，[44-46]で議論されている．

二項係数にもどる．Stein（スタイン）[47]は，
$$f_2(n) = n^\theta F\left(\frac{\ln(n)}{\ln(2)}\right)$$
であることを証明した．ただし，$F(x)$ は周期 1 の連続関数である．比較として，$F(x)$ はほとんどいたるところ微分可能であるが，いたるところ単調でない[48]．しかしながら，この事実は $\lambda_2 = \inf_x F(x)$ の正確な式に関してどんな洞察も与えるとは思えない．$F(x)$ のフーリエ係数はすべて知られている．$F(x)$ の平均値[27]は 0.8636049963…である．再び，基礎にある積分は複雑である．

この分野は，アルゴリズム解析の役に立つ．たとえば，2 進木のレジスター関数（register function）を近似する[49]こと，マージソートの研究[50]，最大値を求めること[51]，その他に分割統治法の漸化式（devide-and-conquer recurrences）[52,53]である．

2.16.2 ウラムの 1 加法数列

デジタル和と Ulam（ウラム）の 1 加法数列[54]とは意外な関連がある．$u < v$ を正の整数とする．基底 u, v をもつ 1 加法数列とは，次のような無限数列 $(u,v) = a_1, a_2, a_3, \cdots$ のことである．$a_1 = u$，$a_2 = v$ であり，a_n は a_{n-1} を超え $a_n = a_i + a_j$（$i < j$，$n \geq 3$）と 1 通りに表現できる最小の整数である．ウラムの原数列
$$(1,2) = 1,2,3,4,6,8,11,13,16,18,26,28,36,38,47,48,53,\cdots$$
は神秘的である．順次の階差数列には，未だにいかなるパターンも見つけられていない．Ulam は，正の整数に対する $(1,2)$ 列の相対密度は 0 であると予想した．この証

明を未だに誰も発見していない.

v が奇数である $(2,v)$ 列の場合および,v が奇数で $v\equiv 1 \bmod 4$ である $(4,v)$ 列の場合については,実質的に多くのことが知られている.Cassaigne & Finch(カサイニュ & フィンチ)[55]は,ウラム1加法数列 $(4,v)$ の階差はいつかは周期的になり,$(4,v)$ 列の密度は

$$d(v)=\frac{1}{2(v+1)}\sum_{k=0}^{(v-1)/2}2^{-b(k)}$$

であることを証明した.$v\to\infty$ のとき $d(v)\to 0$ であることが示される.Stolarsky-Harborth(ストラルスキ-ハルボルス)定数 λ を生みだす技法は,より正確な密度の漸近評価を与えるように修正できる.

$$\frac{1}{4}=\liminf_{\substack{v\to\infty \\ v\equiv 1\bmod 4}}\left(\frac{v}{2}\right)^{2-\theta}d(v)<0.272190<\limsup_{\substack{v\to\infty \\ v\equiv 1\bmod 4}}\left(\frac{v}{2}\right)^{2-\theta}d(v)<0.272191$$

3進2次漸化式のある属とその周期性は,[55]の証明で不思議な役割をする.このアイデアの適用範囲はどこまで広く,技法はどこまで拡張できるのか,を問うことは自然である.

2.16.3 交代ビット集合

n を,$2^{k-1}\leq n<2^k$ を満たす正の整数とする.明らかに,n の2進数展開は k 個のビットをもつ.

n の**交代ビット集合**(alternating bit set)とは,次の性質をもつ n の k ビット位置(k bit positions)の部分集合であると定義する[6, 56-58].

・これらの位置にある n のビットは,1と0を交互にとる
・これらの中で,最も左側(most significant,有効数字のうちの最大桁)のものは1である
・これらの中で,最も右側(least significant,有効数字のうちの最小桁)のものは0である

$c(n)$ を,n のすべての交代ビット集合の個数とする.たとえば,$c(26)=8$ である.なぜならば,26は2進数表示で11010であり,26のすべての交代ビット集合は

$$\{\ \},\ \{5,3\},\ \{5,1\},\ \{4,3\},\ \{4,1\},\ \{2,1\},\ \{5,3,2,1\},\ \{4,3,2,1\}$$

であるからである.$c(n)$ は,$b(n)$ のようにデジタル和ではないけれども,同じような興味ある組み合わせ論的な性質をもつ.$c(n)$ は,各ベキを多くとも2回用いる2のベキの和として n が表される方法の個数である.漸化式

$$c(0)=1,\quad c(n)=\begin{cases}c(m)+c(m-1), & n=2m \text{ のとき} \\ c(m), & n=2m+1 \text{ のとき}\end{cases}\quad (n\geq 1)$$

を満たす.また,巧妙な方法で Fibonacci(フィボナッチ)数列に関連している.

$$0.9588<\limsup_{n\to\infty}\frac{c(n)}{n^{\frac{\ln(\varphi)}{\ln(2)}}}<1.1709$$

が証明できる[57].ただし,φ は黄金比【1.2】である.この上極限の正確な値は何であ

2.16 ストラルスキ-ハルボルス定数

るのだろうか？　正確な値が1であることを疑う根拠があるのだろうか？

[1] J. W. L. Glaisher, On the residue of a binomial-theorem coefficient with respect to a prime modulus, *Quart. J. Math.* 30 (1899) 150-156.
[2] N. J. Fine, Binomial coefficients modulo a prime, *Amer. Math. Monthly* 54 (1947) 589-592; MR 9,331b.
[3] L. Carlitz, The number of binomial coefficients divisible by a fixed power of a prime, *Rend. Circ. Mat. Palermo* 16 (1967) 299-320; MR 40 # 2554.
[4] D. Singmaster, Notes on binomial coefficients III-Any integer divides almost all binomial coefficients, *J. London Math. Soc.* 8 (1974) 555-560; MR 53 # 153.
[5] A. Granville, The arithmetic properties of binomial coefficients, *Organic Mathematics*, Proc. 1995 Burnaby workshop, ed. J. Borwein, P. Borwein, L. Jörgenson, and R. Corless, Amer. Math. Soc., 1997, pp. 253-276; MR 99h:11016.
[6] N. J. A. Sloane, On-Line Encyclopedia of Integer Sequences, A000120, A000788, A001316, A002487, A002858, A003670, A005599, A006046, A006047, A006048, and A006844.
[7] S. Wolfram, Geometry of binomial coefficients, *Amer. Math. Monthly* 91 (1984) 566-571; MR 86d:05007.
[8] D. Flath and R. Peele, Hausdorff dimension in Pascal's triangle, *Applications of Fibonacci Numbers*, v. 5, Proc. 1992 St. Andrews conf., ed. G. E. Bergum, A. N. Philippou, and A. F. Horadam, Kluwer, 1993, pp. 229-244; MR 95a:11068.
[9] K. B. Stolarsky, Power and exponential sums of digital sums related to binomial coefficient parity, *SIAM J. Appl. Math.* 32 (1977) 717-730; MR 55 # 12621.
[10] H. Harborth, Number of odd binomial coefficients, *Proc. Amer. Math. Soc.* 62 (1977) 19-22; MR 55 # 2725.
[11] A. H. Stein, Binomial coefficients not divisible by a prime, *Number Theory: New York Seminar 1985-1988*, ed. D. V. Chudnovsky, G. V. Chudnovsky, H. Cohn, and M. B. Nathanson, Lect. Notes in Math. 1383, Springer-Verlag, 1989, pp. 170-177; MR 91c:11012.
[12] Z. M. Franco, Distribution of binomial coefficients modulo p, *Number Theory: Diophantine, Computational and Algebraic Aspects*, Proc. 1996 Elger conf., ed. K. Györy, A. Pethö, and V. T. Sós, Gruyter, 1998, pp. 199-209; MR 99d:11017.
[13] Z. M. Franco, Distribution of binomial coefficients modulo three, *Fibonacci Quart.* 36 (1998) 272-274.
[14] B.Wilson, Asymptotic behavior of Pascal's triangle modulo a prime, *Acta Arith.* 83 (1998) 105-116; MR 98k:11012.
[15] N. A. Volodin, Number of multinomial coefficients not divisible by a prime, *Fibonacci Quart.* 32 (1994) 402-406; MR 95j:11017.
[16] N. A. Volodin, Multinomial coefficients modulo a prime, *Proc. Amer. Math. Soc.* 127 (1999) 349-353; MR 99c:11019.
[17] Y.-G. Chen and C. Ji, The number of multinomial coefficients not divided by a prime, *Acta Sci. Math. (Szeged)* 64 (1998) 37-48; MR 99j:11020.
[18] J. R. Trollope, An explicit expression for binary digital sums, *Math. Mag.* 41 (1968) 21-25; MR 38 # 2084.
[19] H. Delange, Sur la fonction sommatoire de la fonction "somme des chiffres," *Enseign. Math.* 21 (1975) 31-47; MR 52 # 319.
[20] L. E. Bush, An asymptotic formula for the average sum of the digits of integers, *Amer. Math. Monthly* 47 (1940) 154-156; MR 1 # 199.
[21] R. Bellman and H. N. Shapiro, On a problem in additive number theory, *Annals of Math.* 49 (1948) 333-340; MR 9 # 414.
[22] L. Mirsky, A theorem on representations of integers in the scale of r, *Scripta Math.* 15 (1949) 11-12; MR 11 # 83.

[23] M. P. Drazin and J. S. Griffith, On the decimal representation of integers, *Proc. Cambridge Philos. Soc.* 48 (1952) 555-565; MR 14 #253.
[24] P.-H. Cheo and S.-C. Yien, A problem on the k-adic representation of positive integers (in Chinese), *Acta Math. Sinica* 5 (1955) 433-438; MR 17,828b.
[25] G. F. Clements and B. Lindström, A sequence of (± 1) determinants with large values, *Proc. Amer. Math. Soc.* 16 (1965) 548-550; MR 31 #2259.
[26] M. D. McIlroy, The number of 1's in binary integers: Bounds and extremal properties, *SIAM J. Comput.* 3 (1974) 255-261; MR 55 #9628.
[27] P. Flajolet, P. Grabner, P. Kirschenhofer,H. Prodinger, and R. F. Tichy, Mellin transforms and asymptotics: Digital sums, *Theoret. Comput. Sci.* 123 (1994) 291-314; MR 94m:11090.
[28] J. Coquet, Power sums of digital sums, *J. Number Theory* 22 (1986) 161-176; MR 87d:11070.
[29] P. Kirschenhofer,On the variance of the sum of digits function, *Number-Theoretic Analysis: Vienna 1988-89*, ed. E. Hlawka and R. F. Tichy, Lect. Notes in Math. 1452, Springer-Verlag, 1990, pp. 112-116; MR 92f:11103.
[30] P. J. Grabner, P. Kirschenhofer, H. Prodinger, and R. Tichy, On the moments of the sum-of-digits function, *Applications of Fibonacci Numbers*, v. 5, Proc. 1992 St. Andrews conf., ed. G. E. Bergum, A. N. Philippou, and A. F. Horadam, Kluwer, 1993, pp. 263-271; MR 95d:11123.
[31] C. Cooper and R. E. Kennedy, A generalization of a result by Trollope on digital sums, *J. Inst. Math. Comput. Sci. Math. Ser.* 12 (1999) 17-22; MR 2000e:11008.
[32] J.-P. Allouche and J. Shallit, Sums of digits and the Hurwitz zeta function, *Analytic Number Theory*, Proc. 1988 Tokyo conf., ed. K. Nagasaka and É. Fouvry, Lect. Notes in Math. 1434, Springer-Verlag, 1990, pp. 19-30; MR 91i:11111.
[33] A. H. Osbaldestin, Digital sum problems, *Fractals in the Fundamental and Applied Sciences*, ed. H.-O. Peitgen, J. M. Henriques, and L. F. Penedo, North-Holland, 1991, pp. 307-328; MR 93g: 58102.
[34] T. Okada, T. Sekiguchi, and Y. Shiota, Applications of binomial measures to power sums of digital sums, *J. Number Theory* 52 (1995) 256-266; MR 96d:11084.
[35] J. M. Dumont and A. Thomas, Digital sum problems and substitutions on a finite alphabet, *J. Number Theory* 39 (1991) 351-366; MR 92m:11074.
[36] G. Tenenbaum, Sur la non-dérivabilité de fonctions périodiques associées à certaines formules sommatoires, *The Mathematics of Paul Erdös*, v. 1, ed. R. L. Graham and J. Nesetril, Springer-Verlag, 1997, pp. 117-128; MR 97k:11010.
[37] D. J. Newman, On the number of binary digits in a multiple of three, *Proc. Amer. Math. Soc.* 21 (1969) 719-721; MR 39 #5466.
[38] D. J. Newman and M. Slater, Binary digit distribution over naturally defined sequences, *Trans. Amer. Math. Soc.* 213 (1975) 71-78; MR 52 #5607.
[39] J.-M. Dumont, Discrépance des progressions arithmétiques dans la suite de Morse, *C. R. Acad. Sci. Paris Sér. I Math.* 297 (1983) 145-148; MR 85f:11058.
[40] J. Coquet, A summation formula related to the binary digits, *Invent. Math.* 73 (1983) 107-115; MR 85c:11012.
[41] P. J. Grabner, A note on the parity of the sum-of-digits function, *Séminaire Lotharingien de Combinatoire*, Proc. 1993 Gerolfingen session, ed. R. König and V. Strehl, Univ. Louis Pasteur, 1993, pp. 35-42; MR 95k:11125.
[42] P. J. Grabner, Completely q-multiplicative functions: The Mellin transform approach, *Acta Arith.* 65 (1993) 85-96; MR 94k:11111.
[43] P. J. Grabner, T. Herendi, and R. F. Tichy, Fractal digital sums and codes, *Appl. Algebra Engin. Commun. Comput.* 8 (1997) 33-39; MR 99c:11150.
[44] S. Goldstein, K. A. Kelly, and E. R. Speer, The fractal structure of rarefied sums of the Thue-Morse sequence, *J. Number Theory* 42 (1992) 1-19; MR 93m:11020.
[45] M. Drmota and M. Skalba, Sign-changes of the Thue-Morse fractal function and Dirichlet L-series, *Manuscripta Math.* 86 (1995) 519-541; MR 96b:11027.

[46] M. Drmota and M. Skalba, Rarified sums of the Thue-Morse sequence, *Trans. Amer. Math. Soc.* 352 (2000) 609-642; MR 2000c:11038.

[47] A. H. Stein, Exponential sums of sum-of-digit functions, *Illinois J. Math.* 30 (1986) 660- 675; MR 89a:11014.

[48] G. Larcher, On the number of odd binomial coefficients, *Acta Math. Hungar.* 71 (1996) 183-203; MR 97e:11026.

[49] P. Flajolet, J.-C. Raoult, and J. Vuillemin, The number of registers required for evaluating arithmetic expressions, *Theoret. Comput. Sci.* 9 (1979) 99-125; MR 80e:68101.

[50] P. Flajolet and M. Golin, Mellin transforms and asymptotics: The mergesort recurrence, *Acta Inform.* 31 (1994) 673-696; MR 95h:68035.

[51] P. Flajolet and M. Golin, Exact asymptotics of divide-and-conquer recurrences, *Proc. 1993 Int. Colloq. on Automata, Languages and Programming (ICALP)*, Lund, ed. A. Lingas, R. Karlsson, and S. Carlsson, Lect. Notes in Comp. Sci. 700, Springer-Verlag, 1993, pp. 137-149.

[52] R. Sedgewick and P. Flajolet, *Introduction to the Analysis of Algorithms*, Addison-Wesley, 1996, pp. 62-70.

[53] P. Flajolet and R. Sedgewick, *Analytic Combinatorics*, unpublished manuscript (2001), ch. 7.

[54] R. K. Guy, *Unsolved Problems in Number Theory*, 2nd ed., Springer-Verlag, 1994, sect. C4; MR 96e:11002.

[55] J. Cassaigne and S. R. Finch, A class of 1-additive sequences and quadratic recurrences, *Experim. Math.* 4 (1995) 49-60; MR 96g:11007.

[56] B. Reznick, Some binary partition functions, *Analytic Number Theory*, Proc. 1989 Allerton Park conf., ed. B. C. Berndt, H. G. Diamond,H. Halberstam, and A. Hildebrand, Birkhäuser, 1990, pp. 451-477; MR 91k:11092.

[57] N. J. Calkin and H. S. Wilf, Binary partitions of integers and Stern-Brocot-like trees, unpublished note (1998).

[58] N. J. Calkin and H. S. Wilf, Recounting the rationals, *Amer. Math. Monthly* 107 (2000) 360-363; MR 2001d:11024.

2.17 ガウス-クズミン-ヴィルジング定数

x_0 を区間 $(0,1)$ から一様に選んだ乱数とする．x_0 を（一意に）正則連分数で

$$x_0 = 0 + \cfrac{1}{|a_1|} + \cfrac{1}{|a_2|} + \cfrac{1}{|a_3|} + \cdots$$

と書く．ただし，各 a_k は正の整数である．すべての $n>0$ に対して

$$x_n = 0 + \cfrac{1}{|a_{n+1}|} + \cfrac{1}{|a_{n+2}|} + \cfrac{1}{|a_{n+3}|} + \cdots$$

と定義する．$x_n = \{1/x_{n-1}\}$ であるから，各 n に対して，x_n もまた $(0,1)$ の中にある数である．ただし，$\{y\}$ は y の小数部分である．

1812 年に，Gauss（ガウス）は分布関数[1]

$$F_n(x) = x_n \leq x \text{ となる確率}$$

を調べ，注目すべき極限の結果

$$\lim_{n\to\infty} F_n(x) = \frac{\ln(1+x)}{\ln(2)} \quad (0 \le x \le 1)$$

の証明をもっていると信じられていた．公表された最初の証明は，Kuzmin（クズミン）[2]による．これに続く，誤差の上限を入れた改良が，Lévy（レヴィ）[3]やSzüsz（ジュース）[4]によりされた．Wirsing（ヴィルジング）[5]はさらに詳しく，

$$\lim_{n\to\infty} \frac{F_n(x) - \dfrac{\ln(1+x)}{\ln(2)}}{(-c)^n} = \Psi(x)$$

の証明を与えた．ただし，$c = 0.3036630028\cdots$ で，Ψ は $\Psi(0) = \Psi(1) = 0$ を満たす解析関数である．[6]にあるグラフは，Ψ が凸で，$0 < x < 1$ に対して $-0.1 < \Psi(x) < 0$ であることを示す．定数 c は，一見よく見なれた定数とは関係がなく，ある種の無限次元線型汎関数【2.17.1】（対応する固有関数として $\Psi(x)$ をもつ）の固有値として計算される．この解析の鍵は，等式

$$F_{n+1}(x) = T[F_n](x) = \sum_{k=1}^{\infty} \left[F_n\left(\frac{1}{k}\right) - F_n\left(\frac{1}{k+x}\right) \right]$$

である．

Babenko & Jurev（バベンコ&ジュレフ）[7-9]は，さらに詳しく，ある種の固有値・固有関数展開

$$F_n(x) - \frac{\ln(1+x)}{\ln(2)} = \sum_{k=2}^{\infty} \lambda_k^n \cdot \Psi_k(x) \quad (1 = \lambda_1 > |\lambda_2| = |\lambda_3| \ge \cdots)$$

が，すべての x とすべての $n > 0$ に対して成り立つことを証明した．他の[1,5,6,10,11]の仕事に基づいて，Sebah[12]は，$3 \le k \le 50$ の固有値 λ_k のみならずガウス-クズミン-ヴィルジング定数 c を 100 桁計算した．

いくらかの関連する研究は，[13-19]で示されている．しかし，これらはわれわれには難しすぎて議論することができない．

2.17.1 リュール-マイヤー作用素

ここで調べる作用素は，力学系[20,21]で最初にはじまった．Δ を，中心が 1 で半径 3/2 の開円盤とし，$s > 1$ とする．Δ 上で解析的で，Δ の閉包で連続関数であり，上限ノルム（supremum norm）をもつ，関数 f のバナッハ空間を X とする．線型作用素 $G_s : X \to X$ を次の式[10,11,22,23]

$$G_s[f](z) = \sum_{k=1}^{\infty} \frac{1}{(k+z)^2} f\left(\frac{1}{k+z}\right) \quad (z \in \Delta)$$

により定義する．$s = 2$ の場合のみをここで調べる．$s = 4$ の場合は，【2.19】で必要とされる．

導関数は $T[F]'(x) = G_2[f](x)$ であることに注意せよ．ただし，$F' = f$ である．したがって，G_2 を理解することは，T により理解される．$\lambda_1 = 1$ の後に続く，G_2 の最初の 6 つの固有値[1,6,10-12]は

2.17 ガウス-クズミン-ヴィルジング定数

$$\lambda_2 = -0.3036630028\cdots \quad \lambda_3 = 0.1008845092\cdots \quad \lambda_4 = -0.0354961590\cdots$$
$$\lambda_5 = 0.0128437903\cdots \quad \lambda_6 = -0.0047177775\cdots \quad \lambda_7 = 0.0017486751\cdots$$

である. 一方,

$$\lim_{n\to\infty} \frac{\lambda_{n+1}}{\lambda_n} = -1 - \varphi = -2.6180339887\cdots$$

であると予想される. ただし, φ は黄金比【1.2】である. 他方, G_2 の跡和（トレース）は, [11] により正確に,

$$\tau_1 = \frac{1}{2} - \frac{1}{2\sqrt{5}} + \frac{1}{2} \sum_{k=1}^{\infty} (-1)^{k-1} \binom{2k}{k} (\zeta(2k) - 1) = 0.7711255236\cdots$$

と与えられる. ただし, $\tau_n = \sum_{j=1}^{\infty} \lambda_j^n$ である. G_s とゼータ関数値【1.6】との間の関連は, 驚くに値しない. $f(z) = z^r$ に応用した G_s に注目し, 次に任意の関数 f のマクローリン展開と G_s の線型性とを考えよ.

その他の興味ある跡和公式（トレース公式）は次にある. [24, 25] から

$$\xi_n = 0 + \frac{1|}{|n} + \frac{1|}{|n} + \frac{1|}{|n} + \cdots \quad (n = 1, 2, 3, \cdots)$$

とする. このとき

$$\tau_1 = \int_0^{\infty} \frac{J_1(2u)}{e^u - 1} du = \sum_{n=1}^{\infty} \frac{1}{1 + \xi_n^{-2}}$$

である. ただし,

$$J_1(x) = \sum_{k=0}^{\infty} \frac{(-1)^k}{k!(k+1)!} \left(\frac{x}{2}\right)^{2k+1}$$

は 1 次のベッセル関数である. 同様な方法で,

$$\xi_{m,n} = 0 + \frac{1|}{|m} + \frac{1|}{|n} + \frac{1|}{|m} + \frac{1|}{|n} + \cdots$$

であるとき,

$$\tau_2 = \int_0^{\infty}\int_0^{\infty} \frac{J_1(2\sqrt{uv})^2}{(e^u-1)(e^v-1)} du dv = \sum_{m=1}^{\infty}\sum_{n=1}^{\infty} \frac{1}{(\zeta_{m,n}\xi_{n,m})^{-2} - 1} = 1.1038396536\cdots$$

である. これらの一般化は可能である.

G_s の絶対値が最大な固有値（最大絶対値をもつ）$\lambda_1(s)$ は, 正であり一意的であること, 関数 $s \to \lambda_1(s)$ は解析的で狭義の単調減少関数であること, および [26] より

$$\lim_{s\to 1^+} (s-1)\lambda_1(s) = 1, \quad \lambda_1(2) = 1, \quad \lim_{s\to\infty} \frac{1}{s} \ln(\lambda_1(s)) = -\ln(\varphi)$$

が成り立つことが証明できる. 簡単な議論 [22] で, $\lambda_1'(2) = -\pi^2/(12\ln(2))$ が Lévy（レヴィ）定数【1.8】であることが示される. 後に, ユークリッドの互除法【2.18】の正確な効率性の決定に関連して, $\lambda_1'(2)$ と $\lambda_1''(2)$ のどちらも生じることを見る. 同様に, $\lambda_1'(4)$ は, ある種の比較およびソートアルゴリズムの解析【2.19】に現れる. すべての固有値 $\lambda_j(s)$ は, 実数であることが知られている. しかし, その符号と一意性の問題は, $j > 1$ に対して未解決のままである.

$\lambda_1(s)$ の別な定義がある．正の整数からなる任意の k 次元ベクトル $w=(w_1, w_2, \cdots, w_k)$ に対して，$\langle w \rangle$ を連分数

$$0+\frac{1}{\lfloor w_1} + \frac{1}{\lfloor w_2} + \frac{1}{\lfloor w_3} + \cdots + \frac{1}{\lfloor w_k}$$

の分母とする．$W(k)$ を，そのようなベクトルすべての集合とする．このとき，

$$\lambda_1(s) = \lim_{k \to \infty} \left(\sum_{w \in W(k)} \langle w \rangle^{-s} \right)^{1/k}$$

が，すべての $s>1$ に対して真である．これが，$\lambda_1(s)$ を連分数に付随する擬 (pseudo) ゼータ関数と普通によぶ理由である．

2.17.2 漸近的正規性

はじめに，【1.8】で x に収束する第 n 近似分数の分母 Q_n を研究した．前の節で導入した手続きについて，より詳しいことがいえる．

x を $(0,1)$ から一様に選ぶとき，$\ln(Q_n(x))$ の平均と分散は [22, 26]

$$E(\ln(Q_n(x))) = An + B + O(c^n), \quad \mathrm{Var}(\ln(Q_n(x))) = Cn + D + O(c^n)$$

を満たす．ただし，$c=-\lambda_2(2)=0.3036630028\cdots$，$A=-\lambda_1'(2)=1.1865691104\cdots$ である．【2.18】より

$$C = \lambda_1''(2) - \lambda_1'(2)^2 = 0.8621470373\cdots = (0.9285187329\cdots)^2$$

である．定数 B と D については，数値の評価が待望されている．さらに，$\ln(Q_n(x))$ の分布は漸近的に正規である．すなわち

$$\lim_{n \to \infty} P\left(\frac{\ln(Q_n(x)) - An}{\sqrt{C n}} \leq y \right) = \frac{1}{\sqrt{2\pi}} \int_{-\infty}^{y} \exp\left(-\frac{t^2}{2} \right) dt$$

である．これは，この本に現れるいくつかの中心極限定理の最初のものである．

2.17.3 有界な部分分母

ガウス-クズミン密度の結果は，ほとんどすべての実数は非有界な部分分母 a_k をもつことである．部分分母として 1 と 2 のみをもつ実数全体からなる集合は，何に「似ている」のだろうか？ この集合の Hausdorff（ハウスドルフ）次元は 0.53128049 と 0.53128051 の間である，ことが知られている [27-31]．この数値のさらなる議論は，【8.20】まで後にまわす．

[1] D. E. Knuth, *The Art of Computer Programming*, v. 2, *Seminumerical Algorithms*, 3rd ed., Addison-Wesley, 1998; MR 44 #3531.
[2] R. Kuzmin, Sur un problème de Gauss, *Atti del Congresso Internazionale dei Matematici*, v. 6, Proc. 1928 Bologna conf., ed. N. Zanichelli, Societa Tipografica gia Compositori, 1929, pp. 83-89.
[3] P. Lévy, Sur les lois probabilité dont dépendent les quotients complets et incomplets d'une fraction continue, *Bull. Soc. Math. France* 57 (1929) 178-194; also in Oeuvres, v. 6, ed. D. Dugué, Gauthier-Villars, 1980, pp. 266-282.
[4] P. Szüsz, Über einen Kusminschen Satz, *Acta Math. Acad. Sci. Hungar*. 12 (1961) 447-453; MR 27 #124.

[5] E. Wirsing, On the theorem of Gauss-Kuzmin-Lévy and a Frobenius-type theorem for function spaces, *Acta Arith.* 24 (1974) 507-528; MR 49 #2637.

[6] A. J. MacLeod, High-accuracy numerical values in the Gauss-Kuzmin continued fraction problem, *Comput. Math. Appl. (Oxford)*, v. 26 (1993) n. 3, 37-44; MR 94h:11114.

[7] K. I. Babenko and S. P. Jurev, A problem of Gauss (in Russian), Institut Prikladnoi Mat. Akad. Nauk SSSR preprint n. 63 (1977); MR 58 #19017.

[8] K. I. Babenko, On a problem of Gauss (in Russian), *Dokl. Akad. Nauk SSSR* 238 (1978) 1021-1024; Engl. transl. in *Soviet Math. Dokl.* 19 (1978) 136-140; MR 57 #12436.

[9] K. I. Babenko and S. P. Jurev, On the discretization of a problem of Gauss (in Russian), *Dokl. Akad. Nauk SSSR* 240 (1978) 1273-1276; Engl. transl. in *Soviet Math. Dokl.* 19 (1978) 731-735; MR 81h:65015.

[10] H. Daudé, P. Flajolet, and B.Vallée, An average-case analysis of the Gaussian algorithm for lattice reduction, *Combin. Probab. Comput.* 6 (1997) 397-433; INRIA preprint RR2798; MR 99a:65196.

[11] P. Flajolet and B.Vallée, On the Gauss-Kuzmin-Wirsing constant, unpublished note (1995).

[12] P. Sebah, Computing eigenvalues of Wirsing's operator, unpublished note (2001).

[13] D. H. Mayer, On the thermodynamic formalism for the Gauss map, *Commun. Math. Phys.* 130 (1990) 311-333; MR 91g:58216.

[14] A. Durner, On a theorem of Gauss-Kuzmin-Lévy, *Arch. Math. (Basel)* 58 (1992) 251-256; MR 93c:11056.

[15] C. Faivre, The rate of convergence of approximations of a continued fraction, *J. Number Theory* 68 (1998) 21-28; MR 98m:11083.

[16] M. Iosifescu and S. Grigorescu, *Dependence with Complete Connections and Its Applications*, Cambridge Univ. Press, 1990; MR 91j:60098.

[17] M. Iosifescu, A very simple proof of a generalization of the Gauss-Kuzmin-Lévy theorem on continued fractions, and related questions, *Rev. Roumaine Math. Pures Appl.* 37 (1992) 901-914; MR 94j:40003.

[18] D. Hensley, Metric Diophantine approximation and probability, *New York J. Math.* 4 (1998) 249-257; MR 2000d:11102.

[19] K. Dajani and C. Kraaikamp, A Gauss-Kuzmin theorem for optimal continued fractions, *Trans. Amer. Math. Soc.* 351 (1999) 2055-2079; MR 99h:11089.

[20] D. Ruelle, *Dynamical Zeta Functions for Piecewise Monotone Maps of the Interval*, Amer. Math. Soc., 1994; MR 95m:58101.

[21] D. H. Mayer, Continued fractions and related transformations, *Ergodic Theory, Symbolic Dynamics and Hyperbolic Spaces*, ed. T. Bedford, M. Keane, and C. Series, Oxford Univ. Press, 1991, pp. 175-222; MR 93e:58002.

[22] P. Flajolet and B. Vallée, Continued fraction algorithms, functional operators, and structure constants, *Theoret. Comput. Sci.* 194 (1998) 1-34; INRIA preprint RR2931;MR98j:11061.

[23] P. Flajolet, B. Vallée, and I. Vardi, Continued fractions from Euclid to the present day, École Polytechnique preprint (2000).

[24] D. Mayer and G. Roepstorff, On the relaxation time of Gauss' continued-fraction map. I: The Hilbert space approach (Koopmanism), *J. Stat. Phys.* 47 (1987) 149-171; MR 89a:28017.

[25] D. Mayer and G. Roepstorff, On the relaxation time of Gauss' continued-fraction map. II: The Banach space approach (Transfer operator method), *J. Stat. Phys.* 50 (1988) 331-344; MR 89g:58171.

[26] D. Hensley, The number of steps in the Euclidean algorithm, *J. Number Theory* 49 (1994) 142-182; MR 96b:11131.

[27] I. J. Good, The fractional dimensional theory of continued fractions, *Proc. Cambridge Philos. Soc.* 37 (1941) 199-228; corrigenda 105 (1989) 607;MR3,75b and MR90i:28013.

[28] R. T. Bumby, Hausdorff dimension of sets arising in number theory, *Number Theory: New York 1983-84*, ed. D. V. Chudnovsky, G. V. Chudnovsky, H. Cohn, and M. B. Nathanson, Lect.

Notes in Math. 1135, Springer-Verlag, pp. 1-8, 1985; MR 87a:11074.
[29] D. Hensley, The Hausdorff dimensions of some continued fraction Cantor sets, *J. Number Theory* 33 (1989) 182-198; MR 91c:11043.
[30] D. Hensley, Continued fraction Cantor sets, Hausdorff dimension, and functional analysis, *J. Number Theory* 40 (1992) 336-358; MR 93c:11058.
[31] E. Cesaratto, On the Hausdorff dimension of certain sets arising in number theory, math.NT/9908043.

2.18 ポーター-ヘンズリー定数

与えられた2つの非負整数 m と n に対して，古典的なユークリッドの互除法によって，最大公約数 $\gcd(m,n)$ を計算するときに必要な割り算の回数を，$L(m,n)$ とする．定義から，$m \geq n$ であるとき，

$$L(m,n) = \begin{cases} 1+L(n, m \bmod n) & (n \geq 1 \text{ のとき}) \\ 0 & (n=0 \text{ のとき}) \end{cases}$$

であり，$m<n$ のときには $L(m,n)=1+L(n,m)$ である．すなわち，$L(m,n)$ は，m/n を正則連分数で表示したときの長さと同値である．ユークリッドの互除法の効率性を正確に決定することに興味がある．このことを，3種類の確率変数を調べることにより行う．

$X_n = L(m,n), \quad 0 \leq m < n$ はランダムに選ばれる

$Y_n = L(m,n), \quad 0 \leq m < n$ はランダムに選ばれ，m は n と互いに素

$Z_N = L(m,n), \quad 1 \leq m, \ n < N$ で，m, n ともにランダムに選ばれる

3つのうちで，Y_n の期待値が最もよいふるまいをするので，最初に解析された．これらの平均値を理解するために，その進展に従うことはおもしろい．初版で，Knuth（クヌース）[1]は経験的に $E(Y_n) \sim 0.843 \ln(n) + 1.47$ と述べ，

$$E(Y_n) \sim \frac{12 \ln(2)}{\pi^2} \ln(n) + 1.47, \quad E(Z_N) \sim \frac{12 \ln(2)}{\pi^2} \ln(N) + 0.06$$

にむりやり理由を与えた．ただし，$\ln(n)$ の係数はレヴィ定数【1.8】である．しかしながら，彼はこれらの漸近式の証明において穴がある論理的な欠陥を公然と非難し，「世界でもっとも有名なアルゴリズムを完全に解析する価値がある！」と書いた．

第2版[2]では，注目すべき進歩が Heilbronn（ハイルブロン）[3]，Dixon（ディクソン）[4,5]，Porter（ポーター）[6]により成し遂げられた．すなわち，任意の $\varepsilon > 0$ に対して，次の漸近公式は真である．

$$E(Y_n) \sim \frac{12 \ln(2)}{\pi^2} \ln(n) + C + O(n^{-1/6+\varepsilon})$$

ここで**ポーター定数** C は

2.18 ポーター–ヘンズリー定数

$$C = \frac{6\ln(2)}{\pi^2}\left(3\ln(2) + 4\gamma - \frac{24}{\pi^2}\zeta'(2) - 2\right) - \frac{1}{2} = 1.4670780794\cdots$$

と定義される．ただし，γ はオイラー–マスケローニの定数【1.5】であり，

$$\zeta'(2) = \frac{d}{dx}\zeta(x)\bigg|_{x=2} = -\sum_{k=2}^{\infty}\frac{\ln(k)}{k^2} = -0.9375482543\cdots$$

であり，$\zeta(x)$ は Riemann のゼータ関数【1.6】である．C の表現は Wrench（レンチ）[7] により発見され，彼はさらに $\zeta'(2)$ の値を計算し，C の値を 120 桁求めた[8]．$\zeta'(2)$ のさらなる出現については【2.10】を参照せよ．

他の 2 つの平均値について，いったい何がいえるのだろうか？ Norton（ノートン）[9]は，任意の $\varepsilon > 0$ に対して

$$E(Z_N) \sim \frac{12\ln(2)}{\pi^2}\ln(N) + B + O(N^{-1/6+\varepsilon})$$

を証明した．ただし，

$$B = \frac{12\ln(2)}{\pi^2}\left(-\frac{1}{2} + \frac{6}{\pi^2}\zeta'(2)\right) + C - \frac{1}{2} = 0.0653514259\cdots$$

である．$E(X_n)$ の漸近式は，$E(Y_n)$ から n の約数に基づいた補正項【2.9】を引いた式と類似する．

$$E(X_n) \sim \frac{12\ln(2)}{\pi^2}\left(\ln(n) - \sum_{d|n}\frac{\Lambda(d)}{d}\right) + C + \frac{1}{n}\sum_{d|n}\varphi(d)\cdot O(d^{-1/6+\varepsilon})$$

ただし，φ はオイラーの約数関数【2.7】であり，Λ は von Mangold（フォン・マンゴルド）関数

$$\Lambda(d) = \begin{cases} \ln(p) & （素数 p があり，r \geq 1 に対して，d = p^r のとき）\\ 0 & （その他のとき）\end{cases}$$

である．[9]の証明の中で，Norton は，【2.15】で議論したグレイシャー–キンケリン定数 A について述べた．ポーター定数 C は，A を用いて

$$C = \frac{6\ln(2)}{\pi^2}(48\ln(A) - 4\ln(\pi) - \ln(2) - 2) - \frac{1}{2}$$

と書ける．Knuth[7]は，$(1-2B)/4 = 0.2173242870\cdots$ を含む長い間忘れられていた論文[10]について述べ，C は Lochs–Porter（ロックス–ポーター）定数とよばれるべきであると提案した．

対応する $L(m, n)$ の分散を計算することは，ずっと難しい．Z_N のみに焦点をあてる．Hensley（ヘンズリー）[11]は

$$\mathrm{Var}(Z_N) = H\ln(N) + o(\ln(N))$$

を証明した．ただし，

$$H = -\frac{\lambda_1''(2) - \lambda_1'(2)^2}{\pi^6 \lambda_1'(2)^3} = 0.0005367882\cdots = (0.0231686908\cdots)^2$$

であり，$\lambda_1'(2)$ と $\lambda_1''(2)$ は，【2.17.1】で正確に述べられている．Flajolet & Vallée（フラジョレット & ヴァレ）[12]による数値実験の仕事は，H を評価するために必要な評価 $4\lambda_1''(2) = 9.0803731646\cdots$ を得た．さらに，Z_N の分布は漸近正規分布である．

$$\lim_{n\to\infty} P\left(\frac{Z_N - \frac{12\ln(2)}{\pi^2}\ln(N)}{\sqrt{H\ln(N)}} \leq w\right) = \frac{1}{\sqrt{2\pi}} \int_{-\infty}^{w} \exp\left(-\frac{t^2}{2}\right) dt$$

最近の論文[13]は，連分数表示によるいくつかの実数のソート問題に関連して，ポーター定数に似た定数をいくつか含んでいる．

2.18.1 2進ユークリッドアルゴリズム

m, n を正の奇数とする．$2^{e(m,n)}$ が $m-n$ を割る最大の整数を $e(m,n)$ とする．最大公約数 $\gcd(m,n)$ を計算するために必要な引き算の回数は，2進ユークリッドアルゴリズム（互除法）[14]により，

$$K(m,n) = \begin{cases} 1 + K\left(\frac{m-n}{2^{e(m,n)}}, n\right) & (m > n \text{ のとき}) \\ 0 & (m = n \text{ のとき}) \\ K(n,m) & (m < n \text{ のとき}) \end{cases}$$

である．確率変数を

$$W_N = K(m,n)$$

で定義する．ただし，m, n は $0 < m \leq N$，$0 < n \leq N$ からランダムに選ばれた整数で，m は奇数である．W_N の期待値を計算するのは，Z_N の計算よりもずっと複雑である．【2.17.1】にあるように，関数空間上の線型作用素

$$V_s[f](z) = \sum_{k \geq 1} \sum_{\substack{1 \leq j < 2^k \\ \text{奇数}}} \frac{1}{(j + 2^k z)^s} f\left(\frac{1}{j + 2^k z}\right)$$

の研究[15,16]が必要である．$s = 2$ に対して，Ψ を V_s の（縮尺を除いて）一意的に決まる不動点と定義し，定数

$$\kappa = \frac{2}{\pi^2 \Psi(1)} \sum_{\substack{r \geq 1 \\ \text{奇数}}} 2^{-\lfloor \frac{\ln(r)}{\ln(2)} \rfloor} \int_0^{\frac{1}{r}} \Psi(x)\,dx$$

と定義する．このとき，$E(W_N) \sim \kappa \ln(N)$ である．さらに，Valléeによるある種の予想が真である[15,16]とすると，Brent（ブレント）[17-19]によるある発見的な式が適用でき，

$$\kappa = 1.0185012157\cdots = \ln(2)^{-1} \cdot 0.7059712461\cdots$$

を得る．κ の正確な定義に基づく，直接的な計算は未だにされていない．

他の実行パラメータ（performance parameters）[15,16]および代替アルゴリズム[17]が研究され，多くの定数が得られている．これらの結果の連分数による解釈がある．ユークリッドに類似のアルゴリズムを研究する一般的な枠組み[20,21]は，数論のJacobi（ヤコビ）記号を評価するための解析方法を与える[22]．算術的な操作の回数よりもむしろ平均ビット複雑性を調べるとき，さらにより多くの定数が現れる[23,24]．多くの関連する問題は，未だに未解決のままである．

2.18.2 最悪な場合の考察

Z_N の最大値は，m, n が連続するフィボナッチ数 f_k, f_{k+1} である場合に起こることが知られている[14,25,26]．ただし，k は $f_{k+1} \leq N$ を満たす最大の整数である．したがって，φ を黄金比【1.2】とするとき，

$$\max(Z_N) = k \sim \frac{1}{\ln(\varphi)} \ln(N) = 2.0780869212\cdots \cdot \ln(N)$$

である．比較として[14]，

$$\max(W_N) \sim \frac{1}{\ln(2)} \ln(N) = 1.4426950408\cdots \cdot \ln(N)$$

であり，最大は m と n が $2^{k-1}-1$ と $2^{k-1}+1$ の形であるときに起こる．

[1] D. E. Knuth, *The Art of Computer Programming*, v. 2, *Seminumerical Algorithms*, 1st ed., Addison-Wesley, 1969; MR 44 #3531.
[2] D. E. Knuth, *The Art of Computer Programming*, v. 2, *Seminumerical Algorithms*, 2nd ed., Addison-Wesley, 1981; MR 44 #3531.
[3] H. Heilbronn, On the average length of a class of finite continued fractions, *Number Theory and Analysis*, ed. P. Turán, Plenum Press, 1969, pp. 87-96; MR 41 #3406.
[4] J. D. Dixon, The number of steps in the Euclidean algorithm, *J. Number Theory* 2 (1970) 414-422; MR 42 #1791.
[5] J. D. Dixon, A simple estimate for the number of steps in the Euclidean algorithm, *Amer. Math. Monthly* 78 (1971) 374-376; MR 44 #2697.
[6] J. W. Porter, On a theorem of Heilbronn, *Mathematika* 22 (1975) 20-28; MR 58 #16567.
[7] D. E. Knuth, Evaluation of Porter's constant, *Comput. Math. Appl.* (*Oxford*), v. 2 (1976) n. 1, 137-139; also in *Selected Papers on Analysis of Algorithms*, CSLI, 2000, pp. 189-194; MR 2001c:68066.
[8] B. C. Berndt, *Ramanujan's Notebooks: Part* I, Springer-Verlag, 1985, p. 225; MR 86c:01062.
[9] G. H. Norton, On the asymptotic analysis of the Euclidean algorithm, *J. Symbolic Comput.* 10 (1990) 53-58; MR 91k:11088.
[10] G. Lochs, Statistik der Teilnenner der zu den echten Brüchen gehörigen regelmässigen Kettenbrüche, *Monatsh. Math.* 65 (1961) 27-52; MR 23 #A1622.
[11] D. Hensley, The number of steps in the Euclidean algorithm, *J. Number Theory* 49 (1994) 142-182; MR 96b:11131.
[12] P. Flajolet and B. Vallée, Hensley's constant, unpublished note (1999).
[13] P. Flajolet and B. Vallée, Continued fractions, comparison algorithms, and fine structure constants, *Constructive, Experimental, and Nonlinear Analysis*, Proc. 1999 Limoges conf., ed. M. Théra, Amer. Math. Soc., 2000, pp. 53-82; INRIA preprint RR4072; MR 2001h:11261.
[14] D. E. Knuth, *The Art of Computer Programming*, v. 2, *Seminumerical Algorithms*, 3rd ed., Addison-Wesley, 1997; MR 44 #3531.
[15] B.Vallée, The complete analysis of the binary Euclidean algorithm, *Proc. 1998 Algorithmic Number Theory Sympos. (ANTS-III)*, Portland, ed. J. P. Buhler, Lect. Notes in Comp. Sci. 1423, Springer-Verlag, 1998, pp. 77-94; MR 2000k:11143a.
[16] B. Vallée, Dynamics of the binary Euclidean algorithm: Functional analysis and operators, *Algorithmica* 22 (1998) 660-685; MR 2000k:11143b.
[17] R. P. Brent, Analysis of the binary Euclidean algorithm, *Algorithms and Complexity: New Directions and Recent Results*, ed. J. F. Traub, Academic Press, 1976, pp. 321-355; MR 55 #11701.
[18] R. P. Brent, The binary Euclidean algorithm, *Millennial Perspectives in Computer Science*,

Proc. 1999 Oxford-Microsoft Symp., ed. J. Davies, B. Roscoe, and J.Woodcock, Palgrave, 2000, pp. 41-53.
[19] R. P. Brent, Further analysis of the binary Euclidean algorithm, Oxford Univ. PRG TR-7-99 (1999).
[20] B. Vallée, A unifying framework for the analysis of a class of Euclidean algorithms, *Proc. 2000 Latin American Theoretical Informatics Conf. (LATIN)*, Punta del Este, ed. G. H. Gonnet, D. Panario, and A. Viola, Lect. Notes in Comp. Sci. 1776, Springer-Verlag, 2000, pp. 343-354.
[21] B. Vallée, Dynamical analysis of a class of Euclidean algorithms, *Theoret. Comput. Sci.*, submitted (2000).
[22] B. Vallée and C. Lemée, Average-case analyses of three algorithms for computing the Jacobi symbol, unpublished note (1998).
[23] A. Akhavi and B. Vallée, Average bit-complexity of Euclidean algorithms, *Proc. 2000 Int. Colloq. on Automata, Languages and Programming (ICALP)*, Geneva, ed. U. Montanari, J. D. P. Rolim, and E.Welzl, Lect. Notes in Comp. Sci. 1853, Springer-Verlag, pp. 373-387; MR 2001h: 68052.
[24] B. Vallée, Digits and continuants in Euclidean algorithms, *J. Théorie Nombres Bordeaux* 12 (2000) 531-570; MR 2002b:11105.
[25] J. Shallit, Origins of the analysis of the Euclidean algorithm, *Historia Math.* 21 (1994) 401-419; MR 95h:01015.
[26] P. Schreiber, A supplement to J. Shallit's paper: "Origins of the analysis of the Euclidean algorithm," *Historia Math.* 22 (1995) 422-424; MR 96j:01010.

2.19 ヴァレー定数

x,y を区間 $(0,1)$ から独立に一様に取り出された乱数とする．x と y を**比較する**とは，$x<y$ または $x>y$ うちのどちらが真であるを決めることである．x と y を比較する明らかなアルゴリズムがある．x と y の10進数または2進数の展開で最初に一致しない場所を探す．基底 b での，このアルゴリズムの繰り返し回数 L は，平均値

$$E(L) = \frac{b}{b-1}$$

をもち，確率分布は

$$p_n = P(L \geq n+1) = b^{-n} \quad (n=0,1,2,\cdots)$$

である．明らかに

$$\lim_{n \to \infty} p_n^{\frac{1}{n}} = \frac{1}{b}$$

は，単に基底 b での2つの数の展開において数字が一致する（漸近的）割合を表す方法である．

ここに，x と y を比較するための，[1]で提案された，あまり明らかでないアルゴリズムがある．x と y を正則連分数として（一意に）表す．

$$x = 0 + \frac{1|}{|a_1} + \frac{1|}{|a_2} + \frac{1|}{|a_3} + \cdots, \quad y = 0 + \frac{1|}{|b_1} + \frac{1|}{|b_2} + \frac{1|}{|b_3} + \cdots$$

ただし，各 a_j と b_j は正の整数である．ここで $a_k \neq b_k$ である最初の場所を探す．もし k が偶数であるとき，$x<y$ であるための必要十分条件は $a_k<b_k$ である．もし k が奇数であるとき，$x<y$ であるための必要十分条件は $a_k>b_k$ である．（もし x または y が有理数であるとき，他の必要な対策がある．すなわち，a_j または b_j が 0 であるときであるが，これは議論しない．）

このアルゴリズムの解析は，より多くの難しさがあり，【2.17.1】で議論した技法と考えを用いる．Daudé, Flajolet & Vallée（ダウデ，フラジョレット & ヴァレー）[2-5] は，繰り返し回数の平均値

$$\mathrm{E}(L) = \frac{3}{4} + \frac{180}{\pi^4} \sum_{i=1}^{\infty} \sum_{j=i+1}^{2i} \frac{1}{i^2 j^2} = \frac{17}{4} + \frac{360}{\pi^4} \sum_{i=1}^{\infty} \sum_{j=1}^{i} \frac{(-1)^i}{i^2 j^2}$$
$$= 17 - \frac{60}{\pi} \left[24 \operatorname{Li}_4\left(\frac{1}{2}\right) - \pi^2 \ln(2)^2 + 21 \zeta(3) \ln(2) + \ln(2)^4 \right]$$
$$= 1.3511315744\cdots$$

を示した．ただし，$\operatorname{Li}_4(z)$ は 4 重対数関数（tetralogarithm）【1.6.8】であり，$\zeta(3)$ は Apéry（アペリ）の定数【1.6】である．閉じた式の評価は，[6-8] の結果による．さらに，

$$p_1 = \sum_{i=1}^{\infty} \frac{1}{i^2(i+1)^2} = \frac{\pi^2}{3} - 3 = 0.2898681336\cdots$$
$$p_2 = \sum_{i=1}^{\infty} \sum_{j=1}^{\infty} \frac{1}{(ij+1)^2(ij+i+1)^2} = 0.0484808014\cdots$$
$$= -5 + \frac{2\pi^2}{3} - 2\zeta(3) + 2\sum_{n=0}^{\infty} (-1)^n (n+1) \zeta(n+4) [\zeta(n+2) - 1]$$
$$p_3 = \sum_{i=1}^{\infty} \sum_{j=1}^{\infty} \sum_{k=1}^{\infty} \frac{1}{(ijk+i+k)^2(ijk+ij+i+k+1)^2} = 0.0102781647\cdots$$

である．しかし前のものと違って，簡潔にうまく表す p_n の式は知られていない．p_n のもとになる精巧な漸化式は，後の【2.19.1】にある．下記の v が【2.17.1】で定義される線型作用素 G_4 の最大固有値であるという事実を用いて，

$$v = \lim_{n \to \infty} P_n^{\frac{1}{n}} = 0.1994588183\cdots$$

を得る[2-5]．G_2 がもつように，G_4 の固有値は実数で，符号は交互に正負をもつと思われる（次の固有値は $-0.0757395140\cdots$）．はじめに与えられた斜交座標に対して，2次元空間内の格子の短い基底（short basis）を見つけるためのガウスアルゴリズムの解析に，同じ議論が適用される．連分数表示により，$n>2$ 個の実数のソート問題に関連して，ヴァレー定数がふたたび現れる[9]．

x と y を比較するとき，中心（centerd）連分数を用いるならば，繰り返しの回数 $\hat{L}$ は[2,5]，

$$\mathrm{E}(\hat{L}) = \frac{360}{\pi^4} \sum_{i=1}^{\infty} \sum_{j=\lceil \varphi i \rceil}^{\lfloor (\varphi+1)i \rfloor} \frac{1}{i^2 j^2} = 1.0892214740\cdots$$

$$\hat{v} = \lim_{n\to\infty} \hat{p}_n^{\frac{1}{n}} = 0.0773853773\cdots$$

である．ただし，φ は黄金比【1.2】である．$1/v = 5.01\cdots$，$1/\hat{v} = 12.92\cdots$ であるから，連分数はこの点で，おおよそ基底5と13で表される数のようにふるまう．対応する作用素 $\hat{G}_s$ とそのスペクトルについて，多くは知られていない．Flajolet & Vallée[5] はさらに，モックゼータ関数（mock zeta function，変形ゼータ関数）

$$\zeta_\theta(z) = \sum_{k=1}^{\infty} \frac{1}{\lfloor k\theta \rfloor^z} \quad (\mathrm{Re}(z) > 1, \quad \theta > 1)$$

を数値的に計算した．ただし，$\theta > 1$ は無理数である．たとえば，$\zeta_\varphi(2) = 1.2910603681\cdots$ である．

2.19.1 継続多項式

$k = 2, 3, 4$ に対して，
$$f_k(x_1, x_2, \cdots, x_k) = x_k f_{k-1}(x_1, x_2, \cdots, x_{k-1}) + f_{k-2}(x_1, x_2, \cdots, x_{k-2})$$
で，再帰的に関数を定義する．ただし，
$$f_0 = 1, \quad f_1(x_1) = x_1$$
である．これらを**継続多項式**（continuant polynomials）とよぶ．これはまた隣接した変数 $x_j x_{j+1}$ のすべての可能な組を順次取り除くことによって，$x_1 x_2 \cdots x_k$ から得られる単項式の和をとることによっても定義される．例として，
$$f_2(x_1, x_2) = x_1 x_2 + 1, \quad f_3(x_1, x_2, x_3) = x_1 x_2 x_3 + x_1 + x_3$$
$$f_4(x_1, x_2, x_3, x_4) = x_1 x_2 x_3 x_4 + x_1 x_2 + x_1 x_4 + x_3 x_4 + 1$$
である．興味ある確率は
$$p_k = \sum_{n_1=1}^{\infty} \sum_{n_2=1}^{\infty} \cdots \sum_{n_k=1}^{\infty} \frac{1}{f_k^2 (f_k + f_{k-1})^2}$$
である．各 p_k は，Riemann のゼータ関数の値を含む複雑な級数で表すことができ，多項式時間で計算できる定数のクラスに属する[5]．

[1] M. Beeler, R. W. Gosper, and R. Schroeppel, Continued fraction arithmetic, HAKMEM, MIT AI Memo 239, item 101A.
[2] H. Daudé, P. Flajolet, and B. Vallée, An analysis of the Gaussian algorithm for lattice reduction, *Proc. 1994 Algorithmic Number Theory Sympos.* (ANTS-I), Ithaca, ed. L. M. Adleman and M. -D. Huang, Lect. Notes in Comp. Sci. 877, Springer-Verlag, 1994, pp. 144-158; MR 96a:11075.
[3] H. Daudé, P. Flajolet, and B. Vallée, An average-case analysis of the Gaussian algorithm for lattice reduction, *Combin. Probab. Comput.* 6 (1997) 397-433; INRIA preprint RR2798; MR 99a: 65196.
[4] B. Vallée, Algorithms for computing signs of 2×2 determinants: Dynamics and averagecase analysis, *Proc. 1997 European Symp. on Algorithms (ESA)*, Graz Univ., ed. R. Burkard and G. Woeginger, Lect. Notes in Comp. Sci. 1284, Springer-Verlag, pp. 486-499; MR 99d:68002.
[5] P. Flajolet and B. Vallée, Continued fractions, comparison algorithms, and fine structure constants, *Constructive, Experimental, and Nonlinear Analysis*, Proc. 1999 Limoges conf., ed. M. Théra, Amer. Math. Soc., 2000, pp. 53-82; INRIA preprint RR4072; MR 2001h: 11161.
[6] R. Sitaramachandrarao, A formula of S. Ramanujan, *J. Number Theory* 25 (1987) 1-19; MR 88c:

[7] P. J. de Doelder, On some series containing $\psi(x)-\psi(y)$ and $(\psi(x)-\psi(y))^2$ for certain values of x and y, *J. Comput. Appl. Math.* 37 (1991) 125-141; MR 92m:40002.

[8] G. Rutledge and R. D. Douglass, Evaluation of $\int_0^1 (\log(u)/u) \log(1+u)^2 du$ and related definite integrals, *Amer. Math. Monthly* 41 (1934) 29-36.

[9] P. Flajolet and B. Vallée, Continued fraction algorithms, functional operators, and structure constants, *Theoret. Comput. Sci.* 194 (1998) 1-34; INRIA preprint RR2931; MR 98j: 11061.

2.20 エルデーシュの逆数和定数

2.20.1 A 数 列

無限の正の整数列 $1 \leq a_1 < a_2 < a_3 \cdots$ が **A 数列**であるとは，a_k がそれ以前にある 2 つまたはそれ以上の異なる項の和で表されないときをいう[1]．たとえば，2 の非負のベキの数列は A 数列である．Edrös（エルデーシュ）[2]は

$$S(A) = \sup_{A \, 数列} \sum_{k=1}^{\infty} \frac{1}{a_k} < 103$$

を証明した．したがって，特に最も大きい逆数の和は「有限」でなければならない．Levine & O'Sullivan（レヴィン & オサリバン）[3,4]は，A 数列が **χ 不等式**とよばれる，すべての i,j に対して，

$$(j+1) a_j + a_i \geq (j+1) i$$

を満たすことを示し，結果として，$S(A) < 3.9998$ を証明した．他の方向では，Abbott（アボット）[5]と Zhang（ジャング）[6]は，$S(A) > 2.0649$ が成り立つ特別な例を与えた．これは現在のところ，$S(A)$ の最良の下界である．

χ 不等式は，それ自体おもしろい．Levine & O'Sullivan[3,7]は，欲張り (greedy) アルゴリズムによって具体的な整数列を定義した．$\chi_1 = 1$ で，$i > 1$ に対して

$$\chi_i = \max_{1 \leq j \leq i-1} (j+1)(i - \chi_j)$$

と定義する．すなわち，$1, 2, 4, 6, 9, 12, 15, 18, 21, 24, 28, 32, 36, 40, 45, 50, 55, 60, 65, \cdots$ である．彼らは

$$S(A) \leq \sum_{k=1}^{\infty} \frac{1}{\chi_k} = 3.01 \cdots$$

を予想し，さらに，$\{\chi_k\}$ は χ 不等式を満足する任意の他の整数列の逆数の和で抑えられる，と予想した．Finch（フィンチ）[8-10]は，この後半の予想は任意の（必ずしも整数でない）実数の数列に対しても成り立つのではないか，と想像した．

[3-5]の研究者は，A 数列をさして「無和数列」(sum-free sequence) という表現を使った．これは，「無和」(sum-free) という語がまったく違う種類の数列【2.25】によ

く使われるので，不適切な用語である．ここでは，Guy（ガイ）[1]による言葉「A 数列」を採用した．さらに，違う部分集合の和をもつ集合については，【2.28】も参照せよ．

2.20.2 B_2 数列

正の整数の無限列 $1 \leq b_1 < b_2 < b_3 < \cdots$ が B_2 数列（または Sidon（シドン）数列）であるとは，2つの和 $b_i + b_j$, $i \leq j$ がすべて異なるときをいう[1]．たとえば，欲張りアルゴリズムから，Mian-Chowla（ミアン-チョウラ）[7,11]数列

$$1, 2, 4, 8, 13, 21, 31, 45, 66, 81, 97, 123, 148, 182, 204, 252, 290, \cdots$$

を得る．この数列の逆数の和[12]は，2.158435 と 2.158677 の間にあることが知られている．Zhang[13]は

$$S(B_2) = \sup_{B_2 数列} \sum_{k=1}^{\infty} \frac{1}{b_k} > 2.1597$$

を証明した．これは Mian-Chowla 和よりも大きい．Levine（レヴィン）[1,13]の研究は，$S(B_2)$ が必ず有限であることを示す．実際，$S(B_2) < 2.374$ である．最近の研究 [12,14]により，改良された範囲 $2.16086 < S(B_2) < 2.247327$ を得る．

$b_m \leq n$ である正の整数 $b_1 < b_2 < b_3 < \cdots < b_m$ の有限 B_2 数列が，ある定数 C に対して，$m \leq n^{1/2} + C$ を満たすかどうかを，Erdös-Turán[15-17]は問うた．Lindström（リンドストローム）[18]は，$m < n^{1/2} + n^{1/4} + 1$ を示した．Zhang[19]は，もしそのような C が存在するならば，それは $C > 10.27$ でなければならないことを計算した．Lindström [20]は，C の下界を 13.71 に改良した．最近の論文[21]で，Lindström は，そのような定数 C はたぶん存在しないと結論し，$m \leq n^{1/2} + o(n^{1/4})$ であると予想した．

2.20.3 非平均数列

正の整数の無限列 $1 \leq c_1 < c_2 < c_3 < \cdots$ が非平均（nonaveraging）数列であるとは，どの3つの項も等差数列にはならないことをいう．言いかえると，数列の任意の3つの異なる項に対して，$c + d \neq 2e$ であることである[1]．例として，欲張りアルゴリズムより作った，Szekeres（セケレス）[7,22]数列

$$1, 2, 4, 5, 10, 11, 13, 14, 28, 29, 31, 32, 37, 38, 40, 41, 82, 83, \cdots$$

がある．すなわち，n がこの数列の項であるための必要十分条件は，$n-1$ の3進数展開が0と1だけを含むことである．逆数の和は 3.00793 と 3.00794 の間にあることが知られている．Wróblewski（ロブルフスキー）[23]は，[24,25]を基礎にして，

$$S(C) = \sup_{非平均数列} \sum_{k=1}^{\infty} \frac{1}{c_k} > 3.00849$$

を示すために，特殊な非平均数列を構成した．$S(C)$ が必ず有限であるという証明は，知られていない．c_k の最良の下界[26]が，$O\left(k\sqrt{\frac{\ln(k)}{\ln(\ln(k))}}\right)$ であることが知られているだけである．

別の定式化の規則または違う初期値によって欲張って（greedily）構成された，

$\{c_k\} \cap [1, n]$ の密度のいくつかの関連する研究は，[27-31]にある．ある条件の下で，n が増加するにしたがい，密度は，大雑把には等比数列で（なめらかになるよりもむしろ）山と谷をもち振動する．連続する 2 つの極大の比は，$N \to \infty$ のとき，ある極限値に収束するように見える．この現象はよりよく理解する価値がある．

[1] R. K. Guy, *Unsolved Problems in Number Theory*, 2nd ed., Springer-Verlag, 1994, sect. E10, E28; MR 96e:11002.
[2] P. Erdös, Remarks on number theory. III: Some problems in additive number theory, *Mat. Lapok* 13 (1962) 28-38; MR 26 # 2412.
[3] E. Levine and J. O'Sullivan, An upper estimate for the reciprocal sum of a sum-free sequence, *Acta Arith.* 34 (1977) 9-24; MR 57 # 5900.
[4] E. Levine, An extremal result for sum-free sequences, *J. Number Theory* 12 (1980) 251-257; MR 82d:10078.
[5] H. L. Abbott, On sum-free sequences, *Acta Arith.* 48 (1987) 93-96; MR 88g:11007.
[6] Z. Zhang, A sum-free sequence with larger reciprocal sum, unpublished note (1991).
[7] N. J. A. Sloane, On-Line Encyclopedia of Integer Sequences, A003278, A005282, A014011, and A046185.
[8] S. R. Finch, A convex maximization problem, *J. Global Optim.* 2 (1992) 419.
[9] S. R. Finch, A convex maximization problem: Discrete case, math.OC/9912035.
[10] S. R. Finch, A convex maximization problem: Continuous case, math.OC/9912036.
[11] A. M. Mian and S. Chowla, On the B_2 sequences of Sidon, *Proc. Nat. Acad. Sci. India. Sect. A.* 14 (1944) 3-4; MR 7,243a.
[12] R. Lewis, Mian-Chowla and B_2-sequences, unpublished note (1999).
[13] Z. Zhang, A B_2-sequence with larger reciprocal sum, *Math. Comp.* 60 (1993) 835-839; MR 93m:11012.
[14] G. S. Yovanof and H. Taylor, B_2-sequences and the distinct distance constant, *Comput. Math. Appl.* (*Oxford*) 39 (2000) 37-42; MR 2001j:11007.
[15] P. Erdös and P. Turán, On a problem of Sidon in additive number theory, and on some related problems, *J. London Math. Soc.* 16 (1941) 212-215; addendum 19 (1944) 208; also in *Collected Works of Paul Turán*, v. 1, ed. P. Erdös, Akadémiai Kiadó, pp. 257-261; MR 3,270e.
[16] H. Halberstam and K. F. Roth, Sequences, Springer-Verlag, 1983, pp. 84-88; MR 83m:10094.
[17] P. Erdös, On the combinatorial problems which I would most like to see solved, *Combinatorica* 1 (1981) 25-42; MR 82k:05001.
[18] B. Lindström, An inequality for B_2-sequences, *J. Combin. Theory* 6 (1969) 211-212; MR 38 # 4436.
[19] Z. Zhang, Finding finite B_2-sequences with larger $m - a_m^{1/2}$, *Math. Comp.* 63 (1994) 403-414; MR 94i:11109.
[20] B. Lindström, An Erdös problem studied with the assistance of a computer, *Normat*, v. 45 (1997) n. 4, 145-149, 188; MR 98m:11140.
[21] B. Lindström, Recent results on Sidon sequences-A survey, Royal Institute of Technology report TRITA-MAT-1998-29.
[22] P. Erdös and P. Turán, On some sequences of integers, *J. London Math. Soc.* 11 (1936) 261-264.
[23] J. Wróblewski, A nonaveraging set of integers with a large sum of reciprocals, *Math. Comp.* 43 (1984) 261-262; MR 85k:11006.
[24] F. Behrend, On sets of integers which contain no three terms in an arithmetic progression, *Proc. Nat. Acad. Sci. USA* 32 (1946) 331-332; MR 8,317d.
[25] J. L. Gerver, The sum of the reciprocals of a set of integers with no arithmetic progression of k terms, *Proc. Amer. Math. Soc.* 62 (1977) 211-214; MR 55 # 12678.
[26] J. Bourgain, On triples in arithmetic progression, *Geom. Funct. Anal.* 9 (1999) 968-984; MR 2001h:11132.

[27] A. M. Odlyzko and R. P. Stanley, Some curious sequences constructed with the greedy algorithm, unpublished note (1978).
[28] J. L. Gerver and L. Ramsey, Sets of integers with no long arithmetic progressions generated by the greedy algorithm, *Math. Comp.* 33 (1979) 1353-1360; MR 80k:10053.
[29] J. L. Gerver, Irregular sets of integers generated by the greedy algorithm, *Math. Comp.* 40 (1983) 667-676; MR 84d:10056.
[30] J. Gerver, J. Propp, and J. Simpson, Greedily partitioning the natural numbers into sets free of arithmetic progressions, *Proc. Amer. Math. Soc.* 102 (1988) 765-772; MR 89f:11026.
[31] S. C. Lindhurst, *An Investigation of Several Interesting Sets of Numbers Generated by the Greedy Algorithm*, AB thesis, Princeton Univ., 1990.

2.21 スティルチェス定数

【1.6】で定義されたように，Riemann のゼータ関数 $\zeta(z)$ は，$z=1$ の単純な極の近傍で，次の Laurent（ローラン）展開をもつ．

$$\zeta(z) = \frac{1}{z-1} + \sum_{n=0}^{\infty} \frac{(-1)^n}{n!} \gamma_n (z-1)^n$$

係数 γ_n は，[1-9]

$$\gamma_n = \lim_{m \to \infty} \left(\sum_{k=1}^{m} \frac{\ln(k)^n}{k} - \frac{\ln(m)^{n+1}}{n+1} \right) = \begin{cases} 0.5772156649\cdots & (n=0 \text{ のとき}) \\ -0.0728158454\cdots & (n=1 \text{ のとき}) \\ -0.0096903631\cdots & (n=2 \text{ のとき}) \\ 0.0020538344\cdots & (n=3 \text{ のとき}) \\ 0.0023253700\cdots & (n=4 \text{ のとき}) \\ 0.0007933238\cdots & (n=5 \text{ のとき}) \end{cases}$$

を満たすことが証明できる．とくに，$\gamma_0 = \gamma$ はオイラー-マスケローニの定数【1.5】である．

数論への応用例がある．正の整数 N が**縁取り数**（jagged）とは，最大素因子が $> \sqrt{N}$ であるときをいう．$j(N)$ を，N を超えないそのような整数の個数とする．最初のいくつかの縁取り数は，2, 3, 5, 6, 7, 10, 11, 13, 14, … であり，漸近的に [10, 11]

$$j(N) = \ln(2) N - (1-\gamma_0) \frac{N}{\ln(N)} - (1-\gamma_0-\gamma_1) \frac{N}{\ln(N)^2} + O\left(\frac{N}{\ln(N)^3} \right)$$

である．ただし，$1-\gamma_0 = 0.4227843351\cdots$，$1-\gamma_0-\gamma_1 = 0.4956001805\cdots$ である．【5.4】で，滑らか（smooth）数についての関連する議論を参照せよ．他のところでの γ_n の出現は [12-17] にある．

Stieltjes（スティルチェス）定数 γ_n の符号は，見たところ，ランダムパターンに従うようである．Briggs（ブリッグス）[18] は，無限に多くの γ_n が正であり，無限に多くの γ_n が負であることを証明した．Mitrovic（ミトロヴィチ）[19] は，不等式

$$\gamma_{2n} < 0, \quad \gamma_{2n} > 0, \quad \gamma_{2n-1} < 0, \quad \gamma_{2n-1} > 0$$

の各々が無限に多くの n について成り立つことを示すことによって，結果を拡張した．精密な解析で，Matsuoka（松岡）[20,21]は，任意の $\varepsilon>0$ に対して，$\gamma_n, \gamma_{n+1}, \gamma_{n+2}, \cdots, \gamma_{n+\lfloor(2-\varepsilon)\ln(n)\rfloor}$ のすべてが同じ符号をもつ n が無限に多く存在し，$\gamma_n, \gamma_{n+1}, \gamma_{n+2}, \cdots, \gamma_{n+\lfloor(2+\varepsilon)\ln(n)\rfloor}$ のすべてが同じ符号をもつ n は有限個だけであることを証明した．さらに，

$$f(n)=|\{0\leq k\leq n : \gamma_n>0\}|, \quad g(n)=|\{0\leq k\leq n : \gamma_n<0\}|$$

であるとき，$f(n)=n/2+o(n), \quad g(n)=n/2+o(n)$ である．

はじめの少数のスティルチェス定数は 0 に近い．しかし，これは当てにならない．実際，証明は知られてはいないけれども，その絶対値は $n\to\infty$ のとき $\to\infty$ になるように思われる．$|\gamma_n|$ の上界は，幾人かの研究者[18,22-26]により，得ることに成功した．結果として，

$$|\gamma_n|\leq \frac{(3+(-1)^n)(2n)!}{n^{n+1}(2\pi)^n}$$

を得た．この不等式は，松岡[20,21]の結果にさらに含まれる．彼は，下界

$$\exp(n\ln(\ln(n))-\varepsilon n)<|\gamma_n|$$

が無限に多くの n について成り立ち，一方，上界

$$|\gamma_n|\leq \frac{1}{10000}\exp(n\ln(\ln(n)))$$

がすべての $n\geq 10$ で成り立つことを証明した．

【1.5】で，Vacca（ヴァッカ）による次の式を示した．

$$\gamma_0=\sum_{k=1}^{\infty}\frac{(-1)^k}{k}\left\lfloor\frac{\ln(k)}{\ln(2)}\right\rfloor$$

Hardy[27]は，γ_1 に対する類似の式を与えた．

$$\gamma_1=\sum_{j=1}^{\infty}\frac{(-1)^j\ln(j)}{j}\left\lfloor\frac{\ln(j)}{\ln(2)}\right\rfloor-\frac{\ln(2)}{2}\sum_{k=1}^{\infty}\frac{(-1)^k}{k}\left\lfloor\frac{\ln(2k)}{\ln(2)}\right\rfloor\left\lfloor\frac{\ln(k)}{\ln(2)}\right\rfloor$$

Kluyver（クルイベル）[28]は，高次の定数に対してこのような級数を多数示した．$\{x\}$ を x の小数部分とするとき，さらに[29]

$$\int_1^{\infty}\frac{\{x\}}{x^2}dx=1-\gamma_0, \quad \int_1^{\infty}\int_x^{\infty}\frac{\{y\}}{xy^2}dydx=1-\gamma_0-\gamma_1$$

である．γ_n の補足の式は[7,8,30-32]にある．

ある種の関連する定数を議論する．別の異なる級数

$$\tau_n=\sum_{k=1}^{\infty}(-1)^k\frac{\ln(k)^n}{k}$$

$$=\begin{cases}-\ln(2)=-0.6931471805\cdots & (n=0 \text{ のとき})\\ -\frac{1}{2}\ln(2)^2+\gamma_0\ln(2)=0.1598689037\cdots & (n=1 \text{ のとき})\\ -\frac{1}{3}\ln(2)^3+\gamma_0\ln(2)^2+2\gamma_1\ln(2)=0.0653725925\cdots & (n=2 \text{ のとき})\end{cases}$$

は，式[1,4,8,26]

$$\tau_n = -\frac{\ln(2)^{n+1}}{n+1} + \sum_{k=0}^{n-1}\binom{n}{k}\ln(2)^{n-k}\gamma_k, \quad \gamma_n = \frac{1}{n+1}\sum_{k=0}^{n+1}\binom{n+1}{k}B_{n+1-k}\ln(2)^{n-k}\tau_k$$

によって，スティルチェス定数と関係している．ただし，B_j は j 次のベルヌーイ数である【1.6.1】．さらに，(1 においてでなく) 原点での $\zeta(z)$ のローラン展開を考えて

$$\zeta(z) = \frac{1}{z-1} + \sum_{n=0}^{\infty}\frac{(-1)^n}{n!}\delta_n z^n$$

を得る．Sitaramachandrarao（シタラマチャンドララオ）[33] は，[3,34]

$$\delta_n = \lim_{m\to\infty}\left(\sum_{k=1}^{m}\ln(k)^n - \int_1^m \ln(x)^n dx - \frac{1}{2}\ln(m)^n\right) = (-1)^n(\zeta^{(n)}(0) + n!)$$

$$= \begin{cases} \dfrac{1}{2} = 0.5 & (n=0 \text{ のとき}) \\[2mm] \dfrac{1}{2}\ln(2\pi) - 1 = -0.0810614667\cdots & (n=1 \text{ のとき}) \\[2mm] -\dfrac{\pi^2}{24} - \dfrac{1}{2}\ln(2\pi)^2 + \dfrac{\gamma_0^2}{2} + \gamma_1 + 2 = -0.0063564559\cdots & (n=2 \text{ のとき}) \end{cases}$$

を得た．これらは順に，式[7,26]

$$\sigma_n = \sum_{\rho}\frac{1}{\rho^n} = \begin{cases} -\dfrac{1}{2}\ln(4\pi) + \dfrac{\gamma_0}{2} + 1 = 0.0230957089\cdots & (n=1 \text{ のとき}) \\[2mm] -\dfrac{\pi^2}{8} + \gamma_0^2 + 2\gamma_1 + 1 = -0.0461543172\cdots & (n=2 \text{ のとき}) \\[2mm] -\dfrac{7\zeta(3)}{8} + \gamma_0^3 + 3\gamma_0\gamma_1 + \dfrac{3\gamma_2}{2} + 1 = -0.0001111582\cdots & (n=3 \text{ のとき}) \end{cases}$$

の近似和で Lehmer（レーマー）[35] の研究の役に立つ．各々の和は，$\zeta(z)$ の自明でない零点 ρ すべてをわたる．Keiper（カイパー）[36] と Kreminski（クレミンスキー）[37] は，Lehmer の計算を広く拡張した．

等差数列 $a, a+b, a+2b, a+3b, \cdots$，に対応する類似の γ_n は，Knopfmacher（クノプマッヘル）[38]，Kanemitsu（金光滋）[39]，Dilcher（ディルチャー）[40] により研究された．

$$\gamma_{n,a,b} = \lim_{m\to\infty}\left(\sum_{\substack{0 < k \leq m \\ k \equiv a \bmod b}}\frac{\ln(k)^n}{k} - \frac{1}{b}\frac{\ln(m)^{n+1}}{n+1}\right)$$

たとえば，$\sum_{a=0}^{b-1}\gamma_{n,a,b} = \gamma_n$ で，

$$\gamma_{n,0,2} = \frac{1}{2}\left[\sum_{j=0}^{n}\binom{n}{j}\gamma_{n-j}\ln(2)^j - \frac{\ln(2)^{n+1}}{n+1}\right], \quad \gamma_{1,0,3} = \frac{1}{3}\left[\gamma_1 + \gamma_0\ln(3) - \frac{\ln(3)^2}{2}\right]$$

$$\gamma_{1,1,3} = \frac{1}{6}\left[2\gamma_1 - \gamma_0\ln(3) + \frac{\ln(3)^2}{2} - \frac{(\gamma_0 + \ln(2\pi))}{3} - \ln\left[\Gamma\left(\frac{1}{3}\right)^2\frac{\sqrt{3}}{2\pi}\right]\right]\pi\sqrt{3}$$

である．異なる γ_n の拡張は[23,26,41-46]にある．

読者は，論文のいくつかではスティルチェス定数が γ_n ではなくて $(-1)^n\gamma_n/n!$ として定義されていることに注意を払わなければならない．だから，文献を参照するときには注意が必要である．

2.21.1 一般ガンマ関数

複素数 z に対して，一般ガンマ関数 $\Gamma_n(z)$ は，[47,48] により

$$\Gamma_n(z) = \lim_{m \to \infty} \frac{\exp\left(\frac{\ln(m)^{n+1}}{n+1} z\right) \prod_{k=1}^{m} \exp\left(\frac{\ln(k)^{n+1}}{n+1}\right)}{\prod_{k=0}^{m} \exp\left(\frac{\ln(k+z)^{n+1}}{n+1}\right)}$$

で定義され，負の実軸に沿って切れ目を入れた複素数平面上で解析的である．明らかに，$\Gamma_0(z) = \Gamma(z)$ であり，$\Gamma_n(z)$ は

$$\Gamma_n(1) = 1, \quad \Gamma_n(z+1) = \exp\left(\frac{\ln(z)^{n+1}}{n+1}\right) \Gamma_n(z)$$

を満たす．$\Gamma_n(z)$ と γ_n とは，等式 $\psi_n(1) = -\gamma_n$ で関連している．ただし，

$$\psi_n(x) = \frac{d}{dx} \ln(\Gamma_n(x)) = -\gamma_n - \sum_{k=0}^{\infty} \left(\frac{\ln(x+k)^n}{x+k} - \frac{\ln(k+1)^n}{k+1}\right)$$

は，一般ディガンマ (digamma) 関数である．一般化 Stirling (スターリング) 公式は，特別な場合として

$$\Gamma_0(x) \sim \sqrt{2\pi}\, x^{x-1/2} e^{-x}, \quad \Gamma_1(x) \sim C x^{(1/2)(x-1/2)\ln(x)-x} e^x$$

を含む．ただし，

$$\ln(C) = \ln\left(\Gamma_1\left(\frac{1}{2}\right)\right) - \frac{1}{4}\ln(2)^2 - \frac{1}{2}\ln(2)\ln(2\pi)$$

$$= -\frac{\pi^2}{48} - \frac{1}{4}\ln(2\pi)^2 + \frac{\gamma_0^2}{4} + \frac{\gamma_1}{2} = -1.0031782279\cdots$$

である [48,49]．この種のさらに多くの公式が見つけられるであろう．

[1] J. J. Y. Liang and J. Todd, The Stieltjes constants, *J. Res. Nat. Bur. Standards* B 76 (1972) 161-178; MR 48 #5316.

[2] A. Ivić, *The Riemann Zeta-Function*, Wiley, 1985, pp. 4-6, 49; MR 87d:11062.

[3] B. C. Berndt, *Ramanujan's Notebooks: Part I*, Springer-Verlag, 1985, pp. 164-165, 196-204; MR 86c:01062.

[4] W. E. Briggs and S. Chowla, The power series coefficients of $\zeta(s)$, *Amer. Math. Monthly* 62 (1955) 323-325; MR 16,999f.

[5] D. P. Verma, Laurent's expansion of Riemann's zeta-function, *Indian J. Math.* 5 (1963) 13-16; MR 28 #5046.

[6] R. P. Ferguson, An application of Stieltjes integration to the power series coefficients of the Riemann zeta function, *Amer. Math. Monthly* 70 (1963) 60-61; MR 26 #2408.

[7] M. I. Israilov, The Laurent expansion of the Riemann zeta function (in Russian), *Trudy Mat. Inst. Steklov* 158 (1981) 98-104, 229; Engl. transl. in *Proc. Steklov Inst. Math.* (1983) n. 4, 105-112; MR 83m:10069.

[8] N. Y. Zhang, On the Stieltjes constants of the zeta function (in Chinese), *Beijing Daxue Xuebao* (1981) n. 4, 20-24; MR 84h:10056.

[9] B. K. Choudhury, The Riemann zeta-function and its derivatives, *Proc. Royal Soc. London A* 450 (1995) 477-499; MR 97e:11095.

[10] N. J. A. Sloane, On-Line Encyclopedia of Integer Sequences, A064052.

[11] D. H. Greene and D. E. Knuth, *Mathematics for the Analysis of Algorithms*, 3rd ed., Birkhäuser,

1990, pp. 95-98; MR 92c:68067.
[12] N. G. de Bruijn, On Mahler's partition problem, *Proc. Konink. Nederl. Akad. Wetensch. Sci. Sect.* 51 (1948) 659-669; Indag. Math. 10 (1948) 210-220; MR 10,16d.
[13] W. B. Pennington, On Mahler's partition problem, *Annals of Math.* 57 (1953) 531-546; MR 14, 846m.
[14] A. F. Lavrik, The principal term of the divisor problem and the power series of the Riemann zeta-function in a neighborhood of a pole (in Russian), *Trudy Mat. Inst. Steklov.* 142 (1976) 165-173, 269; Engl. transl. in *Proc. Steklov Inst. Math.* (1979) n. 3, 175-183; MR 58 #27836.
[15] A. F. Lavrik, M. I. Israilov, and Z. Ėdgorov, Integrals containing the remainder term of the divisor problem (in Russian), *Acta Arith.* 37 (1980) 381-389; MR 82e:10078.
[16] Ė. P. Stankus, A remark on the coefficients of Laurent series of the Riemann zeta function (in Russian), Studies in Number Theory, 8, *Zap. Nauchn. Sem. Leningrad Otdel. Mat. Inst. Steklov (LOMI)* 121 (1983) 103-107; MR 85d:11081.
[17] T. W. Cusick, Zaremba's conjecture and sums of the divisor function, *Math. Comp.* 61 (1993) 171-176; MR 93k:11063.
[18] W. E. Briggs, Some constants associated with the Riemann zeta-function, *Michigan Math. J.* 3 (1955-56) 117-121; MR 17,955c.
[19] D. Mitrovic, The signs of some constants associated with the Riemann zeta-function, *Michigan Math. J.* 9 (1962) 395-397; MR 29 #2232.
[20] Y. Matsuoka, On the power series coefficients of the Riemann zeta function, *Tokyo J. Math.* 12 (1989) 49-58; MR 90g:11116.
[21] Y. Matsuoka, Generalized Euler constants associated with the Riemann zeta function, *Number Theory and Combinatorics*, Proc. 1984 Tokyo conf., ed. J. Akiyama, Y. Ito, S. Kanemitsu, T. Kano, T. Mitsui, and I. Shiokawa, World Scientific, pp. 279-295; MR 87e:11105.
[22] E. Lammel, Ein Beweis, dass die Riemannsche Zetafunktion $\zeta(s)$ in $|s-1|\leq 1$ keine Nullstelle besitzt, *Univ. Nacional de Tucumán Rev. Ser.* A 16 (1966) 209-217; MR 36 #5090.
[23] B. C. Berndt, On the Hurwitz zeta-function, *Rocky Mount. J. Math.* 2 (1972) 151-157; MR 44 #6622.
[24] K. Verma, Laurent expansions of Hurwitz and Riemann zeta functions about $s=1$, *Ganita* 42 (1991) 65-70; MR 93i:11106.
[25] B. C. Yang and K.Wu, An inequality for the Stieltjes constants (in Chinese), *J. South China Normal Univ. Natur. Sci. Ed.* (1996) n. 2, 17-20; MR 97m:11163.
[26] N.-Y. Zhang and K. S. Williams, Some results on the generalized Stieltjes constants, *Analysis* 14 (1994) 147-162; MR 95k:11110.
[27] G. H. Hardy, Note on Dr. Vacca's series for γ, *Quart. J. Pure Appl. Math.* 43 (1912) 215-216; also in *Collected Papers*, v. 4, Oxford Univ. Press, 1966, pp. 475-476.
[28] J. C. Kluyver, On certain series of Mr. Hardy, *Quart. J. Pure Appl. Math.* 50 (1927) 185-192.
[29] P. Sebah, Correction to Ellison-Mendès-France example, ch. 1, sect. 5-2, unpublished note (2000).
[30] D. Andrica and L. Tóth, Some remarks on Stieltjes constants of the zeta function, *Stud. Cerc. Mat.* 43 (1991) 3-9; MR 93c:11066.
[31] M.-A. Coppo, Nouvelles expressions des constantes de Stieltjes, *Expos. Math.* 17 (1999) 349-358; MR 2000k:11097.
[32] M.-A. Coppo, Sur les sommes d'Euler divergentes, *Expos. Math.* 18 (2000) 297-308; MR 2001h:11158.
[33] R. Sitaramachandrarao, Maclaurin coefficients of the Riemann zeta function, *Abstracts Amer. Math. Soc.* 7 (1986) 280.
[34] T. M. Apostol, Formulas for higher derivatives of the Riemann zeta function, *Math. Comp.* 44 (1985) 223-232; MR 86c:11063.
[35] D. H. Lehmer, The sum of like powers of the zeros of the Riemann zeta function, *Math. Comp.* 50 (1988) 265-273; MR 88m:11073.

[36] J. B. Keiper, Power series expansions of Riemann's ξ function, Math. Comp. 58 (1992) 765-773; MR 92f:11116.
[37] R. M. Kreminski, Newton-Cotes integration for approximating Stieltjes (generalized Euler) constants, Math. Comp., to appear.
[38] J. Knopfmacher, Generalised Euler constants, Proc. Edinburgh Math. Soc. 21 (1978) 25-32; MR 57 #12432.
[39] S. Kanemitsu, On evaluation of certain limits in closed form, Théorie des nombres, Proc. 1987 Québec conf., ed. J.-M. De Koninck and C. Levesque, Gruyter, 1989, pp. 459-474; MR 90m:11127.
[40] K. Dilcher, Generalized Euler constants for arithmetical progressions, Math. Comp. 59 (1992) 259-282 and S21-S24; MR 92k:11145.
[41] J. R. Wilton, A note on the coefficients in the expansion of $\zeta(s, x)$ in powers of s−1, Quart. J. Pure Appl. Math. 50 (1927) 329-332.
[42] W. E. Briggs and R. G. Buschman, The power series coefficients of functions defined by Dirichlet series, Illinois J. Math. 5 (1961) 43-44; MR 22 #10956.
[43] A. F. Lavrik, Laurent coefficients of a generalized zeta function (in Russian), Theory of Cubature Formulas and Numerical Mathematics, Proc. 1978 Novosibirsk conf., ed. S. L. Sobolev, Nauka Sibirsk. Otdel., 1980, pp. 160-164, 254; MR 82i:10050.
[44] M. I. Israilov, On the Hurwitz zeta function (in Russian), Izv. Akad. Nauk UzSSR Ser. Fiz.-Mat. Nauk (1981) n. 6, 13-18, 78; MR 83g:10031.
[45] U. Balakrishnan, On the Laurent expansion of $\zeta(s, a)$ at $s=1$, J. Indian Math. Soc. 46 (1982) 181-187; MR 88f:11080.
[46] J. Bohman and C.-E. Fröberg, The Stieltjes function-Definition and properties, Math. Comp. 51 (1988) 281-289; MR 89i:11095.
[47] E. L. Post, The generalized gamma functions, Annals of Math. 20 (1919) 202-217.
[48] K. Dilcher, On generalized gamma functions related to the Laurent coefficients of the Riemann zeta function, Aequationes Math. 48 (1994) 55-85; MR 95h:11086.
[49] V. S. Adamchik, $\Gamma_n(1/2)$ and generalized Stirling formulas, unpublished note (2001).

2.22 リューヴィル-ロス定数

他の定数によって定数を研究することがある.与えられた実数 ξ に対して,p, $q>0$ を整数とするとき,不等式

$$0 < \left|\xi - \frac{p}{q}\right| < \frac{1}{q^r}$$

が多くとも有限個の解 (p,q) をもつ,正の実数 r のすべての集合を R とする.Liouville-Roth(リューヴィル-ロス)定数または無理数度を,

$$r(\xi) = \inf_{r \in R} r$$

で定義する.すなわち,有理数で近似できない ξ の臨界閾値である[1-3].

ξ は有理数 　　　$\Rightarrow r(\xi) = 1$
ξ は代数的無理数 $\Rightarrow r(\xi) = 2$ (Thue-Siegel-Roth(トゥエ-ジーゲル-ロス)の定理[4,5])

ξ は超越数 $\Rightarrow r(\xi) \geq 2$

が知られている．もし ξ がリューヴィル数，たとえば

$$\sum_{n=1}^{\infty}\frac{1}{2^{n!}}=\frac{1}{2^1}+\frac{1}{2^2}+\frac{1}{2^6}+\frac{1}{2^{24}}+\frac{1}{2^{120}}+\cdots=0.7656250596\cdots$$

であるならば，$r(\xi)=\infty$ である．同様に，$r(\xi)$ が $2<r(\xi)<\infty$ を満たす任意の値になるように，（適切な速い収束をもつ有理数の級数から）ξ を構成することができる．有名な定数のうちで，

$$r(e)=2$$

が知られている[2]．（実際，さらにずっと正確な不等式が可能である．しかし，e はいくぶん異常である．）さらに

$2 \leq r(\pi) \leq 8.016045\cdots$ (Hata（畑政義）[6,7])

$2 \leq r(\ln(2)) \leq 3.89139978\cdots$ (Rukhadze（ルカジェ）[8,9])

$2 \leq r(\pi^2) \leq 5.441243\cdots$ (Hata[10], Rhin & Viola（リン&ヴィオラ）[11])

$2 \leq r(\zeta(3)) \leq 5.513891\cdots$ (Hata[12], Rhin & Viola[13])

である．ここに，$\zeta(3)$ は Apéry（アペリ）の定数【1.6】である．Catalan（カタラン）の定数 G【1.7】や Khintchine（ヒンチン）の定数 K【1.8】に対応する r の上界は，知られていない．G と K が無理数であるかどうかさえも，未解決である．

π に関する Hata（畑）の仕事の結果より，2つの関数[14,15]

$$C(x)=\inf_{\text{整数}\,n>0} n^x|\sin(n)|, \quad D(x)=\sup_{\text{整数}\,n>0} n^{-x}|\tan(n)|$$

は $C(7.02)>0$，$D(7.02)=0$ を満たす．もし，$r(\pi)=2$ であるという予想[16]が真であるならば，$C(1+\varepsilon)>0$，$D(1+\varepsilon)=0$ がすべての $\varepsilon>0$ で成り立つ．数値的な証拠は $C(1)=0$，$D(1)=\infty$ を示す．

これらの定数の多次元での類似な定数を研究できる．たとえば，$\xi_1,\xi_2,\cdots,\xi_n$ を実数で代数的数とし，$1,\xi_1,\xi_2,\cdots,\xi_n$ は有理数上で線型独立とする．同時不等式

$$0<\left|\xi_i-\frac{p_i}{q}\right|<\frac{1}{q^r}, \quad i=1,2,\cdots,n$$

が，各 p_i，$q>0$ が整数であるような，多くとも有限個の解 $(p_1,p_2,\cdots,p_n,q)$ をもつ正の実数 r のすべての集合を R と定義する．$r(\xi_1,\xi_2,\cdots,\xi_n)$ を上述と同様に正確に定義する．Schmidt（シュミット）[5,17,18]は，

$$r(\xi_1,\xi_2,\cdots,\xi_n)=\frac{n+1}{n}$$

が成り立つことを示し，トゥエ-ジーゲル-ロスの定理を拡張した．

明らかに，結合無理数度 $r(e,\pi)$ は $r(e,\pi) \leq \max\{r(e),r(\pi)\}$ を満たす．しかし，誰もこの上界を改良した人はいない．もちろん，e と π が有理数上で線型独立かどうかさえもわかっていない！

同時ディオファンタス近似定数【2.23】に関する話題には，ここで述べたものとは焦点が異なるものの，類似がある．

[1] G. H. Hardy and E. M. Wright, *An Introduction to the Theory of Numbers*, 5th ed., Oxford Univ. Press, 1985, pp. 154-169; MR 81i:10002.
[2] J. M. Borwein and P. B. Borwein, *Pi and the AGM: A Study in Analytic Number Theory and Computational Complexity*, Wiley, 1987, pp. 351-352, 362-371; MR 99h:11147.
[3] H. M. Stark, *An Introduction to Number Theory*, MIT Press, 1978, pp. 172-180; MR 80a:10001.
[4] K. F. Roth, Rational approximations to algebraic numbers, *Mathematika* 2 (1955) 1-20; corrigendum, 168; MR 17,242d.
[5] A. Baker, *Transcendental Number Theory*, Cambridge Univ. Press, 1975, pp. 66-84; MR 54 # 10163.
[6] M. Hata, Improvement in the irrationality measures of π and π^2, *Proc. Japan Acad. Ser. A. Math. Sci.* 68 (1992) 283-286; MR 94b:11064.
[7] M. Hata, Rational approximations to π and some other numbers, *Acta Arith.* 63 (1993) 335-349; MR 94e:11082.
[8] E. A. Rukhadze, A lower bound for the rational approximation of ln 2 by rational numbers (in Russian), *Vestnik Moskov. Univ. Ser. I Mat. Mekh.* (1987) n. 6, 25-29, 97; Engl. transl. in Moscow Univ. Math. Bull., v. 42 (1987) n. 6, 30-35; MR 89b:11064.
[9] M. Hata, Legendre type polynomials and irrationality measures, *J. Reine Angew. Math.* 407 (1990) 99-125; MR 91i:11081.
[10] M. Hata, A note on Beuker's integral, *J. Austral. Math. Soc.* 58 (1995) 143-153; MR 96c:11081.
[11] G. Rhin and C. Viola, On a permutation group related to $\zeta(2)$, *Acta Arith.* 77 (1996) 23-56; MR 97m:11099.
[12] M. Hata, A new irrationality measure for $\zeta(3)$, *Acta Arith.* 92 (2000) 47-57; MR 2001a:11123.
[13] G. Rhin and C. Viola, The group structure for $\zeta(3)$, *Acta Arith.* 97 (2001) 269-293; MR 2002b:11098.
[14] R. B. Israel, Approximability of π, unpublished note (1996).
[15] I. Rosenholtz, Tangent sequences, world records, π, and the meaning of life: Some applications of number theory to calculus, *Math. Mag.* 72 (1999) 367-376; MR 2000i:11109.
[16] J. M. Borwein, P. B. Borwein, and D. H. Bailey, Ramanujan, modular equations, and approximations to pi, or how to compute one billion digits of pi, *Amer. Math. Monthly* 96 (1989) 201-219; also in *Organic Mathematics*, Proc. 1995 Burnaby workshop, ed. J. Borwein, P. Borwein, L. Jörgenson, and R. Corless, Amer. Math. Soc., 1997, pp. 35-71; MR 90d:11143.
[17] K. B. Stolarsky, *Algebraic Numbers and Diophantine Approximation*, Dekker, 1974, pp. 308-309; MR 51 # 10241.
[18] W. M. Schmidt, Simultaneous approximation to algebraic numbers by rationals, *Acta Math.* 125 (1970) 189-201; MR 42 # 3028.

2.23 ディオファンタス近似定数

リューヴィル-ロス定数の項【2.22】で，1つの無理数 ξ の有理度（無理数度）を議論した．ここでは，少なくとも1つは無理数である n 個の実数 $\xi_1, \xi_2, \cdots, \xi_n$ の同時有理数近似，すなわち，すべて同じ分母をもつ分数近似を研究する．Dirichlet（ディリクレ）の鳩の巣論法[1,2]より，もし $c \geq 1$ ならば，不等式

$$\left|\xi_i - \frac{p_i}{q}\right| < c^{1/n} q^{-(n+1)/n}, \quad i=1,2,\cdots,n$$

は，無限に多くの解 $(p_1, p_2, \cdots, p_n, q)$ をもつ．ただし，$p_1, p_2, \cdots, p_n$ と $q>0$ は整数である．この節での焦点は，右辺の指数 $(n+1)/n$ ではなく，前にもあったように，むしろ 1 次の係数 c である．

いつものように分母を払い，不等式を

$$q \cdot |q\xi_i - p_i|^n < c$$

と書き改め，解の集合 $(p_1, p_2, \cdots, p_n, q)$ が無限集合のままであるすべての $0<c\leq 1$ の下限を c_n と定義する．**n 次元同時ディオファンタス近似定数 γ_n を**，このような，$\xi_1, \xi_2, \cdots, \xi_n$ 上での c_n の上限と定義する．したがって，γ_n は，n 個の数からなる「単独の」集合の，近似度を測る尺度ではなく，「すべての」可能な集合の上で定義される．したがって，次元 n のみに依存する．

近似定数 γ_n について知られているものを要約すると，

$$\gamma_1 = \frac{1}{\sqrt{5}} = 0.4472135955\cdots \text{ (Hurwitz (フルウィツ) [1])}$$

$$0.2857142857\cdots = \frac{2}{7} \leq \gamma_2 \leq \frac{64}{169} = 0.378\cdots$$

(Cassels (カッセル) [2], Nowak (ノバック) [3])

$$0.210\cdots = \frac{2}{5\sqrt{11}} \leq \gamma_3 \leq \delta_2 = \frac{1}{2}\frac{1}{\pi-2} = 0.437\cdots$$

(Cusick (クシック) [4], Spohn (スポーン) [5])

$$0.044\cdots = \frac{16}{9\sqrt{1609}} \leq \gamma_4 \leq \delta_3 = \frac{27}{4}\frac{1}{8\sqrt{3}\pi-27} = 0.408\cdots$$

(Krass (クラス) [6], Spohn[5])

$$0.010\cdots = \frac{16}{207\sqrt{53}} \leq \gamma_5 \leq \delta_4 = 0.390\cdots \quad ([5\text{-}7])$$

$$0.004\cdots = \frac{16}{9\sqrt{184607}} \leq \gamma_6 \leq \delta_5 = 0.379\cdots \quad ([5\text{-}7])$$

ただし，上界[5]は定積分

$$\frac{1}{\delta_k} = k2^{k+1}\int_0^1 \frac{x^{k-1}}{(1+x^k)(1+x)^k}dx$$

で計算される．

$\gamma_2 = 2/7$ であることを示唆する豊富な計算[8]および理論的な証拠がある[9,10]．しかし，まだ定理とは考えられていない．$\xi_1 = 1$ で，ξ_2, ξ_3 が実 3 次体の基底という制約のもとで $\gamma_2 = 2/7$ が正しいことを Adams (アダムス) [9]は証明した．Cusick[10,11] は，$2\cos(2\pi/7)$ の正則連分数展開においてある有限個の部分分母のパターンが無限に多く起こる，という仮定のもとで追加の結果を証明した．[12,13]を参照せよ．

---

†訳注（前頁）：原語は箱論法あるいは抽き出し論法：n 個の箱に，全体で n 個より多くのものを入れれば，必ず複数個が入る箱が生じるという原理である．英語の pigeon hall は区分け巣箱の意味であり，「鳩の巣論法」は慣用の誤訳というべきか．

γ_3 に関して，Szekeres（セケレス）[14]は，真の値は 0.170 ぐらい大きいらしく，ここで与えられている下限よりも十分に大きい値であることを示した．

Nowak[15]は，δ_k の関数を含む，Spohn の上限を改良した．しかし，数値的な評価は現時点では可能でない．

γ_n と数の幾何学との間には顕著な関係がある．まず最初にこれを 2 次元の場合に図示する（図 2.3 を見よ）．$|xy| \leq 1$ (**星体** (star body) とよばれるものの一例である) で決まる平面内の非有界な領域 S を考える．なおその上に，基底ベクトル $(1, 1)$ と $((1+\sqrt{5})/2, (1-\sqrt{5})/2)$ をもつ，格子 L を考える．S の内部に含まれる L の頂点は，原点 $(0, 0)$ のみであることが証明できる．したがって，L は S **許容** (admissible) とよばれる．

L の任意の 1 つの最小の平行四辺形の面積は，明らかに $\sqrt{5}$ である．これは L の**行列式** (determinant) であって，$\det(L)$ と書く．さらに，任意のほかの S 許容な格子 L は $\det(L) \geq \sqrt{5}$ を満たすことが，証明できる．

同様に，
$$|x_{n+1}| \cdot \max\{|x_1|^n, |x_2|^n, \cdots, |x_n|^n\} \leq 1$$
で決まる $(n+1)$ 次元空間の非有界な領域 S を考える．すべての $(n+1)$ 次元 S 許容格子 L を考える．Davenport（ダベンポート）[16,17]は，L の任意の 1 つの最小の平行六面体の体積 $\det(L)$ が $\det(L) \geq 1/\gamma_n$ であることとさらに，等号は L のある頂点の組み合わせで起こることを示した．したがって，

$$\frac{1}{\gamma_n} = \min_{\substack{S \text{許容} \\ \text{格子} L}} \det(L)$$

はまた，星体 S の**臨界行列式** (critical determinant) または**格子定数** (lattice constants) として知られている．この幾何学的な洞察は，残念だが γ_n を計算するとき，限られた助けにしかならない．いくつかの計算例は，[18-24]で与えられている．

数の幾何学からの類似の問題がある（知るかぎりでは γ_n とは関連がない）．再び，2

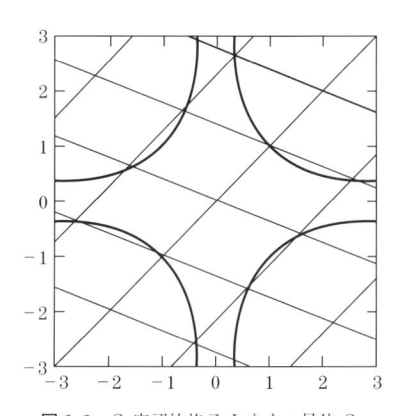

図 2.3　S-容認的格子 L をもつ星体 S．

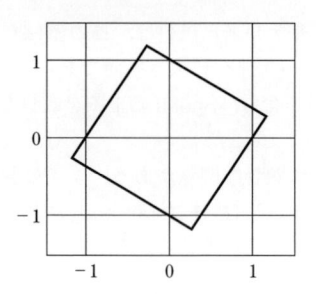

図 2.4 Z-容認的平行四辺形 P.

次元の場合に図示する (図 2.4 を見よ). 平面上の整数の格子点を Z とおく. すなわち, 基底ベクトル $(1,0)$ と $(0,1)$ である. 原点 $(0,0)$ に中心がある, 任意の平行四辺形 P を考える. P が Z **容認的** (allowable) であるとは, P の内部に Z の他の格子点がまったく含まれないときをいう. 平面の任意に与えられた基底 v, w に対して, 明らかに, v と w に垂直な辺をもつ Z 容認的平行四辺形 P が存在する (適切な小さい面積をもつ P をとればよい). このような P のすべてに対して, 面積の上限を, $a(v,w)$ と定義する. このような基底 v, w のすべてに対して, $a(v,w)/4$ の下限を κ_2 と定義する. Szekeres[25] は

$$\kappa_2 = \frac{1}{2}\left(1 + \frac{1}{\sqrt{5}}\right) = 0.7236067977\cdots$$

を証明した.

この場合の「臨界平行四辺形」(critical parallelogram) の傾きは, $(1+\sqrt{5})/2$, $(1-\sqrt{5})/2$ である. 黄金比【1.2】が前出の γ_2 の計算だけでなく, ここにも現れることは, 興味深い.

高次元については, 標準 n 次元整数格子点を Z とし, 与えられた基底 $v_1, v_2, \cdots, v_n$ に垂直な面をもつ n 次元 Z 容認的平行六面体 P を考える. 前と同様に, $2^n \kappa_n$ は P が定めた方向 $v_1, v_2, \cdots, v_n$ と無関係に体積 $2^n \kappa_n$ をもち得るという意味で, P の可能な体積のうちの最大のものである. しかし, 任意の $\varepsilon > 0$ に対して, 体積 $2^n \kappa_n + \varepsilon$ の P に対しては, このことは成り立たない. $\kappa_3 > 1/4$, $\kappa_4 > 1/16$ であることが知られている. たぶん

$$\kappa_3 = \frac{8}{7}\cos\left(\frac{2\pi}{7}\right)\cos\left(\frac{\pi}{7}\right)^2 = 0.5784167628\cdots$$

と思われる理論的な証拠[29]がある. さらに, 漸近的に [28, 30]

$$\frac{n}{(n!)^2}\left(\frac{1}{2}\right)^{n(n+1)/2} < \kappa_n < \left[\frac{1}{2}\left(1+\frac{1}{\sqrt{5}}\right)\right]^{(n-1)/2}$$

が証明されている. 実数 $\kappa_2, \kappa_3, \kappa_4, \cdots$ を Mordell (**モーデル**) **定数**[31] という. さらなる議論は, [32-34] にある.

もう一つ問題がある. K を, 体積が $V(K)$ であり, かつ原点に関して対称な, n 次

元の有界な凸体とする．$\Delta(K)$ を K の臨界行列式とし，
$$\rho_n = \inf_K \frac{V(K)}{\Delta(K)}$$
と定義する．たとえば，$n=2$ で K が円盤のとき，明らかに $V(K)/\Delta(K) = 2\pi/\sqrt{3} = 3.627\cdots$ である．これは最良ではない．というのは[35-38]
$$3.570624\cdots \leq \rho_2 \leq 4 \cdot \frac{8 - 4\sqrt{2} - \ln(2)}{2\sqrt{2} - 1} = 3.6096567319\cdots$$
が知られている．また，ρ_2 はこの上界（各頂点を双曲弧で角を丸めることで得られる，滑らかな 8 角形に対応する）と等しい，と予想されている[39, 40]．さらに，[35, 41, 42]から次のことが知られている．$\rho_3 \geq 4.216$, $\rho_4 \geq 4.721$, $\rho_n \geq r = 4.921553\cdots$ ($n \geq 5$) である．ただし，$r > 1$ は方程式 $r \ln(r) = 2(r-1)$ の一意の解である．しかしながら，Mahler[35]は，$n \to \infty$ のとき $\rho_n \to \infty$ である，と信じた．したがって，改良の余地はかなりある．この定理は古典的な Minkowski-Hlawka（ミンコフスキー-ラウカ）の定理の自然な結果である．有界な星体 S に対応する ρ_n と類似なものを σ_n とすると，平行に考えられる問題群がある．たとえば[43]，$\sigma_2 \leq 3.5128\cdots$（8 つの双曲弧によって縁どられた S に対応する）である．しかし，誰も σ_2 の正確な値を予想しているとは思えない．

[1] G. H. Hardy and E. M. Wright, *An Introduction to the Theory of Numbers*, 5th ed., Oxford Univ. Press, 1985; MR 81i:10002.
[2] G. Szekeres, The n-dimensional approximation constant, *Bull. Austral. Math. Soc.* 29 (1984) 119-125; MR 85c:11056.
[3] W. G. Nowak, A note on simultaneous Diophantine approximation, *Manuscripta Math.* 36 (1981) 33-46; MR 83a:10062.
[4] T. W. Cusick, Estimates for Diophantine approximation constants, *J. Number Theory* 12 (1980) 543-556; MR 82j:10057.
[5] W. G. Spohn, Blichfeldt's theorem and simultaneous Diophantine approximation, *Amer. J. Math.* 90 (1968) 885-894; MR 38 #122.
[6] S. Krass, Estimates for n-dimensional Diophantine approximation constants for $n \geq 4$, *J. Number Theory* 20 (1985) 172-176; MR 86j:11070.
[7] G. Niklasch, Smallest absolute discriminants of number fields, unpublished notes (1997-1998).
[8] G. Szekeres, Computer examination of the 2-dimensional simultaneous approximation constant, *Ars Combin.* 19A (1985) 237-243; MR 86h:11051.
[9] W. W. Adams, The best two-dimensional Diophantine approximation constant for cubic irrationals, *Pacific J. Math.* 91 (1980) 29-30; MR 82i:10038.
[10] T.W. Cusick, The two-dimensional Diophantine approximation constant. II, *Pacific J. Math.* 105 (1983) 53-67; MR 84g:10060.
[11] T. W. Cusick and S. Krass, Formulas for some Diophantine approximation constants, *J. Austral. Math. Soc. Ser. A* 44 (1988) 311-323; MR 89c:11105.
[12] K. M. Briggs, Numbers approximating badly, unpublished note (1998).
[13] K. M. Briggs, On the Furtwängler algorithm for simultaneous rational approximation, BTexact Technologies preprint (2000).
[14] G. Szekeres, Search for the three-dimensional approximation constant, *Diophantine Analysis*, Proc. 1985 Number Theory Sect. Austral. Math. Soc. conf., ed. J. H. Loxton and A. J. van der Poorten, Cambridge Univ. Press, 1986, pp. 139-146; MR 88b:11041.

[15] W. G. Nowak, A remark concerning the s-dimensional simultaneous Diophantine approximation constants, *Österreichisch-Ungarisch-Slowakisches 1992 Kolloquium über Zahlentheorie*, ed. F. Halter-Koch and R. Tichy, Grazer Math. Ber. 318, 1993, pp. 105-110; MR 95f:11047.
[16] H. Davenport, On a theorem of Furtwängler, *J. London Math. Soc.* 30 (1955) 186-195; MR 16, 803a.
[17] P. M. Gruber and C. G. Lekkerkerker, *Geometry of Numbers*, North-Holland 1987, pp. 427-429, 474-498, 546-548; MR 88j:11034.
[18] K. Ollerenshaw, Lattice points in a circular quadrilateral bounded by the arcs of four circles, *Quart. J. Math.* 17 (1946) 93-98; MR 7,506h.
[19] K. Ollerenshaw, On the region defined by $|xy|\leq 1$, $x^2+y^2\leq t$, *Proc. Cambridge Philos.* Soc. 49 (1953) 63-71; MR 14,624d.
[20] A. M. Cohen, Numerical determination of lattice constants, *J. London Math. Soc.* 37 (1962) 185-188; MR 25 #3422.
[21] W. G. Spohn, On the lattice constant for $|x^3+y^3+z^3|\leq 1$, *Math. Comp.* 23 (1969) 141-149; MR 39 #2706.
[22] W. G. Nowak, The critical determinant of the double paraboloid and Diophantine approximation in R^3 and R^4, *Math. Pannonica* 10 (1999) 111-122; MR 2000a:11102.
[23] R. J. Hans, Covering constants of some non-convex domains, *Indian J. Pure Appl. Math.* 1 (1970) 127-141; MR 42 #203.
[24] R. J. Hans-Gill, Covering constant of a star domain, *J. Number Theory* 2 (1970) 298-309; MR 42 #204.
[25] G. Szekeres, On a problem of the lattice plane, *J. London Math. Soc.* 12 (1937) 88-93.
[26] G. Szekeres, Note on lattice points within a parallelepiped, *J. London Math. Soc.* 12 (1937) 36-39.
[27] K. Chao, Note on the lattice points in a parallelepiped, *J. London Math. Soc.* 12 (1937) 40-47.
[28] P. M. Gruber and G. Ramharter, Beiträge zum Umkehrproblem für den Minkowskischen Linearformensatz, *Acta Math. Acad. Sci. Hungar.* 39 (1982) 135-141; MR 83m:10046.
[29] G. Ramharter, Über das Mordellsche Umkehrproblem für den Minkowskischen Linearformensatz, *Acta Arith.* 36 (1980) 27-41; MR 81k:10050.
[30] E. Hlawka, Über Gitterpunkte in Parallelepipeden, *J. Reine Angew. Math.* 187 (1950) 246-252; MR 12,161d.
[31] L. J. Mordell, Note on an arithmetical problem on linear forms, *J. London Math. Soc.* 12 (1937) 34-36.
[32] P. Erdös, P. M. Gruber, and J. Hammer, *Lattice Points*, Longman 1989, pp. 17-18, 97-102; MR 90g:11081.
[33] P. M. Gruber and C. G. Lekkerkerker, *Geometry of Numbers*, North-Holland, 1987, pp. 254-258, 263-266; MR 88j:11034.
[34] R. P. Bambah, V. C. Dumir, and R. J. Hans-Gill, On an analogue of a problem of Mordell, *Studia Sci. Math. Hungar.* 21 (1986) 135-142; MR 88i:11039.
[35] K. Mahler, The theorem of Minkowski-Hlawka, *Duke Math. J.* 13 (1946) 611-621; MR 8,444e.
[36] L. Fejes Tóth, On the densest packing of convex domains, *Proc. Konink. Nederl. Akad. Wetensch. Sci. Sect.* 51 (1948) 544-547; Indag. Math. 10 (1948) 188-192; MR 10,60a.
[37] V. Ennola, On the lattice constant of a symmetric convex domain, *J. London Math. Soc.* 36 (1961) 135-138; MR 28 #1177.
[38] P. Tammela, An estimate of the critical determinant of a two-dimensional convex symmetric domain (in Russian), *Izv. Vyssh. Uchebn. Zaved. Mat.* (1970) n. 12, 103-107;MR44 #2707.
[39] K. Reinhardt, Über die dichteste gitterförmige Lagerung kongruenter Bereiche in der Ebene und eine besondere Art konvexer Kurven, *Abh. Math. Sem. Univ. Hamburg* 10 (1934) 216-230.
[40] K. Mahler, On the minimum determinant and the circumscribed hexagons of a convex domain, *Proc. Konink. Nederl. Akad. Wetensch. Sci. Sect.* 50 (1947) 692-703; Indag. Math. 9 (1947) 326-337; MR 9,10h.

[41] H. Davenport and C. A. Rogers, Hlawka's theorem in the geometry of numbers, *Duke Math.* J. 14 (1947) 367-375; MR 9,11a.
[42] C. G. Lekkerkerker, On the Minkowski-Hlawka theorem, *Proc. Konink. Nederl. Akad. Wetensch. Ser. A.* 59 (1956) 426-434; Indag. Math. 18 (1956) 426-434; MR 18,287a.
[43] K. Ollerenshaw, An irreducible non-convex region, *Proc. Cambridge Philos. Soc.* 49 (1953) 194-200; MR 14,850d.

2.24 自己数密度定数

任意の非負整数 n は一意的に 2 進表示で

$$n=\sum_{k=0}^{\infty} n_k 2^k, \quad n_k=0, 1$$

と表される.この式をすこし動かす.たとえば指数 2^k を 2^k+1 に置きかえると,何が起こるか? 顕著な違いとなる.整数 1, 4, 6 はこの形の表現

$$n_0 \cdot (2^0+1) + n_1 \cdot (2^1+1) + n_2 \cdot (2^2+1) = 2n_0 + 3n_1 + 5n_2, \quad n_k=0,1$$

を「まったく」もたない.しかし,5 はこのような表現として,5 と 2+3 の 2 つをもつ.

存在の問題だけに焦点をあてる.S を,表現

$$n=\sum_{k=0}^{\infty} n_k (2^k+1), \quad n_k=0,1$$

が存在するすべての n の集合 (0 を含めて) と定義する.非負整数に関する S の補集合を T と定義する [1].したがって,$T=\{1,4,6,13,15,18,21,23,30,32,37,39,\cdots\}$ である.これらは **2 進自己数** (binary self-numbers) (Kaprekar (カプレカー) [2, 3]) または **2 進コロンビア数** (binary Columbian numbers) (Recamán (レカマン) [4]) として知られている.

T は無限集合であることが証明される.$\tau(N)$ を,N を超えない 2 進自己数の個数とする.Zannier (ザンニエ) [5] は,極限

$$0 < \lambda = \lim_{N \to \infty} \frac{\tau(N)}{N} < 1$$

が存在し,さらに $\tau(N) = \lambda N + O(\ln(N))^2$ であることを証明した.**自己数密度定数** λ は,式

$$\lambda = \frac{1}{8} \left(\sum_{n \in S} \frac{1}{2^n} \right)^2 = 0.2526602590\cdots$$

で計算され,Troi & Zannier (トロイ&ザンニア) [6,7] によって,λ は超越数であることが最近証明された.

任意の基底 $b>1$ にこの議論を拡張できる.S_b を,表現

$$n=\sum_{k=0}^{\infty}n_k(b^k+1), \quad n_k=0,1,\cdots,b-2,b-1$$

が存在するすべての n の集合と定義する．同様に T_b と $\tau_b(N)$ を定義する．前にあるように，$\tau_b(N)=\lambda_b N+O(\ln(N)^2)$ が成り立つ[5]．近似値は $\lambda_4=0.209\cdots$ であり，$\lambda_{10}=0.097\cdots$ である．しかし，任意の $b>2$ に対して，λ_b に速く収束する無限級数（λ_2 と類似な式）は未だに確立していない．同様に，λ_b, $b\geq 3$ は無理数であることさえ，未だに証明されていない．

さらに，一意性の問題がある．2進数のときのみに焦点をあてる．表現

$$n=\sum_{k=0}^{\infty}n_k(2^k+1), \quad n_k=0,1$$

が存在し，一意的であるすべての n の集合を U と定義する．V を，S に関する U の補集合と定義する．集合 V は明らかに無限集合である．なぜならば，すべての $k>2$ に対して，

$$1\cdot(2^k+1)+1\cdot(2^2+1)=1\cdot(2^k+1)+1\cdot(2^0+1)+1\cdot(2^1+1)$$

だからである．集合 U は明らかに無限集合である．なぜならば，T の要素 t に対して

$$\sum_{k=0}^{t+1}(2^k+1)=(2^{t+2}+1)+t$$

は，他の許される表現をまったくもたないからである．U と V の密度については，いったい何がいえるのだろうか？　等差数列での自己数の密度についてはさらに[8]を，一桁ごとの加法 (digit addition) 級数に関連する議論については[9]を，参照せよ．

[1] N. J. A. Sloane, On-Line Encyclopedia of Integer Sequences, A003052, A010061, A010064, A010067, and A010070.
[2] D. R. Kaprekar, *Puzzles of the Self-Numbers*, unpublished manuscript, 1959; MR 20 # 6381.
[3] M. Gardner, Mathematical games, *Sci. Amer.*, v. 232 (1975) n. 3, 113-114.
[4] B. Recamán and D. W. Bange, Columbian numbers, *Amer. Math. Monthly* 81 (1974) 407.
[5] U. Zannier, On the distribution of self-numbers, *Proc. Amer. Math. Soc.* 85 (1982) 10-14; MR 83i:10007.
[6] G. Troi and U. Zannier, Note on the density constant in the distribution of self-numbers, *Boll. Unione Mat. Ital. A* (7) 9 (1995) 143-148; MR 95k:11093.
[7] G. Troi and U. Zannier, Note on the density constant in the distribution of self-numbers. II, *Boll. Unione Mat. Ital. Sez. B Artic. Ric. Mat.* (8) 2 (1999) 397-399; MR 2000f:11093.
[8] M. B. S. Laporta and E. Laserra, Distribution of self-numbers in arithmetical progressions (in Italian), *Ricerche Mat.* 42 (1993) 307-313; MR 95e:11105.
[9] K. B. Stolarsky, The sum of a digitaddition series, *Proc. Amer. Math. Soc.* 59 (1976) 1-5; MR 53 # 13099.

2.25 カメロンの無和集合定数

　正の整数の集合 S が**無和**（sum-free）であるとは，方程式 $x+y=z$ の解が $x, y, z \in S$ で存在しないときをいう．いいかえると，S が無和であるための必要十分条件は $(S+S) \cap S = \emptyset$ である．ただし，$A+B$ は，$a+b$，$a \in A$，$b \in B$ のすべての和からなる集合である．たとえば，正の奇数すべてからなる集合は無和である．

　無和集合のすべてからなる集合族を考える．Cameron（カメロン）[1-3]はこの集合族に自然な確率測度を定義した．これは，非形式的にランダムな無和集合 S を構成する方法とみなされている．方法は次のとおりである．

- はじめに，集合 $S = \emptyset$ とし，各々の正の整数 n を順に 1 つずつ見ていく．
- もしある $a, b \in S$ で $n = a+b$ が成り立てば，n をとばし，$n+1$ に進む．
- もし $n = x+y$ が解 $x, y \in S$ をもたないとき，正確にできたコインを投げる．もし表が出たら $S = S \cup \{n\}$ とし，$n+1$ に進む．裏が出たら単に $n+1$ に進む．

たとえば，明らかに
$$P(S \text{ は偶数すべてからなる}) = 0$$
である．対照的に Cameron[1]は，定数
$$c = P(S \text{ は奇数すべてからなる})$$
は「正」であるという注目すべき事実を証明した．実際，$0.21759 \leq c \leq 0.21862$ である．これと同値なものは[2]，$N = \{0,1,2,\cdots,n-1\}$ で
$$F(n) = 2^{-2n} \sum_{X \subseteq N} 2^{|(X+X) \cap N|}$$
とおくとき，$F(n)$ は減少関数であり，$\lim_{n \to \infty} F(n) = c$ が成り立つ．和は，N のすべての部分集合 X にわたってとる．$|E|$ は，集合 $|E|$ の個数である．別の証明は Calkin（カルキン）[4]が与えている．

　Cameron[2]はより一般的な結果を証明した．それは S が等差数列のある無和な集合の和に含まれる確率を（下から）評価するものである．彼の一般的な定理を述べるよりもむしろ，単純な一応用例をあげる．
$$P(S \subseteq \{2,7,12,17,22,27,\cdots\} \cup \{3,8,13,18,23,28,\cdots\}) \geq \frac{c^2 d}{2} > 0.0066$$
が成り立つ．ただし，$0.28295 \leq d = \lim_{n \to \infty} G(n) \leq 0.29484$ であり，減少関数 $G(n)$ は
$$G(n) = 2^{-3n} \sum_{X, Y \subseteq N} 2^{|(X+Y) \cap N|}$$
で定義される．しかしながら，これは（コンピュータシミュレーションに基づく）Cameron による近似値の評価 0.022 に近くない．

　Calkin & Cameron[5]は，ランダムな無和集合の理解をさらに先に進めた．再び，

一般的な形では定理を示さず，ただ単に一例だけにとどめる．

P(S は2を含むが，S はその他の偶数を含まない) >0

である．コンピュータシミュレーションによるこの確率の近似値は 0.00016 である．

確率から離れて，かわりに $\{1,2,\cdots,n\}$ の無和部分集合の個数 s_n を考える．はじめの s_n のいくつかの項は[6]，1,2,3,6,9,16,24，…である．Cameron & Erdös[7,8]は，$s_n 2^{-n/2}$ は有界であり，さらに，次の2つの極限が存在し，近似的に

$$\lim_{k\to\infty} s_{2k+1} 2^{-(k+1/2)} = c_o = 6.8\cdots, \quad \lim_{k\to\infty} s_{2k} 2^{-k} = c_e = 6.0\cdots$$

であると予想した．ただし，

$$c_o = \sqrt{2} + \lim_{k\to\infty} H(2k+1), \quad c_e = 1 + \lim_{k\to\infty} H(2k)$$

$$H(n) = 2^{-n/2} \sum_{X \subseteq N'} 2^{-|(X+X) \cap N'|}, \quad N' = \{0,1,\cdots,n\}$$

である．Calkin[9]，Alon（アロン）[10]と Erdös & Granville は，互いに独立に

$$\lim_{n\to\infty} s_n 2^{-(1/2+\varepsilon)n} = 0$$

が，任意の $\varepsilon > 0$ に対して成り立つことを示した．有界性の追加的な証拠は[11]にあり，一般化は[12-15]にある．

もう1つの問題を示す気持ちを抑えることはできない．正の整数の無和集合 S が**完全** (complete) であるとは，十分大きなすべての整数 n に対して，$n \in S$ であるか，または $s+t=n$ を満たす $s, t \in S$ が存在するときをいう．いいかえれば，S が完全であるための必要十分条件は有限集合から欲張って (greedily) 作られていることである．無和集合 S が**周期的** (periodic) であるとは，整数 m が存在して，十分大きなすべての整数 n に対して，$n \in S$ であるための必要十分条件が $n+m \in S$ であることが成り立つときをいう．言いかえると，S が周期的であるための必要十分条件は，昇順に並べた S の要素が（いつかは）周期的な階差数列になることである．

任意の完全無和集合は必然的に周期的であるのか[16]？ Cameron[3]は，最初の潜在的に擬周期的 (aperiodic) な例として，3,4,13,18,24,…からはじまる完全無和集合を与えた．Calkin & Finch（カルキン&フィンチ）[17]は，潜在的に擬周期的なほかの例として，1,3,8,20,26,… と 2,15,16,23,27,… を含む例を与えた．Calkin & Erdös [18]は，「不完全 (incomplete)」な概周期的無和 (aperiodic sum-free) 集合の存在を証明した．実際，彼らは自然に構成されたこのような非可算個の集合を示した．しかし，誰も未だに，完全概周期的無和集合の存在の1つも証明していない．

[1] P. J. Cameron, Cyclic automorphisms of a countable graph and random sum-free sets, *Graphs Combin.* 1 (1985) 129-135; MR 90b:05062.

[2] P. J. Cameron, On the structure of a random sum-free set, *Probab. Theory Relat. Fields* 76 (1987) 523-531; MR 89c:11018.

[3] P. J. Cameron, Portrait of a typical sum-free set, *Surveys in Combinatorics 1987*, ed. C. Whitehead, Cambridge Univ. Press, 1987, pp. 13-42; MR 88k:05138.

[4] N. J. Calkin, On the structure of a random sum-free set of positive integers, *Discrete Math.* 190 (1998) 247-257; MR 99f:11015.

[5] N. J. Calkin and P. J. Cameron, Almost odd random sum-free sets, *Combin. Probab. Comput.* 7 (1998) 27-32; MR 99c:11012.
[6] N. J. A. Sloane, On-Line Encyclopedia of Integer Sequences, A007865.
[7] P. J. Cameron and P. Erdös, On the number of sets of integers with various properties, *Number Theory*, Proc. 1990 Canad. Number Theory Assoc. Banff conf., ed. R. A. Mollin, Gruyter, pp. 61-79; MR 92g:11010.
[8] P. J. Cameron, The Cameron-Erdös constants, unpublished note (2002).
[9] N. J. Calkin, On the number of sum-free sets, *Bull. London Math. Soc.* 22 (1990) 141-144; MR 91b:11015.
[10] N. Alon, Independent sets in regular graphs and sum-free subsets of finite groups, *Israel J. Math.* 73 (1991) 247-256; MR 92k:11024.
[11] G. A. Freiman, On the structure and the number of sum-free sets, *Astérisque* 209 (1992) 195-201; MR 94a:11016.
[12] N. J. Calkin and A. C. Taylor, Counting sets of integers, no k of which sum to another, *J. Number Theory* 57 (1996) 323-327; MR 97d:11015.
[13] Y. Bilu, Sum-free sets and related sets, *Combinatorica* 18 (1998) 449-459; MR 2000j:11035.
[14] N. J. Calkin and J. M. Thomson, Counting generalized sum-free sets, *J. Number Theory* 68 (1998) 151-159; MR 98m:11010.
[15] T. Schoen, A note on the number of (k, l)-sum-free sets, *Elec. J. Combin.* 7 (2000) R30; MR 2001c:11030.
[16] R. K. Guy, *Unsolved Problems in Number Theory*, 2nd ed., Springer-Verlag, 1994, sect. E32; MR 96e:11002.
[17] N. J. Calkin and S. R. Finch, Some conditions on periodicity for a sum-free set, *Experim. Math.* 5 (1996) 131-137; MR 98c:11010.
[18] N. J. Calkin and P. Erdös, On a class of aperiodic sum-free sets, *Math. Proc. Cambridge Philos. Soc.* 120 (1996) 1-5; MR 97b:11030.

2.26 無三重集合定数

正の整数の集合Sが,**無倍**(double free)とは,任意の整数xに対して,集合$\{x, 2x\} \not\subseteq S$であるときをいう.いいかえれば,$x \in S$ならば$2x \notin S$であるとき,$S$は無倍であるという.関数
$$r(n) = \max\{|S| : S \subseteq \{1, 2, \cdots, n\} \text{ が無倍}\}$$
すなわち,nを超える要素がない無倍集合の最大濃度を考える.
$$\lim_{n \to \infty} \frac{r(n)}{n} = \frac{2}{3}$$
すなわち,無倍集合の漸近極大密度が$2/3$である,ことを証明することは難しくない.Wang(ワン)[1]は,$r(n)$の漸化式および閉じた表現のどちらをも得た.さらに,$r(n) = 2n/3 + O(\ln(n))$ ($n \to \infty$) を示した.

より難しい問題を議論する.正の整数の集合Sが,

・**弱無三重**(weakly triple-free)(または**無三重**(triple-free))であるとは,任意の整数xに対して,集合$\{x, 2x, 3x\} \not\subseteq S$であるときをいい,

・**強無三重** (strongly triple-free) であるとは，$x \in S$ ならば $2x \notin S$ かつ $3x \notin S$ であるときをいう．

無倍の場合とは違って，弱と強の意味で2つの無三重は一致しない．関数
$$p(n) = \max\{|S| : S \subseteq \{1,2,\cdots,n\} \text{ は弱無三重}\}$$
$$q(n) = \max\{|S| : S \subseteq \{1,2,\cdots,n\} \text{ は強無三重}\}$$
を考える．定数
$$\lambda = \lim_{n \to \infty} \frac{p(n)}{n}, \quad \mu = \lim_{n \to \infty} \frac{q(n)}{n}$$
を計算したい．無限集合
$$A = \{2^i 3^j : i, j \geq 0\} = \{a_1 < a_2 < a_3 < \cdots\}$$
$$= \{1,2,3,4,6,8,9,12,16,18,24,27,\cdots\}$$
を定義し，A_n を A の最初から n 項までの集合と定義する．このとき，λ と μ は
$$\lambda = \frac{1}{3}\sum_{n=1}^{\infty}(n-f_n)\left(\frac{1}{a_n} - \frac{1}{a_{n+1}}\right), \quad \mu = \frac{1}{3}\sum_{n=1}^{\infty} g_n\left(\frac{1}{a_n} - \frac{1}{a_{n+1}}\right)$$
と表すことができる．ここで，整数列
$$\{f_n\} = \{0,0,1,1,1,1,2,2,2,3,3,4,4,4,4,5,5,5,6,6,7,7,7,8,8,\cdots\}$$
$$\{g_n\} = \{1,1,2,2,3,3,4,4,5,5,6,6,7,7,8,8,9,9,10,11,\cdots\}$$
が，ただちに定義される．

λ に多くの注目があるほどには，定数 μ に関心がない．Eppstein（エプスタイン）[2] は，g_n は格子グラフ (grid graph) A_n における隣り合わない頂点の最大集合の大きさである，ことを証明した（この大きさは**独立数** (independence number) とよばれる）．各 $k=0,1$ に対して，$i+j \equiv k \bmod 2$ を満たす $2^i 3^j$ のすべての要素からなる集合を，$A_{n,k} \subseteq A_n$ と定義する．このとき，$\{A_{n,0}, A_{n,1}\}$ は A_n の分割であり，これらのうちの少なくとも1つは，Cassaigne（カサイニュ）[3] によって発見されたように，極大独立集合 (maximal independent set) である（図 2.5 を見よ）．ここから，Zimmermann（ツィンマーマン）[3] は**無三重集合定数**が $\mu = 0.6134752692\cdots$ であることを計算

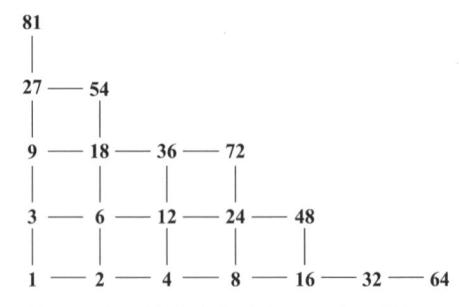

```
81
|
27 ── 54
|     |
9 ── 18 ── 36 ── 72
|     |     |     |
3 ── 6 ── 12 ── 24 ── 48
|     |     |     |     |
1 ── 2 ── 4 ── 8 ── 16 ── 32 ── 64
```

図 2.5 A_{19} に対応した格子グラフ．A_{19} に対して，$g_{19}=10$, $A_{19,0}=\{1, 4, 6, 9, 16, 24, 36, 54, 64, 81\}$, $f_{19}=6=h_{19}$, $B_{19,0}=\{1, 6, 8, 27, 36, 48, 64\}$, $B_{19,0}=\{64\}$ である．

した．

　比較として，定数λは25年以上にわたり人々の興味を引いてきた[4]．Graham, Spencer & Witsenhausen（グラハム，スペンサー&ウィツェンハウゼン）[5]は，$\{1,2,\cdots,n\}$ に含まれる集合上の，一次式 $\sum_{v=1}^{w} c_{uv}x_v$ の値をとらないという，一般的な条件を問題にした．多くの問題の中で，彼らはλが無理数であるかどうかを問題にした．[5]での f_n の値の表から始めて，Cassaigne[6]は $\lambda \geq 4/5$ であることを証明した．Chung, Erdös & Graham（チャン，エルデーシュ&グラハム）[7]は，$\{2^i3^j, 2^{i+1}3^j, 2^i3^{j+1}\} \subseteq A_n$ という形の，すべてのL字形頂点配置（L-shaped vertex configuration）と交差する，A_n の頂点の最小集合の大きさが f_n である，ことを証明した（この大きさを **L-当たり数**（L-hitting number）という）．各 $k=0,1,2$ に対して，$i-j \equiv k \bmod 3$ を満たす 2^i3^j のすべての要素からなる集合 $B_{n,k} \subseteq A_n$ を定義する．このとき，$\{B_{n,0}, B_{n,1}, B_{n,2}\}$ は A_n の分割である．さらに，$1 \leq i \equiv k \bmod 3$ で $2^{i-1}3 \notin A_n$ を満たす 2^i のすべての要素からなる集合を，$\tilde{B}_{n,k} \subseteq B_{n,k}$ と定義する．

$$f_n \leq h_n = \min_{0 \leq k \leq 2} |B_{n,k}| - |\tilde{B}_{n,k}| \leq \left\lfloor \frac{n}{3} \right\rfloor$$

が知られており，その結果 $0.800319 < \lambda < 0.800962$ である．すべての n に対して，$f_n = h_n$ は予想である．もしこれが真であるならば，$\lambda = 0.8003194838\cdots = 1 - 0.1996805161\cdots$ である．

　固定された $s > 1$ に対して，$\{x, 2x, 3x, \cdots sx\} \not\subseteq S$ がすべての整数 x で成り立つような正の整数の集合 S を考える．対応する漸近極大密度を λ_s とする．$s \to \infty$ のとき，λ_s の漸近値について何がいえるのだろうか？ Spencer & Erdös（スペンサー & エルデーシュ）[8]は，定数 c と C が存在し，

$$1 - \frac{C}{s \ln(s)} < \lambda_s < 1 - \frac{c}{s \ln(s)}$$

が適当な大きなすべての s について成り立つことを示した．しかしながら，特定の数値は示されていない．さらに，$\{x, 2x, 3x, 6x\} \not\subseteq T$ がすべての整数 x で成り立つような正の整数の集合 T を考える．対応する漸近極大密度は正確に $11/12$ である[7]．$s=3$ の場合はかなり難しいので，このような値が求められることは驚きである．

　数2と3との相互作用の多くの実例は，3/2のベキ乗の法 $1^\dagger$ に関するもので，【2.30.1】にある．

[1] E. T. H. Wang, On double-free sets of integers, *Ars Combin*. 28 (1989) 97-100; MR 91d:11011.
[2] D. Eppstein, Triple-free set asymptotics and independence numbers of grid graphs, unpublished note (1996).
[3] J. Cassaigne and P. Zimmermann, Numerical evaluation of the strongly triple-free constant, unpublished note (1996).
[4] P. Erdös and R. Graham, *Old and New Problems and Results in Combinatorial Number Theory*, Enseignement Math. Monogr. 28, 1980, p. 20; MR 82j:10001.
[5] R. Graham, J. Spencer, and H.Witsenhausen, On extremal density theorems for linear forms,

---

†訳注：小数部分を考えること．

Number Theory and Algebra, ed. H. Zassenhaus, Academic Press, 1977, pp. 103-109; MR 58 # 569.
[6] J. Cassaigne, Lower bound on triple-free constant λ, unpublished note (1996).
[7] F. Chung, P. Erdös, and R. Graham, On sparse sets hitting linear forms, *Number Theory for the Millennium*, v. 1, Proc. 2000 Urbana conf., ed. M. A. Bennett, B. C. Berndt, N. Boston, H. G. Diamond, A. J. Hildebrand, and W. Philipp, A. K. Peters, to appear.
[8] J. Spencer and P. Erdös, A problem in covering progressions, *Stud. Sci. Math. Hung.* 30 (1995) 149-154; MR 96f:11018.

2.27 エルデーシュ-レーベンソールド定数

正の整数からなる狭義単調数列 $a_1, a_2, a_3, \cdots$ が，**原始的** (primitive) とは[1-3]，任意の $i \neq j$ に対して $a_i \nmid a_j$ が成り立つことである．すなわち，数列のどの項も他の項を割り切らない，ことである．有限の原始的な数列の例は，n を正の整数とするとき，区間 $\lceil (n+1)/2 \rceil \leq m \leq n$ にあるすべての整数 m の集合である．無限の原始的な数列の例は，(固定された) ちょうど r 個の素因子の積から作られるすべての正の整数から構成される．有限と無限の場合を分けて議論する．さらに，関連する注釈については【5.5】を参照せよ．

2.27.1 有限の場合

正の整数 n に対して，最大可能な項の数として
$$M(n) = \sup_{\substack{\text{原始的} \\ A \subseteq \{1,2,\cdots,n\}}} \sum_i 1$$
を，逆数の和の最大可能な値として
$$L(n) = \sup_{\substack{\text{原始的} \\ A \subseteq \{1,2,\cdots,n\}}} \sum_i \frac{1}{a_i}$$
を定義する．明らかに，$M(n) = \lfloor (n+1)/2 \rfloor$ である．だから $\lim_{n\to\infty} M(n)/n = 1/2$ である．
$$\lim_{n\to\infty} \frac{\sqrt{\ln(\ln(n))}}{\ln(n)} L(n) = \frac{1}{\sqrt{2\pi}}$$
を証明することはもっと難しい[4,5]．アルキメデスの定数 (円周率)【1.4】の思いもかけない出現である．

2.27.2 無限の場合

無限の原始的数列は
$$0 = \liminf_{n\to\infty} \frac{1}{n} \sum_{a_i \leq n} 1 \leq \limsup_{n\to\infty} \frac{1}{n} \sum_{a_i \leq n} 1 < \frac{1}{2}$$

を満たす．Besicovitch（ベシコヴィッチ）[1,6]は，任意の $\varepsilon>0$ に対して，
$$\limsup_{n\to\infty}\frac{1}{n}\sum_{a_i\leq n}1>\frac{1}{2}-\varepsilon$$
を満たすような原始的数列が存在する，ことを証明した．とくに，原始的数列は漸近的密度をもつ必要はない！ おそらく極限値 1/2 はそんなに驚くことではない．なぜなら，$M(n)$ についての結果を前に与えているからである．

対比として，Erdös, Sárkozy & Szemerédi（エルデーシュ，サルコジ & セメレディ）[7]は，
$$\lim_{n\to\infty}\frac{\sqrt{\ln(\ln(n))}}{\ln(n)}\sum_{a_i\leq n}\frac{1}{a_i}=0$$
であることを証明した．これは，$L(n)$ についての前の結果とは劇的に異なる．有限と無限の場合は，この点において，お互いに独立にふるまう．

新しい小道を一歩一歩つくりながら，Erdös[1,8]は，級数
$$\sum_i \frac{1}{a_i\ln(a_i)}$$
が収束する（自明な原始的数列 {1} を除いて）ことを証明した．さらに，ある絶対定数により上から有界であることを証明した．Erdös は
$$\sum_i\frac{1}{a_i\ln(a_i)}\leq\sum_i\frac{1}{p_i\ln(p_i)}=1.6366163233\cdots$$
であることを予想した．ただし，最後の和は素数すべてにわたってとる．いくつかの部分的な結果が知られている．Zhang[9,10]は，各項が多くとも 4 つの素因子を含むすべての原始的な数列に対して，不等式が正しいことを証明した．Zhang[11]は同様に，異なるもっと技巧的な条件を仮定して，それを証明した．Erdös & Zhang[12]は，任意の原始的な数列に対して
$$\sum_i\frac{1}{a_i\ln(a_i)}\leq 1.84$$
を証明した．Clark（クラーク）[13]はこれを
$$\sum_i\frac{1}{a_i\ln(a_i)}\leq e^\gamma=1.7810724179\cdots$$
と強めた．γ はオイラー定数【1.5】である．

ちなみに，素数列に対するここで与えられた評価 $1.6366163233\cdots$ は，Cohen（コーヘン）[14]による．

2.27.3 一 般 化

k を正の整数とする．正の整数の狭義単調増加数列 $a_1,a_2,a_3,\cdots$ が **k 原始的**（k primitive）であるとは，数列のどの項も他の k 個の項を割り切らないときをいう．（この用語は新しい．）有限の場合のみを考える．前のように $M(n,k)$ と $L(n,k)$ を定義する．2 原始的数列の例は，区間 $\lceil(n+1)/3\rceil\leq m\leq n$ にあるすべての整数からなる集合である．よって，$\lim_{n\to\infty}M(n,2)/n\geq 2/3$ であるが，ここで改良は可能である．

Lebensold (レーベンソールド) [15]は,

$$0.6725 \leq \lim_{n \to \infty} \frac{M(n,2)}{n} \leq 0.6736$$

を証明し,より正確な範囲が,まったく同じ方法で追加的な計算によって得られることを述べた.Erdösは,極限値が無理数かどうかを問うた[10].$L(n,2)$ または $k>2$ の場合を,知るかぎり,誰も検証していない.

正の整数の狭義単調増加数列 $b_1, b_2, b_3, \cdots$ が**準原始的** (quasi-primitive) [16]であるとは,方程式 $\gcd(b_i, b_j) = b_r$ が $r < i < j$ で解をもたないときをいう.無限準原始的数列は,すべての素数ベキ

$q_1=2, \ q_2=3, \ q_3=2^2, \ q_4=5, \ q_5=7, \ q_6=2^3, \ q_7=3^2, \ q_8=11, \ \cdots$

からなる.Erdös & Zhang[16]は,任意の準原始的数列に対して,

$$\sum_i \frac{1}{b_i \ln(b_i)} \leq \sum_i \frac{1}{q_i \ln(q_i)} = 2.006\cdots$$

であると予想した.Clark[17]は,[16]での間違った主張を修正し,

$$\sum_i \frac{1}{b_i \ln(b_i)} < 4.2022$$

であることを証明した.素数ベキの級数に対するより正確な評価は,未解決問題である.

k-原始的数列と準原始的数列の話題は,無三重定数【2.26】やErdösの逆数和定数【2.20】の関連した話題のように,研究には広く開かれた分野である.

[1] H. Halberstam and K. F. Roth, *Sequences*, Springer-Verlag, 1983, pp. 238-254; MR 83m:10094.
[2] P. Erdös, A. Sárkozy, and E. Szemerédi, On divisibility properties of sequences of integers, *Number Theory*, Proc. 1968 Debrecen conf., ed. P. Turán, Colloq. Math. Soc. János Bolyai 2, North-Holland, 1970, pp. 35-49; MR 43 #4790.
[3] R. Ahlswede and L. H. Khachatrian, Classical results on primitive and recent results on cross-primitive sequences, *The Mathematics of Paul Erdös*, v. 1, ed. R. L. Graham and J. Nesetril, Springer-Verlag, 1997, pp. 104-116; MR 97j:11012.
[4] P. Erdös, On the integers having exactly k prime factors, *Annals of Math.* 49 (1948) 53-66; MR 9,333b.
[5] P. Erdös, A. Sárkozy, and E. Szemerédi, On an extremal problem concerning primitive sequences, *J. London Math. Soc.* 42 (1967) 484-488; MR 36 #1412.
[6] A. S. Besicovitch, On the density of certain sequences of integers, *Math. Annalen* 110 (1934) 336-341.
[7] P. Erdös, A. Sárkozy, and E. Szemerédi, On a theorem of Behrend, *J. Austral. Math. Soc.* 7 (1967) 9-16; MR 35 #148.
[8] P. Erdös, Note on sequences of integers no one of which is divisible by any other, *J. London Math. Soc.* 10 (1935) 126-128.
[9] Z. Zhang, On a conjecture of Erdös on the sum $\sum_{p \leq n} 1/(p \log p)$, *J. Number Theory* 39 (1991) 14-17; MR 92f:11131.
[10] R. K. Guy, *Unsolved Problems in Number Theory*, 2nd ed., Springer-Verlag, 1994, sect. B24, E4; MR 96e:11002.
[11] Z. Zhang, On a problem of Erdös concerning primitive sequences, *Math. Comp.* 60 (1993) 827-834; MR 93k:11120.
[12] P. Erdös and Z. Zhang, Upper bound of $\sum 1/(a_i \log a_i)$ for primitive sequences, *Proc. Amer. Math. Soc.* 117 (1993) 891-895; MR 93e:11018.

[13] D. A. Clark, An upper bound of $\Sigma 1/(a_i \log a_i)$ for primitive sequences, *Proc. Amer. Math. Soc.* 123 (1995) 363-365; MR 95c:11026.
[14] H. Cohen, High precision computation of Hardy-Littlewood constants, unpublished note (1999).
[15] K. Lebensold, A divisibility problem, *Studies in Appl. Math.* 56 (1976-77) 291-294; MR 58 # 21639.
[16] P. Erdös and Z. Zhang, Upper bound of $\Sigma 1/(a_i \log a_i)$ for quasi-primitive sequences, *Comput. Math. Appl. (Oxford)* v. 26 (1993) n. 3, 1-5; MR 94f:11013.
[17] D. A. Clark, An upper bound of $\Sigma 1/(a_i \log a_i)$ for quasi-primitive sequences, *Comput. Math. Appl. (Oxford)*, v. 35 (1998) n. 4, 105-109; MR 99a:11021.

2.28 エルデーシュの和違い集合定数

正の整数の集合 $a_1 < a_2 < a_3 < \cdots < a_n$ が**和違い**（sum-distinct）であるとは，2^n 個の和

$$\sum_{k=1}^{n} \varepsilon_k a_k \quad (\text{各 } \varepsilon_k = 0 \text{ または } 1, \ 1 \leq k \leq n)$$

がすべて異なるときをいう．すなわち，和違い性（sum-distinctness）をもつための必要十分条件は，任意の2つの部分集合の和が決して等しくならない[1-4]ことである．2の非負整数のベキからなる集合は，明らかに和違いであり，比較のための基準として役に立つ．1931年，Erdös（エルデーシュ）は比

$$\alpha_n = \inf_A \frac{a_n}{2^n}$$

を調べた．ただし，下限は濃度 n の和違い集合 A のすべてにわたる．$\alpha = \inf_n \alpha_n$ は正である，と予想されているが，これが真であるかどうか，誰も知らない．しかし，1955年，Erdös と Moser[2,5,6]は，すべての $n \geq 2$ に対して，

$$\alpha_n \geq \max\left(\frac{1}{n}, \frac{1}{4\sqrt{n}}\right)$$

であることを証明した．Elkies（エルキース）[7]は，充分大きな n に対して，

$$\alpha_n \geq \frac{1}{\sqrt{\pi n}}$$

を証明した．Gleason & Elkies（グリーソン&エルキース）[8]は，分散を小さくする技法により係数 π を後で取り除いた．さらに[9]を参照せよ．$\alpha > 1/8 = 0.125$ はおそらく真である．エルデーシュ予想を解決する重要な進歩には，ほとんど確実にまったく新しいアイデア，または，まだ発見されていない洞察力が必要であろう．

いくつかのおもしろい構成によって，α に関する上界が得られる．1986年，Atkinson, Negro & Santoro（アトキンソン，ネグロ & サントロ）[10,11]は，数列

$$u_0 = 0, \quad u_1 = 1, \quad u_{k+1} = 2u_k - u_{k-m}, \quad m = \lfloor (k/2) + 1 \rfloor$$

を定義した．これは，各 n に対して，和違い集合 $a_k = u_n - u_{n-k}, \ 1 \leq k \leq n$ になる．明

らかに，$a_n=u_n$ である．Lunnon（ルノン）[11]は，

$$\lim_{n\to\infty}\frac{u_n}{2^n}=0.3166841737\cdots=\frac{1}{2}\cdot(0.6333683473\cdots)$$

を計算した．より小さい比は，Conway & Guy（コンウェイ&ガイ）[2,11-13] による数列により得られた．

$$v_0=0, \quad v_1=1, \quad v_{k+1}=2v_k-v_{k-m}, \quad m=\lfloor(1/2)+\sqrt{2k}\rfloor$$

つい最近，Bohman（ボーマン）[14]は，この数列が各 n に対して，和違い集合 $a_k=v_n-v_{n-k}$, $1\leq k\leq n$ になることを証明した．(1996 年以前，$n<80$ に対してだけ，この主張が正しいことが知られていた．) Lunnon[11]は

$$\lim_{n\to\infty}\frac{u_n}{2^n}=0.2351252848\cdots=\frac{1}{2}\cdot(0.4702505696\cdots)$$

を計算した．

Atkinson-Negro-Santoro と Conway-Guy（コンウェイ-ガイ）の極限比は，興味ある定数であるけれども，α に関する知られた最良の上界を与えない．よく使う技法は次のとおりである．もし $a_1<a_2<a_3<\cdots<a_n$ が n 個の要素をもつ和違い集合であるならば，明らかに $1<2a_1<2a_2<2a_3<\cdots<2a_n$ は $n+1$ 個の要素をもつ和違い集合である．もちろん，このように拡大することを無限に続けることができる．したがって，n 個の要素と小さい比 ρ をもつ，和違い集合を見つけることができたとき，ただちに上界 $\alpha\leq\rho$ を得る．たとえば，Lunnon[11]は，コンピュータ探索によって，$n=67$ で $\rho=0.22096$ である和違い集合を見つけた．これは，Conway-Guy の上界を改良した．Conway, Guy, Lunnon の仕事を一般化して，Bohman[15]は知られている中で最良の上界 $\alpha\leq 0.22002$ を証明した．さらに，Maltby（マルテビ）[16]は，与えられた和違い集合に対して，より小さい比をもつ，より大きい和違い集合の構成方法を示した．よって，エルデーシュ定数 α は，任意の和違い集合によって実現されていない．すなわち，下限はまったく得られていない！

Bae（バエ）[17]は，与えられた r, q に対して，和違い集合の和が $r \bmod q$ をとらないような集合を研究した．さらに，すべての和違い集合に対して，不等式

$$\sum_{k=1}^{n}\frac{1}{a_k}<2=\sum_{k=1}^{\infty}\frac{1}{2^{k-1}}$$

が真である，ものを考えた．上界 2 は最良のもので，初等的な証明が可能である [9,18,19]ことは，奇妙である．(実際，より多くのことが知られている！) どこかで別にこのような逆数の和【2.20】を議論した．それらを評価するのは，普通ひじょうに難しい．

[1] R. K. Guy, *Unsolved Problems in Number Theory*, 2nd ed., Springer-Verlag, 1994, sect. C8; MR 96e:11002.
[2] R. K. Guy, Sets of integers whose subsets have distinct sums, *Theory and Practice of Combinatorics*, ed. A. Rosa, G. Sabidussi, and J. Turgeon, Annals of Discrete Math. 12, North-Holland, 1982, pp. 141-154; MR 86m:11009.

[3] M. Gardner, Mathematical games: On the fine art of putting players, pills and points into their proper pigeonholes, *Sci. Amer.*, v. 243 (1980) n. 2, 14-18.
[4] P. Smith, Solutions to problem on sum-distinct sets, *Amer. Math. Monthly* 83 (1976) 484 and 88 (1981) 538-539.
[5] P. Erdös, Problems and results in additive number theory, *Colloq. Théorie des Nombres, Bruxelles* 1955, Liege and Paris, 1956, pp. 127-137; MR 18,18a.
[6] N. Alon and J. Spencer, *The Probabilistic Method*, Wiley, 1992, pp. 47-48; MR 93h: 60002.
[7] N. D. Elkies, An improved lower bound on the greatest element of a sum-distinct set of fixed order, *J. Combin. Theory* A 41 (1986) 89-94; MR 87b:05012.
[8] A. M. Gleason and N. D. Elkies, Further improvements on a lower bound, unpublished note (1985).
[9] J. Bae, On subset-sum-distinct sequences, *Analytic Number Theory*, Proc. 1995 Allerton Park conf., v. 1, ed. B. C. Berndt, H. G. Diamond, and A. J. Hildebrand, Birkhäuser, 1996, pp. 31-37; MR 97d:11016.
[10] M.D. Atkinson, A. Negro, and N. Santoro, Sums of lexicographically ordered sets, *Discrete Math.* 80 (1990) 115-122; MR 91m:90136.
[11] W. F. Lunnon, Integer sets with distinct subset-sums, *Math. Comp.* 50 (1988) 297-320; MR 89a: 11019.
[12] J. H. Conway and R. K. Guy, Sets of natural numbers with distinct sums, *Notices Amer. Math. Soc.* 15 (1968) 345.
[13] J. H. Conway and R. K. Guy, Solution of a problem of P. Erdös, *Colloq. Math.* 20 (1969) 307.
[14] T. Bohman, A sum packing problem of Erdös and the Conway-Guy sequence, *Proc. Amer. Math. Soc.* 124 (1996) 3627-3636; MR 97b:11027.
[15] T. Bohman, A construction for sets of integers with distinct subset sums, *Elec. J. Combin.* 5 (1997) R3; MR 98k:11014.
[16] R. Maltby, Bigger and better subset-sum-distinct sets, *Mathematika* 44 (1997) 56-60; MR 98i: 11012.
[17] J. Bae, An extremal problem for subset-sum-distinct sequences with congruence conditions, *Discrete Math.* 189 (1998) 1-20; MR 99h:11018.
[18] R. Honsberger, *Mathematical Gems III*, Math. Assoc. Amer., 1985, pp. 215-223.
[19] P. E. Frenkel, Integer sets with distinct subset sums, *Proc. Amer. Math. Soc.* 126 (1998) 3199-3200; MR 99a:11012.

2.29 高速行列乗法の定数

任意の2つの $n \times n$ 行列の掛算には，掛算の標準的な定義により掛け算をするとき，少なくとも n^3 回の掛算が必要であることは誰もが知っている．

1960年代の半ばに，Pan（パン）とWinograd（ヴィノグラード）[1]が，大きな n について近似的に $n^3/2$ 回の掛け算でできる方法を発見した．数年の間，これが最良の可能な掛け算の縮減であるかもしれない，と信じられていた．

$n \times n$ 行列の掛算が $O(n^\tau)$ 回の掛け算で得られるような，すべての実数 τ の下限として，**行列乗法指数** (exponent of matrix multiplication) ω を定義する．明らかに，$\omega \leq 3$ である．$\omega \geq 2$ が証明できる．

Strassen（ストラッセン）[2]は，7回の掛算だけで2×2行列の積を計算する驚くべき基本アルゴリズムを発見した．その技法は，テンソル積の構成によって大きな行列に帰納的に拡大できる．この場合には，構成はとても簡単である．大きな行列は，帰納的に分割することによって，4等分，16等分などに細かくされる．これから，$\omega \leq \ln(7)/\ln(2) < 2.808$ を得る．

より凝った基本アルゴリズムとテンソル積の構成で，さらに改良される．多くの研究者がこの問題に寄与している．Pan[3,4]は $\omega < 2.781$ を，Strassen[5]は $\omega < 2.479$ を発見した．概観と歴史については[6,7]を参照せよ．

Coppersmith & Winograd（カッパースミス&ヴィノグラード）[8]は，Salem & Spencer（サーレム & スペンサー）の組み合わせ論的定理[9]，すなわちどの3つの項も等差数列にならない整数の稠密な集合を得る定理，に基づく新しい方法を示した．その結果，$\omega < 2.376$ を得た．現在知られている最良の上界である．

$\omega = 2$ であるのか？ Bürgisser（ブルギッサー）[10]は，これを代数的複雑性定理（algebraic complexity theory）の中心問題とよんだ．密接に関連した組み合わせ論の問題がある[8,11]．

与えられた位数 n のアーベル加法群 G に対して，次の性質をもつ最小の整数 $f(n, G)$ を見つけよ．もし G の部分集合 S が濃度 $\geq f(n, G)$ をもつならば，

$$\sum_{a \in A} a = \sum_{b \in B} b = \sum_{c \in C} c$$

が成り立つような，互いに素で空集合でない，S の部分集合 A, B, C が存在する．（明らかに，$n \geq 5$ に対して，$f(n, G)$ は存在する．なぜならば，もし $S = G$ ならば，$A = \{0\}$, $B = \{g, -g\}$, $C = \{h, -h\}$ を考えよ．ただし，非零要素 g と h は，$g \neq h$ および $g \neq -h$ を満たす．）

別の関数

$$F(n) = \max_G f(n, G)$$

を定義する．最大値は，位数 n のアーベル群 G のすべてにわたり考える．比

$$\rho = \lim_{n \to \infty} \frac{\ln(n)}{F(n)}$$

を調べる．Coppersmith & Winograd[8]は，$\rho = 0$ ならば $\omega = 2$ であることを証明した．しかしながら，$\rho = 0$ の証明は，まだ知られていない．どのような（もし可能であれば）数値的な証拠が，$\rho = 0$ を支持するために，存在するのだろうか？

Coppersmith[12]は，さらに定数 $\alpha > 0.294$ であり，任意の $\varepsilon > 0$ に対して，$n \times n$ 行列と $n \times n^\alpha$ 行列との掛算において，複雑度 $O(n^{2+\varepsilon})$ をもつアルゴリズムを得た．α に対する下界の改良は，$\omega = 2$ にさらなる希望を与える．この分野の研究は続いている[13,14]．

[1] S. Winograd, A new algorithm for inner product, *IEEE Trans. Comput.* C-17 (1968) 693-694.
[2] V. Strassen, Gaussian elimination is not optimal, *Numer. Math.* 13 (1969) 354-356; MR 40 # 2223.

[3] V. Pan, Strassen's algorithm is not optimal: Trilinear technique of aggregating, uniting and canceling for constructing fast algorithms for matrix operations, *Proc. 19th Symp. on Foundations of Computer Science (FOCS)*, Ann Arbor, IEEE, 1978, pp. 166-176; MR 80e:68118.

[4] V. Pan, New fast algorithms for matrix operations, *SIAM J. Comput.* 9 (1980) 321-342; MR 81f:65037.

[5] V. Strassen, The asymptotic spectrum of tensors and the exponent of matrix multiplication, *Proc. 27th Symp. on Foundations of Computer Science (FOCS)*, Toronto, IEEE, 1986, pp. 49-54.

[6] D. E. Knuth, *The Art of Computer Programming*, v. 2, *Seminumerical Algorithms*, 2nd ed., Addison-Wesley, 1981, pp. 481-482, 503-505, 654-655; MR 44 #3531.

[7] V. Pan, *How to Multiply Matrices Faster*, Lect. Notes in Comp. Sci. 179; Springer-Verlag, 1984; MR 86g:65006.

[8] D. Coppersmith and S. Winograd, Matrix multiplication via arithmetic progressions, *J. Symbolic Comput*. 9 (1990) 251-280; MR 91i:68058.

[9] R. Salem and D. C. Spencer, On sets of integers which contain no three terms in arithmetical progression, *Proc. Nat. Acad. Sci. USA* 28 (1942) 561-563; MR 4,131e.

[10] P. Bürgisser, Algebraische Komplexitätstheorie II-Schnelle Matrixmultiplikation und Kombinatorik, *Sémin. Lothar. Combin.* 36 (1996) B36b; MR 98d:68110.

[11] G. Chiaselotti, Sums of distinct elements in finite abelian groups, *Boll. Un. Mat. Ital.* A 7 (1993) 243-251; MR 94j:20053.

[12] D. Coppersmith, Rectangular matrix multiplication revisited, *J. Complexity* 13 (1997) 42-49; MR 98b:65028.

[13] X. Huang and V. Pan, Fast rectangular matrix multiplication and applications, *J. Complexity* 14 (1998) 257-299; MR 99i:15002.

[14] L. N. Trefethen, Predictions for scientific computing fifty years from now, *Mathematics Today* (Apr. 2000) 53-57; Oxford Univ. Computing Lab. report NA-98/12.

2.30 ピゾー-ヴィジャヤラハヴァン-サーレム定数

任意に与えられた正の実数 x に対して, $\{x\}=x \bmod 1$ を x の小数部分と定義する. 任意の正の整数 n に対して, 明らかに $\{n+x\}=x$ がすべての x で成り立つ. x が有理数であるとき, 数列 $\{nx\}$ は周期的である. Weyl (ワイル) の判定条件の結果[1-4]により, x が無理数であるとき, 数列 $\{nx\}$ は区間 $[0, 1]$ で稠密である. さらに, 区間 $[0,1]$ で**一様分布** (uniformly distributed) する. この意味は, 任意の部分区間において, その数列の要素が見つかる確率が部分区間の長さに比例する, ことである.

和と積を議論したので, ベキ乗を調べる. 「ほとんどすべて」の実数 $x>1$ について, 数列 $\{x^n\}$ は一様分布をすることが[5,6]で証明されている. (奇妙ではあるが, 一様分布をするような x の特別な値は最近まで1つも知られていなかった[7,8].) $x=3/2$ に対する数列は, 典型的な例だと信じられている【2.30.1】. このふるまい[9-12]に対する例外的な x の測度 0 で非可算集合 E は, $2,3,4,\cdots$ および $1+\sqrt{2}$ などを含む. E について, ほかに何がいえるだろうか?

最初に, 用語を復習する. **モニック多項式** (monic polynomial) とは, 最高次数の

係数が1であるときをいう．αが整数係数のモニック多項式の解であるとき，αを**代数的整数**という．αの**共役数**（conjugate）とは，αの最小多項式のすべての解をいう．実数の代数的整数$\alpha>1$で，そのすべての共役数$\gamma\neq\alpha$が$|\gamma|\leqq 1$を満たような数全体からなる集合を，Uと定義する．$U\subseteqq E$であり，Uは可算集合であることが知られている．例外的なふるまいをより詳しく調べる．

代数的整数$\theta>1$で，その共役数$\gamma\neq\theta$の各々が$|\gamma|<1$を満たすとき，Pisot-Vijayaraghavan（**ピゾー-ヴィジャヤラハヴァン，P-V**）**数**という．P-V数の集合をSと定義する．代数的整数$\tau>1$で，その共役数$\gamma\neq\tau$の各々が$|\gamma|\leqq 1$であり，少なくとも1つが等号を満たすとき，Salem（**サーレム**）**数**という．サーレム数からなる集合をTとする．明らかにSとTは，Uの分割である．さらに，θがP-V数であるとき，
$$\lim_{n\to\infty}\{\theta^n\}=0 \bmod 1$$
である．一方，もしτがサーレム数であるならば，$\{\tau^n\}$は稠密であるが，区間$[0,1]$で一様分布をしない．多くの関連する結果があるが一例をあげる[11]．$\{\lambda\alpha^n\}$が多くとも有限個の集積点を法1（modulo 1）でもつような代数的実数$\alpha>1$と実数$\lambda>0$が与えられていると仮定する．このとき，αは集合Sの要素でなければならない．さらに，集積点の各々は，有理数でなければならない．Eに属するがUに属さない数（すなわち，超越数で例外的なx）を，誰かが明確に示したかどうかは，知られていない．

注意を集合Sに向ける．これは可算かつ閉集合であり，孤立した最小点$\theta_0>1$をもつことが知られている．Salem[13]とSiegel（ジーゲル）[14]は，$\theta_0=1.3247179572\cdots$が多項式$x^3-x-1$の実数の零点，すなわち，
$$\theta_0=\left(\frac{1}{2}+\frac{\sqrt{69}}{18}\right)^{1/3}+\frac{1}{3}\left(\frac{1}{2}+\frac{\sqrt{69}}{18}\right)^{-1/3}=\frac{2\sqrt{3}}{3}\cos\left(\frac{1}{3}\arccos\left(\frac{3\sqrt{3}}{2}\right)\right)$$
であることを証明した．この定数は【1.2.2】にも現れた．

実際，$\varphi+\varepsilon$までのすべてのP-V数の完全なリストは可能である[15]．ただし，$\varphi=1.6180339887\cdots$は黄金比【1.2】であり，$0<\varepsilon<0.0004$である．さらに，$S$の集積点すべての集合を $S^{<1>}$とおく．すなわち，Sの導集合（derived set）である．$S^{<1>}$の最小点は，φであり，孤立点である．さらに一般的に，$S^{<k>}$を，$S^{<k-1>}$（$k\geqq 2$）の導集合とする．$S^{<2>}$の最小点は2であり，$S^{<k>}$の最小点は$\sqrt{k}$と$k+1$の間にある．しかし，$k\geqq 3$に対するこれらの正確な値は，1つも知られていない．

集合Tを調べることは，もっと難しい．Tは可算無限であり，UはTの閉包の真部分集合であることが知られている．最小のサーレム数の存在は，未解決問題のままである．しかし，$\tau_0=1.1762808182\cdots$，これはLehmer（**レーマー**）**の多項式**[16]
$$x^{10}+x^9-x^7-x^6-x^5-x^4-x^3+x+1$$
の零点の1つであると予想されている．たかだか40次の多項式次数をもち1.3より小さいサーレム数は，正確に45個存在することが証明されている[17-20]．（次数が40を超え1.3より小さいサーレム数は2個のみ知られている．しかし，ひょっとするともっと多いかもしれない．）θ_0はTの最小の集積点であるのか？ この問題に対しても答

えが知られていない．

　定数 θ_0 と τ_0 は，0 でない代数的整数 α の Mahler（マーラー）測度についての，Lermer による関連した予想に関して現れる．次数 n の α の共役数を $\alpha_1=\alpha, \alpha_2, \alpha_3, \cdots, \alpha_n$ とする．$|\alpha_j|>1$ を満たすすべての α_j の積の絶対値を $M(\alpha)$ と定義する．Kronecker（クロネッカー）[21,22] は，$M(\alpha)=1$ であるとき α は 1 の累乗根であることを証明した．任意の $\varepsilon>0$ に対して，$1<M(\alpha)<1+\varepsilon$ である α が存在することは真なのだろうか？

　α が**互いに逆数共役でない**（non-reciprocal），すなわち，α と $1/\alpha$ が共役でないならば，答えは否定的であることを，Smyth（スミス）[11,23] が証明した．より正確には，$M(\alpha) \geq \tau_0 = 1.324\cdots$ であるか，α は 1 の累乗根であるか，のどちらかである．

　任意の α に対して，Lehmer[16] は，その答えは否定的のままである，と予想した．より正確には，$M(\alpha) \geq \tau_0 = 1.176\cdots$ であるか，α は 1 の累乗根であるか，のどちらかである．広範囲にわたる研究にもかかわらず，この不等式の反例は見つかっていない．知られているうちで最良の適切な評価は，α が 1 の累乗根でないとき，充分大きな n に対して [24-30]，

$$M(\alpha) > 1 + \left(\frac{9}{4} - \varepsilon\right)\left(\frac{\ln(\ln(n))}{\ln(n)}\right)^3$$

である．マーラー測度に関してはさらに，【3.10】を参照せよ．α の**囲い**（house）とよばれる量を含む関連する不等式 [21,30-32] は

$$\overline{|\alpha|} = \max_{1 \leq k \leq n} |\alpha_k| > 1 + \frac{1}{n}\left(\frac{64}{\pi^2} - \varepsilon\right)\left(\frac{\ln(\ln(n))}{\ln(n)}\right)^3$$

であり，対応する予想は [33]：$\overline{|\alpha|} \geq 1 + (2/3)(\ln(\theta_0)/n) = 1 + (0.4217993614\cdots)/n$ である．さらに [34,35] を参照せよ．

2.30.1 $(3/2)^n$ の小数部分

　Pisot[9] と Vijayaraghavan[36] が，$\{(3/2)^n\}$ は無限に多くの集積点をもつ，すなわち，無限に多くの異なる値に収束する部分列をもつことを証明した．この数列は一様分布すると信じられているが，区間 $[0,1]$ で稠密であることさえ，誰も証明していない．

　いくぶん野心的でない問題がある．$\{(3/2)^n\}$ が，$[0,1/2)$ と $[1/2,1]$ のどちらにも無限に多くの集積点をもつことを証明せよ，という問題である．いいかえれば，数列がある区間を他の区間よりも**好む**（prefer）ことはない，ことを証明することである．この問題は未解決のままである．しかし，Flatto, Lagarias & Pollington（フラットー，ラガリアス & ポリングトン）[37] は最近いくらか進歩させた．$\{(3/2)^n\}$ のすべての，おそらく有限個の乗積点を除いた集積点を含む $[0,1]$ 内の任意の部分区間は，少なくとも長さ $1/3$ でなければならない，ことを証明した．したがって，数列は，任意の $\varepsilon>0$ に対して，$[1/3-\varepsilon,1)$ より $[0,1/3-\varepsilon)$ を好むことはない．同様に，$[0,2/3+\varepsilon)$ より $[2/3+\varepsilon,1]$ を好むこともない．この証明を $[0,1/2)$ と $[1/2,1]$ に拡張することは，意味があるが，難しい仕事である．

Lagarias[38]は，数列 $\{(3/2)^n\}$ と有名な $3x+1$ 問題[†]に関するエルゴード理論的な観点とのゆるい関連について述べた．詳細はとても精密なのでここで議論できない．興味をそそられることに，「さらに」見たところ遠い分野である数論の問題（整数の n 乗の和として整数を表す Waring（ウェアリング）問題）に，この数列は重要である．

すべての正の整数が，非負整数の n 乗の k 個の和として表すことのできる，最小の整数 k を，$g(n)$ とする．Hilbert（ヒルベルト）[39]は，任意の n に対して $g(n)<\infty$ であることを証明した．$2\leq n\leq 6$ に対して，[40-44]

$$g(n)=2^n+\left\lfloor\left(\frac{3}{2}\right)^n\right\rfloor-2$$

であることが知られている．

Dickson（ディクソン）[45,46] と Pillai（ピライ）[47] は独立に，この式がすべての $n>6$ に対して真であることを，条件

$$\left\{\left(\frac{3}{2}\right)^n\right\}\leq 1-\left(\frac{3}{4}\right)^n$$

が成り立つことを仮定して，証明した．したがって，この不等式，すなわちウェアリング問題の解に最後に残る障害，の研究で十分である．

Kubina & Wunderlich（クビナ＆ブンダーリッチ）[48]は，Stemmler（ステムラー）[49]の仕事を拡張して，上記の不等式がすべての $2\leq n\leq 471600000$ で成り立つことをコンピュータで確かめた．Mahler[50]はさらに，代数的数の有理数近似に関するトゥエ-ジーゲル-ロスの定理【2.22】を用いて，この不等式が成り立たないのは，多くとも有限個の n についてだけであることを証明した．しかし，その証明は構成的ではない．したがって，成り立たない数を除外するコンピュータシミュレーションは，まだまったく不可能である．

すべての $n>7$ に対して，不等式が

$$\left(\frac{3}{4}\right)^n<\left\{\left(\frac{3}{2}\right)^n\right\}<1-\left(\frac{3}{4}\right)^n$$

に強められ，ある方法で一般化された[51,52]．再び，Mahler の主張は別にして，証明は知られていない．（最良の有効な結果は，3/4 を 0.577 に置きかえたもので，Beukers（ベウカース）[53]，Dubickas（デュビカス）[54]，Habsieger（ハブジーガー）[55]による．）とても簡単な不等式が，解析のすべての試みを拒む事実は，注目に値する．

$g(n)$ の計算は，時にはウェアリング問題の「究極の目標」とよばれる．「充分大きな」すべての整数が，非負整数の n 乗の k 個の和として表すことのできる，最小の整数 k を $G(n)$ とする．明らかに，$G(n)\leq g(n)$ である．Hurwitz（フルヴィッツ）[56]と Maillet（マイレ）[57]は，$G(n)\geq n+1$ を証明した．いいかえると，n 乗の n

---

[†]訳註：任意の正の整数 n に対して，n が偶数ならば2で割り，奇数ならば $3n+1$ を作る操作を繰り返すと，必ず 1→4→2→1 のループに終わるという予想．いろいろな名でよばれるが，Collatz（コラッツ）の問題とよぶのが適切らしい．

個の和として表すことのできない，任意に大きい整数が存在する．[43,58-60]から，$G(2)=4$, $4 \leq G(3) \leq 7$, $G(4)=16$, $6 \leq G(5) \leq 17$, $9 \leq G(6) \leq 24$ であることが知られている．予想 $G(3)=4$ を支持する数値的な証拠は[61-63]を参照せよ．3乗数の4つの和として n を表現する個数の漸近式については，さらに[64,65]を参照せよ．4つの3乗数は，おもしろいことに $\Gamma(4/3)$ を含むことがわかる．ただし，$\Gamma(x)$ はオイラーガンマ関数【1.5.4】である．

関連のない事実がいくつかある．$x=3/2$ のとき，$\lfloor x^n \rfloor$ の形をした無限に多くの整数は，合成数である[66,67]．さらに，これは $x=4/3$ のときも真である．このような整数が無限に多く素数となるのか？ x のほかの値については，何がいえるのだろうか？

予想は，t が 2^t と 3^t のどちらもが整数となるような実数であるならば，t は有理数である，というものである．これはいわゆる4指数 (four-exponentials) 予想[68,69]から得られる．弱い結果である6指数 (six-exponentials) 定理は，真であることが知られている．

$x_0=1$, $x_n=\lceil (3/2)x_{n-1} \rceil$ ($n \geq 1$) によって，無限数列を定義する．Odlyzko & Wilf (オドリズコ & ヴィルフ) [70]は，すべての n に対して，

$$x_n = \left\lfloor K \cdot \left(\frac{3}{2}\right)^n \right\rfloor$$

を証明した．ただし，定数 $K=1.6222705028\cdots$ である．（実際には，彼らはもっと多くの定理を証明した．）その仕事は，古い Josephus (ヨゼフ) 問題の解に関連している．定数 K は，式が計算（K の正確な値がなんらかの方法で利用可能とならないかぎり）に役に立たないという意味で，Mills (ミルズ) 定数と類似なものである．しかし，そもそも存在するということ自体は注目すべきである．

3円滑数 (3-smooth number) とは，正の整数でかつその素因数が2と3のみである数をいう．正の整数 n が **3円滑表現** (3-smooth representation) をもつとは，n が3円滑数の和で表されるときをいう．ただし，その各項はほかの和を決して割り切らないようにする．$r(n)$ を，n の3円滑表現の個数とする．最近の論文[71-73]で，$r(n)$ の大きさの極大値および平均値の問題の答えを得た．さらに【5.4】を参照せよ．

n を 8 より大きい整数とする．2^n の 3 進数展開は，どこかの桁で数字 2 となるだろうか？ Erdös[74]は，答えは肯定的と予想した．Vardi (ヴァルディ) [75]は $n=2 \cdot 3^{20}$ まで正しいことを確かめた．数字2と3との相互関係のより多くの実例は【2.26】にある．

[1] H. Weyl, Über die Gleichverteilung von Zahlen mod. Eins, *Math. Annalen* 77 (1916) 313-352.
[2] J. W. S. Cassels, *An Introduction to Diophantine Approximation*, Cambridge Univ. Press, 1965; MR 50 # 2084.
[3] A. Miklavc, Elementary proofs of two theorems on the distribution of numbers {nx} (mod 1), *Proc. Amer. Math. Soc.* 39 (1973) 279-280; MR 47 # 4962.
[4] F. M. Dekking and M. Mendès France, Uniform distribution modulo one: A geometrical

viewpoint, *J. Reine Angew. Math.* 329 (1981) 143-153; MR 83b:10062.
- [5] G. H. Hardy and J. E. Littlewood, Some problems of Diophantine approximation, *Acta Math.* 37 (1914) 155-191; also in *Collected Papers of G. H. Hardy*, v. 1, Oxford Univ. Press, 1966, pp. 28-66.
- [6] J. F. Koksma, Ein mengentheoretischer Satz über die Gleichverteilung modulo Eins, *Compositio Math.* 2 (1935) 250-258; Zbl. 12/14.
- [7] M. B. Levin, On the complete uniform distribution of the fractional parts of the exponential function (in Russian), *Trudy Sem. Petrovsk.* 7 (1981) 245-256; Engl. transl. in *J. Soviet Math.* 31 (1985) 3247-3256; MR 83j:10059.
- [8] M. Drmota and R. F. Tichy, *Sequences, Discrepancies and Applications*, Lect. Notes in Math. 1651, Springer-Verlag, 1997; MR 98j:11057.
- [9] C. Pisot, La répartition modulo 1 et les nombres algébriques, *Annali Scuola Norm. Sup.* Pisa 7 (1938) 205-248.
- [10] T. Vijayaraghavan, On the fractional parts of the powers of a number. IV, *J. Indian Math. Soc.* 12 (1948) 33-39; MR 10,433b.
- [11] M.-J. Bertin, A. Decomps-Guilloux, M. Grandet-Hugot, M. Pathiaux-Delefosse, and J.-P. Schreiber, *Pisot and Salem Numbers*, Birkhäuser, 1992; MR 93k:11095.
- [12] M. Mendes France, Book review of "Pisot and Salem Numbers," *Bull. Amer. Math. Soc.* 29 (1993) 274-278.
- [13] R. Salem, A remarkable class of algebraic integers: Proof of a conjecture of Vijayaraghavan, *Duke Math. J.* 11 (1944) 103-108; MR 5,254a.
- [14] C. L. Siegel, Algebraic integers whose conjugates lie in the unit circle, *Duke Math. J.* 11 (1944) 597-602; MR 6,39b.
- [15] J. Dufresnoy and C. Pisot, Etude de certaines fonctions méromorphes bornées sur le cercle unité. Application à un ensemble fermé d'entiers algébriques, *Annales Sci. École Norm. Sup.* 72 (1955) 69-92; MR 17,349d.
- [16] D. H. Lehmer, Factorization of certain cyclotomic functions, *Annals of Math.* 34 (1933) 461-479.
- [17] D. W. Boyd, Small Salem numbers, *Duke Math. J.* 44 (1977) 315-328; MR 56 #11952.
- [18] D. W. Boyd, Pisot and Salem numbers in intervals of the real line, *Math. Comp.* 32 (1978) 1244-1260; MR 58 #10812.
- [19] M. J. Mossinghoff, Polynomials with small Mahler measure, *Math. Comp.* 67 (1998) 1697-1705, S11-S14; MR 99a:11119.
- [20] V. Flammang, M. Grandcolas, and G. Rhin, Small Salem numbers, *Number theory in Progress*, Proc. 1997 Zakopane conf., ed. K. Győry, H. Iwaniec, and J. Urbanowicz, v. 1, de Gruyter, 1999, pp. 165-168; MR 2000e:11132.
- [21] A. Schinzel and H. Zassenhaus, A refinement of two theorems of Kronecker, *Michigan Math. J.* 12 (1965) 81-85; MR 31 #158.
- [22] D. W. Boyd, Variations on a theme of Kronecker, *Canad. Math. Bull.* 21 (1978) 129-133; MR 58 #5580.
- [23] C. J. Smyth, On the product of the conjugates outside the unit circle of an algebraic integer, *Bull. London Math. Soc.* 3 (1971) 169-175; MR 44 #6641.
- [24] E. Dobrowolski, On a question of Lehmer and the number of irreducible factors of a polynomial, *Acta Arith.* 34 (1979) 391-401; MR 80i:10040.
- [25] D. C. Cantor and E. G. Straus, On a conjecture of D. H. Lehmer, *Acta Arith.* 42 (1982) 97-100; correction 42 (1983) 327; MR 84a:12004 and MR 85a:11017.
- [26] R. Louboutin, Sur la mesure de Mahler d'un nombre algébrique, *C. R. Acad. Sci. Paris Sér.* I *Math.* 296 (1983) 707-708; MR 85b:11058.
- [27] U. Rausch, On a theorem of Dobrowolski about the product of conjugate numbers, *Colloq. Math.* 50 (1985) 137-142; MR 87i:11144.
- [28] P. Voutier, An effective lower bound for the height of algebraic numbers, *Acta Arith.* 74 (1996) 81-95. MR 96j:11098.

[29] A. Dubickas, Algebraic conjugates outside the unit circle, *New Trends in Probability and Statistics*, v. 4, *Analytic and Probabilistic Methods in Number Theory*, Proc. 1996 Palanga conf., ed. A. Laurincikas, E. Manstavicius, and V. Stak.enas, VSP, 1997, pp. 11-21; MR 99i:11096.

[30] A. Schinzel, *Polynomials with Special Regard to Reducibility*, Cambridge Univ. Press, 2000; MR 2001h:11135.

[31] A. Dubickas, On a conjecture of A. Schinzel and H. Zassenhaus, *Acta Arith.* 63 (1993) 15-20; MR 94a:11161.

[32] A. Dubickas, The maximal conjugate of a non-reciprocal algebraic integer (in Russian), *Liet. Mat. Rink.* 37 (1997) 168-174; Engl. transl. in *Lithuanian Math. J.* 37 (1997) 129-133; MR 98j:11083.

[33] D.W. Boyd, The maximal modulus of an algebraic integer, *Math.Comp.* 45 (1985) 243-249, S17-S20; MR 87c:11097.

[34] E. Dobrowolski, Mahler's measure of a polynomial in function of the number of its coefficients. *Canad. Math. Bull.* 34 (1991) 186-195; MR 92f:11138.

[35] J. H. Silverman, Small Salem numbers, exceptional units, and Lehmer's conjecture, *Rocky Mount. J. Math.* 26 (1996) 1099-1114; MR 97k:11152.

[36] T. Vijayaraghavan, On the fractional parts of the powers of a number. I, *J. London Math. Soc.* 15 (1940) 159-160; MR 2,33e.

[37] L. Flatto, J. C. Lagarias, and A. D. Pollington, On the range of fractional parts $\{\xi(p/q)^n\}$, *Acta Arith.* 70 (1995) 125-147; MR 96a:11073.

[38] J. C. Lagarias, The $3x+1$ problem and its generalizations, *Amer. Math. Monthly* 92 (1985) 3-23; also in *Organic Mathematics*, Proc. 1995 Burnaby workshop, ed. J. Borwein, P. Borwein, L. Jörgenson, and R. Corless, Amer. Math. Soc., 1997, pp. 305-334; MR 86i:11043.

[39] D. Hilbert, Beweis für die Darstellbarkeit der ganzen Zahlen durch eine feste Anzahl n^{ter} Potenzen (Waringsches Problem), *Nachr. Königlichen Ges. Wiss. Göttingen, Math.-Phys. Klasse* (1909) 17-36; *Math. Annalen* 67 (1909) 281-305; also in *Gesammelte Abhandlungen*, v. 1, Chelsea, 1965, pp. 510-527.

[40] G. H. Hardy and E. M. Wright, *An Introduction to the Theory of Numbers*, 5th ed., Oxford Univ. Press, 1985; MR 81i:10002.

[41] W. J. Ellison, Waring's problem, *Amer. Math. Monthly* 78 (1971) 10-36; MR 54 #2611.

[42] C. Small, Waring's problem, Math. Mag. 50 (1977) 12-16; MR 55 #5561.

[43] P. Ribenboim, *The New Book of Prime Number Records*, Springer-Verlag, 1996, pp. 300-309; MR 96k:11112.

[44] R. Balasubramanian, J.-M. Deshouillers, and F. Dress, Problème de Waring pour les bicarrés 1, 2, *C. R. Acad. Sci. Paris Sér. I Math.* 303 (1986) 85-88, 161-163; MR 87m:11099 and MR 88e:11095.

[45] L. E. Dickson, Proof of the ideal Waring theorem for exponents 7-180, *Amer. J. Math.* 58 (1936) 521-529.

[46] L. E. Dickson, Solution of Waring's problem, *Amer. J. Math.* 58 (1936) 530-535.

[47] S. S. Pillai, On Waring's problem, *J. Indian Math. Soc.* 2 (1936) 16-44; Zbl. 14/294.

[48] J. M. Kubina and M. C.Wunderlich, ExtendingWaring's conjecture to 471,600,000, *Math. Comp.* 55 (1990) 815-820; MR 91b:11101.

[49] R. M. Stemmler, The ideal Waring theorem for exponents 401-200,000, *Math. Comp.* 18 (1964) 144-146; MR 28 #3019.

[50] K. Mahler, On the fractional parts of the powers of a rational number. II, *Mathematica* 4 (1957) 122-124; MR 20 #33.

[51] M. A. Bennett, Fractional parts of powers of rational numbers, *Math. Proc. Cambridge Philos. Soc.* 114 (1993) 191-201; MR 94h:11062.

[52] M. A. Bennett, An ideal Waring problem with restricted summands, *Acta Arith.* 66 (1994) 125-132; MR 95k:11126.

[53] F. Beukers, Fractional parts of powers of rationals, *Math. Proc. Cambridge Philos. Soc.* 90

(1981) 13-20; MR 83g:10028.
[54] A. Dubickas, A lower bound for the quantity $\|(3/2)^n\|$ (in Russian), *Uspekhi Mat. Nauk*, v. 45 (1990) n. 4, 153-154; Engl. transl. in *Russian Math. Survey*, v. 45 (1990) n. 4, 163-164; MR 91k: 11058.
[55] L. Habsieger, Explicit lower bounds for $\|(3/2)^k\|$, unpublished note (1998).
[56] A. Hurwitz, Über die Darstellung der ganzen Zahlen als Summen von n^{ter} Potenzen ganzer Zahlen, *Math. Annalen* 65 (1908) 424-427; also in *Mathematische Werke*, v. 2, Birkhäuser, 1933, pp. 422-426.
[57] E. Maillet, Sur la décomposition d'un entier en une somme de puissances huitièmes d'entiers (Problème de Waring), *Bull. Soc. Math. France* 36 (1908) 69-77.
[58] M. B. Nathanson, *Additive Number Theory: The Classical Bases*, Springer-Verlag, 1996; MR 97e:11004.
[59] R. C. Vaughan and T. D. Wooley, Further improvements in Waring's problem, *Acta Math.* 174 (1995) 147-240; MR 96j:11129a.
[60] R. C. Vaughan and T. D. Wooley, Further improvements in Waring's problem. IV: Higher powers, *Acta Arith.* 94 (2000) 203-285; MR 2001g:11154.
[61] J. Bohman and C.-E. Fröberg, Numerical investigation of Waring's problem for cubes, *BIT* 21 (1981) 118-122; MR 82k:10063.
[62] F. Romani, Computations concerning Waring's problem for cubes, *Calcolo* 19 (1982) 415-431; MR 85g:11088.
[63] J.-M. Deshouillers, F. Hennecart, and B. Landreau, 7,373,170,279,850, *Math. Comp.* 69 (2000) 421-439; MR 2000i:11150.
[64] J. Brüdern and N. Watt, On Waring's problem for four cubes, *Duke Math. J.* 77 (1995) 583-599; MR 96e:11121.
[65] K. Kawada, On the sum of four cubes, *Mathematika* 43 (1996) 323-348; MR 97m:11125.
[66] W. Forman and H. N. Shapiro, An arithmetic property of certain rational powers, *Commun. Pure Appl. Math.* 20 (1967) 561-573; MR 35 #2852.
[67] R. K. Guy, *Unsolved Problems in Number Theory*, 2nd ed., Springer-Verlag 1994, sect. E19; MR 96e:11002.
[68] M.Waldschmidt, *Transcendence Methods*, Queen's Papers in Pure and Appl. Math., n. 52, ed. A. J. Coleman and P. Ribenboim, Queen's Univ., 1979; MR 83a:10068.
[69] M. Waldschmidt, On the transcendence method of Gel'fond and Schneider in several variables, *New Advances in Transcendence Theory*, ed. A. Baker, Cambridge Univ. Press, 1988, pp. 375-398; MR 90d:11089.
[70] A. M. Odlyzko and H. S. Wilf, Functional iteration and the Josephus problem, *Glasgow Math. J.* 33 (1991) 235-240; MR 92g:05006.
[71] R. Blecksmith, M. McCallum, and J. L. Selfridge, 3-smooth representations of integers, *Amer. Math. Monthly* 105 (1998) 529-543; MR 2000a:11019.
[72] M. R. Avidon, On primitive 3-smooth partitions of n, *Elec. J. Combin.* 4 (1997) R2; MR 98a:11136.
[73] P. Erdös and M. Lewin, d-complete sequences of integers, *Math. Comp.* 65 (1996) 837-840; MR 96g:11008.
[74] P. Erdös and R. Graham, *Old and New Problems and Results in Combinatorial Number Theory*, Enseignement Math. Monogr. 28, 1980, p. 80; MR 82j:10001.
[75] I. Vardi, *Computational Recreations in Mathematica*, Addison-Wesley, 1991, pp. 20-25; MR 93e:00002.

2.31 フレイマンの定数

2.31.1 ラグランジュスペクトル, L

ディオファンタス近似定数に関する節【2.23】において，Hurwitz（フルヴィッツ）の定理[1,2]，すなわち任意の無理数 ξ に対して不等式

$$\left|\xi - \frac{p}{q}\right| < \frac{1}{\sqrt{5}} \frac{1}{q^2}$$

が無限に多くの整数解 (p, q) をもつことを議論した．この結果を改良できるのか？ すなわち，$\sqrt{5}$ をより大きい数で置きかえることができるのだろうか？ この答えは，ある特別な数 ξ については，否定的である．しかし，他の場合では肯定的である．詳しく述べる．

各 ξ に対して，

$$\left|\xi - \frac{p}{q}\right| < \frac{1}{c} \frac{1}{q^2}$$

の整数解の集合 (p, q) が無限集合となるような量 c の上限を $\lambda(\xi)$ とする．関数 $\lambda(\xi)$ がとる値 L の集合を，Lagrange（**ラグランジュ**）**スペクトル**という[3]．明らかに，L の最小値は $\sqrt{5}$ である．集合 $L \cap [2,3]$ は，ただ 1 つの極限点（limit point）として 3 をもつ可算無限集合であるが，ある値 $\theta > 4$ に対して $[\theta, \infty) \subseteq L$ であることが証明できる．

2.31.2 マルコフスペクトル, M

実係数をもつ 2 変数 2 次形式 $f(x,y) = \alpha x^2 + \beta xy + \gamma y^2$ が**不定**（indefinite）であるとは，f が正の値と負の値をともにとるときをいう．**判別式**（discriminant）$d(f) = \beta^2 - 4\alpha\gamma$ が正であるならば，xyz 実数空間で $z = f(x,y)$ のグラフを描くと，鞍面（saddle surface，馬の鞍のような曲面）になる．すなわち，最大値も最小値もとる点はない．

このような各 f に対して，

$$\mu(f) = \frac{\sqrt{d(f)}}{\inf_{(m,n) \neq (0,0)} |f(m,n)|}$$

と定義する．ただし，下限はすべての 0 でない整数の組にわたりとる．関数 $\mu(f)$ がとる値 M の集合は，**マルコフスペクトル**（Markov spectrum）といわれる[3]．$L \subseteq M$ であり，さらに $M \cap [2,3] = L \cap [2,3]$ で，L の節で述べたある点 $\theta > 4$ に対して $[\theta, \infty) \subseteq M$ である，ことが証明される．しかしながら，$M \cap [3,\theta] \neq L \cap [3,\theta]$，すなわち L は M の「真」部分集合である．これは永らく未解決問題であった（【2.31.5】

2.31.3 マルコフ-フルヴィッツの方程式

フルヴィッツの定理にもどる．まず，ξ と η が**同値** (equivalent) であるとは，整数 a, b, c, d が存在して，

$$\xi = \frac{a\eta + b}{c\eta + d} \quad (|ad - bc| = 1)$$

となるときをいう．この関係は，数を分割して同値の族へ分ける．2つの無理数 ξ と η が同値であるための必要十分条件は，ある箇所から後，連分数の部分分母の数列が等しくなることである．

黄金比 φ【1.2】と同値なすべての ξ，すなわち部分分母がいつかはすべて 1 となる数，に対して，$\lambda(\xi) = \sqrt{5}$ であることが証明できる．このような数は「最単純（simplest）」と考えることができる．しかし，有理数近似の観点からは，最単純数は「最悪 (worst)」である【1.4】．これらを除くならば，近似の難しさの次のレベルは，ピタゴラスの数 $\sqrt{2}$【1.1】，すなわち部分分母が最後にはすべて 2 となるものと，同値なすべての ξ，に対して，$\lambda(\xi) = \sqrt{8}$ である．同様に，これらを除くと，次のレベルは $\xi = \sqrt{221}/5$ などである．対応する 2 次形式表現 $f(x, y)$ を計算するアルゴリズムのみならず，ラグランジュ・スペクトルにおける最小数の表については [3] を参照せよ．

値 $\sqrt{5}$, $\sqrt{8}$, $\sqrt{221}/5$, $\sqrt{1517}/13$, $\sqrt{7565}/29$, $\cdots$ は，すべて $\sqrt{9w^2 - 4}/w$ の形である．ただし，u, v, w は正の整数であり，ディオファンタス方程式

$$u^2 + v^2 + w^2 = 3uvw, \quad 1 \leq u \leq v \leq w$$

を満たす．これを満足する 3 変数の最初のいくつかの組は

$$(u, v, w) = (1, 1, 1),\ (1, 1, 2),\ (1, 2, 5),\ (1, 5, 13),\ (2, 5, 29),\ (1, 13, 34),\cdots$$

である．w の無限数列

$$1, 2, 5, 13, 29, 34, 89, 169, 194, 233, 433, 610, 985, \cdots$$

は，**マルコフ数**（Markov numbers）[5] とよばれる．すべての w_k が「一意的に」可能な 3 変数の組 (u_k, v_k, w_k) を決定するかどうか，は未解決である [6-12]．明らかに，$k \to \infty$ のとき $\lambda(w_k)$ の極限値は 3 である，ことに注意する．上述のように，これから $L \cap [2, 3]$ の集積点が 3 だけであることが証明される．

副次的な問題がある．取りうる 3 つの組 (u, v, w)，$w \leq n$ の個数 $N(n)$ が，Zagier（ザギエ）[7, 8] によって，

$$N(n) = C \cdot \ln(n)^2 + O[\ln(n) \cdot \ln(\ln(n))^2]$$

を満たすことが証明された．ただし，

$$C = \frac{3}{\pi^2} \cdot \frac{1}{2} \left(\frac{1}{g(1)^2} + \frac{2g(1) - g(2)}{g(1)^2 g(2)} \right) + \frac{3}{\pi^2} \sum_{\substack{\text{取りうる}(u,v,w) \\ u < v < w}} \frac{g(u) + g(v) - g(w)}{g(u)g(v)g(w)}$$

$$= 0.1807171047\cdots$$

$$g(x) = \ln\left(\frac{3x + \sqrt{9x^2 - 4}}{2}\right) = \operatorname{arccosh}\left(\frac{3x}{2}\right), \quad x \geq \frac{2}{3}$$

である．この漸近的結果は，もし一意性の予想が真であれば，
$$w_k = \left(\frac{1}{3} + o(1)\right)\exp\left(\sqrt{\frac{k}{C}}\right) = \left(\frac{1}{3} + o(1)\right)(10.5101504239\cdots)^{\sqrt{k}}$$
となり，
$$N(n) = C \cdot \ln(3n)^2 + o(\ln(n))$$
に強めることができると，Zagier は予想した．

副次的な問題の一般化がある．$m \geq 3$ とする．Markov-Hurwitz（マルコフ-フルヴィッツ）方程式
$$u_1^2 + u_2^2 + \cdots + u_m^2 = m u_1 u_2 \cdots\cdots u_m, \quad 1 \leq u_1 \leq u_2 \leq \cdots \leq u_m$$
を考える．正の整数で方程式の解となる m 個の組 $(u_1, u_2, \cdots, u_m)$ で $u_m \leq n$ を満たすものの個数を，$N_m(n)$ と定義する．$N_m(n)$ の増加率は $O(\ln(n)^{m-1})$ ではなくて，任意の $\varepsilon > 0$ に対して $O(\ln(n)^{\alpha(m)+\varepsilon})$ であることが驚きである．ただし，$\alpha(m)$ は，[13-15]
$$\alpha(3) = 2, \quad 2.430 < \alpha(4) < 2.477, \quad 2.730 < \alpha(5) < 2.798, \quad 2.963 < \alpha(6) < 3.048$$
を満たし，$\lim_{m \to \infty} \alpha(m)/\ln(m) = 1/\ln(2)$ である．$m = 4$ に対するザギエ定数と類似なものは知られていない．

2.31.4 ホール線

$L \cap [3, \infty)$ と $M \cap [3, \infty)$ についての知識は，$L \cap [2, 3]$ に対する前記の情報よりも，あまり完全ではない．L と M それぞれは，実数直線上で閉部分集合である．したがって，各スペクトルの補集合は，開区間の可算和である．すなわち，**空隙** (gap) をもつ．空隙は，もしその端点が考えているスペクトルの中にあれば，**極大** (maximal) である．いくつかの（L と M 両方に関しての）極大な空隙は次のとおりである．
$$(\sqrt{12}, \sqrt{13}) = (3.464101\cdots, 3.605551\cdots),$$
$$\left(\sqrt{13}, \frac{65 + 9\sqrt{3}}{22}\right) = (3.605551\cdots, 3.663111\cdots),$$
$$\left(\frac{480}{7}, \sqrt{10}\right) = (3.129843\cdots, 3.162277\cdots).$$
はじめの 2 つは Perron（ペロン）[16]によって発見された．他の多くは[3]にリストがある．明らかに，左端の点が ≥ 3 である「最初の」空隙は存在しない．

Hall（ホール）[17]は，区間 $[\sqrt{2}-1, 4\sqrt{2}-4]$ の任意の数は，連分数の部分分母が決して 4 を超えない，2 つの数の和で書けることを証明した．このことから，L と M は充分大きな実数をすべて含む，ことが従う．これらのスペクトルのこの一部を，**ホール線** (Hall's ray) とよぶ．Freiman（フレイマン）[18]は，ホール線が始まる正確な点 θ（L と M に共通な点）を計算することに成功した．正確な式は[3,6]
$$\theta = 4 + \frac{253589820 + 283748\sqrt{462}}{491993569} = 4.5278295661\cdots$$
である．実際，右端の点が $< \infty$ であるような「最後の」空隙は，$(4.527829538\cdots,$

4.527829566…) であり，L と M の両方に対して正しい.

比較として，Bumby（バンビ）[3,19]は，$M \cap [3, 3.33437\cdots]$ はルベーグ測度が 0 であることを証明した！ 端点 3.33437…をさらに右にシフトしても，なお測度 0 を保持するのか？ この端点の正確な式を見つけることができるか？

2.31.5 L と M の比較

これはたぶん，この研究の最も謎めいた分野である．ごく簡単に要約する[3]. Freiman[20]は，M には属するが L には属さない，2次無理数 $\xi=3.118120178\cdots$ を構成した．Freiman[21]は後に別な例を見つけた：$\eta=3.293044265\cdots$. 現在では，無限に多くのこのような例が発見されている．Berstein（ベルスタイン）[22,23]は，Freiman の点 ξ と η は含むが L のどの点も含まない，最大の区間を決定した．η に対する区間の長さは，近似的に 2×10^{-7} であるけれども，ξ に対する区間の長さは，近似的に 1.7×10^{-10} である．さらに，Freiman は，これらの区間のそれぞれが M の要素を可算無限個含む，ことを示した．

Cusick & Flahive（クシック&フラヒブ）[3]は，$\sqrt{12}=3.464101\cdots$ より大きな値で L と M が一致する，と予想した．M に属するが L には属さないと知られている最大の数は 3.29304…である．たいへんおもしろいこの分野に関するより多くの論文は，[24]にある．

[1] G. H. Hardy and E. M. Wright, *An Introduction to the Theory of Numbers*, 5th ed., Oxford Univ. Press, 1985, pp. 141-143, 163-169; MR 81i:10002.
[2] J. W. S. Cassels, *An Introduction to Diophantine Approximation*, Cambridge Univ. Press, 1965; MR 50 #2084.
[3] T. W. Cusick and M. E. Flahive, *The Markoff and Lagrange Spectra*, Amer. Math. Soc., 1989; MR 90i:11069.
[4] A. M. Rockett and P. Szüsz, *Continued Fractions*,World Scientific, 1992, pp. 72-110; MR 93m:11060.
[5] N. J. A. Sloane, On-Line Encyclopedia of Integer Sequences, A002559.
[6] J. H. Conway and R. K. Guy, *The Book of Numbers*, Springer-Verlag, 1996, pp. 187-189; MR 98g:00004.
[7] R. K. Guy, *Unsolved Problems in Number Theory*, 2nd ed., Springer-Verlag, 1994, sect. D12; MR 96e:11002.
[8] D. Zagier, On the number of Markoff numbers below a given bound, *Math. Comp.* 39 (1982) 709-723; MR 83k:10062.
[9] P. Schmutz, Systoles of arithmetic surfaces and the Markoff spectrum, *Math. Annalen* 305 (1996) 191-203; MR 97b:11090.
[10] A. Baragar, On the unicity conjecture for Markoff numbers, *Canad. Math. Bull.* 39 (1996) 3-9; MR 97d:11110.
[11] J. O. Button, The uniqueness of the prime Markoff numbers, *J. London Math. Soc.* 58 (1998) 9-17; MR 2000c:11043.
[12] J. O. Button, Markoff numbers, principal ideals and continued fraction expansions, *J. Number Theory* 87 (2001) 77-95; MR 2002a:11074.
[13] Yu. N. Baulina, On the number of solutions of the equation $x_1^1 + \cdots + x_n^2 = nx_1 \cdots x_n$ that do not exceed a given limit (in Russian), *Mat. Zametki* 57 (1995) 297-300; Engl. transl. in *Math. Notes*

57 (1995) 208-210; MR 97b:11049.
[14] A. Baragar, Asymptotic growth of Markoff-Hurwitz numbers, *Compositio Math.* 94 (1994) 1-18; MR 95i:11025.
[15] A. Baragar, The exponent for the Markoff-Hurwitz equations, *Pacific J. Math.* 182 (1998) 1-21; MR 99e:11035.
[16] O. Perron, Über die Approximation irrationaler Zahler durch rationale. II, *Sitzungsberichte der Heidelberger Akademie der Wissenschaften* 8 (1921) 1-12.
[17] M. Hall, On the sum and product of continued fractions, *Annals of Math.* 48 (1947) 966-993; MR 9,226b.
[18] G. A. Freiman, *Diophantine Approximations and the Geometry of Numbers (Markov's Problem)* (in Russian), Kalininskii Gosudarstvennyi Universitet, 1975; MR 58 #5536.
[19] R. T. Bumby, Hausdorff dimensions of Cantor sets, *J. Reine Angew. Math.* 331 (1982) 192-206; MR 83g:10038.
[20] G. A. Freiman, Non-coincidence of the spectra of Markov and of Lagrange (in Russian), Mat. Zametki 3 (1968) 195-200; Engl. transl. in *Math. Notes* 3 (1968) 125-128; MR 37 #2695.
[21] G. A. Freiman, Non-coincidence of the Markov and Lagrange spectra (in Russian), *Number-Theoretic Studies in the Markov Spectrum and in the Structural Theory of Set Addition*, ed. G. A. Freiman, A. M. Rubinov, and E. V. Novoselov, Kalinin. Gos. Univ., 1973, pp. 10-15, 121-125; MR 55 #2777.
[22] A. A. Berstein, The connections between the Markov and Lagrange spectra, *Number-Theoretic Studies in the Markov Spectrum and in the Structural Theory of Set Addition*, Kalinin. Gos. Univ., 1973, pp. 16-49, 121-125; MR 55 #2778.
[23] A. A. Berstein, The structure of the Markov spectrum, *Number-Theoretic Studies in the Markov Spectrum and in the Structural Theory of Set Addition*, Kalinin. Gos. Univ., 1973, pp. 50-78, 121-125; MR 55 #2779.
[24] A. D. Pollington and W. Moran (eds.), *Number Theory with an Emphasis on the Markoff Spectrum*, Proc. 1991 Provo conf., Dekker, 1993; MR 93m:11002.

2.32 ド・ブルイン-ニューマン定数

　本書に収集している他の定数とは似ていない定数を議論する．下記の定数 Λ が正であることは，悪名高いリーマン予想【1.6.2】が偽であることと同値である．これはさらに，精密に数値的な限界がコンピュータ計算でできる方法によって定義される．
　Riemann のゼータ関数 $\zeta(z)$ から始めて[2]，

$$\xi(z) = \frac{1}{2}z(z-1)\pi^{-(1/2)z}\Lambda\left(\frac{1}{2}z\right)\zeta(z), \quad \Xi(z) = \xi\left(iz + \frac{1}{2}z\right), \quad z は複素数$$

と定義する．リーマン予想が真であるための必要十分条件は，$\Xi(z)$ の零点がすべて実数であるのは明らかである．予想のいいかえは次のことに役に立つ．
　複素周波数関数，すなわち時間信号のフーリエ余弦変換として $(1/8)\Xi(z/2)$ を考える．信号は

$$\Phi(t) = \sum_{n=1}^{\infty}(2\pi^2 n^4 e^{9t} - 3\pi n^2 e^{5t})\exp(-\pi n^2 e^{4t}), \quad t は実数で，t \geq 0$$

と計算される. 与えられた実数パラメータ λ に対して, 修正した信号 $\Phi(t)\exp(\lambda t^2)$ を考え, これを周波数領域に逆変換する, すなわち, 最初にあった場所にもどす. フーリエ余弦変換 $H_\lambda(z)$ の結果は, 特別な場合として $H_0(z)=(1/8)\Xi(z/2)$ を含む.

固定された λ に対して, H_λ の零点については何がいえるだろうか? de Bruijn (ド・ブルイン) [3]はとりわけ, H_λ は $\lambda \geq 1/2$ に対して実数の零点のみをもつことを証明した. さらに, H_λ が実数の零点のみをもつための必要十分条件は $\lambda \geq \Lambda$ であるような定数 Λ が存在する, ことを Newman (ニューマン) [4]が証明した. もちろん, de Bruijn の結果からただちに $\Lambda \leq 1/2$ が従う. リーマン予想は, 予想 $\Lambda \leq 0$ と同値である. Newman は, もしもリーマン予想が真であるとするならば, それはきわどく成り立つことをうまく強調して, $\Lambda \geq 0$ と予想した.

Λ の下界は明らかに, 関心のある誰にも非常に興味のあるものである. [1,5-8]での精密な計算により, $\Lambda > -0.0991$ を得た. Csordas, Smith & Varga (チョルダス, スミス & ヴァルガ) [9,10]は, Riemann のゼータ関数のある種の近接して連続する零点 (Lehmer の対として知られる) を含んでいる定理を証明した. これはド・ブルイン-ニューマン定数の評価を劇的によくした. 最新の最良の下界は[11,12], $\Lambda > -2.7 \times 10^{-9}$ である. 知るかぎり, Λ の上界 1/2 を改良した進展は何もない.

余談として, リーマン予想と同値な他の判定条件を1つ述べる. 正の整数 n の各々に対して, 級数

$$\lambda_n = \sum_\rho \left[1 - \left(1 - \frac{1}{\rho}\right)^n\right] = \begin{cases} -\dfrac{1}{2}\ln(4\pi) + \dfrac{\gamma_0}{2} + 1 = 0.0230957089\cdots & (n=1 \text{ のとき}) \\ \dfrac{\pi^2}{8} - \ln(4\pi) + \gamma_0 - \gamma_0^2 - 2\gamma_1 + 1 = 0.0923457352\cdots & (n=2 \text{ のとき}) \\ 0.2076389205\cdots & (n=3 \text{ のとき}) \end{cases}$$

を定義する. ただし, 各和は, $\zeta(z)$ の自明でない零点 ρ のすべてにわたる和であり, γ_k は k 次のスティルチェス定数【2.21】である. Li (リー) [13]は, すべての n に対して $\lambda_n \geq 0$ であるための必要十分条件はリーマン予想が真であることを証明した. 関連する定数 σ_n については【2.21】を, 洞察力のある議論は[14,15]を参照せよ.

[1] R. S. Varga, *Scientific Computation on Mathematical Problems and Conjectures*, SIAM, 1990; MR 92b : 65012.

[2] E. C. Titchmarsh, *The Theory of the Riemann Zeta Function*, 2nd ed., rev. by D. R. Heath-Brown, Oxford Univ. Press, 1986, pp. 16, 255; MR 88c : 11049.

[3] N. G. de Bruijn, The roots of trigonometric integrals, *Duke Math.* J. 17 (1950) 197-226; MR 12, 250a.

[4] C. M. Newman, Fourier transforms with only real zeros, *Proc. Amer. Math. Soc.* 61 (1976) 245-251; MR 55 # 7944.

[5] G. Csordas, T. S. Norfolk, and R. S. Varga, A lower bound for the de Bruijn-Newman constant Λ, *Numer. Math.* 52 (1988) 483-497; MR 89m : 30054.

[6] H. J. J. te Riele, A new lower bound for the de Bruijn-Newman constant, *Numer. Math.* 58 (1991) 661-667; MR 92c : 30030.

[7] T. S. Norfolk, A. Ruttan, and R. S. Varga, A lower bound for the de Bruijn-Newman constant

Λ. II, *Progress in Approximation Theory*, Proc. 1990 Tampa conf., ed. A. A. Gonchar and E. B. Saff, Springer-Verlag, 1992, pp. 403-418; MR 94k：30062.
[8] G. Csordas, A. Ruttan, and R. S. Varga, The Laguerre inequalities with applications to a problem associated with the Riemann hypothesis, *Numer. Algorithms* 1 (1991) 305-329; MR 93c：30041.
[9] G. Csordas, W. Smith, and R. S. Varga, Lehmer pairs of zeros and the Riemann ξ-function, *Mathematics of Computation 1943-1993: A Half-Century of Computational Mathematics*, Proc. 1993 Vancouver conf., ed. W. Gautschi, Amer. Math. Soc., 1994, pp. 553-556; MR 96b：11119.
[10] G. Csordas, W. Smith, and R. S. Varga, Lehmer pairs of zeros, the de Bruijn-Newman constant and the Riemann hypothesis, *Constr. Approx.* 10 (1994) 107-129；MR94k：30061.
[11] G. Csordas, A. Odlyzko, W. Smith, and R. S. Varga, A new Lehmer pair of zeros and a new lower bound for the de Bruijn-Newman constant, *Elec. Trans. Numer. Anal.* 1 (1993) 104-111; MR 94k：11098.
[12] A. M. Odlyzko, An improved bound for the de Bruijn-Newmanconstant, *Numer. Algorithms* 25 (2000) 293-303; MR 2002a：30046.
[13] X.-J. Li, The positivity of a sequence of numbers and the Riemann hypothesis, *J. Number Theory* 65 (1997) 325-333; MR 98d：11101.
[14] E. Bombieri and J. C. Lagarias, Complements to Li's criterion for the Riemann hypothesis, *J. Number Theory* 77 (1999) 274-287; MR 2000h：11092.
[15] P. Biane, J. Pitman, and M. Yor, Probability laws related to the Jacobi theta and Riemann zeta functions, and Brownian excursions, *Bull. Amer. Math. Soc.* 38 (2001) 435-465.

2.33　ホール-モンゴメリー定数

　正の整数上で定義された複素数値関数 f が**完全乗法的**（completely multiplicative）であるとは，任意の m, n に対して $f(mn) = f(m)f(n)$ が成り立つ，ときであると定義する．明らかに，このような関数は $1 \cup \{素数\}$ 上の値によって決まる．簡単な例は，$f(n) = 0$, $f(n) = 1$ またはある固定された $r > 0$ に対して $f(n) = n^r$ である．より複雑な例は，固定された奇素数 p に対して定義される Legendre（ルジャンドル）記号

$$f_p(n) = \left(\frac{n}{p}\right)^\dagger = \begin{cases} 0 & (p \mid n \text{ のとき}) \\ 1 & (p \nmid n \text{ で, } n \text{ が法 } p \text{ で平方剰余であるとき}) \\ -1 & (\text{その他の場合}) \end{cases}$$

である．たとえば，$5^2 \equiv 6 \bmod 19$ であるから $(6/19) = 1$ であり，$x^2 \equiv 39 \bmod 47$ は解をもたないから $(39/47) = -1$ である．

　説明するために，集合 $\{1 \leq n \leq N : f_p(n) = 1\}$ の要素の個数を $g(N)$ と定義する．整数 $\{1, 2, 3, \cdots, p-1\}$ のうち，$(p-1)/2$ 個が平方剰余であり，$(p-1)/2$ 個が平方非剰余であることが知られている[1]．したがって，p の倍数として N をとり，$N \to \infty$ とすると $g(N)/N \to 1/2$ である．N の異なる選び方に対する $g(N)/N$ の可能な極限

---

†訳注：これは分数ではないが印刷の都合で (n/p) とも記す．

値を問題とするのは自然なことである．すぐにこの問題に戻る．

閉区間$[-1,1]$に値をとる完全乗法的関数のすべてからなる集合Fを考える．Fにおける関数の平均値は，どのような数になるのだろうか？ より正確には，fはF上を動き，$N \to \infty$とするとき，

$$\mu_N(f) = \frac{1}{N}\sum_{n=1}^{N} f(n)$$

の集積点の集合Γは何であるか？ 集合Γを$[-1,1]$の**乗法スペクトル**（multiplicative spectrum）という．このスペクトルの理解はつい最近達成された．

Granville & Soundararajan（グランビル & サウンダラランジャン）[2,3]は，Hall & Montgomery（ホール&モンゴメリー）らによる独立な仕事を基礎にして[4]，Γは閉区間であることを証明した．実際，

$$\Gamma = [\delta_1, 1] = [-0.6569990137\cdots, 1]$$

である．ただし，$\delta_1 = 2\delta_0 - 1$であり，

$$\delta_0 = 1 - \frac{\pi^2}{6} - \ln(1+\sqrt{e})\ln\left(\frac{e}{1+\sqrt{e}}\right) + 2\mathrm{Li}_2\left(\frac{e}{1+\sqrt{e}}\right) = 0.1715004931\cdots$$

であり，$\mathrm{Li}_2(x)$は2重対数（dilogarithm）関数【1.6.8】である．解析接続により，δ_0の式は簡単に$1 + \pi^2/6 + 2\mathrm{Li}_2(-\sqrt{e})$となる．この注目すべき式は，より広い理論のほんの先端である．さらに多くのことが$\Gamma(S)$についていえる．ただし，Sは（区間$[-1,1]$ではなくて）複素数平面内の単位円盤Dの任意の部分集合である．証明で重要な役割をするのは，遅れをもつ微積分方程式である【5.4】．

$f_p(n)$の特別な場合に戻る．前に述べた定理によって，

$$g(N) - (N - g(N)) \geq (\delta_1 + o(1))N$$

である．すなわち，$g(N) = (\delta_0 + o(1))N$である．いいかえると，法$p$で平方剰余であ
るNを超えない整数の比率は，pの選び方に依存せずに，少なくともδ_0であり，

$$\delta_0 \leq \liminf_{N \to \infty} \frac{g(N)}{N} \leq \frac{1}{2} \leq \limsup_{N \to \infty} \frac{g(N)}{N} \leq 1$$

が成り立つ．これはHeath-Brown（ヒース-ブラウン）[4]の1994年の予想を証明したことになる．さらに，定数δ_0は最良である．実際，無限に多くの素数pに対して，下極限はδ_0と等しい．

同様に，無限の多くの素数pに対して上極限は1と等しい．これを証明する．固定されたNに対して，素数$p \equiv 1 \bmod M$を選ぶ．ただし，Mは$8 \times (N$以下のすべての奇素数の積）である．このことは，等差数列に存在する素数に関する，ディリクレの定理により可能である．したがって，$(2/p)=1$である．qがN以下の奇素数であるならば，平方剰余の相互法則により$(q/p)=(p/q)=(1/q)=1$が成り立つ．任意の$n \leq N$は，素数$\leq N$の積である．よって，$(n/p)=1$である．したがって，すべての$n \leq N$は法pで平方剰余である．もちろん，このことは無限に多くのpについて可能であるから，結果が得られる．

一般化を調べる．正の整数上で定義された複素数値関数fが，mとnが互いに素で

あるときつねに $f(mn)=f(m)f(n)$ を満たすならば，**乗法的** (multiplicative) であるという．(f が完全乗法的であるときは，明らかに，f は乗法的である．)（前のように）すべての n に対して $-1\leq f(n)\leq 1$ であると仮定する．このとき，平均値は存在し，[5-9]

$$\lim_{N\to\infty}\mu_N(f)=\prod_p\left(1-\frac{1}{p}\right)\left(1+\sum_{k=1}^{\infty}\frac{f(p^k)}{p^k}\right)$$

に等しい．ただし，積はすべての素数 p にわたってとる．たとえば，φ をオイラーの約数関数【2.7】とし，$f(n)=\varphi(n)/n$ とする．このとき，$\lim_{N\to\infty}\mu_N(f)=6/\pi^2$ が成り立つ．この例では，任意の $k\geq 1$ に対して $f(p^k)=f(p)$ であることに注意せよ．$\lim_{N\to\infty}\mu_N(f)$ が存在するための複雑な条件は，仮定をすべての n に対して $f(n)\in D$ だけに弱めるときに生ずる．

かなり人工的な例に対応する（関連のない）漸近的結果がある[10]．乗法的関数 f を，漸化式

$$f(n)=\begin{cases}1 & (n=1\text{ のとき})\\ pf(k) & (\text{任意の素数 }p\text{ に対して }n=p^k\text{ のとき})\end{cases}$$

と定義する．このとき，

$$\lim_{N\to\infty}\frac{1}{N^2}\sum_{k=1}^{N}f(n)=\frac{1}{2}\prod_p\left(1-\frac{1}{p^2}+(p-1)\sum_{n=2}^{\infty}\frac{f(n)}{p^{2n}}\right)$$
$$=\frac{1}{2}(0.8351076361\cdots)$$

である．比較のために，【2.2】で定義された**完全加法的** (completely additive) 関数 $\Omega(n)$ は，任意の素数 p に対して $\Omega(p^k)=k\Omega(p)$ を満し，まったく異なる漸近式となる．

[1] K. Ireland and M. Rosen, *A Classical Introduction to Modern Number Theory*, 2nd ed., Springer-Verlag, 1990, pp. 50-65; MR 92e:11001.

[2] A. Granville and K. Soundararajan, Motivating the multiplicative spectrum, *Topics in Number Theory*, Proc. 1997 Penn. State Univ. conf., ed. S. D. Ahlgren, G. E. Andrews, and K. Ono, Kluwer, 1999, pp. 1-15; math.NT/9909190; MR 2000m:11088.

[3] A. Granville and K. Soundararajan, The spectrum of multiplicative functions, *Annals of Math.* 153 (2001) 407-470; MR 2002g:11127.

[4] R. R. Hall, Proof of a conjecture of Heath-Brown concerning quadratic residues, *Proc. Edinburgh Math. Soc.* 39 (1996) 581-588; MR 97m:11119.

[5] E.Wirsing, Das asymptotische Verhalten von Summen über multiplikative Funktionen. II, *Acta Math. Acad. Sci. Hungar.* 18 (1967) 411-467; MR 36 #6366.

[6] G. Halász, Über die Mittelwerte multiplikativer zahlentheoretischer Funktionen, *Acta Math. Acad. Sci. Hungar.* 19 (1968) 365-403; MR 37 #6254.

[7] A. Hildebrand, On Wirsing's mean value theorem for multiplicative functions, *Bull. London Math. Soc.* 18 (1986) 147-152; MR 87f:11075.

[8] P. D. T. A. Elliott, *Probabilistic Number Theory. I: Mean-Value Theorems*, Springer-Verlag, 1979, pp. 225-256; MR 82h:10002a.

[9] A. Granville and K. Soundararajan, Decay of mean-values of multiplicative functions, math. NT/9911246.

[10] R. A. Gillman and R. Tschiersch, The average size of a certain arithmetic function, *Amer. Math. Monthly* 100 (1993) 296-298; numerical calculation of C due to K. Ford.

3 解析学の不等式に関する定数

3.1 シャピロ-ドリンフェルドの定数

巡回和

$$f_n(x_1, x_2, \cdots, x_n) = \frac{x_1}{x_2+x_3} + \frac{x_2}{x_3+x_4} + \cdots + \frac{x_{n-1}}{x_n+x_1} + \frac{x_n}{x_1+x_2}$$

を考えよう.ここで各 x_j は非負で,各分母は正であるとする.Shapiro(シャピロ)[1]はすべての n に対して $f_n(x_1, x_2, \cdots, x_n) \geq n/2$ が成り立つか否かを問題にした.Lighthill(ライトヒル)[2]は $n=20$ に対する反例を与えた.他の反例も $n=14$ [3,4] および $n=25$ [5,6]に対し次々と発見された.巡回和を理解するにあたりその歴史的経緯については[7-9]を参照してほしい.結果を要約すると,シャピロの不等式は,12以下の偶数と23以下の奇数に対しては正しく(計算機に基づく証明[10])それ以外の整数に対しては誤りである.この結果は偶数の場合には解析的に証明されているが,$13 \leq n \leq 23$ の奇数に対してはまだなされていない.

早くからシャピロの不等式を解明するのに数学者が使ってきた道具を吟味するのも興味深い.そのひとつに注目してみよう.

$$f(n) = \inf_{x \geq 0} f_n(x_1, x_2, \cdots, x_n)$$

とおく.Rankin(ランキン)[12]は次のように定義される式

$$\lambda = \lim_{n \to \infty} \frac{f(n)}{n} = \inf_{n \geq 1} \frac{f(n)}{n}$$

を研究し,$\lambda < 0.49999993 < 1/2$ を証明した.これから彼はただちにシャピロの不等式は十分大きいすべての n に対しては成立しないことを導いた.他の研究者は定数 λ そのものに興味をもち,それを精度を上げて計算していった[7].そのような努力は有限

3.1 シャピロ-ドリンフェルドの定数

の n に対するシャピロの不等式の真偽とは関係がないことに注意しよう．しばしばそうであるように，ある人の用いた道具が他の人にとっては研究の対象になる．

Drinfeld（ドリンフェルド）[13]は，λ を任意の精度で計算する手段を与える λ の「幾何学的」な解釈を見つけた．xy 平面上で2つの曲線を考えよう．

$$y=\frac{1}{\exp(x)}, \quad y=\frac{2}{\exp(x)+\exp(x/2)}$$

$\varphi(x)$ をこれら2つの関数の凸支持線としよう．つまり $\varphi(x)$ は2つの曲線を超えない最大の上に凸な関数である（図3.1参照）．このとき

$$\lambda=\frac{\varphi(0)}{2}=0.4945668172\cdots=\frac{1}{2}(0.9891336344\cdots)$$

シャピロ和の多くの変形が研究されてきた[7]．そのうち2つだけに触れておく．まず初めに次のような巡回和を同じ条件の下で考えよう．

$$g_n(x_1, x_2, \cdots, x_n) = \frac{x_1+x_3}{x_1+x_2} + \frac{x_2+x_4}{x_2+x_3} + \cdots + \frac{x_{n-1}+x_1}{x_{n-1}+x_n} + \frac{x_n+x_2}{x_n+x_1}$$

シャピロの不等式と同じように，不等式 $g_n(x_1,x_2,\cdots,x_n) \geq n$ は一般には成り立たない．Elbert（エルバート）[14]は次の式を研究した．

$$\mu = \lim_{n\to\infty}\frac{g(n)}{n}, \quad \text{ここで } g(n) = \inf_{x\geq 0} g_n(x_1,x_2,\cdots,x_n)$$

Drinfeldの方法を使って，彼は $\mu=\psi(0)=0.9780124781\cdots$ であることを発見した．ここで $y=\psi(x)$ は2つの関数

$$y=\frac{1+\exp(x)}{2}, \quad y=\frac{1+\exp(x)}{1+\exp(x/2)}$$

の凸支持線である．λ と μ の最近の計算結果は[15,16]に含まれている．一般化は[17,18]に見られる．また巡回和の差 $\Delta_n = f_n - h_n$ を考えよう．ここで，f_n は前出と同じもので，

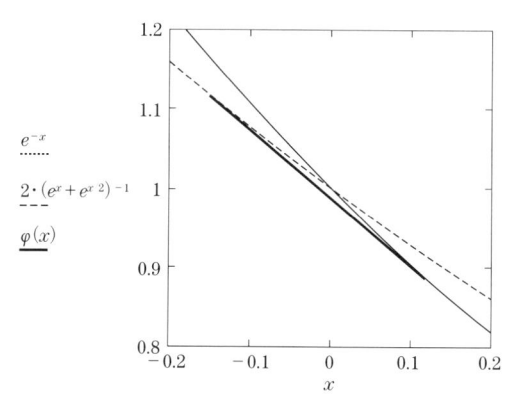

図3.1 $x=0$ の近傍において，$y=\varphi(x)$ のグラフは他の2つの曲線の共通接線となっている．

$$h_n(x_1, x_2, \cdots, x_n) = \frac{x_1}{x_1+x_2} + \frac{x_2}{x_2+x_3} + \cdots + \frac{x_{n-1}}{x_{n-1}+x_n} + \frac{x_n}{x_n+x_1}$$

とする．Gauchman（ガウチマン）[19, 20]は

$$\inf_{n \geq 1} \inf_{x \geq 0} \frac{\Delta_n(x_1 x_2, \cdots, x_n)}{n} = -0.0219875218\cdots$$

を得た．また，対応する2つの曲線は

$$y = \frac{1 - \exp(x/2)}{\exp(x) + \exp(x/2)}, \quad y = \frac{\exp(-x) - 1}{2}$$

である．

Shallit（シャリット）[15, 21]による別の（非巡回）和について触れておこう．

$$s_n(x_1, x_2, \cdots, x_n) = \sum_{i=1}^{n} x_i + \sum_{1 \leq i \leq k \leq n} \prod_{j=i}^{k} \frac{1}{x_j}$$

これについては数値解析的（非幾何学的）手段により

$$\lim_{n \to \infty} \left(\inf_{x > 0} s_n(x_1, x_2, \cdots, x_n) - 3n \right) = -1.3694514039\cdots$$

を満たすことを証明することができる．和 f_n, Δ_n, および s_n の様々な変種が頭に浮かぶ．

3.1.1 ジョコヴィッチの予想

シャピロ予想と同じようにDjokovic（ジョコヴィッチ）予想は"*Monthly* problem"[†]からはじまり，最終的に興味ある定数を生み出すもととなった．$x_1 < x_2 < \cdots < x_n$ を仮定し，

$$P(x_1, x_2, \cdots, x_n) = \frac{1}{M} \int_{x_1}^{x_n} \left(\prod_{k=1}^{n} (t - x_k) \right) dt, \quad \text{ここで } M = \max_{x_1 \leq t \leq x_n} \left| \prod_{k=1}^{n} (t - x_k) \right|$$

と定義しよう．Djokovic（ジョコヴィッチ）[22]は，各 k に対して $(-1)^{n+1-k}(\partial P/\partial x_k) > 0$ であることを予想した．これは現在では，一般的には成立しないことが知られている [23, 24]．$n = 3$ に対してさえ成り立たない．$a_1 = 0.1824878875\cdots$ を3次方程式 $12a^3 - 16a^2 + 8a - 1 = 0$ の一意に存在する実根とし，$a_2 = 1 - a_1 = 0.8175121124\cdots$ としよう．このときジョコヴィッチの不等式は，もし $a_1(x_3 - x_1) < x_2 - x_1 < a_2(x_3 - x_1)$ ならば正しいが，それ以外のときは誤りである．同じように $n \geq 4$ に対して，不等式の妥当性は x たちの分布に依存する．もし x たちが一様に散らばっているならば，そのとき $n \leq 6$ に対して不等式は真であるが，十分大きい n に対しては正しくない．

[1] H. S. Shapiro, Problem 4603, *Amer. Math. Monthly* 61 (1954) 571.
[2] M. J. Lighthill, Note on problem 4603: An invalid inequality, *Amer. Math. Monthly* 63 (1956) 191-192; also Math. Gazette 40 (1956) 266.
[3] A. Zulauf, On a conjecture of L. J. Mordell, *Abh. Math. Sem. Univ. Hamburg* 22 (1958) 240-241;

[†] 訳注：アメリカ数学協会の月刊誌 *American Mathematical Monthly* に毎号掲載されている，読者に解答を求める問題．

MR 23 #A1575.
[4] M. Herschorn and J. E. L. Peck, Partial solution of problem 4603: An invalid inequality, *Amer. Math. Monthly* 67 (1960) 87-88.
[5] D. E. Daykin, Inequalities for functions of a cyclic nature, *J. London Math. Soc.* 3 (1971) 453-462; MR 44 #1622a.
[6] M. A. Malcolm, A note on a conjecture of L. J. Mordell, *Math. Comp.* 25 (1971) 375-377; MR 44 #1622b.
[7] D. S. Mitrinovic, J. E. Pecaric, and A. M. Fink, *Classical and New Inequalities in Analysis*, Kluwer, 1993; MR 94c:00004.
[8] A. Clausing, A review of Shapiro's cyclic inequality, *General Inequalities* 6, Proc. 1990 Oberwolfach conf., ed. W. Walter, Birkhäuser, 1992, pp. 17-31; MR 94g:26030.
[9] P. J. Bushell, Shapiro's cyclic sum, *Bull. London Math. Soc.* 26 (1994) 564-574; MR 96f:26022.
[10] B. A. Troesch, The validity of Shapiro's cyclic inequality, *Math. Comp.* 53 (1989) 657-664; MR 90f:26025.
[11] P. J. Bushell and J. B. McLeod, Shapiro's cyclic inequality for even n, *J. Inequal. Appl.* 7 (2002) 331-348; Univ. of Sussex CMAIA report 2000-08.
[12] R. A. Rankin, An inequality, *Math. Gazette* 42 (1958) 39-40.
[13] V. G. Drinfeld, A certain cyclic inequality (in Russian), *Mat. Zametki* 9 (1971) 113-119; Engl. transl. in *Math. Notes Acad. Sci. USSR* 9 (1971) 68-71; MR 43 #6379.
[14] A. Elbert, On a cyclic inequality, *Period. Math. Hungar.* 4 (1973) 163-168; MR 50 #2424.
[15] A. MacLeod, Three constants, unpublished note (1996).
[16] D. Radcliffe, Calculating Shapiro's constant to 5000 digits, unpublished note (1999).
[17] E. K. Godunova and V. I. Levin, Exactness of a nontrivial estimate in a cyclic inequality (in Russian), *Mat. Zametki* 20 (1976) 203-205; Engl. transl. in *Math. Notes Acad. Sci. USSR* 20 (1976) 673-675; MR 54 #13007.
[18] E. K. Godunova and V. I. Levin, Lower bound for a cyclic sum (in Russian), *Mat. Zametki* 32 (1982) 3-7, 124; Engl. transl. in *Math. Notes Acad. Sci. USSR* 32 (1982) 481-483; MR 84c:26021.
[19] V. C.artoaje, J. Dawson, and H. Volkmer, Solution of problem 10528a: Cyclic sum inequalities, *Amer. Math. Monthly* 105 (1998) 473-474.
[20] H. Gauchman, Solution of problem 10528b: Cyclic sum inequalities, unpublished note (1998).
[21] J. Shallit, C. C. Grosjean, and H. E. De Meyer, Solution of problem 94-15: A minimization problem, *SIAM Rev.* 37 (1995) 451-458.
[22] D. Z. Djokovic, The integral of a normalized polynomial with real roots, *Amer. Math. Monthly* 72 (1965) 794-795; 73 (1966) 788.
[23] G. C. Hu and J. K. Tang, The Djokovic's conjecture on an integral inequality, *J. Math. Res. Expos.* 10 (1990) 271-278; MR 91g:26024.
[24] D. S. Mitrinovic, J. E. Pecaric, and A. M. Fink, *Inequalities Involving Functions and Their Integrals and Derivatives*, Kluwer, 1991; MR 93m:26036.

3.2 カールソン-レヴィンの定数

f を $[0,\infty)$ 上の非負実数値関数としよう．$x^a f(x)^p$ と $x^b f(x)^q$ の積分が存在するという条件が与えられたとき，$f(x)$ の積分値の範囲を決定したい．$a=0$, $b=2$, $p=q=2$ という特別な場合に，Carlson（カールソン）[1-3]は，

$$\int_0^\infty f(x)\,dx \leq \sqrt{\pi}\left(\int_0^\infty f(x)^2 dx\right)^{1/4}\left(\int_0^\infty x^2 f(x)^2 dx\right)^{1/4}$$

および定数 $\sqrt{\pi}$ は最良であることを示した．ここで「最良」とは，$\sqrt{\pi}$ は不等式が成り立つような最小の実係数であるということを意味する．(もし，係数を $\sqrt{\pi}$ よりも小さくして不等式をもっと改良しようとすると，反例となるような所要の条件を満たす関数 f が存在する．)

$p>1$, $q>1$, $\lambda>0$ および $\mu>0$ といった一般の場合に Levin（レヴィン）[2-4]は次の不等式を発見した．

$$\int_0^\infty f(x)\,dx \leq C\left(\int_0^\infty x^{p-1-\lambda} f(x)^p dx\right)^s \left(\int_0^\infty x^{q-1+\mu} f(x)^q dx\right)^t$$

さらに最良定数は

$$C = \frac{1}{(ps)^s}\frac{1}{(qt)^t}\left[\frac{\Gamma\left(\frac{s}{r}\right)\Gamma\left(\frac{t}{r}\right)}{(\lambda+\mu)\,\Gamma\left(\frac{s+t}{r}\right)}\right]^r$$

である．ここで

$$r = 1-s-t, \quad s = \frac{\mu}{p\mu+q\lambda}, \quad t = \frac{\lambda}{p\mu+q\lambda}$$

であり，$\Gamma(x)$ はオイラーのガンマ関数である【1.5.4】．最良の定数に対してこのような厳密な表示が実際に「存在」するというのは興味深い．多くの不等式を，このように完全に決定することはできない．拡張については[5-8]を参照してほしい．

[1] F. Carlson, Une inégalité, *Ark. Mat. Astron. Fysik*, v. 25B (1934) n. 1, 1-5.
[2] E. F. Beckenbach and R. Bellman, *Inequalities*, Springer-Verlag, 1965; MR 33 #236.
[3] D. S. Mitrinovic, J. E. Pecaric, and A. M. Fink, *Inequalities Involving Functions and Their Integrals and Derivatives*, Kluwer, 1991; MR 93m:26036.
[4] V. I. Levin, Exact constants in inequalities of the Carlson type (in Russian), *Dokl. Akad. Nauk SSSR* 59 (1948) 635-638; MR 9,415b.
[5] B. Kjellberg, Ein Momentenproblem, *Ark. Mat. Astron. Fysik*, v. 29A (1943) n. 2, 1-33; MR 6,203a.
[6] B. Kjellberg, A note on an inequality, *Ark. Mat.* 3 (1956) 293-294; MR 17,950a.
[7] V. I. Levin and S. B. Steckin, Inequalities, *Amer. Math. Soc. Transl.* 14 (1960) 1-29; MR 22 #3771.
[8] V. I. Levin and E. K. Godunova, A generalization of Carlson's inequality (in Russian), *Mat. Sbornik* 67 (1965) 643-646; Engl. transl. in *Amer. Math. Soc. Transl.* 86 (1970) 133-136; MR 32 #5824.

3.3 ランダウ-コルモゴロフの定数

関数 f とその導関数 $f^{(k)}$ のノルムを含む不等式については膨大な文献が存在する．ここで4つの別々の場合に出てくるある定数 $C(n,k)$ の定義を述べておこう．これら

の定数は次の不等式に対応している (個々に説明される).
$$\|f^{(k)}\| \leq C(n,k)\|f\|^{1-k/n}\|f^{(n)}\|^{k/n}, \quad 1 \leq k < n$$
これらの不等式を今後「不等式 I」とよぶ.

3.3.1 $L_\infty(0, \infty)$ の場合

$\|f\|$ は, f を $[0, \infty)$ 上で定義された実数値関数とするときの $|f(x)|$ の上限とする. Landau (ランダウ)[1]は, f が2回微分可能で, f と f'' がともに有界であるならば,
$$\|f'\| \leq 2\|f\|^{1/2}\|f''\|^{1/2}$$
が成り立ち, 定数2は最良であることを証明した. この意味は, 任意の正の数 ε に対して 2 を $2-\varepsilon$ に置きかえると必ず反例 f をあげることができるということである.

Schoenberg (シェンベルグ) & Cavaretta (カバレッタ)[2,3]はこの不等式を, f の n 回微分が存在して, f と $f^{(n)}$ がともに有界であるという設定にまで拡張した. 彼らは不等式 I に対する最良定数 $C(n,k)$, $1 \leq k < n$ を決定し, $C(n,k)$ をオイラースプラインのノルムを使って表現した. たとえば,
$$C(3,1) = \left(\frac{243}{8}\right)^{1/3} = 3.14138\cdots^\dagger, \quad C(3,2) = 24^{1/3} = 2.88449\cdots,$$
$$C(4,1) = 4.288\cdots, \quad C(4,2) = 5.750\cdots, \quad C(4,3) = 3.708\cdots$$
すべての n と k に対する具体的な数表示の公式は得られていない[4,5].

3.3.2 $L_\infty(-\infty, \infty)$ の場合

実数値関数 f が $(-\infty, \infty)$ 上で定義されているとき, $|f(x)|$ の上限ノルムを $\|f\|$ としよう. Hadamard (アダマール)[6]は, f が2回微分可能で f と f'' が有界ならば,
$$\|f'\| \leq \sqrt{2}\|f\|^{1/2}\|f''\|^{1/2}$$
であり, かつ定数 $\sqrt{2}$ は最良であることを証明した.

Kolmogorov (コルモゴロフ)[7]は, Favard (ファヴァール) 定数【4.3】を使って不等式 I に対する最良定数 $C(n,k)$, $1 \leq k \leq n$ を次のように決定した.
$$C(n,k) = a_{n-k} a_n^{-1+k/n}, \quad \text{ここで } a_n = \frac{4}{\pi} \sum_{j=0}^{\infty} \left[\frac{(-1)^j}{2j+1}\right]^{n+1}$$
これらの公式は Shilov (シロフ)[8]によって発見された特別な場合を含む.
$$C(3,1) = \left(\frac{9}{8}\right)^{1/3}, \quad C(3,2) = 3^{1/3},$$
$$C(4,1) = \left(\frac{512}{375}\right)^{1/4}, \quad C(4,2) = \left(\frac{6}{5}\right)^{1/2}, \quad C(4,3) = \left(\frac{24}{5}\right)^{1/4},$$
$$C(5,1) = \left(\frac{1953125}{1572864}\right)^{1/5}, \quad C(5,2) = \left(\frac{125}{72}\right)^{1/5}$$
全直線上の関数に関する場合は, 先の半直線上の関数に関する場合よりもやさしいこと

---

†訳注:原著は $\left(\frac{243}{8}\right)^{1/3} = 4.35622\cdots$ となっているが, 数値が合わない. 他の文献により, 新たに計算して修正した.

を見てほしい[4,5].

3.3.3 $L_2(-\infty,\infty)$ の場合

$(-\infty,\infty)$ 上で定義された実数値関数 f に対して
$$\|f\|=\left(\int_{-\infty}^{\infty}f(x)^2dx\right)^{1/2}$$
と定義する．Hardy, Littlewood & Pólya (ハーディ，リトルウッド&ポリヤ) [9]は，f の n 回微分が存在し，f と $f^{(n)}$ がともに2乗可積分であることを仮定したとき，$C(n,k)=1$ が $1\leq k<n$ に対して最良であることを証明した．

3.3.4 $L_2(0,\infty)$ の場合

前述と同じように，半直線の場合は対応する全直線の場合よりも難しい．$(0,\infty)$ 上で定義された実数値関数 f に対して
$$\|f\|=\left(\int_{0}^{\infty}f(x)^2dx\right)^{1/2}$$
と定義しよう．Hardy, Littlewood & Pólya[9]は，f が2回微分可能で，f と f'' がともに2乗可積分であると仮定したとき，
$$\|f'\|\leq\sqrt{2}\|f\|^{1/2}\|f''\|^{1/2}$$
であり，定数 $\sqrt{2}$ が最良であることを証明した．

Ljubic (リュビック) [10]と Kupcov (クプツォフ) [11]はこの不等式を不等式 I にまで拡張し，ある明示的に決まる多項式の零点を使って最良定数 $C(n,k)$ を発見するための注目すべきアルゴリズムを与えた．たとえば[12,13]，
$$C(3,1)=C(3,2)=3^{1/2}[2(2^{1/2}-1)]^{-1/3}=1.84420\cdots$$
$$C(4,1)=C(4,3)=\left[\frac{1}{a}(3^{1/4}+3^{-3/4})\right]^{1/2}=2.27432\cdots$$
$$C(4,2)=\left(\frac{2}{b}\right)^{1/2}=2.97963\cdots$$
ここで，a は $x^8-6x^4-8x^2+1=0$ の最小の正の根，b は $x^4-2x^2-4x+1=0$ の最小の正の根である．また
$$C(5,1)=C(5,4)=2.70247\cdots,\quad C(5,2)=C(5,3)=4.37800\cdots$$

$k=1$ という特別な場合，
$$C(n,1)=\left[\frac{(n-1)^{1/n}+(n-1)^{-1+1/n}}{c}\right]^{1/2}$$
が証明できる．ここで，c は次の等式を満たす最小の正の根である．
$$\int_0^c\int_0^\infty\frac{1}{(x^{2n}-yx^2+1)\sqrt{y}}dxdy=\frac{\pi^2}{2n}$$

$k>1$ に対する類似の公式は知られていない．Ljubic と Kupcov の研究の1つの結果は，この場合に対するすべての $C(n,k)$ が代数的数でなければならないことである．この主張は $L_\infty(0,\infty)$ の場合にも同じく正しいように見える．

ここでとり上げなかった話題には次のようなものがある．
- $p \neq 2$ かつ $p \neq \infty$ であるとき，$L_p(0, \infty)$，$L_p(-\infty, \infty)$ に関連した最良定数，あるいは有限区間上で考えた場合の対応する最良定数[14,15]．
- 離散的な場合，とくに微分を差分で置きかえたとき，l^p ノルムをもつ片側あるいは両側無限の実数列に関連する最良定数[16,17]．

$p = 1, 2, \infty$ は最良定数が正確な公式で表現できる数少ない場合である．他のすべての値 p に対しては数値的近似法が明らかに必要である．

ここで $L_2(0, \infty)$ のわずかな変形に関連する未解決の問題がある．f は 2 回微分可能で，f と f'' が荷重関数 $\omega(x) = x$ に関して 2 乗可積分であることを仮定したとき，Everitt & Guinand（エベリット&グィナンド）[5,18]は次を証明した．

$$\left(\int_0^\infty x f'(x)^2 dx \right)^2 \leq K \cdot \int_0^\infty x f(x)^2 dx \cdot \int_0^\infty x f''(x)^2 dx$$

ここで，最良定数は $2.35070 < K < 2.35075$ である．K の正確な表現は未解決である．

[1] E. Landau, Einige Ungleichungen für zweimal differenzierbare Funktionen, *Proc. London Math. Soc.* 13 (1914) 43-49.
[2] I. J. Schoenberg, The elementary cases of Landau's problem of inequalities between derivatives, *Amer. Math. Monthly* 80 (1973) 121-158; MR 47 #3619.
[3] I. J. Schoenberg and A. Cavaretta, *Solution of Landau's Problem Concerning Higher Derivatives on the Halfline*, MRC TSR 1050 (1970), Univ. of Wisconsin; available through the National Technical Information Service, Springfield VA 22161; MR 51 #5868.
[4] M. K. Kwong and A. Zettl, *Norm Inequalities for Derivatives and Differences*, Lect. Notes in Math 1536, Springer-Verlag, 1992; MR 94f:26011.
[5] D. S. Mitrinovic, J. E. Pecaric, and A. M. Fink, *Inequalities Involving Functions and Their Integrals and Derivatives*, Kluwer, 1991; MR 93m:26036.
[6] J. Hadamard, Sur le module maximum d'une fonctioin et de ses dérivées, *Bull. Soc. Math. de France* 42 (1914) C. R. Séances, pp. 68-72; also in Oeuvres, t. 1, Centre Nat. Recherche Sci., 1968, pp. 379-382.
[7] A. N. Kolmogorov, On inequalities between upper bounds of consecutive derivatives of an arbitrary function defined on an infinite interval (in Russian), *Uchenye Zapiski Moskov. Gos. Univ. Matematika* 30 (1939) 3-16; Engl. transl. in *Amer. Math. Soc. Transl.* 2 (1962) 233-242; also in *Selected Works*, v. 1-, ed. V. M. Tikhomirov, Kluwer, 1991, pp. 277-290; MR 1,298c.
[8] G. E. Shilov, On inequalities between derivatives (in Russian), *Sbornik Rabot Studencheskikh Nauchnykh Kruzhov Moskovskogo Gosudarstvennogo Universiteta* (1937) 17-27.
[9] G. H. Hardy, J. E. Littlewood, and G. Pólya, *Inequalities*, Cambridge Univ. Press, 1934; MR 89d:26016.
[10] J. I. Ljubic, On inequalities between the powers of a linear operator (in Russian), *Izv. Akad. Nauk SSSR Ser. Mat.* 24 (1960) 825-864; Engl. transl. in *Amer. Math. Soc. Transl.* 40 (1964) 39-84; MR 24 #A436.
[11] N. P. Kupcov, Kolmogorov estimates for derivatives in $L_2(0, \infty)$ (in Russian), *Trudy Mat. Inst. Steklov.* 138 (1975) 94-117, 199; Engl. transl. in *Proc. Steklov Inst. Math.* 138 (1975) 101-125; MR 52 #14198.
[12] B. Neta, On determination of best possible constants in integral inequalities involving derivatives, *Math. Comp.* 35 (1980) 1191-1193; MR 81m:26014.
[13] Z. M. Franco, H. G. Kaper, M. K. Kwong, and A. Zettl, Best constants in norm inequalities for derivatives on a half line, *Proc. Royal Soc. Edinburgh* 100A (1985) 67-84; MR 87a:47050.

[14] Z. M. Franco, H. G. Kaper, M. K. Kwong, and A. Zettl, Bounds for the best constant in Landau's inequality on the line, *Proc. Royal Soc. Edinburgh* 95A (1983) 257-262; MR 85m:26018.
[15] B.-O. Eriksson, Some best constants in the Landau inequality on a finite interval, *J. Approx. Theory* 94 (1998) 420-454; MR 99f:41013.
[16] Z. Ditzian, Discrete and shift Kolmogorov type inequalities, *Proc. Royal Soc. Edinburgh* 93A (1983) 307-317; MR 84m:47038.
[17] H. Kaper and B. E. Spellman, Best constants in norminequalities for the difference operator, *Trans. Amer. Math. Soc.* 299 (1987) 351-372; MR 88d:39012.
[18] W. N. Everitt and A. P. Guinand, On a Hardy-Littlewood type integral inequality with a monotonic weight function, *General Inequalities* 5, Proc. 1986 Oberwolfach conf., ed. W. Walter, Birkhäuser, 1987, pp. 29-63; MR 91g:26023.

3.4 ヒルベルト定数

$p>1$ および $q=p/(p-1)$ としよう．もし $\{a_n\}$, $\{b_n\}$ が非負数列で，$f(x)$, $g(x)$ が非負な可積分関数ならば，級数に対する Hilbert（ヒルベルト）の不等式[1-3]とは，すべての a_n が零であるか，あるいはすべての b_n が零である場合を除き，

$$\sum_{m=1}^{\infty}\sum_{n=1}^{\infty}\frac{a_m b_n}{m+n} < \pi \csc\left(\frac{\pi}{p}\right)\left(\sum_{m=1}^{\infty}a_m^p\right)^{1/p}\left(\sum_{n=1}^{\infty}b_n^q\right)^{1/q}$$

が成り立つことをいう[†]．また積分に対するヒルベルトの不等式とは，f が恒等的に零であるか，あるいは g が恒等的に零である場合を除き，

$$\int_0^{\infty}\int_0^{\infty}\frac{f(x)g(y)}{x+y}dxdy < \pi\csc\left(\frac{\pi}{p}\right)\left(\int_0^{\infty}f(x)^p dx\right)^{1/p}\left(\int_0^{\infty}g(y)^q dy\right)^{1/q}$$

が成り立つことをいう．定数 $\pi\csc(\pi/p)$ はそれをより小さな定数で置きかえると反例が存在するという意味で最良である．

ヒルベルトの不等式を次のような2つのパラメータをもつ場合に拡張することを考えてみよう．$p>1$, $q>1$ および

$$\frac{1}{p}+\frac{1}{q}\geq 1, \quad \text{したがって } 0<\lambda=2-\frac{1}{p}-\frac{1}{q}\leq 1$$

とおく．Levin（レヴィン）[4], Steckin（ステキン）[5], および Bonsall（ボンサール）[6]は次を示した．

$$\sum_{m=1}^{\infty}\sum_{n=1}^{\infty}\frac{a_m b_n}{(m+n)^{\lambda}} \leq \left[\pi\csc\left(\frac{\pi(q-1)}{\lambda q}\right)\right]^{\lambda}\left(\sum_{m=1}^{\infty}a_m^p\right)^{1/p}\left(\sum_{n=1}^{\infty}b_n^q\right)^{1/q},$$

$$\int_0^{\infty}\int_0^{\infty}\frac{f(x)g(y)}{(x+y)^{\lambda}}dxdy \leq \left[\pi\csc\left(\frac{\pi(q-1)}{\lambda q}\right)\right]^{\lambda}\left(\int_0^{\infty}f(x)^p dx\right)^{1/p}\left(\int_0^{\infty}g(y)^q dy\right)^{1/q}$$

しかしこの定数が最良であるか否かは知られていない．

最後の点についてはいくらか混乱があるように思われる．Boas（ボアス）[7]は，

[†]訳注：csc は cosec（sin の逆数）の略．

1949 年に Steckin が離散的な場合にその定数が最良であることを証明していると指摘した．1950 年に Boas は自ら訂正し，上界が厳密「ではない」と述べた．Mitrinovic, Pecaric & Fink（ミトリノビッチ，ペツァリツ&フィンク）[1]は，Steckin がその定数が最良であることを確認したと述べた．しかしながら，Levin & Steckin[8]と Walker（ウォーカー）[9]は共に，その問題は依然未解決であると述べた．

理解している限りでは，$\lambda=1/2$ および $p=q=4/3$ に対してさえ最良であることを何人も確かめていない．Copson-de Bruijn（コプソン-ド・ブルイン）定数【3.5】を議論するときと似た計算が可能であろうか？

[1] D. S. Mitrinovic, J. E. Pecaric, and A. M. Fink, *Inequalities Involving Functions and Their Integrals and Derivatives*, Kluwer, 1991; MR 93m:26036.

[2] G. H. Hardy, J. E. Littlewood, and G. Pólya, *Inequalities*, Cambridge Univ. Press, 1934; MR 89d:26016.

[3] K. Oleszkiewicz, An elementary proof of Hilbert's inequality, *Amer. Math. Monthly* 100 (1993) 276-280; MR 94a:51032.

[4] V. I. Levin, On the two-parameter extension and analogue of Hilbert's inequality, *J. London Math. Soc.* 11 (1936) 119-124.

[5] S. B. Steckin, On positive bilinear forms (in Russian), *Dokl. Akad. Nauk SSSR* 65 (1949) 17-20; MR 10,515e and MR 11,870 errata/addenda.

[6] F. F. Bonsall, Inequalities with non-conjugate parameters, *Quart. J. Math.* 2 (1951) 135-150; MR 12,807e.

[7] R. P. Boas, Review of "On positive bilinear forms," MR 10,515e, errata in MR 11,870.

[8] V. I. Levin and S. B. Steckin, Inequalities, *Amer. Math. Soc. Transl.* 14 (1960) 1-29; MR 22 # 3771.

[9] P. L. Walker, A note on an inequality with non-conjugate parameters, *Proc. Edinburgh Math. Soc.* 18 (1973) 293-294; MR 48 # 8723.

3.5 コプソン-ド・ブルイン定数

級数と積分の間の関係はときには非常に自然であるが，またときにはそうでない．$\{a_n\}$ を非負数列，そして $f(x)$ は非負可積分関数としよう．

$$A_n=\sum_{k=1}^{n}a_k, \quad B_n=\sum_{k=n}^{\infty}a_k$$

$$F(x)=\int_0^x f(t)\,dt, \quad G(x)=\int_x^{\infty}f(t)\,dt$$

と定義する．ここでは考察の対象となるすべての無限級数と広義積分は収束して有限な値をとるものと仮定する．2 つの例を調べてみよう．最初の例はすべて期待通りであるが，2 番目の例ではそうならない．$p>1$ のとき，Hardy（ハーディ）の不等式は次の形をとる．a_n のすべてが零でない限り

$$\sum_{n=1}^{\infty}\left(\frac{A_n}{n}\right)^p < \left(\frac{p}{p-1}\right)^p \sum_{n=1}^{\infty} a_n^p$$

が常に成り立つ．積分に対する定理は，f が恒等的に零でない限り

$$\int_0^{\infty}\left(\frac{F(x)}{x}\right)^p dx < \left(\frac{p}{p-1}\right)^p \int_0^{\infty} f(x)^p dx$$

が常に成り立つという形になる．定数 $(p/(p-1))^p$ はそれをより小さな定数で置きかえると反例となる $\{a_n\}$ および $f(x)$ があるという意味で最良である．

$0<p<1$ としよう．コプソンの不等式[2,3]の1つは次の形である．f が恒等的に零でない限り

$$\int_0^{\infty}\left(\frac{G(x)}{x}\right)^p dx > \left(\frac{p}{1-p}\right)^p \int_0^{\infty} f(x)^p dx$$

級数の対応する定理は，奇妙ではあるが，すべての a_n が零でない限り，

$$\left(1+\frac{1}{p-1}\right)\left(\frac{B_1}{1}\right)^p + \sum_{n=2}^{\infty}\left(\frac{B_n}{n}\right)^p > \left(\frac{p}{1-p}\right)^p \sum_{n=1}^{\infty} a_n^p$$

が成り立つという形をとる．Elliott（エリオット）によって発見されたようにここの定数は最良である．驚くべきものは，対応物に到達するに必要な修正項（あるいは[2]で記述されるような注釈[†]）である．

修正項を除けば，次の不等式が現れる[2,4]．すべての a_n が零でない限り，

$$\sum_{n=1}^{\infty}\left(\frac{B_n}{n}\right)^p > p^p \sum_{n=1}^{\infty} a_n^p$$

しかしながら定数 p^p は最良「ではない」．それゆえ，「注釈」を取り除くと不等式の精度を台無しにすることになる．

Levin & Steckin[5]は，$0<p<1/3$ に対して定数 $(p/(1-p))^p$ が最良の定数であることを証明したが，$p>1/3$ に対しては同じようにはいかなかった．

$p=1/2$ という特別な場合を考える．

$$\sum_{n=1}^{\infty}\left(\frac{a_n+a_{n+1}+a_{n+3}+\cdots}{n}\right)^{1/2} \geq C \sum_{n=1}^{\infty} a_n^{1/2}$$

そして a_n を a_n^2 に置きかえて

$$\sum_{n=1}^{\infty} a_n \leq c \sum_{n=1}^{\infty}\left(\frac{a_n^2+a_{n+1}^2+a_{n+3}^2+\cdots}{n}\right)^{1/2}$$

を考える．Steckin[6]は $c \leq 2/\sqrt{3}$ であることを証明し，Boas & de Bruijn[7]はこれを $1.08<c<17/15$ に改良した．より正確に c を評価するために，de Bruijn[8]は複素数列を帰納的に

$$u_1=x, \quad u_n=n^{-1/2}x+(u_{n-1}^2-1)^{1/2} \quad (n \geq 2)$$

と定義した．$c=1.1064957714\cdots$ は，$x \geq c$ ならばすべての $n \geq 1$ に対して $u_n \geq 1$（とくに $\text{Im}(u_m)=0$）が成り立つような最小の実数である．さらに，$x \geq c$ ならば

[†] 訳注：原文は「gloss」，特に古典・古写本などの欄外注の意味．

$$\lim_{n\to\infty} n^{-1/2} u_n = \begin{cases} x + (x^2-1)^{1/2} & (x>c \text{ のとき}) \\ c - (c^2-1)^{1/2} & (x=c \text{ のとき}) \end{cases}$$

が成り立つ．de Bruijn の方法が $p>1/3$ の他の値に対して応用できるかどうかは未解決である．

[1] D. S. Mitrinovic, J. E. Pecaric, and A. M. Fink, *Inequalities Involving Functions and Their Integrals and Derivatives*, Kluwer, 1991; MR 93m:26036.
[2] G. H. Hardy, J. E. Littlewood, and G. Pólya, *Inequalities*, Cambridge Univ. Press, 1934, Thms. 337, 338, 345; MR 89d:26016.
[3] E. T. Copson, Some integral inequalities, *Proc. Royal Soc. Edinburgh* 75A (1975/76) 157-164, Thm. 4; MR 56 # 559.
[4] E. T. Copson, Note on series of positive terms, *J. London Math. Soc.* 3 (1928) 49-51, Thm. 2.3.
[5] V. I. Levin and S. B. Steckin, Inequalities, *Amer. Math. Soc. Transl.* 14 (1960) 1-29; MR 22 # 3771.
[6] S. B. Steckin, On absolute convergence of orthogonal series, I (in Russian), *Mat. Sbornik* 29 (1951) 225-232; MR 13, 229g.
[7] R. P. Boas and N. G. de Bruijn, Solution for problem 83, *Wiskundige Opgaven met de Oplossingen* 20 (1957) 2-4.
[8] N. G. de Bruijn, *Asymptotic Methods in Analysis*, Dover, 1981; MR 83m:41028.

3.6 ソボレフの等周定数

周囲の長さが P である単純閉曲線 C で囲まれる部分の面積 A は $4\pi A \leq P^2$ を満たし，等号が成り立つための必要十分条件は C が円であることである．この**等周問題**をまず 2 次元から n 次元に拡張し，それからそれをある Sobolev（**ソボレフ**）**不等式**に結びつけよう．

Ω を，ユークリッド空間 R^n の区分的に連続的微分可能な境界をもち，その曲面の面積が S である有界な連結開集合の閉包としよう．f を，ある球の外では恒等的に $f=0$ を満たすという意味でコンパクトな台をもつ R^n 上連続的微分可能な関数，また ∇f を f の勾配 (gradient) としよう．R^n の単位球面で囲まれた部分の体積を $\omega_n = \pi^{n/2} \Gamma(n/2+1)^{-1}$ と書くことにする．次の 2 つの命題は同値である[1-4]．

・Ω の体積 V は $n^n \omega_n V^{n-1} \leq S^n$ を満たし，等号が成立する必要十分条件は Ω が球のときである．

・f の $L_{n/(n-1)}$ ノルムは f の勾配ベクトルの L_1 ノルムと

$$\left(\int_{R^n} |f(x)|^{n/(n-1)} dx\right)^{(n-1)/n} \leq \frac{1}{n\omega_n^{1/n}} \int_{R^n} |\nabla f(x)| dx$$

によって結び付けられ，定数 $n^{-1} \omega_n^{-1/n}$ は最良である．

前者はまったく幾何学的である，しかし後者は関数解析学の範囲に入る．1 つの結果として，「等周問題」という語句を，ソボレフ不等式，そしてそれゆえ微分方程式の境

界値問題の固有値を包み込むように拡張した解釈がある．われわれがこのような大きな分野[5-7]を要約することなど望むべくもないが，若干の定数を導入することだけを試みてみよう．

幾人かの著者[8,9]はソボレフ不等式が不確定性原理の役割を果たすと解説している．関数 f の勾配ベクトルの大きさは f の大きさにより下から評価される．定数 ω_n はそれ自体興味のある数である．たとえば，スターリングの公式より $\lim_{n\to\infty} n^{1/2} \omega_n^{1/n} = \sqrt{2\pi e} = 4.1327313541\cdots$ である．物理学から4つの練習問題を例題としてとり上げる．

3.6.1 弦の不等式

もし滑らかな関数 f を $f(0)=f(1)=0$ を満たすように拘束したならば，そのとき
$$\int_0^1 f(x)^2 dx \leq \frac{1}{\pi^2} \int_0^1 \left(\frac{df}{dx}\right)^2 dx$$
であり，定数 $1/\pi^2 = 0.1013211836\cdots$ は最良である[10]．これは，変分学によれば，次の常微分方程式（ODEと略す）の最小の固有値が $\lambda=1$ である事実に対応している．
$$\frac{d^2 g}{dx^2} + \lambda g(x) = 0, \quad g(0) = g(\pi) = 0$$
この ODE は，x 軸上にピンと張られ，両端点が固定された振動する一様な弦の研究から出てくる[11,12]．$\lambda=1$ という値は，弦が弾かれたときに聞こえる音の基本振動として物理学的な解釈ができる．

これの一般化は Talenti（タレンティ）[3]による．
$$\left(\int_0^1 |f(x)|^q dx\right)^{1/q} \leq \frac{q}{2}\left(1+\frac{r}{q}\right)^{1/p}\left(1+\frac{q}{r}\right)^{-1/q} \frac{\Gamma\left(\frac{1}{q}+\frac{1}{r}\right)}{\Gamma\left(\frac{1}{q}\right)\Gamma\left(\frac{1}{r}\right)} \left(\int_0^1 \left|\frac{df}{dx}\right|^p dx\right)^{1/p}$$
ここで $f(0)=f(1)=0$, $p>1$, $q\geq 1$, かつ $r=p/(p-1)$ である．ここに示された定数は最良である．

3.6.2 棒の不等式

「弦の不等式」の2階版がある．適当に滑らかな f が
$$f(0) = \frac{df}{dx}(0) = f(1) = \frac{df}{dx}(1) = 0$$
を満たすならば，
$$\int_0^1 f(x)^2 dx \leq \mu \int_0^1 \left(\frac{d^2 f}{dx^2}\right)^2 dx$$
ここで $\mu=1/\theta^4=0.0019977469\cdots$ であって，$\theta=4.7300407448\cdots$ は方程式
$$\cos(\theta)\cosh(\theta) = 1$$
の最小の正の根である．さらに定数 μ は最良である[12-14]．これは ODE
$$\frac{d^4 g}{dx^4} - \lambda g(x) = 0, \quad g(0) = \frac{dg}{dx}(0) = g(\pi) = \frac{dg}{dx}(\pi) = 0$$

の最小の固有値が $\theta^4/\pi^4 = 5.1387801326\cdots$ であるという事実に対応している．この ODE は，両端が固定された振動する均質な棒や線の研究から出てきた．

3.6.3 膜の不等式

「弦の不等式」の 2 次元版は次のようになる．滑らかな f が単位円板 D の境界で零になるように設定するならば，

$$\int_D f^2 dxdy \leq \mu \int_D \left[\left(\frac{\partial f}{\partial x}\right)^2 + \left(\frac{\partial f}{\partial y}\right)^2\right] dxdy$$

ここで $\mu = 1/\theta^2 = 0.1729150690\cdots$ であって，$\theta = 2.4048255576\cdots$ は 0 次のベッセル関数

$$J_0(z) = \sum_{j=0}^{\infty} \frac{(-1)^j}{(j!)^2} \left(\frac{z}{2}\right)^{2j}$$

の最小の正の零点である．さらに定数 μ は最良である[11,12,15]．これは ODE

$$r^2 \frac{d^2 g}{dr^2} + r \frac{dg}{dr} + \lambda r^2 g(r) = 0, \quad g(0) = 1, \quad g(1) = 0$$

の最小の固有値が $\lambda = \theta^2 = 5.7831859629\cdots$ であるという事実に対応している．この ODE は D 上で一様にピンと張られ，そして境界 C で固定された振動する一様な膜の研究から出てきた．値 $\lambda = \theta^2$ はティンパニを叩いたとき耳にする音の基本振動数である．

面積が一定 A の任意領域 D 上に張られた振動膜で，境界 C で $u = 0$ のラプラスの偏微分方程式（PDE と略す）

$$\frac{\partial^2 u}{\partial x^2} + \frac{\partial^2 u}{\partial y^2} + \Lambda u = 0$$

を考えよう．Rayleigh（レイリー）[16,17]は 1877 年に第 1 固有値 Λ が最小になるのは境界 C が円のときであると予想した．この予想は 1923 年に Faber[18]と Krahn(クラン) [19]によって独立に証明された．$\Lambda \geq (\pi/A)\theta^2$ で等号が成立するのは C が円のときに限る．興味あることに同じことは第 2 固有値に対しては「正しくない」．その場合の臨界境界は円ではなく，8 字形である．

3.6.4 板の不等式

「弦の不等式」の 2 次元の，2 階版は次のとおりである．適当に滑らかな f と外向き法線方向微分 $\partial f/\partial n$ がともに単位円板 D の境界 C 上で 0 になるようにとったと仮定しよう．このとき

$$\int_D f^2 dxdy \leq \mu \int_D \left(\frac{\partial^2 f}{\partial x^2} + \frac{\partial^2 f}{\partial y^2}\right)^2 dxdy$$

ここで $\mu = 1/\theta^4 = 0.0095819302\cdots$ で，$\theta = 3.1962206165\cdots$ は方程式

$$J_0(\theta) I_1(\theta) + I_0(\theta) J_1(\theta) = 0$$

の最小の正の根であり，また $I_0(z)$ は次のように定義される 0 次の変形ベッセル関数である．

$$I_0(z) = \sum_{j=0}^{\infty} \frac{1}{(j!)^2}\left(\frac{z}{2}\right)^{2j}, \quad I_1(z) = \frac{dI_0}{dz}, \quad J_1(z) = -\frac{dJ_0}{dz}$$

さらに，定数 μ は最良である[12,14-16,23]．これは境界 C にしっかりと固定され，振動する均質な板の問題の研究に関係している．

膜の場合と同じようにそれに関連する等周問題を述べてみよう．$u = \partial u/\partial n = 0$ の下で，一定面積 A の任意領域上の振動板に対する PDE

$$\frac{\partial^2}{\partial x^2}\left(\frac{\partial^2 u}{\partial x^2} + \frac{\partial^2 u}{\partial y^2}\right) + \frac{\partial^2}{\partial y^2}\left(\frac{\partial^2 u}{\partial x^2} + \frac{\partial^2 u}{\partial y^2}\right) - \Lambda u = 0$$

を考えよう．Rayleigh[16]は $\Lambda \geq (\pi^2/A^2)\theta^2$ を予想し，Szegö（セゲー）[24-26]はこれが特別な仮定の下で正しいことを証明した．一般的な予想はつい最近証明されたばかりである[27,28]．

3.6.5 その他の変形

$\|f\|$ を $|f(x,y)|$ の上限とする．ただし関数 f は R^2 全体で定義され，2回連続的微分可能とする．そのとき $\|f\|$ は，f のすべての偏微分の平方和の積分と次の不等式で関連づけられる．

$$\|f\| \leq \alpha_{2,2} \Bigl[\int_{R^2}(f^2 + f_x^2 + f_y^2 + f_{xx}^2 + f_{xy}^2 + f_{yy}^2)\,dx dy\Bigr]^{1/2}$$

ここで最良の定数 $\alpha_{2,2} = 0.3187590609\cdots$ が[29]によって与えられている．

$$\alpha_{2,2} = \Bigl(\frac{1}{\pi^2}\int_0^\infty\int_0^\infty \frac{dx dy}{1 + x^2 + y^2 + x^4 + x^2 y^2 + y^4}\Bigr)^{1/2} = \Bigl(\frac{1}{2\pi}\int_0^\infty \frac{dt}{\sqrt{t^2 + 2\sqrt{t^2+3}}}\Bigr)^{1/2}$$

このような公式は $\alpha_{m,n}$ を用いて R^n 全体で定義された m 回連続的に微分可能な関数 f に拡張される．たとえば

$$\alpha_{1,1} = \Bigl(\frac{1}{\pi}\int_0^\infty \frac{dx}{1+x^2}\Bigr)^{1/2} = \frac{\sqrt{2}}{2}, \quad \alpha_{2,3} = 0.23152\cdots, \quad \alpha_{3,3} = 0.142892\cdots$$

もしその代わりに f が R^n の単位立方体上でのみ定義されているなら，対応する定数 $\tilde{\alpha}_{m,n}$ について，われわれは次の結果[30-32]を知っている．

$$\tilde{\alpha}_{1,1} = \tanh(1)^{-1/2} = 1.1458775176\cdots, \quad \tilde{\alpha}_{2,2} = 1.24796\cdots$$

実際，任意の $m \geq 1$ に対して，

$$\alpha_{m,1} = \left[\frac{1}{m+1}\frac{\cos\bigl(\frac{\pi}{2m+2}\bigr)}{\sin\bigl(\frac{3\pi}{2m+2}\bigr)}\right]^{1/2}, \quad \tilde{\alpha}_{m,1} = \left[\frac{2}{m+1}\sum_{k=1}^m \frac{\sin\bigl(\frac{\pi k}{m+1}\bigr)^3}{\tanh\bigl(\sin\bigl(\frac{\pi k}{m+1}\bigr)\bigr)}\right]^{1/2}$$

これらの不等式は数値解析における有限要素法の研究で役に立つ．

関連するアイデアは Friedrichs（フリードリックス）の不等式[33]で，実軸上の閉区間 $[0,1]$ 上の連続的微分可能な関数 f に関連したものである．

$$\Bigl[\int_0^1 (f(x)^2 + f'(x)^2)\,dx\Bigr]^{1/2} \leq \beta \Bigl[f(0)^2 + f(1)^2 + \int_0^1 f'(x)^2 dx\Bigr]^{1/2}$$

最良定数 $\beta = 1.0786902162\cdots$ は $\beta = \sqrt{1 + \theta^{-2}}$ を満たす．ただし $\theta = 2.4725480752\cdots$ は

方程式
$$\cos(\theta) - \theta(\theta^2+1)^{-1}\sin(\theta) = -1 \quad (0 < \theta < \pi)$$
の一意的な解である．さらにたくさんの例が考えられる[34-45]．

もう1つ大きな問題に対して幾何学に戻ってみよう．周囲の長さ P の R^3 における単純閉曲線 C を考えよう．V をその凸包，すなわち，C を含むすべての R^3 の凸集合の共通部分の体積とする．このとき $V \leq \gamma_3 P^3$ の最良定数は $\gamma_3 = 0.0031816877\cdots$ である（ある常微分方程式系の数値解を媒介にして[46,47]に含まれる）．γ_3 に対するどのようなまとまった形の表現も知られていない．もし問題設定が R^3 から R^n（整数 n は偶数とする）に変更されたならば，奇妙なことに最良定数は明確に $\gamma_n = [(\pi n)^{n/2} n!(n/2)!]^{-1}$ で与えられる[48]．$n \geq 5$ が奇数の場合は未解決である．

リーマン多様体（R^n は最も簡単な例である）におけるソボレフ不等式と等周問題のより深い関係はこの本の範囲を超えている．

[1] H. Federer and W. H. Fleming, Normal and integral currents, *Annals of Math.* 72 (1960) 458-520; MR 23 # A588.
[2] V. G. Mazya, Classes of domains and imbedding theorems for function spaces (in Russian), *Dokl. Akad. Nauk SSSR* 133 (1960) 527-530; Engl. transl. in *Soviet Math. Dokl.* 1 (1960) 882-885; MR 23 # A3448.
[3] G. Talenti, Best constant in Sobolev inequality, *Annali Mat. Pura Appl.* 110 (1976) 353-372; MR 57 # 3846.
[4] R. Osserman, The isoperimetric inequality, *Bull. Amer. Math. Soc.* 84 (1978) 1182-1238; MR 58 # 18161.
[5] D. S. Mitrinovic, J. E. Pecaric, and A. M. Fink, *Inequalities Involving Functions and Their Integrals and Derivatives*, Kluwer, 1991; MR 93m:26036.
[6] Y. D. Burago and V. A. Zalgaller, *Geometric Inequalities*, Springer-Verlag, 1988; MR 89b:52020.
[7] P. R. Beesack, Integral inequalities involving a function and its derivative, *Amer. Math. Monthly* 78 (1971) 705-741; MR 48 # 4235.
[8] E. F. Beckenbach and R. Bellman, *Inequalities*, 2nd ed., Springer-Verlag, 1965, pp. 177-188; MR 33 # 236.
[9] E. H. Lieb and M. Loss, *Analysis*, Amer. Math. Soc., 1997, pp. 183-199; MR 98b:00004.
[10] G. H. Hardy, J. E. Littlewood, and G. Pólya, *Inequalities*, Cambridge Univ. Press, 1934; MR 89d:26016.
[11] G. F. Simmons, *Differential Equations with Applications and Historical Notes*, McGraw-Hill, 1972; MR 58 # 17258.
[12] R. Courant and D. Hilbert, *Methods of Mathematical Physics*, v. 1, Interscience, 1953; MR 16, 426a.
[13] N. Anderson, A. M. Arthurs, and R. R. Hall, Extremum principle for a nonlinear problem in magneto-elasticity, *Proc. Cambridge Philos. Soc.* 72 (1972) 315-318; MR 45 # 7151.
[14] C. O. Horgan, A note on a class of integral inequalities, *Proc. Cambridge Philos. Soc.* 74 (1973) 127-131; MR 48 # 9486.
[15] G. Pólya and G. Szegö, *Isoperimetric Inequalities in Mathematical Physics*, Princeton Univ. Press, 1951; MR 13,270d.
[16] J. W. S. Rayleigh, *The Theory of Sound*, v. 1, 2nd rev. ed., Dover, 1945.
[17] C. Bandle, *Isoperimetric Inequalities and Applications*, Pitman, 1980; MR 81e:35095.
[18] G. Faber, Beweis dass unter allen homogenen Membranen von gleicher Fläche und gleicher Spannung die kreisförmige den tiefsten Grundton gibt, *Sitzungsberichte Bayerische Akademie*

der Wissenschaften (1923) 169-172.

[19] E. Krahn, Über eine von Rayleigh formulierte Minimaleigenschaft des Kreises, *Math. Annalen* 94 (1925) 97-100.

[20] E. Krahn, Über Minimaleigenschaften der Kugel in drei und mehr Dimensionen, *Acta et Commentationes Universitatis Tartuensis (Dorpatensis)* A9 (1926) 1-44; Engl. transl. in Edgar Krahn 1894-1961: *A Centenary Volume*, ed. Ü. Lumiste and J. Peetre, IOS Press, 1994, pp. 139-174.

[21] I. Hong, On an inequality concerning the eigenvalue problem of a membrane, *Kōdai Math. Sem. Rep.* (1954) 113-114; MR 16,1116b and MR 16,1337 errata/addenda.

[22] G. Pólya, On the characteristic frequencies of a symmetric membrane, *Math. Z.* 63 (1955) 331-337; also in *Collected Papers*, v. 3, ed. J. Hersch and G.-C. Rota, MIT Press, 1984, pp. 413-419, 519-521; MR 17,372e.

[23] P. M. Morse, *Vibration and Sound*, 2nd ed., McGraw-Hill, 1948.

[24] G. Szegö, On membranes and plates, *Proc. Nat. Acad. Sci. USA* 36 (1950) 210-216; 44 (1958) 314-316; also in *Collected Papers*, v. 3, ed. R. Askey, Birkhäuser, 1982, pp. 185-194 and 479-483; MR 11,757g and MR 20 # 2924.

[25] L. E. Payne, Some comments on the past fifty years of isoperimetric inequalities, *Inequalities: Fifty Years on from Hardy, Littlewood and Pólya*, ed. W. N. Everitt, Dekker, 1991, pp. 143-161; MR 92f:26042.

[26] G. Talenti, On isoperimetric theorems of mathematical physics, *Handbook of Convex Geometry*, ed. P. M. Gruber and J. M. Wills, Elsevier, 1993, pp. 1131-1147; MR 94i:49002.

[27] N. S. Nadirashvili, Rayleigh's conjecture on the principal frequency of the clamped plate, *Arch. Rational Mech. Anal.* 129 (1995) 1-10; MR 97j:35113.

[28] M. S. Ashbaugh and R. D. Benguria, On Rayleigh's conjecture for the clamped plate and its generalization to three dimensions, *Duke Math. J.* 78 (1995) 1-17; MR 97j:35111.

[29] M. Hegland and J. T. Marti, Numerical computation of least constants for the Sobolev inequality, *Numer. Math.* 48 (1986) 607-616; MR 87k:65136.

[30] J. T. Marti, Evaluation of the least constant in Sobolev's inequality for $H^1(0, s)$, *SIAM J. Numer. Anal.* 20 (1983) 1239-1242; MR 85d:46044.

[31] W. Richardson, Steepest descent and the least C for Sobolev's inequality, *Bull. London Math. Soc.* 18 (1986) 478-484; MR 87i:46078.

[32] J. T. Marti, On the norm of the Sobolev imbedding of $H^2(G)$ into $C(G)$ for square domains in $\mathbb{R}^2$, *The Mathematics of Finite Elements and Applications* V (MAFELAP V), Proc. 1984 Uxbridge conf., ed. J. R. Whiteman, Academic Press, 1985, pp. 441-450; MR 87a:46055.

[33] J. T. Marti, The least constant in Friedrichs' inequality in one dimension, *SIAM J. Math. Anal.* 16 (1985) 148-150; MR 86e:46026.

[34] G. Rosen, Minimum value for c in the Sobolev inequality $\|\varphi^3\| \leq c \|\nabla\varphi\|^3$, *SIAM J. Appl. Math* 21 (1971) 30-32; MR 44 # 6927.

[35] G. Rosen, Sobolev-type lower bounds on $\|\nabla\psi\|^2$ for arbitrary regions in two-dimensional Euclidean space, *Quart. Appl. Math.* 34 (1976/77) 200-202; MR 57 # 12803.

[36] G. F. D. Duff, A general integral inequality for the derivative of an equimeasurable rearrangement, *Canad. J. Math.* 28 (1976) 793-804; MR 53 # 13497.

[37] P. S. Crooke, An isoperimetric bound for a Sobolev constant, *Colloq. Math.* 38 (1978) 263-267; MR 58 # 6115.

[38] E. J. M. Veling, Optimal lower bounds for the spectrum of a second order linear differential equation with a p-integrable coefficient, *Proc. Royal Soc. Edinburgh* 92A (1982) 95-101; MR 84h:34050.

[39] P. L. Lions, F. Pacella, and M. Tricarico, Best constants in Sobolev inequalities for functions vanishing on some part of the boundary and related questions, *Indiana Math. J.* 37 (1988) 301-324; MR 89i:46036.

[40] E. H. Lieb, Sharp constants in the Hardy-Littlewood-Sobolev and related inequalities, *Annals*

of Math. 118 (1983) 349-374; MR 86i:42010.
[41] J. T. Marti, New upper and lower bounds for the eigenvalues of the Sturm-Liouville problem, Computing 42 (1989) 239-243; MR 90g:34028.
[42] C. O. Horgan, Eigenvalue estimates and the trace theorem, J. Math. Anal. Appl. 69 (1979) 231-242; MR 80f:49029.
[43] C. O. Horgan and L. E. Payne, Lower bounds for free membrane and related eigenvalues, Rend. Mat. Appl. 10 (1990) 457-491; MR 91h:35237.
[44] S. Waldron, Schmidt's inequality, East J. Approx. 3 (1997) 117-135; MR 98g:41013.
[45] A. A. Ilyin, Best constants in Sobolev inequalities on the sphere and in Euclidean space, J. London Math. Soc. 59 (1999) 263-286; MR 2000f:46043.
[46] Z. A. Melzak, The isoperimetric problem of the convex hull of a closed space curve, Proc. Amer. Math. Soc. 11 (1960) 265-274; MR 22 #7058.
[47] Z. A. Melzak, Numerical evaluation of an isoperimetric constant, Math. Comp. 22 (1968) 188-190; MR 36 #7023.
[48] I. J. Schoenberg, An isoperimetric inequality for closed curves convex in even-dimensional Euclidean spaces, Acta Math. 91 (1954) 143-164; MR 16,508b.

3.7 コーン定数

$u(x)$ を n 次元空間の有界な連結開集合 Ω の閉包上で定義された滑らかなベクトル場としよう．そのとき $\nabla u(x)$ は $u(x)$ の偏微分から作られた $n \times n$ 行列である．行列 M のノルム $|M|$ を M のユークリッドノルムとする，すなわち，すべての要素の平方和の平方根とする．また M^T を M の転置行列とする．

Korn（コーン）の不等式のいわゆる**第2の場合**[1-3]は

$$\int_\Omega |\nabla u(x)|^2 dx \leq K \int_\Omega \left|\frac{\nabla u(x) + \nabla u(x)^T}{2}\right|^2 dx$$

ただし付帯条件

$$\int_\Omega (\nabla u(x) - \nabla u(x)^T) dx = 0$$

のもとで，である．各種の領域 Ω に対する最良定数は線形弾性理論や非圧縮性流体力学において重要である．もし B_n が n 次元球ならば[4,5]，$K(B_2)=4$ かつ $K(B_3)=56/13$ である．$n \geq 4$ に対応する値は知られていない．

P_m を 2 次元の正 m 角形としよう．正方形 P_4 に対して[2]

$$5 \leq K(P_4) \leq 4(2+\sqrt{2})$$

が証明された．また Horgan & Payne（ホーガン&ペイン）[6]は $K(P_4)=7$ を予想した．正三角形 P_3 に対しては，[7-9]のラプラシアンの固有値公式を使って次の結果を得る．

$$6 \leq K(P_3) \leq 8(2+\sqrt{3})$$

任意の m に対して，われわれは次のような上界を知っている[2]．

$$K(P_m) \leq \frac{4}{1-\sin(\pi/m)}$$

また $K(P_6)$ に対する下界が固有値の数値評価を使って得られる[9]．楕円や蝸牛線に対するコーン定数が[2,10]で与えられている．円環や球殻に対しては[11,12]を参照してほしい．

ここに関連する問題（$n=2$ のみに対して）がある．$z=x+iy$，ただし i は虚数単位とし，$f(x,y)$ と $g(x,y)$ を解析関数 $w(z)$ の実数部と虚数部としよう．言いかえると $f(x,y)$ と $g(x,y)$ は**共役調和関数**である．Friedrichs（フリードリックス）の不等式[6,10,13-15]

$$\int_\Omega f(x,y)^2 dxdy \leq \Gamma \int_\Omega g(x,y)^2 dxdy$$

を付帯条件

$$\int_\Omega f(x,y) dxdy = 0$$

の下で考えよう．各種の単連結領域 Ω に対する最良定数 Γ は，Ω が連続的微分可能な境界をもつと仮定したときの $K=2(1+\Gamma)$ によって定義されるコーン定数 K と関連がある．Ω が正方形領域の場合には，Horgan & Payne[6]は，最適な関数が

$$f(x,y) = 2xy, \quad g(x,y) = y^2 - x^2$$

であり，したがって $\Gamma = 5/2$ であることを予想した．これは，もし滑らかさの条件に対して問題がないならば，$K=7$ を直接に示す．

Horgan の概説[2]は貴重な研究の出発点である．関連する話題は[16,17]にもある．

[1] C. O. Horgan, On Korn's inequalities for incompressible media, *SIAM J. Appl. Math.* 28 (1975) 419-430; MR 52 # 2356.

[2] C. O. Horgan, Korn's inequalities and their applications in continuum mechanics, *SIAM Rev.* 37 (995) 491-511; MR 96h:73014.

[3] A. Tiero, On Korn's inequality in the second case, *J. Elasticity* 54 (999) 187-191; MR 2001j:35061.

[4] B. Bernstein and R. A. Toupin, Korn inequalities for the sphere and circle, *Arch. Rational Mech. Anal.* 6 (1960) 51-64; MR 23 # B1719.

[5] L. E. Payne and H. F. Weinberger, On Korn's inequality, *Arch. Rational Mech. Anal.* 8 (1961) 89-98; MR 28 # 1537.

[6] C. O. Horgan and L. E. Payne, On inequalities of Korn, Friedrichs and Babuska-Aziz, *Arch. Rational Mech. Anal.* 82 (1983) 165-179; MR 84d:73014.

[7] M. A. Pinsky, The eigenvalues of an equilateral triangle, *SIAM J. Math. Anal.* 11 (1980) 819-827; 16 (1985) 848-851; MR 82d:35077 and MR 86k:35115.

[8] P. Sjoberg, An investigation of eigenfunctions over the equilateral triangle and square, unpublished note (1995).

[9] L. M. Cureton and J. R. Kuttler, Eigenvalues of the Laplacian on regular polygons and polygons resulting from their dissection, *J. Sound Vibration* 220 (1999) 83-98; MR 99j:35025.

[10] C. O. Horgan, Inequalities of Korn and Friedrichs in elasticity and potential theory, *Z. Angew Math. Phys.* 26 (1975) 155-164; MR 51 # 2399.

[11] C. M. Dafermos, Some remarks on Korn's inequality, *Z. Angew. Math. Phys.* 19 (1968) 913-920; MR 39 # 1154.

[12] E. Andreou, G. Dassios, and D. Polyzos, Korn's constant for a spherical shell, *Quart. Appl.*

Math. 46 (1988) 583-591; MR 89j:73014.
[13] K. O. Friedrichs, On certain inequalities and characteristic value problems for analytic functions and for functions of two variables, *Trans. Amer. Math. Soc.* 41 (1937) 321-364.
[14] K. O. Friedrichs, On the boundary-value problems of the theory of elasticity and Korn's inequality, *Annals of Math.* 48 (1947) 441-471; MR 9,255b.
[15] W. Velte, On inequalities of Friedrichs and Babuska-Aziz, *Meccanica* 31 (1996) 589-596; MR 97k:73022.
[16] V. A. Kondrat'ev and O. A. Oleinik, Boundary value problems for a system in elasticity theory in unbounded domains: Korn's inequalities (in Russian), *Uspekhi Mat. Nauk*, v. 43 (1988) n. 5, 55-98, 239, Engl. transl. in *Russian Math. Surveys* 43 (1988) 65-119; MR 89m:35061.
[17] K. Bhattacharya, Korn's inequality for sequences, *Proc. Royal Soc. London A* 434 (1991) 479-484; MR 92i:73030.

3.8 ウィットニー-ミクリンの拡張定数

$B_{n,r}$ を原点中心，半径 r の n 次元閉球としよう．この節を通して $r>1$ を固定する． n 次元空間全体で定義された関数 F が $B_{n,1}$ 上で定義された関数 f の r **拡張**であるとは，すべての $|x|\leq 1$ に対して $F(x)=f(x)$ で，すべての $|x|\geq r$ に対して $F(x)=0$ を満たすことをいう．

われわれは与えられた f から F を作り上げる方法に関心がある．また「浪費を最小にする」方法でこれをやりたい．ここで「浪費を最小にする」という言葉の解釈を（たくさんある中から）2つとり上げる．

・任意の連続関数 f に対して次の条件を満たす連続 r 拡張 F を作れ．
$$\max_{x\in B_{n,r}}|F(x)|\leq c\max_{x\in B_{n,1}}|f(x)|$$
ただし c は定数（f に依存しない）で，可能なもののうち最小なものとする．

・任意の連続的微分可能な関数 f に対して次の不等式を満たす連続的微分可能な r 拡張 F を構成せよ．
$$\left[\int_{B_{n,r}}\left(F(x)^2+\sum_{k=1}^{n}\left(\frac{\partial F}{\partial x_k}\right)^2\right)dx\right]^{1/2}\leq \chi\left[\int_{B_{n,1}}\left(f(x)^2+\sum_{k=1}^{n}\left(\frac{\partial f}{\partial x_k}\right)^2\right)dx\right]^{1/2}$$
ここで（再び）χ は定数で，可能なもののうち最小なものである．

この言葉のいま1つの解釈は次のようなものである．$B_{n,1}$ と $B_{n,r}$ 上で定義された関数の2つのバナッハ空間に対して，一方から他方への r 拡張作用素で最小の（作用素）ノルムをもつものを決定せよ．最初の場合では，バナッハ空間ノルムは L_∞ あるいは最大ノルムであり，2番目の場合では，ソボレフ空間 W_2^1 の積分ノルムである，ただし2番目の場合はとらえがたい微分の取り扱いが難しい．

Whitney（ウィットニー）[1]は最初の場合1の分割の手法をもちいて $c=1$ であることを証明した．変分学は2番目の場合（r に依存することに注意）に対しては $n=1$

のとき
$$\chi=\sqrt{1+\coth(1)\coth(r-1)}$$
という結果をもたらした[2,3].

Mikhlin（ミクリン）[4-6]は2番目の場合に対して$n\geq 2$のとき最良定数$\chi=\chi(n,r)$を決定した．初期の関連する結果には Hestenes（ヘステンス）[7], Calderón（カルデロン）[8], Stein（スタイン）[9]がある．便宜上$\nu=(n-2)/2$と定義し，変形ベッセル関数を次のように定義する．

$$I_\nu(r)=\left(\frac{r}{2}\right)^\nu \sum_{j=0}^\infty \frac{1}{j!\Gamma(\nu+j+1)}\left(\frac{r}{2}\right)^{2j}, \quad K_\nu(r)=\frac{\pi}{2}\frac{I_{-\nu}(r)-I_\nu(r)}{\sin(\nu\pi)}$$

$I_\nu(r)$ や $K_\nu(r)$ を含む代数公式に基づく, $\chi(n,r)$ の数値評価の一覧表については[4]を参照してほしい．われわれの関心事はもっぱら漸近値

$$\chi_x=\lim_{r\to\infty}\chi(n,r)=\sqrt{1+\frac{I_\nu(1)}{I_{\nu+1}(1)}\frac{K_{\nu+1}(1)}{K_\nu(1)}}$$

であり，明らかに奇数次元 n に対して

$$\chi_1=\sqrt{\frac{2e^2}{e^2-1}}, \quad \chi_3=e, \quad \chi_5=\sqrt{\frac{e^2}{e^2-7}}, \quad \chi_7=\sqrt{\frac{2}{7}}\sqrt{\frac{e^2}{37-5e^2}}, \quad \chi_9=\sqrt{\frac{1}{37}}\sqrt{\frac{e^2}{18e^2-133}}$$

となる．ここに思いがけず自然対数の底 e が出現する．偶数次元の n に対しては e ではなく，$I_0(1)$, $I_1(1)$, $K_0(1)$, および $K_1(1)$ を用いて同じような公式を記述することができる．

[1] H. Whitney, Analytic extensions of differentiable functions defined in closed sets, *Trans. Amer. Math. Soc.* 36 (1934) 63-89.
[2] R. Johnson, Sharp Sobolev constants, unpublished note (2000).
[3] D. H. Luecking, Extension operator of minimal norm, unpublished note (2000).
[4] S. G. Mikhlin, *Konstanten in einigen Ungleichungen der Analysis*, Teubner, 1981; *Constants in Some Inequalities of Analysis*, Wiley, 1986; MR 84a:46076 and MR 87g:46057.
[5] S. G. Mikhlin, Equivalent norms in Sobolev spaces and norms of extension operators (in Russian), *Sibirskii Mat. Z.* 19 (1978) 1141-1153; Engl. transl. in *Siberian Math. J.* 19 (1978) 804-813; MR 81g:46043.
[6] S. G. Mikhlin, On the minimal extension constant for functions of Sobolev classes (in Russian), Numerical Methods and Questions in the Organization of Calculations, 3, *Zap. Nauchn. Sem. Leningrad Otdel. Mat. Inst. Steklov (LOMI)* 90 (1979) 150-185, 300; MR 81k:46034.
[7] M. R. Hestenes, Extension of the range of a differentiable function, *Duke Math. J.* 8 (1941) 183-292; MR 2,219c.
[8] A.-P. Calderón, Lebesgue spaces of differentiable functions and distributions, *Partial Differential Equations*, ed. C. B. Morrey, Proc. Symp. Pure Math. 4, Amer. Math. Soc., 1961, pp. 33-49; MR 26 #603.
[9] E. M. Stein, *Singular Integrals and Differentiability Properties of Functions*, Princeton Univ. Press, 1970; MR 44 #7280.

3.9　ゾロタリョーフ-シュア定数

nを正の整数としよう．S_nをn次実係数の多項式で，すべての$-1 \leq x \leq 1$に対して$|p(x)| \leq 1$を満たすものの全体としよう．

Markov（マルコフ）[1,2]は，もし$p \in S_n$ならばすべての$-1 \leq x \leq 1$に対して$|p'(x)| \leq n^2$であることを証明した．ただしp'はpの導関数である．等号が成立するのは$x = \pm 1$かつ$p(x) = \pm T_n(x)$のときである．ここでT_nはn次のチェビシェフ多項式である【4.9】．

$-1 \leq \xi \leq 1$を実数とし，$n \geq 3$を整数としよう．$S_{n,\xi}$を付加的な条件$p''(\xi) = 0$によって特徴づけられたS_nの部分集合と定義する．$T_n \notin S_{n,\pm 1}$に注意すると，集合$S_{n,\pm 1}$上で$|p'(\pm 1)|$を最大にするような多項式は，前のものとは全く異なる解になる．

Schur（シュア）[3,4]は，もし$p \in S_{n,\xi}$ならば$|p'(\xi)| < n^2/2$であることを証明した．さらに

$$s_n = \sup_{-1 \leq \xi \leq 1} \sup_{p \in S_{n,\xi}} \frac{|p'(\xi)|}{n^2} \text{ かつ } \sigma = \limsup_{n \to \infty} s_n$$

とおいて，$0.217 \leq \sigma \leq 0.465$という評価を得た．

定数σを確定することは，Zolotarev（ゾロタリョーフ）によってなされた研究の1つの結果[5-12]であることがわかる．ちょうど$T_n(x)$がマルコフの定理における極値多項式として現れたように，新しい多項式の集合$Z_n(x)$がシュアの定理を十二分に理解するために必要とされる．Zolotarevは1877年に，その時代に先んじた研究の中で，楕円関数を使って色々な多項式近似の厳密な解を決定した．

Erdös & Szegö（エルデーシュ&セゲー）[4]はシュアの定理とゾロタリョーフの多項式との関係をはっきりさせた．彼らは

$$\sigma = \frac{1}{c^2}\left(1 - \frac{E(c)}{K(c)}\right)^2 = 0.3110788667\cdots$$

を証明した．ただし$K(x)$と$E(x)$は第1種および第2種の完全楕円積分【1.4.6】で，cは方程式

$$[K(c) - E(c)]^3 + (1-c^2)K(c) - (1+c^2)E(c) = 0, \quad 0 < c < 1$$

の一意的な解である．極値$s_n n^2$には，$n > 3$に対しては，$\xi = 1$および$p(x) = \pm Z_n(x)$，あるいは，$\xi = -1$および$p(x) = \pm Z_n(-x)$のとき到達する．ゾロタリョーフの多項式と関連する微分方程式を議論するのはあまりにも本題を外れるので，ここで終わることにする．

3.9.1　楕円に関するシーウェル問題

ここにマルコフの問題の拡張がある．$p(z)$を$z = x + iy$に関するn次の複素多項式

とし，$x^2+(y/g)^2 \leq 1$, ただし $0 < g \leq 1$, によって与えられる楕円領域 E 上で $|p(z)| \leq 1$ と仮定しよう．E 上すべてで $|p'(z)| \leq n \cdot K(g)$ が成り立つような n に無関係な最小の定数 $K(g)$ は何か？

$K(1)=1$ かつ $K(g) \leq 1/g$ であることが知られている [13-16]．2次多項式 $p(z) = (8z^2-3)/5$ の例から，van Delden（ファン・デルデン）[17] は $K(1/2) \geq 8/5$ を導いた．彼はさらに一般化されたチェビシェフ多項式列 [4.9]

$$T_n(z,g) = \cos(n \arccos(\tilde{z})) = \frac{(\tilde{z}+\sqrt{\tilde{z}^2-1})^n + (\tilde{z}-\sqrt{\tilde{z}^2-1})^n}{2}, \quad \tilde{z} = \frac{z}{\sqrt{1-g^2}}$$

を利用して，$K(g)$ がその上界 $1/g$ に等しいことを示唆した．類似の定数が他の境界曲線に対しても同様に定義される [18-20]．また [21-25] も参照してほしい．

[1] A. A. Markov, On a certain problem of D. I. Mendeleev (in Russian), *Zapiski Imperatorskoi Akademii Nauk* 62 (1877) 1-24.
[2] R. J. Duffin and A. C. Schaeffer, A refinement of an inequality of the brothers Markoff, *Trans. Amer. Math. Soc.* 50 (1941) 517-528; MR 3,235c.
[3] I. Schur, Über das Maximum des absoluten Betrages eines Polynoms in einem gegebenen Intervall, *Math. Z.* 4 (1919) 271-287.
[4] P. Erdös and G. Szegö, On a problem of I. Schur, *Annals of Math.* 43 (1942) 451-470; correction 74 (1961) 628; also in *Gabor Szegö: Collected Papers*, v. 2, ed. R. Askey, Birkhäuser, 1982, pp. 805 -828; MR 4,41d and MR 24 # A1341.
[5] E. I. Zolotarev, Applications of elliptic functions to questions about functions deviating least or most from zero (in Russian), *Zapiski Imperatorskoi Akademii Nauk*, v. 30 (1877) n. 5; also in *Collected Works*, v. 2, Izdat. Akad. Nauk SSSR, 1932, pp. 1-59.
[6] N. I. Achieser, *Theory of Approximation*, F. Ungar, 1956; MR 3,234d and MR 10,33b.
[7] B. C. Carlson and J. Todd, Zolotarev's first problem-The best approximation by polynomials of degree $\leq n-2$ to $x^n - \sigma x^{n-1}$ in $[-1, 1]$, *Aequationes Math.* 26 (1983) 1-33; MR 85c:41012.
[8] J. Todd, Applications of transformation theory: A legacy from E. I. Zolotarev (1847-1878), *Approximation Theory and Spline Functions*, Proc. 1983 St. John's conf., ed. S. P. Singh, J. W. H. Burry, and B. Watson, Reidel, 1984, pp. 207-245; MR 86g:41044.
[9] J. Todd, A legacy from E. I. Zolotarev (1847-1878), *Math. Intellig.* 10 (1988) 50-53; MR 89m: 01057.
[10] P. P. Petrushev and V. A. Popov, *Rational Approximation of Real Functions*, Cambridge Univ. Press, 1987; MR 89i:41022.
[11] V. M. Tikhomirov, Approximation theory, *Analysis II*, ed. R. V. Gamkrelidze, Springer-Verlag, 1990, pp. 93-255; MR 91e:00001.
[12] F. Peherstorfer and K. Schiefermayr, Description of extremal polynomials on several intervals and their computation. II, *Acta Math. Hungar.* 83 (1999) 59-83; MR 99m:41010.
[13] S. N. Bernstein, Sur l'ordre de la meilleure approximation des fonctions continues par des polynômes de degré donné, *Mémoires de l'Académie Royale de Belgique. Classe des Sciences. Collection*, 2nd ser., 4 (1912) 1-103; Russian transl. in *Collected Works*, v. 1, *Constructive Theory of Functions (1905-1930)*, Izdat. Akad. Nauk SSSR, 1952; MR 14,2c.
[14] M. Riesz, Eine trigonometrische Interpolationsformel und einige Ungleichungen für Polynome, *Jahresbericht Deutsch. Math.-Verein.* 23 (1914) 354-368.
[15] W. E. Sewell, On the polynomial derivative constant for an ellipse, *Amer. Math. Monthly* 44 (1937) 577-578.
[16] Q. I. Rahman, Some inequalities for polynomials, *Amer. Math. Monthly* 67 (1960) 847-851; 68 (1961) 349; MR 23 A1011.

[17] J. van Delden, Polynomial optimization over an ellipse, unpublished note (2001).
[18] W. E. Sewell, The derivative of a polynomial on various arcs of the complex domain, *National Math. Mag.* 12 (1938) 167-170.
[19] W. E. Sewell, The derivative of a polynomial on further arcs of the complex domain, *Amer. Math. Monthly* 46 (1939) 644-645.
[20] W. E. Sewell, The polynomial derivative at a zero angle, *Proc. Amer. Math. Soc.* 12 (1961) 224-228; MR 22 # 12206.
[21] A. C. Schaeffer and G. Szegö, Inequalities for harmonic polynomials in two and three dimensions, *Trans. Amer. Math. Soc.* 50 (1941) 187-225; also in *Gabor Szegö: Collected Papers*, v. 2, ed. R. Askey, Birkhäuser, 1982, pp. 755-794; MR 3,111b.
[22] J. H. B. Kemperman, Markov type inequalities for the derivatives of a polynomial, *Aspects of Mathematics and Its Applications*, ed. J. A. Barroso, North-Holland, 1986, pp. 465-476; MR 87j:26020.
[23] A. Jonsson, On Markov's and Bernstein's inequalities in the unit ball in $\mathbb{R}^k$, *J. Approx. Theory* 78 (1994) 151-154; MR 95h:41026.
[24] G. V. Milovanovic, D. S. Mitrinovic, and T. M. Rassias, *Topics in Polynomials: Extremal Problems, Inequalities, Zeros*, World Scientific, 1994, pp. 404-407, 448-449, 527-723; MR 95m:30009.
[25] W. Plesniak, Recent progress in multivariate Markov inequality, *Approximation Theory: In Memory of A. K. Varma*, ed. N. K. Govil, R. N. Mohapatra, Z. Nashed, A. Sharma, and J. Szabados, Dekker, 1998, pp. 449-464; MR 99g:41016.

3.10 クネーザー-マーラー多項式定数

多項式が与えられたとき，その因子の大きさについて何がいえるか？ $\|p\|$を閉区間$[-1,1]$上で定義された複素係数のn次多項式$p(x)$の最大値ノルムとしよう．$p(x)=q(x)r(x)$，ただし$q(x)$はk次で，$r(x)$は$n-k$次である，と仮定しよう．このときKneser（クネーザー）[1]は，Aumann（アウマン）[2]の仕事を足がかりに，

$$\|q\|\|r\| \le \frac{1}{2}C_{n,k}C_{n,n-k}\cdot\|p\|$$

を証明した[3-5]．ただし

$$C_{n,k} = 2^k\prod_{j=1}^{k}\left[1+\cos\left(\frac{(2j-1)\pi}{n}\right)\right]$$

である．さらに任意のnと$k\le n$に対して，この定数は最良である．ここで右辺は，$q(x)$の次数が「わかっている」とした場合なのに注意してほしい．

$q(x)$の次数kに関する情報が利用できないと仮定したとき，Borwein（ボーウィン）[4,5]はKneserの結果の1つの系として$k=\lfloor n/2\rfloor$が$C_{n,k}$を最大にすること，したがって$n\to\infty$のとき漸近的に

$$\|q\|\|r\| \le \delta^{2n}\|p\|$$

を示した．ここで

$$\delta = \exp\left(\frac{2G}{\pi}\right) = 1.7916228120\cdots$$

はダイマーに関する定数【5.23】であり，G は Catalan（カタラン）の定数である【1.7】．さらに，その不等式は

$$\limsup_{n\to\infty}\left(\frac{\|q\|\|r\|}{\|p\|}\right)^{1/n} = \delta^2 = 3.2099123007\cdots$$

という意味で最良である．ここで上限（sup）はすべての n 次の多項式とその因子 q と r についてとるものとする．

この表現における δ の注目すべき登場は，別の分野で研究していた Boyd（ボイド）[6]によって数年前から予想されていた．以後 $\|p\|$ を複素数平面の単位円板 D 上で定義された $p(z)$ の最大値ノルムとする．Boyd は，もし $p(z) = q(z)r(z)$ ならば漸近的に

$$\|q\|\|r\| \leq \delta^n \|p\|$$

であること，そしてこれが最良であることを証明した．$[-1,1]$ の場合に δ^2 が現れ，D の場合は δ が現れるのは興味深い．

この不等式から $\|r\|$ を取り去ってみよう．大きな定数によるつまらない掛け算を避けるため，p と q，それゆえ r がモニックであると仮定する．Boyd[6]は漸近的に

$$\|q\| \leq \beta^n \|p\|$$

が成り立って，これが最良であることを証明した．ここで

$$\beta = \exp\left(\frac{1}{\pi}I\left(\frac{2}{3}\pi\right)\right) = 1.3813564445\cdots$$

かつ

$$I(\theta) = \int_0^\theta \ln\left(2\cos\left(\frac{x}{2}\right)\right)dx$$

この積分は，$\mathrm{Cl}(\theta)$ を Clausen（クラウゼン）積分[7,8]とすれば，単に $\mathrm{Cl}(\pi-\theta)$ である．同じような表現[6,9]

$$\delta = \exp\left(\frac{2}{\pi}I\left(\frac{1}{2}\pi\right)\right)$$

さらに2つの級数[10,11]

$$\ln(\delta) = \frac{2}{\pi}\left(1 - \frac{1}{3^2} + \frac{1}{5^2} - \frac{1}{7^2} + \frac{1}{9^2} - \frac{1}{11^2} + \cdots\right) = 0.5831218080\cdots,$$

$$\ln(\beta) = \frac{3\sqrt{3}}{4\pi}\left(1 - \frac{1}{2^2} + \frac{1}{4^2} - \frac{1}{5^2} + \frac{1}{7^2} - \frac{1}{8^2} + \cdots\right) = 0.3230659472\cdots$$

に注目する．定数 β は文献中にいくつかの場面に現れている．最初は Mahler[12]の中で一見関係のない多項式にかかわる不等式の中に現れている．[13,14]の中で二項巡回行列式と呼ばれる漸近式の中で出現している．[15]の中で，$\ln(\beta)$ は単純2次元ずらし変換のエントロピーであり，[16]において，$\pi \log(\beta) = 1.0149416064\cdots$ は双曲型四面体の最大の可能な体積である．さらに，【5.23】，【8.9】を参照してほしい．$\pi\log(\beta)$ に関する面白い最近の記述は[17]に見られる．そこではこの定数は Gieseking（**ギーゼキング**）**の定数**と呼ばれている．

3.10 クネーザー-マーラー多項式定数

さらに，δ は文献の中にいたるところで現れる．2次元整数格子のダイマーの充填問題にすでに言及したところである．[18,19]において $\ln(\delta)$ は Schmidt のガウス整数の連分数に関連して現れる．他方 δ は，[20,21]に記述されたものを含む数理物理学においてある役割を果たしている．

Boyd[9]はこの議論を2つの因子から m 個の因子の場合へと拡張した．m を固定して，$p(z) = p_1(z) p_2(z) \cdots p_m(z)$ とすると，漸近的に

$$\|p_1\| \cdot \|p_2\| \cdots \|p_m\| \leq c_m^n \cdot \|p\|$$

が成り立ち，これは最良である．ただし，

$$c_m = \exp\left(\frac{m}{\pi} I\left(\frac{1}{m}\pi\right)\right)$$

である．$c_2 = \delta$，また $I(\pi/3) = (2/3) I(2\pi/3)$ だから $c_3 = \beta^2 = 1.9081456268\cdots$ であることを確認しよう[8]．われわれは，さらに $c_4 = 1.9484547890\cdots$，$c_5 = 1.9670449011\cdots$ および $c_6 = 1.9771268308\cdots$ をも知っている．

Boyd[9]は，$p(z)$ およびすべての $p_i(z)$ が実係数であるが，D 上で定義されている場合を考えた．このとき，定数 c_m は単に δ に置きかえられて，かつこれが最良である．すなわち，実係数の場合，最良定数は m に依存しない．Borwein[4,5]は $p(z)$ と $p_i(z)$ が区間 $[-1,1]$ 上で定義された複素係数の場合を考えた．このとき c_m は単純に δ^2 に置きかえられ，再びこの不等式は最良である．Pritsker（プリッカー）[22,23]は，区間 $[-a,a]$ 上の Boyd の不等式[6]に対して β の類似物 $B(a)$ に対する一般的な公式を得た．たとえば，$B(2) = \beta^2 = 1.90815\cdots$ かつ $B(1) = \sqrt{2}\,\delta = 2.53373\cdots$．なお，[24,25]をも参照されたい．

【2.30】で我々は代数的整数 α に対する Mahler（マーラー）の尺度（measure）$M(\alpha)$ をとり上げた．これは1変数の多項式[26]

$$f(z) = a_0 \prod_{j=1}^{n} (z - \alpha_j)$$

に対するマーラーの尺度 $M(f)$ と本質的に同値である．ここでマーラーの尺度とは，Jensen（イェンセン）の公式[27]の1つの結果として

$$M(f) = \exp\left(\int_0^1 \ln(|f(e^{2\pi i\theta})|)\,d\theta\right) = |a_0| \prod_{j=1}^{n} \max(|\alpha_j|, 1)$$

によって与えられるものである．

多変数関数 $f(z_1, z_2, \cdots, z_m)$ への重要な一般化が

$$M(f) = \exp\left(\int_0^1 \int_0^1 \cdots \int_0^1 \ln(|f(e^{2\pi i\theta_1}, e^{2\pi i\theta_2}, \cdots, e^{2\pi i\theta_m})|)\,d\theta_1 d\theta_2 \cdots d\theta_m\right)$$

で与えられる．いくつかの例について計算すると

$$M(1+x) = 1, \quad M(1+x+y) = \beta = M(\max(1, |x+1|)),$$

$$M(1+x+y+z) = \exp\left(\frac{7\zeta(3)}{2\pi^2}\right),$$

$$M(1+x+y-xy) = \delta = M(\max(|x-1|, |x+1|))$$

ただし，$\zeta(3)$ は Apéry（アペリ）の定数である【1.6】．次の2つの漸近的な結果があ

る[10]. それらはオイラー-マスケロニの定数 γ【1.5】を含む

$$\lim_{m\to\infty} \frac{M(z_1+z_2+\cdots+z_m)}{\sqrt{m}} = \exp\left(-\frac{1}{2}\gamma\right) = 0.7493060013\cdots$$

および

$$\lim_{m\to\infty} M(z_1+(1+z_2)(1+z_3)\cdots(1+z_m))^{1/\sqrt{m}} = \exp\left(\sqrt{\frac{\pi}{24}}\right)$$

最後に,Bombieri(ボンビエリ)の最大値ノルムをとり上げる. $p(z) = \sum_{j=0}^{n} a_j z^j$ のとき

$$[p] = \max_{0 \leq j \leq n} |a_j| \frac{n!}{j!(n-j)!}$$

と定義する. $p(z)$ と $q(z)$ が D 上のモニックな多項式で,$\deg(p) = n$ かつ q が p の因子であるとき,$\|q\|$ の $[p]$ との相対的大きさに興味がある.漸近的に次の式が成り立つことが知られている[28-31].

$$\|q\| \leq K^n [p]$$

ただし

$$K = M(1+|x+1|) = M((1+x+x^2+y)^2) = 2.1760161352\cdots$$

しかし K が最良であることの証明はいまだ見つかっていない.

[1] H. Kneser, Das Maximum des Produkts zweier Polynome, *Sitzungsber. Preuss. Akad. Wiss.* (1934) 426-431.
[2] G. Aumann, Satz über das Verhalten von Polynomen auf Kontinuen, *Sitzungsber. Preuss. Akad. Wiss.* (1933) 924-931.
[3] D. S. Mitrinovic, *Analytic Inequalities*, Springer-Verlag, 1970; MR 43 # 448.
[4] P. B. Borwein, Exact inequalities for the norms of factors of polynomials, *Canad. J. Math.* 46 (1994) 687-698; MR 95k:26015.
[5] P. Borwein andT. Erdélyi, *Polynomials and Polynomial Inequalities*, Springer-Verlag, 1995; MR 97e:41001.
[6] D. W. Boyd, Two sharp inequalities for the norm of a factor of a polynomial, *Mathematika* 39 (1992) 341-349; MR 94a:11162.
[7] M. Abramowitz and I. A. Stegun, *Handbook of Mathematical Functions*, Dover, 1972;MR 94b: 00012.
[8] L. Lewin, *Polylogarithms and Associated Functions*, North-Holland, 1981;MR83b:33019.
[9] D. W. Boyd, Sharp inequalities for the product of polynomials, *Bull. London Math. Soc.* 26 (1994) 449-454; MR 95m:30008.
[10] C. J. Smyth, On measures of polynomials in several variables, *Bull. Austral. Math. Soc.* 23 (1981) 49-63; G. Myerson and C. J. Smyth, corrigendum 26 (1982) 317-319; MR 82k:10074 and MR 84g: 10088.
[11] D. W. Boyd, Speculations concerning the range of Mahler's measure, *Canad. Math. Bull.* 24 (1981) 453-469; MR 83h:12002.
[12] K. Mahler, A remark on a paper of mine on polynomials, *Illinois J. Math.* 8 (1964) 1-4; MR 28 #2194.
[13] J. S. Frame, Factors of the binomial circulant determinant, *Fibonacci Quart.* 18 (1980) 9-23; MR 81j:10007.
[14] D. W. Boyd, The asymptotic behaviour of the binomial circulant determinant, *J. Math. Anal. Appl.* 86 (1982) 30-38; MR 83f:10007.

[15] D. Lind, K. Schmidt, and T. Ward, Mahler measure and entropy for commuting automorphisms of compact groups, *Invent. Math.* 101 (1990) 593-629; MR 92j:22013.
[16] J. Milnor, Hyperbolic geometry: The first 150 years, *Bull. Amer. Math. Soc.* 6 (1982) 9-24; MR 82m:57005.
[17] C. C. Adams, The newest inductee in the Number Hall of Fame, *Math. Mag.* 71 (1998) 341-349.
[18] A. L. Schmidt, Ergodic theory of complex continued fractions, *Number Theory with an Emphasis on the Markoff Spectrum*, Proc. 1991 Provo conf., ed. A. D. Pollington and W. Moran, Dekker, 1993, pp. 215-226; MR 95f:11055.
[19] H. Nakada, The metrical theory of complex continued fractions, *Acta Arith.* 56 (1990) 279-289; MR 92e:11081.
[20] P. Sarnak, Spectral behavior of quasiperiodic potentials, *Commun. Math. Phys.* 84 (1982) 377-401; MR 84g:35136.
[21] D. J. Thouless, Scaling for the discrete Mathieu equation, *Commun. Math. Phys.* 127 (1990) 187-193; MR 90m:82004.
[22] I. E. Pritsker, An inequality for the norm of a polynomial factor, *Proc. Amer. Math. Soc.* 129 (2001) 2283-2291; math.CV/0001124; MR 2002f:30005.
[23] I. E. Pritsker, Norms of products and factors polynomials, *Number Theory for the Millennium*, v. 2, Proc. 2000 Urbana conf., ed. M. A. Bennett, B. C. Berndt, N. Boston, H. G. Diamond, A. J. Hildebrand, and W. Philipp, A. K. Peters, to appear; math.CV/0101164.
[24] A. Kroó and I. E. Pritsker, A sharp version of Mahler's inequality for products of polynomials, *Bull. London Math. Soc.* 31 (1999) 269-278; MR 99m:30008.
[25] I. E. Pritsker, Products of polynomials in uniform norms, *Trans. Amer. Math. Soc.* 353 (2001) 3971-3993; MR 2002f:30006.
[26] G. Everest, Measuring the height of a polynomial, *Math. Intellig.*, v. 20 (1998) n. 3, 9-16; MR 2000b:11115.
[27] R. M. Young, On Jensen's formula and $\int_0^{2\pi} \log|1-e^{i\theta}|\,d\theta$, *Amer. Math. Monthly* 93 (1986) 44-45.
[28] D. W. Boyd, Bounds for the height of a factor of a polynomial in terms of Bombieri's norms. I: The largest factor, *J. Symbolic Comput.* 16 (1993) 115-130; MR 94m:11032a.
[29] D. W. Boyd, Bounds for the height of a factor of a polynomial in terms of Bombieri's norms. II: The smallest factor, *J. Symbolic Comput.* 16 (1993) 131-145; MR 94m:11032b.
[30] D. W. Boyd, Large factors of small polynomials, *The Rademacher Legacy to Mathematics*, ed. G. E. Andrews, D. M. Bressoud, and L. Alayne Parson, Contemp. Math. 166, Amer. Math. Soc., 1994, pp. 301-308; MR 95e:11034.
[31] D. W. Boyd, Mahler's measure and special values of L-functions, *Experim. Math.* 7 (1998) 37-82; MR 99d:11070.

3.11 グロタンディークの定数

任意の整数 $n \geq 2$ に対して次の性質を満たす定数 $k(n)$ が存在する. A を $m \times m$ 行列で,

$$\left|\sum_{i=1}^m \sum_{j=1}^m a_{ij}s_i t_j\right| \leq 1$$

が $|s_i| \leq 1$, $|t_j| \leq 1$ を満たすすべてのスカラー $s_1, s_2, \cdots, s_m, t_1, t_2, \cdots, t_m$ に対して満たさ

れるものとする．このとき $\|x_i\| \leq 1$, $\|y_j\| \leq 1$ を満たす n 次元ヒルベルト空間のすべてのベクトル $x_1, x_2, \cdots, x_m$, $y_1, y_2, \cdots, y_m$ に対して

$$\left| \sum_{i=1}^{m} \sum_{j=1}^{m} a_{ij} \langle x_i, y_j \rangle \right| \leq k(n)$$

が成り立つ．いつものとおり，$\langle x, y \rangle$ は x と y の内積で $\|x\| = \sqrt{\langle x, x \rangle}$ である．定数 $k(n)$ は最小なものが採用される．

この定義は，実際には2つの可能性のある場合を扱っている．
・スカラーと行列が実数で，ベクトルが実ヒルベルト空間に属する場合．
・スカラーと行列が複素数で，ベクトルが複素ヒルベルト空間に属する場合．

対応する定数をそれぞれ $k_R(n)$, $k_C(n)$ と書くことにする．次のことがわかっている [3-6]．

$$k_R(2) = \sqrt{2}, \quad k_R(3) < 1.517, \quad k_R(4) \leq \pi/2$$

これに反して

$$1.1526 \leq k_C(2) \leq 1.2157, \quad 1.2108 \leq k_C(3) \leq 1.2744, \quad 1.2413 \leq k_C(4) \leq 1.3048$$

各数列は n に関して明らかに単調増加である．実数の場合も複素数の場合も $\kappa = \lim_{n \to \infty} k(n)$ と定義する．極限に関して次の式を示すのは困難ではない [2]．

$$\frac{1}{2} \kappa_R \leq \kappa_C \leq 2 \kappa_R$$

もっともよく知られた数値的限界は次のとおりである [3,4,7-9]．

$$1.67696 \leq \kappa_R \leq \frac{\pi}{2 \ln(1 + \sqrt{2})} = 1.7822139781 \cdots,$$

$$1.33807 \leq \kappa_C \leq \frac{8}{\pi \cdot (x_0 + 1)} = 1.40491 \cdots$$

ここで x_0 は第1種と第2種の完全楕円積分 $K(x)$, $E(x)$ を含む次の方程式の解である【1.4.6】．

$$\psi(x) = \frac{\pi}{8}(x+1) \quad (-1 < x < 1)$$

ただし

$$\psi(x) = x \int_0^{\pi/2} \frac{\cos^2 \theta}{\sqrt{1 - x^2 \sin^2 \theta}} \, d\theta = \frac{1}{x} [E(x) - (1-x^2) K(x)]$$

κ_R の上方評価は，Krivine（クリビン）[3,4,10] によって正確な値であろうと予想された．対照的に，Haagerup（ハーゲラップ）[7] は 1.40491 が κ_C に対する正確な値であるかどうかを疑い，

$$\frac{1}{|\psi(i)|} = \left(\int_0^{\pi/2} \frac{\cos^2 \theta}{\sqrt{1 + \sin^2 \theta}} \right)^{-1} = 1.4045759346 \cdots$$

がそれよりもっともらしい候補であると考えた．彼の推論は類推によるものであった．複素形の場合の関数 $\psi(x)$ は実数形の場合の Krivine によって用いられた関数 $\varphi(x)$ と似ている．ここで，

$$\varphi(x) = \frac{2}{\pi}\arcsin(x)$$

であり，そして

$$\frac{1}{|\varphi(i)|} = \frac{\pi}{2\operatorname{arcsinh}(1)} = \frac{\pi}{2\ln(1+\sqrt{2})}$$

であることもわかっている．

κ_R の範囲を求めるための異なる研究が[11]に与えられている．

[1] J. Lindenstrauss and A. Pelczynski, Absolutely summing operators in L_p spaces and their applications, *Studia Math.* 29 (1968) 275-326; MR 37 # 6743.

[2] G. L.O. Jameson, *Summing and Nuclear Norms in Banach Space Theory*, Cambridge Univ. Press, 1987; MR 89c:46020.

[3] J.-L. Krivine, Sur la constante de Grothendieck, *C. R. Acad. Sci. Paris* 284 (1977) 445-446; MR 55 # 1435.

[4] J.-L. Krivine, Constantes de Grothendieck et fonctions de type positif sur les spheres, *Adv. Math.* 31 (1979) 16-30; MR 80e:46015.

[5] H. König, On the complex Grothendieck constant in the n-dimensional case, *Geometry of Banach Spaces*, Proc. 1989 Strobl conf., ed. P. F. X. Müller and W. Schachermayer, Cambridge Univ. Press, 1990, pp. 181-198; MR 92g:46011.

[6] H. König, Some remarks on the Grothendieck inequality, *General Inequalities* 6, Proc. 1990 Oberwolfach conf., ed. W. Walter, Birkhäuser, 1992, pp. 201-206; MR 94h:46035.

[7] U. Haagerup, A new upper bound for the complex Grothendieck constant, *Israel J. Math.* 60 (1987) 199-224; MR 89f:47029.

[8] A. M. Davies, Lower bound for κ_C, unpublished note (1984).

[9] J. A. Reeds, A new lower bound on the real Grothendieck constant, unpublished note (1991).

[10] F. Le Lionnais, *Les Nombres Remarquables*, Hermann 1983.

[11] P. C. Fishburn and J. A. Reeds, Bell inequalities, Grothendieck's constant, and root two, *SIAM J. Discrete Math.* 7 (1994) 48-56; MR 95e:05013.

3.12 デュボア・レイモン定数

微積分学で Abel（アーベル）の定理は，実数項級数 $\sum_{n=0}^{\infty} a_n$ が収束するならば，対応するベキ級数が

$$\lim_{r \to 1-} \sum_{n=0}^{\infty} a_n r^n = \sum_{n=0}^{\infty} a_n$$

をみたすことを主張する．これは区間 $[0,1]$ 上一様収束することの1つの結果である．われわれは1つの疑問から始めることにする．もし $\sum_{n=0}^{\infty} a_n$ が「発散」したら何が起こるだろうか？

部分和 $s_n = \sum_{k=0}^{n} a_k$ の列を定義し，

$$s = \liminf_{n \to \infty} s_n, \quad S = \limsup_{n \to \infty} s_n$$

がともに有限だと仮定する．すなわち級数は有界で，2つの有限な限界の間を振動している．ここで，

$$s \leq \liminf_{r \to 1-} \sum_{n=0}^{\infty} a_n r^n \leq \limsup_{r \to 1-} \sum_{n=0}^{\infty} a_n r^n \leq S$$

と考えるのは自然であり，実際それは正しい[1].

実際，もっと多くのことが成り立つ．$\varphi(x)$ を $x>0$ で連続的微分可能で，次の条件をみたすものとする．

$$\lim_{x \to 0+} \varphi(x) = 1, \quad \lim_{x \to \infty} \varphi(x) = 0, \quad I = \int_0^{\infty} \left| \frac{d}{dx} \varphi(x) \right| dx < \infty$$

そして

$$f(x) = \sum_{n=0}^{\infty} a_n \varphi(nx) \quad \text{はすべての } x>0 \text{ に対して収束する．}$$

そのとき次のことが証明できる[1,2].

$$\frac{1}{2}(S+s) - \frac{1}{2}(S-s)I \leq \liminf_{x \to 0+} f(x) \leq \limsup_{x \to 0+} f(x) \leq \frac{1}{2}(S+s) + \frac{1}{2}(S-s)I$$

さらに，これは前に議論したものを真に拡張している．$r = \varphi(x) = \exp(-x)$ とおいてみれば，その理由がわかる．

ある整数 $m \geq 2$ に対して $\varphi(x) = (\sin x/x)^m$ とおけば，もう1つの重要な場合が出てくる．m 次の du Bois Reymond（**デュボア・レイモン**）**定数**を次のように定義する．

$$c_m = I - 1 = \int_0^{\infty} \left| \frac{d}{dx} \left(\frac{\sin x}{x} \right)^m \right| dx - 1$$

Watson（ワトソン）[2-6]は

$$c_2 = \frac{1}{2}(e^2 - 7) = 0.1945280495\cdots, \quad c_4 = \frac{1}{8}(e^4 - 4e^2 - 25) = 0.0052407047\cdots,$$

$$c_6 = \frac{1}{32}(e^6 - 6e^4 + 3e^2 - 98) = 0.0002206747\cdots$$

そして c_{2k} は有理係数の e^2 に関する k 次の多項式として表されることを証明した．c_{2k+1} に対してはそのような表現は知られていないが，すべての c_m に対して役立つ興味深い級数が存在する．$\xi_1, \xi_2, \xi_3, \cdots$ を方程式 $\tan x = x$ のすべての正の解とする．そのとき，

$$c_m = 2 \sum_{j=1}^{\infty} \frac{1}{(1+\xi_j^2)^{m/2}}$$

および，とくに，$c_3 = 0.0282517642\cdots$ である．同様に $c_5, c_7, \cdots$ を数値的に評価することができる．Watson は

$$c_3 = -\frac{2}{\pi} \int_1^{\infty} \frac{1}{\sqrt{x^2-1}} \frac{d}{dx} \left(\frac{\tanh^2 x}{x - \tanh x} \right) dx$$

であることを決定したが，この積分をもっと単純化したものはないように思える．

数列，$\xi_1, \xi_2, \xi_3, \cdots$ は最近の "*Monthly* problem"

$$\sum_{n=1}^{\infty}\frac{1}{\xi_j^2}=\frac{1}{10}$$

に出てきて，大きな注目をひいた[7]．この公式はちょうど今議論したものと平行しており，Watson の他の結果，すなわち

$$b_m=2\sum_{j=1}^{\infty}\frac{(-1)^{j+1}}{(1+\xi_j^2)^{m/2}}, \quad b_3=-\frac{1}{4}(e^3-3e-12)=0.0173271405\cdots$$

および b_{2k+1} は，e に関する有理係数の $2k+1$ 次多項式として表現できる．e に関する類似の表現が【3.8】にも出てきたことに注意しておく．

ここに正接関数（tan）にかかわる方程式を含む別の定数がある．制約条件 $\sum_{k=1}^{n}x_k^2\leq 1$ の下での関数

$$\left(\sum_{k=1}^{n}\frac{x_k}{k}\right)^2+\sum_{k=1}^{n}\left(\frac{x_k}{k}\right)^2$$

の最大値 $M(n)$ は次の漸近的な結果を満たす．

$$\lim_{n\to\infty}M(n)=\left(\frac{\pi}{\xi}\right)^2=2.3979455861\cdots$$

ここで，$\xi=2.0287578381\cdots$ は方程式 $x+\tan x=0$ の正の最小解である．【3.14】に述べられているもう１つの例[9]は，方程式 $\pi+x=\tan x$ に関係している．

[1] E. W. Hobson, *The Theory of Functions of a Real Variable and the Theory of Fourier's Series*, v. 2, Dover, 1957, pp. 221-225; MR 19,1166b.
[2] G. N. Watson, Du Bois Reymond's constants, *Quart. J. Math.* 4 (1933) 140-146.
[3] A. Fletcher, J. C. P. Miller, L. Rosenhead, and L. J. Comrie, *An Index of Mathematical Tables*, 2nd ed., v. 1, Addison-Wesley, 1962, p. 129; MR 26 # 365a.
[4] H. P. Robinson and E. Potter, *Mathematical Constants*, UCRL-20418, Univ. of Calif. at Berkeley, 1971; available through the National Technical Information Service, Springfield VA 22151.
[5] F. Le Lionnais, *Les Nombres Remarquables*, Hermann, 1983.
[6] S. Plouffe, 2nd Du Bois Reymond constant (Plouffe's Tables).
[7] R. M. Young, A Rayleigh popular problem, *Amer. Math. Monthly* 93 (1986) 660-664.
[8] G. Szegö, Über das Maximum einer quadratischen Form von unendlich-vielen Veränderlichen, *Jahresbericht Deutsch. Math.-Verein.* 31 (1922) 85-88; also *in Collected Papers*, v. 1, ed. R. Askey, Birkhaüser, 1982; pp. 589-593.
[9] G. Brown and K.-Y. Wang, An extension of the Fejér-Jackson inequality, *J. Austral. Math. Soc. Ser. A* 62 (1997) 1-12; MR 98e:42003.

3.13 シュタイニッツ定数

3.13.1 動　　機

もし $\sum x_i$ が実数項の絶対収束級数ならば，級数の項 x_i を並べ換えてもその和にどん

な影響も与えない.

それと比較して, $\sum x_i$ が実数項の条件収束級数ならば, どんな和 (∞ や $-\infty$ でさえ) をもつ級数をも作りだすように項 x_i を並べ替えられる. これは Riemann によるよく知られた定理である.

話変わって, 項 x_i が有限次元実ノルム空間の元, すなわち, x_i が実ベクトルであるが, 異なった長さの概念 (計量の選択) をもつものとする. $\sum x_i$ の性質については何も仮定しない. C を収束するように項 x_i を並べ替えたときの和の全体とする. Steinitz (シュタイニッツ) [1-3] は C が空集合であるか, あるベクトル y とある線形部分空間 L に対して $y+L$ という形であるかのいずれかであることを示した. ($L=\{0\}$ (零空間) も可能であることに注意してほしい.)

この定理を証明するために, Steinitz は次の節で定義されるある定数 $K(0,0)$ に関する評価を必要とした. 正確な関係についての詳細は, [4-6] を参照してほしい.

3.13.2 定　義

a と b を非負実数とする. m 次元実ノルム空間において, $S=\{u, v_1, v_2 \cdots, v_{n-1}, v_n, w\}$ を $|u| \leq a$, $|v_j| \leq 1$ (各 $1 \leq j \leq n$ に対して), $|w| \leq b$ および $u+\sum_{j=1}^{n} v_j + w = 0$ (図 3.2) を満足する $n+2$ 個のベクトルの集合と定義する.

π を添え数 $\{1, 2, \cdots, n\}$ の1つの並べ替え (順列) とし, 関数

$$F(\pi, n, S) = \max_{1 \leq k \leq n} \left| u + \sum_{j=1}^{k} v_{\pi(j)} \right|$$

を定義する. 要するに, F は, 0 を中心, $u, v_{\pi(1)}, v_{\pi(2)}, \cdots, v_{\pi(n)}$ を辺とする多角形を取り囲む最小の球の半径である. あらゆる可能な π によって決定されるベクトルの並びのうちで, 球の半径を最小にするものが (少なくとも1つ) 存在する. そこで

$$K_m(a, b) = \max_{n, S} \min_{\pi} F(\pi, n, S)$$

と定義する. すなわち, $K_m(a, b)$ は, すべての整数 n と集合 S に対して, さらにある順列 π に対して, $|u + \sum_{j=1}^{k} v_{\pi(j)}| \leq K_m(a, b)$ を満たす最小数である.

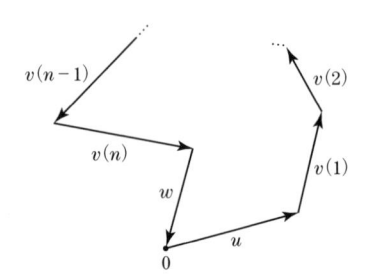

図 3.2　$u + \sum_{j=1}^{n} v_j + w = 0$ を満たすベクトル達の集合.

3.13.3 結　果

一般にいま述べられた設定（ノルムに制限のない）の下で，m 次元シュタイニッツ定数の上からの評価で最もよく知られているのは，Banaszczyk（バナシェジク）[7] によるもので，[8]の研究で改良された

$$K_m(0,0) \leq m-1+\frac{1}{m}$$

である．さらに，Grinberg & Sevastyanov（グリンベルグ&セバスチャノフ）[8]は，$m=2$ に対し上からの評価 3/2 が最良であることを考察した．言いかえると，等式が成り立つような例が存在する．この考察がより大きな m に対して成り立つか否かは知られていない．

以後ノルムはユークリッドノルムであると仮定する．Banaszczyk[7]は，

$$K_2(a,b) = \sqrt{1+\max(a^2,b^2,1/4)}$$

を証明した．これは早くから研究者に知られていた結果 $K_2(1,0) = K_2(1,1) = \sqrt{2}$，$K_2(0,0) = \sqrt{5}/2 = 1.1180339887\cdots$ の拡張である．Damsteeg & Halperin（ダムステック&ハルペリン）[4]は，

$$K_m(0,0) \geq \frac{1}{2}\sqrt{m+3}$$

そして $m \geq 2$ に対して，

$$K_m(1,1) \geq K_m(1,0) \geq \frac{1}{2}\sqrt{m+6}$$

を証明した．Behrend（ベーレント）[10]は，

$$K_m(1,0) \leq K_m(1,1) < m, \quad K_3(1,0) \leq K_3(1,1) < \sqrt{5+2\sqrt{3}} = 2.9093129112\cdots$$

を証明したが，任意の $m>2$ に対する正確な値は依然としてわかっていない．（注：[11]において $K(0,0)$ と $K(1,0)$ の間に混乱があるように見える．しかしそれ以前の文献[12]には見られない．）Behrend は，これらの定数の真のオーダーが $\sqrt{m}$ らしいと信じていた．また関連するアイデアについては[13-18]を参照してほしい．

---

[1] E. Steinitz, Bedingt konvergente Reihen und konvexe Systeme, *J. Reine Angew. Math.* 143 (1913) 128-175; 144 (1914) 1-40.

[2] P. Rosenthal, The remarkable theorem of Lévy and Steinitz, *Amer. Math. Monthly* 94 (1987) 342-351; MR 88d:40005.

[3] M. I. Kadets and V. M. Kadets, *Series in Banach Spaces: Conditional and Unconditional Convergence*, Birkhäuser, 1997; MR 98a:46016.

[4] I. Damsteeg and I. Halperin, The Steinitz-Gross theorem on sums of vectors, *Trans. Royal Soc. Canada Ser. III* 44 (1950) 31-35; MR 12,419a.

[5] I. Halperin and N. Miller, An inequality of Steinitz and the limits of Riemann sums, *Trans. Royal Soc. Canada Ser. III* 48 (1954) 27-29l; MR 16,596a.

[6] I. Halperin, Sums of a series, permitting rearrangements, *C. R. Math. Rep. Acad. Sci. Canada* 8 (1986) 87-102; MR 87m:40004.

[7] W. Banaszczyk, The Steinitz constant of the plane, *J. Reine Angew. Math.* 373 (1987) 218-220; MR 88e:52016.

[8] V. S. Grinberg and S. V. Sevastyanov, O velicine konstanty Steinica, *Funkcional. Anal. i Prilozen.*, v. 14 (1980) n. 2, 56-57; Engl. transl. in *Functional Anal. Appl.* 14 (1980) 125-126; MR 81h:52008.
[9] W. Banaszczyk, A note on the Steinitz constant of the Euclidean plane, *C. R. Math. Rep. Acad. Sci. Canada* 12 (1990) 97-102; MR 91g:52005.
[10] F. A. Behrend, The Steinitz-Gross theorem on sums of vectors, *Canad. J. Math.* 6 (1954) 108-124; MR 15,551c.
[11] D. S. Mitrinovic, J. E. Pecaric, and A. M. Fink, *Classical and New Inequalities in Analysis*, Kluwer, 1993; MR 94c:00004.
[12] D. S. Mitrinovic, *Analytic Inequalities*, Springer-Verlag, 1970; MR 43 # 448.
[13] W. Banaszczyk, The Dvoretzky-Hanani lemma for rectangles, *Period. Math. Hungar.* 31 (1995) 1-3; MR 96j:52011.
[14] S. Sevastjanov and W. Banaszczyk, To the Steinitz lemma in coordinate form, *Discrete Math.* 169 (1997) 145-152; MR 97m:05004.
[15] I. Bárány, M. Katchalski, and J. Pach, Quantitative Helly-type theorems, *Proc. Amer. Math. Soc.* 86 (1982) 109-114; MR 84h:52016.
[16] I. Bárány and A. Heppes, On the exact constant in the quantitative Steinitz theorem in the plane, *Discrete Comput. Geom.* 12 (1994) 387-398; MR 95g:52011.
[17] P. Brass, On the quantitative Steinitz theorem in the plane, *Discrete Comput. Geom.* 17 (1997) 111-117; MR 97h:52005.
[18] I. Halperin and T. Ando, *Bibliography: Series of Vectors and Riemann Sums*, Hokkaido Univ., Sapporo, Japan, 1989.

3.14 ヤング-フェイエール-ジャクソン定数

3.14.1 余弦和の非負性

以下において n は正の整数, $0 \leq \theta \leq \pi$ と a は考察の対象となるパラメータとする. Young (ヤング) [1] は余弦和について

$$C(\theta, a, n) = \frac{1}{1+a} + \sum_{k=1}^{n} \frac{\cos(k\theta)}{k+a} \geq 0 \quad (-1 < a \leq 0 \text{ において})$$

を証明した. Rogosinski & Szegö (ロゴシンスキ&セゲー) [2] はこの結果を $-1 < a \leq 1$ に拡張し, 次の意味で最良の上方限界 A, $1 \leq A \leq 2(1+\sqrt{2})$ が存在することを証明した.

・$-1 < a \leq A$ に対して, すべての n とすべての θ に対し, $C(\theta, a, n) \geq 0$.
・$a > A$ に対して, ある n とある θ に対して, $C(\theta, a, n) < 0$.

Gasper (ガスパー) [3,4] は $A = 4.5678018826\cdots$ であること, および A は最小多項式

$$9x^7 + 55x^6 - 14x^5 - 948x^4 - 3247x^3 - 5013x^2 - 3780x - 1134$$

をもつことを証明した. 実際, もし $a > A$ ならば, ある θ に対して $C(\theta, a, 3) < 0$ である. これで余弦和に対する話は完結する.

3.14.2 正弦和の正値性

ここでは n は正の整数，$0<\theta<\pi$ および b は研究の対象となるパラメータである．Fejér（フェイエール）[5]，Gronwall（グロンウォール）[6,7]および Jackson（ジャクソン）[8]は，対応する正弦級数が $b=0$ に対して

$$S(\theta,b,n)=\sum_{k=1}^{n}\frac{\sin(k\theta)}{k+b}>0$$

という結果を得た．手短な証明に対しては[9]を参照してほしい．また，[10-13]も参照してほしい．Brown & Wang（ブラウン&ワン）[14]は，この結果を奇数 n に対して $-1<b\leq B$ まで拡張した．ここで，B は最良の上方限界である．偶数の n に対しては，この話はより複雑になり，後ほど説明することになるだろう．

B が定義されるために2つの媒介的な定数が必要となる．
- $\lambda=0.4302966531\cdots$，方程式 $(1+\lambda)\pi=\tan(\lambda\pi)$ の解．
- $\mu=0.8128252421\cdots$，方程式 $(1+\lambda)\sin(\mu\pi)=\mu\sin(\lambda\pi)$ の解．

これらを使って次の方程式の解を B と定義する[14,15]．

$$(1+\lambda)\pi\left((B-1)\psi\left(1+\frac{B-1}{2}\right)-2B\psi\left(1+\frac{B}{2}\right)+(B+1)\psi\left(1+\frac{B+1}{2}\right)\right)=2\sin(\lambda\pi)$$

ただし，$\psi(x)$ はディガンマ関数である【1.5.4】．B は代数的数だろうか？　その答えは未知である．

さて n が偶数の場合を議論しよう．$c_n(x)=1-2x/(4n+1)$ と定義する．もし $-1<b\leq B$ で，n が偶数ならば，$0<\theta\leq\pi c_n(\mu)$ に対して $S(\theta,b,n)>0$ が成り立つ．さらに，定数 μ は，$0<\nu<\mu$ ならばある $b<B$ と無限に多くの n に対して $S(\pi c_n(\nu),b,n)<0$ が成り立つという意味で最良である．

Wilson（ウィルソン）[16]は，Belov（ベロフ）の研究[17]に基づいて $S<0$ が期待され得ることを指摘した．

3.14.3 一様有界性

パラメータの値を $0<r<1$ に固定しよう．関数列

$$F_n(\theta,r)=\sum_{k=1}^{n}k^{-r}\cos(k\theta),\quad n=1,2,3,\cdots$$

を考えよう．もしすべての θ とすべての n に対して $m<F_n(\theta,r)$ が成り立つような定数 $m>-\infty$ が存在するならば，この数列は**下から一様有界**であるということにする．m は r の選び方に依存することに注意する．

Zygmund（ジグムント）[11]は，次の意味で，r に対して最良の下方限界 $0<R<1$ が存在することを証明した．
- $F_n(\theta,r)$ は $r\geq R$ に対して下に一様有界で，
- $F_n(\theta,r)$ は $r<R$ に対して下に一様有界でない．

定数 $R=0.3084437795\cdots$ は次の方程式の一意解である[15,18-22]．

$$\int_0^{3\pi/2} x^{-R}\cos x\, dx = 0$$

そしてこれは Belov の論文 [17,23] においても同様にある役割を果たしている. 興味深いことに, 関数列

$$G_n(\theta, r) = \sum_{k=1}^{n} k^{-r}\sin(k\theta), \quad n=1,2,3,\cdots$$

はすべての $r>0$ に対して下に一様有界である. それゆえ, 関数列 $G_n(\theta, r)$ に対する R の類似物というものは存在しない.

[1] W. H. Young, On a certain series of Fourier, *Proc. London Math. Soc.* 11 (1912) 357-366.
[2] W. Rogosinski and G. Szegö, Über die Abschnitte von Potenzreihen, die in einem Kreise beschränkt bleiben, *Math. Z.* 28 (1928) 73-94; also in *Gabor Szegö: Collected Papers*, v. 2, ed. R. Askey, Birkhäuser, 1982, pp. 87-111.
[3] G. Gasper, Nonnegative sums of cosine, ultraspherical and Jacobi polynomials, *J. Math. Anal. Appl.* 26 (1969) 60-68; MR 38 # 6130.
[4] D. S. Mitrinovic, J. E. Pecaric, and A. M. Fink, *Classical and New Inequalities in Analysis*, Kluwer, 1993, p. 615; MR 94c:00004.
[5] L. Fejér, Lebesguesche Konstanten und divergente Fourierreihen, *J. Reine Angew. Math.* 138 (1910) 22-53.
[6] T. H. Gronwall, Über die Gibbssche Erscheinung und die trigonometrischen Summen $\sin(x) + 1/2\sin(2x) + \cdots + 1/n\sin(nx)$, *Math. Annalen* 72 (1912) 228-243.
[7] E. Hewitt and R. E. Hewitt, The Gibbs-Wilbraham phenomenon: An episode in Fourier analysis, *Arch. Hist. Exact Sci.* 21 (1979) 129-160; MR 81g:01015.
[8] D. Jackson, Über eine trigonometrische Summe, *Rend. Circ. Mat. Palermo* 32 (1911) 257-262.
[9] R. Askey, J. Fitch, and G. Gasper, On a positive trigonometric sum, *Proc. Amer. Math. Soc.* 19 (1968) 1507; MR 37 # 6662.
[10] P. Turán, On a trigonometrical sum, *Annales Soc. Polonaise Math.* 25 (1952) 155-161; also in *Collected Works*, v. 1, ed. P. Erdös, Akadémiai Kiadó, pp. 661-667; MR 14, 1080c.
[11] A. G. Zygmund, *Trigonometric Series*, v. 1, 2nd ed., Cambridge Univ. Press, 1959, pp. 8, 191-192; MR 89c:42001.
[12] R. Askey and J. Steinig, Some positive trigonometric sums, *Trans. Amer. Math. Soc.* 187 (1974) 295-307; MR 49 # 3245.
[13] R. Askey and J. Steinig, A monotonic trigonometric sum, *Amer. J. Math.* 98 (1976) 357-365; MR 53 # 11288.
[14] G. Brown and K.-Y.Wang, An extension of the Fejér-Jackson inequality, *J. Austral. Math. Soc. Ser. A* 62 (1997) 1-12; MR 98e:42003.
[15] J. Keane, Estimating Brown-Wang B and Zygmund R constants, unpublished note (2000).
[16] D. C. Wilson, Review of "An extension of the Fejér-Jackson inequality," MR 98e:42003.
[17] A. S. Belov, On the coefficients of trigonometric series with nonnegative partial sums (in Russian), *Mat. Zametki* 41 (1987) 152-158, 285; Engl. transl. in *Math. Notes* 41 (1987) 88-92; MR 88f:42001.
[18] R. P. Boas and V. C. Klema, A constant in the theory of trigonometric series, *Math. Comp.* 18 (1964) 674; MR 31 # 558.
[19] R. F. Church, On a constant in the theory of trigonometric series, *Math. Comp.* 19 (1965) 501.
[20] Y. L. Luke, W. Fair, G. Coombs, and R. Moran, On a constant in the theory of trigonometric series, *Math. Comp.* 19 (1965) 501-502.
[21] R. D. Halbgewachs and S. M. Shah, Trigonometric sums and Fresnel-type integrals, *Proc. Indian Acad. Sci. Sect. A* 65 (1967) 227-232; MR 35 # 3855.

[22] S. M. Shah, Trigonometric series with nonnegative partial sums, *Entire Functions and Related Parts of Analysis*, ed. J. Korevaar, S. S. Chern, L. Ehrenpreis, W. H. J. Fuchs, and L. A. Rubel, Proc. Symp. Pure Math. 11, Amer. Math. Soc., 1968, pp. 386-391; MR 38 #6293.

[23] A. S. Belov, Coefficients of trigonometric cosine series with nonnegative partial sums, *Theory of Functions* (in Russian), Proc. 1987 Amberd conf., ed. S. M. Nikolskii, *Trudy Mat. Inst. Steklov*. 190 (1989) 3-21; Engl. transl. in *Proc. Steklov Inst. Math*. 190 (1992) 1-19; MR 90m:42005.

3.15 ファン・デル・コルプト定数

f を区間 $[a,b]$ 上 2 回連続的微分可能な実関数で，すべての x に対して $|f''(x)| \geq r$ という性質をもつものとする．f ばかりでなく，a にも b にも依存しない最小の定数 m が存在し，

$$\left|\int_a^b \exp(if(x))\,dx\right| \leq \frac{m}{\sqrt{r}}$$

が成り立つ．ただし，i は虚数単位である[1-3]．この不等式は最初 van der Corput (ファン・デル・コルプト) によって証明され，解析的数論でいくつかの応用がある．Kershner (カーシュナ) [4,5] は，Wintner の示唆にしたがって，最大となる関数 f が

$$-a = b = \sqrt{(\pi-2c)/r}$$

を端点とする区間上の放物線 $f(x) = rx^2/2 + c$ であることを証明した．ここで，係数 $c = -0.7266432468\cdots$ は方程式

$$\int_0^{\sqrt{\pi/2-c}} \sin(x^2+c)\,dx = 0, \quad -\frac{\pi}{2} \leq c \leq \frac{\pi}{2}$$

の唯一の解である．これから，ファン・デル・コルプト定数 m は次のようになる．

$$m = 2\sqrt{2}\int_0^{\sqrt{\pi/2-c}} \cos(x^2+c)\,dx = 3.3643175781\cdots$$

[1] J. G. van der Corput, Zahlentheoretische Abschätzungen, *Math. Annalen* 84 (1921) 53-70.
[2] E. Landau, *Vorlesungen über Zahlentheorie*, v. 2, Verlag von S. Hirzel, 1927, p. 60.
[3] E. C. Titchmarsh, On van der Corput's method and the zeta-function of Riemann, *Quart. J. Math*. 2 (1932) 161-173.
[4] R. Kershner, Determination of a van der Corput-Landau absolute constant, *Amer. J. Math*. 57 (1935) 840-846.
[5] R. Kershner, Determination of a van der Corput absolute constant, *Amer. J. Math*. 60 (1938) 549-554.

3.16 トゥラーンの指数和定数

固定した複素数 $z_1, z_2, \cdots, z_n$ に対して，次数が n 以下のベキ和の絶対値の最大値を

$$S(z) = \max_{1 \leq k \leq n} \left| \sum_{j=1}^{n} z_j^k \right|$$

と定義する．また，$(n-1)$ 次元複素領域 K_n を

$$K_n = \{ z \in C^n : z_1 = 1 \text{ かつ } |z_j| \leq 1 \, (2 \leq j \leq n) \}$$

と定義する．

$z \in K_n$ という条件の下で $S(z)$ を最小化する問題を考えよう．$S(z)$ の最小の値 σ_n は，

$n=2$ ならば $\dfrac{\sqrt{5}-1}{\sqrt{2}} = 0.8740320488\cdots$, $n=3$ ならば $x = 0.824780309\cdots$

である[2-4]．ただし，x は次のような最小多項式をもつ数である．

$x^{30} - 81x^{28} + 2613x^{26} - 43629x^{24} + 417429x^{22} - 2450985x^{20} + 9516137x^{18}$
$- 26203659x^{16} + 53016480x^{14} - 83714418x^{12} + 112601340x^{10} - 140002992x^8$
$+ 156204288x^6 - 124361568x^4 + 55427328x^2 - 10077696$

$n \geq 4$ に対して σ_n の正確な値は知られていない．しかし範囲については，十分大きい n に対して $0.3579 < \sigma_n < 1 - (250n)^{-1}$ がわかっている[1,6,7]．$\lim_{n \to \infty} \sigma_n$ が存在すると予想されているが，私の知る限りこの問題を数値的に探った人はいない．

代わりに

$$T(z) = \max_{2 \leq k \leq n+1} \left| \sum_{j=1}^{n} z_j^k \right|$$

を定義し[1,8], $z \in K_n$ という条件の下で $T(z)$ を最小化する問題を考えよう．驚いたことに $T(z)$ の最小値 τ_n は十分大きな n に対して，$\tau_n < 1.321^{-n}$ を満たす．これは，σ_n とは非常に異なった動きである．もし指数の範囲 $2 \leq k \leq n+1$ を $3 \leq k \leq n+2$ に置きかえるならば，定数 1.321 は 1.473 に置きかわる．

Turán（トゥラーン）の本[1]は関連する理論と応用の宝庫とよぶべきものである．

[1] P. Turán, *Über eine neue Methode der Analysis und ihre Anwendungen*, Akadémiai Kiadó, 1953 (original ed.); *On a New Method of Analysis and Its Applications*, Wiley, 1984 (revised ed.); MR 15, 688c and MR 86b : 11059.
[2] J. Lawrynowicz, Remark on a problem of P. Turán, *Bull. Soc. Sci. Lettres Lódz*, v. 11 (1960) n. 1, 1-4; MR 23 # A3959.
[3] J. Lawrynowicz, Calculation of a minimum maximorum of complex numbers, *Bull. Soc.Sci. Lettres Lódz*, v. 11 (1960) n. 2, 1-9; MR 23 # A3958.
[4] J. Lawrynowicz, Remark on power-sums of complex numbers, *Acta Math. Acad. Sci. Hungar.* 18 (1967) 279-281; MR 36 # 379.

[5] P. Sebah, Minimal polynomial for Turán's constant, unpublished note (2000).
[6] F. V. Atkinson, Some further estimates concerning sums of powers of complex numbers, *Acta Math. Acad. Sci. Hungar.* 20 (1969) 193-210; MR 39 #443.
[7] J. Komlós, A. Sárkozy, and E. Szemerédi, On sums of powers of complex numbers (in Hungarian), *Mat. Lapok* 15 (1964) 337-347; MR 34 #2534.
[8] E. Makai, An estimation in the theory of diophantine approximations, *Acta Math. Acad. Sci. Hungar.* 9 (1958) 299-307; MR 21 #1293.

4 関数の近似に関する定数

4.1 ギッブス-ウィルブラハム定数

f を半開区間 $[-\pi, \pi)$ 上で定義され，周期的に実直線全体に拡張された関数で，もとの区間でたかだか有限個の不連続点（すべて有限な跳躍の形の）をもつ区分的に滑らかな関数としよう．

$$a_k = \frac{1}{\pi}\int_{-\pi}^{\pi} f(t)\cos(kt)\,dt, \quad b_k = \frac{1}{\pi}\int_{-\pi}^{\pi} f(t)\sin(kt)\,dt$$

を f のフーリエ係数といい，

$$S_n(f, x) = \frac{a_0}{2} + \sum_{k=1}^{n}(a_k\cos(kx) + b_k\sin(kx))$$

を f のフーリエ級数の n 部分和という．$x = c$ を不連続点の 1 つとする．

$$\delta = \left(\lim_{x\to c-} f(x)\right) - \left(\lim_{x\to c+} f(x)\right), \quad \mu = \frac{1}{2}\left[\left(\lim_{x\to c-} f(x)\right) + \left(\lim_{x\to c+} f(x)\right)\right]$$

と定義する．また一般性を失わないで $\delta > 0$ と仮定する．$x_n < c$ を c の左側で $S_n(f, x)$ が極大値をとる最初の点とし，$\xi_n > c$ を c の右側で $S_n(f, x)$ が極小値をとる最初の点とする．このとき

$$\lim_{n\to\infty} S_n(f, x_n) = \mu + \frac{\delta}{\pi}G, \quad \lim_{n\to\infty} S_n(f, \xi_n) = \mu - \frac{\delta}{\pi}G$$

ただし

$$G = \int_0^{\pi} \frac{\sin\theta}{\theta}\,d\theta = \sum_{n=0}^{\infty} \frac{(-1)^n \pi^{2n+1}}{(2n+1)(2n+1)!} = 1.8519370519\cdots$$

$$= \frac{\pi}{2}(1.1789797444\cdots)$$

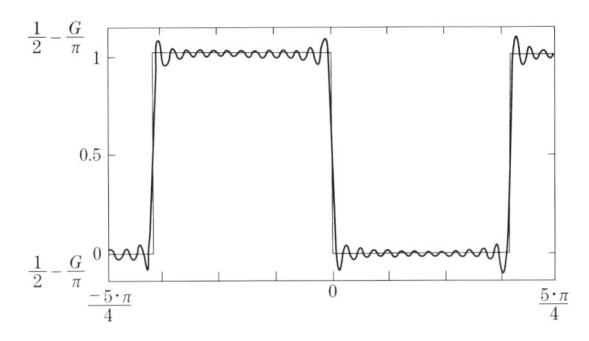

図 4.1 方形波のフーリエ級数近似は「上がりすぎ」と「下がりすぎ」の両方を見せている.

は Gibbs-Wilbraham (**ギッブス-ウィルブラハム**) 定数である [1-5].

図 4.1 の $-\pi \leq x < 0$ に対して $f(x) = 1$, $0 \leq x < \pi$ に対して $f(x) = 0$ という関数 f のグラフを考えよう. 振動の最も高い頂上の極限は 1 ではなく $1/2 + G/\pi = 1.0894898722\cdots$ に収束している. 同じように最も深い溝は 0 ではなく, $1/2 - G/\pi = -0.0894898722\cdots$ に収束している. 言いかえれば, ギッブス-ウィルブラハム定数は, 関数のフーリエ級数が跳躍不連続点でそこでの関数値を上回ったり, 下回ったりする程度を定量化したものである.

これらの現象は最初 Wilbraham[6] および Gibbs[7] によって観察された, Bôcher (ボーシェル)[8] は任意の関数 f にそのような考察を拡張した.

さらに一般に, $x_{n,2r-1} < c$ が $S_n(f, x)$ の c の左側にあって第 r 番目の極大値をとる点とし, $x_{n,2r} < c$ が $S_n(f, x)$ の c の左側にあって第 r 番目の極小値をとる点とし, さらに右側に同じように $\xi_{n,2r}$ と $\xi_{n,2r-1}$ をとると,

$$\lim_{n\to\infty} S_n(f, x_{n,s}) = \mu + \frac{\delta}{\pi}\int_0^{s\pi} \frac{\sin\theta}{\theta}\,d\theta, \quad \lim_{n\to\infty} S_n(f, \xi_{n,s}) = \mu - \frac{\delta}{\pi}\int_0^{s\pi} \frac{\sin\theta}{\theta}\,d\theta$$

この正弦積分は, $s = 2r - 1$ の整数値が増加するとき $\pi/2$ に減少するが, $s = 2r$ が増加するとき $\pi/2$ に増加する. 十分大きな r に対してその極限は $\mu \pm \delta/2$ になり, これは直感とも一致する.

フーリエ級数は L_2 (最小 2 乗ノルム) による最良三角多項式である. ギッブス-ウィルブラハム現象は, スプライン [5,9-11], ウェーブレット [5,12], および一般化された Padé (パデ) 近似 [13] に関連して同様に現れる. だからたくさんのギッブス-ウィルブラハム定数がある! Moskona, Petrushev & Staff (モスコナ, ペトルシェフ&スタッフ)[5,14] は L_1 における最良三角多項式に対応するものを研究し, この設定における $2G/\pi - 1 = 0.1789797444\cdots$ に類似する量を決定した. その値は $\max_{x \geq 1} g(x) = 0.0657838882\cdots$ である. ただし, $x > 0$ に対して

$$g(x) = -\frac{\sin(\pi x)}{\pi}\int_0^1 t^{x-1}\frac{1-t}{1+t}\,dt = -\frac{\sin(\pi x)}{\pi x}\sum_{k=1}^{\infty}\frac{k! \cdot 2^{-k}}{(x+1)(x+2)\cdots(x+k)}$$

である．$1<p\neq 2$ のときの L_p 近似の場合はつい最近研究されるようになったばかりである[15]．

[1] T. H. Gronwall, Über die Gibbssche Erscheinung und die trigonometrischen Summen $\sin(x)$ $+(1/2)\sin(2x)+\cdots+(1/n)\sin(nx)$, *Math. Annalen* 72 (1912) 228-243.
[2] H. S. Carslaw, *Introduction to the Theory of Fourier's Series and Integrals*, 3rd ed., Dover, 1930.
[3] A. G. Zygmund, *Trigonometric Series*, v. 1, 2nd ed., Cambridge Univ. Press, 1959; MR 89c:42001.
[4] E. Hewitt and R. E. Hewitt, The Gibbs-Wilbraham phenomenon: An episode in Fourier analysis, *Arch. Hist. Exact Sci.* 21 (1979) 129-160; MR 81g:01015.
[5] A. J. Jerri, *The Gibbs Phenomenon in Fourier Analysis, Splines and Wavelet Approximations*, Kluwer, 1998; MR 99i:42001.
[6] H. Wilbraham, On a certain periodic function, *Cambridge and Dublin Math. J.* 3 (1848) 198-201.
[7] J. W. Gibbs, Fourier's series, *Nature* 59 (1898) 200 and 59 (1899) 606; also in *Collected Works*, v. 2, ed. H. A. Bumstead and R. G. Van Name, Longmans and Green, 1931, pp. 258-260.
[8] M. Bôcher, Introduction to the theory of Fourier's series, *Annals of Math.* 7 (1906) 81-152.
[9] J. Foster and F. B. Richards, The Gibbs phenomenon for piecewise-linear approximation, *Amer. Math. Monthly* 98 (1991) 47-49; MR 91m:41037.
[10] F. B. Richards, A Gibbs phenomenon for spline functions, *J. Approx. Theory* 66 (1991) 334-351; MR 92h:41025.
[11] J. Foster and F. B. Richards, Gibbs-Wilbraham splines, *Constr. Approx.* 11 (1995) 37-52; MR 96b:41018.
[12] S. E. Kelly, Gibbs phenomenon for wavelets, *Appl. Comput. Harmon. Anal.* 3 (1996) 72-81; MR 97f:42056.
[13] G. Németh and G. Páris, The Gibbs phenomenon in generalized Padé approximation, *J. Math. Phys.* 26 (1985) 1175-1178; MR 87a:41021.
[14] E. Moskona, P. Petrushev, and E. B. Saff, The Gibbs phenomenon for best L_1-trigonometric polynomial approximation. *Constr. Approx.* 11 (1995) 391-416; MR 96f:42004.
[15] E. B. Saff and S. Tashev, Gibbs phenomenon for best L_p approximation by polygonal lines, *East J. Approx.* 5 (1999) 235-251; MR 2000i:41025.

4.2 ルベーグ定数

4.2.1 三角フーリエ級数

関数 f が区間 $[-\pi,\pi]$ 上可積分ならば，f のフーリエ係数を

$$a_k=\frac{1}{\pi}\int_{-\pi}^{\pi}f(t)\cos(kt)\,dt,\quad b_k=\frac{1}{\pi}\int_{-\pi}^{\pi}f(t)\sin(kt)\,dt$$

と定義し，f のフーリエ級数の n 部分和を

$$S_n(f,x)=\frac{a_0}{2}+\sum_{k=1}^{n}(a_k\cos(kx)+b_k\sin(kx))$$

と定義する．さらに $|f(x)|\leq 1$ と仮定すると，すべての x に対して

$$|S_n(f,x)| \leq \frac{1}{\pi}\int_0^\pi \frac{\left|\sin\left(\frac{2n+1}{2}\theta\right)\right|}{\sin\left(\frac{\theta}{2}\right)}d\theta = L_n$$

ここで L_n は n 次 Lebesgue（**ルベーグ**）**定数**である[1,2]．はじめのいくつかのルベーグ定数の値は次のとおりである．

$$L_0 = 1, \quad L_1 = \frac{1}{3} + \frac{2\sqrt{3}}{\pi} = 1.4359911241\cdots,$$
$$L_2 = 1.6421884352\cdots, \quad L_3 = 1.7783228615\cdots$$

いくつかの別な公式は Fejér（フェイエール）[3,4]や Szegö（セゲー）[5]によるものである．

$$L_n = \frac{1}{2n+1} + \frac{2}{\pi}\sum_{k=1}^n \frac{1}{k}\tan\left(\frac{\pi k}{2n+1}\right) = \frac{16}{\pi^2}\sum_{k=1}^\infty \sum_{j=1}^{(2n+1)k} \frac{1}{4k^2-1}\frac{1}{2j-1}$$

後者の表現は $\{L_n\}$ が単調に増加することを明示している．

ルベーグ定数は $L_n = \sup_f |S_n(f,0)|$ という意味で最良である．ただし，上限（sup）はすべての x に対して $|f(x)| \leq 1$ を満たすすべての連続関数についてとるものとする．

$$\frac{4}{\pi^2}\ln(n) < L_n < 3 + \frac{4}{\pi^2}\ln(n)$$

が成り立つことは容易に示される[6,7]．これは $L_n \to \infty$ を意味しており，結果的に，f が連続であっても f に対するフーリエ級数が非有界であり得ることを意味している[8-10]．また f の**連続度**，

$$\omega(f,\delta) = \sup_{|x-y|<\delta} |f(x) - f(y)|$$

が $\lim_{\delta \to 0}\omega(f,\delta)\ln(\delta) = 0$ を満たすならば，f のフーリエ級数は f に一様収束することも意味している．これは Dini-Lipschitz（ディニ-リプシッツ）の定理として知られている[2,7]．いいかえれば，単に連続性だけでは十分でないが，連続性に「付加」された条件（たとえば，微分可能性）が一様収束性を保証する．

ルベーグ定数をもっと精密に評価することもできる．Watson（ワトソン）[11]は，

$$\lim_{n\to\infty}\left(L_n - \frac{4}{\pi^2}\ln(2n+1)\right) = c$$

を示した．ここで

$$c = \frac{8}{\pi^2}\left(\sum_{k=1}^\infty \frac{\ln(k)}{4k^2-1}\right) - \frac{4}{\pi^2}\psi\left(\frac{1}{2}\right) = \frac{8}{\pi^2}\left(\sum_{j=0}^\infty \frac{\lambda(2j+2)-1}{2j+1}\right) + \frac{4}{\pi^2}(2\ln(2) + \gamma)$$

$$= 0.9894312738\cdots = \frac{4}{\pi^2}(2.4413238136\cdots)$$

γ はオイラー-マスケローニの定数【1.5】，$\psi(x)$ はディガンマ関数【1.5.4】であり，さらに $\lambda(x)$ は【1.7】に現れた関数である．L_n の漸近展開で生ずる高次の係数は，ベルヌーイ数【1.6.1】の有限の組み合わせとして書くことができる．Galkin（ガルキン）[12]はさらに $n \to \infty$ のとき $L_n - (4/\pi^2)\ln(2n+1)$ は c に単調減少し，一方 $L_n - (4/\pi^2)\ln(2n+2)$ は c に単調増加することを示した．さらなる漸近性については[13,14]にある．

Hardy（ハーディ）[15]によって発見された2つの積分公式に触れておこう．

$$L_n = 4\int_0^\infty \frac{\tanh((2n+1)x)}{\tanh(x)}\frac{1}{\pi^2+4x^2}dx$$
$$= \frac{4}{\pi^2}\int_0^\infty \frac{\sinh((2n+1)x)}{\sinh(x)}\ln\left(\coth\left(\frac{2n+1}{2}x\right)\right)dx$$

関連する議論についてはFavard（ファヴァール）定数に関するわれわれの小論【4.3】を参照されたい．

L_nの可能な拡張はたくさんある．フーリエ級数のどのような性質が当面の問題に移行するかを確かめることは興味深い．たとえば，Legendre（ルジャンドル）級数に対するルベーグ定数の単調性が，Szegöの予想を裏付けるかたちで証明されている[16]．

ここに関連するアイデアがある．fが単位円板の内部で複素解析的で境界上では連続，そしてすべての$|z|<1$に対して$|f(z)|<1$を満たすものとすると，そのとき[17]

$$f(z) = \sum_{k=0}^\infty a_k z^k \text{ とすると } \left|\sum_{k=0}^n a_k\right| \leq G_n$$

が成立する．ただし，

$$G_n = \sum_{m=0}^n \frac{1}{2^{4m}}\binom{2m}{m}^2 = 1 + \left(\frac{1}{2}\right)^2 + \left(\frac{1\cdot 3}{2\cdot 4}\right)^2 + \cdots + \left(\frac{1\cdot 3\cdots(2n-1)}{2\cdot 4\cdots(2n)}\right)^2$$

はn次Landau（ランダウ）の定数（【1.5.4】との類似性に注意）である．定数G_nは各nに対して最良である．次のことが知られている[11]．

$$\lim_{n\to\infty}\left(G_n - \frac{1}{\pi}\ln(n+1)\right) = \frac{1}{\pi}(4\ln(2)+\gamma) = 1.0662758532\cdots,$$

$$G_{2n} \leq L_n < \frac{4}{\pi}G_{2n}$$

そして数列$\{G_n\}$と$\{L_n/G_{2n}\}$はともに単調増加である．もっと詳細なことが[18-21]に見られる．

4.2.2 ラグランジュ補間

ここでは同じ「ルベーグ定数」という言葉が違った意味で使われる．実数値データ$X=\{x_1, x_2, \cdots, x_n\}$, $Y=\{y_1, y_2, \cdots, y_n\}$で，$-1\leq x_1 \leq x_2 \leq \cdots \leq x_n \leq 1$を満たすものが与えられたとき，$n-1$次の多項式$p_{X,Y}(x)$で

$$p_{X,Y}(x_i) = y_i, \quad i=1,2,\cdots,n$$

を満たすものが一意的に存在し，この多項式をXとYが与えられたときのLagrange（ラグランジュ）の補間多項式とよぶ．$p_{X,Y}(x)$の公式は

$$p_{X,Y}(x) = \sum_{k=1}^n \left(y_k \prod_{j\neq k}\frac{x-x_j}{x_k-x_j}\right)$$

である．$\{x_i\}$の空間内の配置が変化したり，nが増加したりするときの補間多項式の近似力の意味を考えてみよう．次の表現

$$\Lambda_n(X) = \max_{-1\leq x\leq 1}\sum_{k=1}^n\left|\prod_{j\neq k}\frac{x-x_j}{x_k-x_j}\right|$$

4.2 ルベーグ定数

はこの目的のために有用であり，X に対する n 次の**ルベーグ定数**とよばれる．Λ_n は Y に依存しないことに注意しよう．X の選択に無関係に，すべての n に対して

$$\Lambda_n > \frac{4}{\pi^2} \ln(n) - 1$$

が容易に示される．したがって $\lim_{n\to\infty} \Lambda_n = \infty$ である．これは，任意の X に対して n が大きくなるとき，$p_{X,f(X)}(x)$ が f に一様収束「しない」ような連続関数が存在することを意味している．いいかえれば，すべての連続関数 f に対して一様収束を保証するような「普遍的」な集合 X は存在しない．

Erdös[23]はルベーグ定数の下界をもっと厳しく詰めた．彼はすべての n と任意の X に対して

$$\Lambda_n > \frac{2}{\pi} \ln(n) - C$$

を満たす定数 C が存在することを証明した．C の最小の可能な値を簡潔に説明しよう．Erdös の結果（主要項）を改良することはできない．というのは，もし T を n 次 Chebyshev（チェビシェフ）多項式の n 個の零点

$$x_j = -\cos\left(\frac{(2j-1)\pi}{2n}\right) \quad (j=1,2,\cdots,n)$$

からなる集合とすると，次の不等式が成り立つからである．

$$\Lambda_n(T) = \frac{1}{n} \sum_{j=1}^{n} \cot\left(\frac{(2j-1)\pi}{4n}\right) \leq \frac{2}{\pi} \ln(n) + 1$$

実際，$\{\Lambda_n(T) - (2/\pi)\ln(n)\}$ は単調減少で[24-26]

$$\lim_{n\to\infty}\left(\Lambda_n(T) - \frac{2}{\pi}\ln(n)\right) = \frac{2}{\pi}(3\ln(2) - \ln(\pi) + \gamma) = 0.9625228267\cdots$$

である．ある完全な漸近展開（またしてもベルヌーイ数を含む）が[27-30]で得られている．

Λ_n を最小とするとするような最適集合 X^* はどんなものだろうか[22]？ 確かにチェビシェフ多項式の零点は X^* のよい候補ではあるが，他の X の方がさらによいことを示すことができる．Kilgore（キルゴア）[31]や de Boor & Pinkus（ド・ブーア＆ピンクス）[32]は，そのような X^* に関して Bernstein（ベルンシュテイン）の同程度振動予想[33]を証明した．X^* に関する，より正確な解析的表現は知られていない．

必ずしも絶望的とはいえない問題は，$\Lambda_n = \Lambda_n(X^*)$ を評価するものである．Vertesi（ヴェルテシ）[34-36]は Erdös[23]の仕事を足場に，

$$\lim_{n\to\infty}\left(\Lambda_n^* - \frac{2}{\pi}\ln(n)\right) = \frac{2}{\pi}(2\ln(2) - \ln(\pi) + \gamma) = 0.5212516264\cdots$$

を証明した．これは C の正体を決定したが，高次の漸近性や単調性の問題については未解決のままである．

[1] H. Lebesgue, Sur la représentation trigonométrique approchée des fonctions satisfaisant à une condition de Lipschitz, *Bull. Soc. Math. France* 38 (1910) 184-210; also in *Oeuvres Scientifiques*,

v. 3, L'Enseignement Math., 1972, pp. 363-389.
[2] A. G. Zygmund, *Trigonometric Series*, v. 1, 2nd ed., Cambridge Univ. Press, 1959; MR 89c:42001.
[3] L. Fejér, Sur les singularités de la série de Fourier des fonctions continues, *Annales Sci. École Norm. Sup.* 28 (1911) 64-103; also in *Gesammelte Arbeiten*, v. 1, Birkhäuser, 1970, pp. 654-689.
[4] L. Carlitz, Note on Lebesgue's constants, *Proc. Amer. Math. Soc.* 12 (1961) 932-935; MR 24 # A2791.
[5] G. Szegö, Über die Lebesgueschen Konstanten bei den Fourierschen Reihen, *Math. Z.* 9 (1921) 163-166; also in *Collected Papers*, v. 1, ed. R. Askey, Birkhäuser, 1982, pp. 307-313.
[6] T. J. Rivlin, *An Introduction to the Approximation of Functions*, Blaisdell, 1969; MR 83b:41001.
[7] E. W. Cheney, *Introduction to Approximation Theory*, McGraw-Hill, 1966; MR 99f:41001.
[8] R. L. Wheeden and A. G. Zygmund, *Measure and Integral: An Introduction to Real Analysis*, Dekker, 1977, pp. 222-229; MR 58 # 11295.
[9] V. M. Tikhomirov, Approximation theory, *Analysis II*, ed. R. V. Gamkrelidze, Springer-Verlag, 1990, pp. 93-255; MR 91e:00001.
[10] N. Korneichuk, *Exact Constants in Approximation Theory*, Cambridge Univ. Press, 1991; MR 92m:41002.
[11] G. N. Watson, The constants of Landau and Lebesgue, *Quart. J. Math.* 1 (1930) 310-318.
[12] P. V. Galkin, Estimates for Lebesgue constants (in Russian), *Trudy Mat. Inst. Steklov.* 109 (1971) 3-5; Engl. transl. in *Proc. Steklov Inst. Math.* 109 (1971) 1-4; MR 46 # 2338.
[13] L. Lorch, On Fejér's calculation of the Lebesgue constants, *Bull. Calcutta Math. Soc.* 37 (1945) 5-8; MR 7,59f.
[14] L. Lorch, The principal term in the asymptotic expansion of the Lebesgue constants, *Amer. Math. Monthly* 61 (1954) 245-249; MR 15,788d.
[15] G. H. Hardy, Note on Lebesgue's constants in the theory of Fourier series, *J. London Math. Soc.* 17 (1942) 4-13; also in *Collected Papers*, v. 3, Oxford Univ. Press, 1966, pp. 89-98; MR 4,36f.
[16] C. K. Qu and R. Wong, Szegö's conjecture on Lebesgue constants for Legendre series, *Pacific J. Math.* 135 (1988) 157-188; MR 89m:42025.
[17] E. Landau, Abschätzung der Koeffizientensumme einer Potenzreihe, *Archiv Math. Phys.* 21 (1913) 42-50, 250-255; also in *Collected Works*, v. 5-6, ed. L. Mirsky, I. J. Schoenberg, W. Schwarz, and H. Wefelscheid, Thales Verlag, 1985, pp. 432-440, 11-16.
[18] L. Brutman, A sharp estimate of the Landau constants, *J. Approx. Theory* 34 (1982) 217-220; MR 84m:41058.
[19] L. P. Falaleev, Inequalities for the Landau constants (in Russian), *Sibirskii Mat. Z.* 32 (1991) 194-195; Engl. transl. in *Siberian Math. J.* 32 (1992) 896-897; MR 93c:30001.
[20] J. E. Wilkins, The Landau constants, *Progress in Approximation Theory*, Proc. 1990 Tampa conf., ed. A. A. Gonchar and E. B. Saff, Academic Press, 1991, pp. 829-842; MR 92k:41045.
[21] D. Cvijovic and J. Klinowski, Inequalities for the Landau constants, *Math. Slovaca* 50 (2000) 159-164; MR 2001e:11125.
[22] L. Brutman, Lebesgue functions for polynomial interpolation-A survey, *Annals of Numer. Math.* 4 (1997) 111-127; MR 97m:41003.
[23] P. Erdös, Problems and results on the theory of interpolation. II, *Acta Math. Acad. Sci. Hungar.* 12 (1961) 235-244; MR 26 # 2779.
[24] F. W. Luttmann and T. J. Rivlin, Some numerical experiments in the theory of polynomial interpolation, *IBM J. Res. Develop.* 9 (1965) 187-191; MR 31 # 4147.
[25] H. Ehlich and K. Zeller, Auswertung der Normen von Interpolationsoperatoren, *Math. Annalen* 164 (1966) 105-112; MR 33 # 3005.
[26] T. J. Rivlin, The Lebesgue constants for polynomial interpolation, *Functional Analysis and Its Applications*, Proc. 1973 Madras conf., ed. H. G. Garnir, K. R. Unni, and J. H. Williamson, Lect. Notes in Math. 399, Springer-Verlag, 1974, pp. 422-437; MR 53 # 3549.
[27] R. Günttner, Evaluation of Lebesgue constants, *SIAM J. Numer. Anal.* 17 (1980) 512-520; MR 81i:41003.

- [28] A. K. Pokalo and L. I. Sloma,Ona representation of the Lebesgue constant by an asymptotic series (in Russian), *Dokl. Akad. Nauk BSSR* 25 (1981) 204-205, 284; MR 82i:41036.
- [29] P. N. Shivakumar and R.Wong, Asymptotic expansion of the Lebesgue constants associated with polynomial interpolation, *Math. Comp.* 39 (1982) 195-200; MR 83i:41006.
- [30] V. K. Dzjadik and V. V. Ivanov, On asymptotics and estimates for the uniform norms of the Lagrange interpolation polynomials corresponding to the Chebyshev nodal points, *Anal. Math.* 9 (1983) 85-97; MR 85d:41001.
- [31] T. A. Kilgore, A characterization of the Lagrange interpolating projection with minimal Tchebycheff norm, *J. Approx. Theory* 24 (1978) 273-288; MR 80d:41002.
- [32] C. de Boor and A. Pinkus, Proof of the conjectures of Bernstein and Erdös concerning the optimal nodes for polynomial interpolation, *J. Approx. Theory* 24 (1978) 289-303; MR 80d:41003.
- [33] S. N. Bernstein, Sur la limitation des valeurs d'un polynome $P_n(x)$ de degré n sur tout un segment par ses valeurs en $n+1$ points du segment, *Izv. Akad. Nauk SSSR* 8 (1931) 1025-1050; Russian transl. in *Collected Works*, v. 2, *Constructive Theory of Functions (1931-1953)*, Izdat. Akad. Nauk SSSR, 1954, pp. 107-126; MR 16,433o.
- [34] P. Vértesi, On the optimal Lebesgue constants for polynomial interpolation, *Acta Math. Hungar.* 47 (1986) 165-178; MR 87i:41005.
- [35] P. Vértesi, Optimal Lebesgue constant for Lagrange interpolation, *SIAM J. Numer. Anal.* 27 (1990) 1322-1331; MR 91k:41010.
- [36] R. Günttner, Note on the lower estimate of optimal Lebesgue constants, *Acta Math. Hungar.* 65 (1994) 313-317; MR 95d:41051.

4.3 アキェゼル-クレイン-ファヴァール定数

本節ではルベーグ定数【4.2】の知識を前提とする．関数 f は区間 $[-\pi,\pi]$ 上可積分そして $S_n(f,x)$ を f のフーリエ級数の n 次部分和とする．すべての x に対して $|f(x)|\leq 1$ ならば

$$|S_n(f,x)|\leq L_n=\frac{4}{\pi^2}\ln(n)+O(1)$$

であること，さらに L_n は最良（これは最大値である）であることは既知である．可積分関数の集合の部分集合である連続関数に制限しても，L_n は依然として最良である（このときは上限（supremum）にしかならないが）．

これは次の Kolmogorov（コルモゴロフ）[1-3] の結果の極限的な場合（$r=0$）と考えることもできる．ある整数 $r\geq 1$ を固定する．関数 f が r 回微分可能ですべての x に対して $|f^{(r)}(x)|\leq 1$ であるならば，

$$|f(x)-S_n(f,x)|\leq L_{n,r}=\frac{4}{\pi^2}\frac{\ln(n)}{n^r}+O\left(\frac{1}{n^r}\right)$$

である．ただし

$$L_{n,r} = \begin{cases} \dfrac{1}{\pi} \int_{-\pi}^{\pi} \left| \sum_{k=n+1}^{\infty} \dfrac{\sin(k\theta)}{k^r} \right| d\theta & (r \geq 1 \text{ が奇数のとき}) \\ \dfrac{1}{\pi} \int_{-\pi}^{\pi} \left| \sum_{k=n+1}^{\infty} \dfrac{\cos(k\theta)}{k^r} \right| d\theta & (r \geq 2 \text{ が偶数のとき}) \end{cases}$$

は最良である.

これはすべて Achiser-Krein-Favard(**アキェゼル-クレイン-ファヴァール**)**定数**を導入するための多少遠回りな方法である.この定数はしばしば単に**ファヴァール定数**と呼ばれる.これまで f のフーリエ評価 $S_n(f,x)$ の特性のみに焦点を絞ってきた.$S_n(f,x)$ を任意の三角多項式

$$P_n(x) = \frac{a_0}{2} + \sum_{k=1}^{n} (a_k \cos(kx) + b_k \sin(kx))$$

によって置きかえたとしよう.ここで係数にいかなる条件も課さない(実数であることは別として)ものとする.以前と同様に f の r 階導関数が有界で -1 と 1 の間の値をとるものとするならば,すべての x に対して

$$|f(x) - P_n(x)| \leq \frac{K_r}{(n+1)^r}$$

を満たす多項式 $P_n(x)$ が「存在する」.ここで r 次のファヴァール定数

$$K_r = \frac{4}{\pi} \sum_{j=0}^{\infty} \left[\frac{(-1)^j}{2j+1} \right]^{r+1}$$

は可能な最小の分子である.いいかえれば,ルベーグ定数は最小2乗法の意味(フーリエ級数)で最良である近似に関連しているが,ファヴァール定数は各点ごとの意味で最良であるような近似に関係している.

次のことを見ておこう.

$$K_r = \begin{cases} \dfrac{4}{\pi} \lambda(r+1) & (r \text{ が奇数のとき}) \\ \dfrac{4}{\pi} \beta(r+1) & (r \text{ が偶数のとき}) \end{cases}$$

ただし,λ と β 関数は【1.7】で議論したものである.各ファヴァール定数はそれゆえ π^r の有理数倍である.たとえば,

$$K_0 = 1, \quad K_1 = \frac{\pi}{2}, \quad K_2 = \frac{\pi^2}{8}, \quad K_3 = \frac{\pi^3}{24}$$

また $1 = K_0 < K_2 < \cdots < 4/\pi < \cdots < K_3 < K_1 = \pi/2$.

これは,いろいろな関数族と近似の方法において,定数 K_r を含む多くの最良の結果の中で最初に得られたものである.この定理はかなり技巧的であって,ここでは議論することができない.しかしながら Bohr-Favard(ボーア-ファヴァール)の不等式[7-9]と Landau-Kolmogorov(ランダウ-コルモゴロフ)の定数【3.3】に触れておこう.また,[10,11]も参照してほしい.

ここに未解決な問題がある.任意の三角不等式 $P_n(\theta)$ に対して

$$\max_{-\pi \le \theta \le \pi} |P_n(\theta)| \le C \frac{n}{2\pi} \int_{-\pi}^{\pi} |P_n(\theta)| d\theta$$

が知られている[12,13]．さらに最良の定数は漸近的に $n \to \infty$ のとき $0.539 \le C \le 0.58$ を満たす．C の正確な表現は知られていない．

[1] A. N. Kolmogorov, Zur Grössenordnung des Restgliedes Fourierscher reihen differenzierbarer Funktionen, *Annals of Math.* 36 (1935) 521-526; Engl. transl. in *Selected Works*, v. 1., ed. V. M. Tikhomirov, Kluwer, 1991, pp. 196-201; MR 93d:01096.
[2] S. Nikolsky, Approximations of periodic functions by trigonometrical polynomials (in Russian), *Travaux Inst. Math. Stekloff* 15 (1945) 1-76; MR 7,435c.
[3] A. G. Zygmund, *Trigonometric Series*, v. 1, 2nd ed., Cambridge Univ. Press, 1959; MR 89c:42001.
[4] J. Favard, Sur les meilleurs procédés d'approximation de certaines classes de fonctions par des polynomes trigonométriques, *Bull. Sci. Math.* 61 (1937) 209-224, 243-256.
[5] N. Achieser and M. Krein, Sur la meilleure approximation des fonctions périodiques dérivables au moyen de sommes trigonométriques, *C. R. (Dokl.) Acad. Sci. URSS* 15 (1937) 107-111.
[6] N. I. Achieser, *Theory of Approximation*, F. Ungar, 1956, pp. 187-199, 300-301; MR 3,234d and MR 10,33b.
[7] V. M. Tikhomirov, Approximation theory, *Analysis II*, ed. R. V. Gamkrelidze, Springer-Verlag, 1990, pp. 93-255; MR 91e:00001.
[8] D. S. Mitrinovic, J. E. Pecaric, and A. M. Fink, *Inequalities Involving Functions and Their Integrals and Derivatives*, Kluwer, 1991; MR 93m:26036.
[9] N. Korneichuk, *Exact Constants in Approximation Theory*, Cambridge Univ. Press, 1991; MR 92m:41002.
[10] B. C. Yang and Y. H. Zhu, Inequalities for the Hurwitz zeta-function on the real axis (in Chinese), *Acta Sci. Natur. Univ. Sunyatseni*, v. 36 (1997) n. 3, 30-35; MR 99h: 11101.
[11] B. C. Yang and K. Wu, Inequalities for Favard's constant (in Chinese), *J. South China Normal Univ. Natur. Sci. Ed.* (1998) n. 1, 12-17; MR 2001a:11031.
[12] L. V. Taikov, A group of extremal problems for trigonometric polynomials (in Russian), *Uspekhi Mat. Nauk*, v. 20 (1965) n. 3, 205-211; MR 32 #2829.
[13] G. V. Milovanovic, D. S. Mitrinovic, and T. M. Rassias, *Topics in Polynomials: Extremal Problems, Inequalities, Zeros*, World Scientific, 1994, pp. 495-496; MR 95m:30009.

4.4 ベルンシュテイン定数

区間 $[-1,1]$ で定義された任意の実関数 $f(x)$ に対して $E_n(f)$ をたかだか n 次の実多項式による f の最良一様近似の誤差とする．すなわち，

$$E_n(f) = \inf_{p \in P_n} \sup_{-1 \le x \le 1} |f(x) - p(x)|$$

ただし $P_n = \{\sum_{k=0}^{n} a_k x^k : a_k$ は実数$\}$．特別な場合 $\alpha(x) = |x|$ を考えよう．これに対しては Jackson（ジャクソン）の定理[1,2]は $E_n(\alpha) \le 6/n$ であることを示している．$|x|$ が $[-1,1]$ 上の連続偶関数であることから，$[-1,1]$ 上の P_n からの（一意的に存在する）最良一様近似もそうである．そのことより $E_{2n}(\alpha) = E_{2n+1}(\alpha)$ が従う．それゆえ，

偶数次の場合のみを考える．Bernstein（ベルンシュテイン）[3]はジャクソンの不等式
$$2nE_{2n}(a) \leq 6$$
を，チェビシェフの多項式【4.9】を利用して，
$$2nE_{2n}(a) \leq \frac{4n}{\pi(2n+1)} < \frac{2}{\pi} = 0.636\cdots$$
まで改良した．彼は次のような極限の存在を証明し，そして次のような上界と下界を得た．
$$0.278\cdots < \beta = \lim_{n\to\infty} 2nE_{2n}(a) < 0.286\cdots$$
Bernstein は $\beta = 1/(2\sqrt{\pi}) = 0.2821\cdots$ を予想した．この予想は，70年ほど未解決のままであった．その理由は，大きい n に対して $E_{2n}(a)$ を計算することの難しさと，$2nE_{2n}(a)$ の β への収束がゆっくりしていることによる．

Varga（ヴァルガ）と Carpenter（カーペンター）[4,5]は $\beta = 0.2801694990\cdots$ を小数50位まで計算し，Bernstein の予想が誤っていることを立証した．95桁の精度の計算といくつかの他の技法を使った $n=52$ までの $2nE_{2n}(a)$ の計算を必要とした．[4]の終わりに，彼らは，β が古典的な超幾何級数あるいは他のよく知られた定数を用いて簡潔に表現できることを信ずる方がまだしももっともらしいのではないかと指摘した．

$|x|$ への最良一様「多項式」近似の問題を議論したところでもあるので，同じように最良一様「有理式」近似の問題を考察するのは自然である．$[-1,1]$ 上の任意の関数 f に対して，
$$E_{m,n}(f) = \inf_{r \in R_{m,n}} \sup_{-1\leq x\leq 1} |f(x) - r(x)|$$
とおく．ただし，$R_{m,n} = \{p(x)/q(x) : p \in P_m, q \in P_n, q \neq 0\}$ である．Newman（ニューマン）[6]は
$$\frac{1}{2} e^{-9\sqrt{n}} \leq E_{n,n}(a) \leq 3e^{-\sqrt{n}} \quad (n \geq 4)$$
を，またこれと同値であるが，E_n とは比較にならない速さで $E_{n,n} \to 0$ であることを証明した．Newman のこの結果は研究者の間に大きな興奮を引き起こした[5,7]．Bulanov（ブラノフ）[8]は Gonchar（ゴンチャール）[9]の結果を拡張して，その下界が
$$e^{-\sqrt{n+1}} \leq E_{n,n}(a)$$
に改良できることを証明し，Vjacheslavov（ヴャチェスラボフ）[10]は正の定数 m と M が存在して，
$$m \leq e^{\pi\sqrt{n}} E_{n,n}(a) \leq M$$
であることを証明した．(Petrushev & Popov（ペトルシェフ＆ポポフ）[7]は一見関わりのないように見える定数 e と π が興味深くも並び立っているのに注目した．) 多項式の場合と同じように $E_{2n,2n}(a) = E_{2n+1,2n+1}(a)$ であるから，添え数が偶数の場合に焦点を合わせる．Varga, Ruttan & Carpenter（ヴォルガ，ルッタン＆カーペンター）[11]は，注意深い計算をもとに

$$\lim_{n\to\infty} e^{\pi\sqrt{2n}} E_{2n,2n}(a) = 8$$

を予想したが，最近 Stahl（スタール）[12,13]が証明した．多項式と有理式の対比はなんと魅惑的なことか！

Gonchar[9]は Zolotarev の結果【3.9】がこの一連の研究と関連していることを指摘した．

[1] T. J. Rivlin, *An Introduction to the Approximation of Functions*, Blaisdell, 1969; MR 83b:41001.
[2] E. W. Cheney, *Introduction to Approximation Theory*, McGraw-Hill, 1966; MR 99f: 41001.
[3] S. N. Bernstein, Sur la meilleure approximation de $|x|$ par les polynomes de degrés donnés, *Acta Math.* 37 (1913) 1-57; Russian transl. in *Collected Works*, v. 1, *Constructive Theory of Functions (1905-1930)*, Izdat. Akad. Nauk SSSR, 1952; MR 14,2c.
[4] R. S. Varga and A. J. Carpenter, A conjecture of S. Bernstein in approximation theory (in Russian), *Mat. Sbornik* 129 (1986) 535-548, 591-592; Engl. transl. in *Math. USSR Sbornik* 57 (1987) 547-560; MR 87g:41066.
[5] R. S. Varga, *Scientific Computation on Mathematical Problems and Conjectures*, SIAM, 1990; MR 92b:65012.
[6] D. J. Newman, Rational approximation to $|x|$, *Michigan Math. J.* 11 (1964) 11-14; MR 30 # 1344.
[7] P. P. Petrushev and V. A. Popov, *Rational Approximation of Real Functions*, Cambridge Univ. Press, 1987; MR 89i:41022.
[8] A. P. Bulanov, Asymptotics for least deviations of $|x|$ from rational functions (in Russian), *Mat. Sbornik* 76 (1968) 288-303; Engl. transl. in *Math. USSR Sbornik* 5 (1968) 275-290; MR 37 # 4468.
[9] A. A. Gonchar, Estimates of the growth of rational functions and some of their applications (in Russian), *Mat. Sbornik* 72 (1967) 489-503; Engl. transl. in *Math. USSR Sbornik* 1 (1967) 445-456; MR 35 # 4652.
[10] N. S. Vjacheslavov, The uniform approximation of $|x|$ by rational functions (in Russian), *Dokl. Akad. Nauk SSSR* 220 (1975) 512-515; Engl. transl. in *Soviet Math. Dokl.* 16 (1975) 100-104; MR 52 # 1114.
[11] R. S. Varga, A. Ruttan, and A. J. Carpenter, Numerical results on best uniform rational approximations to $|x|$ on $[-1, +1]$ (in Russian), *Mat. Sbornik* 182 (1991) 1523-1541; Engl. transl. in *Math. USSR Sbornik* 74 (1993) 271-290; MR 92i:65040.
[12] H. Stahl, Best uniform rational approximation of $|x|$ on $[-1, 1]$ (in Russian), *Mat. Sbornik* 183 (1992) 85-118; Engl. transl. in *Russian Acad. Sci. Sbornik Math.* 76 (1993) 461-487; MR 93i:41019.
[13] H. Stahl, Uniform rational approximation of $|x|$, *Methods of Approximation Theory in Complex Analysis and Mathematical Physics*, ed. A. A. Gonchar and E. B. Saff, Lect. Notes in Math. 1550, Springer-Verlag, 1993, pp. 110-130; MR 95m:41031.

4.5 「1/9」予想の定数

ここで半直線 $[0,\infty)$ 上の関数 $\exp(-x)$ の有理式近似を問題にする．$\lambda_{m,n}$ を最良一様近似の誤差

$$\lambda_{m,n} = \inf_{r \in R_{m,n}} \sup_{x \geq 0} |e^{-x} - r(x)|$$

とする．ただし，【4.4】で定義したように $R_{m,n}$ は有理関数 $p(x)/q(x)$ ($\deg(p) \leq m, \deg(q) \leq n$ かつ $q \neq 0$)) の集合である．

明らかに

$$0 < \lambda_{n,n} \leq \lambda_{n-1,n} \leq \lambda_{n-2,n} \leq \cdots \leq \lambda_{2,n} \leq \lambda_{1,n} \leq \lambda_{0,n}$$

となるから，$m=0$ と $m=n$ のときが最も興味ある 2 つの場合である．多くの研究者[1-4]は**チェビシェフ定数**に関連させてこれらの定数 $\lambda_{m,n}$ を研究してきた[4]．ここではほんの少しの結果について述べるにとどめる．Schönhage（シェーンハーゲ）[5]は

$$\lim_{n \to \infty} \lambda_{0,n}^{1/n} = \frac{1}{3}$$

を証明し，このことが

$$\lim_{n \to \infty} \lambda_{n,n}^{1/n} = \frac{1}{9}$$

という予想に何人かの研究者を導いた．Schönhage[6]と Trefethen & Gutknecht（トレフェテン&グトクネヒト）[7]によって発見された数値的証拠は，予想が間違いであることを示唆した．その極限が存在するという証明は発見されてはいなかったが，Carpenter, Ruttan & Varga（カーペンター，ルッタン&ヴァルガ）[8]はチェビシェフ定数を $n=30$ まで 200 桁の精度で計算し，注意深く

$$\lim_{n \to \infty} \lambda_{n,n}^{1/n} = \frac{1}{9.2890254919\cdots} = 0.1076539192\cdots$$

を得た．

Opitz & Scherer（オピッツ&シュラー）[9]と Magnus（マグナス）[10-12]の結果を基礎に Gonchar & Rakhmanov（ゴンチャール&ラクマノフ）[4,13]は，極限が存在して，それは

$$\Lambda = \exp\left(\frac{-\pi K(\sqrt{1-c^2})}{K(c)}\right)$$

に等しいことを示した．ただし $K(x)$ は第 1 種完全楕円積分【1.4.6】で，定数 c は次のように得られる定数である．$E(x)$ を第 2 種完全楕円積分【1.4.6】とするとき，$0 < c < 1$ は方程式 $K(c) = 2E(c)$ の一意的な解である．

Gonchar と Rakhmanov の「1/9」予想の精密な反証では，複素ポテンシャル論からのアイデアを使ったが，それは $\exp(-x)$ の有理式近似からあまりにもはずれすぎるように思われる！ 彼らはまた「1/9」定数 Λ の数論的特徴付けも得た．もし

$$f(z) = \sum_{j=1}^{\infty} a_j z^j, \quad \text{ここで } a_j = \left|\sum_{d|j} (-1)^d d\right|$$

とするならば，f は単位開円板で複素解析的である．方程式 $f(z) = 1/8$ の一意に存在する正の解が定数 Λ である．a_j を書き下すもう 1 つの方法は次のようなものである[14]．もし

$$j = 2^m p_1^{m_1} p_2^{m_2} \cdots p_k^{m_k}$$

が整数 j の素因数分解で，$p_1<p_2<\cdots<p_k$ が奇素数で，$m\geqq 0$ および $m_i\geqq 1$ であるならば，

$$a_j=|2^{m+1}-3|\frac{p_1^{m_1+1}-1}{p_1-1}\frac{p_2^{m_2+1}-1}{p_2-1}\cdots\frac{p_k^{m_k+1}-1}{p_k-1}$$

Carpenter[4]はこの方程式を使って Λ を 101 桁まで計算した.
ここに Magnus[10]によるもう1つの表現がある.「1/9」定数 Λ は，方程式

$$\sum_{k=0}^{\infty}(2k+1)^2(-x)^{k(k+1)/2}=0 \quad (0<x<1)$$

の一意的な解であり，このことは既に Halphen（ハルフェン）[15]によって 100 年前から研究されてきたものであることがわかっている．Halphen はテータ関数に関心があり，明らかにこの定数が 1 世紀後に有名になるとは気づかず，Λ を 6 桁まで計算した．Λ はハルフェン定数と改名されるべきことを Varga[4]は提案した．このようにこの近似問題の解決に多くの研究者たちが貢献してきた．しかしながらおかしくも不正確な「1/9」の名称を使い続けるのが最も簡単なのかもしれない．

Λ を定義する定数 $c=0.9089085575\cdots$ は，まったく関連のない分野である「Euler の弾性学」の研究[16-18]に出てくる．ここで議論されたような楕円関数の商は【7.8】に現れる．

[1] W. J. Cody, G. Meinardus, and R. S. Varga, Chebyshev rational approximations to e^{-x} in $[0, +\infty)$ and applications to heat-conduction problems, *J. Approx. Theory* 2 (1969) 50-65; MR 39 # 6536.
[2] R. S. Varga, *Topics in Polynomial and Rational Interpolation and Approximation*, Les Presses de l'Université de Montréal, 1982; MR 83h:30041.
[3] P. P. Petrushev and V. A. Popov, *Rational Approximation of Real Functions*, Cambridge Univ. Press, 1987; MR 89i:41022.
[4] R. S. Varga, *Scientific Computation on Mathematical Problems and Conjectures*, SIAM, 1990; MR 92b:65012.
[5] A. Schönhage, Zur rationalen Approximierbarkeit von e^{-x} über $[0,\infty)$, *J. Approx. Theory* 7 (1973) 395-398; MR 49 # 3391.
[6] A. Schönhage, Rational approximation to e^{-x} and related L_2-problems, *SIAM J. Numer. Anal.* 19 (1982) 1067-1080; MR 83k:41016.
[7] L. N. Trefethen and M. H. Gutknecht, The Carathéodory-Fejér method for real rational approximation, *SIAM J. Numer. Anal.* 20 (1983) 420-436; MR 85g:41024.
[8] A. J. Carpenter, A. Ruttan, and R. S. Varga, Extended numerical computations on the "1/9" conjecture in rational approximation theory, *Rational Approximation and Interpolation*, ed. P. R. Graves-Morris, E. B. Saff, and R. S. Varga, Lect. Notes in Math. 1105, Springer-Verlag, 1984, pp. 383-411.
[9] H.-U. Opitz and K. Scherer, On the rational approximation of e^{-x} on $[0,\infty)$, *Constr. Approx.* 1 (1985) 195-216; MR 88f:41027.
[10] A. P. Magnus, On the use of the Carathéodory-Fejér method for investigating "1/9" and similar constants, *Nonlinear Numerical Methods and Rational Approximation*, ed. A. Cuyt, Reidel, 1988, pp. 105-132, MR 90j:65035.
[11] A. P. Magnus, Asymptotics and super asymptotics for best rational approximation error norms to the exponential function (the "1/9" problem) by the Carathéodory-Fejér method, *Nonlinear Numerical Methods and Rational Approximation II*, ed. A. Cuyt, Kluwer, 1994, pp. 173-185; MR

96b:41023.
[12] A. P. Magnus and J. Meinguet, The elliptic functions and integrals of the "1/9" problem, *Numer. Algorithms* 24 (2000) 117-139; MR 2001f:41017.
[13] A. A. Gonchar and E. A. Rakhmanov, Equilibrium distributions and degree of rational approximation of analytic functions (in Russian), *Mat. Sbornik* 134 (1987) 306-352, 447; Engl. transl. in *Math. USSR Sbornik* 62 (1989) 305-348; MR 89h:30054.
[14] A. A. Gonchar, Rational approximations of analytic functions (in Russian), *International Congress of Mathematicians*, v. 1, Proc. 1986 Berkeley conf., ed. A. M. Gleason, Amer. Math. Soc., 1987, pp. 739-748; Engl. transl. in *Amer. Math. Soc. Transl.* 147 (1990) 25-34; MR 89e:30066.
[15] G.-H. Halphen, *Traité des fonctions elliptiques et de leurs applications*, v. 1, Gauthier-Villars, 1886, p. 287.
[16] D. A. Singer, Curves whose curvature depends on distance from the origin, *Amer. Math. Monthly* 106 (1999) 835-841; MR 2000j:53005.
[17] T. A. Ivey and D. A. Singer, Knot types, homotopies and stability of closed elastic rods, *Proc. London Math. Soc.* 79 (1999) 429-450; MR 2000g:58015.
[18] C. Truesdell, The influence of elasticity on analysis: The classic heritage, *Bull. Amer. Math. Soc.* 9 (1983) 293-310; MR 85f:01004.

4.6 フランセン-ロビンソン定数

x が増加するとき,ガンマ関数の逆数 $1/\Gamma(x)$ は,任意の定数 c に対して $\exp(-cx)$ よりもはるかに急激に減少する.かくして,ある確率モデルに対する片側密度関数として役立っている.1つの結果として,次の値

$$I = \int_0^\infty \frac{1}{\Gamma(x)} dx = 2.8077702420\cdots$$

が正規化のために必要である.

この積分を計算する1つの方法はリーマン和 I_n の $n \to \infty$ のときの極限によるものである,ただし[1]

$$I_n = \frac{1}{n}\sum_{k=1}^\infty \frac{1}{\Gamma\left(\dfrac{k}{n}\right)} = \begin{cases} e = 2.7182818284\cdots & (n=1 \text{ のとき}) \\ \dfrac{1}{2}\left(\dfrac{1}{\sqrt{\pi}} + e\,\mathrm{erfc}(-1)\right) = 2.7865848321 & (n=2 \text{ のとき}) \end{cases}$$

そして

$$\mathrm{erf}(x) = \frac{2}{\sqrt{\pi}}\int_0^x \exp(-t^2)\,dt = 1 - \mathrm{erfc}(x)$$

は誤差関数である.しかしながら,これは高い精度で I を計算するための方法としてはあまりに遅すぎる.

Fransén (フランセン)[2]は Euler-Maclaurin (オイラー-マクローリン) の総和法と次の公式を使って I を小数65位まで計算した.その公式とは

4.6 フランセン-ロビンソン定数

$$\Gamma(x) = \frac{e^{-\gamma x}}{x} \prod_{n=1}^{\infty} \left(1+\frac{x}{n}\right)^{-1} e^{x/n} = \frac{1}{x} \exp\left[\sum_{k=1}^{\infty} \frac{(-1)^k s_k}{k} x^k\right]$$

である．ただし，$s_1 = \gamma$ および $s_k = \zeta(k)$，$k \geq 2$ である．オイラー-マスケロニ定数 γ の背景は【1.5】に，そして Riemann のゼータ関数 $\zeta(x)$ の背景は【1.6】に見られる．

Robinson（ロビンソン）[2]は独立に I の評価を11点の Newton-Cotes（ニュートン-コーツ）近似公式を使って36桁まで得た．Fransén & Wrigge（フランセン&リッジュ）[3,4]はテイラー級数と他の解析的な手段により80桁を，そして続いて Johnson（ジョンソン）[5]は300桁を成し遂げた．

Sebah（セバ）[6]は（チェビシェフ多項式に基づく）Clenshaw & Curtis（クレンショウ&カーティス）の方法を利用してフランセン-ロビンソン定数を600桁以上計算した．彼はまた

$$I = \int_1^2 \frac{f(x)}{\Gamma(x)} dx$$

という基本的事実に着目した．ただし $f(x)$ は高速に収束する級数

$$f(x) = x + \sum_{k=0}^{\infty} \left(\prod_{j=0}^{k} \frac{1}{x+j}\right) = x + e \sum_{k=0}^{\infty} \frac{(-1)^k}{k!(x+k)}$$

によって定義されて，$f(1) = f(2) = e$，$f(3/2) = (1+e\sqrt{\pi}\,\text{erf}(1))/2$ である．これを使って，I は現在1025桁まで知られている．

Ramanujan（ラマヌジャン）[7,8]は $w=1$ のときの値として $2.2665345077\cdots$ をもつ公式

$$\int_0^{\infty} \frac{w^x}{\Gamma(1+x)} dx = e^w - \int_{-\infty}^{\infty} \frac{\exp(-we^y)}{y^2+\pi^2} dy$$

を考察した．w について微分すると，I を一般化した類似の表現が得られる．

$$\frac{1}{w} \int_0^{\infty} \frac{w^x}{\Gamma(x)} dx = e^w + \int_{-\infty}^{\infty} \frac{\exp(-we^y+y)}{y^2+\pi^2} dy$$

このような公式は，ガンマ関数の逆数分布に対するモーメントを計算するときに有用である．

関数 x^x は $\Gamma(x)$ より速く大きくなる．そして[10]は次のような計算をした．

$$\int_0^{\infty} \frac{1}{x^x} dx = 1.994559575\cdots, \quad \int_1^{\infty} \frac{1}{x^x} dx = 0.7041699604\cdots$$

反復指数関数（iterated exponentials）に関してはもっとたくさんのことが【6.11】に見られる．逆分布は多重 Barnes（バーンズ）関数【2.15】あるいは一般ガンマ関数【2.21】に基づいても作られる．

[1] J. Spanier and K. B. Oldham, *An Atlas of Functions*, Hemisphere, 1987, p. 415, formula 43:5:12.
[2] A. Fransén, Accurate determination of the inverse gamma integral, *BIT* 19 (1979) 137-138; MR 80c:65047.
[3] A. Fransén and S. Wrigge, High-precision values of the gamma function and of some related coefficients, *Math. Comp.* 34 (1980) 553-566; addendum and corrigendum, 37 (1981) 233-235; MR 81f:65004 and MR 82m:65002.

[4] S. Plouffe, Fransén-Robinson constant (Plouffe's Tables).
[5] A. Fransén and S. Wrigge, Calculation of the moments and the moment generating function for the reciprocal gamma distribution, *Math. Comp.* 42 (1984) 601-616; MR 86f:65042a.
[6] P. Sebah, Several computations of the Fransén-Robinson constant, unpublished notes (2000-2001).
[7] G. H. Hardy, Another formula of Ramanujan, *J. London Math. Soc.* 12 (1937) 314-318; also in *Collected Papers*, v. 4, Oxford Univ. Press, 1966, pp. 544-548.
[8] G. H. Hardy, *Ramanujan: Twelve Lectures on Subjects Suggested by His Life and Work*, Chelsea, 1940, p. 196; MR 21 # 4881.
[9] S. Wrigge, A note on the moment generating function for the reciprocal gamma distribution, *Math. Comp.* 42 (1984) 617-621; MR 86f:65042b.
[10] G. N. Watson, Theorems stated by Ramanujan. VIII: Theorems on divergent series, *J. London Math. Soc.* 4 (1929) 82-86.

4.7 ベリー-エッセン定数

$X_1, X_2, \cdots, X_n$ を独立な確率変数で，モーメントについて，各 $1 \leq k \leq n$ に対して
$$E(X_k)=0, \quad E(X_k^2)=\sigma_k^2>0, \quad E(|X_k|^3)=\beta_k<\infty$$
を満たすものとする．Φ_n を次の確率変数

$$X=\frac{1}{\sigma}\sum_{k=1}^{n}X_k, \quad \text{ただし} \quad \sigma^2=\sum_{k=1}^{n}\sigma_k^2$$

の分布関数とする．また Lyapunov（リャプノフ）比を次のように定義する．

$$\lambda=\frac{\beta}{\sigma^3}, \quad \text{ここに} \quad \beta=\sum_{k=1}^{n}\beta_k$$

Φ を標準正規分布関数としよう．Berry（ベリー）[1] と Esseen（エッセン）[2,3] は次の条件を満たす定数 C が存在することを証明した．

$$\sup_n \sup_{F_k} \sup_x |\Phi_n(x)-\Phi(x)| \leq C\lambda$$

ここで，すべての k に対して F_k は X_k の分布関数とする．そのような定数 C の最小なものは，ここで与えられた条件のもとで，下に示す範囲にある [4-12]．

$$0.4097321837\cdots=\frac{3+\sqrt{10}}{6\sqrt{2\pi}} \leq C < 0.7915$$

$X_1, X_2, \cdots, X_n$ が同一分布ならば，C の上界は 0.7655 に改良できる．さらに，C がここに示した下界に等しいことの漸近的証拠が存在する．

[13-22] は関連する研究を含む．要するに，ベリー-エッセンの不等式は中心極限定理における収束の速さを，すなわち，正規分布が独立確率変数の和の分布にどれほど近いかを定式化するものである [23-26]．Hall & Barbour（ホール&バーブール）[27] は，これと反対に，2つの分布がどれくらい離れていなければならないかを記述する不等式を示した．ここにもまたもう 1 つの定数が派生してくるが，それについては少ししかわ

かっていない.

[1] A. C. Berry, The accuracy of the Gaussian approximation to the sum of independent variates, *Trans. Amer. Math. Soc.* 49 (1941) 122-136; MR 2,228i.
[2] C.-G. Esseen, On the Liapounoff limit of error in the theory of probability, *Ark. Mat. Astron. Fysik*, v. 28A (1942) n. 9, 1-19; MR 6,232k.
[3] C.-G. Esseen, Fourier analysis of distribution functions: A mathematical study of the Laplace -Gaussian law, *Acta Math.* 77 (1945) 1-125; MR 7,312a.
[4] H. Bergström, On the central limit theorem, *Skand. Aktuarietidskr.* 27 (1944) 139-153; 28 (1945) 106-127; 32 (1949) 37-62; MR 7,458e, MR 7,459a, and MR 11,255b.
[5] C.-G. Esseen, A moment inequality with an application to the central limit theorem, *Skand. Aktuarietidskr* 39 (1956) 160-170; MR 19,777f.
[6] S. Ikeda, A note on the normal approximation to the sum of independent random variables, *Annals Inst. Statist. Math. Tokyo* 11 (1959) 121-130; MR 24 # A570.
[7] S. Zahl, Bounds for the central limit theorem error, *SIAM J. Appl. Math.* 14 (1966) 1225-1245; MR 35 # 1077.
[8] V. M. Zolotarev, Absolute estimate of the remainder in the central limit theorem (in Russian), *Teor. Verojatnost. i Primenen.* 11 (1966) 108-119; Engl. transl. in *Theory Probab. Appl.* 11 (1966) 95-105; MR 33 # 6686.
[9] V. M. Zolotarev, A sharpening of the inequality of Berry-Esseen, *Z. Wahrsch. Verw. Gebiete* 8 (1967) 332-342; MR 36 # 4622.
[10] V. M. Zolotarev, Some inequalities from probability theory and their application to a refinement of A. M. Ljapunov's theorem (in Russian), *Dokl. Akad. Nauk SSSR* 177 (1967) 501-504; Engl. transl. in *Soviet Math. Dokl.* 8 (1967) 1427-1430; MR 36 # 3398.
[11] P. van Beek, An application of Fourier methods to the problem of sharpening the Berry-Esseen inequality, *Z. Wahrsch. Verw. Gebiete* 23 (1972) 187-196; MR 48 # 7342.
[12] I. S. Shiganov, Refinement of the upper bound of a constant in the remainder term of the central limit theorem (in Russian), *Problemy Ustoi Chivosti Stokhasticheskikh Modelei*, Proc. 1982 Moscow seminar, ed. V. M. Zolotarev and V. V. Kalashnikov, Vsesoyuz. Nauchno-Issled. Inst. Sistem. Issled., 1982, pp. 109-115; MR 85d:60008.
[13] U. V. Linnik, On the accuracy of the approximation to a Gaussian distribution of sums of independent random variables (in Russian), *Izv. Akad. Nauk SSSR* 11 (1947) 111-138; Engl. transl. in *Selected Transl. Math. Statist. and Probab.*, v. 2, Amer. Math. Soc., pp. 131-158; MR 8,591c.
[14] A. N. Kolmogorov, Some recent work on limit theorems in probability theory (in Russian), *Vestn. Moskov. Gos. Univ.* 10 (1953) 29-38; Engl. transl. in *Selected Works*, v. 2, ed. A. N. Shiryayev, Kluwer, 1992, pp. 406-418; MR 92j:01071.
[15] V. V. Petrov, On precise estimates in limit theorems (in Russian), *Vestnik Leningrad. Univ.*, v. 10 (1955) n. 11, 57-58; MR 17,753g.
[16] B. A. Rogozin, A remark on the paper "A moment inequality with an application to the central limit theorem" by C. G. Esseen (in Russian), *Teor. Verojatnost. i Primenen.* 5 (1960) 125-128; Engl. transl. in *Theory Probab. Appl.* 5 (1960) 114-117; MR 24 # A3683.
[17] H. Prawitz, On the remainder in the central limit theorem. I: One dimensional independent variables with finite absolute moments of third order, *Scand. Actuar. J.*, 1975, 145-156; MR 53 # 1695.
[18] R. Michel, On the constant in the nonuniform version of the Berry-Esseen theorem, *Z. Wahrsch. Verw. Gebiete* 55 (1981) 109-117; MR 82c:60042.
[19] L. Paditz, On the analytical structure of the constant in the nonuniform version of the Esseen inequality, *Statistics* 20 (1989) 453-464; MR 90k:60046.
[20] A. Mitalauskas, On the calculation of the constant in the Berry-Esseen inequality for a class

of distributions (in Russian), *Liet. Mat. Rink.* 32 (1992) 526-531; Engl. transl. in *Lithuanian Math. J.* 32 (1992) 410-413; MR 94i:60031.
[21] V. Bentkus, On the asymptotical behavior of the constant in the Berry-Esseen inequality, *J. Theoret. Probab.* 7 (1994) 211-224; MR 95d:60042.
[22] G. P. Chistyakov, Asymptotically proper constants in Lyapunov's theorem (in Russian), Probability and Statistics, 1, *Zap. Nauchn. Sem. S.-Peterburg. Otdel. Mat. Inst. Steklov (POMI)* 228 (1996) 349-355, 363-364; Engl. transl. in *J. Math. Sci.* 93 (1999) 480-483; MR 98g:60038.
[23] V. V. Petrov, *Sums of Independent Random Variables*, Springer-Verlag, 1975; MR 52 # 9335.
[24] V. V. Petrov, *Limit Theorems of Probability Theory: Sequences of Independent Random Variables*, Oxford Univ. Press, 1995; MR 96h:60048.
[25] P. Hall, *Rates of Convergence in the Central Limit Theorem*, Pitman, 1982; MR 84k:60032.
[26] R. N. Bhattacharya and R. Ranga Rao, *Normal Approximation and Asymptotic Expansions*, Wiley, 1976; MR 55 # 9219.
[27] P. Hall and A. D. Barbour, Reversing the Berry-Esseen inequality, *Proc. Amer. Math. Soc.* 90 (1984) 107-110; MR 86a:60028.

4.8 ラプラス限界定数

実数 M と ε, $|\varepsilon|\leq 1$ が与えられたとき，Kepler（ケプラー）の方程式

$$M = E - \varepsilon \sin(E)$$

の厳密解は，天体力学では決定的に重要である[1-4]．それは太陽の周りを楕円軌道を描いて運動する惑星の**平均近点離角** M が，惑星の**離心近点離角** E と楕円の離心率 ε とに関係していることを述べている．それは超越方程式である．すなわち M と ε を使って代数的に解くことはできない．E を計算するのは，時間の関数として惑星の位置を計算するとき共通に使われる中間的段階である．それゆえ，なぜ Newton から現在まで多くの数学者がこの問題に没頭してきたのかを理解することは難しいことではない．

ケプラーの方程式の基礎をなす軌道力学について説明することはしないが，簡単な幾何学的な動機付けを与える例について述べることにしよう．まず任意の点 F を単位円の中にとる．P を F に最も近い円周上の点とし，もう1つの点 Q を円周上に別にとる．E と ε を図 4.2 に描かれたように定義する．M を網を掛けた扇形 PFQ の面積の 2 倍とする．このとき，

$$\frac{M}{2} = (\text{扇形 } POQ \text{ の面積}) - (\text{三角形 } FOQ \text{ の面積}) = \frac{1}{2}E - \frac{1}{2}\varepsilon \sin(E)$$

だから，面積 M と長さ ε を与えたとき，ケプラーの方程式の解から角度 E を計算することができる．

ケプラーの方程式は，(Lagrange（ラグランジュ）の逆転法（inversion method）によって) ε のベキ級数で与えられた一意解をもつ．

$$E = M + \sum_{n=1}^{\infty} a_n \varepsilon^n$$

4.8 ラプラス限界定数

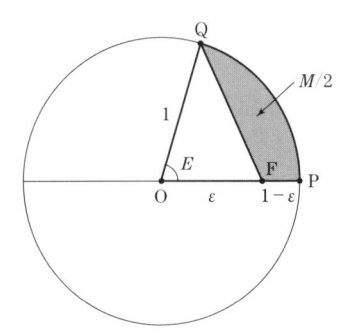

図 4.2 ケプラーの方程式のための幾何学的動機付けを与える例.

ただし [1,5-7],

$$a_n = \frac{1}{n!2^{n-1}} \sum_{k=0}^{\lfloor n/2 \rfloor} (-1)^k \binom{n}{k} (n-2k)^{n-1} \sin((n-2k)M)$$

このようなベキ級数解は，コンピュータのなかった 19 世紀において計算を実行するために選ばれた方法であった．だから Laplace（ラプラス）によって明白にはじめて発見されたように，この級数が $|\varepsilon| > 0.662$ に対して発散するという結果はおそらくショックだったであろう．Arnold[8]は，「これは数学の歴史において重要な役割を演じている… この神秘的な定数の起源の研究が Cauchy（コーシ）を複素解析学の創造に駆り立てた」と述べている.

実際，E に対するベキ級数は，公比

$$f(\varepsilon) = \frac{\varepsilon}{1+\sqrt{1+\varepsilon^2}} \exp(\sqrt{1+\varepsilon^2})$$

の等比級数のように収束する．$f(\lambda) = 1$ となる値 $\lambda = 0.6627434193\cdots$ は**ラプラスの限界値**と呼ばれている．初等関数で λ をまとまった形で表現したものは知られていない．λ に対する無限級数，あるいは定積分で表現することも同じように知られていない.

物語はここで終わりではない．E に対するベッセル関数級数は次の式である [5,6,9].

$$E = M + \sum_{n=1}^{\infty} \frac{2}{n} J_n(n\varepsilon) \sin(nM)$$

ここで

$$J_p(x) = \sum_{k=0}^{\infty} \frac{(-1)^k}{k!(p+k)!} \left(\frac{x}{2}\right)^{p+2k}$$

この級数は，ベキ級数よりもよい．というのもそれはすべての $|\varepsilon| \leq 1$ に対して $|g(\varepsilon)| \leq 1$ を満たす公比

$$g(\varepsilon) = \frac{\varepsilon}{1+\sqrt{1-\varepsilon^2}} \exp(\sqrt{1-\varepsilon^2})$$

をもつ等比級数のように収束するからである．

しかしながら反復法はこれらの級数展開による方法のいずれをもしのぐ．関数
$$T(E)=M+\varepsilon\sin(E) \quad (\text{固定された } M \text{ と } \varepsilon \text{ に対して})$$
は縮小写像であることに注目しよう．このとき，逐次近似法
$$E_0=0, \quad E_{i+1}=T(E_i)=M+\varepsilon\sin(E_i)$$
が有効に働く．ニュートン法
$$E_0=0, \quad E_{i+1}=E_i+\frac{M+\varepsilon\sin(E_i)-E_i}{1-\varepsilon\cos(E_i)}$$
はさらにより急速に収束する．これらの変形はたくさんある．実用性を別にすれば，ケプラーの方程式を解く定積分表現がいくつかある[10-13]．これらは現在知られている限りでは，高精度を追求する競争における対抗者とはみなされていない．

λ の別な種類の表現は次のようなものである[7,14,15]．$\mu=1.1996786402\cdots$ を $\coth(\mu)=\mu$ の正の一意的な解とすると，$\lambda=\sqrt{\mu^2-1}$ である．

[1] F. R. Moulton, *An Introduction to Celestial Mechanics*, 2nd ed., MacMillan, 1914.
[2] H. Dörrie, *100 Great Problems of Elementary Mathematics: Their History and Solution*, Dover, 1965; MR 84b:00001.
[3] J. B. Marion, *Classical Dynamics of Particles and Systems*, 2nd ed., Academic Press, 1970.
[4] J. M. A. Danby, *Fundamentals of Celestial Mechanics*, 2nd ed., Willmann-Bell, 1988; MR 90b:70017.
[5] A. Wintner, *The Analytical Foundations of Celestial Mechanics*, Princeton Univ. Press, 1941; MR 3,215b.
[6] P. Henrici, *Applied and Computational Complex Analysis*, v. 1, *Power Series-Integration-Conformal Mapping-Location of Zeros*, Wiley, 1974; MR 90d:30002.
[7] P. Colwell, *Solving Kepler's Equation over Three Centuries*, Willmann-Bell, 1993, p. 72.
[8] V. I. Arnold, *Huygens and Barrow, Newton and Hooke*, Birkhäuser, 1990, p. 85; MR 91h:01014.
[9] G. N. Watson, *A Treatise on the Theory of Bessel Functions*, 2nd ed., Cambridge Univ. Press, 1966; MR 96i:33010.
[10] C. E. Siewert and E. E. Burniston, An exact analytical solution of Kepler's equation, *Celest. Mech.* 6 (1972) 294-304; MR 51 #9596.
[11] N. I. Ioakimidis and K. E. Papadakis, A new simple method for the analytical solution of Kepler's equation, *Celest. Mech.* 35 (1985) 305-316; MR 86g:70001.
[12] N. I. Ioakimidis and K. E. Papadakis, A new class of quite elementary closed-form integral formulae for roots of nonlinear systems, *Appl. Math. Comput.*, v. 29 (1989) n. 2, 185-196; MR 89m:65050.
[13] S. Ferraz-Mello, The convergence domain of the Laplacian expansion of the disturbing function, *Celest. Mech. Dynam. Astron.* 58 (1994) 37-52; MR 94j:70011.
[14] E. Goursat, *A Course in Mathematical Analysis*, v. 2, Dover, 1959; MR 21 #4889.
[15] F. Le Lionnais, *Les Nombres Remarquables*, Hermann, 1983.

4.9 整係数チェビシェフ定数

P_n を n 次のすべての実係数のモニックな（最高次の係数が 1 の）多項式の全体と考える．この族の零でない多項式で区間 $[0,1]$ において零関数との距離が最小であるのは何か？ すなわち，次の最適化問題の解は何か？

$$\min_{p \in P_n, p \neq 0} \max_{0 \leq x \leq 1} |p(x)| = f(n)$$

この一意的な解は $p_n(x) = 2^{1-2n} T_n(2x-1)$ である．ただし [1,2]

$$T_n(x) = \cos(n \arccos(x)) = \frac{(x+\sqrt{x^2-1})^n + (x-\sqrt{x^2-1})^n}{2}$$

および

$$\lim_{n \to \infty} f(n)^{1/n} = \frac{1}{4}$$

である．$p_n(x)$ を（伝統を無視して）**実チェビシェフ多項式**とよぶことにするが，初めのいくつかの多項式は表 4.1 にあげられている．（「チェビシェフ多項式」という語は通例多項式 $T_n(x)$ を表すのに使われる．）$f(n)$ の定義においては「モニック」という言葉を「最高次の係数が少なくとも 1」という言葉に置きかえた方がよいことに注意する．

代わりに最高次の係数が正であるような n 次の整数係数多項式の族 Q_n を考えてみる．再び，この族の零でない多項式で区間 $[0,1]$ において零関数との距離が最小であるのは何か？ すなわち，次の解は何か？

$$\min_{q \in Q_n, q \neq 0} \max_{0 \leq x \leq 1} |q(x)| = g(n)$$

明らかにこれは先の問題の制限を強くした 1 つの変形である．この場合は完全な解をもつこともないし，一意性すらもたない．$q_n(x)$ は，**整係数チェビシェフ多項式**とよばれるものになるが，最初のいくつかは表 4.2 に一覧表となっている．**整係数チェビシェ**

表 4.1 実チェビシェフ多項式

n	$p_n(x)$	$f(n)$	$f(n)^{1/n}$
1	$x - \dfrac{1}{2}$	$\dfrac{2}{4^1} = \dfrac{1}{2}$	0.500
2	$x^2 - x + \dfrac{1}{8}$	$\dfrac{2}{4^2} = \dfrac{1}{8}$	0.353
3	$x^3 - \dfrac{3}{2}x^2 + \dfrac{9}{16}x - \dfrac{1}{32}$	$\dfrac{2}{4^3} = \dfrac{1}{32}$	0.314
4	$x^4 - 2x^3 + \dfrac{5}{4}x^2 - \dfrac{1}{4}x + \dfrac{1}{128}$	$\dfrac{2}{4^4} = \dfrac{1}{128}$	0.297
5	$x^5 - \dfrac{5}{2}x^4 + \dfrac{35}{16}x^3 - \dfrac{25}{32}x^2 + \dfrac{25}{256}x - \dfrac{1}{512}$	$\dfrac{2}{4^5} = \dfrac{1}{512}$	0.287

表4.2 整係数チェビシェフ多項式

n	$q_n(x)$	$g(n)$	$g(n)^{1/n}$
1	x または $x-1$ または $2x-1$	1	1.000
2	$x(x-1)$	$\dfrac{1}{4}$	0.500
3	$x(x-1)(2x-1)$	$\dfrac{\sqrt{3}}{18}$	0.458
4	$x^2(x-1)^2$ または $x(x-1)(2x-1)^2$ または $x(x-1)(5x^2-5x+1)$	$\dfrac{1}{16}$	0.500
5	$x^2(x-1)^2(2x-1)$	$\dfrac{\sqrt{5}}{125}$	0.447
6	$x^2(x-1)^2(2x-1)^2$	$\dfrac{1}{108}$	0.458
7	$x^3(x-1)^3(2x-1)$	$\dfrac{27\sqrt{7}}{19208}$	0.449

フ定数（あるいは**整数超越直径，整数対数直径【4.9.1】**）を次のように定義する．

$$\chi=\lim_{n\to\infty}g(n)^{1/n}$$

χ について何が言えるのか？ 一方，我々はある下界を知っている[3-5]．

$$\exp(-0.8657725922\cdots)=\frac{1}{2.3768417063\cdots}=0.4207263771\cdots=\alpha\leqq\chi$$

ただし

$$a_0=2,\quad a_k=a_{k-1}+\frac{1}{a_{k-1}},\quad k\geqq 1,\quad \alpha=\frac{1}{2}\prod_{j=0}^{\infty}\left(1+\frac{1}{a_j^2}\right)^{-1/(2^{j+1})}$$

この漸化式は Gorshkov & Wirsing（ゴルシュコフ&ワーシング）多項式[3,6]として知られているもののおかげで得られた．Borwein & Erdélyi（ボーウィン&エルデリ）[4]が $\chi>\alpha$ ということを証明して，誰もを驚愕させるまでは，$\chi=\alpha$ が予想されていた[5]．一方，我々は上界を知っている．

$$\chi\leqq\beta=0.42347945=\frac{1}{2.36138964}=\exp(-0.85925028)$$

これは Habsieger & Salvy（ハブシーガー&サルビ）[7]によるものであるが，かれらは 75 までの各次数に対して整係数チェビシェフ多項式を計算することに成功した．そのようなかなり高次までの多項式を発見するため，そしてこの方法で β を決定するためにはもっとよいアルゴリズムが必要とされるであろう．しかしながら最近 Pritsker（プリッカー）[8]は，異なる方法によって改良された上界・下界 $0.4213<\chi<0.4232$ を得た．

これまですべての注意を区間 $[0,1]$ に，すなわち定数 $\chi=\chi(0,1)$ に集中してきた．他の区間 $[a,b]$ については何がいえるか？

$$\chi(-1,1)^4=\chi(0,1)^2=\chi\left(0,\frac{1}{4}\right)$$

が知られている[4,9]．それゆえ先に求めた範囲が適用できる．任意の $0<b-a<4$ に

対して $\chi(a,b)$ の正確な値は未解決のままである[4]．しかしながら，もし $b-a\geq 4$ ならば，$\chi(a,b)=1$，およびすべての $1-0.17^2\leq c<1+\varepsilon$（ある $\varepsilon>0$ に対して）である c について，$\chi(0,c)=\chi(0,1)$，すなわち $\chi(0,c)$ は $c=1$ の近傍で定数である．また，[10]

$$\chi(0,1)=\chi(1,2)>0.42$$

しかし，初等的な考察から

$$\chi(0,2)\leq \frac{1}{\sqrt{2}}<0.71<0.84=2(0.42)$$

すなわち，$\chi(0,2)$ は $2\chi(0,1)$ とも $\chi(0,1)+\chi(1,2)$ とも同じではない．また $\chi(0,1)=\chi(d,d+1)$ という関係も整数でない d に対しては成り立たない．だから，スケーリングや加法そして平行移動に関する不変性は（実チェビシェフの場合と違って）整数係数チェビシェフの場合には成り立たない．

Gel'fond（ゲルフォンド）と Schnirelmann（シュニレルマン）によれば $\chi(0,1)$ の計算と素数論とに興味ある関係がある[3,5]．もし $\chi=1/e=0.36\cdots$ が真ならば，有名な素数定理の新しい証明を手にすることができたのだが，あいにくにもこれは（われわれの上界・下界が明らかに示すように）正しくなかった．

最終的に区間 $[0,1]$ において，Aparicio & Bernardo（アパリチオ&ベルナルド）[11]は，整数係数チェビシェフ多項式 $q_n(x)$ が常に因子

$$x(x-1),\quad 2x-1,\quad \text{および}\quad 5x^2-5x+1$$

をもち，n が大きくなるにつれてそれらの累乗を含み，その指数が増加する傾向があることを考察した．これが出現する相対的な比率，すなわち多項式 $q_n(x)$ の漸近的な構造は一層興味のある定数を生み出している[4,6,8]．

4.9.1 超越直径

われわれは以前ポテンシャル論の用語を利用して話を精密にした．E を複素数平面のコンパクト集合としよう．**(実) 超越直径**あるいは**(実) 対数容量**は

$$\gamma(E)=\lim_{n\to\infty}\max_{z_2,\cdots,z_n\in E}\left(\prod_{j<k}|z_j-z_k|\right)^{2/n(n-1)}$$

つまり，E の n 個の点に対する相異なる 2 点間の距離の相乗平均の最大値を $n\to\infty$ としたときの極限である．たとえば，

$$\gamma([0,1])=\frac{1}{4}=\lim_{n\to\infty}f(n)^{1/4}$$

そしてこの等式は偶然の一致ではない．任意の E に対して超越直径，対数容量および(実) チェビシェフ定数という語は，専門用語として相互に入れ替えが可能である[1,12]．実例の計算については[13-15]を参照してほしい．Robin（ロバン）の定数として知られているものに関連する議論は[16-18]にある．

[1] P. Borwein and T. Erdélyi, *Polynomials and Polynomial Inequalities*, Springer-Verlag, 1995;

MR 97e:41001.
[2] G. F. Simmons, *Differential Equations with Applications and Historical Notes*, McGraw-Hill, 1972; MR 58 #17258.
[3] H. L. Montgomery, *Ten Lectures on the Interface between Analytic Number Theory and Harmonic Analysis*, CBMS v. 84, Amer. Math. Soc., 1994, pp. 179-193; MR 96i:11002.
[4] P. Borwein and T. Erdélyi, The integer Chebyshev problem, *Math. Comp.* 65 (1996) 661-681; MR 96g:11077.
[5] G. V. Chudnovsky, Number theoretic applications of polynomials with rational coefficients defined by extremality conditions, *Arithmetic and Geometry: Papers Dedicated to I. R. Shafarevich*, v. 1, ed. M. Artin and J. Tate, Birkhäuser, 1983, pp. 61-105; MR 86c:11052.
[6] I. E. Pritsker, Chebyshev polynomials with integer coefficients, *Analytic and Geometric Inequalities and Applications*, ed. Th. M. Rassias and H. M. Srivastava, Kluwer, 1999, pp. 335 -348; MR 2001h:30007.
[7] L. Habsieger and B. Salvy, On integer Chebyshev polynomials, *Math. Comp.* 66 (1997) 763-770; MR 97f:11053.
[8] I. E. Pritsker, Small polynomials with integer coefficients, math.CA/0101166.
[9] V. Flammang, Sur le diamètre transfini entier d'un intervalle à extrémités rationnelles, *Annales Inst. Fourier (Grenoble)* 45 (1995) 779-793; MR 96i:11083.
[10] V. Flammang, G. Rhin, and C. J. Smyth, The integer transfinite diameter of intervals and totally real algebraic integers, *J. Théorie Nombres Bordeaux* 9 (1997) 137-168; MR 98g:11119.
[11] E. Aparicio Bernardo, On the asymptotic structure of the polynomials of minimal Diophantic deviation from zero, J. *Approx. Theory* 55 (1988) 270-278; MR 90b:41010.
[12] E. Hille, *Methods in Classical and Functional Analysis*, Addison-Wesley, 1972; MR 57 #3802.
[13] N. S. Landkoff, *Foundations of Modern Potential Theory*, Springer-Verlag, 1972; MR 50 #2520.
[14] T. J. Ransford, *Potential Theory in the Complex Plane*, Cambridge Univ. Press, 1995; MR 96e: 31001.
[15] J. Rostand, Computing logarithmic capacity with linear programming, *Experim. Math.* 6 (1997) 221-238; MR 98f:65032.
[16] L. V. Ahlfors, *Conformal Invariants: Topics in Geometric Function*, McGraw-Hill, 1973; MR 50 #10211.
[17] J. B. Conway, *Functions of One Complex Variable II*, Springer-Verlag, 1995; MR96i:30001.
[18] P. K. Kythe, *Computational Conformal Mapping*, Birkhäuser, 1998; MR 99k:65027.

5 離散構造数え上げに関する定数

5.1 アーベル群数え上げに関する定数

　すべての有限アーベル群は巡回部分群の直和である．この基本定理の系は次のとおりである．与えられた正の整数 n に対して，位数 n の互いに同型でない（non-isomorphic）アーベル群の個数 $a(n)$ は，[1,2]

$$a(n) = P(a_1)P(a_2)P(a_3)\cdots P(a_r)$$

で与えられる．ただし，$n = p_1^{a_1} p_2^{a_2} p_3^{a_3} \cdots p_r^{a_r}$ は n の素因数分解であり，$p_1, p_2, p_3, \cdots, p_r$ は相異なる素数であり，各 a_k は正の数で，$P(a_k)$ は制限を付けない a_k の分割の個数である．たとえば，任意の素数 p に対して $a(p^4) = 5$ である．なぜならば，4 には 5 個の分割

$$4 = 1+3 = 2+2 = 1+1+2 = 1+1+1+1$$

があるからである．他の例は，異なる素数 p と q に対して $a(p^4 q^4) = 25$ である．しかし，$a(p^8) = 22$ である．明らかに，

$$\liminf_{n\to\infty} a(n) = 1$$

である．だが，[3-6]

$$\limsup_{n\to\infty} \ln(a(n)) \frac{\ln(\ln(n))}{\ln(n)} = \frac{\ln(5)}{4}$$

を示すことは，もっと難しい．

　多くの研究者が，すべての正の整数にわたる $a(n)$ の和の平均値のふるまいを研究している．最も精密に知られている結果は[7-10]，

$$\sum_{n=1}^{N} a(n) = A_1 N + A_2 N^{1/2} + A_3 N^{1/3} + O(N^{50/199+\varepsilon})$$

である．ただし，$\varepsilon>0$ は任意に小さく，

$$A_k = \sum_{\substack{j=1 \\ j \neq k}}^{\infty} \zeta\left(\frac{j}{k}\right) = \begin{cases} 2.2948565916\cdots & (k=1 \text{ のとき}) \\ -14.6475663016\cdots & (k=2 \text{ のとき}) \\ 118.6924619727\cdots & (k=3 \text{ のとき}) \end{cases}$$

であり，$\zeta(x)$ は Riemann のゼータ関数【1.6】である．次の評価

$$\sum_{n=1}^{N} a(n) \sim \sum_{k=1}^{\infty} A_k N^{-1/k} + \varDelta(N)$$

について思いをめぐらせずにはいられない．しかし，誤差 $\varDelta(N)$ の理解は，見たところいまだに成し遂げられていない[11,12]．有限半単純結合環 (finite semisimple associative ring) の，同様な数え上げは【5.1.1】にある．

$a(n)$ の逆数の和に焦点を移すならば，[13,14]

$$\sum_{n=1}^{N} \frac{1}{a(n)} = A_0 N + O(N^{1/2} \ln(N)^{-1/2})$$

が成り立つ．ただし，A_0 はすべての素数 p にわたる無限積で

$$A_0 = \prod_p \left[1 - \sum_{k=2}^{\infty} \left(\frac{1}{P(k-1)} - \frac{1}{P(k)}\right)\frac{1}{p^k}\right] = 0.7520107423\cdots$$

である．要約すると，任意の与えられた位数の互いに同型でないアーベル群の平均個数は，「平均」を相加平均の意味で理解すると $A_1=2.2948$ であり，調和平均の意味で理解すると $A_0^{-1}=1.3297$ である．一般の場合（必ずしもアーベル群でない）に対する類似の統計量を得ることは，現在望むことさえできない．いくつかの興味ある限界が知られ[15-19]，その限界は有限単純群の分類定理 (classification theorem) に基づいて得られる．

定数 A_1 は，二次体の類数の算術的性質と関連して[20]にもある．

Erdös & Szekeres（エルデーシュ&セケレス）[21,22]は，$a(n)$ および次のような一般化を調べた．$p^j, j \geq i$ の形の因数の積（項の数は任意であり，順序は無視する）として，n を表す表現の個数を $a(n,i)$ とする．彼らは

$$\sum_{n=1}^{N} a(n,i) = C_i N^{1/i} + O(N^{1/(i+1)}), \quad C_i = \prod_{k=1}^{\infty} \zeta\left(1 + \frac{k}{i}\right)$$

を証明した．確実にすでに誰かがこの評価をよくしているだろう．充平方 (square-full) と充立方 (cube-full) 整数の議論については，さらに【2.6.1】を参照せよ．

5.1.1 半単純結合環

恒等元 $1 \neq 0$ をもつ有限環（結合法則をもつ）R が**単純** (simple) であるとは，R が真部分（両側）イデアルをもたないときをいい，**半単純** (semisimple) であるとは，R が単純イデアルの直和であるときをいう．

単純環は体を一般化している．同様に，半単純環は単純環を一般化している．すべての（有限）半単純環は，実際，有限体上の全 (full) 行列環の直和である．したがって，与えられた正の整数 n に対して，位数 n の同型でない半単純環の個数 $s(n)$ は，

$$s(n) = Q(\alpha_1) Q(\alpha_2) Q(\alpha_3) \cdots Q(\alpha_r)$$

で与えられる．ただし，$n = p_1^{a_1} p_2^{a_2} p_3^{a_3} \cdots p_r^{a_r}$ は n の素因数分解であり，$p_1, p_2, p_3, \cdots, p_r$ は相異なる素数であり，各 α_k は正の数であり，$Q(\alpha_k)$ は整数の組 (r_j, m_j) の（順序を問わない）集合で，

$$\alpha_k = \sum_j r_j m_j^2, \quad \text{かつ，すべての } j \text{ に対して } r_j m_j^2 > 0$$

を満たすものの個数である．例として，任意の素数 p に対して $s(p^5) = 8$ が成り立つのは，

$$5 = 1 \cdot 1^2 + 1 \cdot 2^2 = 5 \cdot 1^2 = 2 \cdot 1^2 + 3 \cdot 1^2 = 1 \cdot 1^2 + 4 \cdot 1^2$$
$$= 1 \cdot 1^2 + 1 \cdot 1^2 + 3 \cdot 1^2 = 1 \cdot 1^2 + 2 \cdot 1^2 + 2 \cdot 1^2$$
$$= 1 \cdot 1^2 + 1 \cdot 1^2 + 1 \cdot 1^2 + 2 \cdot 1^2 = 1 \cdot 1^2 + 1 \cdot 1^2 + 1 \cdot 1^2 + 1 \cdot 1^2 + 1 \cdot 1^2$$

だからである．漸近的に，極端な結果[23, 24]として

$$\liminf_{n \to \infty} s(n) = 1$$

$$\limsup_{n \to \infty} \ln(s(n)) \frac{\ln(\ln(n))}{\ln(n)} = \frac{\ln(6)}{4}$$

があり，平均値に関する結果は[25-30]

$$\sum_{n=1}^{N} s(n) = A_1 B_1 N + A_2 B_2 N^{1/2} + A_3 B_3 N^{1/3} + O(N^{50/199 + \varepsilon})$$

である．ただし，$\varepsilon > 0$ は任意に小さく，A_k は前に定義した値であり，

$$B_k = \prod_{r=1}^{\infty} \prod_{m=2}^{\infty} \zeta\left(\frac{rm^2}{k}\right)$$

である．とくに，平均（相加平均の意味での「平均」）として，任意に与えられた位数の，同型でない半単純環が

$$A_1 B_1 = \prod_{rm^2 > 1} \zeta(rm^2) = 2.4996161129 \cdots$$

個存在する．

[1] D. G. Kendall and R. A. Rankin, On the number of abelian groups of a given order, *Quart. J. Math.* 18 (1947) 197-208; MR 9,226c.
[2] N. J. A. Sloane, On-Line Encyclopedia of Integer Sequences, A000688 and A004101.
[3] E. Krätzel, Die maximale Ordnung der Anzahl der wesentlich verschiedenen abelschen Gruppen n^{ter} Ordnung, *Quart. J. Math.* 21 (1970) 273-275; MR 42 #3171.
[4] W. Schwarz and E. Wirsing, The maximal number of non-isomorphic abelian groups of order n, *Arch. Math. (Basel)* 24 (1973) 59-62; MR 47 #4953.
[5] A. Ivić, The distribution of values of the enumerating function of non-isomorphic abelian groups of finite order, *Arch. Math. (Basel)* 30 (1978) 374-379; MR 58 #16562.
[6] E. Krätzel, The distribution of values of $a(n)$, *Arch. Math. (Basel)* 57 (1991) 47-52; MR 92e:11097.
[7] H.-E. Richert, Über die Anzahl Abelscher Gruppen gegebener Ordnung. I, *Math. Z.* 56 (1952) 21-32; MR 14,349e.
[8] B. R. Srinivasan, On the number of abelian groups of a given order, *Acta Arith.* 23 (1973) 195-205; MR 49 #2610.
[9] G. Kolesnik, On the number of abelian groups of a given order, *J. Reine Angew. Math.* 329 (1981) 164-175; MR 83b:10055.

[10] H.-Q. Liu, On the number of abelian groups of a given order, *Acta Arith.* 59 (1991) 261-277; supplement 64 (1993) 285-296; MR 92i:11103 and MR 94g:11073.
[11] E. Krätzel, *Lattice Points*, Kluwer, 1988, pp. 293-303; MR 90e:11144.
[12] A. Ivić, *The Riemann Zeta-Function*, Wiley, 1985, pp. 37-38, 413-420, 438-439; MR 87d:11062.
[13] J.-M. DeKoninck and A. Ivić, *Topics in Arithmetical Functions: Asymptotic Formulae for Sums of Reciprocals of Arithmetical Functions and Related Fields*, North-Holland, 1980; MR 82a: 10047.
[14] P. Zimmermann, Estimating the infinite product A_0, unpublished note (1996).
[15] P. M. Neumann, An enumeration theorem for finite groups, *Quart. J. Math.* 20 (1969) 395-401; MR 40 #7344.
[16] L. Pyber, Enumerating finite groups of given order, *Annals of Math.* 137 (1993) 203-220; MR 93m:11097.
[17] M. Ram Murty, Counting finite groups, *Analysis, Geometry and Probability*, ed. R. Bhatia, Hindustan Book Agency, 1996, pp. 161-172; MR 98j:11076.
[18] L. Pyber, Group enumeration and where it leads us, *European Congress of Mathematics*, v. 2, Budapest, 1996, ed. A. Balog, G. O. H. Katona, A. Recski, and D. Szász, Birkhäuser, 1998, pp. 187-199; MR 99i:20037.
[19] B. Eick and E. A. O'Brien, Enumerating p-groups, *J. Austral. Math. Soc. Ser. A* 67 (1999) 191-205; MR 2000h:20033.
[20] H. Cohen, *A Course in Computational Algebraic Number Theory*, Springer-Verlag, 1993, pp. 290-293; MR 94i:11105.
[21] P. Erdös and G. Szekeres, Über die Anzahl der Abelschen Gruppen gegebener Ordnung und über ein verwandtes zahlentheoretisches Problem, *Acta Sci. Math. (Szeged)* 7 (1934-35) 95-102.
[22] I. Z. Ruzsa, Erdös and the integers, *J. Number Theory* 79 (1999) 115-163; MR 2002e: 11002.
[23] J. Knopfmacher, Arithmetical properties of finite rings and algebras, and analytic number theory. IV, *J. Reine Angew. Math.* 270 (1974) 97-114; MR 51 #389.
[24] J. Knopfmacher, Arithmetical properties of finite rings and algebras, and analytic number theory. VI, *J. Reine Angew. Math.* 277 (1975) 45-62; MR 52 #3088.
[25] J. Knopfmacher, Arithmetical properties of finite rings and algebras, and analytic number theory. I, *J. Reine Angew. Math.* 252 (1972) 16-43; MR 47 #1769.
[26] J. Knopfmacher, *Abstract Analytic Number Theory*, North-Holland, 1975; MR 91d:11110.
[27] J. Duttlinger, Eine Bemerkung zu einer asymptotischen Formel von Herrn Knopfmacher, *J. Reine Angew. Math.* 266 (1974) 104-106; MR 49 #2605.
[28] C. Calderón and M. J. Zárate, The number of semisimple rings of order at most x, *Extracta Math.* 7 (1992) 144-147; MR 94m:11111.
[29] M.Kühleitner, Comparing the number of abelian groups and of semisimple rings of a given order, *Math. Slovaca* 45 (1995) 509-518; MR 97d:11143.
[30] W. G. Nowak, On the value distribution of a class of arithmetic functions, *Comment. Math. Univ. Carolinae* 37 (1996) 117-134; MR 97h:11104.

5.2 ピタゴラス3数定数

正の整数 a, b, c が**原始ピタゴラス数** (primitive Pythagorean triple) であるとは、$a \leq b$, $\gcd(a, b, c) = 1$, $a^2 + b^2 = c^2$ であるときをいう。明らかに、任意のこのような組は、比例した辺の長さをもつ直角三角形の辺の長さとして幾何学的に解釈できる。

$P_h(n)$, $P_p(n)$, $P_a(n)$ を，それぞれ n を超えない斜辺，周および面積をもつ原始ピタゴラス数の個数と定義する．D. N. Lehmer（レーマー）[1]は，

$$\lim_{n\to\infty}\frac{P_h(n)}{n}=\frac{1}{2\pi}, \quad \lim_{n\to\infty}\frac{P_p(n)}{n}=\frac{\ln(2)}{\pi^2}$$

を証明した．Lambek & Moser（ランベック&モザー）[2]は，$\Gamma(x)$ を Euler（オイラー）のガンマ関数【1.5.4】とするとき，

$$\lim_{n\to\infty}\frac{P_a(n)}{\sqrt{n}}=C=\frac{1}{\sqrt{2\pi^5}}\Gamma\left(\frac{1}{4}\right)^2=0.5313399499\cdots$$

を証明した．

誤差項について何がいえるだろうか？ D. H. Lehmer（レーマー）[3]は，

$$P_p(n)=\frac{\ln(2)}{\pi^2}n+O(n^{1/2}\ln(n))$$

を証明した．Lambek & Moser[2]と Wild（ワイルド）[4]はさらに

$$P_h(n)=\frac{1}{2\pi}n+O(n^{1/2}\ln(n)), \quad P_a(n)=Cn^{1/2}-Dn^{1/3}+O(n^{1/4}\ln(n))$$

を証明した．ただし，

$$D=-\frac{1+2^{-1/3}\zeta\left(\frac{1}{3}\right)}{1+4^{-1/3}\zeta\left(\frac{4}{3}\right)}=0.2974615529\cdots$$

であり，$\zeta(x)$ は Riemann のゼータ関数である【1.6】．より精密な $P_a(n)$ の評価は[5-8]で得られている．

原始ピタゴラス数 a,b,c の斜辺 c と周の長さ $a+b+c$ は，ともに整数でなければならない．もし ab が奇数であれば，a と b はともに奇数である．したがって，$c^2\equiv 2$ mod 4 であり，これは不可能である．よって，面積 $ab/2$ もまた整数でなければならない．$P'_a(n)$ を，面積 $\leq n$ が整数である原始ピタゴラス数の個数とする．このとき，$P'_a(n)=P_a(n)$ である．もちろん，このような等式は，直角三角形でない三角形に対して成り立たない．

やや関係する問題は，大昔の**合同数問題**（congruent number problem）[9]である．解は，Tunnel（タンネル）[10]が楕円曲線理論から Birch-Swinnerton-Dyer（バーチ-スウィナートン-ダイヤー）予想の弱い形にしたもので与えられる．合同数問題では，直角三角形において（ちょうど整数ではなく）有理数の辺をとることを認める．与えられた n に対して，面積が n であり有理数を辺とする直角三角形は存在するか？[†]

さらに，**原始ヘロン3数**（primitive Heronian triples），すなわち，（面積が整数で）互いに素な整数 $a\leq b\leq c$ に比例した辺をもつ**任意の**三角形の辺の長さ，の数え上げの問題がある．（同じように定義された）数 $H_h(n)$, $H_p(n)$, $H_a(n)$, $H'_a(n)$ について，漸近的に何がいえるのか？ この問題に答える出発論文は，[11,12]である．

[†]訳注：$n\equiv 5,6,7 \pmod 8$ で無平方な場合，つねに存在するかという予想．

[1] D. N. Lehmer, Asymptotic evaluation of certain totient sums, *Amer. J. Math.* 22 (1900) 293-335.
[2] J. Lambek and L. Moser, On the distribution of Pythagorean triples, *Pacific J. Math.* 5 (1955) 73-83; MR 16,796h.
[3] D. H. Lehmer, A conjecture of Krishnaswami, *Bull. Amer. Math. Soc.* 54 (1948) 1185-1190; MR 10,431c.
[4] R. E. Wild, On the number of primitive Pythagorean triangles with area less than n, *Pacific J. Math.* 5 (1955) 85-91; MR 16,797a.
[5] J. Duttlinger and W. Schwarz, Über die Verteilung der pythagoräischen Dreiecke, *Colloq. Math.* 43 (1980) 365-372; MR 83e:10018.
[6] H. Menzer, On the number of primitive Pythagorean triangles, *Math. Nachr.* 128 (1986) 129-133; MR 87m:11022.
[7] W. Müller, W. G. Nowak, and H. Menzer, On the number of primitive Pythagorean triangles, *Annales Sci. Math. Québec* 12 (1988) 263-273; MR 90b:11020.
[8] W. Müller and W. G. Nowak, Lattice points in planar domains: Applications of Huxley's 'Discrete Hardy-Littlewood method,' *Number-Theoretic Analysis: Vienna 1988-89*, ed. E. Hlawka and R. F. Tichy, Lect. Notes in Math. 1452, Springer-Verlag, 1990, pp. 139-164; MR 92d:11113.
[9] N. Koblitz, *Introduction to Elliptic Curves and Modular Forms*, Springer-Verlag, 1984; MR 94a:11078.
[10] J. B. Tunnell, A classical Diophantine problem and modular forms of weight 3/2, *Invent. Math.* 72 (1983) 323-334; MR 85d:11046.
[11] C. P. Popovici, Heronian triangles (in Russian), *Rev. Math. Pures Appl.* 7 (1962) 439-457; MR 31 #121.
[12] D. Singmaster, Some corrections to Carlson's "Determination of Heronian triangles," *Fibonacci Quart.* 11 (1973) 157-158; MR 45 #156 and MR 47 #4922.

5.3 レーニィの駐車定数

$x>1$ である1次元区間 $[0,x]$ を考える．道に一方の側だけに駐車が許されると仮定する．単位長さの車が，道に「**完全にランダム**」に1台ずつ駐車する．明らかにすでに駐車しているところには駐車が許されない．駐車できる車の平均数 $M(x)$ は，いくらだろうか？

Rényi（レーニィ）[1-3] は，$M(x)$ が次の積分方程式を満たすことを証明した．

$$M(x) = \begin{cases} 0 & (0 \leq x < 1 \text{ のとき}) \\ 1 + \dfrac{2}{x-1} \int_0^{x-1} M(t)\,dt & (x \geq 1 \text{ のとき}) \end{cases}$$

ラプラス変換から，Rényi は，区間 $[0,x]$ の車の極限平均密度 m が，

$$m = \lim_{x \to \infty} \frac{M(x)}{x} = \int_0^\infty \beta(x)\,dx = 0.7475979202\cdots$$

であることを証明した．ただし，

$$\beta(x) = \exp\left(-2 \int_0^x \frac{1-e^{-t}}{t}\,dt\right) = e^{-2(\ln(x) - \operatorname{Ei}(-x) + \gamma)}, \quad \alpha(x) = m - \int_0^x \beta(t)\,dt$$

であり，γ は Euler-Mascheroni（オイラー-マスケローニ）の定数【1.5】であり，$\mathrm{Ei}(x)$ は指数積分【6.2.1】である．いくつかの別証明が[4,5]にある．

道に駐車できる車の数の分散 $V(x)$ について，何がいえるだろうか？ Mackenzie（マッケンジー）[6]，Dvoretzky & Robbins（ドボレツキー＆ロビンス）[7]と Mannion（マニオン）[8,9]らは，独立にこの問題に取り組み，

$$v = \lim_{x\to\infty} \frac{V(x)}{x} = 4\int_0^\infty \left[e^{-x}(1-e^{-x})\frac{\alpha(x)}{x} - e^{-2x}(x+e^{-x}-1)\frac{\alpha(x)^2}{\beta(x)x^2} \right]dx - m$$
$$= 0.0381563991\cdots$$

を証明した．さらに，中心極限定理が成り立つ[7]．すなわち，車が駐車できる総数は漸近的に，十分大きな x に対して，平均 mx で分散 vx の正規分布になる．

駐車問題をより高い次元のもとで考えるのは自然である．長さ $x>1$ で幅 $y>1$ の2次元長方形を考える．駐車場の長方形の2辺と平行に，単位正方形の車を駐車すると仮定する．駐車できる車の数の平均 $M(x,y)$ はいくらなのだろうか？ Palasti（パラスティ）[10-12]は

$$\lim_{x\to\infty}\lim_{y\to\infty} \frac{M(x,y)}{xy} = m^2 = (0.7475979202\cdots)^2 = 0.558902\cdots$$

と予想した．解析において議論の余地のある試みをすでにいくつか試したにもかかわらず[13,14]，予想は証明されないままである．極限駐車密度自体の存在も，つい最近証明されたばかりである[15]．しかしながら，徹底的なコンピュータシミュレーション[16-18]により，予想は間違いであり，真の極限値は $0.562009\cdots$ であるという推定がなされた．

1次元の設定を変えたものがある．レーニィ問題においては，以前に停められた車と駐車位置が重複する車は，無視した．Solomon（ソロモン）[14,19-21]は，後からきた車が左または右に停められた車のどちらか近いほうを，ただちに「動かす」というルールに変えて研究した．その後で，十分なスペースがあれば駐車し，そうでなければ駐車しないとする．このとき，駐車台数の平均値は大きく

$$m = \int_0^\infty (2x+1)\exp[-2(x+e^{-x}-1)]\beta(x)dx = 0.8086525183\cdots$$

である．なぜならば，バンパーとバンパーをくっつけて駐車するために，車はより大きな自由度を与えられるからである．レーニィ問題を3次元体積を球で充填するモデルとして考える．このとき，Solomon の変形問題は，「揺さぶり」ながら充填することに対応する，それは追加の球のための空間をつくることに対応する．

別の問題は，ランダムな長さをもつ車を含むものである[22,23]．k 番目に到着した車の，右と左の端の点を $[0,x]$ から独立かつ一様にとられた2つの数とする．このとき，うまく駐車に成功する車の漸近的期待値は，$C\cdot k^{(\sqrt{17}-3)/4}$ である．ただし，[24,25]

$$C = \left(1 - \frac{1}{2^{(\sqrt{17}-1)/4}}\right)\sqrt{\pi}\,\frac{\Gamma\left(\frac{\sqrt{17}}{2}\right)}{\Gamma\left(\frac{\sqrt{17}+1}{4}\right)\Gamma\left(\frac{\sqrt{17}+3}{4}\right)^2} = 0.9848712825\cdots$$

であり，Γはガンマ関数である【1.5.4】．この場合にはxは単なる計量因子であり，結果には現れていない．

駐車場問題の応用（または，もっと一般的に，一列駐車（sequential parking）または空間充塡問題）は，次のような広く広がる諸分野を含んでいる．

- 物理学：液体構造のモデル[26-29]
- 化学：結晶表面の薄い流体膜の吸収[5.3.1]
- モンテカルロ法：定積分の評価[30]
- 言語学：1シラブル，長さnの英語の単語の頻度[31]
- 社会学：選挙のモデルおよび駐車問題で発生する空隙の長さ[32-35]
- 材料科学：鉄筋コンクリートの多数の破損後の内部の割れ目の距離[36]
- コンピュータ科学：CD上の最適なデータの配置[37]および線型探査ハッシング（linear probing hashing）[38]

さらに[39-41]を参照せよ．Golomb-Dickman（ゴロム-ディックマン）定数【5.4】とレーニイ定数との定式化における類似性に注意せよ．

5.3.1 ランダム逐次吸着

区間$[0, x]$を離散有限線型格子点$1, 2, 3, \cdots, n$に置きかえた場合を考える．各々の車は単位長さをもつ線分であり，それが駐車するとき2つの格子点を覆う．車はすでに覆われている点に接触することは許されない．覆われていない隣接した2つの格子点が残っていないとき，この操作を停止する．$n \to \infty$のとき[19,42-45]，

$$m = \frac{1-e^{-2}}{2} = 0.4323323583\cdots, \quad v = e^{-4} = 0.0183156388\cdots$$

が，証明される．どちらも連続な場合に対応したものよりは，値は小さい．2次元離散的な類似は，4つの格子点を単位正方形が覆う問題であり，連続な場合と同様に解析的に手におえない．Palasti（パラスティ）予想はここでも誤りのようである．平面の極限平均密度は，$m^2 = 0.186911\cdots$ではなく，むしろ$0.186985\cdots$である[46-48]．

簡単のために，無限線形格子点$1, 2, 3, \cdots$を$1 \times \infty$帯（strip）とよぶ．$2 \times \infty$帯は，2つの平行線と橋桁（横断線）をもつ無限の梯子格子である．$3 \times \infty$の帯は，3つの平行線をもつものと同じである．$\infty \times \infty$帯は，当然，無限の正方形格子である．したがって，$1 \times \infty$におけるmおよびvの閉じた式はある．しかし，$\infty \times \infty$では，Palastiの評価の数値的な補正のみしかない．

車は，$2 \times \infty$帯上において，単位長さの線分（dimer，**ダイマー**）をもつとする．このとき，車の平均密度は$1/2 \cdot (0.91556671\cdots)$である．車が，$\infty \times \infty$帯上にあれば，対応する平均密度は$1/2 \cdot (0.90682\cdots)$である[49-55]．これらの2つの量について，厳密な式を得ることができるだろうか？

車は，$1 \times \infty$帯上において，長さ2の線分（**線型トリマー**，trimer）をもつとする．このとき，空きの平均密度は$\mu(3) = 0.1763470368\cdots$である．ただし，[6,56-58]

$$\mu(r) = 1 - r\int_0^1 \exp\left(-2\sum_{k=1}^{r-1}\frac{1-x^k}{k}\right)dx$$

である.さらに一般的に,$\mu(r)$ は任意の $r \geq 2$ に対して,$1\times\infty$ 帯上において,線型 r-マーに対する空きの平均密度である.分散に対応する空きの式は知られていない.

車が単一粒子であり,隣接する格子点には駐車できないとする.すなわち最も近くの隣を排除する(**最近接排除**)モノマー(monomer with nearest neighbor exclusion)と仮定する.$1\times\infty$ 帯に対する,車の平均密度は前のように,もちろん $m_1 = (1-e^{-2})/2$ である.$2\times\infty$ および $3\times\infty$ 帯に対する,平均密度は[59-61]

$$m_2 = \frac{2-e^{-1}}{4} = 0.4080301397\cdots, \quad m_3 = \frac{1}{3} = 0.3333333333\cdots$$

である.$\infty\times\infty$ 帯に対応する密度は,$m_\infty = 0.364132\cdots$ [47,48,50,53,55,62] である.再び,m_4 または m_∞ に対する,厳密な式を得ることができるだろうか?

連続な場合は,適当な極限操作の議論により,離散的な場合から得ることができる [6,58,63].ランダム逐次吸着モデル(random sequential adsorption models)の包括的な概論は[64-66]にある.

[1] A. Rényi, On a one-dimensional problem concerning random space-filling (in Hungarian), *Magyar Tud. Akad. Mat. Kutató Int. Közl.* 3 (1958) 109-127; Engl. transl. in *Selected Transl. Math. Statist. and Probab.*, v. 4, Inst. Math. Stat. and Amer. Math. Soc., 1963, pp. 203-224; also in *Selected Papers of Alfréd Rényi*, v. 2, Akadémiai Kiadó, 1976, pp. 173-188; MR 21 #3039.

[2] M. Lal and P. Gillard, Evaluation of a constant associated with a parking problem, *Math. Comp.* 28 (1974) 561-564; MR 49 #6560.

[3] G. Marsaglia, A. Zaman, and J. C. W. Marsaglia, Numerical solution of some classical differential-difference equations, *Math. Comp.* 53 (1989) 191-201; MR 90h:65124.

[4] P. C. Hemmer, The random parking problem, *J. Stat. Phys.* 57 (1989) 865-869; MR 92e:82054.

[5] P. L. Krapivsky, Kinetics of random sequential parking on a line, *J. Stat. Phys.* 69 (1992) 135-150; MR 93h:82063.

[6] J. K. Mackenzie, Sequential filling of a line by intervals placed at random and its application to linear adsorption, *J. Chem. Phys.* 37 (1962) 723-728.

[7] A. Dvoretzky and H. Robbins, On the parking problem, *Magyar Tud. Akad. Mat. Kutató Int. Közl.* 9 (1964) 209-224; MR 30 #3488.

[8] D. Mannion, Random space-filling in one dimension, *Magyar Tud. Akad. Mat. Kutató Int. Közl.* 9 (1964) 143-154; MR 31 #1698.

[9] D. Mannion, Random packing of an interval, *Adv. Appl. Probab.* 8 (1976) 477-501; 11 (1979) 591-602; MR 55 #1401 and MR 80i:60014.

[10] I. Palasti, On some random space filling problems, *Magyar Tud. Akad. Mat. Kutató Int. Közl.* 5 (1960) 353-359; MR 26 #4466.

[11] I. Palasti, On a two-dimensional random space filling problem, *Studia Sci. Math. Hungar.* 11 (1976) 247-252; MR 81d:52011.

[12] I. Palasti, A two-dimensional case of random packing and covering, *Random Fields*, v. 2, ed. J. Fritz, J. L. Lebowitz, and D. Szász, Colloq. Math. Soc. János Bolyai 27, North-Holland, 1976, pp. 821-834; MR 84j:60022.

[13] H. J. Weiner, Sequential random packing in the plane, *J. Appl. Probab.* 15 (1978) 803-814; letters to the editor and replies, 16 (1979) 697-707 and 17 (1980) 878-892;MR80a:60015, MR 81b:60013, and MR 82m:60021.

[14] H. Solomon and H. J. Weiner, A review of the packing problem, *Commun. Statist. Theory*

Methods 15 (1986) 2571-2607; MR 88a:60028.
[15] M. D. Penrose, Random parking, sequential adsorption, and the jamming limit, *Commun. Math. Phys.* 218 (2001) 153-176; MR 2002a:60015.
[16] B. E. Blaisdell and H. Solomon, On random sequential packing in the plane and a conjecture of Palasti, *J. Appl. Probab.* 7 (1970) 667-698; MR 43 #8101.
[17] B. J. Brosilow, R. M. Ziff, and R. D. Vigil, Random sequential adsorption of parallel squares, *Phys. Rev. A* 43 (1991) 631-638.
[18] B. Bonnier, M. Hontebeyrie, and C. Meyers, On the random filling of $\mathbb{R}^d$ by nonoverlapping d-dimensional cubes, *Physica A* 198 (1993) 1-10; MR 94g:82045.
[19] H. Solomon, Random packing density, *Proc. Fifth Berkeley Symp. Math. Stat. Probab.*, v. 3, ed. L. M. Le Cam and J. Neyman, Univ. of Calif. Press, 1967, pp. 119-134; MR 41 #1187.
[20] J. Talbot and S. M. Ricci, Analytic model for a ballistic deposition process, *Phys. Rev. Lett.* 68 (1992) 958-961.
[21] P. Viot, G. Tarjus, and J. Talbot, Exact solution of a generalized ballistic-deposition model, *Phys. Rev. E* 48 (1993) 480-488.
[22] P. E. Ney, A random interval filling problem, *Annals of Math. Statist.* 33 (1962) 702-718; MR 25 #1561.
[23] J. P. Mullooly, A one dimensional random space-filling problem, *J. Appl. Probab.* 5 (1968) 427-435; MR 38 #5261.
[24] E. G. Coffman, C. L. Mallows, and B. Poonen, Parking arcs on the circle with applications to one-dimensional communication networks, *Annals of Appl. Probab.* 4 (1994) 1098-1111; MR 95k:60245.
[25] E. G. Coffman, L. Flatto, P. Jelenkovic, and B. Poonen, Packing random intervals on-line, *Algorithmica* 22 (1998) 448-476; MR 2000h:60012.
[26] J. D. Bernal, A geometrical approach to the structure of liquids, *Nature* 183 (1959) 141-147.
[27] J. D. Bernal, Geometry of the structure of monatomic liquids, *Nature* 185 (1960) 68-70.
[28] J. D. Bernal and J. Mason, Coordination of randomly packed spheres, *Nature* 188 (1959) 910-911.
[29] J. D. Bernal, J. Mason, and K. R. Knight, Radial distribution of the random close packing of equal spheres, *Nature* 194 (1962) 956-958.
[30] G. Bánkövi, Evaluation of integrals by Monte Carlo methods based on the one-dimensional random space filling, *Magyar Tud. Akad. Mat. Kutató Int. Közl.* 5 (1960) 339-352; MR 26 #4465.
[31] J. Dolby and H. Solomon, Information density phenomena and random packing, *J. Appl. Probab.* 12 (1975) 364-370.
[32] Y. Itoh and S. Ueda, A random packing model for elections, *Annals Inst. Statist. Math.* 31 (1979) 157-167.
[33] G. Bánkövi, On gaps generated by a random space filling procedure, *Magyar Tud. Akad. Mat. Kutató Int. Közl.* 7 (1962) 395-407; MR 27 #6314.
[34] Y. Itoh, On the minimum of gaps generated by one-dimensional random packing, *J. Appl. Probab.* 17 (1980) 134-144; MR 81a:60019.
[35] J. A. Morrison, The minimum of gaps generated by random packing of unit intervals into a large interval, *SIAM J. Appl. Math.* 47 (1987) 398-410; MR 88g:60033.
[36] A. C. Kimber, An application of random packing to the multiple fracture of composite materials, *J. Appl. Probab.* 31 (1994) 564-569; MR 94m:73093.
[37] E. G. Coffman and G. S. Lueker, *Probabilistic Analysis of Packing and Partitioning Algorithms*, Wiley, 1991; MR 92h:68038.
[38] P. Flajolet, P. Poblete, and A. Viola, On the analysis of linear probing hashing, *Algorithmica* 22 (1998) 490-515; MR 2000h:68056.
[39] J. M. F. Chamayou, On two cascade models, *Stochastic Process. Appl.* 7 (1978) 153-163; MR 58 #24621.
[40] M. Sibuya and Y. Itoh, Random sequential bisection and its associated binary tree, *Annals Inst. Statist. Math.* 39 (1987) 69-84; MR 88e:60120.

[41] F. Komaki and Y. Itoh, A unified model for Kakutani's interval splitting and Rényi's random packing, *Adv. Appl. Probab.* 24 (1992) 502-505; MR 93c:60009.

[42] P. J. Flory, Intramolecular reaction between neighboring substituents of vinyl polymers, *J. Amer. Chem. Soc.* 61 (1939) 1518-1521.

[43] E. S. Page, The distribution of vacancies on a line, *J. Royal Statist. Soc. Ser. B* 21 (1959) 364-374; MR 22 # 9984.

[44] F. Downton, A note on vacancies on a line, *J. Royal Statist. Soc. Ser. B* 23 (1961) 207-214; MR 24 # A3672.

[45] J. Texter, Alternative solution to a discrete car parking problem, *J. Chem. Phys.* 91 (1989) 6295-6301; MR 92c:82118.

[46] V. Privman, J.-S.Wang, and P. Nielaba, Continuum limit in random sequential adsorption, *Phys. Rev. B* 43 (1991) 3366-3371.

[47] R. Dickman, J.-S. Wang, and I. Jensen, Random sequential adsorption: Series and virial expansions, *J. Chem. Phys.* 94 (1991) 8252-8257.

[48] A. Baram and M. Fixman, Random sequential adsorption: Long time dynamics, *J. Chem. Phys.* 103 (1995) 1929-1933.

[49] R. S. Nord and J. W. Evans, Irreversible immobile random adsorption of dimers, trimers, ... on 2D lattices, *J. Chem. Phys.* 82 (1985) 2795-2810.

[50] J. W. Evans and R. S. Nord, Random and cooperative sequential adsorption on infinite ladders and strips, *J. Stat. Phys.* 69 (1992) 151-162.

[51] M. J. de Oliveira, T. Tomé, and R. Dickman, An isotropic random sequential adsorption of dimers on a square lattice, *Phys. Rev. A* 46 (1992) 6294-6299.

[52] J.-S. Wang and R. B. Pandey, Kinetics and jamming coverage in a random sequential adsorption of polymer chains, *Phys. Rev. Lett.* 77 (1996) 1773-1776.

[53] C. K. Gan and J.-S. Wang, Extended series expansions for random sequential adsorption, *J. Chem. Phys.* 108 (1998) 3010-3012.

[54] C. Fusco, P. Gallo, A. Petri, and M. Rovere, Random sequential adsorption and diffusion of dimers and k-mers on a square lattice, *J. Chem. Phys.* 114 (2001) 7563-7569.

[55] J.-S. Wang, Series expansion and computer simulation studies of random sequential adsorption, *Colloids and Surfaces A* 165 (2000) 325-343.

[56] M. Gordon and I. H. Hillier, Statistics of random placement, subject to restrictions, on a linear lattice, *J. Chem. Phys.* 38 (1963) 1376-1380.

[57] E. A. Boucher, Kinetics and statistics of occupation of linear arrays: A model for polymer reactions, *J. Chem. Phys.* 59 (1973) 3848-3852.

[58] J. J. González, P. C. Hemmer, and J. S. Hoye, Cooperative effects in random sequential polymer reactions, *Chem. Phys.* 3 (1974) 228-238.

[59] A. Baram and D. Kutasov, Random sequential adsorption on a quasi-one-dimensional lattice: An exact solution, *J. Phys. A* 25 (1992) L493-L498.

[60] Y. Fan and J. K. Percus, Random sequential adsorption on a ladder, *J. Stat. Phys.* 66 (1992) 263-271; MR 92j:82075.

[61] A. Baram and D. Kutasov, Random sequential adsorption on a $3 \times \infty$ lattice: An exact solution, *J. Phys. A* 27 (1994) 3683-3687.

[62] P. Meakin, J. L. Cardy, E. Loh, and D. J. Scalapino, Maximal coverage in random sequential adsorption, *J. Chem. Phys.* 86 (1987) 2380-2382.

[63] M. C. Bartelt, J. W. Evans, and M. L. Glasser, The car-parking limit of random sequential adsorption: Expansions in one dimension, *J. Chem. Phys.* 99 (1993) 1438-1439.

[64] J. W. Evans, Random and cooperative sequential adsorption, *Rev. Mod. Phys.* 65 (1993) 1281-1329.

[65] J. Talbot, G. Tarjus, P. R. van Tassel, and P.Viot, From car parking to protein adsorption: An overview of sequential adsorption processes, *Colloids and Surfaces A* 165 (2000) 287-324.

[66] V. Privman, Recent theoretical results for nonequilibrium deposition of submicron particles, *J.*

Adhesion 74 (2000) 421-440; cond-mat/0003062.

5.4 ゴロム-ディックマン定数

n 個の記号のすべての置換は，互いに素な巡回置換の積で表すことができる．たとえば，$\pi(x)=3x \bmod 10$ で定義される $\{0,1,2,\cdots,9\}$ の置換は，巡回構造

$$\pi = (0)(1397)(2684)(5)$$

をもつ．この場合，置換 π は，長さ 1 の巡回置換 $a_1(\pi)=2$，長さ 2 の巡回置換 $a_2(\pi)=0$，長さ 3 の巡回置換 $a_3(\pi)=0$，長さ 4 の巡回置換 $a_4(\pi)=2$ 個からなる．π における巡回置換の総数 $\sum_{j=1}^{\infty}a_j$ は，この例では 4 に等しい．

n を固定し，$\{0,1,2,\cdots,n-1\}$ の $n!$ 個の置換が等しい確率で起こると仮定する．ランダムに π を選ぶと，古典的な結果[1-4]

$$\mathrm{E}\left(\sum_{j=1}^{\infty}a_j\right)=\sum_{i=1}^{n}\frac{1}{i}=\ln(n)+\gamma+O\left(\frac{1}{n}\right)$$

$$\mathrm{Var}\left(\sum_{j=1}^{\infty}a_j\right)=\sum_{i=1}^{n}\frac{i-1}{i^2}=\ln(n)+\gamma-\frac{\pi^2}{6}+O\left(\frac{1}{n}\right)$$

$$\lim_{n\to\infty}\mathrm{P}(a_j=k)=\frac{1}{k!}\exp\left(-\frac{1}{j}\right)\left(\frac{1}{j}\right)^k \quad \text{(漸近ポアソン分布)}$$

$$\lim_{n\to\infty}\mathrm{P}\left(\frac{\sum_{j=1}^{\infty}a_j-\ln(n)}{\sqrt{\ln(n)}}\leq x\right)=\frac{1}{\sqrt{2\pi}}\int_{-\infty}^{x}\exp\left(-\frac{t^2}{2}\right)dt \quad \text{(漸近正規分布)}$$

がある．ただし，γ はオイラー-マスケローニの定数【1.5】である．

与えられたランダム置換 π について，**最長の巡回置換の長さ**および**最短の巡回置換の長さ**

$$M(\pi)=\max\{j\geq 1: a_j>0\}, \quad m(\pi)=\min\{j\geq 1: a_j>0\}$$

の極限分布については，何が言えるだろうか？ Goncharov（ゴンチャロフ）[1,2]と Golomb（ゴロム）[5-7]の両者は，$M(\pi)$ の平均値を調べた．Golomb は定数[8-10]

$$\lambda=\lim_{n\to\infty}\frac{\mathrm{E}(M(\pi))}{n}=1-\int_{1}^{\infty}\frac{\rho(x)}{x^2}dx=0.6243299885\cdots$$

を証明した．ただし，$\rho(x)$ は次の遅れをもつ微分方程式の一意の連続な解である．

$$\rho(x)=1 \quad (0\leq x\leq 1), \quad x\rho'(x)+\rho(x-1)=0 \quad (x>1)$$

（実際には，彼は関数 $\rho(x-1)$ で研究した．）Shepp & Lloyd（シェップ&ロイド）[11]その他[6]は，別の表現式

$$\lambda=\int_{0}^{\infty}e^{-x+\mathrm{Ei}(-x)}dx=\int_{0}^{1}e^{\mathrm{Li}(x)}dx=G(1,1)$$

を発見した．ただし，

$$G(a,r)=\frac{1}{a}\int_0^\infty\Bigl(1-\exp(a\mathrm{Ei}(-x))\sum_{k=0}^{r-1}\frac{(-a)^k}{k!}\mathrm{Ei}(-x)^k\Bigr)dx$$

であり，$\mathrm{Ei}(x)$ は指数積分【6.2.1】であり，$\mathrm{Li}(x)$ は対数積分【6.2.2】である．Gourdon（グールドン）[12]は $\mathrm{E}(M(\pi))$ に対する完全な漸近展開を決定した．

$$\mathrm{E}(M(\pi))=\lambda n+\frac{\lambda}{2}-\frac{e^\gamma}{24}\frac{1}{n}+\left[\frac{e^\gamma}{48}-\frac{(-1)^n}{8}\right]\frac{1}{n^2}$$
$$+\left[\frac{17e^\gamma}{3840}+\frac{(-1)^n}{8}+\frac{e^{2(2n+1)\pi i/3}}{6}+\frac{e^{2(n+2)\pi i/3}}{6}\right]\frac{1}{n^3}+O\!\left(\frac{1}{n^4}\right)$$

1の累乗根を含む周期的な変動に注意せよ．

$\lim_{n\to\infty}(\mathrm{Var}(M(\pi))/n^2)=0.0369078300\cdots=H(1,1)$ に対して，同様な積分公式が成り立つ．ただし[12]，

$$H(a,r)=\frac{2}{a(a+1)}\int_0^\infty\Bigl(1-\exp(a\mathrm{Ei}(-x))\sum_{k=0}^{r-1}\frac{(-a)^k}{k!}\mathrm{Ei}(-x)^k\Bigr)xdx-G(a,r)^2$$

である．後に，$a\neq 1\neq r$ のときの $G(a,r)$ と $H(a,r)$ の値が必要になる．λ に類似した量は，多項式の因数分解に関連して[13,14]にある．

$m(\pi)$ の漸近平均値を導く議論は，より複雑である．Shepp & Lloyd[11]は，より高次のモーメントの式だけでなく，

$$\lim_{n\to\infty}\frac{\mathrm{E}(m(\pi))}{\ln(n)}=e^{-\gamma}=0.5614594835\cdots$$

を証明した．しかしながら，$\mathrm{E}(m(\pi))$ の完全な漸近展開は，未解決のままである．

（$n\to\infty$ のとき，n と n^2 で正規化された）r 番目に長い巡回置換の平均と分散は，それぞれ $G(1,r)$ と $H(1,r)$ で与えられる．たとえば，$G(1,2)=0.2095808742\cdots$，$H(1,2)=0.0125537906\cdots$，$G(1,3)=0.0883160988\cdots$，$H(1,3)=0.0044939231\cdots$ である[11,12]．

λ と素因数分解のアルゴリズムには，すばらしい関連がある[15,16]．$f(n)$ を n の最大素因数とする．1と N の間にランダムに整数 n を選ぶ．Dickman（ディックマン）[17-20]は，$0<x\leq 1$ に対して，

$$\lim_{N\to\infty}\mathrm{P}(f(n)\leq n^x)=\rho\!\left(\frac{1}{x}\right)$$

であることを証明した．これを考慮して，$f(n)=n^x$ であるような x の平均値は何であるのか？ Dickman は，数値的に

$$\mu=\lim_{N\to\infty}\mathrm{E}(x)=\lim_{N\to\infty}\mathrm{E}\!\left(\frac{\ln(f(n))}{\ln(n)}\right)=\int_0^1 x\,d\rho\!\left(\frac{1}{x}\right)=1-\int_1^\infty\frac{\rho(y)}{y^2}dy=\lambda$$

を得た．これは，本当に驚くべきことである！ ディックマン定数 μ とゴロム定数 λ は，同一である！ Knuth & Trabb Pardo（クヌース&トラップパルド）[15]はこの結果を次のように述べた．λn は，n 桁の数がもつ最大素因数における**漸近的平均桁数**である．さらに一般的に，n 桁のランダム整数を素因数分解するとき，その素因数の平均桁数は，近似的に n 個の要素のランダム置換における巡回置換の長さの分布と同じ

である．この注目すべき意外な事実は，[21,22]で広く徹底的に調べられている．

最大素因数関数 $f(n)$ を含む他の漸近式は，[15,23,24]

$$\mathrm{E}(f(n)^k) \sim \frac{\zeta(k+1)}{k+1} \frac{N^k}{\ln(N)}, \quad \mathrm{E}(\ln(f(n))) \sim \lambda \ln(N) - \lambda(1-\gamma)$$

である．ただし，$\zeta(x)$ はゼータ関数である【1.6】．さらに[25-29]を参照せよ．積分と和

$$\int_0^\infty \rho(x)\,dx = e^\gamma = \sum_{n=1}^\infty n\rho(n)$$

を含む不思議な偶然の一致[15]に注意せよ．Dickman の関数は y-円滑数[24,30-32]，すなわち素因子が y を決して超えない整数の研究に重要である．確率論では，

$$X_1 + X_1 X_2 + X_1 X_2 X_3 + \cdots, \quad X_j \text{ は } [0,1] \text{ の独立一様ランダム変数}$$

の（e^γ で正規化された）密度関数として[33,34]に現れる．

$\rho(x)$ の他の応用については[35-40]を参照せよ．密接に関連する Buchstab（ブッシュスタブ）による関数は[24,34,41-45]

$$\omega(x) = \frac{1}{x} \quad (1 \leq x \leq 2), \quad x\omega'(x) + \omega(x) - \omega(x-1) = 0 \quad (x > 2)$$

を満たす．これは，「最小の」素因子 $\geq n^x$ であるような整数 n の頻度を評価する場合に現れる．関数は両方とも，いたるところ正であり，特殊な値[46]

$$\rho\left(\frac{3+\sqrt{5}}{2}\right) = 1 - \ln\left(\frac{3+\sqrt{5}}{2}\right) + \ln\left(\frac{1+\sqrt{5}}{2}\right)^2 - \frac{\pi^2}{60} \quad \lim_{x \to \infty} \rho(x) = 0$$

$$\frac{5+\sqrt{5}}{2} \omega\left(\frac{5+\sqrt{5}}{2}\right) = 1 + \ln\left(\frac{3+\sqrt{5}}{2}\right) + \ln\left(\frac{1+\sqrt{5}}{2}\right)^2 - \frac{\pi^2}{60} \quad \lim_{x \to \infty} \omega(x) = e^{-\gamma}$$

を含む．$\rho(x)$ は非減少であるのに反して，差 $\omega(x) - e^{-\gamma}$ は，長さ1のすべての区間で（多くとも2回）符号を変える．振動する動きは，素数の分布の不規則さを理解するときその役割を果たす．

ゴロム-ディックマン定数とレーニィの駐車定数【5.3】との定式化には類似性があることに注意せよ．

5.4.1 対 称 群

いくつかの関連する問題がある．与えられた n 個の記号の置換 π がある．$\pi^m =$ 単位元となるような最小の正の整数を，π の**位数** $\theta(\pi)$ という．明らかに，$1 \leq \theta(\pi) \leq n!$ である．平均値 $\mathrm{E}(\theta(\pi))$ の値は何であるのか？ Goh & Schmutz（ゴー&シュムツ）[47]は，Erdös & Turán（エルデーシュ&トゥラン）[48]の仕事を基礎にして，

$$\ln(\mathrm{E}(\theta(\pi))) = B\sqrt{\frac{n}{\ln(n)}} + o(1)$$

を証明した．ただし，$B = 2\sqrt{2b} = 2.9904703993\cdots$ であり，

$$b = \int_0^\infty \ln(1 - \ln(1 - e^{-x}))\,dx = 1.1178641511\cdots$$

である．Stong（ストング）[49]は，$o(1)$ の評価を改良し，b の別な表現

$$b=\int_0^\infty \frac{xe^{-x}}{(1-e^{-x})(1-\ln(1-e^{-x}))}dx=\int_0^\infty \frac{\ln(x+1)}{e^x-1}dx=-\sum_{k=1}^\infty \frac{e^k}{k}\text{Ei}(-k)$$

を得た．

代表的な置換 π は $\ln(\theta(\pi)) \sim (1/2)\ln(n)^2$ を満たす，ことが証明できる．したがって，少数の例外的な置換が平均値に重大な寄与をする．$\theta(\pi)$ の分散については何がいえるだろうか？

さらに，すべての n 置換 π の最大位数 $\theta(\pi)$ を，$g(n)$ と定義する．Landau[50, 51]は，$\ln(g(n)) \sim \sqrt{n\ln(n)}$ を証明した．とても緻密な $g(n)$ の評価は，[52]にある．

自然な同値関係は，共役関係によって対称群 S_n 上で定義される．$n \to \infty$ とすると，ほとんどすべての共役類 C に対して，C の要素は，$\exp(\sqrt{n}(A+o(1)))$ に等しい位数をもつ．ただし[48,53,54]，

$$A=\frac{2\sqrt{6}}{\pi}\sum_{j\neq 0}\frac{(-1)^{j+1}}{3j^2+j}=4\sqrt{2}-\frac{6\sqrt{6}}{\pi}$$

である．和は，零でない5角数の逆数を含むことに注意せよ．

ランダムに（復元抽出で）選んだ対称群 S_n の2つの要素が，実際に S_n を生成する確率を s_n とする．最初のいくつかの値は，$s_1=1$, $s_2=3/4$, $s_3=1/2$, $s_4=3/8$, … である[55]．s_n の漸近的な値について何が言えるだろうか？ Dixon（ディクソン）[56]は，Netto（ネットー）[57]による1892番の予想，$n\to\infty$ のとき $s_n \to 3/4$ を証明した．Babai（ババイ）[58]はより緻密な評価を得た．

5.4.2 ランダム写像の統計量

n 記号の置換（全単射関数）を任意の n 記号の写像に，議論を一般化する．たとえば，$\varphi(x)=2x \bmod 10$ で定義された $\{0,1,2,\cdots,9\}$ 上の関数 φ は，サイクル (0) と (2 4 8 6) をもつ．残りのシンボル1,3,5,7,9は，3から出発すると，サイクル (2 4 8 6) に吸収され，決して3には戻らない，という意味で一時的である．それにもかかわらず，前のようにサイクルの長さ a_j を定義することができる．この単純な場合，$a_1(\varphi)=1$, $a_2(\varphi)=a_3(\varphi)=0$, $a_4(\varphi)=1$ である．

最長サイクルと最短サイクルの長さ，それぞれ $M(\varphi)$, $m(\varphi)$, は明らかに，擬似乱数の生成に重要である．Purdom & Williams（プルドン＆ウィリアムズ）[59-61]は

$$\lim_{n\to\infty}\frac{\text{E}(M(\varphi))}{\sqrt{n}}=\lambda\sqrt{\frac{\pi}{2}}=0.7824816009\cdots, \quad \lim_{n\to\infty}\frac{\text{E}(m(\varphi))}{\ln(n)}=\frac{1}{2}e^{-\gamma}$$

を発見した．$\text{E}(M(\varphi))$ は，以前にあった n ではなく $\sqrt{n}$ のオーダーで増加することに注意せよ．

他の例として，$\psi(x)=x^2+2 \bmod 20$ で定義された $\{0,1,2,\cdots,19\}$ 上の関数 ψ を考える．図 5.1 から，明らかに $a_2(\psi)=2$ である．他に興味ある量がある[62]．一時的な記号 0, 5, 10, 15 の各々は，サイクルに達するまで2段階かかることに注意せよ．これは，このような距離の最大値である．だから，**最長の裾** (longest tail) $L(\psi)=2$ と定義する．4 は，0,5,10,15 の各々に対して繰り返しのない軌道での頂点の個数である

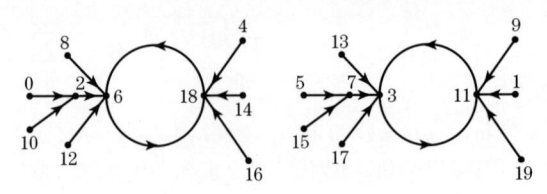

図 5.1 $\psi(x)=x^2+2 \bmod 20$ の関数グラフ (functional graph) は2つの成分 (component) をもつ. 各々は長さ2の閉路 (cycle) をもつ.

ことにも注意せよ. これは, このような長さの最大値である. だから, **最長ロー パス** (longest rho-path) を $R(\psi)=4$ と定義する. 明らかに, 前の例に対して, $L(\varphi)=1$ と $R(\varphi)=5$ である. 任意の n 写像 φ に対して[61],

$$\lim_{n\to\infty}\frac{\mathrm{E}(L(\varphi))}{\sqrt{n}}=\sqrt{2\pi}\ln(2)=1.7374623212\cdots$$

$$\lim_{n\to\infty}\frac{\mathrm{E}(R(\varphi))}{\sqrt{n}}=\sqrt{\frac{\pi}{2}}\int_0^\infty(1-e^{\mathrm{EI}(-x)-I(x)})\,dx=2.4149010237\cdots$$

である. ただし,

$$I(x)=\int_0^x\frac{e^{-y}}{y}\left(1-\exp\left(\frac{-2y}{e^{x-y}-1}\right)\right)dy$$

である.

写像 φ に関連する他の量は, **最大木** (largest tree) $P(\varphi)$ である. φ の各サイクルにおける各頂点は, 一意の極大木 (maximal tree) の根 (root) である【5.6】. 頂点が最大数の木を選ぶ. この数を $P(\varphi)$ とする. 2つの例に対して, 明らかに, $P(\varphi)=2$ と $P(\psi)=6$ である. 任意の n 写像 φ に対して[12,61],

$$\nu=\lim_{n\to\infty}\frac{\mathrm{E}(P(\varphi))}{n}=2\int_0^\infty[1-(1-F(x))^{-1}]\,dx=0.4834983471\cdots$$

$$\lim_{n\to\infty}\frac{\mathrm{Var}(P(\varphi))}{n^2}=\frac{8}{3}\int_0^\infty[1-(1-F(x))^{-1}]x\,dx-\nu^2=0.0494698522\cdots$$

である. ただし,

$$F(x)=\frac{-1}{2\sqrt{\pi}}\int_x^\infty e^{-t}t^{-3/2}dt=1-\frac{1}{\sqrt{\pi x}}\exp(-x)-\mathrm{erf}(\sqrt{x})$$

であり, $\mathrm{erf}(x)$ は誤差関数である【4.6】. Gourdon[63]は, 前の2つの定数を生む硬貨投げゲームの解析について調べた.

最後に, 写像の連結成分の構造を調べる. サイクルとして写像に関係する成分は置換と関連するので, ある意味で, 全サイクルになる. 2つの例に対して, 数え上げ関数 (counting function) は, $\beta_2(\varphi)=1$, $\beta_8(\varphi)=1$ であるが, $\beta_{10}(\varphi)=2$ である. 類似した興味ある対象には, より詳細なものがある. (連結)成分の総数 $\sum_{j=1}^\infty \beta_j$ は, どちらの場合も2に等しい. φ をランダムに選ぶ. このとき[64-67],

$$\mathrm{E}\Big(\sum_{j=1}^{\infty}\beta_j\Big)=\sum_{i=1}^{n}c_{n,0,i}=\frac{1}{2}\ln(n)+\frac{1}{2}(\ln(2)+\gamma)+o(1)$$

$$\mathrm{Var}\Big(\sum_{j=1}^{\infty}\beta_j\Big)=\sum_{i=1}^{n}c_{n,0,i}-\Big(\sum_{i=1}^{n}c_{n,0,i}\Big)^2+\sum_{i=1}^{n}c_{n,0,i}\sum_{j=1}^{n-i}c_{n,i,j}$$

$$=\frac{1}{2}\ln(n)+o(\ln(n))$$

$$\lim_{n\to\infty}\mathrm{P}(\beta_j=k)=\frac{1}{k!}\exp(-d_j)\,d_j^k \quad \text{(漸近ポアソン分布)}$$

である.ただし,

$$c_{n,p,q}=\binom{n-p}{q}\frac{(q-1)!}{n^q},\quad d_j=\frac{e^{-j}}{j}\sum_{i=0}^{j-1}\frac{j^i}{i!}$$

であり,対応するガウス極限も成り立つ.**最大成分** $Q(\varphi)=\max\{j\geq 1:\beta_j>0\}$ を定義する.このとき[12,61,68],

$$\lim_{n\to\infty}\frac{\mathrm{E}(Q(\varphi))}{n}=G\Big(\frac{1}{2},1\Big)=0.7578230112\cdots$$

$$\lim_{n\to\infty}\frac{\mathrm{Var}(Q(\varphi))}{n^2}=H\Big(\frac{1}{2},1\Big)=0.0370072165\cdots$$

である.この結果は[69-71]で取り上げた問題の答えである.重要性から $0.75782\cdots$ を,Flajolet-Odlyzko(**フラジョレット-オドリズコ**)定数とよぶのがふさわしい.r 番目に大きい成分(再び,$n\to\infty$ のとき,それぞれ n と n^2 で正規化された)の平均値と分散は,それぞれ $G(1/2,r)$ と $H(1/2,r)$ で与えられる.たとえば,$G(1/2,2)=0.1709096198\cdots$ であり,$H(1/2,2)=0.0186202233\cdots$ である.最小成分の議論は[72]にある.

[1] W. Goncharov, Sur la distribution des cycles dans les permutations, *C. R. (Dokl.) Acad. Sci. URSS* 35 (1942) 267-269; MR 4,102g.
[2] W. Goncharov, On the field of combinatory analysis (in Russian), *Izv. Akad. Nauk SSSR* 8 (1944) 3-48; Engl. transl. in *Amer. Math. Soc. Transl.* 19 (1962) 1-46; MR 6,88b and MR 24 A1221.
[3] R. E. Greenwood, The number of cycles associated with the elements of a permutation group, *Amer. Math. Monthly* 60 (1953) 407-409; MR 14,939b.
[4] H. S. Wilf, *generatingfunctionology*, Academic Press, 1990; MR 95a:05002.
[5] S. W. Golomb, Random permutations, *Bull. Amer. Math. Soc.* 70 (1964) 747.
[6] S. W. Golomb, *Shift Register Sequences*, Holden-Day, 1967; MR 39 #3906.
[7] S. W. Golomb and P. Gaal, On the number of permutations on n objects with greatest cycle length k, *Adv. Appl. Math.* 20 (1998) 98-107; *Probabilistic Methods in Discrete Mathematics*, Proc. 1996 Petrozavodsk conf., ed. V. F. Kolchin, V. Ya. Kozlov, Yu. L. Pavlov, and Yu. V. Prokhorov, VSP, 1997, pp. 211-218; MR 98k:05003 and MR 99j:05001.
[8] W. C. Mitchell, An evaluation of Golomb's constant, *Math. Comp.* 22 (1968) 411-415.
[9] D. E. Knuth, *The Art of Computer Programming*, v. 1, *Fundamental Algorithms*, Addison-Wesley, 1969, pp. 180-181, 519-520; MR 51 #14624.
[10] A. MacLeod, Golomb's constant to 250 decimal places, unpublished note (1997).
[11] L. A. Shepp and S. P. Lloyd, Ordered cycle lengths in a random permutation, *Trans. Amer. Math. Soc.* 121 (1966) 350-557; MR 33 #3320.
[12] X. Gourdon, *Combinatoire, Algorithmique et Géométrie des Polynômes*, Ph.D. thesis, École

Polytechnique, 1996.
- [13] P. Flajolet, X. Gourdon, and D. Panario, Random polynomials and polynomial factorization, *Proc. 1996 Int. Colloq. on Automata, Languages and Programming (ICALP)*, Paderborn, ed. F. Meyer auf der Heide and B. Monien, Lect. Notes in Comp. Sci. 1099, Springer-Verlag, pp. 232-243; MR 98e:68123.
- [14] P. Flajolet, X. Gourdon, and D. Panario, The complete analysis of a polynomial factorization algorithm over finite fields, *J. Algorithms* 40 (2001) 37-81; INRIA report RR3370; MR 2002f:68193.
- [15] D. E. Knuth and L. Trabb Pardo, Analysis of a simple factorization algorithm, *Theoret. Comput. Sci.* 3 (1976) 321-348; also in *Selected Papers on Analysis of Algorithms*, CSLI, 2000, pp. 303-339; MR 58 #16485.
- [16] D. E. Knuth, *The Art of Computer Programming*, v. 2, *Seminumerical Algorithms*, 2nd ed., Addison-Wesley, 1981, pp. 367-368, 395, 611; MR 83i:68003.
- [17] K. Dickman, On the frequency of numbers containing prime factors of a certain relative magnitude, *Ark. Mat. Astron. Fysik*, v. 22A (1930) n. 10, 1-14.
- [18] S. D. Chowla and T. Vijayaraghavan, On the largest prime divisors of numbers, *J. Indian Math. Soc.* 11 (1947) 31-37; MR 9,332d.
- [19] V. Ramaswami, On the number of positive integers less than x and free of prime divisors greater than x^c, *Bull. Amer. Math. Soc.* 55 (1949) 1122-1127; MR 11,233f.
- [20] N. G. de Bruijn, On the number of positive integers $\leq x$ and free of prime factors $> y$, *Proc. Konink. Nederl. Acad. Wetensch. Ser.* A 54 (1951) 50-60; *Indag. Math.* 13 (1951) 50-60; MR 13, 724e.
- [21] R. Arratia, A. D. Barbour, and S. Tavaré, Random combinatorial structures and prime factorizations, *Notices Amer. Math. Soc.* 44 (1997) 903-910; MR 98i:60007.
- [22] R. Arratia, A. D. Barbour, and S. Tavaré, *Logarithmic Combinatorial Structures: A Probabilistic Approach*, unpublished manuscript (2001).
- [23] K. Alladi and P. Erdös, On an additive arithmetic function, *Pacific J. Math.* 71 (1977) 275-294; MR 56 #5401.
- [24] G. Tenenbaum, *Introduction to Analytic and Probabilistic Number Theory*, Cambridge Univ. Press, 1995, pp. 365-366, 393, 399-400; MR 97e:11005b.
- [25] P. Erdös and A. Ivić, Estimates for sums involving the largest prime factor of an integer and certain related additive functions, *Studia Sci. Math. Hungar.* 15 (1980) 183-199; MR 84a:10046.
- [26] T. Z. Xuan, On a result of Erdös and Ivić, *Arch. Math. (Basel)* 62 (1994) 143-154; MR 94m:11109.
- [27] H. Z. Cao, Sums involving the smallest prime factor of an integer, *Utilitas Math.* 45 (1994) 245-251; MR 95d:11126.
- [28] S. R. Finch, Moments of the Smarandache function, *Smarandache Notions J.* 10 (1999) 95-96; 11 (2000) 140-141; MR 2000a:11129.
- [29] K. Ford, The normal behavior of the Smarandache function, *Smarandache Notions J.* 10 (1999) 81-86; MR 2000a:11130.
- [30] K. K. Norton, Numbers with small prime factors, and the least k^{th} power non-residue, *Memoirs Amer. Math. Soc.* 106 (1971) 1-106; MR 44 #3948.
- [31] A. Hildebrand and G. Tenenbaum, Integers without large prime factors, *J. Théorie Nombres Bordeaux* 5 (1993) 411-484; MR 95d:11116.
- [32] P. Moree, *On the Psixyology of Diophantine Equations*, Ph.D. thesis, Univ. of Leiden, 1993.
- [33] J.-M.-F. Chamayou, A probabilistic approach to a differential-difference equation arising in analytic number theory, *Math. Comp.* 27 (1973) 197-203; MR 49 #1725.
- [34] G. Marsaglia, A. Zaman, and J. C. W. Marsaglia, Numerical solution of some classical differential-difference equations, *Math. Comp.* 53 (1989) 191-201; MR 90h:65124.
- [35] H. Davenport and P. Erdös, The distribution of quadratic and higher residues, *Publ. Math. (Debrecen)* 2 (1952) 252-265; MR 14,1063h.
- [36] L. I. Pál and G. Németh, A statistical theory of lattice damage in solids irradiated by high-

energy particles, *Nuovo Cimento* 12 (1959) 293-309; MR 21 #7630.
[37] G. A.Watterson, The stationary distribution of the infinitely-many neutral alleles diffusion model, *J. Appl. Probab.* 13 (1976) 639-651; correction 14 (1977) 897; MR 58 #20594a and b.
[38] D. Hensley, The convolution powers of the Dickman function, *J. London Math. Soc.* 33 (1986) 395-406; MR 87k:11097.
[39] D. Hensley, Distribution of sums of exponentials of random variables, *Amer. Math. Monthly* 94 (1987) 304-306.
[40] C. J. Lloyd and E. J. Williams, Recursive splitting of an interval when the proportions are identical and independent random variables, *Stochastic Process. Appl.* 28 (1988) 111-122; MR 89e:60025.
[41] A. A. Buchstab, Asymptotic estimates of a general number theoretic function (in Russian), *Mat. Sbornik* 2 (1937) 1239-1246.
[42] S. Selberg, The number of cancelled elements in the sieve of Eratosthenes (in Norwegian), *Norsk Mat. Tidsskr.* 26 (1944) 79-84; MR 8,317a.
[43] N. G. de Bruijn, On the number of uncancelled elements in the sieve of Eratosthenes, *Proc. Konink. Nederl. Acad. Wetensch. Sci. Sect.* 53 (1950) 803-812; Indag. Math. 12 (1950) 247-256; MR 12,11d.
[44] A. Y. Cheer and D. A. Goldston, A differential delay equation arising from the sieve of Eratosthenes, *Math. Comp.* 55 (1990) 129-141; MR 90j:11091.
[45] A. Hildebrand and G. Tenenbaum, On a class of differential-difference equations arising in analytic number theory, *J. d'Analyse Math.* 61 (1993) 145-179; MR 94i:11069.
[46] P. Moree, A special value of Dickman's function, *Math. Student* 64 (1995) 47-50.
[47] W. M. Y. Goh and E. Schmutz, The expected order of a random permutation, *Bull. London Math. Soc.* 23 (1991) 34-42; MR 93a:11080.
[48] P. Turán, Combinatorics, partitions, group theory, *Colloquio Internazionale sulle Teorie Combinatorie*, t. 2, Proc. 1973 Rome conf., Accad. Naz. Lincei, 1976, pp. 181-200; also in *Collected Works*, v. 3, ed. P. Erdös, Akadémiai Kiadó, pp. 2302-2321; MR 58 #21978.
[49] R. Stong, The average order of a permutation, *Elec. J. Combin.* 5 (1998) R41; MR 99f:11122.
[50] W. Miller, The maximum order of an element of a finite symmetric group, *Amer. Math. Monthly* 94 (1987) 497-506; MR 89k:20005.
[51] J.-L. Nicolas, On Landau's function $g(n)$, *The Mathematics of Paul Erdös*, v. 1, ed. R. L. Graham and J. Nesetril, Springer-Verlag, 1997, pp. 228-240; MR 98b:11096.
[52] J.-P. Massias, J.-L. Nicolas, and G. Robin, Effective bounds for the maximal order of an element in the symmetric group, *Math. Comp.* 53 (1989) 665-678; MR 90e:11139.
[53] D. H. Lehmer, On a constant of Turán and Erdös, *Acta Arith.* 37 (1980) 359-361; MR 82b:05014.
[54] W. Schwarz, Another evaluation of an Erdös-Turán constant, *Analysis* 5 (1985) 343-345; MR 86m:40003.
[55] N. J. A. Sloane, On-Line Encyclopedia of Integer Sequences, A040173 and A040174.
[56] J. D. Dixon, The probability of generating the symmetric group, *Math. Z.* 110 (1969) 199-205; MR 40 #4985.
[57] E. Netto, *The Theory of Substitutions and Its Applications to Algebra*, 2nd ed., Chelsea, 1964; MR 31 #184.
[58] L. Babai, The probability of generating the symmetric group, *J. Combin. Theory Ser. A* 52 (1989) 148-153; MR 91a:20007.
[59] P. W. Purdom and J. H. Williams, Cycle length in a random function, *Trans. Amer. Math. Soc.* 133 (1968) 547-551; MR 37 #3616.
[60] D. E. Knuth, *The Art of Computer Programming*, v. 2, *Seminumerical Algorithms*, 2nd ed., Addison-Wesley, 1981; pp. 7-8, 517-520; MR 83i:68003.
[61] P. Flajolet and A. M. Odlyzko, Random mapping statistics, *Advances in Cryptology - EURO-CRYPT '89*, ed. J.-J. Quisquater and J. Vandewalle, Lect. Notes in Comp. Sci. 434, Springer-Verlag, 1990, pp. 329-354; MR 91h:94003.

[62] R. Sedgewick and P. Flajolet, *Introduction to the Analysis of Algorithms*, Addison-Wesley, 1996, pp. 357-358, 454-465.
[63] X. Gourdon, Largest components in random combinatorial structures, *Discrete Math.* 180 (1998) 185-209; MR 99c:60013.
[64] M. D. Kruskal, The expected number of components under a random mapping function, *Amer. Math. Monthly* 61 (1954) 392-397; MR 16,52b.
[65] S. M. Ross, A random graph, *J. Appl. Probab.* 18 (1981) 309-315; MR 81m:05124.
[66] P. M. Higgins and E. J.Williams, Random functions on a finite set, *Ars Combin.* 26 (1988) A, 93-102; MR 90g:60008.
[67] V. F. Kolchin, *Random Mappings*, Optimization Software Inc., 1986, pp. 46, 79, 164; MR 88a:60022.
[68] D. Panario and B. Richmond, Exact largest and smallest size of components, *Algorithmica* 31 (2001) 413-432; MR 2002j:68065.
[69] G. A.Watterson and H. A. Guess, Is the most frequent allele the oldest?, *Theoret. Populat. Biol.* 11 (1977) 141-160.
[70] P. J. Donnelly,W. J. Ewens, and S. Padmadisastra, Functionals of random mappings: Exact and asymptotic results, *Adv. Appl. Probab.* 23 (1991) 437-455; MR 92k:60017.
[71] P. M. Higgins, *Techniques of Semigroup Theory*, Oxford Univ. Press, 1992;MR93d:20101.
[72] D. Panario and B. Richmond, Smallest components in decomposable structures: exp-log class, *Algorithmica* 29 (2001) 205-226.

5.5 カルマールの合成定数

整数 n の**加法分解** (additive composition) とは，ある $k≧1$ と整数の列 $x_1, x_2, \cdots, x_k$ で

$$n = x_1 + x_2 + \cdots + x_k, \quad すべての j, \ 1≦j≦k で，\ x_j≧1$$

を満たすときをいう．

整数 n の**乗法分解** (multiplicative composition) とは，ある $k≧1$ と整数の列 $x_1, x_2, \cdots, x_k$ で

$$n = x_1 x_2 \cdots x_k, \quad すべての j, \ 1≦j≦k で，\ x_j≧2$$

を満たすときをいう．

n の加法分解の個数 $a(n)$ が，2^{n-1} であることは自明である．乗法分解の個数 $m(n)$ には，閉じた式がない．しかし，漸近的には，

$$\sum_{n=1}^{N} m(n) \sim \frac{-1}{\rho \zeta'(\rho)} N^\rho = (0.3181736521 \cdots) \cdot N^\rho$$

を満たす．ただし，$\rho = 1.7286472389 \cdots$ は $\zeta(x) = 2 \ (x>1)$ の解であり，$\zeta(x)$ は Riemann のゼータ関数である【1.6】．この結果は最初 Kalmár (カルマール) [1,2] によって得られ，[3-8]で洗練された．

整数 n の**加法分割** (additive partition) とは，ある $k≧1$ と整数の列 $x_1, x_2, \cdots, x_k$ が

$$n = x_1 + x_2 + \cdots + x_k, \quad 1 ≦ x_1 ≦ x_2 ≦ \cdots ≦ x_k$$

を満たすときをいう．並べ替えのもとで，分割は自然に加法分解の同値類を表している．n の加法分割の個数 $A(n)$ は，【1.4.2】で述べた．一方，n の**乗法分割**（multiplicative partition）の個数 $M(n)$ は，漸近的に[9,10]

$$\sum_{n=1}^{N} M(n) \sim \frac{1}{2\sqrt{\pi}} N \exp(2\sqrt{\ln(N)}) \ln(N)^{-3/4}$$

を満たす．

これまでは，「制限なし」の分解と分割を扱った．変形したものが多くある．各 x_j を素数に制限した場合に焦点を当てる．たとえば，**素数乗法分割**（prime multiplicative partitions）の個数 $M_p(n)$ は，明らかに，$n \geq 2$ に対して 1 である．**素数加法分解**（prime additive compositions）の個数 $a_p(n)$ は[11]，

$$a_p(n) \sim \frac{1}{\xi f'(\xi)} \left(\frac{1}{\xi}\right)^n = (0.3036552633\cdots) \cdot (1.4762287836\cdots)^n$$

である．ただし，$\xi = 0.6774017761\cdots$ は方程式

$$f(x) = \sum_p x^p = 1, \quad x > 0$$

の一意の解である．和は，すべての素数 p にわたる．**素数乗法分解**（prime multiplicative compositions）の個数 $m_p(n)$ は[12]，

$$\sum_{n=1}^{N} m_p(n) \sim \frac{-1}{\eta g'(\eta)} N^{-\eta} = (0.4127732370\cdots) \cdot N^{-\eta}$$

を満たす．ただし，$\eta = -1.3994333287\cdots$ は方程式

$$g(y) = \sum_p p^y = 1, \quad y < 0$$

の一意の解である．$n \geq 8$ に対して $A_p(n+1) > A_p(n)$ を除いて，**素数加法分割**（prime additive partition）の個数について多くは知られていない[13-16]．

関連のあるいくぶん人工的な話題がある．p_n を n 番目の素数とする．$p_1 = 2$ である．形式的ベキ級数

$$P(z) = 1 + \sum_{n=1}^{\infty} p_n z^n, \quad Q(z) = \frac{1}{P(z)} = \sum_{n=0}^{\infty} q_n z^n$$

を定義する．係数が次の漸近式[17]

$$q_n \sim \frac{1}{\theta P'(\theta)} \left(\frac{1}{\theta}\right)^n = (-0.6223065745\cdots) \cdot (-1.4560749485\cdots)^n$$

を満たすことを知って，ある人は驚くだろう．ただし，$\theta = -0.6867778344\cdots$ は，円板 $|z| < 3/4$ の内部にある $P(z)$ の一意の零点である．対比として，素数定理により $p_n \sim n \ln(n)$ である．同様な精神で，データセット[18]

$$(1,2), (2,3), (3,5), (4,7), (5,11), (6,13), \cdots, (n, p_n)$$

を通る $n-1$ 次の多項式

$$c_0 + c_1(x-1) + c_2(x-1)(x-2) + \cdots + c_{n-1}(x-1)(x-2)(x-3)\cdots(x-n+1)$$

の係数を考える．$n \to \infty$ のとき，和 $\sum_{k=0}^{n-1} c_k$ は $3.4070691656\cdots$ に収束する．

分解と分割の数え上げに戻る．各 x_j は無平方[12]または x たちがすべて異なる[8]

という制約をもつ変形だけを注意しておく．さらに，n の分解/分割 $x_1, x_2, \cdots, x_k$ と $y_1,$ $y_2, \cdots, y_k$ が **独立** (independent) であるとは，x と y の真部分数列 (proper subsequence) の和/積が決して一致しないときをいう．このような組は，いったいどのくらい（n の関数として）あるのだろうか？ 漸近的な解答については，[19] を参照せよ．

$i < j$ であるときつねに $z_i | z_j$ であるような数列 $1 \leq z_1 < z_2 < \cdots < z_k = n$ の個数は，$2m(n)$ であることを，Cameron & Erdös (カメロン&エルデーシュ) [20] が指摘した．数列に1を含めるか含めないかの選び方ができるので，係数2が現れる．$i \neq j$ であるときはつねに $w_i \nmid w_j$ であるような数列 $1 \leq w_1 < w_2 < \cdots < w_k \leq n$ の個数 $c(n)$ については，何がいえるだろうか？ $\lim_{n \to \infty} c(n)^{1/n}$ は存在すると予想されている．充分大きな n に対して，$1.55967^n \leq c(n) \leq 1.59^n$ であることが知られている．**原始的数列** (primitive sequences) として知られているこのような数列については，さらに【2.27】を参照せよ．

最後に，集合 $\{2^{-i} : i \geq 0\}$ の $n+1$ 個の要素の和として，1を表す方法の個数を $h(n)$ と定義する．ただし，繰り返しを許し，順序は問わない．Flajolet & Prodinger (フラジョレット&プロディンガー) [21] は
$$h(n) \sim (0.2545055235 \cdots) \kappa^n$$
を証明した．ただし，$\kappa = 1.7941471875 \cdots$ は方程式
$$\sum_{j=1}^{\infty} (-1)^{j+1} \frac{x^{2^{j+1}-2-j}}{(1-x)(1-x^3)(1-x^7) \cdots (1-x^{2^j-1})} - 1 = 0$$
の最小の正の解 x の逆数である．

これは，2分木 (binary trees)【5.6】に結びついた数え上げレベル数列 (enumerating level number sequences) と関連している．

[1] L. Kalmár, A "factorisatio numerorum" problémájáról, *Mat. Fiz. Lapok* 38 (1931) 1-15.
[2] L. Kalmár, Über die mittlere Anzahl Produktdarstellungen der Zahlen, *Acta Sci. Math. (Szeged)* 5 (1930-32) 95-107.
[3] E. Hille, A problem in "Factorisatio Numerorum," *Acta Arith.* 2 (1936) 136-144.
[4] P. Erdös, On some asymptotic formulas in the theory of the "factorisatio numerorum," *Annals of Math.* 42 (1941) 989-993; MR 3,165b.
[5] P. Erdös, Corrections to two of my papers, *Annals of Math.* 44 (1943) 647-651;MR5,172c.
[6] S. Ikehara, On Kalmár's problem in "Factorisatio Numerorum." II, *Proc. Phys.-Math. Soc. Japan* 23 (1941) 767-774; MR 7,365h.
[7] R. Warlimont, Factorisatio numerorum with constraints, *J. Number Theory* 45 (1993) 186-199; MR 94f:11098.
[8] H.-K. Hwang, Distribution of the number of factors in random ordered factorizations of integers, *J. Number Theory* 81 (2000) 61-92; MR 2001k:11183.
[9] A. Oppenheim, On an arithmetic function. II, *J. London Math. Soc.* 2 (1927) 123-130.
[10] G. Szekeres and P. Turán, Über das zweite Hauptproblem der "Factorisatio Numerorum," *Acta Sci. Math. (Szeged)* 6 (1932-34) 143-154; also in *Collected Works of Paul Turán*, v. 1, ed. P. Erdös, Akadémiai Kiadó, pp. 1-12.
[11] P. Flajolet, Remarks on coefficient asymptotics, unpublished note (1995).
[12] A. Knopfmacher, J. Knopfmacher, and R. Warlimont, Ordered factorizations for integers and arithmetical semigroups, *Advances in Number Theory*, Proc. 1991 Kingston conf., ed. F. Q.

Gouvêa and N. Yui, Oxford Univ. Press, 1993, pp. 151-165; MR 97e:11118.
[13] P. T. Bateman and P. Erdös, Monotonicity of partition functions, *Mathematika* 3 (1956) 1-14; MR 18,195a.
[14] P. T. Bateman and P. Erdös, Partitions into primes, *Publ. Math. (Debrecen)* 4 (1956) 198-200; MR 18,15c.
[15] J. Browkin, Sur les décompositions des nombres naturels en sommes de nombres premiers, *Colloq. Math.* 5 (1958) 205-207; MR 21 # 1956.
[16] N. J. A. Sloane, On-Line Encyclopedia of Integer Sequences, A000041, A000607, A001055, A002033, A002572, A008480, and A023360.
[17] N. Backhouse, Formal reciprocal of a prime power series, unpublished note (1995).
[18] F. Magata, Newtonian interpolation and primes, unpublished note (1998).
[19] P. Erdös, J.-L. Nicolas, and A. Sárközy, On the number of pairs of partitions of n without common subsums, *Colloq. Math.* 63 (1992) 61-83; MR 93c:11087.
[20] P. J. Cameron and P. Erdös, On the number of sets of integers with various properties, *Number Theory*, Proc. 1990 Canad. Number Theory Assoc. Banff conf., ed. R. A. Mollin, Gruyter, pp. 61-79; MR 92g:11010.
[21] P. Flajolet and H. Prodinger, Level number sequences for trees, *Discrete Math.* 65 (1987) 149-156; MR 88e:05030.

5.6 オッターの木の数え上げ定数

位数（order）nの**グラフ**（graph）は，**辺**（edge）（異なる点の順序をつけない対）の集合をもったn個の**頂点**（vertex，点）の集合からなる．ループと多重並行辺（頂点の対を2つ以上の辺で結ぶ）は，自動的に許されないことに注意する．辺で結ばれた2つの頂点は，**隣接**（adjacent）しているとよばれる．

林（forest，森ともいう）とは，**無閉路的**（acyclic）グラフのことである．無閉路とは，隣接頂点$v_0, v_1, \cdots, v_m$の列で，すべての$i<j<m$に対して$v_i \neq v_j$で，$v_0 = v_m$となるものは存在しないものである．

木（tree）（または**自由木**（free tree））が林であるとは，任意の2点が**連結している**（connected）ときをいう．任意の2つの異なる頂点uとvが連結（connected）しているとは，隣接頂点$v_0, v_1, \cdots, v_m$で$v_0 = u$で$v_m = w$である点列が存在するときをいう．

2つの木σとτが**同型**（isomorphic）とは，σの頂点からτの頂点へ，辺の隣接関係を保つ，1対1対応の写像が存在するときをいう（図5.2）．位数<11のすべての非同型な木の図は，[1]にある．応用は[2]にある．

位数nの非同型な木の個数t_nの漸近式について何がいえるだろうか？ Cayley（ケーリー）とPólya（ポリヤ）の仕事を基礎にして，Otter（オッター）[3-6]は

$$\lim_{n \to \infty} \frac{t_n n^{5/2}}{\alpha^n} = \beta$$

を証明した．ただし，$\alpha = 2.9557652856\cdots = (0.3383218568\cdots)^{-1}$は，すぐ後に定義され

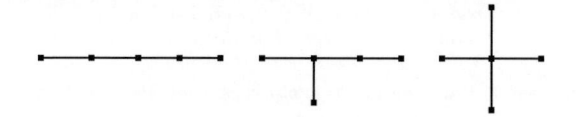

図5.2 次数5の非同型木は3個ある.

るある種の関数 T を含む, 方程式 $T(x^{-1})=1$ の一意の正の解であり,

$$\beta = \frac{1}{\sqrt{2\pi}}\Big(1+\sum_{k=2}^{\infty} T'\Big(\frac{1}{\alpha^k}\Big)\Big)^{3/2} = 0.5349496061\cdots$$

である. ただし, T' は T の導関数である. α と β は高い精度で効率的に計算できるけれども, それらが代数的数であるか, または超越数であるか, どちらとも知られていない[6,7].

根付き木 (rooted tree) とは, **根** (root) とよばれるちょうど1つの頂点が, 他のものから区別されている木のことである (図5.3). 木の一番上の頂点として根 (root) を描く. 根付き木の同型写像は, 根を根に写像する. 位数 n の非同型な根付き木の個数 T_n の漸近式について何がいえるだろうか? Otter の対応する結果は

$$\lim_{n\to\infty} \frac{T_n n^{3/2}}{\alpha^n} = \Big(\frac{\beta}{2\pi}\Big)^{1/3} = 0.4399240125\cdots = \Big(\frac{1}{4\pi\alpha}\Big)^{1/2}(2.6811281472\cdots)$$

である. 実際, 母関数

$$t(x) = \sum_{n=1}^{\infty} t_n x^n$$
$$= x + x^2 + x^3 + 2x^4 + 3x^5 + 6x^6 + 11x^7 + 23x^8 + 47x^9 + 106x^{10} + \cdots$$

$$T(x) = \sum_{n=1}^{\infty} T_n x^n$$
$$= x + x^2 + 2x^3 + 4x^4 + 9x^5 + 20x^6 + 48x^7 + 115x^8 + 286x^9 + \cdots$$

は, 式 $t(x) = T(x) - (T(x)^2 - T(x^2))/2$ で関連づけられ, 定数 α^{-1} は $t(x)$, $T(x)$ 両方の収束半径であり, 係数 T_n は

$$T(x) = x \exp\Big(\sum_{k=1}^{\infty} \frac{T(x^k)}{k}\Big), \quad T_{n+1} = \frac{1}{n}\sum_{k=1}^{n}\Big(\sum_{d|k} dT_d\Big)T_{n-k+1}$$

を用いて計算できる.

木を変形したものが数多くある. それらを1つ1つ丹念に詳細に記述することは,

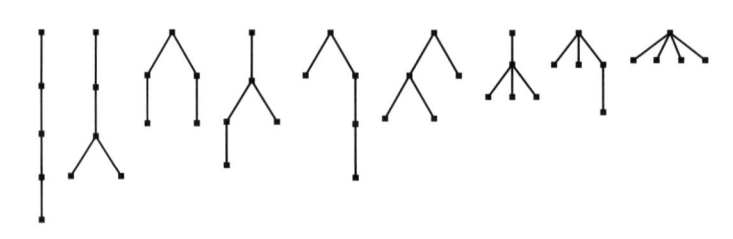

図5.3 次数5の非同型根付き木は9個ある.

5.6 オッターの木の数え上げ定数

[4,5]に残しておくことが最適である．多くの例のうち最初のものは，**弱2分木**(weakly binary tree) である．これは，根が多くとも2つの頂点と隣接していて，根でない頂点のすべては，多くとも3つの頂点と隣接している，木のことである．たとえば，位数5の同型でない弱2分木は6個ある．位数nの同型でない弱2分木の個数，B_nの漸近式は，Otter[3,8-10]によって，

$$\lim_{n\to\infty} \frac{B_n n^{3/2}}{\xi^n} = \eta$$

である．ただし，$\xi^{-1} = 0.4026975036\cdots = (2.4832535361\cdots)^{-1}$ は，級数

$$B(x) = \sum_{n=0}^{\infty} B_n x^n = 1 + x + x^2 + 2x^3 + 3x^4 + 6x^5 + 11x^6 + 23x^7 + 46x^8 + 98x^9 + \cdots$$

の収束半径であり，

$$\eta = \sqrt{\frac{\xi}{2\pi}} \left(1 + \frac{1}{\xi} B\left(\frac{1}{\xi^2}\right) + \frac{1}{\xi^3} B'\left(\frac{1}{\xi^2}\right)\right)^{1/2}$$

$$= 0.7916031835\cdots = (0.3187766258\cdots)\xi$$

である．級数の係数は

$$B(x) = 1 + \frac{1}{2} x (B(x)^2 + B(x^2))$$

$$B_k = \begin{cases} \dfrac{B_i(B_i+1)}{2} + \sum_{j=0}^{i-1} B_{k-j-1} B_j & (k=2i+1 \text{ のとき}) \\ \sum_{j=0}^{i-1} B_{k-j-1} B_j & (k=2i \text{ のとき}) \end{cases}$$

から出てくる．Otterは，この特別な場合について，$\xi = \lim_{n\to\infty} c_n^{2^{-n}}$ であることを示した．ただし，数列 $\{c_n\}$ は2次漸化式

$$c_0 = 2, \quad c_n = c_{n-1}^2 + 2, \quad n \geq 1$$

で定義される．したがって，

$$\eta = \frac{1}{2} \sqrt{\frac{\xi}{\pi}} \sqrt{3 + \frac{1}{c_1} + \frac{1}{c_1 c_2} + \frac{1}{c_1 c_2 c_3} + \frac{1}{c_1 c_2 c_3 c_4} + \cdots}$$

である．

前の概念を少し特殊化したものがある．**強2分木** (strongly binary tree) とは，根 (root, ルート) が0個または2個の頂点と隣接していて，根でない頂点は1つまたは3つの頂点と隣接しているときをいう（図5.4）．このような木は**2分木** (binary trees) ともよばれ，さらに【5.6.9】および【5.13】で議論される．位数$2n+1$の同型でな

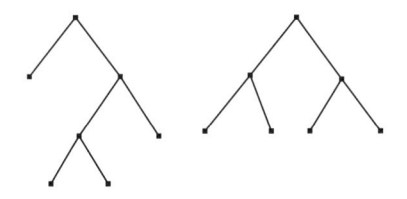

図5.4 次数7の非同型な強2分木は2個ある．

い強2分木の個数は，ちょうど B_n であることがわかる．1対1対応は，前の方向に向かって，強2分木の**葉**（leaves）（終端頂点）のすべてを取り除くことによって得られる．反対側に進むために，弱2分木から始める．任意の次数1の頂点（またはもし次数0であれば根）に2つの葉を付け加える，任意の次数2の頂点（またはもし次数1であれば根）に1つの葉を付け加える．したがって，同様な漸近式が，弱（weak）と強（strong）の両方の場合に適用される．

さらに，可換であるが結合法則が成り立たない代数においては，x^4 という表現は不明確で，xx^3 または x^2x^2 として解釈できる．式 x^5 は同様に，xxx^3，xx^2x^2，x^2x^3 の意味にとられる．明らかに，B_{n-1} は x^n の可能な解釈の数である．だから，$\{B_n\}$ を Wedderburn-Etherington（ウェダーバーン-エセリントン）数列[11-15]とよぶこともある．

5.6.1 化学での異性体

弱3分木（weakly ternary tree）とは，根が多くとも3つの頂点と隣接し，根でない頂点のすべては多くとも4つの頂点と隣接している，根付き木のことである．たとえば，位数5の非同型な弱3分木は，8個ある．位数 n の非同型な弱3分木の個数，R_n の漸近式は，再び Otter [3,15-17] により

$$\lim_{n\to\infty}\frac{R_n n^{3/2}}{\xi_R^n}=\eta_R$$

である．ただし，$\xi_R^{-1}=0.3551817423\cdots=(2.8154600332\cdots)^{-1}$ は，級数

$$R(x)=\sum_{n=0}^{\infty}R_n x^n$$
$$=1+x+x^2+2x^3+4x^4+8x^5+17x^6+39x^7+89x^8+211x^9+\cdots$$

の収束半径であり，

$$\eta_R=\sqrt{\frac{\xi_R}{2\pi}}\left(-1+\rho+\frac{1}{\xi_R^3}R'\!\left(\frac{1}{\xi_R^2}\right)+\frac{1}{\xi_R^4}R'\!\left(\frac{1}{\xi_R^3}\right)\right)^{1/2}\rho^{-1/2}$$
$$=0.5178759064\cdots$$

であり，$\rho=R(\xi_R^{-1})$ である．級数の係数は

$$R(x)=1+\frac{1}{6}x(R(x)^3+3R(x)R(x^2)+2R(x^3))$$

から得られる．

この分野の応用は，有機化学[18-21]，たとえば分子式 $C_nH_{2n+1}OH$（アルコール類，図5.5）の**構造異性体**（constitutional isomers）の個数 R_n に関係する．2つの構造異性体は，原子的結合状態が違う．水酸基（OH）の立体的な相対的位置は関係ない．

さらに[18,19,22,23]，

$$r(x)=\frac{1}{24}x(R(x)^4+6R(x)^2R(x^2)+8R(x)R(x^3)+3R(x^2)^2+6R(x^4))$$
$$-\frac{1}{2}(R(x)^2-R(x^2))+R(x)$$

5.6 オッターの木の数え上げ定数

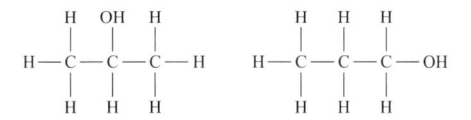

図 5.5 化学式 C_3H_7OH（プロパノール）の 2 つの異性体.

と定義すると，
$$r(x) = \sum_{n=0}^{\infty} r_n x^n$$
$$= 1 + x + x^2 + x^3 + 2x^4 + 3x^5 + 5x^6 + 9x^7 + 18x^8 + 35x^9 + 75x^{10} + \cdots$$

であり，r_n は分子式 C_nH_{2n+2}（アルカン基 (alkanes)，図 5.6）の構造異性体の個数である．級数 $t(x)$ が $T(x)$ と関連しているように，級数 $r(x)$ は $R(x)$ に関連している（r, t が自由 (free) であり R, T が根付き木であるという意味で）．$r(x)$ の収束半径は同様に ξ_R^{-1} であり，

$$\lim_{n \to \infty} \frac{r_n n^{5/2}}{\xi_R^n} = 2\pi \frac{\eta_R^3}{\xi_R} \rho = 0.6563186958 \cdots$$

が成り立つ．

炭素原子が 4 つの異なった対象（原子または官能基）と結合しているとき，**キラル** (chiral) または **非対称** (asymmetric, **不斉**) である，という．$C_nH_{2n+1}OH$ の構造異性体の個数 Q_n は，キラル炭素 (C) 原子なしで[18, 24]，

$$\lim_{n \to \infty} \frac{Q_n}{\xi_Q^n} = \eta_Q$$

が成り立つ．ただし，$\xi_Q^{-1} = 0.5947539639 \cdots = (1.6813675244 \cdots)^{-1}$ は，級数

$$Q(x) = \sum_{n=0}^{\infty} Q_n x^n$$
$$= 1 + x + x^2 + 2x^3 + 3x^4 + 5x^5 + 8x^6 + 14x^7 + 23x^8 + 39x^9 + \cdots$$

の収束半径である．係数は，$Q(x) = 1 + xQ(x)Q(x^2)$ から得られる．よって，

$$Q(x) = \cfrac{1}{1} - \cfrac{x}{1} - \cfrac{x^2}{1} - \cfrac{x^4}{1} - \cfrac{x^8}{1} - \cfrac{x^{16}}{1} - \cdots$$

であり，これは興味ある連分数である．これから，$Q(x) = \psi(x^2)/\psi(x)$ は，（ψ は解析的関数で，$\psi(0) = 1$ を仮定すると）一意であることが容易に従う．したがって，

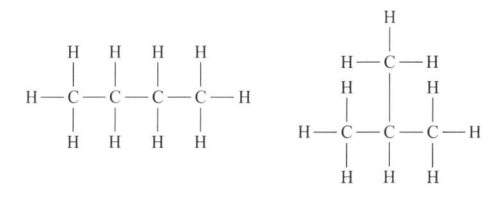

図 5.6 化学式 C_4H_{10}（ブタン）の 2 つの異性体.

$$\eta_Q = -\xi_Q \psi\left(\frac{1}{\xi_Q^2}\right)\left(\psi'\left(\frac{1}{\xi_Q}\right)\right)^{-1} = 0.3607140971\cdots$$

を得る.

$C_nH_{2n+1}OH$ の**立体異性体** (stereoisomers) の個数を S_n とする. 今度は上述の注意の他に, 水酸基の相対的位置も重要である[18,19,25]. たとえば, 描かれた立体異性体の対 (2つの4面体, 図5.7) は, (鏡像体だが) 移動して重ね合わせることはできない. S_n の母関数は,

$$S(x) = \sum_{n=0}^{\infty} S_n x^n$$
$$= 1 + x + x^2 + 2x^3 + 5x^4 + 11x^5 + 28x^6 + 74x^7 + 199x^8 + 551x^9 + \cdots$$
$$S(x) = 1 + \frac{1}{3}x(S(x)^3 + 2S(x^3)), \quad S_n \sim \eta_S n^{-3/2} \xi_S^n$$

であり, 収束半径は, $\xi_S^{-1} = 0.3042184090\cdots = (3.2871120555\cdots)^{-1}$ である. 簡単のために η_S の値は除く.

5.6.2 木のいろいろな種類

恒等木 (identity tree) とは, 自己同型写像が恒等写像のみである木のことである. 明らかに, 位数7と8には一意的な恒等木がある. しかし, 位数≦6には自明な場合はない. 恒等木の母関数は[4,26],

$$u(x) = \sum_{n=1}^{\infty} u_n x^n$$
$$= x + x^7 + x^8 + 3x^9 + 6x^{10} + 15x^{11} + 29x^{12} + 67x^{13} + 139x^{14} + \cdots$$

である. **根付き恒等木** (rooted identity tree) とは, 根を固定する自己同型写像が恒等写像のみである根付き木のことである. この追加の条件の下で, 根付き恒等木はすべての位数に対して存在する. 母関数は

$$U(x) = \sum_{n=1}^{\infty} U_n x^n = x + x^2 + x^3 + 2x^4 + 3x^5 + 6x^6 + 12x^7 + 25x^8 + 52x^9 + \cdots$$

である. 根付き恒等木の図は【6.11】にあるので参照せよ. このような木は, またすべての頂点と辺は一意的, すなわち, 同型なものを許さないという意味で, **非対称** (asym-

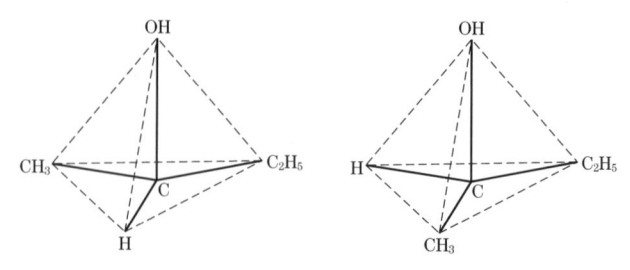

図5.7 (自明でない) 立体異性体が存在する最も簡単なアルコール類 C_4H_9OH.

5.6 オッターの木の数え上げ定数

metric) であるといわれる．[5,27]で，

$$\lim_{n\to\infty}\frac{U_n n^{3/2}}{\xi_U^n}=\eta_U=\frac{1}{\sqrt{2\pi}}\left(1-\sum_{k=2}^{\infty}\frac{(-1)^k}{\xi_U^k}U'\left(\frac{1}{\xi_U^k}\right)\right)^{1/2}=0.3625364234\cdots$$

$$\lim_{n\to\infty}\frac{u_n n^{5/2}}{\xi_U^n}=2\pi\eta_U^3=0.2993882877\cdots$$

が証明された．ただし，$\xi_U^{-1}=0.3972130965\cdots=(2.5175403550\cdots)^{-1}$ は，$U(x)$ および $u(x)$ の収束半径である．さらに，

$$U(x)=x\exp\left(\sum_{k=1}^{\infty}(-1)^{k+1}\frac{U(x^k)}{k}\right), \quad u(x)=U(x)-\frac{1}{2}(U(x)^2+U(x^2))$$

である．

木が，**同相的既約** (homeomorphically irreducible)，または**級数既約** (series-reduced) とは，ちょうど2つの他の頂点に隣接する頂点が，まったく存在しないときをいう．明らかに，位数3には，このような木は存在しない．母関数は[4,26,28]，

$$h(x)=\sum_{n=1}^{\infty}h_n x^n$$
$$=x+x^2+x^4+x^5+2x^6+2x^7+4x^8+5x^9+10x^{10}+14x^{11}+\cdots$$

である．**同相的既約植木** (planted homeomorphically irreducible tree) とは，同相的既約であり，根が他の頂点とちょうど1つだけに隣接している根付き木のことである．対応する母関数は

$$H(x)=\sum_{n=1}^{\infty}H_n x^n$$
$$=x^2+x^4+x^5+2x^6+3x^7+6x^8+10x^9+19x^{10}+35x^{11}+\cdots=x\widetilde{H}(x)$$

である．[5,29]で，

$$\lim_{n\to\infty}\frac{H_n n^{3/2}}{\xi_H^n}=\eta_H=\frac{1}{\xi_H\sqrt{2\pi}}\left(\frac{\xi_H}{\xi_H+1}+\sum_{k=2}^{\infty}\frac{1}{\xi_H^k}\widetilde{H}'\left(\frac{1}{\xi_H^k}\right)\right)^{1/2}=0.1924225474\cdots$$

$$\lim_{n\to\infty}\frac{h_n n^{5/2}}{\xi_H^n}=2\pi\xi_H^2(\xi_H+1)\eta_H^3=0.6844472720\cdots$$

が証明された．ただし，$\xi_H^{-1}=0.4567332095\cdots=(2.1894619856\cdots)^{-1}$ は，$H(x)$ および $h(x)$ の収束半径である．さらに，

$$\widetilde{H}(x)=\frac{x}{x+1}\exp\left(\sum_{k=1}^{\infty}\frac{\widetilde{H}(x^k)}{k}\right),$$
$$h(x)=(x+1)\widetilde{H}(x)-\frac{x+1}{2}\widetilde{H}(x)^2-\frac{x-1}{2}\widetilde{H}(x^2)$$

である．

任意の頂点の部分木の（左側から右側への）順序付けを考慮に入れると，**順序木** (ordered trees) が得られ，数え上げの異なる問題が生ずる．たとえば，順序付けられた2つの木 σ と τ が**巡回同型** (cyclically isomorphic) とは，σ と τ が根付き木として同型であり，もし τ が任意の頂点または同様にいくつかの頂点のそれぞれに対して，すべての部分木を巡回的に置きかえて得られる，ときをいう．この関係の下での同値類

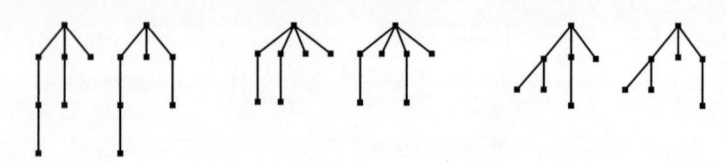

図 5.8 根付き木として同一である（次数 7 の）異なる 3 つのモビール．

は，**モビール**（mobile）といわれる．位数 7 には，51 個のモビールが存在するが，位数 7 には，根付き木は 48 個しかない（図 5.8）．

モビールの母関数は [22, 26, 30]

$$M(x)=\sum_{n=1}^{\infty}M_n x^n$$
$$=x+x^2+2x^3+4x^4+9x^5+20x^6+51x^7+128x^8+345x^9+\cdots$$
$$M(x)=x\Big(1-\sum_{k=1}^{\infty}\frac{\varphi(k)}{k}\ln(1-M(x^k))\Big), \quad M_n\sim\eta_M n^{-3/2}\xi_M^n$$

である．ただし，φ はオイラーの約数関数【2.7】であり，$\xi_M^{-1}=0.3061875165\cdots=(3.2659724710\cdots)^{-1}$ である．

頂点をはっきりさせるために，整数 $1,2,3,\cdots,n$ でグラフの頂点に**ラベル**（label，名札）をつける．このとき，対応する数え上げの問題の多くは簡単になる．たとえば，ちょうど n^{n-2} 個のラベルつき自由木（labeled free tree）があり，ちょうど n^{n-1} 個のラベルつき根付き木（labeled rooted tree）がある．ラベルつきモビールに対して，指数的母関数 [31]

$$\hat{M}(x)=\sum_{n=1}^{\infty}\hat{M}_n x^n$$
$$=x+\frac{2}{2!}x^2+\frac{9}{3!}x^3+\frac{68}{4!}x^4+\frac{730}{5!}x^5+\frac{10164}{6!}x^6+\frac{173838}{7!}x^7+\cdots$$
$$\hat{M}(x)=x\Big(1-\ln(1-\hat{M}(x))\Big), \quad \hat{M}_n\sim\hat{\eta}\hat{\xi}^n n^{n-1}$$

があり，この問題はかなりおもしろくなる．ただし，$\hat{\xi}=e^{-1}(1-\mu)^{-1}=1.1574198038\cdots$，$\hat{\eta}=\sqrt{\mu(1-\mu)}=0.4656386467\cdots$ であり，$\mu=0.6821555671\cdots$ は方程式 $\mu(1-\mu)^{-1}=1-\ln(1-\mu)$ の一意の解である．

増加木（increasing tree）とは，根から始まる任意の枝に沿ってラベル（についている数）が増加するようにラベルがつけられた木のことである．根には 1 をつけなければならない．増加モビール（increasing mobiles）に対して，数え上げの問題は再び，興味ある定数を生む [32]．

$$\widetilde{M}(x)=\sum_{n=1}^{\infty}\widetilde{M}_n x^n$$
$$=x+\frac{1}{2!}x^2+\frac{2}{3!}x^3+\frac{7}{4!}x^4+\frac{36}{5!}x^5+\frac{245}{6!}x^6+\frac{2076}{7!}x^7+\cdots$$

$$\tilde{M}'(x)=1-\ln(1-\tilde{M}(x)), \quad \tilde{M}_n \sim \tilde{\xi}^{n-1} n!\left(\frac{1}{n^2}-\frac{1}{n^2\ln(n)}+O\left(\frac{1}{n^2\ln(n)^2}\right)\right)$$

が成り立つ．ただし，$\tilde{\xi}^{-1}=-e\mathrm{Ei}(-1)=0.5963473623\cdots=e^{-1}(0.6168878482\cdots)^{-1}$ は Euler-Gompertz（オイラー–ゴンパーツ）定数【6.2】である．これらの漸近式を強化したものは[31,33]を参照せよ．

5.6.3 属　　性

これまで数え上げの問題のみを議論してきた．本来のオッター定数 α と β は，しかしながら，木のもつ他の属性を決定するいくつかの漸近式に現れている．頂点の**次数** (degree，または valency（**価数**））とは，それに隣接する頂点の個数を意味する．n 個の頂点をもつ，ランダムな根付き木が与えられているとする．根の期待次数は，[34]，$n\to\infty$ のとき

$$\theta=1+\sum_{i=1}^{\infty}T\left(\frac{1}{\alpha^i}\right)=2+\sum_{j=1}^{\infty}T_j\frac{1}{\alpha^j(\alpha^j-1)}=2.1918374031\cdots$$

である．根の次数の分散は

$$\sum_{i=1}^{\infty}iT\left(\frac{1}{\alpha^i}\right)=1+\sum_{j=1}^{\infty}T_j\frac{2\alpha^j-1}{\alpha^j(\alpha^j-1)^2}=1.4741726868\cdots$$

である．

2つの頂点の**距離** (distance) とは，それらを結ぶもっとも短い道の辺の個数である．頂点と根との平均の距離は，$n\to\infty$ のとき，

$$\frac{1}{2}\left(\frac{2\pi}{\beta}\right)^{1/3}n^{1/2}=(1.1365599187\cdots)n^{1/2}$$

であり，距離の分散は

$$\frac{4-\pi}{4\pi}\left(\frac{2\pi}{\beta}\right)^{2/3}n=(0.3529622229\cdots)n$$

である．

n 個の頂点をもつランダム自由木の，任意の頂点を v とする．$n\to\infty$ のとき，v が次数 m である確率を p_m とする．このとき[35]，

$$p_1=\frac{\alpha^{-1}+\sum_{k=1}^{\infty}D_k\dfrac{\alpha^{-2k}}{1-\alpha^{-k}}}{1+\sum_{k=1}^{\infty}kT_k\dfrac{\alpha^{-2k}}{1-\alpha^{-k}}}=0.4381562356\cdots$$

が成り立つ．ただし，$D_1=1$, $D_{k+1}=\sum_{j=1}^{n}(\sum_{d|j}D_d)T_{k-j+1}$ である．明らかに，$m\to\infty$ のとき $p_m\to 0$ である．さらに，正確には

$$\omega=\prod_{i=1}^{\infty}\left(1-\frac{1}{\alpha^i}\right)^{-T_{i+1}}=\exp\left(\sum_{j=1}^{\infty}\frac{1}{j}\left[\alpha^jT\left(\frac{1}{\alpha^j}\right)-1\right]\right)=7.7581602911\cdots$$

のとき，$\lim_{m\to\infty}\alpha^m p_m$ は，[36,37]によって，

$$(2\pi\beta^2)^{-1/3}\omega=(1.2160045618\cdots)^{-1}\omega=6.3800420942\cdots$$

である．θ と ω ともに後で必要になる．さらに[38,39]を参照せよ．

G をグラフとし，G の自己同型写像の群を $A(G)$ とする．G の頂点 v は，任意の $\phi \in A(G)$ に対して，$\phi(v)=v$ であるとき，**不動点** (fixed point) であるという．$n \to \infty$ のとき，位数 n のランダム木の任意の頂点が不動点である確率を q とする．Harary & Palmer（ハラリー&パルマー）[7,40] は

$$q = (2\pi\beta^2)^{-1/3}\left(1 - E\left(\frac{1}{\alpha^2}\right)\right) = 0.6995388700\cdots$$

を証明した．ただし，$E(x) = T(x)(1+F(x)-F(x^2))$ である．おもしろいことに，同じ値 q は，根付き木に対しても同様に当てはまる．

紙数の関係で，被覆および充填問題に関する定数は除く[41-43]．同様に，頂点の極大な独立集合の数え上げ問題[44-47]，ゲームの問題[48]，同等に塗り分けられる木の問題[49]も除く．

5.6.4 林

位数 n の非同型な林の個数を f_n とする．母関数[26]

$$f(x) = \sum_{n=1}^{\infty} f_n x^n$$
$$= x + 2x^2 + 3x^3 + 6x^4 + 10x^5 + 20x^6 + 37x^7 + 76x^8 + 153x^9 + 329x^{10} + \cdots$$

は，

$$1 + f(x) = \exp\left(\sum_{k=1}^{\infty} \frac{t(x^k)}{k}\right), \quad f_n = \frac{1}{n}\sum_{k=1}^{n}\left(\sum_{d|k} d t_d\right) f_{n-k}$$

を満たす．$f_0 = 1$ としたのは後の式のためだけである．Palmer & Schwenk（パルマー&シュウェンク）[50] は

$$f_n \sim c t_n = \left(1 + f\left(\frac{1}{\alpha}\right)\right) t_n = (1.9126258077\cdots) t_n$$

を示した．もし林をランダムに選ぶと，$n \to \infty$ のとき，林にある木の期待個数は，

$$1 + \sum_{i=1}^{\infty} t\left(\frac{1}{\alpha^i}\right) = \frac{3}{2} + \frac{1}{2} T\left(\frac{1}{\alpha^2}\right) + \sum_{j=1}^{\infty} t_j \frac{1}{\alpha^j(\alpha^j-1)} = 1.7555101394\cdots$$

である．根付き木に対応する個数は，$\theta = 2.1918374031\cdots$ である．この定数は【5.6.3】にあるので意外なものではない．ランダムな林にある，ちょうど k 個の根付き木の確率は，漸近的に $\omega \alpha^{-k} = (7.7581602911\cdots)\alpha^{-k}$ である．自由木に対して，類似の確率は，係数

$$\frac{\alpha}{c} \prod_{i=1}^{\infty} \left(1 - \frac{1}{\alpha^i}\right)^{-t_{i+1}} = \frac{\alpha}{c} \exp\left(\sum_{j=1}^{\infty} \frac{1}{j}\left[\alpha^j t\left(\frac{1}{\alpha^j}\right) - 1\right]\right) = 3.2907434386\cdots$$

をもち，α^{-k} と同じように幾何級数的に減少する．

さらに，位数 n の2つの根付き林 (rooted forest) が共通の木をもたない，漸近的確率は[51]，

$$\prod_{i=1}^{\infty} \left(1 - \frac{1}{\alpha^{2i}}\right)^{T_i} = \exp\left(-\sum_{j=1}^{\infty} \frac{1}{j} T\left(\frac{1}{\alpha^{2j}}\right)\right) = 0.8705112052\cdots$$

である．

5.6.5 カクタスと 2-木

木でなく，それにもかかわらず木のようなグラフを調べる．**カクタス**（cactus, **サボテン**）とは，各辺が 1 つの（最小の）サイクル以外には含まれない，連結したグラフである[52-54]．図 5.9 を見よ．

さらに，すべての辺が 1 つのサイクル上にだけあり，すべてのサイクルが固定された m に対して，m 辺多角形であると仮定する．このとき，カクタスは m-**カクタス**（m-cactus）とよばれる．慣例により，2-カクタスは単なる木である．3-カクタスの議論は [4] にあり，4-カクタスは [55]，塗り分けした頂点をもつ m-カクタスは [56] にある．このような特別な場合については議論しない．カクタスと根付きカクタスに対する母関数は，それぞれ [57]

$$c(x) = \sum_{n=1}^{\infty} c_n x^n$$
$$= x + x^2 + 2x^3 + 4x^4 + 9x^5 + 23x^6 + 63x^7 + 188x^8 + 596x^9 + 1979x^{10} + \cdots$$
$$C(x) = \sum_{n=1}^{\infty} C_n x^n$$
$$= x + x^2 + 3x^3 + 8x^4 + 26x^5 + 84x^6 + 297x^7 + 1066x^8 + 3976x^9 + \cdots$$

である．これらは，[58-60]

$$C(x) = x \exp\left[-\sum_{k=1}^{\infty} \frac{1}{k}\left(\frac{C(x^k)^2 - 2 + C(x^{2k})}{2(C(x^k) - 1)(C(x^{2k}) - 1)} + 1\right)\right]$$

$$c(x) = C(x) - \frac{1}{2}\sum_{k=1}^{\infty} \frac{\phi(k)}{k} \ln(1 - C(x_k)) + \frac{(C(x)+1)(C(x)^2 - 2C(x) + C(x^2))}{4(C(x)-1)(C(x^2)-1)}$$

を満たし，収束半径は $0.2221510651\cdots$ である．ラベルをもつ場合は，

$$\hat{c}(x) = \sum_{n=1}^{\infty} \frac{\hat{c}_n}{n!} x^n$$
$$= x + \frac{1}{2!}x^2 + \frac{4}{3!}x^3 + \frac{31}{4!}x^4 + \frac{362}{5!}x^5 + \frac{5676}{6!}x^6 + \frac{111982}{7!}x^7 + \cdots$$

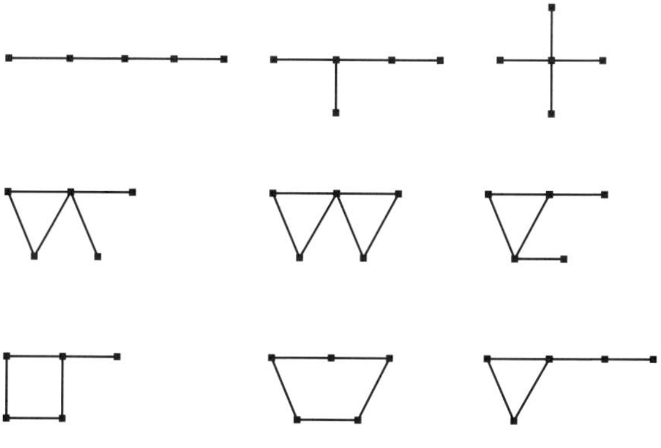

図 5.9 次数 5 の非同型な 9 個のカクタス．

$$\widehat{C}(x) = \sum_{n=1}^{\infty} \frac{\widehat{C}_n}{n!} x^n$$
$$= x + \frac{2}{2!}x^2 + \frac{12}{3!}x^3 + \frac{124}{4!}x^4 + \frac{1810}{5!}x^5 + \frac{34056}{6!}x^6 + \frac{783874}{7!}x^7 + \cdots$$

であり，これらは，収束半径 $0.2387401436\cdots$ をもち，
$$\widehat{C}(x) = x \exp\left(\frac{\widehat{C}(x)}{2} \frac{2-\widehat{C}(x)}{1-\widehat{C}(x)}\right), \quad x\widehat{c}'(x) = \widehat{C}(x)$$
を満たす．

2-木（2-tree）は，次のように再帰的に定義される[4]．ランク1の2-木は，三角形（3つの頂点と3つの辺をもつ）である．ランク $n \geq 2$ の2-木は，ランク $n-1$ の2-木から，隣接する頂点の2つの各々に隣接する，次数2の新しい頂点を付け加えることによって作られる．したがって，ランク n の2-木は，$n+2$ 個の頂点と $2n+1$ 個の辺をもつ．2-木に対する母関数は[61]，

$$w(x) = \sum_{n=0}^{\infty} w_n x^n$$
$$= 1 + x + x^2 + 2x^3 + 5x^4 + 12x^5 + 39x^6 + 136x^7 + 529x^8 + 2171x^9 + \cdots$$
$$w(x) = \frac{1}{2}\left[W(x) + \exp\left(\sum_{k=0}^{\infty} \frac{1}{2k}\left(2x^k W(x^{2k}) + x^{2k} W(x^{2k})^2 - x^{2k} W(x^{4k})\right)\right)\right]$$
$$+ \frac{1}{3}x(W(x^3) - W(x)^3)$$

である．ただし，$W(x)$ は，区別される有向辺をもつ2-木に対する母関数であり，

$$W(x) = \sum_{n=0}^{\infty} W_n x^n$$
$$= 1 + x + 3x^2 + 10x^3 + 39x^4 + 160x^5 + 702x^6 + 3177x^7 + 14830x^8 + \cdots$$
$$W(x) = \exp\left(\sum_{k=1}^{\infty} \frac{x^k W(x^k)^2}{k}\right), \quad w_n \sim \eta_w n^{-5/2} \xi_w^n$$

を満たす．さらに，$w(x)$ は収束半径 $\xi_w^{-1} = 0.1770995223\cdots = (5.6465426162\cdots)^{-1}$ をもち，

$$\eta_w = \frac{1}{16\xi\sqrt{\pi}}\left(\xi + 2\widetilde{W}'\left(\frac{1}{\xi}\right)\widetilde{W}\left(\frac{1}{\xi}\right)^{-1}\right)^{3/2} = 0.0948154165\cdots$$
$$\widetilde{W}(x) = e^{-xW(x)^2} W(x)$$

を満たす．

5.6.6 写像パターン

【5.4】で n 個の頂点をもつラベル付き関数グラフ（labeled functional graphs）を調べた．ラベルを取り除き，グラフの同型類のみ，すなわち**写像パターン**（mapping patterns）というものを考える．本来のオッター定数 α と β は，ここで決定的な役割をすることを述べる．写像パターンの母関数は[57, 62]，

$$P(x) = \sum_{n=1}^{\infty} P_n x^n$$
$$= x + 3x^2 + 7x^3 + 19x^4 + 47x^5 + 130x^6 + 343x^7 + 951x^8 + 2615x^9 + \cdots$$
$$1 + P(x) = \prod_{k=1}^{\infty} (1 - T(x^k))^{-1}, \quad P_n \sim \eta_P n^{-1/2} \alpha^n$$

である．ただし，

$$\eta_P = \frac{1}{2\pi} \left(\frac{2\pi}{\beta}\right)^{1/3} \prod_{i=2}^{\infty} \left(1 - T\left(\frac{1}{\alpha^i}\right)\right)^{-1} = 0.4428767697\cdots$$
$$= (1.2241663491\cdots)(4\pi^2 \beta)^{-1/3}$$

である．これから，ランダムな写像パターンにある，任意のサイクルの長さの期待値は，

$$\frac{1}{2}\left(\frac{2\pi}{\beta}\right)^{1/3} n^{1/2} = (1.1365599187\cdots) n^{1/2}, \quad n \to \infty$$

(偶然の一致で，【5.6.3】で見た式）であり，写像パターンが連結である漸近確率は，

$$\frac{1}{2\eta_P} n^{1/2} = (1.1289822228\cdots) n^{-1/2}$$

である．さらに，連結写像パターン（connected mapping pattern）だけを考慮すると，母関数は

$$K(x) = \sum_{n=1}^{\infty} K_n x^n$$
$$= x + 2x^2 + 4x^3 + 9x^4 + 20x^5 + 51x^6 + 125x^7 + 329x^8 + 862x^9 + \cdots$$
$$K(x) = -\sum_{j=1}^{\infty} \frac{\varphi(j)}{j} \ln(1 - T(x^j)), \quad K_n \sim \frac{1}{2} n^{-1} \alpha^n$$

である．ただし，φ はオイラーの約数関数である．よって，ランダムな連結写像パターンに含まれる（一意の）サイクルの長さの期待値は，

$$\frac{1}{\pi}\left(\frac{2\pi}{\beta}\right)^{1/3} n^{1/2} = (0.7235565167\cdots) n^{1/2}, \quad n \to \infty$$

であり，これは前のものより小さい．ラベル付きでない場合とラベル付きの両方の場合（数値的な結果は，確かに少し異なる）に対するこのような統計量の比較は[62]にある．この分野のさらに最近の成果は[63,64]を参照せよ．

5.6.7　種々のグラフ

グラフ G が**区間グラフ**（interval graph）とは，次のように表現できるときをいう．2つの頂点が隣接するための必要十分条件はこれらの頂点に対応する区間の共通部分が空集合でない，という方法で，G の頂点の各々が実数直線の部分区間に対応しているグラフのことである．**単位区間グラフ**（unit interval graph）とは，区間の長さがすべて1であるような区間グラフのことである．単位区間グラフの母関数は，たとえば[65, 66]，

$$I(x) = \sum_{n=1}^{\infty} I_n x^n$$
$$= x + 2x^2 + 4x^3 + 9x^4 + 21x^5 + 55x^6 + 151x^7 + 447x^8 + 1389x^9 + \cdots$$
$$1 + I(x) = \exp\Bigl(\sum_{k=1}^{\infty} \frac{\psi(x^k)}{\beta}\Bigr), \quad \psi(x) = \frac{1 + 2x - \sqrt{1-4x}\sqrt{1-4x^2}}{4\sqrt{1-4x^2}}$$

であり，漸近式

$$I_n \sim \frac{1}{8\kappa\sqrt{\pi}} n^{-3/2} 4^n, \quad \kappa = \exp\Bigl(-\frac{\sqrt{3}}{4}\Bigr) \exp\Bigl(-\sum_{j=2}^{\infty} \frac{\psi(4^{-j})}{j}\Bigr) = 0.6231198963\cdots$$

が成り立つ．区間グラフは，遺伝学および他の分野での応用がある[67, 68]．

グラフが **2-正則** (2-regular) とは，すべての頂点が次数2であるときをいう．n個の頂点をもつ2-正則グラフの個数J_nは，nを3個以上の部分に分ける分割の個数と等しい．一方，2-正則なラベル付きグラフの指数的母関数は[69]

$$\tilde{J}(x) = \sum_{n=0}^{\infty} \frac{\tilde{J}_n}{n!} x^n = 1 + \frac{1}{3!} x^3 + \frac{3}{4!} x^4 + \frac{12}{5!} x^5 + \frac{70}{6!} x^6 + \cdots$$
$$= \frac{1}{\sqrt{1-x}} \exp\Bigl(-\frac{1}{2}x - \frac{1}{4}x^2\Bigr)$$

である．したがって，

$$J_N \sim \frac{\pi^2}{12\sqrt{3}n^2} \exp\Bigl(\pi\sqrt{\frac{2n}{3}}\Bigr), \quad \tilde{J}_n \sim \sqrt{2} e^{-3/4} \Bigl(\frac{n}{e}\Bigr)^n$$

である．最後の式には，おもしろい幾何学的な解釈がある[14, 70]．平面上の一般的な位置にあるn個の直線を与えると，$\binom{n}{2}$個の交点を生じる．大きさnの**雲** (cloud) とは，どの3つの直線も1点で交わらない[†] n個の直線の交点の（極大）集合である．大きさnの雲の個数は，明らかに$\tilde{J}_n$である．

有向グラフ (directed graph または digraph) とは，辺が（単なる対ではなくて）異なる頂点の順序対であるグラフのことである．ループは定義から明らかに許されない．さらに，**無閉路有向グラフ** (acyclic digraph) とは，有向閉路を含んでいないものをいう．とくに，多重並行な辺をもたない．ラベル付き無閉路有向グラフの（変形された）指数的母関数は[65, 71-74]，

$$A(x) = \sum_{n=0}^{\infty} \frac{A_n}{n! \, 2^{\binom{n}{2}}} x^n = 1 + x + \frac{3}{2! \cdot 2} x^2 + \frac{25}{3! \cdot 2^3} x^3 + \frac{543}{4! \cdot 2^6} x^4 + \frac{29281}{5! \cdot 2^{10}} x^5 + \cdots$$
$$A'(x) = A(x)^2 A\Bigl(\frac{1}{2}x\Bigr)^{-1}, \quad A_n \sim \frac{n! \, 2^{\binom{n}{2}}}{\eta_A \xi_A^n}$$

である．ただし，$\xi_A = 1.4880785456\cdots$ は，関数

$$\lambda(x) = \sum_{n=0}^{\infty} \frac{(-1)^n}{n! \, 2^{\binom{n}{2}}} x^n = A(x)^{-1}, \quad \lambda'(x) = -\lambda\Bigl(\frac{1}{2}x\Bigr)$$

の最小の正の零点であり，$\eta_A = \xi_A \lambda(\xi_A/2) = 0.5743623733\cdots = (1.7410611252\cdots)^{-1}$である．関数$\lambda(-x)$が2の累乗による整数の分割に関するMahler（マーラー）[75]で，先行研究されたことは不思議である．ラベル付きでない無閉路有向グラフの類似なものの議論は[76, 77]を参照せよ．

[†] 訳注：交わるとき共線性をもつ（collinear）という．

5.6.8 データ構造

組み合わせ論の学者にとっては，言葉「$2n+1$ 個の頂点をもつ強2分木」(strongly binary tree) は，木の同型写像の同値類を意味する．しかしながら，コンピュータ科学者にとっては，同じ言葉は事実上つねに，正確に述べていようといなくとも，「向き」という語を含む．したがって，「ランダム2分木」という言葉は文献では，時にはあいまいである．標本空間は，組み合わせ論の学者にとっては B_n の要素をもつが，コンピュータ科学者にとっては $\binom{2n}{n}/(n+1)$ の要素をもつ！ コンピュータアルゴリズムにおける木の役割をここで概観することはできない．いくつかの定数を述べるだけにする．

大きさ n の**左翼の木** (leftist tree) とは，任意の部分木 σ において，σ の根に一番近い葉が σ の右側の部分木にあるような，n 枚の葉をもつ順序付けられた2分木のことである．左翼の木の母関数は[6, 65, 78, 79]，

$$L(x) = \sum_{n=0}^{\infty} L_n x^n$$
$$= x + x^2 + x^3 + 2x^4 + 4x^5 + 8x^6 + 17x^7 + 38x^8 + 87x^9 + 203x^{10} + \cdots$$
$$L(x) = x + \frac{1}{2}L(x)^2 + \frac{1}{2}\sum_{m=1}^{\infty} l_m(x)^2 = \sum_{m=1}^{\infty} l_m(x)$$

である．ただし，補助母関数 $l_m(x)$ は，

$$l_1(x) = x, \quad l_2(x) = xL(x), \quad l_{m+1}(x) = l_m(x)\Big(L(x) - \sum_{k=1}^{m-1} l_k(x)\Big), \quad m \geq 2$$

を満たす．難しいが

$$L_n \sim (0.2503634293\cdots) \cdot (2.7494879027\cdots)^n n^{-3/2}$$

が証明できる．左翼の木はある種のソート問題およびマージアルゴリズムで役に立つ．

大きさ n の **2,3-木**とは，次の条件を満たす n 枚の葉がある根付き順序木のことである．

・葉でない頂点の各々には，2個または3個の子孫（後続の頂点，successors）がある．
・根から葉への道（root-to-leaf path）の長さは，すべて同じである．

2,3-木の母関数（2-木とは関係ない）は[65, 80, 81]，

$$Z(x) = \sum_{n=0}^{\infty} Z_n x^n$$
$$= x + x^2 + x^3 + x^4 + 2x^5 + 2x^6 + 3x^7 + 4x^8 + 5x^9 + 8x^{10} + 14x^{11} + \cdots$$
$$Z(x) = x + Z(x^2 + x^3), \quad Z_n = \sum_{k=\lceil n/3 \rceil}^{\lfloor n/2 \rfloor} \binom{k}{3k-n} Z_k \sim \varphi^n n^{-1} f(\ln(n))$$

である．ただし，φ は黄金比【1.2】である．$f(x)$ は，定数ではない正の連続関数で，周期 $\ln(4-\varphi) = 0.867\cdots$ をもつ周期関数である．この関数 $f(x)$ は，平均 $(\varphi \ln(4-\varphi))^{-1}$

$=0.712\cdots$ をもち，$0.682\cdots$ と $0.806\cdots$ の間を振動する．さらに，**B木**（B-tree）という特別な型がある．同様の解析[82]は，AVL-木（または高さ平衡木（height-balanced trees））として知られている木の漸近式を明らかにした．このような木は，データベースの検索，削除，挿入に効果的な助けとなる．他の種類についてはたくさんありすぎて，述べることはできない．

τ が順序付き2分木のとき，その**高さ**（height）と**レジスター関数**（register functions）は，[83]から，次のように帰納的に定義される．

$$\mathrm{ht}(\tau) = \begin{cases} 0 & (\tau \text{ が点のとき}) \\ 1+\max(\mathrm{ht}(\tau_L), \mathrm{ht}(\tau_R)) & (\text{その他のとき}) \end{cases}$$

$$\mathrm{rg}(\tau) = \begin{cases} 0 & (\tau \text{ が点のとき}) \\ 1+\mathrm{rg}(\tau) & (\mathrm{rg}(\tau_L)=\mathrm{rg}(\tau_R) \text{ のとき}) \\ \max(\mathrm{rg}(\tau_L), \mathrm{rg}(\tau_R)) & (\text{その他のとき}) \end{cases}$$

ただし，τ_L と τ_R はそれぞれ，その根の左と右の部分木である．すなわち，$\mathrm{ht}(\tau)$ は，根から最長の長さをもつ枝に沿っての辺の数である．一方，$\mathrm{rg}(\tau)$ は，（数表現とみなした）木を評価するために必要なレジスターの最小個数である．$2n+1$ 個の頂点をもつ2分木 τ をランダムに選ぶ．このとき，$E(\mathrm{ht}(\tau))$ の漸近式は，【1.4】で述べたように，$2\sqrt{\pi n}$ を含む．$E(\mathrm{rg}(\tau))$ の漸近式は，$\ln(n)/\ln(4)$ に平均0で振動する関数【2.16】を加えたものを含む．さらに，$\mathrm{ym}(\tau)$ を，$\mathrm{rg}(\tau)$ よりちょうど1だけ少ないレジスター関数をもつ τ の極大部分木の個数と定義する．Prodinger（プロディンガー）[84]は，Yekutiele & Mandelbrot（イェクティエル＆マンデルブロ）[85]の仕事を基礎にして，$E(\mathrm{ym}(\tau))$ は，漸近的に

$$\frac{2G}{\pi \ln(2)} + \frac{5}{2} = 3.3412669407\cdots$$

に平均0で振動する関数を加えたものである，ことを証明した．ただし，G は Catalan（カタラン）の定数である【1.7】．さらに，これは根での**分岐比**（bifurcation ratio）として知られている．より一般的な分岐構造の階層的な複雑性の量である．

5.6.9 ゴルトン-ワトソン分岐過程

ここまで，「ランダム2分木」は，一様な確率分布をもつ母集団から n 個の頂点をもつ2分木を選ぶことを意味していた．整数 n は固定されている．

しかしながら，（単に選ぶのではなく）2分木が「生長」することも可能でもある．確率 p，$0<p<1$ を固定する．次のように根の左と右の部分木によって，（強）2分木 τ を帰納的に定義する．確率 $1-p$ で $\tau_L=\emptyset$ を選び，これと独立に，確率 $1-p$ で $\tau_R=\emptyset$ を選ぶ．[86-88]で，この過程は終了することが示されている．すなわち τ は，$p \leq 1/2$ のとき**絶滅確率**（extinction probability）は1であり，$p>1/2$ のとき絶滅確率は $1/p-1$ である有限木である．もちろん，頂点の個数 N は，ここでは確率変数であり，**全子孫**（total progeny）とよばれる．

（実際ここで述べていること以上に一般的な）Bienaymé-Galton-Watson（ビエナイ

メーゴルトン-ワトソン）過程について，多くのことがいえる．ちょうど1つの項目について焦点をあてる．根から距離 k にある頂点の個数を，N_k とする．すなわち，k 世代の大きさである．劣臨界 (subcritical) な場合 $p<1/2$ を考える．$N_k=0$ である確率を a_k とする．このとき，数列 $a_0, a_1, a_2, \cdots$ は2次の漸化式【6.10】

$$a_0=0, \quad a_k=(1-p)+pa_{k-1}^2 \ (k\geq 1), \quad \lim_{k\to\infty} a_k=1$$

を満たす．$\{a_k\}$ の収束速度について，何がいえるだろうか？

$$C(p)=\lim_{k\to\infty}\frac{1-a_k}{(2p)^k}=\prod_{l=0}^{\infty}\frac{1+a_l}{2}$$

が証明できる．知るかぎり，p に関して閉じた式はない．これで終わりであり，またここで最大の興味のある事実，$0<p<1/2$ に対して $P(N_k>0) \sim C(p)(2p)^k$ であること，を越えている．他の興味あるパラメータは，**絶滅のモーメント** (monent of extinction) $\min\{k: N_k=0\}$ すなわち木の高さと，**極大世代サイズ** (maximal generation size) $\max\{N_k: k\geq 0\}$ すなわち木の幅である．

5.6.10 エルデーシュ-レーニィの進化過程

はじめに n 個の非連結な頂点からはじめて，$\binom{n}{2}$ 個の候補から元に戻さないで（非復元抽出），異なる点の対の間に辺を付け加えることを連続的に行うことによって，ランダムグラフを定義する．この操作を候補の辺がなくなるまで，続ける[89-92]．

操作のある段階で，**複体成分** (complex component) が発生する．すなわち，1つより多くのサイクルをもつ最初の成分である．この複体成分は，全過程を通して，ふつう一意のままである．これが真である確率は，$n\to\infty$ のとき $5\pi/18=0.8726\cdots$ である．いいかえると，頂点よりも辺を多くとる最初の成分は，ランダムグラフの**巨大成分** (giant component) にかなりなりやすい．ちょうど2つの複体成分になる確率は，$50\pi/1296=0.1212\cdots$ である．しかし，任意の時点に，2つの複体成分よりも多くの成分を決してもたない発展グラフの確率（$>0.9938\cdots$）は，正確には知られていない[93]．

多くの関連する結果があるが，1つだけ述べる．各部屋に4つの壁がある，部屋が $m\times n$ の長方形の格子の状態から始める．操作のある段階で，（すべての mn 個の部屋を頂点とし，開いた通路をもつ，隣り合う部屋の対のすべてを辺として）関連づけたグラフが木となるような，ランダムな方法で室内の壁を連続的に取り除く．条件が満たされたときに停止する．結果は**ランダム迷路** (random maze) となる[94]．難しさは，新しい辺を付け加えることが，望まないサイクルを作るかどうかを検出することにある．（時がたてば変わる同値類を保持しながら）これを行う効果的な方法は，QF および QFW で発見された．これは，コンピュータ科学における**合併発見アルゴリズム** (union-find algorithm) の2種類である．QF および QFW の厳密な実行解析は，ランダムグラフ理論および Erdös-Rényi（エルデーシュ-レーニィ）過程を変形したものを用いて，[95-97]にある．

[1] F. Harary, *Graph Theory*, Addison-Wesley, 1969; MR 41 #1566.
[2] R. Sedgewick and P. Flajolet, *Introduction to the Analysis of Algorithms*, Addison-Wesley, 1996.
[3] R. Otter, The number of trees, *Annals of Math.* 49 (1948) 583-599; MR 10,53c.
[4] F. Harary and E. M. Palmer, *Graphical Enumeration*, Academic Press, 1973;MR50 #9682.
[5] F. Harary, R. W. Robinson, and A. J. Schwenk, Twenty-step algorithm for determining the asymptotic number of trees of various species, *J. Austral. Math. Soc. Ser. A* 20 (1975) 483-503; corrigenda 41 (1986) 325; MR 53 #10644 and MR 87j:05091.
[6] A. M. Odlyzko, Asymptotic enumeration methods, *Handbook of Combinatorics*, ed. R. L. Graham, M. Groetschel, and L. Lovasz, MIT Press, 1995, pp. 1063-1229; MR 97b: 05012.
[7] D. J. Broadhurst and D. Kreimer, Renormalization automated by Hopf algebra, *J. Symbolic Comput.* 27 (1999) 581-600; hep-th/9810087; MR 2000h:81167.
[8] E. A. Bender, Asymptotic methods in enumeration, *SIAM Rev.* 16 (1974) 485-515; errata 18 (1976) 292; MR 51 #12545 and MR 55 #10276.
[9] D. E. Knuth, *The Art of Computer Programming*, v. 1, *Fundamental Algorithms*, Addison-Wesley, 1997, pp. 396-397, 588-589; MR 51 #14624.
[10] J. N. Franklin and S. W. Golomb, A function-theoretic approach to the study of nonlinear recurring sequences, *Pacific J. Math.* 56 (1975) 455-468; MR 51 #10212.
[11] J. H. M.Wedderburn, The functional equation $g(x^2)=2ax+[g(x)]^2$, *Annals of Math.* 24 (1922-23) 121-140.
[12] I. M. H. Etherington, Non-associate powers and a functional equation, *Math. Gazette* 21 (1937) 36-39, 153.
[13] H. W. Becker, Genetic algebra, *Amer. Math. Monthly* 56 (1949) 697-699.
[14] L. Comtet, *Advanced Combinatorics: The Art of Finite and Infinite Expansions*, Reidel, 1974, pp. 52-55, 273-279; MR 57 #124.
[15] N. J. A. Sloane, On-Line Encyclopedia of Integer Sequences, A000055, A000081, A000598, A000602, A000621, A000625, and A001190.
[16] C. Bailey, E. Palmer, and J. Kennedy, Points by degree and orbit size in chemical trees. II, *Discrete Appl. Math.* 5 (1983) 157-164; MR 85b:05098.
[17] E. A. Bender and S. Gill Williamson, *Foundations of Applied Combinatorics*, Addison-Wesley, 1991, pp. 372-373, 395-396.
[18] G. Pólya and R. C. Read, *Combinatorial Enumeration of Groups, Graphs, and Chemical Compounds*, Springer-Verlag, 1987, pp. 4-8, 78-86, 130-134; MR 89f:05013.
[19] A. T. Balaban, *Chemical Applications of Graph Theory*, Academic Press, 1976; MR 58 #21870.
[20] N. Trinajstic, *Chemical Graph Theory*, v. 2, CRC Press, 1992; MR 93g:92034.
[21] D. H. Rouvray, Combinatorics in chemistry, *Handbook of Combinatorics*, ed. R. L. Graham, M. Groetschel, and L. Lovasz, MIT Press, 1995, pp. 1955-1981; MR 97e:92017.
[22] F. Bergeron, G. Labelle, and P. Leroux, *Combinatorial Species and Tree-Like Structures*, Cambridge Univ. Press, 1998; MR 2000a:05008.
[23] E. M. Rains and N. J. A. Sloane, On Cayley's enumeration of alkanes (or 4-valent trees), *J. Integer Seq.* 2 (1999) 99.1.1; math.CO/0207176; MR 99j:05010.
[24] P. Flajolet and R. Sedgewick, *Analytic Combinatorics*, unpublished manuscript, 2001.
[25] R. W. Robinson, F. Harary, and A. T. Balaban, The numbers of chiral and achiral alkanes and monosubstituted alkanes, *Tetrahedron* 32 (1976) 355-361.
[26] N. J. A. Sloane, On-Line Encyclopedia of Integer Sequences, A000014, A000220, A001678, A004111, A005200, A005201, A029768, A032200, A038037, A005195, and A059123.
[27] A. Meir, J.W. Moon, and J. Mycielski, Hereditarily finite sets and identity trees, *J. Combin. Theory Ser. B* 35 (1983) 142-155; MR 85h:05009.
[28] F. Harary and G. Prins, The number of homeomorphically irreducible trees, and other species, *Acta Math.* 101 (1959) 141-162; MR 21 #653.
[29] S. R. Finch, Counting topological trees, unpublished note (2001).

[30] C. G. Bower, Mobiles (cyclic trees), unpublished note (2001).
[31] I. M. Gessel, B. E. Sagan, and Y. N. Yeh, Enumeration of trees by inversions, *J. Graph Theory* 19 (1995) 435-459; MR 97b:05070.
[32] F. Bergeron, P. Flajolet, and B. Salvy, Varieties of increasing trees, *17th Colloq. on Trees in Algebra and Programming (CAAP)*, Proc. 1992 Rennes conf., ed. J.-C. Raoult, Lect. Notes in Comp. Sci. 581, Springer-Verlag, 1992, pp. 24-48; MR 94j:68233.
[33] W. Y. C. Chen, The pessimistic search and the straightening involution for trees, *Europ. J. Combin.* 19 (1998) 553-558; MR 99g:05103.
[34] A. Meir and J. W. Moon, On the altitude of nodes in random trees, *Canad. J. Math.* 30 (1978) 997 -1015; MR 80k:05043.
[35] R. W. Robinson and A. J. Schwenk, The distribution of degrees in a large random tree, *Discrete Math.* 12 (1975) 359-372; MR 53 # 10629.
[36] A. J. Schwenk, An asymptotic evaluation of the cycle index of a symmetric group, *Discrete Math.* 18 (1977) 71-78; MR 57 # 9598.
[37] A. Meir and J.W. Moon, On an asymptotic evaluation of the cycle index of the symmetric group, *Discrete Math.* 46 (1983) 103-105; MR 84h:05071.
[38] C. K. Bailey, Distribution of points by degree and orbit size in a large random tree, *J. Graph Theory* 6 (1982) 283-293; MR 83i:05037.
[39] W. M. Y. Goh and E. Schmutz, Unlabeled trees: Distribution of the maximum degree, *Random Structures Algorithms* 5 (1994) 411-440; MR 95c:05111.
[40] F. Harary and E. M. Palmer, The probability that a point of a tree is fixed, *Math. Proc. Cambridge Philos. Soc.* 85 (1979) 407-415; MR 80f:05020.
[41] A. Meir and J. W. Moon, Packing and covering constants for certain families of trees. I, *J. Graph Theory* 1 (1977) 157-174; MR 57 # 2965.
[42] A. Meir and J. W. Moon, Packing and covering constants for certain families of trees. II, *Trans. Amer. Math. Soc.* 233 (1977) 167-178; MR 57 # 157.
[43] A. Meir and J.W. Moon, Path edge-covering constants for certain families of trees, *Utilitas Math.* 14 (1978) 313-333; MR 80a:05073.
[44] A. Meir and J. W. Moon, On maximal independent sets of nodes in trees, *J. Graph Theory* 12 (1988) 265-283; MR 89f:05064.
[45] P. Kirschenhofer, H. Prodinger, and R. F. Tichy, Fibonacci numbers of graphs. II, *Fibonacci Quart.* 21 (1983) 219-229; MR 85e:05053.
[46] P. Kirschenhofer, H. Prodinger, and R. F. Tichy, Fibonacci numbers of graphs. III: Planted plane trees, *Fibonacci Numbers and Their Applications*, Proc. 1984 Patras conf., ed. A. N. Philippou, G. E. Bergum, and A. F. Horadam, Reidel, 1986, pp. 105-120, MR 88k: 05065.
[47] M. Drmota, On generalized Fibonacci numbers of graphs, *Applications of Fibonacci Numbers*, v. 3, Proc. 1988 Pisa conf., ed. G. E. Bergum, A. N. Philippou, and A. F. Horadam, Kluwer, 1990, pp. 63-76; MR 92e:05032.
[48] A. Meir and J. W. Moon, Games on random trees, Proc. 15th Southeastern *Conf. on Combinatorics, Graph Theory and Computing*, Baton Rouge, 1984, ed. F. Hoffman, K. B. Reid, R. C. Mullin, and R. G. Stanton, Congr. Numer. 44, Utilitas Math., 1984, pp. 293-303; MR 86h:05045.
[49] N. Pippenger, Enumeration of equicolorable trees, *SIAM J. Discrete Math.* 14 (2001) 93-115; MR 2001j:05073.
[50] E. M. Palmer and A. J. Schwenk, On the number of trees in a random forest, *J. Combin. Theory Ser. B* 27 (1979) 109-121; MR 81j:05071.
[51] S. Corteel, C. D. Savage, H. S. Wilf, and D. Zeilberger, A pentagonal number sieve, *J. Combin. Theory Ser. A* 82 (1998) 186-192; MR 99d:11111.
[52] K. Husimi, Note on Mayers' theory of cluster integrals, *J. Chem. Phys.* 18 (1950) 682-684; MR 12,467i.
[53] G. E. Uhlenbeck and G. W. Ford, *Lectures in Statistical Mechanics*, Lect. in Appl. Math., v. 1, Amer. Math. Soc., 1963; MR 27 # 1241.

[54] G. E. Uhlenbeck and G.W. Ford, The theory of linear graphs with applications to the theory of the virial development of the properties of gases, *Studies in Statistical Mechanics*, v. 1, ed. J. de Boer and G. Uhlenbeck, North-Holland, 1962, pp. 119-211; MR 24 # B2420.

[55] F. Harary and G. E. Uhlenbeck, On the number of Husimi trees. I, *Proc. Nat. Acad. Sci. USA* 39 (1953) 315-322; MR 14,836a.

[56] M. Bóna, M. Bousquet, G. Labelle, and P. Leroux, Enumeration of m-ary cacti, *Adv. Appl. Math.* 24 (2000) 22-56; MR 2001c:05072.

[57] N. J. A. Sloane, On-Line Encyclopedia of Integer Sequences, A000083, A000237, A000314, A001205, A001372, A002861, A005750, A035351, and A054581.

[58] C. G. Bower, Cacti (mixed Husimi trees), unpublished note (2001).

[59] G. W. Ford and G. E. Uhlenbeck, Combinatorial problems in the theory of graphs. III, *Proc. Nat. Acad. Sci. USA* 42 (1956) 529-535; MR 18,326c.

[60] C. Domb, Graph theory and embeddings, *Phase Transitions and Critical Phenomena*, v. 3, ed. C. Domb and M. S. Green, Academic Press, 1974, pp. 33-37; MR 50 # 6393.

[61] T. Fowler, I. Gessel, G. Labelle, and P. Leroux, Specifying 2-trees, *Formal Power Series and Algebraic Combinatorics (FPSAC)*, Proc. 2000 Moscow conf., ed. D. Krob, A. A. Mikhalev, and A. V. Mikhalev, Springer-Verlag, 2000, pp. 202-213; MR 2001m:05132.

[62] A. Meir and J. W. Moon, On random mapping patterns, *Combinatorica* 4 (1984) 61-70; MR 85g:05125.

[63] L. R. Mutafchiev, Large trees in a random mapping pattern, *Europ. J. Combin.* 14 (1993) 341-349; MR 94f:05072.

[64] L. R. Mutafchiev, Limit theorem concerning random mapping patterns, *Combinatorica* 8 (1988) 345-356; MR 90a:05170.

[65] N. J. A. Sloane, On-Line Encyclopedia of Integer Sequences, A001205, A003024, A003087, A005217, A006196, A008483, A015701, A014535, and A037026.

[66] P. Hanlon, Counting interval graphs, *Trans. Amer. Math. Soc.* 272 (1982) 383-426; MR 83h:05050.

[67] F. S. Roberts, *Discrete Mathematical Models*, Prentice-Hall, 1976, pp. 111-140.

[68] P. C. Fishburn, *Interval Orders and Interval Graphs: A Study of Partially Ordered Sets*, Wiley, 1985; MR 86m:06001.

[69] H. S. Wilf, *generatingfunctionology*, Academic Press, 1990, pp. 76-77, 151; MR 95a:05002.

[70] R. M. Robinson, A new absolute geometric constant, *Amer. Math. Monthly* 58 (1951) 462-469; addendum 59 (1952) 296-297; MR 13,200b.

[71] R. W. Robinson, Counting labeled acyclic digraphs, *New Directions in the Theory of Graphs*, Proc. 1971 Ann Arbor conf., ed. F. Harary, Academic Press, 1973, pp. 239-273; MR 51 # 249.

[72] R. P. Stanley, Acyclic orientations of graphs, *Discrete Math.* 5 (1973) 171-178; MR 47 # 6537.

[73] V. A. Liskovec, The number of maximal vertices of a random acyclic digraph (in Russian), *Teor. Veroyatnost. i Primenen.* 20 (1975) 412-421; Engl. transl. in *Theory Probab. Appl.* 20 (1975) 401-421; MR 52 # 1822.

[74] E. A. Bender, L. B. Richmond, R. W. Robinson, and N. C. Wormald, The asymptotic number of acyclic digraphs. I, *Combinatorica* 6 (1986) 15-22; MR 87m:05102.

[75] K. Mahler, On a special functional equation, *J. London Math. Soc.* 15 (1940) 115-123; MR 2,133e.

[76] R. W. Robinson, Counting unlabeled acyclic digraphs, *Combinatorial Mathematics V*, Proc. 1976 Melbourne conf., ed. C. H. C. Little, Lect. Notes in Math. 622, Springer-Verlag, 1977, pp. 28-43; MR 57 # 16129.

[77] E. A. Bender and R. W. Robinson, The asymptotic number of acyclic digraphs. II, *J. Combin. Theory Ser.* B 44 (1988) 363-369; MR 90a:05098.

[78] R. Kemp, A note on the number of leftist trees, *Inform. Process. Lett.* 25 (1987) 227-232.

[79] R. Kemp, Further results on leftist trees, *Random Graphs '87*, Proc. 1987 Poznan conf., ed. M. Karonski, J. Jaworski, and A. Rucinski,Wiley, 1990, pp. 103-130; MR 92e:05034.

[80] R. E. Miller, N. Pippenger, A. L. Rosenberg, and L. Snyder, Optimal 2,3-trees, *SIAM J. Comput.* 8 (1979) 42-59; MR 80c:68050.

[81] A. M. Odlyzko, Periodic oscillations of coefficients of power series that satisfy functional equations, *Adv. Math.* 44 (1982) 180-205; MR 84a:30042.

[82] A. M. Odlyzko, Some new methods and results in tree enumeration, *Proc. 13*[th] *Manitoba Conf. on Numerical Mathematics and Computing*, Winnipeg, 1983, ed. D. S. Meek and G. H. J. van Rees, Congr. Numer. 42, Utilitas Math., 1984, pp. 27-52; MR 85g:05061.

[83] H. Prodinger, Some recent results on the register function of a binary tree, *Random Graphs '85*, Proc. 1985 Poznan conf, ed. M. Karonski and Z. Palka, Annals of Discrete Math. 33, North-Holland, 1987, pp. 241-260; MR 89g:68058.

[84] H. Prodinger, On a problem of Yekutieli and Mandelbrot about the bifurcation ratio of binary trees, *Theoret. Comput. Sci.* 181 (1997) 181-194; also in *Proc. 1995 Latin American Theoretical Informatics Conf. (LATIN)*, Valparáiso, ed. R. A. Baeza-Yates, E. Goles Ch., and P. V. Poblete, Lect. Notes in Comp. Sci. 911, Springer-Verlag, 1995, pp. 461-468; MR 98i:68212.

[85] I.Yekutieli and B. B. Mandelbrot, Horton-Strahler ordering of random binary trees, *J. Phys. A* 27 (1994) 285-293; MR 94m:82022.

[86] T. E. Harris, *The Theory of Branching Processes*, Springer-Verlag, 1963; MR 29 #664.

[87] K. B. Athreya and P. Ney, *Branching Processes*, Springer-Verlag, 1972; MR 51 #9242.

[88] G. Sankaranarayanan, *Branching Processes and Its Estimation Theory*, Wiley, 1989; MR 91m:60156

[89] P. Erdös and A. Rényi, On random graphs. I, *Publ. Math. (Debrecen)* 6 (1959) 290-297; also in *Selected Papers of Alfréd Rényi*, v. 2, Akadémiai Kiadó, 1976, pp. 308-315; MR 22 #10924.

[90] B. Bollobás, The evolution of random graphs, *Trans. Amer. Math. Soc.* 286 (1984) 257-274; MR 85k:05090.

[91] B. Bollobás, *Random Graphs*, Academic Press, 1985; MR 87f:05152.

[92] S. Janson, T. Luczak, and A. Rucinski, *Random Graphs*, Wiley, 2000; MR 2001k:05180.

[93] S. Janson, D. E. Knuth, T. Luczak, and B. Pittel, The birth of the giant component, *Random Structures Algorithms* 4 (1993) 231-358; MR 94h:05070.

[94] M. A. Weiss, *Data Structures and Algorithm Analysis in C^{++}*, 2nd ed., Addison-Wesley, 1999, pp. 320-322.

[95] A. C. C. Yao, On the average behavior of set merging algorithms, *8*[th] *ACM Symp. on Theory of Computing (STOC)*, Hershey, ACM, 1976, pp. 192-195; MR 55 #1819.

[96] D. E. Knuth and A. Schönhage, The expected linearity of a simple equivalence algorithm, *Theoret. Comput. Sci.* 6 (1978) 281-315; also in *Selected Papers on Analysis of Algorithms*, CSLI, 2000, pp. 341-389; MR 81a:68049.

[97] B. Bollobás and I. Simon, Probabilistic analysis of disjoint set union algorithms, *SIAM J. Comput.* 22 (1993) 1053-1074; also in *17*[th] *ACM Symp. on Theory of Computing (STOC)*, Providence, ACM, 1985, pp. 124-231; MR 94j:05110.

5.7 レンジェル定数

5.7.1 スターリングの分割数

S は n 個の要素をもつ集合とする。S のすべての部分集合の集合は、2^n 個の要素をもつ。S の**分割** (partition) とは、互いに素な空集合でない部分集合 (**区塊**, **ブロック** (blocks) という) で、その和集合が S となるものをいう。ちょうど k 個のブロッ

クをもつ，S の分割の集合には，$S_{n,k}$ 個の要素がある．ただし，$S_{n,k}$ は**第2種の** Stirling（**スターリング**）**数**である．S の「すべて」の分割の集合は，B_n 個の要素をもつ．ただし，B_n は Bell（**ベル**）**数**

$$B_n = \sum_{k=1}^{n} S_{n,k} = \frac{1}{e}\sum_{j=0}^{\infty}\frac{j^n}{j!} = \frac{d^n}{dx^n}\exp(e^x-1)\Big|_{x=0}$$

である．たとえば，$S_{4,1}=1$，$S_{4,2}=7$，$S_{4,3}=6$，$S_{4,4}=1$，$B_4=15$ である．より一般的に，$S_{n,1}=1$，$S_{n,2}=2^{n-1}-1$，$S_{n,3}=(3^{n-1}+1)/2-2^{n-1}$ である．次の漸化式は役に立つ[1-4]．

$$S_{n,0}=\begin{cases} 1 & (n=0\text{ のとき}) \\ 0 & (n\geq 1\text{ のとき}) \end{cases} \qquad S_{n,k}=kS_{n-1,k}+S_{n-1,k-1} \quad (1\leq k\leq n)$$

$$B_0=1, \quad B_n=\sum_{k=0}^{n-1}\binom{n-1}{k}B_k$$

漸近式は[5-9]で議論されている．

5.7.2 S の部分集合束における鎖

U，V を S の集合とする．U が V の真部分集合であるとき，$U\subset V$ と書く．これにより，S のすべての部分集合の族に，**半順序**（partial ordering）が入る．実際，それは最大元 S と最小元 $\emptyset$ をもつ**束**（lattice）である．長さ k の**鎖**（chain）$\emptyset = U_0 \subset U_1 \subset \cdots \subset U_{k-1} \subset U_k = S$ の個数は，$k!\,S_{n,k}$ である．したがって，$\emptyset$ から S までのすべての鎖の個数は[1,6,10]，

$$\sum_{k=0}^{n}k!\,S_{n,k} = \sum_{j=0}^{\infty}\frac{j^n}{2^{j+1}} = \frac{1}{2}\operatorname{Li}_{-n}\left(\frac{1}{2}\right) = \frac{d^n}{dx^n}\frac{1}{2-e^x}\Big|_{x=0} \sim \frac{n!}{2}\left(\frac{1}{\ln(2)}\right)^{n+1}$$

である．ただし，$\operatorname{Li}_m(x)$ は多重対数関数（polylogarithm）である．Wilf（ヴィルフ）[10]は，この漸近近似がなんと正確であるか，と驚いた．

さらに，鎖が**極大**（maximal），すなわち鎖に真部分集合を入れることができないとする．このとき，このような鎖の個数は $n!$ である．Doubilet, Rota & Stanley（ドゥビレ，ロタ&スタンレイ）[11]による一般的な技法は，いわゆる「結合代数」（incidence algebras）を含む，前記の2つの結果を得るために使われる．同様に，より複雑なポセット（poset，半順序集合）にある鎖を数えるために使われる[12]．

別なものとして，結合代数のより深い応用を述べる．有限ベクトル空間の線型部分空間の鎖を数えることである[6]．q-**二項係数**（q-binomial coefficient）と q-**階乗**（q-factorial）を，$q>1$ としてそれぞれ

$$\binom{n}{k}_q = \frac{\prod_{j=1}^{n}(q^j-1)}{\prod_{j=1}^{k}(q^j-1)\cdot\prod_{j=1}^{n-k}(q^j-1)}$$

$$[n!]_q = (1+q)(1+q+q^2)\cdots(1+q+q^2\cdots+q^{n-1})$$

で定義する．$q\to 1+0$ とした極限という特別な場合に注意せよ．q を素数の累乗とし，有限体 $\mathbb{F}_q$ 上の n 次元ベクトル空間 $\mathbb{F}_q^n$ を考える[12-16]．$\mathbb{F}_q^n$ の k 次元線型部分空間の

個数は，$\binom{n}{k}_q$ である．$\mathbb{F}_q^n$ の線型部分空間の全個数は，漸近的に，n が偶数のとき $c_e q^{n^2/4}$ であり，奇数のとき $c_o q^{n^2/4}$ である．ただし，係数は[17,18]

$$c_e = \frac{\sum_{k=-\infty}^{\infty} q^{-k^2}}{\prod_{j=1}^{\infty}(1-q^{-j})}, \quad c_o = \frac{\sum_{k=-\infty}^{\infty} q^{-(k+1/2)^2}}{\prod_{j=1}^{\infty}(1-q^{-j})}$$

である．真部分空間（再び，包含関係により順序づけられた）の鎖の個数 χ_n に対する漸化式を与えると，

$$\chi_1 = 1, \quad \chi_n = 1 + \sum_{k=1}^{n-1}\binom{n}{k}_q \chi_k \quad (n \geq 2)$$

である．漸近式は，[6,17]から，

$$\chi_n \sim \frac{1}{\zeta_q'(r) r}\left(\frac{1}{r}\right)^n \prod_{j=1}^{n}(q^j-1) = \frac{A}{r^n}(q-1)(q^2-1)(q^3-1)\cdots(q^n-1)$$

である．ただし，$\zeta_q(x)$ は部分空間のポセットに対するゼータ関数

$$\zeta_q(x) = \sum_{k=1}^{\infty} \frac{x^k}{(q-1)(q^2-1)(q^3-1)\cdots(q^k-1)}$$

で，$r>0$ は方程式 $\zeta_q(r)=1$ の一意の解である．とくに，$q=2$ のとき，$c_e = 7.3719688014\cdots$，$c_o = 7.3719494907\cdots$ であり，

$$\chi_n \sim \frac{A}{r^n} \cdot Q \cdot 2^{n(n+1)/2}$$

である．ただし，$r = 0.7759021363\cdots$，$A = 0.8008134543\cdots$ であり，

$$Q = \prod_{k=1}^{\infty}\left(1 - \frac{1}{2^k}\right) = 0.2887880950\cdots$$

は，デジタル検索木定数【5.14】の1つである．さらに強調するなら，鎖が極大であれば，このような鎖の個数は $[n!]_q$ である．

5.7.3 S を分割した束における鎖

集合 S の部分集合のポセットに関する鎖を議論した．しかしながら，あまり知られていないし，かつ研究するにはより難しい，S に自然に関連する別のポセットがある．それは S の**分割のポセット**（poset of partitions）である．P と Q が2つの S の分割と仮定する．このとき，$P<Q$ とは，$P \neq Q$ であり，かつ $p \in P$ は，ある $q \in Q$ が存在し，p が q の部分集合であるときとした半順序を考える．いいかえれば，P は，p の区塊の各々が Q のある区塊に含まれるという意味で，Q の「細分」（refinement）である．任意の n に対して，このポセットは実際，最小元 $m = \{\{1\}, \{2\}, \cdots, \{n\}\}$ と最大元 $M = \{\{1, 2, \cdots, n\}\}$ をもつ束である．

S の分割束における長さ k の鎖 $m = P_0 < P_1 < P_2 < \cdots < P_{k-1} < P_k = M$ の個数は何個か？ $n=3$ の場合には，$k=1$ のとき，明らかに $m<M$ に対して鎖が1個だけ存在する．$k=2$ に対して，図5.10で描かれているように，このような鎖は3個存在する．

任意の長さをもつ，m から M までのすべての鎖の個数を Z_n とする．明らかに，Z_1

```
          {{1,2,3}}
{{1,2},{3}}  {{1,3},{2}}  {{2,3},{1}}
          {{1},{2},{3}}
```

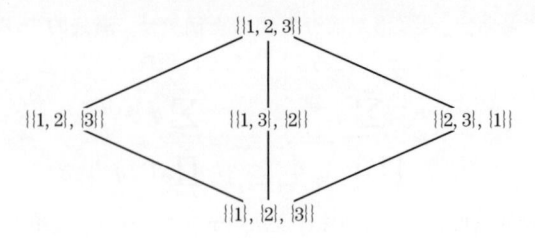

図 5.10 集合 $\{1,2,3\}$ の分割束の鎖 $m < P_1 < M$ は3個ある.

$= Z_2 = 1$ であり,前述のことから,$Z_3 = 4$ である.漸化式

$$Z_n = \sum_{k=1}^{n-1} S_{n,k} Z_k$$

が成り立つ.指数的母関数は,

$$Z(x) = \sum_{n=1}^{\infty} \frac{Z_n}{n!} x^n, \quad 2Z(x) = x + Z(e^x - 1)$$

である.しかし,Doubilet, Rota & Stanley と Bender (ベンダー) の技法は,Z_n の漸近式を得るためには適用できない.分割束は,「二項束」(binomial lattice) の構造がない,最初の束である.これは,よく知られた母関数の技法が,もはや役に立たないことを意味する.

Lengyel (レンジェル) [19]は,商

$$r_n = \frac{Z_n}{(n!)^2 (2\ln(2))^{-n} n^{-1-\ln(2)/3}}$$

が,$n \to \infty$ のとき2つの正の定数の間になければならないことを証明するために,違う研究法を組み立てた.彼は r_n が一意の値に収束することを支持する数値的な証拠を示した.Babai & Lengyel (ババイ&レンジェル) [20]は,かなり一般的な収束判定条件を証明した.その判定条件により,$\Lambda = \lim_{n\to\infty} r_n$ が存在し,$\Lambda = 1.09\cdots$ であると結論することができる.[19]の解析は,Stirling (スターリング) 数の複雑な評価を含んでいる.[20]では,有限遅れ (finite retardation) と能動的先祖 (active predecessors) をもつ,近凸線型漸化式に焦点をあてている.

Flajolet & Salvy (フラジョレット&サルヴィ) [21]は,大がかりな計算で $\Lambda = 1.0986858055\cdots$ を得た.その方法は,$\exp(x) - 1$ の (複素関数の) 解析的反復およびそれ以上多くのものに基づいている.しかし,あいにく彼らの論文は現在,不完全のままである.Takeuchi-Prellberg (竹内-プレルバーグ) 定数の関連する議論については,【5.8】を参照せよ.

対比として,「極大」鎖の個数は,正確に $n!(n-1)!/2^{n-1}$ で与えられる.Lengyel [19]は,Z_n は指数関数的に増大する係数倍だけこれを超えると述べた.

5.7.4 ランダム鎖

Van Cutsem & Ycart (ファン・カットセン&ユカート) [22]は,部分集合と分割束

の両方における，ランダム鎖を調べた．これらを調べることによって，共通の枠組みが存在すること，および鎖の両方の型の極限分布は「同一」(identical) であること，は驚くべきことである．1つだけ結果を述べる．ランダム鎖の正規化された長さを $\kappa_n = k/n$ とする．このとき，

$$\lim_{n \to \infty} E(\kappa_n) = \frac{1}{2 \ln(2)} = 0.7213475204\cdots$$

であり，対応する中心極限定理も成り立つ．

[1] L. Lovász, *Combinatorial Problems and Exercises*, 2nd ed., North-Holland, 1993, pp. 16-18, 162-173;MR 94m:05001.
[2] L. Comtet, *Advanced Combinatorics: The Art of Finite and Infinite Expansions*, Reidel, 1974, pp. 59-60, 204-211;MR 57 #124.
[3] N. J. A. Sloane, On-Line Encyclopedia of Integer Sequences, A000110, A000670, A005121, and A006116.
[4] G.-C. Rota, The number of partitions of a set, *Amer. Math. Monthly* 71 (1964) 498-504;MR 28 #5009.
[5] N. G. de Bruijn, *Asymptotic Methods in Analysis*, Dover, 1958, pp. 102-109;MR 83m:41028.
[6] E. A. Bender, Asymptotic methods in enumeration, SIAM Rev. 16 (1974) 485-515;errata 18 (1976) 292;MR 51 #12545 and MR 55 #10276.
[7] A. M. Odlyzko, Asymptotic enumeration methods, *Handbook of Combinatorics*, v. I, ed.R. Graham, M. Grötschel, and L. Lovász, MIT Press, 1995, pp. 1063-1229;MR 97b:05012.
[8] B. Salvy and J. Shackell, Symbolic asymptotics: multiseries of inverse functions, *J. Symbolic Comput*. 27 (1999) 543-563;MR 2000h:41039.
[9] B. Salvy and J. Shackell, Asymptotics of the Stirling numbers of the second kind, *Studies in Automatic Combinatorics*, v. 2, Algorithms Project, INRIA, 1997.
[10] H. S. Wilf, *generatingfunctionology*, Academic Press, 1990, pp. 21-24, 146-147;MR 95a:05002.
[11] P. Doubilet, G.-C. Rota, and R. Stanley, On the foundations of combinatorial theory. VI: The idea of generating function, *Proc. Sixth Berkeley Symp. Math. Stat. Probab*., v. 2, ed. L. M. Le Cam, J. Neyman, and E. L. Scott, Univ. of Calif. Press, 1972, pp. 267-318;MR 53 #7796.
[12] M. Aigner, *Combinatorial Theory*, Springer-Verlag, 1979, pp. 78-79, 142-143;MR 80h:05002.
[13] J. Goldman and G.-C. Rota, The number of subspaces of a vector space, *Recent Progress in Combinatorics*, Proc. 1968 Waterloo conf., ed. W. T. Tutte, Academic Press, 1969, pp. 75-83;MR 40 #5453.
[14] G. E. Andrews, *The Theory of Partitions*, Addison-Wesley, 1976;MR 99c:11126.
[15] H. Exton, *q-Hypergeometric Functions and Applications*, Ellis Horwood, 1983;MR 85g:33001.
[16] M. Sved, Gaussians and binomials, *Ars Combin*. 17 A (1984) 325-351;MR 85j:05002.
[17] T. Slivnik, Subspaces of $\mathbb{Z}_2^n$, unpublished note (1996).
[18] M. Wild, The asymptotic number of inequivalent binary codes and nonisomorphic binary matroids, *Finite Fields Appl*. 6 (2000) 192-202;MR 2001i:94077.
[19] T. Lengyel, On a recurrence involving Stirling numbers, *Europ. J. Combin*. 5 (1984) 313-321;MR 86c:11010.
[20] L. Babai and T. Lengyel, A convergence criterion for recurrent sequences with application to the partition lattice, *Analysis* 12 (1992) 109-119;MR 93f:05005.
[21] P. Flajolet and B. Salvy, Hierarchical set partitions and analytic iterates of the exponential function, unpublished note (1990).
[22] B. Van Cutsem and B. Ycart, Renewal-type behavior of absorption times in Markov chains, *Adv. Appl. Probab*. 26 (1994) 988-1005;MR 96f:60118.

5.8 竹内-プレルバーグ定数

1978年，Takeuchi（竹内郁雄）は3変数の帰納的関数[1,2]

$$t(x,y,z)=\begin{cases} y & (x\leq y \text{ のとき}) \\ t(t(x-1,y,z),\ t(y-1,z,x),\ t(z-1,x,y)) & (\text{その他のとき}) \end{cases}$$

を定義した．これはプログラミング言語のベンチマークテストに役立つ．$t(x,y,z)$ の値自体に実用的な意味はない．実際，McCarthy（マッカーシー）[1,2]は，この関数はもっと簡単に

$$t(x,y,z)=\begin{cases} y & (x\leq y \text{ のとき}) \\ \begin{cases} z, & y\leq z \text{ のとき} \\ x, & \text{その他のとき} \end{cases} & (\text{その他のとき}) \end{cases}$$

となることを示した．

興味ある量は $t(x,y,z)$ ではなく，むしろ「その他のとき」の文が繰り返し実行される回数 $T(x,y,z)$ である．以前に計算された結果が未来の任意の時に利用できないという意味で，プログラムにはメモリーがないと仮定する．Knuth（クヌース）[1,3]は，Takeuchi（竹内）数 $T_n=T(n,0,n+1)$，

$$T_0=0,\quad T_1=1,\quad T_2=4,\quad T_3=14,\quad T_4=53,\quad T_5=223,\quad \cdots$$

を研究し，十分大きなすべての n に対して，

$$e^{n\ln(n)-n\ln(\ln(n))-n} < T_n < e^{n\ln(n)-n+\ln(n)}$$

を得た．彼は T_n の増大について，より正確な漸近式の情報を問題とした．

竹内数に対する Knuth の再帰関数から始めて，

$$T_{n+1}=\sum_{k=0}^{n}\left[\binom{n+k}{n}-\binom{n+k}{n+1}\right]T_{n-k}+\sum_{k=1}^{n-1}\binom{2k}{k}\frac{1}{k+1}$$

であり，Bell（ベル）数【5.7】

$$B_{n+1}=\sum_{k=0}^{n}\binom{n}{k}B_{n-k},\quad B_0=1,\quad B_1=1,\quad B_2=2,\quad B_3=5,\quad B_4=15,\quad B_5=52,\quad \cdots$$

といくぶん関連している．

Prellberg（プレルバーグ）[4]は，次の極限値が存在することを述べた．

$$c=\lim_{n\to\infty}\frac{T_n}{B_n\exp\left(\frac{1}{2}W_n^2\right)}=2.2394331040\cdots$$

ただし，$W_n\exp(W_n)=n$ は Lambert（ランベルト）W 関数の特別な値である【6.11】．

ベル数と W 関数はともに，よくわかっているから，Knuth の問題に答えることができる．基礎理論は今もなお発展している．しかし，Prellberg の数値的な証拠は説得力がある．最近の理論的な成果[5]は，Lengyel の定数【5.7】を得た方法と平行した手法

で，関連する関数方程式

$$T(z)=\sum_{n=0}^{\infty}T_n z^n, \quad T(z)=\frac{T(z-z^2)}{z}-\frac{1}{(1-z)(1-z+z^2)}$$

に定数 c を関連付けている．

[1] D. E. Knuth, Textbook examples of recursion, *Artificial Intelligence and Mathematical Theory of Computation*, ed.V. Lifschitz, Academic Press, 1991, pp. 207-229; also in *Selected Papers on Analysis of Algorithms*, CSLI, 2000, pp. 391-414; MR 93a:68093.
[2] I. Vardi, *Computational Recreations in Mathematica*, Addison-Wesley, 1991, pp. 179-199; MR 93e: 00002.
[3] N. J. A. Sloane, On-Line Encyclopedia of Integer Sequences, A000651.
[4] T. Prellberg, On the asymptotics of Takeuchi numbers, *Symbolic Computation, Number Theory, Special Functions, Physics and Combinatorics*, Proc. 1999 Gainesville conf., ed. F. G. Garvan and M. Ismail, Kluwer, 2001, pp. 231-242; math.CO/0005008.
[5] T. Prellberg, On the asymptotic analysis of a class of linear recurrences, presentation at *Formal Power Series and Algebraic Combinatorics (FPSAC)* conf., Univ. of Melbourne, 2002.

5.9 ポリヤの酔歩定数

d 次元空間内の座標すべてが整数値である頂点からなる d 次元立方体格子を L とする．L 上の**歩行** (walk) ω とは，原点から始まる，頂点の無限列 $\omega_0, \omega_1, \omega_2, \omega_3, \cdots$ である．ただし，$\omega_0=0$ で，すべての j に対して $|\omega_{j+1}-\omega_j|=1$ である．各段階において，可能なすべての方向 $2d$ 個は等しい確率で起こるという意味で，歩行はランダムで対称であると仮定する．ある $n>0$ で，$\omega_n=0$ であることは，どれほど起きやすいのか？すなわち，**回帰確率** (return probability) p_d はどれほどか？

Pólya（ポリヤ）[1-4]は，$p_1=p_2=1$ であるが，$d>2$ に対して $p_d<1$ である，という驚くべき事実を証明した．McCrea & Whipple（マックレア&ウィッペル）[5]，Watson（ワトソン）[6]，Domb（ドンブ）[7]および Glasser & Zucker（グラッサー&ザッカー）[8]の各々は，評価 $p_3=1-1/m_3=0.3405373295\cdots$ に多面的に貢献した．ただし，原点へ回帰する数に 1 を加えた期待値数 m_3 は，

$$\begin{aligned}
m_3 &= \frac{3}{(2\pi)^3}\int_{-\pi}^{\pi}\int_{-\pi}^{\pi}\int_{-\pi}^{\pi}\frac{1}{3-\cos(\theta)-\cos(\phi)-\cos(\psi)}d\theta d\phi d\psi \\
&= \frac{12}{\pi^2}(18+12\sqrt{2}-10\sqrt{3}-7\sqrt{6})K[(2-\sqrt{3})(\sqrt{3}-\sqrt{2})]^2 \\
&= 3(18+12\sqrt{2}-10\sqrt{3}-7\sqrt{6})\left[1+2\sum_{k=1}^{\infty}\exp(-\sqrt{6}\pi k^2)\right]^4 \\
&= \frac{\sqrt{6}}{32\pi^3}\Gamma\left(\frac{1}{24}\right)\Gamma\left(\frac{5}{24}\right)\Gamma\left(\frac{7}{24}\right)\Gamma\left(\frac{11}{24}\right)=1.5163860591\cdots
\end{aligned}$$

である．したがって，3次元立方格子における酔歩（ランダムウォーク，乱歩）の**脱出確率** (escape probability) は，$1-p_3=0.6594626704\cdots$ である．これらの式において，K は第1種の完全楕円積分【1.4.6】であり，Γ はガンマ関数【1.5.4】である．回帰確率および脱出確率は，（単純な立方格子とは異なる）体心立方格子または面心立方格子に対しても計算できる．しかし，これらの場合や他の一般化[9]は議論しない．

$d>3$ に対して，p_d については何がいえるのだろうか？ 閉じた式は，ここでは存在するとは思えない．Montroll（モントロル）[10-12]は $p_d=1-1/m_d$ を示した．ただし，

$$m_d = \frac{d}{(2\pi)^d} \int_{-\pi}^{\pi} \int_{-\pi}^{\pi} \cdots \int_{-\pi}^{\pi} \Big(d - \sum_{k=1}^{d} \cos(\theta_k)\Big)^{-1} d\theta_1 d\theta_2 \cdots d\theta_d$$

$$= \int_0^{\infty} e^{-t} \Big(I_0\Big(\frac{t}{d}\Big)\Big)^d dt$$

であり，$I_0(x)$ は0次の変形ベッセル関数【3.6】である．d の関数として，対応する数値近似は，表5.1に示した[10,13-17]．

戻るために必要な道の長さは，どれくらいであるのか？ 原点から出発し，格子点 l で終わる，d 次元の n 段階の酔歩の個数を $U_{d,l,n}$ とする．原点から出発し，格子点 $l\neq 0$ に達することが最後に「初めて」起こる（$l=0$ のときは2度目である）d 次元の n 段階の酔歩の個数を $V_{d,l,n}$ とする．このとき，母関数

$$U_{d,l}(x) = \sum_{n=0}^{\infty} \frac{U_{d,l,n}}{(2d)^n} x^n, \quad V_{d,l}(x) = \sum_{n=0}^{\infty} \frac{V_{d,l,n}}{(2d)^n} x^n$$

は，$l\neq 0$ のとき $V_{d,l}(x)=U_{d,l}(x)/U_{d,0}(x)$ であり，$l=0$ のとき $V_{d,l}(x)=1-1/U_{d,0}(x)$ であり，$U_{d,0}(1)=m_d$, $V_{d,0}(1)=p_d$ である．たとえば，

$$U_{1,l}(x) = \sum_{n=0}^{\infty} \frac{1}{2^n} \binom{n}{\frac{l+n}{2}} x^n, \quad U_{2,l}(x) = \sum_{n=0}^{\infty} \frac{1}{4^n} \binom{n}{\frac{l_1+l_2+n}{2}} \binom{n}{\frac{l_1-l_2+n}{2}} x^n$$

である．ただし，$d=1$ で $l+n$ が奇数であるか，または $d=2$ で l_1+l_2+n が奇数であるときには，二項係数を0に等しいとする．$d=3$ のとき，$a_n=U_{3,0,2n}$ は[18]，

$$a_n = \binom{2n}{n} \sum_{k=0}^{n} \binom{n}{k}^2 \binom{2k}{k} = \sum_{k=0}^{n} \frac{(2n)!(2k)!}{(n-k)!^2 k!^4}, \quad \sum_{n=0}^{\infty} \frac{a_n}{(2n)!} y^{2n} = I_0(2y)^3$$

$$(n+2)^3 a_{n+2} - 2(2n+3)(10n^2+30n+23) a_{n+1} + 36(n+1)(2n+1)(2n+3) a_n = 0$$

を満たす．$d=4$ のとき，$b_n=U_{4,0,2n}$ は[19]，

$$(n+2)^4 b_{n+2} - 4(2n+3)^2 (5n^2+15n+12) b_{n+1}$$
$$+ 256(n+1)^2 (2n+1)(2n+3) b_n = 0$$

表5.1 回帰期待値と回帰確率

d	m_d	p_d
4	1.2394671218⋯	0.1932016732⋯
5	1.1563081248⋯	0.1351786098⋯
6	1.1169633732⋯	0.1047154956⋯
7	1.0939063155⋯	0.0858449341⋯
8	1.0786470120⋯	0.0729126499⋯

を満たす.

任意の d に対して,任意の格子点 l に到着する,平均**初通過時間**(**初到達時間**,first-passage time)は,(初通過する確率は $d=1,2$ のとき $V_{d,l}(1)=1$ であるにもかかわらず)無限である.必要な歩く長さを得るためのいくつかの別な方法がある.$V_{d,l}(x)$ の式を用いると,初通過時間の中央値は,$d=1$ のとき $l=0$, 1, 2, 3 に対してそれぞれ 2-4, 1-3, 6-8, 17-19 段であり,$d=2$ のとき $l=(0,0)$, $(1,0)$, $(1,1)$ に対してそれぞれ 2-4, 25-27, 520-522 段である.Hughes(ヒューズ)[3,20]は,(いつかは回帰が起こるという条件で)原点へ回帰する条件付平均時間を調べた.さらに,$d=1$ に対して,独立に歩する3点のうち,原点へ回帰する最初の酔歩の平均時間は,有限であり,その値は[6,21-23]

$$2\sum_{n=0}^{\infty}\frac{1}{2^{6n}}\binom{2n}{n}^3=\frac{2}{\pi^3}\int_0^\pi\int_0^\pi\int_0^\pi\frac{1}{1-\cos(\theta)\cos(\phi)\cos(\psi)}d\theta d\phi d\psi$$
$$=\frac{8}{\pi^2}K\left(\frac{1}{\sqrt{2}}\right)^2=\frac{1}{2\pi^3}\Gamma\left(\frac{1}{4}\right)^4=2(1.3932039296\cdots)$$

である.一方,$d=2$ のとき,「任意個」の互いに独立な酔歩に対して,原点へ回帰する最初の酔歩の平均時間は,無限である.好意的(friendly)と悪意的(vicious)をともに含む,多重酔歩に関する多くの結果は,[24]にある.

$$U_{d,l}(x)=\frac{d}{(2\pi)^d}\int_{-\pi}^{\pi}\int_{-\pi}^{\pi}\cdots\int_{-\pi}^{\pi}\left(d-x\sum_{k=1}^d\cos(\theta_k)\right)^{-1}\exp\left(i\sum_{k=1}^d\theta_k l_k\right)d\theta_1 d\theta_2\cdots d\theta_d$$

が知られている.小さい d に対しては,数値的に評価できる.3次元酔歩がある点 l に到着する,いくつかの標本確率[11,16]がある.

$$V_{3,l}(1)=\frac{U_{3,l}(1)}{m_3}=\begin{cases}0.3405373295\cdots & (l=(1,0,0))\\ 0.2183801414\cdots & (l=(1,1,0))\\ 0.1724297877\cdots & (l=(1,1,1))\end{cases}$$

である.これらの確率の漸近展開は[11,12],

$$V_{3,l}(1)=\frac{3}{2\pi m_3|l|}\left[1+\frac{1}{8|l|^2}\left(-3+\frac{5(l_1^4+l_2^4+l_3^4)}{|l|^2}+\cdots\right)\right]\sim\frac{0.3148702313\cdots}{|l|}$$

である.$|l|^2=l_1^2+l_2^2+l_3^2\to\infty$ のとき,これは妥当である.

d 次元の n 段階の酔歩をする間に訪れる,異なる格子点の平均個数を $W_{d,n}$ とする.[25-28]より,$n\to\infty$ のとき,

$$W_d(x)=\sum_{n=0}^{\infty}W_{d,n}x^n=\frac{1}{(1-x)^2 U_{d,0}(x)},\quad W_{d,n}\sim\begin{cases}\sqrt{\dfrac{8n}{\pi}} & (d=1)\\ \dfrac{\pi n}{\ln(n)} & (d=2)\\ (1-p_3)n & (d=3)\end{cases}$$

が成り立つ.$W_{3,n}$ の高い次数の漸近式は,展開式[11,12,29-31]を用いることで可能で,$x\to 1-0$ のとき,

$$U_{3,0}=m_3-\frac{3\sqrt{3}}{2\pi}(1-x^2)^{1/2}+c(1-x^2)-\frac{3\sqrt{3}}{4\pi}(1-x^2)^{2/3}+\cdots$$

である. ただし,
$$c=\frac{9}{32}\left(m_3+\frac{6}{\pi^2 m_3}\right)=0.5392381750\cdots$$
である. 他のパラメータ, たとえば原点からの距離の平均増大度[32]について

$$\lim_{n\to\infty}\frac{1}{\ln(n)}\sum_{j=1}^{n}\frac{j^{-1/2}}{1+|\omega_j|}=\lambda_1 \text{ が } d=1 \text{ のとき, 確率1で成り立つ.}$$

$$\lim_{n\to\infty}\frac{1}{\ln(n)^2}\sum_{j=1}^{n}\frac{1}{1+|\omega_j|^2}=\lambda_2 \text{ が } d=2 \text{ のとき, 確率1で成り立つ.}$$

$$\lim_{n\to\infty}\frac{1}{\ln(n)}\sum_{j=1}^{n}\frac{1}{1+|\omega_j|^2}=\lambda_d \text{ が } d\geq 3 \text{ のとき, 確率1で成り立つ.}$$

を, 解析するのはより難しい. 定数 λ_d は, 有限でかつ正であることのみが知られている.

1次元 n 段階の酔歩 ω に対して, ω_j の最大値を M_n^+, $-\omega_j$ の最大値を M_n^- と定義する. このとき, M_n^+ と M_n^- の各々は, $n\to\infty$ のとき, 半正規分布【6.2】に従い[33,34],

$$\lim_{n\to\infty}\mathrm{E}(n^{-1/2}M_n^+)=\sqrt{\frac{2}{\pi}}=\lim_{n\to\infty}\mathrm{E}(n^{-1/2}M_n^-)$$

が成り立つ. さらに, $\omega_j=M_n^+$ となる最小の j を T_n^+, $-\omega_k=M_n^-$ となる最小の k を T_n^- と定義する. このとき, **逆正弦法則** (arcsine law) が成り立つ.

$$\lim_{n\to\infty}\mathrm{P}(n^{-1}T_n^+<x)=\frac{2}{\pi}\arcsin\sqrt{x}=\lim_{n\to\infty}\mathrm{P}(n^{-1}T_n^-<x)$$

これは1次元酔歩が, 高い確率で位置が負または正のいずれかである (両方は成り立たない) ことを意味する. d 次元酔歩についてのこのような詳しい情報は, 未だにない. さらに, 任意の正の整数 r に対して, $|\omega_j|\geq r$ となる最小の j を $\tau_{d,r}$ と定義する. このとき[35],

$$\tau_{1,r}=r^2, \quad \tau_{2,2}=\frac{9}{2}, \quad \tau_{2,3}=\frac{135}{13}, \quad \tau_{2,4}=\frac{11791}{668}$$

である. しかし, そのパターンは明らかになっていない. $r\to\infty$ のとき, $\tau_{d,r}$ については, 正確に何がいえるだろうか？

計算は別にして, Odlyzko (オドリズコ) の結果[36-38]を述べる. 正確に M_n^+ (または M_n^-) を決定するための任意のアルゴリズムは, ω_j の値を平均で, 少なくとも $(A+o(1))\sqrt{n}$ 回調べなければならない. ただし, $A=\sqrt{8/\pi}\ln(2)=1.1061028674\cdots$ である.

他方, 1次元酔歩が, 最初の n 段階では訪れていない新しい頂点に達する, 待ち時間 N_n については[39],

$$\text{確率1で,} \quad \limsup_{n\to\infty}\frac{N_n}{n\ln(\ln(n))^2}=\frac{1}{\pi^2}$$

が成り立つ. 一方, n 段階までに酔歩によって, 最大回数訪れた頂点の集合を F_n とする. F_n は**お気に入りの場所** (favorite sites) とよばれる. このとき, $|F_n|\geq 4$ は, 確率1で有限回だけ起こる[40].

2次元酔歩に対して，類似な F_n を定義する．最初の n 段階以内で，F_n 内のある選んだ点を訪れる回数は，$n \to \infty$ のとき確率 1 で $\sim \ln(n)^2/\pi$ である．これは大酒飲みが彼のお気に入りのバーにちょっと立ち寄る漸近的回数として，いいかえができる[41, 42]．双対的に $r \times r$ トーラス（輪環面，反対側の辺を同一視した正方形）のすべての頂点を覆うのに必要な時間の長さ C_r は，すべての $\varepsilon > 0$ に対して，（確率収束の意味で）[43]

$$\lim_{r \to \infty} P\left(\left|\frac{C_r}{r^2(\ln r)^2} - \frac{4}{\pi}\right| < \varepsilon\right) = 1$$

を満たす．これは「白色画面問題（white screen problem）」として知られているものの解である．

3次元酔歩 ω を，領域 $z \leq y \leq x$ に制限すると，類似な級数の係数は

$$\bar{a}_n = \sum_{k=0}^{n} \frac{(2n)!(2k)!}{(n-k)!(n+1-k)!k!^2(k+1)!^2}$$

であり，これから[45]，

$$\bar{m}_3 = \sum_{n=0}^{\infty} \frac{\bar{a}_n}{6^{2n}} = 1.0693411205\cdots, \quad \bar{p}_3 = 1 - \frac{1}{\bar{m}_3} = 0.0648447153\cdots$$

が回帰を特徴づける．回帰を特徴づけている他の領域，たとえば，半空間，4分空間や8分空間に関しては何がいえるだろうか？

変形が1つある．$X_1, X_2, X_3, \cdots$ を平均 μ で分散 1 の正規分布からの独立な変数とする．（1次元格子点ではなくて）実数直線上で，（ベルヌーイ増加ではなくて）ガウス増加をもって酔歩する部分和 $S_j = \sum_{k=1}^{j} X_k$ を考える．$\{S_j\}$ に関しては多くの文献があるが，1つの結果のみを述べる．H を，S_j がとる初めての正の値とする．H は，過程の**はしごの最初の高さ**（first ladder height）とよばれる．$\mu = 0$ のとき，H のモーメントは[46]，

$$E_0(H) = \frac{1}{\sqrt{2}}, \quad E_0(H^2) = -\frac{\zeta\left(\frac{1}{2}\right)}{\sqrt{\pi}} = \sqrt{2}\,\rho = \sqrt{2}\,(0.5825971579\cdots)$$

であり，0 の近傍にある任意の μ に対して，

$$E_\mu(H) = \frac{1}{\sqrt{2}} \exp\left[-\frac{\mu}{\sqrt{2\pi}} \sum_{k=0}^{\infty} \frac{\zeta\left(\frac{1}{2} - k\right)}{k!(2k+1)} \left(-\frac{\mu^2}{2}\right)^k\right]$$

が成り立つ．ただし，$\zeta(x)$ は Riemann のゼータ関数【1.6】である．統計学の文献にあるおもしろい定数 ρ の他の出現については，[47-50]にある．

他の変形がある．$Y_1, Y_2, Y_3, \cdots$ を区間 $[-1,1]$ からの独立な一様乱数とする．$S_0 = 0$, $S_j = \sum_{k=1}^{j} Y_j$ とする．このとき，$\{S_0, S_1, \cdots, S_n\}$ の最大期待値は $n \to \infty$ のとき[51]，

$$E\left(\max_{0 \leq j \leq n} S_j\right) = \sqrt{\frac{2}{3\pi}}\, n^{1/2} + \sigma + \frac{1}{5}\sqrt{\frac{2}{3\pi}}\, n^{-1/2} + O(n^{-3/2})$$

である．ただし，$\sigma = -0.2979521902\cdots$ は

$$\sigma = \frac{\zeta\left(\frac{1}{2}\right)}{\sqrt{6\pi}} + \frac{\zeta\left(\frac{3}{2}\right)}{20\sqrt{6\pi}} + \sum_{k=1}^{\infty}\left(\frac{t_k}{k} - \frac{k^{-1/2}}{\sqrt{6\pi}} - \frac{k^{-3/2}}{20\sqrt{6\pi}}\right)$$

であり，

$$t_k = \frac{2(-1)^k}{(k+1)} \sum_{k/2 \leq j \leq k} (-1)^j \binom{k}{j}\left(j - \frac{k}{2}\right)^{k+1}$$

である．$\zeta(x)$ と酔歩との深い関連は，[52]で議論されている．

5.9.1 交差と捕獲

格子点上の歩行 ω は，ある $i<j$ で $\omega_i = \omega_j$ になるとき，**自己交差**（self-intersecting）であるという．これが起こる最小の j を**自己交差時間**（self-intersection time）という．歩行の折に通った場所をすべて記憶しておかなければならないので，自己交差時間を計算することは，初通過時間よりももっと難しい．$d=1$ のとき，明らかに，平均自己交差時間は 3 である．$d=2$ のとき，平均自己交差時間は[53]，

$$\frac{2\cdot 4}{4^2} + \frac{3\cdot 12}{4^3} + \frac{4\cdot 44}{4^4} + \frac{5\cdot 116}{4^5} + \cdots = \sum_{n=2}^{\infty} \frac{n(4c_{n-1} - c_n)}{4^n}$$
$$= \frac{c_1}{2} + \sum_{n=2}^{\infty} \frac{c_n}{4^n} = 4.5860790989\cdots$$

である．ただし，数列 $\{c_n\}$ は【5.10】で定義される．n が大きいときには，前に議論した数列 $\{a_n\}$，$\{\bar{a}_n\}$，$\{b_n\}$ とは違って，c_n を評価するための正確な式は知られていない．この例は，後に出てくる難しさを事前に示している．一般化は[54,55]を参照せよ．

ある k が存在し，すべての $i<j\leq k$ に対して $\omega_i \neq \omega_j$ であり，ω_k は以前に訪れた頂点によって完全に取り囲まれているとき，歩行は**自己捕獲**（self-trapping）であるという．

$d=2$ とする．$k=7$ のとき 8 個の自己捕獲歩行がある．$k=8$ のとき，16 個の自己捕獲歩行がある．[56,57]のモンテカルロシミュレーションでは，近似的に $70.7\cdots$ の平均自己捕獲時間を得ている．

2 つの歩行が ω と ω' が**交差**（intersect）するとは，0 でないある i と j に対して $\omega_i = \omega'_j$ となる，ときをいう．2 つの n 段階の独立な酔歩が決して交差しない確率 q_n は [58-61]，$n \to \infty$ のとき，

$$\ln(q_n) \sim \begin{cases} -\dfrac{5}{8}\ln(n) & (d=2) \\ -\xi \ln(n) & (d=3) \\ -\dfrac{1}{2}\ln(\ln(n)) & (d=4) \end{cases}$$

である．ただし，指数 ξ は近似的に $0.29\cdots$ である．（これもまたシミュレーションにより得られた．）各 $d\geq 5$ に対して，[62]より，$\lim_{n\to\infty} q_n$ は，正確に 0 と 1 の間にあることが示されている．さらにシミュレーション[63]より，$q_5 = 0.708\cdots$，$q_6 = 0.822\cdots$ を得る．これらを【5.10】で言及する．

5.9.2 ホロノミー性

ホロノーム関数 (holonomic function) (Zeilberger (ザイルバーガー) [45,64,65] の意味で) は，線型同次微分方程式

$$f^{(n)}(z) + r_1(z)f^{(n-1)}(z) + \cdots + r_{n-1}(z)f'(z) + r_n(z)f(z) = 0$$

の解 $f(z)$ である．ただし，各 $r_k(z)$ は有理数係数をもつ有理関数である．**正則ホロノーム** (regular holonomic) **定数**とは，各 r_k が解析的であるときの，代数的数 z_0 での f の値のことをいう．そのとき，f は z_0 で解析的であることが証明できる．**特異ホロノーム** (singular holonomic) **定数**とは，各 r_k が z_0 で，最悪の場合でも，k 次の極 (フックス的すなわち「確定」特異点 ("regular" singularities) [66-68]) をもつ代数的数 z_0 での f の値のことをいう．前者には π，$\ln(2)$ および4重対数 $\text{Li}_4(1/2)$ があり，後者には Apéry (アペリ) 定数 $\zeta(3)$，Catalan (カタラン) の定数 G および Pólya (ポリヤ) の定数 p_d，$d > 2$ がある．どの型のホロノーム定数も，多項式時間で計算できる定数の族である[69]．Chow (チョウ) [70]による EL 数の，少々関連する理論だけを注意する．

---

[1] G. Pólya, Über eine Aufgabe der Wahrscheinlichkeitsrechnung betreffend die Irrfahrt im Stassennetz, *Math. Annalen* 84 (1921) 149-160; also in *Collected Papers*, v. 4, ed. G.-C. Rota, MIT Press, 1984, pp. 69-80, 609.

[2] P. G. Doyle and J. L. Snell, *Random Walks and Electric Networks*, Math. Assoc. Amer., 1984; math.PR/0001057; MR 89a:94023.

[3] B. D. Hughes, *Random Walks and Random Environments*, v. 1, Oxford Univ. Press, 1995; MR 96i:60070.

[4] F. Spitzer, *Principles of Random Walks*, 2nd ed, Springer-Verlag, 1976, p. 103; MR 52 #9383.

[5] W. H. McCrea and F. J. W. Whipple, Random paths in two and three dimensions, *Proc. Royal Soc. Edinburgh* 60 (1940) 281-298; MR 2,107f.

[6] G. N. Watson, Three triple integrals, *Quart. J. Math.* 10 (1939) 266-276; MR 1,205b.

[7] C. Domb, On multiple returns in the random-walk problem, *Proc. Cambridge Philos. Soc.* 50 (1954) 586-591; MR 16,148f.

[8] M. L. Glasser and I. J. Zucker, Extended Watson integrals for the cubic lattices, *Proc. Nat. Acad. Sci. USA* 74 (1977) 1800-1801; MR 56 #686.

[9] D. J. Daley, Return probabilities for certain three-dimensional random walks, *J. Appl. Probab.* 16 (1979) 45-53; MR 80e:60083.

[10] E. W. Montroll, Random walks in multidimensional spaces, especially on periodic lattices, *J. SIAM* 4 (1956) 241-260; MR 19,470d.

[11] E. W. Montroll, Random walks on lattices, *Stochastic Processes in Mathematical Physics and Engineering*, ed. R. Bellman, Proc. Symp. Appl. Math. 16, Amer. Math. Soc., 1964, pp. 193-220; MR 28 #4585.

[12] E. W. Montroll and G. H. Weiss, Random walks on lattices. II, *J. Math. Phys.* 6 (1965) 167-181; MR 30 #2563.

[13] K. Kondo and T. Hara, Critical exponent of susceptibility for a class of general ferromagnets in $d > 4$ dimensions, *J. Math. Phys.* 28 (1987) 1206-1208; MR 88k:82052.

[14] D. I. Cartwright, Some examples of random walks on free products of discrete groups, *Annali Mat. Pura Appl.* 151 (1988) 1-15; MR 90f:60018.

[15] P. Griffin, Accelerating beyond the third dimension: Returning to the origin in simple random

walk, *Math. Sci.* 15 (1990) 24-35; MR 91g:60083.
[16] T. Hara, G. Slade, and A.D. Sokal, New lower bounds on the self-avoiding-walk connective constant, *J. Stat. Phys.* 72 (1993) 479-517; erratum 78 (1995) 1187-1188; MR 94e:82053.
[17] J. Keane, Pólya's constants p_d to 80 digits for $3 \le d \le 64$, unpublished note (1998).
[18] N. J. A. Sloane, On-Line Encyclopedia of Integer Sequences, A002896, A039699, A049037, and A063888.
[19] M. L. Glasser and A. J. Guttmann, Lattice Green function (at 0) for the 4D hypercubic lattice, *J. Phys. A* 27 (1994) 7011-7014; MR 95i:82047.
[20] B. D. Hughes, On returns to the starting site in lattice random walks, *Physica* 134A (1986) 443-457; MR 87c:60057.
[21] K. Lindenberg, V. Seshadri, K. E. Shuler, and G. H. Weiss, Lattice random walks for sets of random walkers: First passage times, *J. Stat. Phys.* 23 (1980) 11-25; MR 83c:60100.
[22] W. D. Fryer and M. S. Klamkin, Comment on problem 612, *Math. Mag.* 40 (1967) 52-53.
[23] B. C. Berndt, *Ramanujan's Notebooks: Part II*, Springer-Verlag, 1989, p. 24; MR 90b:01039.
[24] M. E. Fisher, Walks, walls, wetting, and melting, *J. Stat. Phys.* 34 (1984) 667-729; MR 85j:82022.
[25] A. Dvoretzky and P. Erdös, Some problems on randomwalk in space, *Proc. Second Berkeley Symp. Math. Stat. Probab.*, ed. J. Neyman, Univ. of Calif. Press, 1951, pp. 353-367; MR 13,852b.
[26] G. H. Vineyard, The number of distinct sites visited in a random walk on a lattice, *J. Math. Phys.* 4 (1963) 1191-1193; MR 27 #4610.
[27] M. N. Barber and B.W. Ninham, *Random and Restricted Walks: Theory and Applications*, Gordon and Breach, 1970.
[28] E.W. Montroll and B. J.West, On an enriched collection of stochastic processes, *Studies in Statistical Mechanics*, v. 7, *Fluctuation Phenomena*, ed. E. W. Montroll and J. L. Lebowitz, North-Holland, 1979, pp. 61-175; MR 83g:82001.
[29] A. A. Maradudin, E. W. Montroll, G. H. Weiss, R. Herman, and H. W. Milnes, Green's functions for monatomic simple cubic lattices, *Acad. Royale Belgique Classe Sci. Mém. Coll. 4°, 2ᵉ Sér.*, v. 14 (1960) n. 7, 1-176; MR 22 #7440.
[30] G. S. Joyce, Lattice Green function for the simple cubic lattice, *J. Phys. A* (1972) L65-L68.
[31] G. S. Joyce, On the simple cubic lattice Green function, *Philos. Trans. Royal Soc. London Ser. A* 273 (1973) 583-610; MR 58 #28708.
[32] P. Erdös and S. J. Taylor, Some problems concerning the structure of random walk paths, *Acta Math. Acad. Sci. Hungar.* 11 (1960) 137-162; MR 22 #12599.
[33] P.Révész, *RandomWalk in Random and Nonrandom Environments*, World Scientific, 1990; MR 92c:60096.
[34] W. Feller, *An Introduction to Probability Theory and Its Applications*, v. 1, 3rd ed., Wiley, 1968; MR 37 #3604.
[35] I. Kastanas, Simple random walk barriers, unpublished note (1997).
[36] A. M. Odlyzko, Search for the maximum of a random walk, *Random Structures Algorithms* 6 (1995) 275-295; MR 97b:60117.
[37] H.-K. Hwang, A constant arising from the analysis of algorithms for determining the maximum of a random walk, *Random Structures Algorithms* 10 (1997) 333-335; MR 98i:05007.
[38] P. Chassaing, How many probes are needed to compute the maximum of a random walk?, *Stochastic Process. Appl.* 81 (1999) 129-153; MR 2000k:60087.
[39] E. Csáki, A note on: "Three problems on the random walk in Z^{d+1}" by P. Erdös and P. Révész, *Studia Sci. Math. Hungar.* 26 (1991) 201-205; MR 93k:60172.
[40] B. Tóth, No more than three favorite sites for simple random walk, *Annals of Probab.* 29 (2001) 484-503; MR 2002c:60076.
[41] A. Dembo, Y. Peres, J. Rosen, and O. Zeitouni, Thick points for planar Brownian motion and the Erdös-Taylor conjecture on random walk, *Acta Math.* 186 (2001) 239-270; math.PR/0105107.
[42] I. Stewart, Where drunkards hang out, *Nature* 413 (2001) 686-687.
[43] A. Dembo, Y. Peres, J. Rosen, and O. Zeitouni, Cover times for Brownian motion and random

walks in two dimensions, math.PR/0107191.
[44] H. S. Wilf, The white screen problem, *Amer. Math. Monthly* 96 (1989) 704-707.
[45] J. Wimp and D. Zeilberger, How likely is Pólya's drunkard to stay in $x>y>z$?, *J. Stat. Phys.* 57 (1989) 1129-1135; MR 90m:82083.
[46] J. T. Chang and Y. Peres, Ladder heights, Gaussian random walks and the Riemann zeta function, *Annals of Probab.* 25 (1997) 787-802; MR 98c:60086.
[47] H. Chernoff, Sequential tests for the mean of a normal distribution. IV. (Discrete case), *Annals of Math. Statist.* 36 (1965) 55-68; MR 30 # 681.
[48] T. L. Lai, Asymptotic moments of random walks with applications to ladder variables and renewal theory, *Annals of Probab.* 4 (1976) 51-66; MR 52 # 12086.
[49] D. Siegmund, Corrected diffusion approximations in certain random walk problems, *Adv. Appl. Probab.* 11 (1979) 701-719; MR 80i:60096.
[50] S. Asmussen, P. Glynn, and J. Pitman, Discretization error in simulation of one-dimensional reflecting Brownian motion, *Annals of Appl. Probab.* 5 (1995) 875-896; MR 97e:65156.
[51] E. G. Coffman, P. Flajolet, L. Flatto, and M. Hofri, The maximum of a random walk and its application to rectangle packing, *Probab. Engin. Inform. Sci.* 12 (1998) 373-386; MR 99f:60127.
[52] P. Biane, J. Pitman, and M. Yor, Probability laws related to the Jacobi theta and Riemann zeta functions, and Brownian excursions, *Bull. Amer. Math. Soc.* 38 (2001) 435-465.
[53] A. J. Guttmann, Padé approximants and SAW generating functions, unpublished note (2001).
[54] D. J. Aldous, Self-intersections of random walks on discrete groups, *Math. Proc. Cambridge Philos. Soc.* 98 (1985) 155-177; MR 86j:60157.
[55] D. J. Aldous, Self-intersections of 1-dimensional random walks, *Probab. Theory Relat. Fields* 72 (1986) 559-587; MR 88a:60125.
[56] R. S. Lehman and G. H. Weiss, A study of the restricted random walk, *J. SIAM* 6 (1958) 257-278; MR 20 # 4891.
[57] S. Hemmer and P. C. Hemmer, An average self-avoiding random walk on the square lattice lasts 71 steps, *J. Chem. Phys.* 81 (1984) 584-585; MR 85g:82099.
[58] K. Burdzy, G. F. Lawler, and T. Polaski, On the critical exponent for random walk intersections, *J. Stat. Phys.* 56 (1989) 1-12; MR 91h:60073.
[59] G. F. Lawler, *Intersections of Random Walks*, Birkhäuser, 1991; MR 92f:60122.
[60] G. Slade, Random walks, *Amer. Scientist*, v. 84 (1996) n. 2, 146-153.
[61] G. F. Lawler, O. Schramm, and W. Werner, Values of Brownian intersection exponents. II: Plane exponents, *Acta Math.* 187 (2001) 275-308; math.PR/0003156.
[62] G. F. Lawler, A self-avoiding random walk, *Duke Math. J.* 47 (1980) 655-693; MR 81j:60081.
[63] T. Prellberg, Intersection probabilities for high dimensional walks, unpublished note (2002).
[64] D. Zeilberger, A holonomic systems approach to special functions identities, *J. Comput. Appl. Math.* 32 (1990) 321-348; MR 92b:33014.
[65] H. S.Wilf and D. Zeilberger, Towards computerized proofs of identities, *Bull. Amer. Math. Soc.* 23 (1990) 77-83; MR 91a:33003.
[66] W. Wasow, *Asymptotic Expansions for Ordinary Differential Equations*, Wiley, 1965, pp. 1-29; MR 34 # 3401.
[67] E. L. Ince, *Ordinary Differential Equations*, Dover, 1956, pp. 356-365; MR 6,65f.
[68] G. F. Simmons, *Differential Equations with Applications and Historical Notes*, McGraw-Hill, 1972, pp. 153-174; MR 58 # 17258.
[69] P. Flajolet and B. Vallée, Continued fractions, comparison algorithms, and fine structure constants, *Constructive, Experimental, and Nonlinear Analysis*, Proc. 1999 Limoges conf., ed. M. Théra, Amer. Math. Soc., 2000, pp. 53-82; INRIA preprint RR4072; MR 2001h:11161.
[70] T. Y. Chow, What is a closed-form number?, *Amer. Math. Monthly* 106 (1999) 440-448; math.NT/9805045; MR 2000e:11156.

5.10 自己回避歩行定数

d 次元空間内のすべての座標が整数である点全体を頂点とする d 次元立方格子を L とする．L 上の n 段の**自己回避歩行**（self-avoiding walk）$\omega^\dagger$ とは，原点からはじまり，$\omega_0=0$ で，すべての j に対して $|\omega_i-\omega_j|=1$ で，かつ，すべての $i \neq j$ で $\omega_i \neq \omega_j$ であるような頂点の列 $\omega_0, \omega_1, \omega_2, \cdots, \omega_n$ のことである．このような歩行（walk）の個数を c_n とする．たとえば，$c_0=1$，$c_1=2d$，$c_2=2d(2d-1)$，$c_3=2d(2d-1)^2$，$c_4=2d(2d-1)^3-2d(2d-2)$ である．普通の歩行よりも，自己回避歩行を研究することは非常に難しい[1-6]．歴史的には，化学での線型ポリマーのモデルとして現れた[7,8]．大きな n に対して，正確な組合せ論的数え上げは不可能である．したがって，解析の方法は，有限の級数展開とモンテカルロシミュレーションである．

簡単のために，d に関する c_n の従属性を伏せておく．いつでも可能なときは関連する定数に対しても，従属性を伏せておく．

c_n の漸近式については何がいえるだろうか？ Fekete（フェケテ）の劣乗法性定理[9-12]の基礎である $c_{n+m} \leq c_n c_m$ が成り立っているから，**結合**（connective）**定数**

$$\mu_d = \lim_{n \to \infty} c_n^{1/n} = \inf_n c_n^{1/n}$$

が存在し，0 でない．$\mu=\mu_d$ を推定する先行研究は，[13-15]にある．詳しい概観については[2]を参照せよ．現在の μ の最も厳密な下界および上界，および知られている限りの最良（best-known）推定値は表 5.2 に与えられている[16-24]．

我々が無知である範囲については，かなり驚きである．すなわち，すべての n とすべての d に対して，$\mu^2=\lim_{n \to \infty} c_{n+2}/c_n$ および $c_{n+1} \geq c_n$ であることを知っているけれども，$2 \leq d \leq 4$ に対する $\mu = \lim_{n \to \infty} c_{n+1}/c_n$ の存在証明は，未解決問題のままである[25,26]．

次の極限が存在し，0 でない正の定数 $\gamma=\gamma_d$ が存在すると信じられている．

表 5.2 結合定数 μ の推定値

d	下界	μ の最良推定値	上界
2	2.6200	2.6381585303	2.6792
3	4.5721	4.68404	4.7114
4	6.7429	6.77404	6.8040
5	8.8285	8.83854	8.8602
6	10.8740	10.87809	10.8886

---

†訳注：自分自身と交わらない歩行のこと．

5.10 自己回避歩行定数

$$A = \begin{cases} \lim_{n\to\infty} \dfrac{c_n}{\mu^n n^{\gamma-1}} & (d \neq 4) \\ \lim_{n\to\infty} \dfrac{c_n}{\mu^n n^{\gamma-1} \ln(n)^{1/4}} & (d=4) \end{cases}$$

臨界指数(critical exponent) γ は,

$$\gamma_2 = \frac{43}{32} = 1.34375, \quad \gamma_3 = 1.1575\cdots, \quad \gamma_4 = 1$$

であると予想されている[27-29]. $d>4$ に対しては,1 に等しいことが証明されている [1,30]. 小さい数 d に対して,上界[1,25,31]は

$$c_n \leq \begin{cases} \mu^n \exp(C n^{1/2}) & (d=2) \\ \mu^n \exp(C n^{2/(d+2)} \ln(n)) & (3 \leq d \leq 4) \end{cases}$$

である. A の存在を証明するにはほど遠い. $d=5$ に対して $1 \leq A \leq 1.493$ であり,充分大きな d に対して $A = 1 + 1/2d + 1/d^2 + O(1/d^3)$ であることが知られている[32].

別のおもしろい研究対象は,**端から端までの平均平方距離**(mean square end-to-end distance)

$$r_n = \mathrm{E}(|\omega_n|^2) = \frac{1}{c_n} \sum_\omega |\omega_n|^2$$

である.ただし,和は L 上のすべての n 段自己回避歩行 ω にわたってとる. c_n と同様に,次の極限が存在し,0 でない正の定数 $v = v_d$ が存在すると信じられている.

$$B = \begin{cases} \lim_{n\to\infty} \dfrac{r_n}{n^{2v}} & (d \neq 4) \\ \lim_{n\to\infty} \dfrac{r_n}{n^{2v} \ln(n)^{1/4}} & (d=4) \end{cases}$$

前のように,

$$v_2 = \frac{3}{4} = 0.75, \quad v_3 = 0.5877\cdots, \quad v_4 = \frac{1}{2} = 0.5$$

と予想されている[27,33,34]. $d>4$ に対して, $v=1/2$ であることが証明されている [1,30]. この最後の値は,ポリヤの酔歩の値と同じである.すなわち,自己回避の制約は,高い次元ではほとんど影響しない. $d=5$ に対して $1.098 \leq B \leq 1.803$ であり,充分大きな d に対して, $B = 1 + 1/d + 2/d^2 + O(1/d^3)$ であることが知られている[32]. したがって,自己回避歩行は,ポリヤの酔歩よりも速く原点から遠ざかっていく.ただしそれは振幅のレベルについてだけであり,指数のレベルについてではない.

予想漸近式 $c_n \sim A\mu^n n^{\gamma-1}$ および $r_n \sim Bn^{2v}$ が真 ($d \neq 4$ に対して) であることを認めると,表 5.3 にある計算が可能となる[23,24,33,35-37].

($d=4$ に対する対数補正 (logarithmic correction) は, A または B の信頼できる推

表5.3 振幅 A と B の推定値

d	A の推定値	B の推定値	d	A の推定値	B の推定値
2	1.177043	0.77100	5	1.275	1.4767
3	1.205	1.21667	6	1.159	1.2940

定値をとても難しくする．）応用がある．2つの歩行 ω と ω' が**交差** (intersect) するとは，0でないある i と j に対して $\omega_i = \omega'_j$ となるときをいう．2つの n 段の独立なランダム自己回避歩行が決して交差しない確率は[1,38]，$n \to \infty$ のとき，

$$\frac{c_{2n}}{c_n^2} \sim \begin{cases} A^{-1} 2^{\gamma-1} n^{1-\gamma} \to 0 & (2 \leq d \leq 3) \\ A^{-1} \ln(n)^{-1/4} \to 0 & (d=4) \\ A^{-1} > 0 & (d \geq 5) \end{cases}$$

である．予想されたふるまいは，直感と一致する．自己回避歩行は空間内でよりまばらに散らばるので，c_{2n}/c_n^2 は普通の歩行に対応する確率 q_n よりも（ほんの少し）大きい【5.9.1】．

歩行の大きさの，他のおもしろい測度は，**平均平方回転半径** (mean square radius of gyration)

$$s_n = \mathrm{E}\left(\frac{1}{n+1}\sum_{i=0}^{n}\left|\omega_i - \frac{1}{n+1}\sum_{j=0}^{n}\omega_j\right|^2\right) = \mathrm{E}\left(\frac{1}{2(n+1)^2}\sum_{i=0}^{n}\sum_{j=0}^{n}|\omega_i - \omega_j|^2\right)$$

と，**端点からのモノマーの平均平方距離** (mean square distance of a monomer from the endpoints)

$$t_n = \mathrm{E}\left(\frac{1}{n+1}\sum_{i=0}^{n}\frac{|\omega_i|^2 + |\omega_n - \omega_i|^2}{2}\right)$$

である．回転半径は，たとえば弱い散乱によって希薄溶液におけるポリマーに対して実験的に測ることができる．しかし，端から端までの距離は，理論的な単純さが好まれる[33,39-41]．ν を r_n に関する同じ指数とする．このとき，$s_n \sim En^{2\nu}$，$t_n \sim Fn^{2\nu}$ および，$d=2$ のとき $E/B = 0.14026\cdots$，$F/B = 0.43961\cdots$，$d=3$ のとき $E/B = 0.1599\cdots$，と予想されている．

この議論を d 次元空間の任意の格子 L に一般化できる．たとえば，$d=2$ の場合，正三角形格子に対して，厳密な上界 $\mu < 4.278$ およびある推定値 $\mu = 4.1507951\cdots$ がある[17,35,42-45]．正六角（蜂の巣）格子に対して，$\mu = \sqrt{2+\sqrt{2}} = 1.8477590650\cdots$ である，と予想されている[46-48]．しかしながら，臨界指数 γ，ν と振幅比 E/B，F/B については，これらは格子の構造に独立（次元には従属）である，という意味で「普遍的」(universal) であると考えられている．だから，重要な挑戦は，このような指数と比の性質をよりよく理解することであり，また低次元においてこれらの存在を疑いなく厳密に証明することである．

5.10.1 多角形と小道

以前に与えた結合定数 μ の値は，自己回避歩行の個数の漸近的増大に適用できるだけでなく，あらかじめ決められた端点をもつ，**自己回避多角形** (self-avoiding polygons) およびあらかじめ定めた端点をもつ自己回避歩行の個数の漸近的増大にも適用できる[2,49]．自己回避多角形に関連した，格子動物すなわちポリオミノの議論に関しては【5.19】を参照せよ．

自己回避歩行においては，1度より多く訪れる場所 (site) や枝 (bond) はない．対

比として，**自己回避小道**（self-avoiding trail）は，場所を再び訪れてもよいが，枝は再訪しない．だから，歩行は小道の真部分集合である[50-55]．小道の個数 h_n は，$h_n \sim G\lambda^n n^{\gamma-1}$ を満たすと予想されている．ただし，γ は c_n に関する同じ指数である．結合定数 λ は前のようにたぶん存在する．実際，$\lambda \geq \mu$ を満たす．正方格子に対して，厳密な範囲 $2.634 < \lambda < 2.851$ および推定値 $\lambda = 2.72062\cdots$ がある．振幅は近似的に $G = 1.272\cdots$ である．立方格子に対して，上界が $\lambda < 4.929$ であり，推定値は $\lambda = 4.8426\cdots$ である．多くの関連する問題が可能である．

5.10.2 チェス盤上のルークの通り道

$m \times n$ チェス盤上の固定された端から，チェス盤から決して離れないで反対側の端へ行くとき，ルーク[†]はどのくらい多くの数の自己回避歩行をすることができるだろうか？ このような道の個数を $p_{m-1,n-1}$ とする．明らかに，$p_{k,1} = 2^k$，$p_{2,2} = 12$ であり[56-58]，$k \to \infty$ のとき，

$$p_{k,2} \sim \frac{4+\sqrt{13}}{2\sqrt{13}} \left(\sqrt{\frac{3+\sqrt{13}}{2}} \right)^{2k} = 1.0547001962\cdots (1.8173540210\cdots)^{2k}$$

である．もっと一般的に，数列 $\{p_{k,l}\}_{k=1}^{\infty}$ の母関数は任意の整数 $l \geq 1$ に対して有理関数である．したがって，関連のある漸近式の係数はすべて代数的数である．$k \to \infty$ のときの $p_{k,k}$ の漸近式については何が言えるだろうか？ Whittington & Guttmann（ウィティングトン&ガットマン）[59]は，

$$p_{k,k} \sim (1.756\cdots)^{k^2}$$

を証明し，次のことを予想した[60,61]．$\pi_{j,k}$ を，母関数

$$P_k(x) = \sum_{j=1}^{\infty} \pi_{j,k} x^j, \quad P_k(1) = p_{k,k}$$

をもつ j 段の道の個数とする．このとき，

$0 < \lim_{k \to \infty} P_k(x)^{1/k} < 1$ が，$0 < x < \mu^{-1} = 0.3790522777\cdots$ に対して存在する．

$\lim_{k \to \infty} P_k(\mu^{-1})^{1/k} = 1$

$1 < \lim_{k \to \infty} P_k(x)^{1/k^2} < \infty$ が，$x > \mu^{-1}$ に対して存在する．

という意味で，「相転移」(phase transition) が存在する．

証明は Madras（マドラス）[62]により与えられた．これは，結合定数 $\mu = \mu_2$ のおもしろい出現である．d 次元チェス盤上の類似の定理もまた成り立ち，当然 μ_d を使用する．

5.10.3 うねり図形とスタンプの折り曲げ

位数 n の**メアンダー**[††]（うねり図形，meander）とは，無限の直線（川）が $2n$ 回

[†]訳注：飛将．将棋の飛車と同じ動きをする．
[††]訳注：ギリシャのメアンドロス川が蛇行していたことから，蛇行する川あるいは曲がりくねった道の意味．ここではうねり図形と仮訳する．

図 5.11 次数 3 のうねり図形は 8 個あり，次数 5 の半うねり図形は 10 個ある．川を横切る反転は除外した．

($2n$ の橋で）交わる，平面上の自己交差しない閉路（self-avoiding loop）のことである．2 つのうねり図形は，橋（交点）を固定しながら，他に連続的に変形できるとき，同値であると定義する．位数 n の同値でないうねり図形の個数 M_n は，$M_1=1$, $M_2=2$, $M_3=8$, $M_4=42$, $M_5=262$, ⋯ である．

位数 n の**半うねり図形**（semi-meander）とは，半無限直線（源流をもった川）が n 回（n 個の橋で）交わるような平面上の自己交差しない閉路のことである．半うねり図形の同値関係は，同じように定義される．位数 n の同値でない半うねり図形の個数 $\tilde{M}_n$ は，$\tilde{M}_1=1$, $\tilde{M}_2=1$, $\tilde{M}_3=2$, $\tilde{M}_4=4$, $\tilde{M}_5=10$, ⋯ である．

うねり図形および半うねり図形を数えることに，多くの注意が払われている[63-73]．図 5.11 を見よ．

前のように，漸近的ふるまい

$$M_n \sim C \frac{R^{2n}}{n^\alpha}, \quad \tilde{M}_n \sim \tilde{C} \frac{R^n}{n^{\tilde{\alpha}}}$$

が成り立つことを期待する．ただし，$R=3.501838\cdots$，すなわち $R^2=12.262874\cdots$ である．結合定数 R に対する正確な式は知られていない．対照的に，臨界指数は，

$$\alpha = \sqrt{29}\,\frac{\sqrt{29}+\sqrt{5}}{12} = 3.4201328816\cdots$$

$$\tilde{\alpha} = 1+\sqrt{11}\,\frac{\sqrt{29}+\sqrt{5}}{24} = =2.0531987328\cdots$$

だろうという予想[74-76]がある．しかし，半うねり図形の臨界指数の値については，疑問があがっている[77-79]．さらに，数列 $\tilde{M}_n$ と M_n は，線型または環状の切手の列を 1 枚の切手の上に折りたたむ方法の数，を数える問題に関連している[80-87]．

[1] N. Madras and G. Slade, *The Self-Avoiding Walk*, Birkhäuser, 1993; MR 94f:82002.
[2] B. D. Hughes, *Random Walks and Random Environments*, v. 1, Oxford Univ. Press, 1995; MR 96i:60070.

[3] G. Slade, Self-avoiding walks, *Math. Intellig.* 16 (1994) 29-35.
[4] G. Slade, Random walks, *Amer. Scientist*, v. 84 (1996) n. 2, 146-153.
[5] B. Hayes, How to avoid yourself, *Amer. Scientist*, v. 86 (1998) n. 4, 314-319.
[6] N. J. A. Sloane, On-Line Encyclopedia of Integer Sequences, A001334, A001411, A001412, and A001668.
[7] P. J. Flory, The configuration of real polymer chains, *J. Chem. Phys.* 17 (1949) 303-310.
[8] K. F. Freed, Polymers as self-avoiding walks, *Annals of Probab.* 9 (1981) 537-556;MR 82j:60123.
[9] M. Fekete, Über die Verteilung der Wurzeln bei gewissen algebraischen Gleichungen mit ganzzahligen Koeffizienten, *Math. Z.* 17 (1923) 228-249.
[10] G. Pólya and G. Szegö, *Problems and Theorems in Analysis*, v. 1, Springer-Verlag, 1972, ex. 98, pp. 23, 198;MR 81e:00002.
[11] J. M. Hammersley and K. W. Morton, Poor man's Monte Carlo, *J. Royal Statist. Soc. Ser. B*. 16 (1954) 23-38;discussion 61-75;MR 16,287i.
[12] J. M. Hammersley, Percolation processes. II: The connective constant, *Proc. Cambridge Philos. Soc*. 53 (1957) 642-645;MR 19,989f.
[13] M. E. Fisher and M. F. Sykes, Excluded-volume problem and the Ising model of ferromagnetism, *Phys. Rev.* 114 (1959) 45-58;MR 22 # 10274.
[14] W. A. Beyer and M. B. Wells, Lower bound for the connective constant of a self-avoiding walk on a square lattice, *J. Combin. Theory Ser. A* 13 (1972) 176-182;MR 46 # 1605.
[15] F. T. Wall and R. A. White, Macromolecular configurations simulated by random walks with limited orders of non-self-intersections, *J. Chem. Phys.* 65 (1976) 808-812.
[16] T. Hara, G. Slade, and A.D. Sokal, New lower bounds on the self-avoiding-walk connective constant, *J. Stat. Phys.* 72 (1993) 479-517;erratum 78 (1995) 1187-1188;MR 94e:82053.
[17] S. E. Alm, Upper bounds for the connective constant of self-avoiding walks, *Combin. Probab. Comput.* 2 (1993) 115-136;MR 94g:60126.
[18] A. R. Conway and A. J. Guttmann, Lower bound on the connective constant for square lattice self-avoiding walks, *J. Phys. A* 26 (1993) 3719-3724.
[19] J. Noonan, New upper bounds for the connective constants of self-avoiding walks, *J. Stat. Phys.* 91 (1998) 871-888;MR 99g:82030.
[20] J. Noonan and D. Zeilberger, The Goulden-Jackson cluster method: Extensions, applications and implementations, *J. Differ. Eq. Appl.* 5 (1999) 355-377;MR 2000f:05005.
[21] I. Jensen and A. J. Guttmann, Self-avoiding polygons on the square lattice, *J. Phys. A* 32 (1999) 4867-4876.
[22] A. Pönitz and P. Tittmann, Improved upper bounds for self-avoiding walks in Z^d, *Elec. J. Combin.* 7 (2000) R21;MR 2000m:05015.
[23] D. MacDonald, S. Joseph, D. L. Hunter, L. L. Moseley, N. Jan, and A. J. Guttmann, Self-avoiding walks on the simple cubic lattice, *J. Phys. A* 33 (2000) 5973-5983;MR 2001g:82051.
[24] A. L. Owczarek and T. Prellberg, Scaling of self-avoiding walks in high dimensions, *J. Phys. A* 34 (2001) 5773-5780;cond-mat/0104135.
[25] H. Kesten, On the number of self-avoiding walks, *J. Math. Phys.* 4 (1963) 960-969;5 (1964) 1128-1137;MR 27 # 2006 and 29 # 4118.
[26] G. L. O'Brien, Monotonicity of the number of self-avoiding walks, *J. Stat. Phys.* 59 (1990) 969-979;MR 91k:8007.
[27] P. Butera and M. Comi, N-vector spin models on the simple-cubic and the body-centeredcubic lattices: A study of the critical behavior of the susceptibility and of the correlation length by high-temperature series extended to order 21, *Phys. Rev. B* 56 (1997) 8212-8240;hep-lat/9703018.
[28] S. Caracciolo, M. S. Causo, and A. Pelissetto, Monte Carlo results for three-dimensional self-avoiding walks, *Nucl. Phys. Proc. Suppl.* 63 (1998) 652-654;hep-lat/9711051.
[29] S. Caracciolo, M. S. Causo, and A. Pelissetto, High-precision determination of the critical exponent gamma for self-avoiding walks, *Phys. Rev. E* 57 (1998) R1215-R1218;condmat/9703250.

[30] T. Hara and G. Slade, Self-avoiding walk in five or more dimensions. I: The critical behaviour, *Commun. Math. Phys.* 147 (1992) 101-136;MR 93j:82032.
[31] J. M. Hammersley and D. J. A. Welsh, Further results on the rate of convergence to the connective constant of the hypercubical lattice, *Quart. J. Math.* 13 (1962) 108-110;MR 25 # 2967.
[32] T. Hara and G. Slade, The self-avoiding-walk and percolation critical points in high dimensions, *Combin. Probab. Comput.* 4 (1995) 197-215;MR 96i:82081.
[33] B. Li,N. Madras, and A.D. Sokal, Critical exponents, hyperscaling and universal amplitude ratios for two- and three-dimensional self-avoiding walks, *J. Stat. Phys.* 80 (1995) 661-754;hep-lat/9409003;MR 96e:82046.
[34] T. Prellberg, Scaling of self-avoiding walks and self-avoiding trails in three dimensions, *J. Phys. A* 34 (2001) L599-L602.
[35] I. G. Enting and A. J. Guttmann, Self-avoiding rings on the triangular lattice, *J. Phys. A* 25 (1992) 2791-2807;MR 93b:82030.
[36] A. R. Conway, I. G. Enting, and A. J. Guttmann, Algebraic techniques for enumerating self-avoiding walks on the square lattice, *J. Phys. A* 26 (1993) 1519-1534;MR 94f: 82040.
[37] A. R. Conway and A. J. Guttmann, Square lattice self-avoiding walks and corrections-toscaling, *Phys. Rev. Lett.* 77 (1996) 5284-5287.
[38] G. F. Lawler, *Intersections of Random Walks*, Birkhäuser, 1991;MR 92f:60122.
[39] S. Caracciolo, A. Pelissetto, and A. D. Sokal, Universal distance ratios for two-dimensional self-avoiding walks: Corrected conformal-invariance predictions, *J. Phys. A* 23 (1990) L969-L974;MR 91g:81130.
[40] J. L. Cardy and A. J. Guttmann, Universal amplitude combinations for self-avoiding walks, polygons and trails, *J. Phys. A* 26 (1993) 2485-2494;MR 94f:82039.
[41] K. Y. Lin and J. X. Huang, Universal amplitude ratios for self-avoiding walks on the Kagomé lattice, *J. Phys. A* 28 (1995) 3641-3643.
[42] A. J. Guttmann, On two-dimensional self-avoiding random walks, *J. Phys. A* 17 (1984) 455-468; MR 85g:82096.
[43] T. Ishinabe, Critical exponents and corrections to scaling for the two-dimensional selfavoiding walk, *Phys. Rev. B* 37 (1988) 2376-2379.
[44] T. Ishinabe, Reassessment of critical exponents and corrections to scaling for self-avoiding walks, *Phys. Rev. B* 39 (1989) 9486-9495.
[45] A. J. Guttmann and J. Wang, The extension of self-avoiding random walk series in two dimensions, *J. Phys. A* 24 (1991) 3107-3109;MR 92b:82073.
[46] B. Nienhuis, Exact critical point and critical exponents of O(n) models in two dimensions, *Phys. Rev. Lett.* 49 (1982) 1062-1065.
[47] I. G. Enting and A. J. Guttmann, Polygons on the honeycomb lattice, *J. Phys. A* 22 (1989) 1371-1384;MR 90c:82050.
[48] D. MacDonald, D. L. Hunter, K. Kelly, and N. Jan, Self-avoiding walks in two to five dimensions: Exact enumerations and series study, *J. Phys. A* 25 (1992) 1429-1440.
[49] N. J. A. Sloane, On-Line Encyclopedia of Integer Sequences, A006817, A006818, A006819, and A006851.
[50] A. Malakis, The trail problem on the square lattice, *J. Phys. A* 9 (1976) 1283-1291.
[51] A. J. Guttmann and S. S. Manna, The geometric properties of planar lattice trails, *J. Phys. A* 22 (1989) 3613-3619.
[52] A. J. Guttmann, Lattice trails. I: Exact results, *J. Phys. A* 18 (1985) 567-573;MR87f:82069.
[53] A. J. Guttmann, Lattice trails. II: Numerical results, *J. Phys. A* 18 (1985) 575-588;MR 87f:82070.
[54] A. R. Conway and A. J. Guttmann, Enumeration of self-avoiding trails on a square lattice using a transfer matrix technique, *J. Phys. A* 26 (1993) 1535-1552;MR 94f:82041.
[55] N. J. A. Sloane, On-Line Encyclopedia of Integer Sequences, A001335, A001413, A002931, and A036418.
[56] N. J. A. Sloane, On-Line Encyclopedia of Integer Sequences, A006192, A007764, A007786, and

A007787.
[57] H. L. Abbott and D. Hanson, A lattice path problem, *Ars Combin.* 6 (1978) 163-178;MR 80m:05009.
[58] K. Edwards, Counting self-avoiding walks in a bounded region, *Ars Combin.* 20 B (1985) 271-281;MR 87h:05015.
[59] S. G. Whittington and A. J. Guttmann, Self-avoiding walks which cross a square, *J. Phys. A* 23 (1990) 5601-5609;MR 92e:82038.
[60] T.W. Burkhardt and I. Guim, Self-avoiding walks that cross a square, *J. Phys. A* 24 (1991) L1221-L1228.
[61] J. J. Prentis, Renormalization theory of self-avoiding walks which cross a square, *J. Phys. A* 24 (1991) 5097-5103.
[62] N. Madras, Critical behaviour of self-avoiding walks that cross a square, *J. Phys. A* 28 (1995) 1535-1547;MR 96d:82032.
[63] H. Poincaré, Sur un théorème de géométrie, *Rend. Circ. Mat. Palmero* 33 (1912) 375-407;also in Oeuvres, t. 6, Gauthier-Villars, 1953, pp. 499-538.
[64] J. Touchard, Contribution à l'étude du problème des timbres poste, *Canad. J. Math.* 2 (1950) 385-398;MR 12,312i.
[65] V. I. Arnold, The branched covering $CP^2 \to S^4$, hyperbolicity and projective topology (in Russian), *Sibirskii Mat. Zh.*, v. 29 (1988) n. 5, 36-47, 237;Engl. transl. in *Siberian Math. J*. 29 (1988) 717-726;MR 90a:57037.
[66] S. K. Lando and A. K. Zvonkin, Meanders, *Selecta Math. Soviet*. 11 (1992) 117-144;MR 93k:05013.
[67] S. K. Lando and A. K. Zvonkin, Plane and projective meanders, *Theoret. Comput. Sci*. 117 (1993) 227-241;MR 94i:05004.
[68] P. Di Francesco, O. Golinelli, and E. Guitter, Meanders: A direct enumeration approach, *Nucl. Phys. B* 482 (1996) 497-535;hep-th/9607039;MR 97j:82074.
[69] P. Di Francesco, O. Golinelli, and E. Guitter, Meander, folding and arch statistics, *Math. Comput. Modelling* 26 (1997) 97-147;hep-th/9506030;MR 99f:82029.
[70] P. Di Francesco, Meander determinants, *Commun. Math. Phys*. 191 (1998) 543-583;hepth/9612026;MR 99e:05007.
[71] M. G. Harris, A diagrammatic approach to the meander problem, hep-th/9807193.
[72] J. Reeds and L. Shepp, An upper bound on the meander constant, unpublished note (1999).
[73] O. Golinelli, A Monte-Carlo study of meanders, *Europ. Phys. J. B* 14 (2000) 145-155;cond-mat/9906329.
[74] P. Di Francesco, O. Golinelli, and E. Guitter, Meanders: Exact asymptotics, *Nucl. Phys. B* 570 (2000) 699-712;cond-mat/9910453;MR 2001f:82032.
[75] P. Di Francesco, Folding and coloring problems in mathematics and physics, *Bull. Amer. Math. Soc*. 37 (2000) 251-307;MR 2001g:82004.
[76] P. Di Francesco, E. Guitter, and J. L. Jacobsen, Exact meander asymptotics: A numerical check, *Nucl. Phys. B* 580 (2000) 757-795;MR 2002b:82024.
[77] I. Jensen, Enumerations of plane meanders, cond-mat/9910313.
[78] I. Jensen, A transfer matrix approach to the enumeration of plane meanders, *J. Phys. A* 33 (2000) 5953-5963;cond-mat/0008178;MR 2001e:05008.
[79] I. Jensen and A. J. Guttmann, Critical exponents of plane meanders, *J. Phys. A* 33 (2000) L187-L192;MR 2001b:82024.
[80] W. F. Lunnon, A map-folding problem, *Math. Comp*. 22 (1968) 193-199;MR 36 #5009.
[81] J. E. Koehler, Folding a strip of stamps, *J. Combin. Theory* 5 (1968) 135-152;errata *J. Combin. Theory Ser. A* 19 (1975) 367;MR 37 #3945 and MR 52 #2910.
[82] W. F. Lunnon, Multi-dimensional map-folding, *Comput. J*. 14 (1971) 75-80;MR44 #2627.
[83] W. F. Lunnon, Bounds for the map-folding function, unpublished note (1981).
[84] M. Gardner, The combinatorics of paper-folding, *Wheels, Life, and Other Mathematical Amusements*, W. H. Freeman, 1983, pp. 60-73.

[85] M. Gardner, *The Sixth Book of Mathematical Games from Scientific American*, Univ. of Chicago Press, 1984, pp. 21, 26-27.
[86] N. J. A. Sloane, My favorite integer sequences, *Sequences and Their Applications (SETA)*, Proc. 1998 Singapore conf., ed. C. Ding, T. Helleseth, and H. Niederreiter, Springer-Verlag, 1999, pp. 103-130;math.CO/0207175.
[87] N. J. A. Sloane, On-Line Encyclopedia of Integer Sequences, A000682, A001011, A005315, and A005316.

5.11 フェラーの硬貨投げ定数

理想的な硬貨を n 回独立に投げるとき，表が連続して 3 回出ない確率を w_n とする．明らかに，$w_0=w_1=w_2=1$ であり，$n \geq 3$ に対して $w_n=w_{n-1}/2+w_{n-2}/4+w_{n-3}/8$ であり，$\lim_{n \to \infty} w_n=0$ である．Feller（フェラー）[1]は，より正確な次の漸近式を証明した．

$$\lim_{n \to \infty} w_n \alpha^{n+1} = \beta$$

である．ただし，

$$\alpha = \frac{(136+24\sqrt{33})^{1/3} - 8(136+24\sqrt{33})^{-1/3} - 2}{3} = 1.0873780254\cdots$$

であり，

$$\beta = \frac{2-\alpha}{4-3\alpha} = 1.2368398446\cdots$$

である．
まず，これらの式を一般化する．$k>1$ とし，k 回続けて表が出ることはないとする．このとき，類似の定数は[1,2]

α は，$1-x+\left(\dfrac{x}{2}\right)^{k+1}=0$ の最小の正の解であり，$\beta=\dfrac{2-\alpha}{k+1-k\alpha}$

である．同値なことであるが，k 回続けて表が出ない，コイン投げの数列を数える母関数は[3]，

$$S_k(z) = \frac{1-z^k}{1-2z+z^{k+1}}, \quad \frac{1}{n!} \frac{d^n}{dz^n} S_k(z) \bigg|_{z=0} \sim \frac{\beta}{\alpha} \left(\frac{2}{\alpha}\right)^n$$

である．組合せ論的な性質の，さらなる資料については[4-8]を参照せよ．
硬貨が理想的でないとする．すなわち，$P(H)=p$, $P(T)=q$, $p+q=1$ であるが，p と q は等しくないとする．このとき，w_n の漸近的ふるまいは，

α は，$1-x+qp^kx^{k+1}=0$ の最小の正の解であり，$\beta=\dfrac{1-p\alpha}{(k+1-k\alpha)q}$

によって，決定される．
さらなる一般化は，斉時 2 状態（time-homogeneous two-state）のマルコフ連鎖を

含む．ここでは硬貨投げという表現をするのは，あまり意味がない．だから，異なる応用に注意を向けなければならない．大気を通しての観測線は，どんよりした曇 (0) または快晴 (1) かどうかを，1時間に1回地面に置かれたセンサーで決定されると仮定する．気象事象はふつう時間に関して頑健性があるので，センサーの観測値は独立でない．観測値の時系列 $X_1, X_2, X_3, \ldots$ の簡単なモデルは，推移確率

$$\begin{pmatrix} P(X_{j+1}=0 \mid X_j=0) & P(X_{j+1}=1 \mid X_j=0) \\ P(X_{j+1}=0 \mid X_j=1) & P(X_{j+1}=1 \mid X_j=1) \end{pmatrix} = \begin{pmatrix} \pi_{00} & \pi_{01} \\ \pi_{10} & \pi_{11} \end{pmatrix}$$

をもつマルコフ連鎖と考えられる．ただし，条件付き確率変数は $\pi_{00}+\pi_{01}=1=\pi_{10}+\pi_{11}$ を満たす．$\pi_{00}=\pi_{10}$ かつ $\pi_{01}=\pi_{11}$ の特別な場合は，硬貨投げに関連して議論した Bernoulli（ベルヌーイ）試行のシナリオと同値である．長さ $k>1$ の曇った区間が起こらない確率を $w_{n,k}$ とする．初めに $P(X_0=1)=\theta_1$ と仮定する．漸近的ふるまいは，前と同様である．α は方程式 [9, 10]

$$1-(\pi_{11}+\pi_{00})x+(\pi_{11}-\pi_{01})x^2+\pi_{10}\pi_{01}\pi_{11}^{k-1}x^{k+1}=0$$

の最小の正の解であり，

$$\beta = \frac{[-1+(2\pi_{11}-\pi_{01})\alpha-(\pi_{11}-\pi_{01})\pi_{11}\alpha^2][\theta_1+(\pi_{01}-\theta_1)\alpha]}{\pi_{10}\pi_{01}[-1-k+(\pi_{11}+\pi_{00})k\alpha+(\pi_{11}-\pi_{01})(1-k)\alpha^2]}$$

である．分子生物学に応用されるパターン統計の解析のための，一般的な技法については [11] を参照せよ．

この問題に関して，多くの変形があるうちの1つを議論する．すべての女の子の隣には，少なくとも1人の他の女の子がいるように，1列になった n 人の子供のパターンはどれだけあるのか？ この答えを Y_n とすると，$Y_1=1$，$Y_2=2$，$Y_3=4$ であり，$n \geq 4$ に対して $Y_n = 2Y_{n-1}-Y_{n-2}+Y_{n-3}$ が成り立つ．したがって，

$$\lim_{n \to \infty} \frac{Y_{n+1}}{Y_n} = \frac{(100+12\sqrt{69})^{1/3}+4(100+12\sqrt{69})^{-1/3}+4}{6} = 1.7548776662\cdots$$

が成り立つ．この一般化，すなわち女の子が少なくとも k 人のグループで現れなければならない，という場合は [12, 13] で与えられる．類似の3次無理数は【1.2.2】に現れている．

硬貨投げに戻ろう．n 回の理想的な硬貨投げの数列において，連続して表が出る最長の連の長さの期待値は何になるのか？ 答えは驚くほど複雑である [14-21]．$n \to \infty$ のとき，

$$\sum_{k=1}^{n}(1-w_{n,k}) = \frac{\ln(n)}{\ln(2)}-\left(\frac{3}{2}-\frac{\gamma}{\ln(2)}\right)+\delta(n)+o(1)$$

が成り立つ．ただし，γ はオイラー–マスケロニの定数であり，

$$\delta(n) = \frac{1}{\ln(2)}\sum_{\substack{k=-\infty \\ k \neq 0}}^{n} \Gamma\left(\frac{2\pi i k}{\ln(2)}\right) \exp\left(-2\pi i k \frac{\ln(n)}{\ln(2)}\right)$$

である．すなわち，期待長さは，$\ln(n)/\ln(2)-0.6672538227\cdots$ プラス「小さく振幅する修正項」である．関数 $\delta(n)$ は，周期的（$\delta(n)=\delta(2n)$）で，平均値0で，「無視できる」（すべての n に対して $|\delta(n)|<1.574\times 10^{-6}$）関数である．対応する分散は

$C+c+\varepsilon(n)+o(1)$ である．ただし，$\varepsilon(n)$ は別の小さい振幅をもつ関数であり，

$$C=\frac{1}{12}+\frac{\pi^2}{6\ln(2)^2}=3.5070480758\cdots$$

$$c=\frac{2}{\ln(2)}\sum_{k=0}^{n}\ln\left[1-\exp\left(-\frac{2\pi^2}{\ln(2)}(2k+1)\right)\right]=(-1.237412\cdots)\times10^{-12}$$

である．$\delta(n)$ や $\varepsilon(n)$ と類似した関数は，【2.3】，【2.16】，【5.6】，【5,14】にある．

さらに，n 個の理想的な硬貨を投げるとき，最初に投げて裏が出た硬貨だけを次に投げる．2番目に投げて裏が出た硬貨だけを再び投げる，以下同様に続ける．最後に投げた硬貨がちょうど1枚である確率は何であるのか？ 再び，答えは複雑である[22-25]．$n\to\infty$ のとき

$$\frac{n}{2}\sum_{j=0}^{\infty}2^{-j}(1-2^{-j})^{n-1}\sim\frac{1}{2\ln(2)}+\rho(n)+o(1)$$

が成り立つ．ただし，

$$\rho(n)=\frac{1}{2\ln(2)}\sum_{\substack{k=-\infty\\k\neq0}}^{n}\Gamma\left(1-\frac{2\pi ik}{\ln(2)}\right)\exp\left(2\pi ik\frac{\ln(n)}{\ln(2)}\right)$$

である．すなわち，最終に1枚だけ残る（引き分けがない）確率は，$1/(2\ln(2))=0.7213475204\cdots$ プラス「すべての n に対して $|\rho(n)|<7.131\times10^{-6}$ を満たす振動関数」である．n 個の硬貨投げ数列の最も長い連の，期待長さは $\sum_{j=0}^{\infty}[1-(1-2^{-j})^n]$ であり，同じように解析される[26]．関連する議論は[27-31]にある．

[1] W. Feller, *An Introduction to Probability Theory and Its Applications*, v. 1, 3rd ed., Wiley, 1968, pp. 278-279, 322-325; MR 37 #3604.

[2] W. Feller, Fluctuation theory of recurrent events, *Trans. Amer. Math. Soc.* 67 (1949) 98-119; MR 11,255c.

[3] R. Sedgewick and P. Flajolet, *Introduction to the Analysis of Algorithms*, Addison-Wesley, 1996, pp. 366-373.

[4] R. A. Howard, *Dynamic Probabilistic Systems: Markov Models*, v. 1, Wiley, 1971.

[5] A. N. Philippou and F. S. Makri, Successes, runs and longest runs, *Statist. Probab. Lett.* 4 (1986) 211-215; MR 88c:60035b.

[6] K. Hirano, Some properties of the distributions of order *k*, *Fibonacci Numbers and Their Applications*, Proc. 1984 Patras conf., ed. A. N. Philippou, G. E. Bergum, and A. F. Horadam, Reidel, 1986; MR 88b:62031.

[7] A. P. Godbole, Specific formulae for some success run distributions, *Statist. Probab. Lett.* 10 (1990) 119-124; MR 92e:60021.

[8] K. A. Suman, The longest run of any letter in a randomly generated word, *Runs and Patterns in Probability: Selected Papers*, ed. A. P. Godbole and S. G. Papastavnidis, Kluwer, 1994, pp. 119-130; MR 95k:60028.

[9] M. B. Rajarshi, Success runs in a two-state Markov chain, *J. Appl. Probab.* 11 (1974) 190-192; correction 14 (1977) 661; MR 51 #4402 and MR 57 #7777.

[10] C. Banderier and M. Vandenbogaert, A Markovian generalization of Feller's coin tossing constants, unpublished note (2000).

[11] P. Nicodème, B. Salvy, and P. Flajolet, Motif statistics, *Proc. 1999 European Symp. on Algorithms (ESA)*, Prague, ed. J. Nesetril, Lect. Notes in Comp. Sci. 1643, Springer-Verlag, 1999, pp. 194-211; INRIA research report RR-3606.

[12] J. H. Conway and R. K. Guy, *The Book of Numbers*, Springer-Verlag, 1996, pp. 205-206; MR 98g:00004.

[13] R. Austin and R. Guy, Binary sequences without isolated ones, *Fibonacci Quart.* 16 (1978) 84-86; MR 57 #5778.

[14] D. W. Boyd, Losing runs in Bernoulli trials, unpublished note (1972).

[15] P. Erdös and P. Révész, On the length of the longest head-run, *Topics in Information Theory*, Proc. 1975 Keszthely conf., ed. I. Csiszár and P. Elias, Colloq. Math. Soc. János Bolyai 16, North-Holland, 1977, pp. 219-228; MR 57 #17788.

[16] D. E. Knuth, The average time for carry propagation, *Proc. Konink. Nederl. Akad.Wetensch. Ser. A* 81 (1978) 238-242; *Indag. Math.* 40 (1978) 238-242; also in *Selected Papers on Analysis of Algorithms*, CSLI, 2000, pp. 467-471; MR 81b:68030.

[17] L. J. Guibas and A. M. Odlyzko, Long repetitive patterns in random sequences, *Z. Wahrsch. Verw. Gebiete* 53 (1980) 241-262; MR 81m:60047.

[18] L. Gordon, M. F. Schilling, and M. S. Waterman, An extreme value theory for long head runs, *Probab. Theory Relat. Fields* 72 (1986) 279-287; MR 87i:60023.

[19] H. Prodinger, Über längste 1-Teilfolgen in 0-1-Folgen, *Zahlentheoretische Analysis II*, Wien Seminar 1984-86, ed. E. Hlawka, Lect. Notes in Math. 1262, Springer-Verlag, 1987, pp. 124-133; MR 90g:11106.

[20] P. Hitczenko and G. Stengle, Expected number of distinct part sizes in a random integer composition, *Combin. Probab. Comput.* 9 (2000) 519-527; MR 2002d:05008.

[21] M. F. Schilling, The longest run of heads, *College Math J.* 21 (1990) 196-206.

[22] F. T. Bruss and C. A. O'Cinneide, On the maximum and its uniqueness for geometric random samples, *J. Appl. Probab.* 27 (1990) 598-610; MR 92a:60096.

[23] L. Räde, P. Griffin, and O. P. Lossers, Tossing coins until all show heads, *Amer. Math. Monthly* 101 (1994) 78-80.

[24] P. Kirschenhofer and H. Prodinger, The number of winners in a discrete geometrically distributed sample, *Annals of Appl. Probab.* 6 (1996) 687-694; MR 97g:60016.

[25] N. J. Calkin, E. R. Canfield, and H. S. Wilf, Averaging sequences, deranged mappings, and a problem of Lampert and Slater, *J. Combin. Theory Ser. A* 91 (2000) 171-190; MR 2002c:05019.

[26] C. F. Woodcock, On the asymptotic behaviour of a function arising from tossing coins, *Bull. London Math. Soc.* 28 (1996) 19-23; MR 96i:26001.

[27] B. Eisenberg, G. Stengle, and G. Strang, The asymptotic probability of a tie for first place, *Annals of Appl. Probab.* 3 (1993) 731-745; MR 95d:60044.

[28] J. J. A. M. Brands, F.W. Steutel, and R. J. G.Wilms, On the number of maxima in a discrete sample, *Statist. Probab. Lett.* 20 (1994) 209-217; MR 95e:60010.

[29] Y. Baryshnikov, B. Eisenberg, and G. Stengle, A necessary and sufficient condition for the existence of the limiting probability of a tie for first place, *Statist. Probab. Lett.* 23 (1995) 203-209; MR 96d:60015.

[30] Y. Qi, A note on the number of maxima in a discrete sample, *Statist. Probab. Lett.* 33 (1997) 373-377; MR 98f:60061.

[31] D. E. Lampert and P. J. Slater, Parallel knockouts in the complete graph, *Amer. Math. Monthly* 105 (1998) 556-558.

5.12 強正方形エントロピー定数

$n \times n$ の 2 値行列 (binary matrix) のすべての集合を考える. 1 が隣り合わないこのような行列の個数 $F(n)$ について何がいえるだろうか？ 2 つの 1 が隣接する (adjacent) とは, ある i, j が存在して, 行列成分 (i, j) と $(i+1, j)$, または行列成分 (i, j) と $(i, j+1)$ にともに 1 があるときをいう. いいかえると, $F(n)$ は, $n \times n$ のチェス盤上にいるプリンスたちが互いに攻撃しない, 位置の個数である. ただし, 「プリンス」[†] は上下左右に隣接する 4 つのマス目に動くが, 対角方向には動かない駒である. $N = n^2$ とする. このとき [1-3],

$$\kappa = \lim_{n \to \infty} F(n)^{1/N} = 1.5030480824\cdots = \exp(0.4074951009\cdots)$$

は, **強正方形** (hard square) **エントロピー定数**である. 初期の推定値は, 物理学者 [4-9] および数学者 [10-13] によって得られた. いくつかの関連する組合せ的数え上げの問題は, [14-16] にある.

$n \times n$ の 2 値行列の代わりに, $n \times n$ の 2 値配列

$$\begin{pmatrix} a_{11} & & a_{23} & \\ & a_{22} & & a_{34} \\ a_{21} & & a_{33} & \\ & a_{32} & & a_{44} \\ a_{31} & & a_{43} & \\ & a_{42} & & a_{54} \\ a_{41} & & a_{53} & \\ & a_{52} & & a_{64} \end{pmatrix}$$

(ここでは $n = 4$ である) を考える. 1 が隣り合わないこのような配列の個数 $G(n)$ について, 何がいえるだろうか？ ただし, 2 つの 1 が隣接しているとは, ある i, j が存在して, 1 が行列成分 (i, j) と $(i+1, j)$, (i, j) と $(i, j+1)$, または (i, j) と $(i+1, j+1)$ にあるときをいう. いいかえると, $G(n)$ は, 正六角形のマス目をもつ $n \times n$ チェス盤上にいるキング (隣の 6 方向に動く) たちが互いに攻撃しない位置の個数である. **強六角形** (hard hexagon) **エントロピー定数**

$$\kappa = \lim_{n \to \infty} G(n)^{1/N} = 1.3954859724\cdots = \exp(0.3332427219\cdots)$$

が「代数的数」(実際, 根号で解ける [17-22]) であることは, 驚きである. それが満足する最小整数係数多項式は [23]

$$25937424601 x^{24} + 2013290651222784 x^{22} + 2505062311720673792 x^{20}$$

[†] 訳注: 動きを制限したチェスの駒につけた仮の名である. 王より位が下という気持であり, デューク (公爵) ということもある. 中国将棋の将, 帥も同じ動きをする.

$$+797726698866658379776x^{18}+7449488310131083100160x^{16}$$
$$+2958015038376958230528x^{14}-7240567028564916161 7408x^{12}$$
$$+1071554481504433880432 64x^{10}-71220809441400405884928x^{8}$$
$$-73347491183630103871488x^{6}+9714313527737757 5190528x^{4}$$
$$-32751691810479015985152$$

である.

これは, 楕円テータ関数および数論からの Rogers-Ramanujan（ロジャー-ラマヌジャン）恒等式を使った強六角形モデルの Baxter（バクスター）の正確な解の結果である[28-31]! κ は, 実際, より一般的な式

$$\kappa(z)=\lim_{n\to\infty}Z_n(z)^{1/N}$$

から導かれる. ただし, $Z_n(z)$ はモデルの**分割関数**（partition function）として知られている. $G(n)=Z_n(1)$, $\kappa=\kappa(1)$ である. 格子ガスモデルでの相転移の物理に関するさらなるものは,【5.12.1】にある.

McKay（マッケイ）と Calkin（カルキン）はそれぞれ独立に, 正方形のマス目をもつチェス盤で, プリンスをキングに変えると, 対応する定数 κ は 1.3426439511… となることを示した. さらに[32-34]を参照せよ. 正六角形のマス目のチェス盤上のプリンスとキングとの相違は, 重要でない.（解明. プリンスがマス目 c にいると, c と辺を共有する任意のマス目が攻撃を受ける. 対照的に, キングがマス目 c にいると, c と辺または対角線を共有する任意のマス目が攻撃を受ける.）

チェス盤の代わりに, 正三角形のマス目にすると, プリンスに対して, $\kappa=$ 1.5464407087… である[3]. これを**強三角形**（hard triangle）**エントロピー**とよぶ. プリンスをキングに置きかえたときの κ の値は, 知られていない.

正方形マス目のチェス盤上で, ナイトまたはクイーンの攻撃しない場合に対する定数 κ は何であるのか? ナイトの解析は, プリンスやキングの解析と同じである. しかし, クイーンに対しては, 交互作用がもはや局所的ではないので, すべて異なる[35].

強正方形エントロピー定数は, さらに, $\ln(\kappa)/\ln(2)=0.5878911617\cdots$ の形で, 符号理論に関するいくつかの論文[36-41]に現れる. それらはホログラフィックデータの格納および検索を含む応用をもつ.

5.12.1 気体モデル格子における相転移

統計力学は, 粒子の大きなシステムの平均的な性質に関する理論である. たとえば, 温度が下がるかまたは密度が増加するにつれて, 無秩序な流体から秩序ある固体への相転移を考える.

この現象の簡単なモデルは, **格子ガス**（lattice gas）である. 粒子が, 正則な格子の位置に置かれ, 隣接する粒子とのみ交互作用する. 剛体の分子はこのような完全な対称性の要求を満たす可能性がないので, 理想化は絶望的である. しかしながら, このモデルは, 物質のミクロとマクロとの間のつながりを理解するには役立つ.

広く研究されている格子ガスモデルの2つの型は，**強正方形** (hard square) モデルと**強六角形** (hard hexagon) モデルである．ひとたび，1つの粒子が格子点の上に置かれると，他の粒子は，図5.12に示したように，同じ場所またはその場所の隣に置くことができない．いいかえると，指し示している正方形および六角形は，重なることができない．したがって，隣接「強」（adjective "hard"）という言葉が生まれる．

N 個の場所をもつ（正方または三角（六角））格子が与えられている．各 i，$1 \leq i \leq N$ に対して，場所 i が塞がれているとき，変数 $\sigma_i = 1$ とし，そうでないとき $\sigma_i = 0$ とする．**分割関数** (partition function)

$$Z_n(z) = \sum_{\sigma} \left(z^{\sigma_1 + \sigma_2 + \sigma_3 + \cdots + \sigma_N} \cdot \prod_{(i,j)} (1 - \sigma_i \sigma_j) \right)$$

を調べる．ただし，和はベクトル $\sigma = (\sigma_1, \sigma_2, \sigma_3, \cdots, \sigma_N)$ の 2^N のすべての可能な値にわたるものであり，積は格子のすべての辺（場所 i と j が異なり隣接する）にわたるものである．積は最も近い隣点を排除していることに注意せよ．すなわち，配置がお互い隣り合っている2つの粒子とすると，分割関数への寄与はなくなる．

習慣的に，トーラスを形作るように格子の周りを巻きつけることによって境界の影響を扱う．より詳しくいうと，正方格子について，$2n$ の新しい辺は，n 個の最も右および上の各点を，n 個の最も左および下の点にそれぞれ対応させて繋げることにより作られる．したがって，各場所がすべて「同じように見える」．正方格子においては，全部で $2N$ の辺がある．三角格子では，$4n-1$ の新しい辺が作られ，全部で $3N$ の辺がある．どちらの場合も，境界の場所の個数は，N に対して，$n \to \infty$ のとき，0になるくらいに小さい．だから，この扱いによって，いかなる誤差も生じない．

明らかに，次の組み合わせ論的な式は正しい[4, 42, 43]．正方格子に対して，

$$Z_n = \sum_{k=0}^{\lfloor \frac{N}{2} \rfloor} f_{k,n} z^k, \quad f_{0,n} = 1, \quad f_{1,n} = N, \quad f_{2,n} = \begin{cases} 2 & (n=2 \text{ のとき}) \\ \dfrac{1}{2} N(N-5) & (n \geq 3 \text{ のとき}) \end{cases}$$

$$f_{3,n} = \begin{cases} 6 & (n=3 \text{ のとき}) \\ \dfrac{1}{6}(N(N-10)(N-13) + 4N(N-9) + 4N(N-8)) & (n \geq 4 \text{ のとき}) \end{cases}$$

ただし，$f_{k,n}$ は k 個の正方形をもつ，N 点格子（N-site lattice with k squares）の可

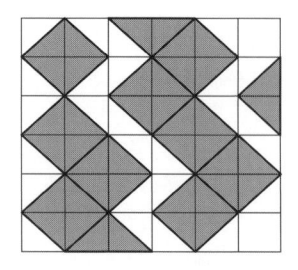

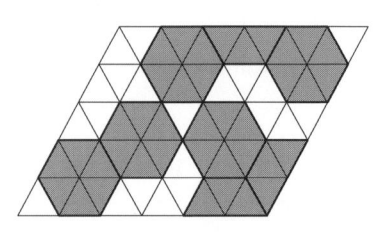

図5.12 正方格子と三角格子上にそれぞれ置かれた強正方形と強六角形の場所．

5.12 強正方形エントロピー定数

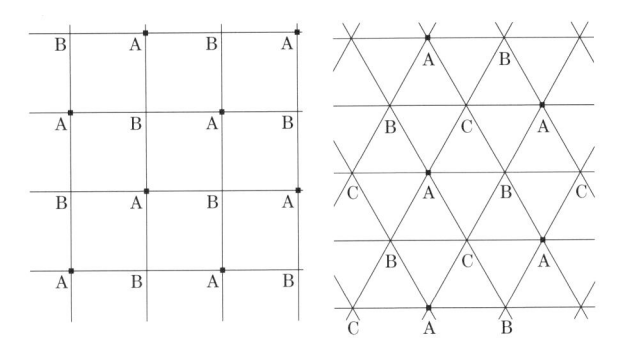

図 5.13 正方格子の2つの部分格子および三角格子の3つの部分格子.

能なタイル張りの個数である. 三角格子に対しては,

$$Z_n = \sum_{k=0}^{\lfloor N/3 \rfloor} g_{k,n} z^k, \quad g_{0,n}=1, \quad g_{1,n}=N, \quad g_{2,n}=\frac{1}{2}N(N-7)$$

$$g_{3,n} = \begin{cases} 0 & (n=3\text{ のとき}) \\ \frac{1}{6}(N(N-14)(N-19)+6N(N-13)+6N(N-12)) & (n\geq 4\text{ のとき}) \end{cases}$$

であり, $g_{k,n}$ は六角形タイル張りに対応する個数である.

物理に戻る. 分割関数は確率計算における「分母」の役割をするので, 分割関数は重要である. たとえば, 図 5.13 で示すような場所をもつ正方格子の, 2つの部分格子 A, B を考える. 部分格子 A において, 任意の格子点 α が占める確率は,

$$\rho_A(z) = \lim_{n\to\infty} \frac{1}{Z_n} \sum_\sigma \left(\sigma_\alpha \cdot z^{\sigma_1+\sigma_2+\sigma_3+\cdots+\sigma_N} \prod_{(i,j)}(1-\sigma_i\sigma_j) \right)$$

である. これは, さらに α における局所密度 (local density at α) とよばれる. 三角格子の3つの部分格子 A, B, C に対しても, 類似の確率を定義することができる.

アクティビティ (活性度, activity) として知られる正の変数 z の関数として, これらのモデルのふるまいに興味がある. 図 5.14 は, たとえば, 強六角形の場合に対する平均密度のグラフである. [18] で与えられる正確な式を用いて,

$$\rho(z) = z\frac{d}{dz}\ln(\kappa(z)) = \frac{\rho_A(z)+\rho_B(z)+\rho_C(z)}{3}$$

を得る.

相転移の存在は, 図より明らかである. 極端な場合を調べる. ぎっしり詰め込まれた配置 (closely-packed configurations) (大きな z) とまばらに散らばった配置 (sparsely-distributed configulations) (小さい z) である. 無数の z に対して, すべての可能な部分格子 (sublattice) のうちの1つは, 完全に使われる. それを A の部分格子とする. 他方はまったく空になる. すなわち,

$$\rho_A=1, \quad \rho_B=0 \quad (\text{正方形モデル})$$

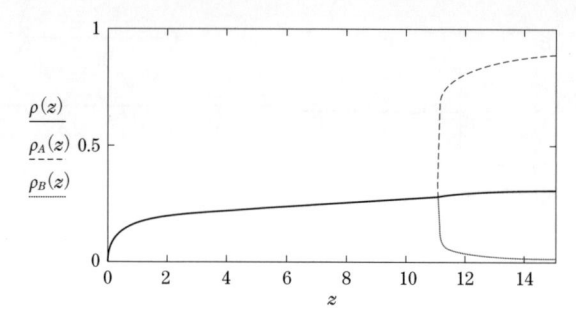

図5.14 六角形モデルの平均密度および部分格子密度を z の関数とみたグラフ．

$$\rho_A=1, \quad \rho_B=\rho_C=0 \quad (六角形モデル)$$

である．z が 0 に近いとき，部分格子に特別な好みの順序はない．すなわち，

$$\rho_A=\rho_B \;(正方格子のとき), \quad \rho_A=\rho_B=\rho_C \;(六角格子のとき)$$

である．低い活性（activity）は均一性に対応し，高い活性は不均一に対応する．だから，相転移が起こるような臨界値 z_c が存在する．**秩序パラメータ**（order parameter）

$$R=\rho_A-\rho_B\;(正方形のとき), \quad R=\rho_A-\rho_B=\rho_A-\rho_C\;(六角形のとき)$$

を定義する．このとき，$z<z_c$ で $R=0$，$z>z_c$ で $R>0$ である．

緻密な数値計算[7,4,45]によって，$n\to\infty$ のとき，位置 α が格子の中で無限に深い (infinitely deep) と仮定して，

$$z_c=3.7962\cdots\;(正方形のとき), \quad z_c=11.09\cdots\;(六角形のとき)$$

を示した．計算は，R の高い精度の級数展開および角転送行列（corner transfer-matrices）として知られているものを使った．ここでは，紙数の関係で議論することはできない．

すばらしい進展があり，Baxter[24,25]は六角形モデルの厳密な解を得た．この業績のすべての範囲をここでは伝えることはできない．しかし，多くの系のうちの1つである六角形モデルに対する厳密な式は

$$z_c=\frac{11+5\sqrt{5}}{2}=\left(\frac{1+\sqrt{5}}{2}\right)^5=11.0901699437\cdots$$

である．正方形モデルに対して，同じような理論的な大発展はない．だから，3.7962… の正体は，我々の理解から覆い隠されたままである．六角形または蜂の巣格子（hexagonal or honeycomb lattice）上の三角形モデルに対して，臨界値 $z_c=7.92\cdots$ も正確に知られていない[46]．

強六角形に対しては，臨界値での $\rho(z)$ と $R(z)$ のふるまいは，重要である[24,26,27]．

$$\rho \sim \rho_c - 5^{-3/2}\left(1-\frac{z}{z_c}\right)^{2/3} \quad (z\to z_c-0), \quad \rho_c=\frac{5-\sqrt{5}}{10}=0.2763932022\cdots$$

5.12 強正方形エントロピー定数

$$R \sim \frac{3}{\sqrt{5}}\left[\frac{1}{5\sqrt{5}}\left(\frac{z}{z_c}-1\right)\right]^{1/9} \quad (z \to z_c+0)$$

であり，指数 1/3 と 1/9 は普遍的（universal）であると予想されている．強正方形と強三角形に対して，それぞれ数値的な推定値 $\rho_c=0.368\cdots$ と $0.422\cdots$ だけが知られている．臨界値から遠い $z=1$ での計算は難しくない [3, 47]．

$$\rho(1)=\begin{cases} 0.1624329213\cdots & (\text{強六角形 (hard hexagons) のとき}) \\ 0.2265708154\cdots & (\text{強正方形 (hard squares) のとき}) \\ 0.2424079763\cdots & (\text{強三角形 (hard triangles) のとき}) \end{cases}$$

この最初の $0.1624329213\cdots$ は，次数 12 の代数的数である [18, 22]．$\rho(1)$ の一般化は，任意の点 a と近傍の点 a' の特別な位置がすべて占有されたときの確率である．実例計算は [3] にある．

いうまでもないが，ここで議論したモデルの 3 次元の類似なものは，厳密な解を求めるいかなる試みも拒んでいる [44]．

[1] N. J. Calkin and H. S. Wilf, The number of independent sets in a grid graph, *SIAM J. Discrete Math.* 11 (1998) 54-60; MR 99e:05010.

[2] B. D. McKay, On Calkin and Wilf's limit theorem for grid graphs, unpublished note (1996).

[3] R. J. Baxter, Planar lattice gases with nearest-neighbour exclusion, *Annals of Combin.* 3 (1999) 191-203; cond-mat/9811264; MR 2001h:82010.

[4] D. S. Gaunt and M. E. Fisher, Hard-sphere lattice gases. I: Plane-square lattice, *J. Chem. Phys.* 43 (1965) 2840-2863; MR 32 #3716.

[5] L. K. Runnels and L. L. Combs, Exact finite method of lattice statistics. I: Square and triangular lattice gases of hard molecules, *J. Chem. Phys.* 45 (1966) 2482-2492.

[6] B. D. Metcalf and C. P. Yang, Degeneracy of antiferromagnetic Ising lattices at critical magnetic field and zero temperature, *Phys. Rev. B* 18 (1978) 2304-2307.

[7] R. J. Baxter, I. G. Enting, and S. K. Tsang, Hard-square lattice gas, *J. Stat. Phys.* 22 (1980) 465-489; MR 81f:82037.

[8] P. A. Pearce and K. A. Seaton, A classical theory of hard squares, *J. Stat. Phys.* 53 (1988) 1061-1072; MR 89j:82076.

[9] S. Milosevic, B. Stosic, and T. Stosic, Towards finding exact residual entropies of the Ising antiferromagnets, *Physica A* 157 (1989) 899-906.

[10] N. G. Markley and M. E. Paul, Maximal measures and entropy for $Z\nu$ subshifts of finite type, *Classical Mechanics and Dynamical Systems*, ed. R. L. Devaney and Z. H. Nitecki, Dekker, 1981, pp. 135-157; MR 83c:54059.

[11] N. J. Calkin, *Sum-Free Sets and Measure Spaces*, Ph. D. thesis, Univ. of Waterloo, 1988.

[12] K. Weber, On the number of stable sets in an $m \times n$ lattice, *Rostock. Math. Kolloq.* 34 (1988) 28-36; MR 89i:05172.

[13] K. Engel, On the Fibonacci number of an $m \times n$ lattice, *Fibonacci Quart.* 28 (1990) 72-78; MR 90m:11033.

[14] H. Prodinger and R. F. Tichy, Fibonacci numbers of graphs, *Fibonacci Quart.* 20 (1982) 16-21; MR 83m:05125.

[15] Y. Kong, General recurrence theory of ligand binding on three-dimensional lattice, *J. Chem. Phys.* 111 (1999) 4790-4799.

[16] N. J. A. Sloane, On-Line Encyclopedia of Integer Sequences, A000045, A001333, A006506, A051736, A051737, A050974, A066863, and A066864.

[17] G. S. Joyce, Exact results for the activity and isothermal compressibility of the hardhexagon

model, *J. Phys. A* 21 (1988) L983-L988; MR 90a:82070.
[18] G. S. Joyce, On the hard hexagon model and the theory of modular functions, *Philos. Trans. Royal Soc. London A* 325 (1988) 643-702; MR 90c:82055.
[19] G. S. Joyce, On the icosahedral equation and the locus of zeros for the grand partition function of the hard hexagon model, *J. Phys. A* 22 (1989) L237-L242; MR 90e: 82101.
[20] M. P. Richey and C. A. Tracy, Equation of state and isothermal compressibility for the hard hexagon model in the disordered regime, *J. Phys. A* 20 (1987) L1121-L1126; MR 89b:82121.
[21] C. A. Tracy, L. Grove, and M. F. Newman, Modular properties of the hard hexagon model, *J. Stat. Phys.* 48 (1987) 477; MR 89b:82125.
[22] M. P. Richey and C. A. Tracy, Algorithms for the computation of polynomial relationships for the hard hexagon model, *Nucl. Phys. B* 330 (1990) 681-704; MR 91c:82029.
[23] P. Zimmermann, The hard hexagon entropy constant, unpublished note (1996).
[24] R. J. Baxter, Hard hexagons: Exact solution, *J. Phys. A*13 (1980) L61-L70;MR80m:82052.
[25] R. J. Baxter, Rogers-Ramanujan identities in the hard hexagon model, *J. Stat. Phys.* 26 (1981) 427-452; MR 84m:82104.
[26] R. J. Baxter, *Exactly Solved Models in Statistical Mechanics*, Academic Press, 1982; MR 90b: 82001.
[27] R. J. Baxter, Ramanujan's identities in statistical mechanics, *Ramanujan Revisited*, Proc. 1987 Urbana conf., ed. G. E. Andrews, R. A. Askey, B. C. Berndt, K. G. Ramanathan, and R. A. Rankin, Academic Press, 1988, pp. 69-84; MR 89h:82043.
[28] G. E. Andrews, The hard-hexagon model and Rogers-Ramanujan type identities, *Proc. Nat. Acad. Sci. USA* 78 (1981) 5290-5292; MR 82m:82005.
[29] G. E. Andrews, *q-Series: Their Development and Application in Analysis, Number Theory, Combinatorics, Physics and Computer Algebra*, Amer. Math. Soc., 1986; MR 88b:11063.
[30] G. E. Andrews, The reasonable and unreasonable effectiveness of number theory in statistical mechanics, *The Unreasonable Effectiveness of Number Theory*, Proc. 1991 Orono conf., ed. S. A. Burr, Amer. Math. Soc., 1992, pp. 21-34; MR 94c:82021.
[31] G. E. Andrews and R. J. Baxter, A motivated proof of the Rogers-Ramanujan identities, *Amer. Math. Monthly* 96 (1989) 401-409; 97 (1990) 214-215; MR 90e:11147 and MR 91j:11086.
[32] D. E. Knuth, Nonattacking kings on a chessboard, unpublished note (1994).
[33] H. S. Wilf, The problem of the kings, *Elec. J. Combin.* 2 (1995) R3; MR 96b:05012.
[34] M. Larsen, The problem of kings, *Elec. J. Combin.* 2 (1995) R18; MR 96m:05011.
[35] I. Rivin, I. Vardi, and P. Zimmermann, The n-queens problem, *Amer. Math. Monthly* 101 (1994) 629-638; MR 95d:05009.
[36] W. Weeks and R. E. Blahut, The capacity and coding gain of certain checkerboard codes, *IEEE Trans. Inform. Theory* 44 (1998) 1193-1203; MR 98m:94061.
[37] S. Forchhammer and J. Justesen, Entropy bounds for constrained two-dimensional random fields, *IEEE Trans. Inform. Theory* 45 (1999) 118-127; MR 99k:94023.
[38] A. Kato and K. Zeger, On the capacity of two-dimensional run-length constrained channels, *IEEE Trans. Inform. Theory* 45 (1999) 1527-1540; MR 2000c:94012.
[39] Z. Nagy and K. Zeger, Capacity bounds for the 3-dimensional (0, 1) runlength limited channel, *Applied Algebra, Algebraic Algorithms and Error-Correcting Codes (AAECC-13)*, Proc. 1999 Honolulu conf., ed. M. Fossorier, H. Imai, S. Lin, and A. Poli, Lect. Notes in Comp. Sci. 1719, Springer-Verlag, 1999, pp. 245-251.
[40] H. Ito, A. Kato, Z. Nagy, and K. Zeger, Zero capacity region of multidimensional run length constraints, *Elec. J. Combin.* 6 (1999) R33; MR 2000i:94036.
[41] R. M. Roth, P. H. Siegel, and J. K. Wolf, Efficient coding schemes for the hard-square model, *IEEE Trans. Inform. Theory* 47 (2001) 1166-1176; MR 2002b:94017.
[42] N. J. A. Sloane, On-Line Encyclopedia of Integer Sequences, A007197, A027683, A066866, and A067967.
[43] R. Hardin, Computation of $f_{k,n}$ and $g_{k,n}$, $1 \leq n \leq 9$, unpublished note (2002).

[44] D. S. Gaunt, Hard-sphere lattice gases. II: Plane-triangular and three-dimensional lattices, *J. Chem. Phys.* 46 (1967) 3237.

[45] R. J. Baxter and S. K. Tsang, Entropy of hard hexagons, *J. Phys. A* 13 (1980) 1023-1030.

[46] L. K. Runnels, L. L. Combs, and J. P. Salvant, Exact finite method of lattice statistics. II: Honeycomb lattice gas of hard molecules, *J. Chem. Phys.* 47 (1967) 4015-4020.

[47] C. Richard, M.Höffe, J. Hermisson, and M. Baake, Random tilings: Concepts and examples, *J. Phys. A* 31 (1998) 6385-6408; MR 99g:82033.

5.13 2分検索木定数

最初にある種の関数 f を定義する．その定式化は少し難解と思われる．しかし，f は，弱2分木（weakly binary tree）（後で議論する応用の1つ）のある型に沿った道の長さ（pass length）として，自然に解釈される【5.6】．

k 個の異なる整数のベクトル $V=(v_1,v_2,\cdots,v_k)$ が与えられているとする．2つの部分ベクトル V_L, V_R を
$$V_L=(v_j : v_j<v_1,\ 2\leqq j\leqq k), \quad V_R=(v_j : v_j>v_1,\ 2\leqq j\leqq k)$$
で定義する．添え字 L と R は「左」と「右」を意味する．部分リスト V_L と V_R は，V にリストされた要素の順序を保存することを強調しておく．

さて，すべての整数 x 上で，帰納的関数
$$f(x,V)=\begin{cases} 0 & (V=\emptyset\ (\emptyset\text{は空ベクトル})) \\ \begin{cases} 1 & x=v_1 \\ 1+f(x,V_L) & x<v_1 \\ 1+f(x,V_R) & x>v_1 \end{cases} & (\text{その他}\ (v_1\text{はベクトルの第1成分})) \end{cases}$$
を定義する．明らかに，つねに $0\leqq f(x,V)\leqq k$ であり，$v_1,v_2,\cdots,v_k$ の順序は，$f(x,V)$ の値を決定するために欠くことができない．たとえば，$f(7,(3,9,5,1,7))=4$ であり，$f(4,(3,9,5,1,7))=3$ である．

V を $(1,3,5,\cdots,2n-1)$ のランダムな置換とする．2つの規則
- $1\leqq x\leqq 2n-1$ を満たすランダムな奇数 x（検索に成功）
- $0\leqq x\leqq 2n$ を満たすランダムな偶数 x（検索に失敗）

のもとでの $f(x,V)$ の確率分布に興味がある．

V と x ともにランダムであることに注意せよ．それらは一様サンプリングで独立に選ばれると仮定する．$f(x,V)$ の期待値は，コンピュータ科学の言葉では[1-3]，
- n 個のレコードをもつデータ構造内に，ランダムに存在するレコード x を「見つける」ために必要な比較の平均回数
- n 個のレコードをもつデータ構造内に，新しいランダムレコード x を「挿入する」ために必要な比較の平均回数

$$V = \begin{pmatrix} 3 \\ 9 \\ 5 \\ 1 \\ 7 \end{pmatrix}$$

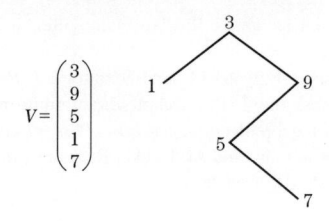

図 5.15 V で構成される 2 分検索木.

である.ただし,データ構造は **2 分検索木** (binary search tree) の構造に従う,と仮定する.図 5.15 は,指示された V から始まるこのような木の作り方を示したものである.さらに,$g(l,V) = |\{x : f(x,V) = l,\ 1 \leq x \leq 2n-1,\ x \text{ は奇数}\}|$ と定義する.すなわち,木の l 番目のレベルを占めている頂点の個数 ($l=1$ は根 (root level) である) を定義する.たとえば,$g(2,(3,9,5,1,7))=2$ であり,$g(3,(3,9,5,1,7))=1$ である.

平均的な場合の 2 つのパラメータに加えて,(与えられた V に対して,レコード x を見つけるための最悪の場合のシナリオを捕まえる) 木の**高さ** (height)
$$h(V) = \max\{f(x,V) : 1 \leq x \leq 2n-1,\ x \text{ は奇数}\} - 1$$
および木の**飽和レベル** (saturation level) (木にある頂点の全レベルの個数 -1 である)
$$s(V) = \max\{l : g(l,V) = 2^{l-1}\} - 1$$
の確率分布が知りたい.したがって,$h(V)$ は木の根から葉への一番長い道の長さである.一方,$s(V)$ はこのような一番短い道の長さである.たとえば,$h(3,9,5,1,7)=3$,$s(3,9,5,1,7)=1$ である.

いつものように,調和数列
$$H_n = \sum_{k=1}^{n} \frac{1}{k} = \ln(n) + \gamma + \frac{1}{n} + O\left(\frac{1}{n^2}\right),\quad H_n^{(2)} = \sum_{k=1}^{n} \frac{1}{k^2} = \frac{\pi^2}{6} - \frac{1}{n} + O\left(\frac{1}{n^2}\right)$$
を定義する.ただし,γ はオイラー-マスケロニの定数【1.5】である.このとき,ランダム木の検索成功 (ランダムで奇数 $1 \leq x \leq 2n-1$) における比較の期待回数は [2-4],
$$\mathrm{E}(f(x,V)) = 2\left(1 + \frac{1}{n}\right)H_n - 3 = 2\ln(n) + 2\gamma - 3 + O\left(\frac{\ln(n)}{n}\right)$$
であり,ランダム木の検索失敗 (ランダムで偶数 $0 \leq x \leq 2n$) における比較の期待回数は,
$$\mathrm{E}(f(x,V)) = 2(H_{n+1} - 1) = 2\ln(n) + 2\gamma - 2 + O\left(\frac{1}{n}\right)$$
である.奇数 x に対応する分散は
$$\mathrm{Var}(f(x,V)) = \left(2 + \frac{10}{n}\right)H_n - 4\left(1 + \frac{1}{n}\right)\left(H_n^{(2)} + \frac{H_n^2}{4}\right) + 4$$
$$\sim 2\left(\ln(n) + \gamma - \frac{\pi^3}{3} + 2\right)$$
であり,偶数 x に対応する分散は

$$\mathrm{Var}(f(x,V)) = 2(H_{n+1} - 2H_{n+1}^{(2)} + 1) \sim 2\left(\ln(n) + \gamma - \frac{\pi^3}{3} + 1\right)$$

である.

　$h(V)$ と $s(V)$ の完全な解析は, Robson (ロブソン) [8] と Pittel (ピッテル) [9] の仕事を基礎にして, Devroye (デブロイ) [3,5-7] が $n \to \infty$ のとき, ほとんど確実 (almost surely) に

$$\frac{h(V)}{\ln(n)} \to c, \quad \frac{s(V)}{\ln(n)} \to d$$

が成り立つことを証明する 1985 年まで未解決であった. ただし, $c=4.3110704070\cdots$, $d=0.3733646177\cdots$ は, 方程式

$$\frac{2}{x}\exp\left(1-\frac{1}{x}\right) = 1$$

の 2 つだけ存在する実数の解である. $h(V)/\ln(n)$ と $s(V)/\ln(n)$ が収束する速度は, 遅いことに注意せよ. だから, 数値的検証には効率的なシミュレーションが必要である [10]. かなりの努力が, これらの漸近式をより精密にするために払われている [11-14]. Reed (リード) [15,16] と Drmota (ドロモタ) [17-19] は最近, $n \to \infty$ のとき,

$$\mathrm{E}(h(V)) = c\ln(n) - \frac{3c}{2(c-1)}\ln(\ln(n)) + O(1)$$

$$\mathrm{E}(s(V)) = d\ln(n) + O(\sqrt{\ln(n)}\ln(\ln(n)))$$

および $\mathrm{Var}(h(V)) = O(1)$ であることを証明した. 後者の数値的な推定値は, 未だにない. さらに [20] を参照せよ.

　2 分探索木よりもいくぶんより複雑な, デジタル検索木 (digital search trees) 【5.14】に対して, 類似の極限

$$\frac{h(V)}{\ln(n)} \to \frac{1}{\ln(2)}, \quad \frac{s(V)}{\ln(n)} \to \frac{1}{\ln(2)}$$

が, 新しい定数を含まないことは奇妙である. $h(V)/\ln(n)$ と $s(V)/\ln(n)$ の極限値が等しいという事実は, 木がほとんど完全 ($\log_2(n)$ の周りの小さい「変動」だけをもつ) であることを意味する. これは, デジタル検索木での検索/挿入アルゴリズムが, 平均的に, 2 分木よりも効率的である, という微かな徴候である.

　関連する 1 つの話題がある [21-23]. 長さ r の棒をランダムに 2 つの部分に折る. 独立に, 2 つの折った棒それぞれを同様にランダムに 2 つに折る. 帰納的にこれを続ける. n 回後には, 2^n 個の断片になる. すべての断片の長さが <1 となる確率を $P_n(r)$ とする. 固定した r に対して, 明らかに $n \to \infty$ のとき $P_n(r) \to 1$ である. さらに面白いことに,

$$\lim_{n\to\infty} P_n(r^n) = \begin{cases} 0 & (r > e^{1/c} \text{ のとき}) \\ 1 & (0 < r < e^{1/c} \text{ のとき}) \end{cases}$$

が成り立つ. ただし, $e^{1/c} = 1.2610704868\cdots$ であり, c は以前に定義した. これを証明する技法は, 【5.3】で利用したものと同様である.

　2 分検索木の一般化である **4 分木** (quadtrees) を単に述べる [24-30]. これはさらに

魅力ある漸近定数をもつ．4分木は，多次元の実際のデータ，たとえば地図作成・コンピュータグラフィックスや画像処理などを格納したり検索したりすることに役立つ[31-33]．

[1] R. Sedgewick and P. Flajolet, *Introduction to the Analysis of Algorithms*, Addison-Wesley, 1996, pp. 236-240, 246-250, 261.
[2] D. E. Knuth, *The Art of Computer Programming*, v. 3, *Sorting and Searching*, Addison-Wesley, 1973, pp. 427, 448, 671-672; MR 56 # 4281.
[3] H. M. Mahmoud, *Evolution of Random Search Trees*, Wiley 1992, pp. 68-85, 92-95, 177-206; MR 93f:68045.
[4] G. G. Brown and B. O. Shubert, On random binary trees, *Math. Oper. Res.* 9 (1984) 43-65; MR 86c:05099.
[5] L. Devroye, A note on the height of binary search trees, *J. ACM* 33 (1986) 489-498; MR 87i:68009.
[6] L. Devroye, Branching processes in the analysis of the heights of trees, *Acta Inform.* 24 (1987) 277-298; MR 88j:68022.
[7] J. D. Biggins, How fast does a general branching random walk spread?, *Classical and Modern Branching Processes*, Proc. 1994 Minneapolis conf., ed. K. B. Athreya and P. Jagers, Springer-Verlag, 1997, pp. 19-39; MR 99c:60186.
[8] J. M. Robson, The height of binary search trees, *Austral. Comput. J.* 11 (1979) 151-153; MR 81a:68071.
[9] B. Pittel, On growing random binary trees, *J. Math. Anal. Appl.* 103 (1984) 461-480; MR 86c:05101.
[10] L. Devroye and J. M. Robson, On the generation of random binary search trees, *SIAM J. Comput.* 24 (1995) 1141-1156; MR 96j:68040.
[11] L. Devroye and B. Reed, On the variance of the height of random binary search trees, *SIAM J. Comput.* 24 (1995) 1157-1162; MR 96k:68033.
[12] J. M. Robson, On the concentration of the height of binary search trees, *Proc. 1997 Int. Colloq. on Automata, Languages and Programming (ICALP)*, Bologna, ed. P. Degano, R. Gorrieri, and A. Marchettio-Spaccamela, Lect. Notes in Comp. Sci. 1256, Springer-Verlag, pp. 441-448; MR 98m:68047.
[13] J. M. Robson, Constant bounds on the moments of the height of binary search trees, *Theoret. Comput. Sci.* 276 (2002) 435-444.
[14] M. Drmota, An analytic approach to the height of binary search trees, *Algorithmica* 29 (2001) 89-119.
[15] B. Reed, How tall is a tree?, *Proc. 32nd ACM Symp. on Theory of Computing (STOC)*, Portland, ACM, 2000, pp. 479-483.
[16] B. Reed, The height of a random binary search tree, *J. ACM*, to appear.
[17] M. Drmota, The variance of the height of binary search trees, *Theoret. Comput. Sci.* 270 (2002) 913-919.
[18] M. Drmota, An analytic approach to the height of binary search trees. II, *J. ACM*, to appear.
[19] M. Drmota, The saturation level in binary search trees, *Mathematics and Computer Science. Algorithms, Trees, Combinatorics and Probabilities*, Proc. 2000 Versailles conf., ed. D. Gardy and A. Mokkadem, Birkhäuser, 2000, pp. 41-51; MR 2001j:68026.
[20] C. Knessl and W. Szpankowski, The height of a binary search tree: The limiting distribution perspective, *Theoret. Comput. Sci.*, to appear.
[21] M. Sibuya and Y. Itoh, Random sequential bisection and its associated binary tree, *Annals Inst. Statist. Math.* 39 (1987) 69-84; MR 88e:60120.
[22] Y. Itoh, Binary search trees and 1-dimensional random packing, INRIA Project Algorithms Seminar 1997-1998, lecture summary.

[23] T. Hattori and H. Ochiai, A note on the height of binary search trees, random sequential bisection, and successive approximation for differential equation with moving singularity, unpublished note (1998).
[24] L. Laforest, *Étude des arbes hyperquaternaires*, LaCIM Tech. Report 3, Université du Québec à Montréal, 1990.
[25] L. Devroye and L. Laforest, An analysis of random d-dimensional quad trees, *SIAM J. Comput.* 19 (1990) 821-832; MR 91f:68018.
[26] P. Flajolet, G. Gonnet, C. Puech, and J. M. Robson, Analytic variations on quadtrees, *Algorithmica* 10 (1993) 473-500; MR 94i:68052.
[27] P. Flajolet, G. Labelle, L. Laforest, and B. Salvy, Hypergeometrics and the cost structure of quadtrees, *Random Structures Algorithms*, 7 (1995) 117-144; MR 96m:68034.
[28] G. Labelle and L. Laforest, Combinatorial variations on multidimensional quadtrees, *J. Combin. Theory Ser. A* 69 (1995) 1-16; MR 95m:05018.
[29] G. Labelle and L. Laforest, Sur la distribution de l'artié de la racine d'une arborescence hyperquaternaire à d dimensions, *Discrete Math.* 139 (1995) 287-302; MR 96f:05093.
[30] G. Labelle and L. Laforest, Etude de constantes universelles pour les arborescences hyperquaternaires de recherche, *Discrete Math.* 153 (1996) 199-211; MR 97c:05003.
[31] H. Samet, The quadtree and related hierarchical structures, *ACM Comput. Surveys* 16 (1984) 187-206.
[32] H. Samet, *The Design and Analysis of Spatial Data Structures*, Addison-Wesley, 1990.
[33] H. Samet, *Applications of Spatial Data Structures*, Addison-Wesley, 1990.

5.14 デジタル検索木定数

この節を読む前に，前節の2分検索木の知識【5.13】を勧める．k 個の異なる行からなる，2値 $k \times n$ 行列 $M = (m_{i,j}) = (m_1, m_2, \cdots, m_k)$ が与えられているとする．任意の整数 $1 \leq p \leq n$ に対して，2つの部分行列 $M_{L,p}$ と $M_{R,p}$ を，
$$M_{L,p} = (m_i : m_{i,p} = 0, \ 2 \leq i \leq k), \quad M_{R,p} = (m_i : m_{i,p} = 1, \ 2 \leq i \leq k)$$
と定義する．すなわち，$M_{L,p}$ の第 p 列はすべて0であり，$M_{R,p}$ の第 p 列はすべて1である．添え字 L と R は，左と右を意味する．部分リスト $M_{L,p}$ と $M_{R,p}$ は M でリストされているように行の順序を保存する，ことを強調しておく．

さて，すべての2値 n 次元ベクトル (binary n-vectors) x 上で，再帰的に関数
$$f(x, M, p) = \begin{cases} 0 & (M = \emptyset \text{のとき}) \\ \begin{cases} 1, & x = m_1 \text{のとき} \\ 1 + f(x, M_{L,p}, p+1), & x \neq m_1 \text{で} x_p = 0 \text{のとき} \\ 1 + f(x, M_{R,p}, p+1), & x \neq m_1 \text{で} x_p = 1 \text{のとき} \end{cases} & (\text{その他}) \end{cases}$$
を定義する．明らかに，つねに $0 \leq f(x, M, p) \leq k$ であり，$m_1, m_2, \cdots, m_k$ の順序は，p の値と同様に，$f(x, M, p)$ の値を決定するには欠かすことはできない．

$M = (m_1, m_2, \cdots, m_k)$ を，n 個の異なる行をもつランダムな2値 $n \times n$ 行列とし，x を2値 n ベクトルとする．2つの規則における $f(x, M, 1)$ の確率分布に興味がある．

- ある i, $1 \leq i \leq n$ に対して, $x=m_i$ を満たす乱数 x（検索成功）
- すべての i, $1 \leq i \leq n$ に対して, $x \neq m_i$ を満たす乱数 x（検索失敗）

である．ここで2分検索木【5.13】がもつように，2重のランダム性がある．しかし，x は，前よりも取り扱いが難しい M に従属していることに注意する．$f(x, M, 1)$ の期待値は，コンピュータ科学の言葉では[1-6]，

- n 個のレコードをもつデータ構造内に，ランダムに存在するレコード x を「見つける」ために必要な比較の平均回数
- n 個のレコードをもつデータ構造内に，新しいランダムレコード x を「挿入する」ために必要な比較の平均回数

である．ここでデータ構造は**デジタル検索木**（digital search tree）の構造に従うと仮定されている．図5.16は，規定された M から始めて，木の作り方を示している．

興味ある別のパラメータは，次数1の根（root）でない頂点の個数 A_n である．すなわち，子（children）をもたない節（node）のことである．2分検索木【3.7】に対して，$\mathrm{E}(A_n) = (n+1)/3$ が知られている．デジタル検索木に対して，対応する結果は，すぐ後で見るように，より複雑である．デジタル検索木は通例，2分検索木よりもうまく「バランス」しているので，1次の係数が 1/3 よりも 1/2 に近いと期待される．

γ をオイラー-マスケローニの定数【1.5】とし，新しい定数

$$\alpha = \sum_{k=1}^{n} \frac{1}{2^k - 1} = 1.6066951524\cdots$$

を定義する．このとき，ランダム木の検索成功（ランダムで，ある i で $x = m_i$）での比較の期待値は，[3-6, 8, 9]

$$\mathrm{E}(f(x,M,1)) = \frac{1}{\ln(2)}\ln(n) + \frac{3}{2} + \frac{\gamma-1}{\ln(2)} - \alpha + \delta(n) + O\left(\frac{\ln(n)}{n}\right)$$
$$\sim \log_2(n) - 0.716644\cdots + \delta(n)$$

であり，検索失敗（ランダムで，すべての i で $x \neq m_i$）での比較の期待値は，

$$\mathrm{E}(f(x,M,1)) = \frac{1}{\ln(2)}\ln(n) + \frac{1}{2} + \frac{\gamma}{\ln(2)} - \alpha + \delta(n) + O\left(\frac{\ln(n)}{n}\right)$$
$$\sim \log_2(n) - 0.273948\cdots + \delta(n)$$

である．ただし，

$$\delta(n) = \frac{1}{\ln(2)} \sum_{\substack{k=-\infty \\ k \neq 0}}^{\infty} \Gamma\left(-1 - \frac{2\pi i k}{\ln(2)}\right) \exp\left(2\pi i k \frac{\ln(n)}{\ln(2)}\right)$$

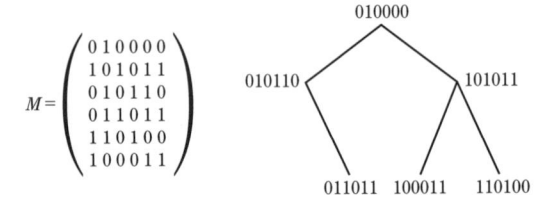

$$M = \begin{pmatrix} 0 & 1 & 0 & 0 & 0 & 0 \\ 1 & 0 & 1 & 0 & 1 & 1 \\ 0 & 1 & 0 & 1 & 1 & 0 \\ 0 & 1 & 1 & 0 & 1 & 1 \\ 1 & 1 & 0 & 1 & 0 & 0 \\ 1 & 0 & 0 & 0 & 1 & 1 \end{pmatrix}$$

図5.16　M を使って構成されるデジタル検索木．

5.14 デジタル検索木定数

である．関数 $\delta(n)$ は，振動し（$\delta(n) = \delta(2n)$），平均 0 であり，「無視できる」（すべての n で $|\delta(n)| < 1.726 \times 10^{-7}$）関数である．類似した関数 $\varepsilon(n)$, $\rho(n)$, $\sigma(n)$, $\tau(n)$ は後で必要になる．これらは多くのアルゴリズム解析[3,4,6]に現れる．同様に，【2.3】，【2.16】，【5.6】，【5.11】で議論した問題にも現れる．このような関数は，実際の目的には安全に無視できるけれども，理論的な厳密性のためにはある種の取り扱いが必要になる．

検索に対応する分散は

$$\mathrm{Var}(f(x,M,1)) \sim \frac{1}{12} + \frac{\pi^2+6}{6\ln(2)^2} - \alpha - \beta + \varepsilon(n) \sim 2.844383\cdots + \varepsilon(n)$$

であり，挿入に対応する分散は，

$$\mathrm{Var}(f(x,M,1)) \sim \frac{1}{12} + \frac{\pi^2}{6\ln(2)^2} - \alpha - \beta + \varepsilon(n) \sim 0.763014\cdots + \varepsilon(n)$$

である．ただし，新しい定数 β は

$$\beta = \sum_{k=1}^{\infty} \frac{1}{(2^k-1)^2} = 1.1373387363\cdots$$

で与えられる．

Flajolet & Sedgewick（フラジョレット&セジェウィック）[3,8,10]は，パラメータ A_n に関する Knuth（クヌース）の未解決な問題に答えた．

$$\mathrm{E}(A_n) = \left[\theta + 1 - \frac{1}{Q}\left(\frac{1}{\ln(2)} + \alpha^2 - \alpha\right) + \rho(n)\right]n + O(n^{1/2})$$

が成り立つ，である．ただし，新しい定数 Q と θ は

$$Q = \prod_{k=1}^{\infty}\left(1 - \frac{1}{2^k}\right) = 0.2887880950\cdots = (3.4627466194\cdots)^{-1}$$

$$\theta = \sum_{k=1}^{\infty} \frac{k2^{k(k-1)/2}}{1\cdot 3\cdot 7\cdots(2^k-1)} \sum_{j=1}^{k} \frac{1}{2^j-1} = 7.7431319855\cdots$$

で与えられる．$\mathrm{E}(A_n)$ の 1 次の係数は，

$$c = \theta + 1 - \frac{1}{Q}\left(\frac{1}{\ln(2)} + \alpha^2 - \alpha\right) = 0.3720486812\cdots$$

の周りで変動している．前に予想していたほど，1/2 に近くない！ さらに，[11]により，c には積分表示がある．

$$c = \frac{1}{\ln(2)} \int_0^{\infty} \frac{x}{1+x}\left(1+\frac{x}{1}\right)^{-1}\left(1+\frac{x}{2}\right)^{-1}\left(1+\frac{x}{4}\right)^{-1}\left(1+\frac{x}{8}\right)^{-1}\cdots dx$$

m 分検索木の主な型が 3 つある．すなわち，デジタル検索木（digital search trees）・基数検索トライ（radix search tries，単にトライともいう）・パトリシアトライ（Patricia tries）である．今まで $m=2$ と仮定していた．たとえば，パトリシアトライに対応する検索の分散は何であるのか？ 変動項を省略すると，残りの項

$$v = \frac{1}{12} + \frac{\pi^2}{6\ln(2)^2} + \frac{2}{\ln(2)} \sum_{k=1}^{\infty} \frac{(-1)^k}{k(2^k-1)}$$

は面白い．なぜならば，一見すると，正確に 1 に等しいと思われるからである．実際，

$v > 1+10^{-12}$ で，デデキントのエータ関数[12,13]により，さらに綿密に説明できる．

5.14.1 他との関連

数論では，約数関数 $d(n)$ は，$1 \leq d \leq n$ を満たす n を割る整数 d の個数である．母関数の特別な値[4,14,15]

$$\sum_{n=1}^{\infty} d(n) q^n = \sum_{k=1}^{\infty} \frac{q^k}{1-q^k} = \sum_{k=1}^{\infty} \frac{q^{k^2}(1+q^k)}{1-q^k}$$

は，$q=1/2$ のとき α である．Erdös（エルデーシュ）[16,17]は，α が無理数であることを証明した．

$$\sum_{n=1}^{\infty} \frac{1}{2^n-3}, \quad \sum_{n=1}^{\infty} \frac{1}{2^n+1}$$

のような定数について，あれこれ思いをめぐらし 40 年が過ぎた．（前者は[18]にある．一方，後者は基数検索トライ[6]とマージソートの漸近式[19, 20]に関連している．）$|a| \geq 2$ が整数であり，$b \neq 0$ が有理数であり，すべての n に対して $b \neq -a^n$ であるとき，級数の和

$$\sum_{n=1}^{\infty} \frac{1}{a^n+b}, \quad \sum_{n=1}^{\infty} \frac{(-1)^n}{a^n+b}$$

はともに無理数であることを，Borwein（ボーウィン）[21,22]は証明した．同じ条件のもとで，積

$$\prod_{n=1}^{\infty}\left(1+\frac{b}{a^n}\right)$$

は無理数である[23,24]．したがって，Q も無理数である．コンピュータを利用した無理性の最近の証明は，[25]を参照せよ．

他方，整数の分割の組み合わせ論からは，オイラーの5角数定理[14,26-28]

$$\prod_{n=1}^{\infty}(1-q^n) = \sum_{n=-\infty}^{\infty}(-1)^n q^{(3n+1)n/2} = 1+\sum_{n=1}^{\infty}(-1)^n(q^{(3n-1)n/2}+q^{(3n+1)n/2})$$

$$\prod_{n=1}^{\infty}(1-q^n)^{-1} = 1+\sum_{n=1}^{\infty}\frac{q^n}{(1-q)(1-q^2)(1-q^3)\cdots(1-q^n)}$$

$$= 1+\sum_{n=1}^{\infty}\frac{q^{n^2}}{(1-q)^2(1-q^2)^2(1-q^3)^2\cdots(1-q^n)^2} = 1+\sum_{n=1}^{\infty}p(n)q^n$$

が成り立つ．ただし，$p(n)$ は n に制限を付けない分割の個数である．$q=1/2$ のとき，これらは Q と $1/Q$ に特殊化される．他方，有限ベクトル空間の理論において，Q は，$q=2$ のときの線型部分空間 $\mathbb{F}_{q,n}$ の個数に対する漸近式【5.7】に現れる．

実際の理論では，さまざまの古典的な数学の対象の q-アナログ（q-analog）を含んだものに現れる．たとえば，定数 α はオイラー--マスケローニの定数[11]の 1/2-アナログ（1/2-analog）と見なされる．他の定数（たとえば，アペリ定数 $\zeta(3)$ やカタラン定数 G）も同様に一般化される．

Q の多くの表現可能な式のうち，3 つを述べる[4,14,26,29]．

$$Q = \frac{1}{3} - \frac{1}{3\cdot 7} + \frac{1}{3\cdot 7\cdot 15} - \frac{1}{3\cdot 7\cdot 15\cdot 31} + -\cdots$$
$$= \exp\left(-\sum_{n=1}^{\infty}\frac{1}{n(2^n-1)}\right)$$
$$= \sqrt{\frac{2\pi}{\ln(2)}}\exp\left(\frac{\ln(2)}{24}-\frac{\pi^2}{6\ln(2)}\right)\prod_{n=1}^{\infty}\left[1-\exp\left(\frac{-4\pi^2 n}{\ln(2)}\right)\right]$$

2番目の式は,Q と α の間に簡単な関係式があるのかと思わせる.Q は,ランダムな $n\times n$ の2値行列の行列式が,奇数となる漸近確率であることが示される.Q に類似の定数 P は,【2.8】にある.P での指数は奇数にかぎる.

反復単位 (repunits)[†] の逆数の和[30]
$$9\sum_{n=1}^{\infty}\frac{1}{10^n-1}=\frac{1}{1}+\frac{1}{11}+\frac{1}{111}+\frac{1}{1111}+\cdots = 1.1009181908\cdots$$
は,ボーウィンの定理より,無理数である.フィボナッチ数の逆数の級数は
$$\sum_{k=1}^{\infty}\frac{1}{f_k}=\sqrt{5}\sum_{n=0}^{\infty}\frac{(-1)^n}{\varphi^{2n+1}-(-1)^n}=3.3598856662\cdots$$
として表される[31-33]. ただし, φ は黄金比であり,この和は無理数であることが知られている[34-37]. 添え字が偶数の項の部分級数は,簡単に評価されることに注意せよ[26,31].
$$\sum_{k=1}^{\infty}\frac{1}{f_{2k}}=\sqrt{5}\left(\sum_{n=1}^{\infty}\frac{1}{\lambda^n-1}-\sum_{n=1}^{\infty}\frac{1}{\mu^n-1}\right)=1.5353705088\cdots$$
である.ただし,$2\lambda=\sqrt{3}+5$, $2\mu=7+3\sqrt{5}$ である.フィボナッチ数(このときは定数 Q に似ている)と完全に異なる関係は【1.2】にある.

ある正規化された定数[38-40]
$$K=\sqrt{\prod_{n=0}^{\infty}\left(1+\frac{1}{2^{2n}}\right)}=1.6467602581\cdots$$
は,2次元ベクトルの回転を効率的に「2進コーディック」(binary cordic) によって実装するときに起こる.しかしながら,Q や K のような積は,$q=\exp(-\pi\xi)$ ($\xi>0$ は代数的数)を除いて,閉じた式の表現は知られていない[26, 41].

$2^{n+1}-1$ は,2^i, $i\geq 0$ の形の n 個の整数の和として表すことのできない最小の正の整数であることに注意せよ.$2^i 3^j$, $i\geq 0$, $j\geq 0$ の形の n 個の整数の和として,表すことのできない最小の正の整数を,h_n と定義する.すなわち,$h_0=1$, $h_1=5$, $h_2=23$, $h_3=431$, …である[42,43]. $n\to\infty$ として,h_n の増大する正確な速度は何であるのか? (定数 α の「2-3 アナログ」(2-3 analog) とよんでよいのかもしれない.) これは【2.26】【2.30.1】の議論に漠然と関連している.

5.14.2 近似数え上げ

コンピュータ科学に戻る.Morris (モリス) [44] によるアルゴリズム,**近似数え上**

[†]訳注:1, 11, 111, ...のように1のみからなる10進数のこと.

げ（approximate counting）を議論する．近似数え上げは，わずか $\log_2(\log_2(N))$ ビットだけの記憶装置によって，大きな数 N 個の事象を見失わないようにすることと関係する．ただし，精度は優先されない．整数の時系列 $X_0, X_1, \cdots, X_N$ を，帰納的に

$$X_n = \begin{cases} 1 & (n=0 \text{ のとき}) \\ \begin{cases} 1+X_{n-1}, & \text{確率 } 2^{-X_{n-1}} \text{ で} \\ X_{n-1}, & \text{確率 } 1-2^{-X_{n-1}} \text{ で} \end{cases} & (\text{その他}) \end{cases}$$

により定義する．

$$\mathrm{E}(2^{X_N} - 2) = N, \quad \mathrm{Var}(2^{X_N}) = \frac{1}{2}N(N+1)$$

を証明することは難しくない．したがって，この体系による確率的な更新は，N の不偏推定量を与える．Flajolet[45-50]は，より広く詳細に X_n の分布を調べた．

$$\mathrm{E}(X_N) = \frac{1}{\ln(2)}\ln(N) + \frac{1}{2} + \frac{\gamma}{\ln(2)} - \alpha + \sigma(n) + O\left(\frac{\ln(N)}{N}\right)$$
$$\sim \log_2(N) - 0.273948\cdots + \sigma(N)$$
$$\mathrm{Var}(X_N) \sim \frac{1}{12} + \frac{\pi^2}{6\ln(2)^2} - \alpha - \beta - \chi + \tau(n) \sim 0.763014\cdots + \tau(n)$$

である．α と β は前のとおりであり，新しい定数 χ は

$$\chi = \frac{1}{\ln(2)}\sum_{n=1}^{\infty}\frac{1}{n}\mathrm{csch}\left(\frac{2\pi^2 n}{\ln(2)}\right) = (1.237412\cdots) \times 10^{-12}$$

であり，$\sigma(n)$ と $\tau(n)$ は振動し，「無視できる」関数である．とくに，$\chi > 0$ であるから，$\mathrm{Var}(X_N)$ の定数項は，以前に与えた $\mathrm{Var}(f(x, M, 1))$ の定数項よりも（少しだけ）小さい．確率的数え上げアルゴリズムにおける類似の考えは，【6.8】にある．

[1] E. G. Coffman and J. Eve, File structures using hashing functions, *Commun. ACM* 13 (1970) 427-432.

[2] A. G. Konheim and D. J. Newman, A note on growing binary trees, *Discrete Math.* 4 (1973) 57-63; MR 47 #1650.

[3] P. Flajolet and R. Sedgewick, Digital search trees revisited, *SIAM Rev.* 15 (1986) 748-767; MR 87m:68014.

[4] D. E. Knuth, *The Art of Computer Programming*, v. 3, *Sorting and Searching*, Addison-Wesley, 1973, pp. 21, 134, 156, 493-502, 580, 685-686; MR 56 #4281.

[5] G. Louchard, Exact and asymptotic distributions in digital and binary search trees, *RAIRO Inform. Théor. Appl.* 21 (1987) 479-495; MR 89h:68031.

[6] H. M. Mahmoud, *Evolution of Random Search Trees*, Wiley, 1992, pp. 226-227, 260-291; MR 93f:68045.

[7] G. G. Brown and B. O. Shubert, On random binary trees, *Math. Oper. Res.* 9 (1984) 43-65; MR 86c:05099.

[8] P. Kirschenhofer and H. Prodinger, Further results on digital search trees, *Theoret. Comput. Sci.* 58 (1988) 143-154; MR 89j:68022.

[9] P. Kirschenhofer, H. Prodinger, and W. Szpankowski, Digital search trees again revisited: The internal path length perspective, *SIAM J. Comput.* 23 (1994) 598-616; MR 95i:68034.

[10] H. Prodinger, External internal nodes in digital search trees via Mellin transforms, *SIAM J. Comput.* 21 (1992) 1180-1183; MR 93i:68047.

[11] P. Flajolet and B. Richmond, Generalized digital trees and their difference-differential equations, *Random Structures Algorithms* 3 (1992) 305-320; MR 93f:05086.
[12] P. Kirschenhofer and H. Prodinger, Asymptotische Untersuchungen über charakteristische Parameter von Suchbäumen, *Zahlentheoretische Analysis II*, ed. E. Hlawka, Lect. Notes in Math. 1262, Springer-Verlag, 1987, pp. 93-107; MR 90h:68028.
[13] P. Kirschenhofer, H. Prodinger, and J. Schoissengeier, Zur Auswertung gewisser numerischer Reihen mit Hilfe modularer Funktionen, *Zahlentheoretische Analysis II*, ed. E. Hlawka, Lect. Notes in Math. 1262, Springer-Verlag, 1987, pp. 108-110; MR 90g:11057.
[14] G. H. Hardy and E. M. Wright, *An Introduction to the Theory of Numbers*, 5th ed., Oxford Univ. Press, 1985; Thms. 310, 350, 351, 353; MR 81i:10002.
[15] B. C. Berndt, *Ramanujan's Notebooks: Part I*, Springer-Verlag, 1985, p. 147; MR 86c:01062.
[16] P. Erdős, On arithmetical properties of Lambert series, *J. Indian Math. Soc.* 12 (1948) 63-66; MR 10,594c.
[17] D. H. Bailey and R. E. Crandall, Random generators and normal numbers, Perfectly Scientific Inc. preprint (2001).
[18] P. Erdős and R. L. Graham, *Old and New Problems and Results in Combinatorial Number Theory*, Enseignement Math. Monogr. 28, 1980, p. 62; MR 82j:10001.
[19] P. Flajolet and M. Golin, Exact asymptotics of divide-and-conquer recurrences, *Proc. 1993 Int. Colloq. on Automata, Languages and Programming (ICALP)*, Lund, ed. A. Lingas, R. Karlsson, and S. Carlsson, Lect. Notes in Comp. Sci. 700, Springer-Verlag, 1993, pp. 137-149.
[20] P. Flajolet and M. Golin, Mellin transforms and asymptotics: The mergesort recurrence, *Acta Inform.* 31 (1994) 673-696; MR 95h:68035.
[21] P.B. Borwein, On the irrationality of $\Sigma(1/(q^n+r))$, *J. Number Theory* 37 (1991) 253-259; MR 92b:11046.
[22] P. B. Borwein, On the irrationality of certain series, *Math. Proc. Cambridge Philos. Soc.* 112 (1992) 141-146; MR 93g:11074.
[23] J. Lynch, J. Mycielski, P. A. Vojta, and P. Bundschuh, Irrationality of an infinite product, *Amer. Math. Monthly* 87 (1980) 408-409.
[24] R.Wallisser, Rational approximation of the q-analogue of the exponential function and irrationality statements for this function, *Arch. Math. (Basel)* 44 (1985) 59-64;MR86i:11036.
[25] T. Amdeberhan and D. Zeilberger, q-Apéry irrationality proofs by q-WZ pairs, *Adv. Appl. Math.* 20 (1998) 275-283; MR 99d:11074.
[26] J. M. Borwein and P. B. Borwein, *Pi and the AGM: A Study in Analytic Number Theory and Computational Complexity*, Wiley, 1987, pp. 62-75, 91-101; MR 99h:11147.
[27] G. Gasper and M. Rahman, *Basic Hypergeometric Series*, Cambridge Univ. Press, 1990; MR 91d:33034.
[28] G. E. Andrews, *q-Series: Their Development and Application in Analysis, Number Theory, Combinatorics, Physics and Computer Algebra*, CBMS, v. 66, Amer. Math. Soc., 1986; MR 88b:11063.
[29] R. J. McIntosh, Some asymptotic formulae for q-hypergeometric series, *J. London Math. Soc.* 51 (1995) 120-136; MR 95m:11112.
[30] A. H. Beiler, *Recreations in the Theory of Numbers*, Dover, 1966.
[31] A. F. Horadam, Elliptic functions and Lambert series in the summation of reciprocals in certain recurrence-generated sequences, *Fibonacci Quart.* 26 (1988) 98-114; MR 89e:11013.
[32] P. Griffin, Acceleration of the sum of Fibonacci reciprocals, *Fibonacci Quart.* 30 (1992) 179-181; MR 93d:11024.
[33] F.-Z. Zhao, Notes on reciprocal series related to Fibonacci and Lucas numbers, *Fibonacci Quart.* 37 (1999) 254-257; MR 2000d:11020.
[34] R. André-Jeannin, Irrationalité de la somme des inverses de certaines suites récurrentes, *C. R. Acad. Sci. Paris Sér. I Math.* 308 (1989) 539-541; MR 90b:11012.
[35] P. Bundschuh and K. Väänänen, Arithmetical investigations of a certain infinite product,

Compositio Math. 91 (1994) 175-199; MR 95e:11081.
[36] D. Duverney, Irrationalité de la somme des inverses de la suite de Fibonacci, *Elem. Math.* 52 (1997) 31-36; MR 98c:11069.
[37] M. Prevost, On the irrationality of $\sum t^n/(A\alpha^n+B\beta^n)$, *J. Number Theory* 73 (1998) 139-161; MR 2000b:11085.
[38] J. E. Volder, The CORDIC trigonometric computing technique, *IRE Trans. Elec. Comput.* EC-8 (1959) 330-334; also in *Computer Arithmetic*, v. 1, ed. E. E. Swartzlander, IEEE Computer Soc. Press, 1990, pp. 226-230.
[39] J. S. Walther, A unified algorithm for elementary functions, *Spring Joint Computer Conf. Proc.* (1971) 379-385; also in *Computer Arithmetic*, v. 1, ed. E. E. Swartzlander, IEEE Computer Soc. Press, 1990, pp. 272-278.
[40] H. G. Baker, Complex Gaussian integers for 'Gaussian graphics,' *ACM Sigplan Notices*, v. 28 (1993) n. 11, 22-27.
[41] R. W. Gosper, Closed forms for $\prod_{n=1}^{\infty}(1+\exp(-\pi\xi n))$, unpublished note (1998).
[42] V. S. Dimitrov, G. A. Jullien, and W. C. Miller, Theory and applications for a doublebase number system, *IEEE Trans. Comput.* 48 (1999) 1098-1106; also in *Proc. 13th IEEE Symp. on Computer Arithmetic (ARITH)*, Asilomar, 1997, ed. T. Lang, J.-M. Muller, and N. Takagi, IEEE, 1997, pp. 44-51.
[43] N. J. A. Sloane, On-Line Encyclopedia of Integer Sequences, A018899.
[44] R. Morris, Counting large numbers of events in small registers, *Commun. ACM* 21 (1978) 840-842.
[45] P. Flajolet, Approximate counting: A detailed analysis, *BIT* 25 (1985) 113-134; MR 86j:68053.
[46] P. Kirschenhofer and H. Prodinger, Approximate counting:An alternative approach, *RAIRO Inform. Théor. Appl.* 25 (1991) 43-48; MR 92e:68070.
[47] H. Prodinger, Hypothetical analyses: Approximate counting in the style of Knuth, path length in the style of Flajolet, *Theoret. Comput. Sci.* 100 (1992) 243-251; MR 93j:68076.
[48] P. Kirschenhofer and H. Prodinger, A coin tossing algorithm for counting large numbers of events, *Math. Slovaca* 42 (1992) 531-545; MR 93j:68073.
[49] H. Prodinger, Approximate counting via Euler transform, *Math. Slovaca* 44 (1994) 569-574; MR 96h:11013.
[50] P. Kirschenhofer, A note on alternating sums, *Elec. J. Combin.* 3 (1996) R7;MR97e:05025.

5.15 最適停止定数

よく知られた**秘書問題**（secretary problem）[†]を考える．**応募者**（applicants）（異なる実数）$s_1, s_2, \cdots, s_n$が順序に関係なく並んだ数列があり，1度に1つずつ，あなたによって値を尋ねられる．sについて前もって何らの情報をもっていないが，nの値はわかっている．s_kの値を尋ねているので，s_kを採用してこの操作を終わるか，またはs_kを不採用にしてs_{k+1}の値を尋ねる．採用か不採用かの判断は，単に，すべての$1 \leq j < k$に対して$s_k > s_j$（すなわち，s_kが**候補者**（candidate））であるかどうかに基づかなけれ

[†]訳注：海辺の美女の問題ともよばれる．以下の記述のように一度調べた乱数を棄却して先へ進み，なるべく値の大きな候補を選ぶ確率的算法である．

ばならない．一度不採用とした応募者には，後で不採用を取り消すことができない．

目標は最も高い質をもつ応募者（最大の s_k）を選び出すことである．このとき，最適戦略は，最初から $m-1$ の応募者はすべて不採用とし，最初から m 以降の最初の条件を満たす候補者を採用することである．ただし[1-4]，$n \to \infty$ のとき

$$m = \min\left\{k \geq 1 : \sum_{j=k+1}^{n} \frac{1}{j-1} \leq 1\right\} \sim \frac{n}{e}$$

とする．だから，この戦略で最もよい応募者を得る漸近的確率は，$1/e = 0.3678794411\cdots$ である．ただし，e は自然対数の底である【1.3】．この一般化は[5-7]にあるので参照せよ．

目標を変える．選んだ応募者の期待ランク R_n（最大の s_k はランク1であり，2番目に大きなものはランク2である，以下同様）を最小にすることに変えると，異なる式となる．Lindley（リンドレイ）[8]とChowら[9]は，この場合の最適戦略を導き[10]

$$\lim_{n \to \infty} R_n = \prod_{k=1}^{\infty}\left(1 + \frac{2}{k}\right)^{1/(k+1)} = 3.8695192413\cdots = C$$

を証明した．

変形したものには，前もって $s_1, s_2, \cdots, s_n$ が独立で，区間 $[0,1]$ 上の一様分布である，と知っている場合を含むものでもよい．これは，（上に議論した，情報のない問題とは対照的に）**完全情報問題**（full-information problem）として知られている．分布の情報がどれほど成功の機会を改善するのか？「最上だけ」（nothing but the best）という目標に対して，Gilbert & Mosteller（ギルバート＆モステレ）[11]は，成功の確率を[12,13]

$$e^{-a} - (e^a - a - 1)\operatorname{Ei}(-a) = 0.5801642239\cdots$$

と計算した．ただし，$a = 0.8043522628\cdots$ は方程式 $\operatorname{Ei}(a) - \gamma - \ln(a) = 1$ の一意の実数解である．Ei は指数積分【6.2】であり，γ はオイラー-マスケローニの定数【1.5】である．

$\lim_{n \to \infty} R_n$ に対する，完全情報問題と類似なものは，未解決問題として[14-16]にある．しかしながら，また1つ別の目標は，雇用される者の期待品質 Q_n それ自身（k 番目の応募者は品質 s_k をもつ）を最大にすることかもしれない．明らかに，

$$Q_0 = 0, \quad Q_n = \frac{1}{2}(1 + Q_{n-1}^2) \quad (n \geq 1)$$

で，$n \to \infty$ のとき $Q_n \to 1$ である．Moser[11,17-19]は

$$Q_n \sim 1 - \frac{2}{n + \ln(n) + b}$$

を導いた．定数 b は，$1.76799378\cdots$ であると[10]で推定されている．

密接に関連する問題がある．$s_1, s_2, \cdots, s_n$ を，独立で，区間 $[0, N]$ 上に一様分布する変数と仮定する．目標は，期待する品質が $\geq N-1$ の応募者を選び出すために必要な質問の回数 T_N を最小にすることである．Gum[20]は，$N \to \infty$ のとき $T_N = 2N - O(\ln(N))$ であることの証明の概要を述べた．代わりに，$s'_1, s'_2, \cdots, s'_n$ は集合 $\{1, 2, \cdots, N\}$ から復元抽出で選ばれることを除いて，すべて前のように仮定する．このとき，$T'_N = cN$

$+O(\sqrt{N})$ であることが証明できる. ただし[10]

$$c=2\sum_{k=3}^{\infty}\frac{\ln(k)}{k^2-1}-\frac{\ln(2)}{3}=1.3531302722\cdots=\ln(C)$$

である.

秘書問題とそれから派生した問題は，**最適停止**（optimal stopping）の理論分野であり[19]，具体的な演習問題がある．正しくできた硬貨を繰り返し投げることを観察し，いつでも観察を停止することができるとする．停止したとき表が観察できた平均回数を，支払いとする．このとき，期待支払いを最大にする最上の戦略は何であるのか？ Chow & Robbins[21,22]は，期待支払い$>0.79=(0.59+1)/2$を達成できる戦略を述べた．

[1] T. S. Ferguson, Who solved the secretary problem?, *Statist. Sci.* 4 (1989) 282-296; MR 91k:01011.
[2] P. R. Freeman, The secretary problem and its extensions: A review, *Int. Statist. Rev.* 51 (1983) 189-206; MR 84k:62115.
[3] F. Mosteller, *Fifty Challenging Problems in Probability with Solutions*, Addison-Wesley, 1965, pp. 12, 74-79; MR 53 # 1666.
[4] M. Gardner, *New Mathematical Diversions from Scientific American*, Simon and Schuster, 1966, pp. 35-36, 41-43.
[5] A.Q. Frank and S. M. Samuels, On an optimal stopping problem of Gusein-Zade, *Stochastic Process. Appl.* 10 (1980) 299-311; MR 83f:60067.
[6] M. P. Quine and J. S. Law, Exact results for a secretary problem, *J. Appl. Probab.* 33 (1996) 630-639; MR 97k:60127.
[7] G. F. Yeo, Duration of a secretary problem, *J. Appl. Probab.* 34 (1997) 556-558; MR 98a:60056.
[8] D. V. Lindley, Dynamic programming and decision theory, *Appl. Statist.* 10 (1961) 39-51; MR 23 # A740.
[9] Y. S. Chow, S. Moriguti, H. Robbins, and S. M. Samuels, Optimal selection based on relative rank (the "secretary problem"), *Israel J. Math.* 2 (1964) 81-90; MR 31 # 855.
[10] P. Sebah, Computations of several optimal stopping constants, unpublished note (2001).
[11] J. P. Gilbert and F. Mosteller, Recognizing the maximum of a sequence, *J. Amer. Statist. Assoc.* 61 (1966) 35-73; MR 33 # 6792.
[12] S. M. Samuels, Exact solutions for the full information best choice problem, Purdue Univ. Stat. Dept. report 82-17 (1982).
[13] S. M. Samuels, Secretary problems, in *Handbook of Sequential Analysis*, ed. B. K. Ghosh and P. K. Sen, Dekker, 1991, pp. 381-405; MR 93g:62102.
[14] F. T. Bruss and T. S. Ferguson, Minimizing the expected rank with full information, *J. Appl. Probab.* 30 (1993) 616-626; MR 94m:60090.
[15] F. T. Bruss and T. S. Ferguson, Half-prophets and Robbins' problem of minimizing the expected rank, *Proc. 1995 Athens Conf. on Applied Probability and Time Series Analysis*, v. 1, ed. C. C. Heyde, Yu. V. Prohorov, R. Pyke, and S. T. Rachev, Lect. Notes in Statist. 114, Springer-Verlag, 1996, pp. 1-17; MR 98k:60066.
[16] D. Assaf and E. Samuel-Cahn, The secretary problem: Minimizing the expected rank with i.i.d. random variables, *Adv. Appl. Probab.* 28 (1996) 828-852; MR 97f:60089.
[17] L. Moser, On a problem of Cayley, *Scripta Math.* 22 (1956) 289-292.
[18] I. Guttman, On a problem of L. Moser, *Canad. Math. Bull.* 3 (1960) 35-39;MR23 # B2064.
[19] T. S. Ferguson, *Optimal Stopping and Applications*, unpublished manuscript (2000).
[20] B. Gum, The secretary problem with a uniform distribution, presentation at *Tenth SIAM Conf. on Discrete Math.*, Minneapolis, 2000.

[21] Y. S. Chow and H. Robbins, On optimal stopping rules for s_n/n, *Illinois J. Math.* 9 (1965) 444-454; MR 31 #4134.

[22] T. S. Ferguson, *Mathematical Statistics: A Decision Theoretic Approach*, Academic Press, 1967, p. 314; MR 35 #6231.

5.16 極値定数

$X_1, X_2, \cdots, X_n$ は，連続確率密度関数 $f(x)$ をもつ母集団からのランダムサンプルとする．多くのおもしろい結果が，**順序統計量**（order statistics）
$$X^{<1>} < X^{<2>} < \cdots < X^{<n>}$$
の分布に関して存在する．ただし，$X^{<1>} = \min\{X_1, X_2, \cdots, X_n\} = m_n$, $X^{<n>} = \max\{X_1, X_2, \cdots, X_n\} = M_n$ である．簡略のために，最大値 M_n だけに焦点をあてる．

$X_1, X_2, \cdots, X_n$ を，一様分布 $[0,1]$ からとられているとする（すなわち，$0 \leq x \leq 1$ のとき $f(x) = 1$ であり，その他の場合は $f(x) = 0$ である）．このとき，M_n の確率分布は
$$P(M_n < x) = \begin{cases} 0 & (x < 0 \text{ のとき}) \\ x^n & (0 \leq x \leq 1 \text{ のとき}) \\ 1 & (x > 1 \text{ のとき}) \end{cases}$$
と書け，そのモーメントは
$$\mu_n = E(M_n) = \frac{n}{n+1}, \quad \sigma_n^2 = \text{Var}(M_n) = \frac{n}{(n+1)^2(n+2)}$$
で与えられる．これらはすべて厳密な結果である [1-3]．明らかに，
$$\lim_{n \to \infty} P(n(M_n - 1) < y) = \lim_{n \to \infty} P\left(M_n < 1 + \frac{1}{n}y\right) = \begin{cases} e^y & (y < 0 \text{ のとき}) \\ 1 & (y \geq 0 \text{ のとき}) \end{cases}$$
が成り立つことに注意せよ．この漸近結果は，Fisher & Tippett（フィッシャー&ティペット）[4] と Gnedenko（グネデンコ）[5] による，内容のより深い一般的な定理の特別な場合である．広い状況の下で，（うまく正規化された）M_n の漸近分布は，ちょうど 3 つの可能な分布族のうちの 1 つに属さなければならない．別なものでそれほど自明でない，次の例を調べる．

$X_1, X_2, \cdots, X_n$ を，正規分布 $N(0,1)$ から取られているとする．すなわち密度関数 $f(x)$ および分布関数 $F(x)$ は，
$$f(x) = \frac{1}{\sqrt{2\pi}} \exp\left(-\frac{x^2}{2}\right), \quad F(x) = \int_{-\infty}^{x} f(\xi) \, d\xi = \frac{1}{2} \text{erf}\left(\frac{x}{\sqrt{2}}\right) + \frac{1}{2}$$
である．このとき，M_n の確率分布は，
$$P(M_n < x) = F(x)^n = n \int_{-\infty}^{x} F(\xi)^{n-1} f(\xi) \, d\xi$$
であり，モーメントは

$$\mu_n = n\int_{-\infty}^{\infty} xF(x)^{n-1}f(x)\,dx, \qquad \sigma_n^2 = n\int_{-\infty}^{\infty} x^2 F(x)^{n-1}f(x)\,dx - \mu_n^2$$

である．小さな n に対して，厳密な式は可能であり [2,3,6-11]

$$\mu_2 = \frac{1}{\sqrt{\pi}} = 0.564\cdots, \qquad \sigma_2^2 = 1 - \mu_2^2 = 0.681\cdots$$

$$\mu_3 = \frac{3}{2\sqrt{\pi}} = 0.846\cdots, \qquad \sigma_3^2 = 1 + \frac{\sqrt{3}}{2\pi} - \mu_3^2 = 0.559\cdots$$

$$\mu_4 = \frac{3}{\sqrt{\pi}}(1 - 2S_2) = 1.029\cdots, \qquad \sigma_4^2 = 1 + \frac{\sqrt{3}}{\pi} - \mu_4^2 = 0.491\cdots$$

$$\mu_5 = \frac{5}{\sqrt{\pi}}(1 - 3S_2) = 1.162\cdots, \qquad \sigma_5^2 = 1 + \frac{5\sqrt{3}}{2\pi}(1 - 2S_3) - \mu_5^2 = 0.447\cdots$$

$$\mu_6 = \frac{15}{2\sqrt{\pi}}(1 - 4S_2 + 2T_2) = 1.267\cdots, \qquad \sigma_6^2 = 1 + \frac{5\sqrt{3}}{\pi}(1 - 3S_3) - \mu_6^2 = 0.415\cdots$$

$$\mu_7 = \frac{21}{2\sqrt{\pi}}(1 - 5S_2 + 5T_2) = 1.352\cdots, \qquad \sigma_7^2 = 1 + \frac{35\sqrt{3}}{4\pi}(1 - 4S_3 + 2T_3) - \mu_7^2 = 0.391\cdots$$

である．ただし，

$$S_k = \frac{\sqrt{k}}{\pi}\int_0^{\pi/4} \frac{dx}{\sqrt{k+\sec(x)^2}} = \frac{1}{\pi}\arcsin\sqrt{\frac{k}{2(1+k)}}$$

$$T_k = \frac{\sqrt{k}}{\pi^2}\int_0^{\pi/4}\int_0^{\pi/4} \frac{dxdy}{\sqrt{k+\sec(x)^2+\sec(y)^2}} = \frac{1}{\pi^2}\int_0^{\pi S(x)} \arcsin\sqrt{\frac{1}{2}\frac{k(k+1)}{k(k+2)-\tan(z)^2}}\,dz$$

である．$\mu_8 = 1.423\cdots$ と $\sigma_8^2 = 0.372\cdots$ に対する類似の式を発見することが残っている．Ruben[12]は，順序統計量のモーメントとある種の超球面単体 (hyperspherical simplices, 一般化された球面三角形) の体積との関係を示した．Calkin[13]は，極限の場合，μ_3 の厳密な式を得る二項恒等式 (binomial identity) を発見した．

M_n の漸近分布に戻る．

$$a_n = \sqrt{2\ln(n)} - \frac{1}{2}\frac{\ln(\ln(n)) + \ln(4\pi)}{\sqrt{2\ln(n)}}$$

とする．[14-18]から，

$$\lim_{n\to\infty} \mathrm{P}(\sqrt{2\ln(n)}\,(M_n - a_n) < y) = \exp(-e^{-y})$$

と，2重指数密度関数 $g(y) = \exp(-y - e^{-y})$ は右へ歪曲している (Gumbel (ガンベル) またはFisher-Tippett (フィッシャー-ティペット) I型極値密度とよばれている), ことが証明できる．Gumbel の式に従い分布するランダム変数 Y は[4]

$$\mathrm{E}(Y) = \gamma = 0.577215\cdots, \quad \mathrm{Skew}(Y) = \frac{\mathrm{E}[Y - \mathrm{E}(Y))^3]}{\mathrm{Var}(Y)^{3/2}} = \frac{12\sqrt{6}}{\pi^3}\zeta(3) = 1.139547\cdots$$

$$\mathrm{Var}(Y) = \frac{\pi^2}{6} = 1.644934\cdots, \quad \mathrm{Kurt}(Y) = \frac{\mathrm{E}[Y - \mathrm{E}(Y))^4]}{\mathrm{Var}(Y)^2} - 3 = \frac{12}{5} = 2.4$$

を満たす．ただし，γ はオイラー-マスケローニの定数【1.5】であり，$\zeta(3)$ は Apéry (アペリ) の定数【1.6】である．(ある研究者は，歪度 (skewness) の「平方」を報告している．[2]では1.2986と見積もり，[19]では1.3と見積もっている．) 定数 $\zeta(3)$ は，

さらに[20]にもある．$g(y)$のような2重指数関数は他の場所でも現れる（【2.13】，【5.7】，【6.10】を参照せよ）．

よく知られた中心極限定理により，独立で同一分布に従う多くの変数の「和」は，共通のもとの分布がどのようであろうとも，漸近的に正規分布になる．よく似た状況が，極値理論で成り立つ．（正規化された）M_nの漸近分布は，次の分布族のうちの1つに属さなければならない[2, 14-17]．

$$G_{1,a}(y) = \begin{cases} 0 & (y \leq 0 \text{ のとき}) \\ \exp(-y^{-a}) & (y > 0 \text{ のとき}) \end{cases} \quad \text{「Fréchet（フレッシェ）」またはII型}$$

$$G_{2,a}(y) = \begin{cases} \exp(-(-y)^{-a}) & (y \leq 0 \text{ のとき}) \\ 1 & (y > 0 \text{ のとき}) \end{cases} \quad \text{「Weibull（ワイブル）」またはIII型}$$

$$G_3(y) = \exp(-e^{-y}) \qquad \text{「Gumbel（ガンベル）」またはI型}$$

ただし，$a>0$は任意の形（shape）の変数である．$G_{2,1}(y)$は一様分布するXに，$G_3(y)$は正規分布しているXに関する議論に現れた．「吸収領域」(domain of attraction) に属することを確かめるために，Xの分布Fについて多くを知る必要がないことがわかる．Fの裾（tail）のふるまいは，本質的な部分である．これらの3つの分布族は，さらに1つにまとめられる．

$$H_\beta(y) = \exp(-(1+\beta y)^{-1/\beta}) \quad (1+\beta y > 0), \quad H_0(y) = \lim_{\beta \to 0} H_\beta(y)$$

とすると，$\beta>0$，$\beta<0$，$\beta=0$それぞれに対応して，3つの場合に帰着する．

上に述べたこととランダム行列理論（RTMと略記）との間には，すばらしい関連がある．ランダムな対角要素$X_1, X_2, \cdots, X_n$をもつ$n \times n$対角行列を，はじめに考える．もちろん，最大固有値はMである．ランダムな$n \times n$複素エルミート行列を考える．これは$X_{ij} = \overline{X}_{ij}$を意味するから，対角成分は実数であり，対角成分でない要素は共役対称性の条件を満たす．さらに，すべての固有値は実数である．このような行列を生成する「自然な」方法は，いわゆるガウスユニタリー集合 (Gaussian unitary ensemble, GUE) の確率分布に従うことである[21]．最大固有値の厳密なモーメントの式は，前に議論した対角正規分布の場合と同様に，小さいnに対して，ここでも存在する[22]．固有値は，対角行列の場合には独立である．しかし，一般のエルミート行列では強い従属性がある．RTMはいくつかの方面で重要である．第1に，Riemannのゼータ関数の自明でない零点の間の間隔の分布は，GUEの固有値分布と密接に関係しているようである【2.15.3】．第2に，RTMは【5.20】で議論する，最長増加部分数列問題 (longest increasing subsequence problem) を解くときにきわめて重要であり，その手法は2次元イジングモデル (Ising model) を理解する上で役に立つ【5.22】．最後に，RTMは原子エネルギー準位の物理学に関連している．しかし，これに関しては複雑であるので，ここでは述べることができない．

[1] J. D. Gibbons, *Nonparametric Statistical Inference*, McGraw-Hill, 1971; MR 86m:62067.
[2] H. A. David, *Order Statistics*, 2nd ed., Wiley, 1981; MR 82i:62073.

[3] N. Balakrishnan and A. Clifford Cohen, *Order Statistics and Inference*, Academic Press, 1991; MR 92k:62098.
[4] R. A. Fisher and L. H. C. Tippett, Limiting forms of the frequency distribution of the largest or smallest member of a sample, *Proc. Cambridge Philos. Soc.* 24 (1928) 180-190.
[5] B. Gnedenko, Sur la distribution limite du terme maximum d'une série aléatoire, *Annals of Math.* 44 (1943) 423-453; MR 5,41b.
[6] H. L. Jones, Exact lower moments of order statistics in small samples from a normal distribution, *Annals of Math. Statist.* 19 (1948) 270-273; MR 9,601d.
[7] H. J. Godwin, Some low moments of order statistics, *Annals of Math. Statist.* 20 (1949) 279-285; MR 10,722f.
[8] H. Ruben, On the moments of the range and product moments of extreme order statistics in normal samples, *Biometrika* 43 (1956) 458-460; MR 18,607d.
[9] J. K. Patel and C. B. Read, *Handbook of the Normal Distribution*, Dekker, 1982, pp. 238-241; MR 83j:62002.
[10] Y. Watanabe, M. Isida, S. Taga, Y. Ichijo, T. Kawase, G. Niside, Y. Takeda, A. Horisuzi, and I. Kuriyama, Some contributions to order statistics, *J. Gakugei, Tokushima Univ.* 8 (1957) 41-90; MR 20 #5545.
[11] Y. Watanabe, T. Yamamoto, T. Sato, T. Fujimoto, M. Inoue, T. Suzuki, and T. Uno, Some contributions to order statistics (continued), *J. Gakugei, Tokushima Univ.* 9 (1958) 31-86; MR 21 #2330.
[12] H. Ruben, On the moments of order statistics in samples from normal populations, *Biometrika* 41 (1954) 200-227; also in *Contributions to Order Statistics*, ed. A. E. Sarhan and B. G. Greenberg, Wiley, 1962, pp. 165-190; MR 16,153c.
[13] N. J. Calkin, A curious binomial identity, *Discrete Math.* 131 (1994) 335-337; MR 95i:05002.
[14] J. Galambos, *The Asymptotic Theory of Extreme Order Statistics*, 2nd ed., Krieger, 1987; MR 89a:60059.
[15] R.-D. Reiss, *Approximate Distributions of Order Statistics*, Springer-Verlag, 1989; MR 90e:62001.
[16] J. Galambos, Order statistics, *Handbook of Statistics*, v. 4, *Nonparametric Methods*, ed. P. R. Krishnaiah and P. K. Sens, Elsevier Science, 1984, pp. 359-382; MR 87g:62001.
[17] M. R. Leadbetter, G. Lindgren, and H. Rootzén, *Extremes and Related Properties of Random Sequences and Processes*, Springer-Verlag, 1983; MR 84h:60050.
[18] P. Hall, On the rate of convergence of normal extremes, *J. Appl. Probab.* 16 (1979) 433-439; MR 80d:60025.
[19] M. Abramowitz and I. A. Stegun, *Handbook of Mathematical Functions*, Dover, 1972, p. 930; MR 94b:00012.
[20] P. C. Joshi and S. Chakraborty, Moments of Cauchy order statistics via Riemann zeta functions, *Statistical Theory and Applications*, ed. H. N. Nagaraja, P. K. Sen, and D. F. Morrison, Springer-Verlag, 1996, pp. 117-127; MR 98f:62149.
[21] C. A. Tracy and H. Widom, Universality of the distribution functions of random matrix theory, *Integrable Systems: From Classical to Quantum*, Proc. 1999 Montréal conf., ed. J. Harnad, G. Sabidussi, and P. Winternitz, Amer. Math. Soc, 2000, pp. 251-264; math-ph/9909001; MR 2002f:15036.
[22] J. Gravner, C. A. Tracy, and H. Widom, Limit theorems for height fluctuations in a class of discrete space and time growth models, *J. Stat. Phys.* 102 (2001) 1085-1132; math.PR/0005133; MR 2002d:82065.

5.17 パターンをもたない語の定数[†]

$a, b, c, \cdots$ を有限種類のアルファベットの文字とする．**語**（word）とは文字の有限列のことである．例を2つあげると，$abcacbacbc$ と $abcacbabcb$ である．**平方**（square）とは，空集合でない x に対して，xx の形の語のことである．**無平方**（square-free）とは，因子として平方を含まないときをいう．最初の例は平方 $acbacb$ を含む．他方，2番目は無平方である．次の質問をする：長さ n の無平方の語は何個存在するのか？

2種類の文字の場合は，無平方の語は a, b, ab, ba, aba, bab だけである．だから，**2値**（binary）無平方語はおもしろくない．しかしながら，任意に長い **3値**（ternary）無平方語が，3種類のアルファベット上で存在する．この事実は，現在 Prouhet-Thue-Morse（プルーエ-トゥエ-モース）数列【6.8】とよばれるものを用いて，Thue [1,2]によって初めて証明された．このような語の，正確な漸近的な数え上げは，複雑である[3-7]．Brandenburg（ブランデンブルグ）[8]は，長さ $n>24$ の3値無平方な語の個数 $s(n)$ が

$$6 \cdot 1.032^n < 6 \cdot 2^{n/22} \leq s(n) \leq 6 \cdot 1172^{(n-2)/22} < 3.157 \cdot 1.379^n$$

を満たすことを証明した．Brinkhuis（ブリンクハウス）[9]は，ある定数 $A>0$ に対して，$s(n) \leq A \cdot 1.316^n$ を証明した．Noonan & Zeilberger（ヌーナン&ザイルバーガー）[10]は，ある定数 $A'>0$ に対して，上界を $A' \cdot 1.302128^n$ に改良し，極限の厳密でない推定値

$$S = \lim_{n \to \infty} s(n)^{1/n} = 1.302 \cdots$$

を得た．独立な計算では[11]，$k>3$ での，k 文字アルファベットの S の推定値のみではなく，$S = \exp(0.263719 \cdots) = 1.301762 \cdots$ を得ている．Ekhad & Zeilberger（エッカド&ザイルバーガー）[12]は最近15年間で下界の最初の改良，$1.041^n < 2^{n/17} \leq s(n)$ を証明した．自己回避歩行【5.10】に関連するある種の定数 μ と同じ意味で，S は結合定数（connective constant）であることに注意せよ．実際，Noonan & Zeilberger の S の計算は，μ と結びつけて使う同じ Goulden-Jackson の技術に基づいている．

無立方語（cube-free word）とは，x が空集合でなく，xxx の形の因子を含まないことをいう．Prouhet-Thue-Morse 数列は，任意に長い2値無立方語の例を与える．Brandenburg[8]は，長さ $n>18$ の2値無立方語の個数 $c(n)$ が，

$$2 \cdot 1.080^n < 2 \cdot 2^{n/9} \leq c(n) \leq 2 \cdot 1251^{(n-1)/17} < 1.315 \cdot 1.522^n$$

を満たすことを証明した．Edlin（エドリン）[13]は，ある定数 $B>0$ に対して，上界

[†]訳注：この節の原題 Pattern-Free Word Constants は，ある種のパターン，たとえば同じ語の繰り返しを含まない語などに関する諸定数を意味する．

を $B \cdot 1.45757921^n$ に改良した．さらに，Edlin は極限の厳密でない推定値
$$C=\lim_{n\to\infty} c(n)^{1/n}=1.457\cdots$$
を得た．

x が空集合でなく，$xyxyx$ の形の因子が含まれないとき，語は**無重複**（**オーバーラップフリー**，overlap-free）という．プルーエ-トゥエ-モース数列は，再び，任意に長い無重複語の例を与える．無平方語は無重複語でなければならないことと，無重複語は無立方語でなければならないこと，に注意せよ．実際，重複は，任意に長い2値語において，避けられる最も短いパターンである．長さ n の2値無重複語の個数 $t(n)$ [14,15] は，ある定数 p, q に対して，
$$p \cdot n^{1.155} \leq t(n) \leq q \cdot n^{1.587}$$
を満たす．したがって，$t(n)$ の増大は，単に多項式オーダーであり，$s(n)$ や $c(n)$ とは異なる．Cassaigne（カサイニュ）[16]は，興味ある事実として，極限値 $\lim_{n\to\infty} \ln(t(n))/\ln(n)$ は存在しないが，
$$1.155 < T_L = \liminf_{n\to\infty} \frac{\ln(t(n))}{\ln(n)} < 1.276 < 1.332 < T_U = \limsup_{n\to\infty} \frac{\ln(t(n))}{\ln(n)} < 1.587$$
が成り立つことを証明した．（実際，彼はもっと多くのことを証明している．）【2.16】で，同様なよくない漸近的ふるまいを調べた．

アーベル平方（abelian square）とは，空集合でない x と x の文字を置換した語 x' による，語 xx' のことである．語が**無アーベル平方**（abelian square-free）とは，因子としてアーベル平方を含まない，ときをいう．語 $abcacbabcb$ はアーベル平方 $abcacb$ を含む．実際，少なくとも長さ8の3値語はアーベル平方を含む．Pleasants（プレサンツ）[17]は，5つの文字を基礎にした，任意に長い無アーベル平方語が存在することを証明した．4文字の場合は，最近まで未解決問題のままであったが，Keränen（ケレーネン）[18]は，任意に長い4値無アーベル平方語も存在することを証明した．Carpi（カルピ）[19]は，さらにこれらの個数 $h(n)$ は
$$\liminf_{n\to\infty} h(n)^{1/n} > 1.000021$$
を満たさなければならないことを証明した．彼は，「…右辺の値が1に近いということは，たぶん，最適な値からほど遠いと思われる」と書いている．

3値語 w が**部分アーベル平方**（partially abelian square）とは，空集合でない語 x と文字 b を固定し，隣接した文字 a と c の交換のみを許す x の置換 x' で，$w=xx'$ と書けるときをいう．たとえば，語 $bacbca$ は部分アーベル平方である．語が**無部分アーベル平方**（partially abelian square-free）とは，因子として部分アーベル平方を含まないときをいう．Cori & Formisano（コリ&フォルミサノ）[20]は，無部分アーベル平方語の個数の範囲を導くために，$t(n)$ に Kobayashi（小林良和）の不等式を用いた．

Kolpakov & Kucherov（コルパコフ&クシェロフ）[21,22]は次のことを問題にした．無限の無平方3値語（infinite square-free ternary words）において，1文字の最小の比率は何であるのか？ Tarannikov（タランニコフ）は，彼の仕事の続き[23]

で，答えは $0.2746\cdots$ であることを示唆した．

k 文字アルファベット上の語が**原始的** (primitive) とは，ある部分語のベキ（繰り返し）でないときをいう．長さ n の原始的な語の個数は，$\sum_{d|n} \mu(d) k^{n/d}$ である．ただし，$\mu(d)$ はメビウス関数である【2.2】．したがって，原始的な語の比率は，$n \to \infty$ のとき，1 に近づくことが簡単に示される．一方，ベキを「含ま」ない語をすべて数える問題は，おそらく無平方語や無立方語などを数えるのと同程度に難しい．

長さ n の 2 値語 $w_1 w_2 w_3 \cdots w_n$ が**偽造不可能** (unforgeable) とは，それ自身の左または右へのシフトとは決して一致しないときをいう．すなわち，u_i または v_j と任意の $1 \leq m \leq n-1$ の任意の可能な選び方に対して，$u_1 u_2 \cdots u_m w_1 w_2 \cdots w_{n-m}$ または $w_{m+1} w_{m+2} \cdots w_n v_1 v_2 \cdots v_m$ のどれかに決して一致しないことである．たとえば，$m=n-1$ のときを考えると，$w_1 = w_n$ であることはない．長さ n の偽造不可能語の個数を $f(n)$ とする．例からただちに

$$0 \leq \rho = \lim_{n \to \infty} \frac{f(n)}{2^n} \leq \frac{1}{2}$$

を得る．さらに，母関数によって [7, 25-27]，

$$\rho = \sum_{n=1}^{\infty} (-1)^{n-1} \frac{2}{2^{(2^{n+1}-1)}-1} \prod_{m=2}^{n} \frac{2^{(2^{m-1})}}{2^{(2^{m-1})}-1} = 0.2677868402\cdots$$
$$= 1 - 0.7322131597\cdots$$

である．この級数は極端に速く収束する．

[1] A. Thue, Über die gegenseitige Lage gleicher Teile gewisser Zeichenreihen, *Videnskapsselskapets Skrifter I, Matematisk-Naturvidenskapelig Klasse, Kristiania*, n. 1, Dybwad, 1912, pp. 1-67; also in *Selected Mathematical Papers*, ed. T. Nagell, A. Selberg, S. Selberg, and K. Thalberg, Universitetsforlaget, 1977, pp. 413-478; MR 57 #46.

[2] J.-P. Allouche and J. Shallit, The ubiquitous Prouhet-Thue-Morse sequence, *Sequences and Their Applications (SETA)*, Proc. 1998 Singapore conf., ed. C. Ding, T. Helleseth, and H. Niederreiter, Springer-Verlag, 1999, pp. 1-16; MR 2002e:11025.

[3] M. Lothaire, *Combinatorics on Words*, Addison-Wesley, 1983; MR 98g:68134.

[4] J. Berstel, Some recent results on squarefree words, *Proc. 1984 Symp. on Theoretical Aspects of Computer Science (STACS)*, Paris, ed. M. Fontet and K. Mehlhorn, Lect. Notes in Comp. Sci. 166, Springer-Verlag, 1984, pp. 14-25; MR 86e:68056.

[5] J. Currie, Open problems in pattern avoidance, *Amer. Math. Monthly* 100 (1993) 790-793.

[6] M. Lothaire, *Algebraic Combinatorics on Words*, Cambridge Univ. Press, 2002.

[7] N. J. A. Sloane, On-Line Encyclopedia of Integer Sequences, A003000, A006156, A028445, and A007777.

[8] F.-J. Brandenburg, Uniformly growing k th power-free homomorphisms, *Theoret. Comput. Sci.* 23 (1983) 69-82; MR 84i:68148.

[9] J. Brinkhuis, Non-repetitive sequences on three symbols, *Quart. J. Math.* 34 (1983) 145-149; MR 84e:05008.

[10] J. Noonan and D. Zeilberger, The Goulden-Jackson cluster method: Extensions, applications and implementations, *J. Differ. Eq. Appl.* 5 (1999) 355-377; MR 2000f:05005.

[11] M. Baake, V. Elser, and U. Grimm, The entropy of square-free words, *Math. Comput. Modelling* 26 (1997) 13-26; math-ph/9809010; MR 99d:68202.

[12] S. B. Ekhad and D. Zeilberger, There are more than $2^{n/17}$ n-letter ternary square-free words,

J. Integer Seq. 1 (1998) 98.1.9.
[13] A. E. Edlin, The number of binary cube-free words of length up to 47 and their numerical analysis, *J. Differ. Eq. Appl.* 5 (1999) 353-354.
[14] A. Restivo and S. Salemi, Overlap-free words on two symbols, *Automata on Infinite Words*, ed. M. Nivat and D. Perrin, Lect. Notes in Comp. Sci. 192, Springer-Verlag, 1985, pp. 198-206; MR 87c:20101.
[15] Y. Kobayashi, Enumeration of irreducible binary words, *Discrete Appl. Math.* 20 (1988) 221-232; MR 89f:68036.
[16] J. Cassaigne, Counting overlap-free binary words, *Proc. 1993 Symp. on Theoretical Aspects of Computer Science (STACS)*, Würzburg, ed. P. Enjalbert, A. Finkel, and K. W. Wagner, Lect. Notes in Comp. Sci. 665, Springer-Verlag, 1993, pp. 216-225; MR 94j:68152.
[17] P. A. B. Pleasants, Non-repetitive sequences, *Proc. Cambridge Philos. Soc.* 68 (1970) 267-274; MR 42 #85.
[18] V.Keränen, Abelian squares are avoidable on 4 letters, *Proc. 1992 Int. Colloq. on Automata, Languages and Programming (ICALP)*, Vienna, ed. W. Kuich, Lect. Notes in Comp. Sci. 623, Springer-Verlag, 1992, pp. 41-52; MR 94j:68244.
[19] A. Carpi, On the number of abelian square-free words on four letters, *Discrete Appl. Math.* 81 (1998) 155-167; MR 98j:68139.
[20] R. Cori and M. R. Formisano, On the number of partially abelian square-free words on a three-letter alphabet, *Theoret. Comput. Sci.* 81 (1991) 147-153; MR 92i:68129.
[21] R. Kolpakov and G. Kucherov, Minimal letter frequency in n^{th} power-free binary words, *Mathematical Foundations of Computer Science (MFCS)*, Proc. 1997 Bratislava conf., ed. I. Prívara and P. Ruzicka, Lect. Notes in Comp. Sci. 1295, Springer-Verlag, pp. 347-357; MR 99d:68206.
[22] R. Kolpakov, G. Kucherov, and Y. Tarannikov, On repetition-free binary words of minimal density, *Theoret. Comput. Sci.* 218 (1999) 161-175; MR 2000b:68180.
[23] Y. Tarannikov, Minimal letter density in infinite ternary square-free word is 0.2746 . . . , *Proc. Two Joint French-Russian Seminars on Combinatorial and Algorithmical Properties of Discrete Structures*, Moscow State Univ., 2001, pp. 51-56.
[24] H. Petersen, On the language of primitive words, *Theoret. Comput. Sci.* 161 (1996) 141-156; MR 97e:68065.
[25] P. T. Nielsen, A note on bifix-free sequences, *IEEE Trans. Inform. Theory* 19 (1973) 704-706; MR 52 #2724.
[26] G. Blom and O. P. Lossers, Overlapping binary sequences, *SIAM Rev.* 37 (1995) 619-620.
[27] D. J. Greaves and S. J. Montgomery-Smith, Unforgeable marker sequences, unpublished note (2000).

5.18 浸透のクラスター密度定数

浸透理論(パーコレーション理論,percolation)はランダム媒質での流体の流れに関する理論である.たとえば,多孔性固体(porus solid)を通り抜ける分子,または野火が森を焼き尽くすような場合である.Broadbent & Hammersley(ブロードベント&ハマースレイ)[1-3]は流体の通路に関する媒質の開いた経路の,起こりそうな個数および構造について考えた.彼らの疑問の答えは,研究の新しい分野を創った[4-

5.18 浸透のクラスター密度定数

10]．分野は広大なので，いくつかの定数を提示するのみにする．

$M=(m_{ij})$ を，ランダムな $n \times n$ の2値行列（binary matrix）で，次の条件を満たすものとする．

- 各 i, j に対して，確率 p で $m_{ij}=1$，確率 $1-p$ で $m_{ij}=0$
- すべての $(i,j) \neq (k,l)$ に対して，m_{ij} と m_{kl} は独立である．

隣接（adjacency）とは，横と縦の隣り（対角線方向でない）であるときとする．**s-クラスター**とは，M における s 個の1たちが隣接している孤立したグループ分けである．例として，4×4 行列

$$m = \begin{pmatrix} 1 & 0 & 1 & 1 \\ 1 & 1 & 0 & 0 \\ 0 & 1 & 0 & 1 \\ 1 & 0 & 0 & 1 \end{pmatrix}$$

は，1個の1-クラスター，2個の2-クラスター，1個の4-クラスターをもつ．この場合，クラスターの総個数 K_4 は4である．任意の n に対して，全クラスターの個数 K_n はランダムな変数である．正規化された期待値 $E(K_n)/n^2$ の $n \to \infty$ での極限 $\kappa_S(p)$ が存在する．$\kappa_S(p)$ は，**位置浸透**（site percolation）**モデル**に対する**平均クラスター密度**（mean cluster density）とよばれる．$\kappa_S(p)$ は区間 $[0,1]$ で2回連続微分可能であることが知られている．さらに，$\kappa_S(p)$ は，もし存在するとしても1点 $p=p_c$ を除いて，区間 $[0,1]$ で解析的である．モンテカルロシミュレーションと数値的 Padé（パデ）近似は，$\kappa_S(p)$ を計算することに使える．たとえば[11]より，$\kappa_S(1/2)=0.065770\cdots$ であると知られている．

$n \times n$ の2値行列 M の代わりに，$2n(n-1)$ 個の要素をもつ2値配列で，次のようなもの

$$A = \begin{pmatrix} & a_{12} & & a_{14} & & a_{16} & \\ a_{11} & & a_{13} & & a_{15} & & a_{17} \\ & a_{22} & & a_{24} & & a_{26} & \\ a_{21} & & a_{23} & & a_{25} & & a_{27} \\ & a_{32} & & a_{34} & & a_{36} & \\ a_{31} & & a_{33} & & a_{35} & & a_{37} \\ & a_{42} & & a_{44} & & a_{46} & \end{pmatrix}$$

を考える（ここでは，$n=4$ である）．（m_{ij} に対してするように）a_{ij} を $n \times n$ 正方格子（square lattice）の位置（site）とは結びつけないが，ある枝（bond）と結びつける．このとき，s-クラスターは1とすべて結びついているグラフの孤立した連結な部分グラフである．たとえば，配列

$$A = \begin{pmatrix} & 1 & & 0 & & 0 & \\ 1 & & 0 & & 0 & & 0 \\ & 0 & & 1 & & 0 & \\ 0 & & 1 & & 1 & & 0 \\ & 0 & & 1 & & 0 & \\ 1 & & 0 & & 0 & & 0 \\ & 0 & & 0 & & 0 & \end{pmatrix}$$

は，1つの1-クラスター，1つの2-クラスター，1つの4-クラスターをもつ．このような**枝の浸透モデル**（bond percolation model）に対して，総個数にはなおその上0-クラスターをも含める．すなわち1に結びついたものがない孤立した位置である．この例では，7つの0-クラスターがある．したがって，クラスターの総個数 K_4 は10である．平均クラスター密度 $\kappa_B(p) = \lim_{n \to \infty} \mathrm{E}(K_n)/n^2$ が存在し，同様な滑らかな性質が成り立つ．しかしながら，注目に値する厳密な積分表示が，平均クラスター密度に関して $p = 1/2$ の場合に発見された[13,14]．

$$\kappa_B\left(\frac{1}{2}\right) = -\frac{1}{8}\cot(y) \cdot \frac{d}{dy}\left\{\frac{1}{y}\int_{-\infty}^{\infty} \mathrm{sech}\left(\frac{\pi x}{2y}\right) \ln\left(\frac{\cosh(x) - \cos(2y)}{\cosh(x) - 1}\right) dx \right\}\bigg|_{y = \pi/3}$$

Adamchik（アダムチック）[11,12]は，最近これを

$$\kappa_B\left(\frac{1}{2}\right) = -\frac{3\sqrt{3} - 5}{2} = 0.0980762113\cdots$$

と簡単化した．この値は，ときには，0.0355762113… と記されるが，これは，0-クラスターを総個数に含めない，$\kappa_B(1/2) - 1/16$ の値である．代わりに，0.0177881056… とも書かれるが，これは，位置の数 n^2 によって正規化せずに，枝の個数 $2n(n-1)$ によって正規化した場合に起こる．文献を参照するときは，注意が必要である．この積分の他のところでの出現は，[15-18]にある．

枝クラスター密度の分散の極限の式は，知られていない．しかし，モンテカルロ法による推定では 0.164… であり，関連がある議論は[11]にある．「三角」格子上の枝の浸透モデルでは，特別な値 $p = 0.347\cdots$ における，平均クラスター密度の極限値は 0.111… である．（より精密なものについては，次の項を参照せよ．）対応する分散は 0.183… であり，再び，式は知られていない．

5.18.1 臨界確率

平均クラスター密度 $\kappa(p)$ から離れて，代わりに**平均クラスターの大きさ** $\sigma(p)$ に注意を向ける．前に与えた例では，位置の場合は $S_4 = (1 + 2 + 2 + 4)/4 = 9/4$ であり，枝の場合は $S_4 = (1 + 2 + 4)/3 = 7/3$ であり，$n \to \infty$ のとき，$\sigma(p)$ は $\mathrm{E}(S_n)$ の極限値である．**臨界確率**（critical probability）または**浸透閾値**（percolation threshold）p_c は

$$p_c = \inf_{\substack{0 < p < 1 \\ \sigma(p) = \infty}} p$$

と定義される[5,6,10]．すなわち，∞-クラスターが無限格子（infinite lattice）に現れ

5.18 浸透のクラスター密度定数

る凝縮確率である．たいていの条件の下では同値であることがわかる，他の可能な定義がある．たとえば，$\theta(p)$ を**浸透確率**とする．すなわち，∞-クラスターが所定の位置または枝を含む確率とする．このとき，p_c は，$p<p_c$ のとき $\theta(p)=0$ で，$p>p_c$ のとき $\theta(p)>0$ となる，一意の点である．臨界確率はシステムにおける相転移を示す，【5.12】と【5.22】で調べたものと類似なものである．

正方格子上での位置の浸透に対する，厳密な範囲[19-24]は

$$0.566 < p_c < 0.679492$$

であり，広範なシミュレーションに基づく推定値[25,26]は $p_c = 0.5927460\cdots$ である．Ziff（ジフ）[11]は，シミュレーションにより $\kappa_S(p_c) = 0.0275981\cdots$ であることを，さらに計算した．立方格子および，より高い次元におけるパラメータの範囲は，[27-30]にある．

対比として，正方格子および三角格子上の枝の浸透に対しては，Sykes & Essam（サイケス&エッサム）[31,32]による正確な結果がある．Kesten（ケステン）[33]は，前の項での表現 $\kappa_B(1/2)$ に対応する，正方格子上で $p_c = 1/2$ を証明した．三角格子上で，Wierman（ウィルマン）[34]は，

$$p_c = 2\sin\left(\frac{\pi}{18}\right) = 0.3472963553\cdots$$

を証明した．これは，別の厳密な式[11,35-37]

$$\kappa_B(p_c) = -\frac{3}{8}\csc(2y)\cdot\frac{d}{dy}\left\{\int_{-\infty}^{\infty}\frac{\sinh((\pi-y)x)\sinh\left(\frac{2}{3}yx\right)}{x\sinh(\pi x)\cosh(yx)}dx\right\}\bigg|_{y=\pi/3} + \frac{3}{2} - \frac{2}{1+p_c}$$

$$= \frac{35}{4} - \frac{3}{p_c} - \frac{23}{4} - \frac{3}{2}\cdot\left\{\sqrt[3]{4(1+i\sqrt{3})} + \sqrt[3]{4(1-i\sqrt{3})}\right\}$$

$$= 0.1118442752\cdots$$

に対応する．双対性により，類似の結果が六角形格子（蜂の巣）に対して成り立つ．

三角格子上の位置の浸透に対して，$p_c = 1/2$ である[10]．この場合，シミュレーションにより[11,38]，$\kappa_S(1/2) = 0.0176255\cdots$ であることが知られている．六角形格子上の位置の浸透に対して，範囲は[39]

$$0.6527 < 1 - 2\sin\left(\frac{\pi}{18}\right) \leq p_c \leq 0.8079$$

であり，推定値は $p_c = 0.6962\cdots$ [40,41]である．

5.18.2 級数展開

どのようにして関数 $\kappa_S(p)$，$\kappa_B(p)$ が計算されるか，に関して詳細に記述する[6,42,43]．位置の浸透にほとんどすべて焦点を当て，正方格子を研究する．面積 s と周の長さ t をもつ格子動物（lattice animals）【5.19】の個数を g_{st} とし，$q = 1-p$ とする．固定した位置が，1-クラスターである確率は明らかに pq^4 である．2-クラスターは縦または横のどちらかに向くことができるので，位置ごとの2-クラスターの個数の平均は，

$2p^2q^6$ である．3-クラスターは，線型（2つの方向）またはL-形（4つの方向）となることができる．したがって，位置ごとの平均3-クラスターの個数は，$p^3(2q^8+4q^7)$ である．一般的には，平均 s-クラスター密度は $\sum_t g_{st} p^s q^t$ である．表5.4[44,45]における左の列の要素を加えることにより，要素の個数 $\to \infty$ として $\kappa_S(p)$ が得られる．

$$\kappa_S(p) = p - 2p^2 + p^4 + p^8 - p^9 + 2p^{10} - 4p^{11} + 11p^{12} + - \cdots$$
$$\sim \kappa_S(p_c) + a_S(p-p_c) + b_S(p-p_c)^2 + c_S|p-p_c|^{2-\alpha}$$

同様に，表での右の列を加えることにより，$\kappa_B(p)$ を得る．

$$\kappa_B(p) = q^4 + 2p - 6p^2 + 4p^3 + 2p^6 - 2p^7 + 7p^8 - 12p^9 + 28p^{10} + \cdots$$
$$\sim \kappa_B\left(\frac{1}{2}\right) + a_B\left(p - \frac{1}{2}\right) + b_B\left(p - \frac{1}{2}\right)^2 + c_B\left|p - \frac{1}{2}\right|^{2-\alpha}$$

である．ただし，$a_B = -0.50\cdots$, $b_B = 2.8\cdots$, $c_B = -8.48\cdots$ である[46]．指数 α は $-2/3$ であると予想されている．すなわち，$2-\alpha = 8/3$ である．

$\sum_{s,t} g_{st} p^s q^t$ の代わりに，$\sum_{s,t} s^2 g_{st} p^{s-1} q^t$ を調べる．このとき，位置モデルに対して，

$$\sigma_S(p) = 1 + 4p + 12p^2 + 24p^3 + 52p^4 + 108p^5 + 224p^6 + 412p^7 + \cdots$$
$$\sim C|p-p_c|^{-\gamma}$$

は，(低い密集 $p < p_c$ に対する) 平均クラスターの大きさに関する級数である．指数 γ は $43/18$ であると予想されている．

式 $1 - \sum_{s,t} s g_{st} p^{s-1} q^t$ は，q について展開すると，

$$\theta_S(p) = 1 - q^4 - 4q^6 - 8q^7 - 23q^8 - 28q^9 - 186q^{10} + 48q^{11} - + \cdots$$
$$\sim D|p-p_c|^\beta$$

であり，これは (高い密集 $p > p_c$ に対する) 位置の浸透確率級数である．指数 β は $5/36$ であると予想されている．

Smirnov & Werner（スミルノフ&ウェルナー）[47]は最近，三角格子上の位置の浸透に対して，α, β, γ は確かに存在し，それらの予想された値に等しいことを証明した．一般的な証明は，正方格子上の位置と枝の両方の場合を含むだろう．しかし，これはまだ成し遂げられていない．

5.18.3 変　　種

無限格子の位置に，それぞれ独立に，確率 p でラベル A を，確率 $1-p$ でラベル B をつける．普通の位置の浸透理論は，A のクラスターを含む．代わりに，「反対の」ラベルをもつ隣接位置を結び，同じラベルをもつ非連結な隣接位置をそのままにする．こ

表5.4　平均 s-クラスター密度

s	位置モデルの平均 s クラスター密度	枝モデルの平均 s クラスター密度
0	0	q^4
1	pq^4	$2pq^6$
2	$2p^2q^6$	$6p^2q^8$
3	$p^3(2q^8+4q^7)$	$p^3(18q^{10}+4q^9)$
4	$p^4(2q^{10}+8q^9+9q^8)$	$p^4(55q^{12}+32q^{11}+q^8)$

れは**AB 浸透**（percolation）または**反浸透**（antipercolation）として知られている．無限 AB クラスターがあらかじめ定められた位置を含む，確率 $\theta(p)$ について何がいえるかを知りたい．無限正方格子に対して，すべての p で $\theta(p)=0$ であるとわかる [48]．しかし，無限三角格子に対しては，1/2 を含むある空でない部分区間内にあるすべての p に対して，$\theta(p)>0$ である [49]．この区間の正確な範囲は知られていない．Mai & Halley（メイ&ハリー）[50] は，モンテカルロシミュレーションによって，$[0.2145, 0.7855]$ を与えたが，Wierman（ウィルマン）[51] は $[0.4031, 0.5969]$ を与えた．関数 $\theta(p)$ は，三角格子に対して，区間 $[0, 1/2]$ で非減少である．したがって，Appel [52] により，$[0,1]$ で単峰であると考えられた．

普通の枝の浸透理論（bond percolation theory）は，任意の選んだ枝が開いている（open）(1) か，または閉じている（closed）(0) か，のどちらかであるようなモデルに関するものである．初通過浸透（first-passage percolation）[53] は，各枝に 2 値ランダム変数を与えるのではなく，非負の「実」ランダム変数を長さとして考えて与える．各枝が，独立に，$[0,1]$ での一様な確率分布からの長さとして与えられている正方格子を考える．原点 $(0,0)$ から出発し，点 $(n,0)$ で終わる，すべての格子道の長さの最小を T_n とする．このとき，極限

$$\tau = \lim_{n\to\infty} \frac{\mathrm{E}(T_n)}{n} = \inf_n \frac{\mathrm{E}(T_n)}{n}$$

が存在することが証明される．先行研究 [54-58] を基礎にして，Alm & Parviainen（アルム&パルヴィアイネン）[59] は，厳密な範囲 $0.243666 \leq \tau \leq 0.403141$，およびシミュレーションによって推定値 $\tau = 0.312\cdots$ を得た．代わって，長さが平均 1 の指数分布からとられたとする．このとき，範囲 $0.300282 \leq \tau \leq 0.503425$，およびシミュレーションによって推定値 $\tau = 0.402$ を得る．Godsil, Grötschel & Welsh（ゴッドシル，グレートシェル&ウェルシュ）[9] は，τ の正確な評価は「望みがないほど手に負えない問題」である，と示唆した．

最後に，**連続体浸透**（continuum percolation, 連続媒質中の浸透）[5,60] において現れる定数 $\lambda_c = 0.359072\cdots$ を述べる．平面上における母数 λ の同次ポアソン過程を考える．すなわち点が平面内に，

- 測度 μ の部分集合 S に，正確に n 個の点をもつ確率は $e^{-\lambda\mu}(\lambda\mu)^n/n!$ である．
- 互いに素である可測な部分集合 S_i の，任意の集合における点の個数 n_i が，独立なランダム変数である．

を満たすように一様分布していると仮定する．

各点の周りに，単位円板を描く．円は重なってもよい．すなわち，これらは完全に進入可能である（fully penetrable）．円板の非有界連結クラスターが，$\lambda > \lambda_c$ のとき確率 1 で，$\lambda < \lambda_c$ のとき確率 0 で，広がる一意の臨界量 λ_c が存在する．Hall（ホール）[61] は，最良の厳密な範囲は $0.174 < \lambda_c < 0.843$ である，と証明した．数値的な推定値 $0.359072\cdots$ は [62-64] にある．いくつかの別の表現のうちで，$\phi_c = 1 - \exp(-\pi\lambda_c) = 0.676339\cdots$ [65] および $\pi\lambda_c = 1.128057\cdots$ [66] を述べる．後者は，円板が重なるかどうか

に注意を払わないで，ただ単にすべての円板の正規化された全面積である．だが，ϕ_c は重複した部分を考慮に入れる．連続体浸透は，格子浸透とともに多くの数学的性質を共有する．けれども，多くの方法において，物理的無秩序（physical disorder）のより精密なモデルである．おもしろいことには，さらに最近，純粋数学それ自体において，ガウス素数の集合における空隙の研究に応用された[67]．

[1] S. R. Broadbent and J. M. Hammersley, Percolation processes. I: Crystals and mazes, *Proc. Cambridge Philos.* Soc. 53 (1957) 629-641; MR 19,989e.

[2] J. M. Hammersley, Percolation processes. II: The connective constant, *Proc. Cambridge Philos.* Soc. 53 (1957) 642-645; MR 19,989f.

[3] J. M. Hammersley, Percolation processes: Lower bounds for the critical probability, *Annals of Math. Statist.* 28 (1957) 790-795; MR 21 #374.

[4] D. Stauffer, Scaling theory of percolation clusters, *Phys. Rep.* 54 (1979) 1-74.

[5] G. Grimmett, *Percolation*, Springer-Verlag, 1989; MR 90j:60109.

[6] D. Stauffer and A. Aharony, *Introduction to Percolation Theory*, 2nd ed., Taylor and Francis, 1992; MR 87k:82093.

[7] M. Sahimi, *Applications of Percolation Theory*, Taylor and Francis, 1994.

[8] S. Havlin and A. Bunde, Percolation, *Contemporary Problems in Statistical Physics*, ed. G. H. Weiss, SIAM, 1994, pp. 103-146.

[9] C. Godsil, M. Grötschel, and D. J. A.Welsh, Combinatorics in statistical physics, *Handbook of Combinatorics*, v. II, ed. R. Graham, M. Grötschel, and L. Lovász, MIT Press, 1995, pp. 1925-1954; MR 96h:05001.

[10] B. D. Hughes, *Random Walks and Random Environments*, v. 2, Oxford Univ. Press, 1996; MR 98d:60139.

[11] R. M. Ziff, S. R. Finch, and V. S. Adamchik, Universality of finite-size corrections to the number of critical percolation clusters, *Phys. Rev. Lett.* 79 (1997) 3447-3450.

[12] V. S. Adamchik, S. R. Finch, and R. M. Ziff, The Potts model on the square lattice, unpublished note (1996).

[13] H. N. V. Temperley and E. H. Lieb, Relations between the 'percolation' and 'colouring' problem and other graph-theoretical problems associated with regular planar lattices: Some exact results for the 'percolation' problem, *Proc. Royal Soc. London A* 322 (1971) 251-280; MR 58 #16425.

[14] J. W. Essam, Percolation and cluster size, *Phase Transitions and Critical Phenomena*, v. II, ed. C. Domb and M. S. Green, Academic Press, 1972, pp. 197-270; MR 50 #6392.

[15] M. L. Glasser, D. B. Abraham, and E. H. Lieb, Analytic properties of the free energy for the "ice" models, *J. Math. Phys.* 13 (1972) 887-900; MR 55 #7227.

[16] R. J. Baxter, Potts model at the critical temperature, *J. Phys.* C 6 (1973) L445-L448.

[17] H. N. V. Temperley, The shapes of domains occurring in the droplet model of phase transitions, *J. Phys. A* 9 (1976) L113-L117.

[18] R. J. Baxter, *Exactly Solved Models in Statistical Mechanics*, Academic Press, 1982, pp. 322-341; MR 90b:82001.

[19] B. Tóth, A lower bound for the critical probability of the square lattice site percolation, *Z. Wahr. Verw. Gebiete* 69 (1985) 19-22; MR 86f:60118.

[20] S. A. Zuev, Bounds for the percolation threshold for a square lattice (in Russian), *Teor. Veroyatnost. i Primenen.* 32 (1987) 606-609; Engl. transl. in *Theory Probab. Appl.* 32 (1987) 551-553; MR 89h:60171.

[21] S. A. Zuev, A lower bound for a percolation threshold for a square lattice (in Russian), *Vestnik Moskov. Univ. Ser. I Mat. Mekh.* (1988) n. 5, 59-61; Engl. transl. in *Moscow Univ. Math. Bull.*, v. 43 (1988) n. 5, 66-69; MR 91k:82035.

[22] M. V. Menshikov and K. D. Pelikh, Percolation with several defect types: An estimate of critical probability for a square lattice, *Math. Notes Acad. Sci. USSR* 46 (1989) 778-785; MR 91h: 60116.
[23] J. C. Wierman, Substitution method critical probability bounds for the square lattice site percolation model, *Combin. Probab. Comput.* 4 (1995) 181-188; MR 97g:60136.
[24] J. van den Berg and A. Ermakov, A new lower bound for the critical probability of site percolation on the square lattice, *Random Structures Algorithms* 8 (1996) 199-212; MR 99b: 60165.
[25] R. M. Ziff and B. Sapoval, The efficient determination of the percolation threshold by a frontier-generating walk in a gradient, *J. Phys. A* 19 (1986) L1169-L1172.
[26] R. M. Ziff, Spanning probability in 2D percolation, *Phys. Rev. Lett.* 69 (1992) 2670-2673.
[27] M. Campanino and L. Russo, An upper bound on the critical percolation probability for the three-dimensional cubic lattice, *Annals of Probab.* 13 (1985) 478-491; MR 86j: 60222.
[28] J. Adler, Y. Meir, A. Aharony, and A. B. Harris, Series study of percolation moments in general dimension, *Phys. Rev. B* 41 (1990) 9183-9206.
[29] B. Bollobás and Y. Kohayakawa, Percolation in high dimensions, *Europ. J. Combin.* 15 (1994) 113-125; MR 95c:60092.
[30] T. Hara and G. Slade, The self-avoiding-walk and percolation critical points in high dimensions, *Combin. Probab. Comput.* 4 (1995) 197-215; MR 96i:82081.
[31] M. F. Sykes and J. W. Essam, Exact critical percolation probabilities for site and bond problems in two dimensions, *J. Math. Phys.* 5 (1964) 1117-1121; MR 29 # 1977.
[32] J.W. Essam and M. F. Sykes, Percolation processes. I: Low-density expansion for the mean number of clusters in a random mixture, *J. Math. Phys.* 7 (1966) 1573-1581;MR34 # 3952.
[33] H. Kesten, The critical probability of bond percolation on the square lattice equals 1/2, *Commun. Math. Phys.* 74 (1980) 41-59; MR 82c:60179.
[34] J. C.Wierman, Bond percolation on honeycomb and triangular lattices, *Adv. Appl. Probab.* 13 (1981) 298-313; MR 82k:60216.
[35] R. J. Baxter, H. N. V. Temperley, and S. E. Ashley, Triangular Potts model at its transition temperature, and related models, *Proc. Royal Soc. London A* 358 (1978) 535-559; MR 58 # 20108.
[36] V. S. Adamchik, S. R. Finch, and R. M. Ziff, The Potts model on the triangular lattice, unpublished note (1996).
[37] V. S. Adamchik, A class of logarithmic integrals, *Proc. 1997 Int. Symp. on Symbolic and Algebraic Computation (ISSAC)*, ed. W. W. Küchlin, Maui, ACM, 1997, pp. 1-8; MR 2001k:33043.
[38] D. W. Erbach, Advanced Problem 6229, *Amer. Math. Monthly* 85 (1978) 686.
[39] T. Luczak and J. C. Wierman, Critical probability bounds for two-dimensional site percolation models, *J. Phys. A* 21 (1988) 3131-3138; MR 89i:82053.
[40] Z.V. Djordjevic, H. E. Stanley, and A. Margolina, Site percolation threshold for honeycomb and square lattices, *J. Phys. A* 15 (1982) L405-L412.
[41] R. P. Langlands, C. Pichet, P. Pouliot, and Y. Saint-Aubin, On the universality of crossing probabilities in two-dimensional percolation, *J. Stat. Phys.* 67 (1992) 553-574; MR 93e:82028.
[42] J. Adler, Series expansions, *Comput. in Phys.* 8 (1994) 287-295.
[43] A. R. Conway and A. J. Guttmann, On two-dimensional percolation, *J. Phys. A* 28 (1995) 891-904.
[44] M. F. Sykes and M. Glen, Percolation processes in two dimensions. I: Low-density series expansions, *J. Phys. A* 9 (1976) 87-95.
[45] M. F. Sykes, D. S. Gaunt, and M. Glen, Perimeter polynomials for bond percolation processes, *J. Phys. A* 14 (1981) 287-293; MR 81m:82045.
[46] C. Domb and C. J. Pearce, Mean number of clusters for percolation processes in two dimensions, *J. Phys. A* 9 (1976) L137-L140.
[47] S. Smirnov and W. Werner, Critical exponents for two-dimensional percolation, *Math. Res. Lett.* 8 (2001) 729-744; math.PR/0109120.

[48] M. J. Appel and J. C. Wierman, On the absence of infinite AB percolation clusters in bipartite graphs, *J. Phys. A* 20 (1987) 2527-2531; MR 89b:82061.

[49] J. C. Wierman and M. J. Appel, Infinite AB percolation clusters exist on the triangular lattice, *J. Phys. A* 20 (1987) 2533-2537; MR 89b:82062

[50] T. Mai and J.W. Halley, AB percolation on a triangular lattice, *Ordering in Two Dimensions*, ed. S. K. Sinha, North-Holland, 1980, pp. 369-371.

[51] J. C. Wierman, AB percolation on close-packed graphs, *J. Phys. A* 21 (1988) 1939-1944; MR 89m:60258.

[52] M. J. B. Appel, AB percolation on plane triangulations is unimodal, *J. Appl. Probab.* 31 (1994) 193-204; MR 95e:60101.

[53] R. T. Smythe and J. C. Wierman, *First-Passage Percolation on the Square Lattice*, Lect. Notes in Math. 671, Springer-Verlag, 1978; MR 80a:60135.

[54] J. M. Hammersley and D. J. A. Welsh, First-passage percolation, subadditive processes, stochastic networks, and generalized renewal theory, *Bernoulli, 1713; Bayes, 1763; Laplace, 1813. Anniversary Volume*, Proc. 1963 Berkeley seminar, ed. J. Neyman and L. M. Le Cam, Springer-Verlag, 1965, pp. 61-110; MR 33 # 6731.

[55] S. Janson, An upper bound for the velocity of first-passage percolation, *J. Appl. Probab.* 18 (1981) 256-262; MR 82c:60178.

[56] D. J. A.Welsh, An upper bound for a percolation constant, *Z. Angew Math. Phys.* 16 (1965) 520-522.

[57] R. T. Smythe, Percolation models in two and three dimensions, *Biological Growth and Spread*, Proc. 1979 Heidelberg conf., ed. W. Jäger, H. Rost, and P. Tautu, Lect. Notes in Biomath. 38, Springer-Verlag, 1980, pp. 504-511; MR 82i:60183.

[58] R. Ahlberg and S. Janson, Upper bounds for the connectivity constant, unpublished note (1981).

[59] S. E. Alm and R. Parviainen, Lower and upper bounds for the time constant of first-passage percolation, *Combin. Probab. Comput*, to appear.

[60] R. Meester and R. Roy, *Continuum Percolation*, Cambridge Univ. Press, 1996; MR 98d:60193.

[61] P. Hall,Oncontinuum percolation, *Annals of Probab.* 13 (1985) 1250-1266;MR87f:60018.

[62] S. W. Haan and R. Zwanzig, Series expansions in a continuum percolation problem, *J. Phys. A* 10 (1977) 1547-1555.

[63] E. T. Gawlinski and H. E. Stanley, Continuum percolation in two dimensions: Monte Carlo tests of scaling and universality for noninteracting discs, *J. Phys. A* 14 (1981) L291-L299; MR 82m:82026.

[64] J. Quintanilla, S. Torquato, and R. M. Ziff, Efficient measurement of the percolation threshold for fully penetrable discs, *J. Phys. A* 33 (2000) L399-L407.

[65] C. D. Lorenz and R. M. Ziff, Precise determination of the critical percolation threshold for the three-dimensional "Swiss cheese" model using a growth algorithm, *J. Chem. Phys.* 114 (2001) 3659-3661.

[66] P. Hall, *Introduction to the Theory of Coverage Processes*, Wiley, 1988; MR 90f:60024.

[67] I. Vardi, Prime percolation, *Experim. Math.* 7 (1998) 275-289; MR 2000h:11081.

5.19 クラナーのポリオミノ定数

ドミノ (domino) は隣接した正方形の対のことである．これを拡張して，次数 n のポリオミノ (polyomino) または格子動物 (lattice animal) とは，n 個の隣接した正

5.19 クラナーのポリオミノ定数

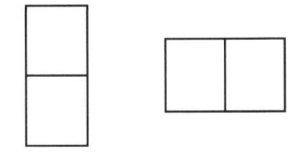

図 5.17 すべてのドミノ (次数2のポリオミノ): $A(2)=2$.

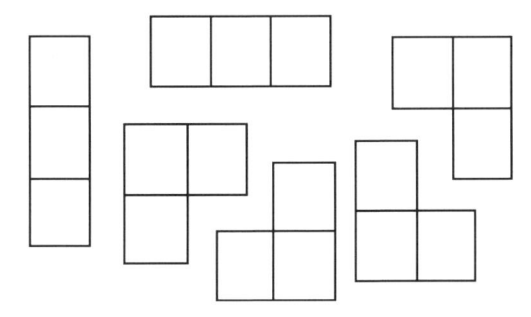

図 5.18 次数3のすべてのポリオミノ: $A(3)=6$.

方形の連結集合のことである[1-7]. 図5.17および図5.18を見よ.

次数 n のポリオミノの個数を $A(n)$ とする. ただし, 2つのポリオミノが異なるとは, それらが違った形 (different shape) または異なる向き (different orientation) をもつときに限る[†].

$$A(1)=1, \quad A(2)=2, \quad A(3)=6, \quad A(4)=19, \quad A(5)=63,$$
$$A(6)=216, \quad A(7)=760, \cdots$$

ポリオミノの定義には異なる意味をもつものがいくつかある. たとえば, 自由対固定, 枝対位置, 単連結対必ずしもそうでないもの, その他である. 簡単のために, 固定された位置について, 多重連結も許した場合だけに焦点をあてる.

Redelmeier (レデルマイヤ) [8]は $A(n)$ を $n=24$ まで計算した. Conway & Guttmann (コンウェイ&ガットマン) [9]は $A(25)$ を発見した. 最近の相次ぐ研究で, Oliveira e Silva[10]は $n=28$ まで $A(n)$ を計算した. Jensen & Guttmann (イェンセン&ガットマン) [11,12]は $A(46)$ まで拡張した. Knuth[13]は $A(47)$ を発見した. Klarner (クラナー) [14,15]は, 極限

$$\alpha = \lim_{n \to \infty} A(n)^{1/n} = \sup_n A(n)^{1/n}$$

が存在し, 0でないことを証明した. しかしながら, Eden (エデン) [16]は数年前に,

[†]訳注: ここではポリオミノの向きの差を考慮に入れ, たとえばドミノを縦と横に置いた場合を区別している. 裏返しも許して形が合同なものを同一とみなすと, 個数は順次1,1,2,5,12,35,...となる.

数値的に a を研究していた。a の最良範囲は，[17-20]で議論されているように，$3.903184 \leq a \leq 4.649551$ である。新しい値 $A(47)$ を用いて，改良は可能である。差分近似 (differential approximants) [11]による級数展開解析で得られた最良の推定値は，$a=4.062570\cdots$ である。$A(n)$ に対する，より精密な漸近式は

$$A(n) \sim \left(\frac{0.316915\cdots}{n} - \frac{0.276\cdots}{n^{3/2}} + \frac{0.335\cdots}{n^2} - \frac{0.25\cdots}{n^{5/2}} + O\left(\frac{1}{n^3}\right)\right) a^n$$

である。しかし，このような実験結果は，厳密に証明するにはかなり遠い結果である。

Satterfield（サターフィールド）[5,21]は，Klarner と Shende（シェンデ）とともに開発した，いくつかのアルゴリズムのうちの１つを用いて，a の下界 3.91336 を報告した。あいにく彼らの仕事の詳細は，未発表のままである。

平行した解析が，三角格子および六角格子で実行できることを述べる[7,22]。

任意の自己回避多角形【5.10】はポリオミノを定める。しかし，ポリオミノは穴をもつことができるので，逆は正しくない。ポリオミノが**行凸** (row-convex) とは，すべての（水平の）列が正方形の１つの帯 (strip) からなるときをいう。**凸** (convex) とは，すべての列に対しても同様の性質が成り立つときをいう。凸ポリオミノは，一般的に普通の意味において，凸多角形を定めないことに注意せよ。行凸ポリオミノの個数は，3次の線型漸化式に従う[23-28]。しかし，凸ポリオミノの個数 $\tilde{A}(n)$ を解析するにはいくぶん難しい[29,30]。

$$\tilde{A}(1)=1, \quad \tilde{A}(2)=2, \quad \tilde{A}(3)=6, \quad \tilde{A}(4)=19, \quad \tilde{A}(5)=59,$$
$$\tilde{A}(6)=176, \quad \tilde{A}(7)=502, \quad \cdots$$
$$\tilde{A}(n) \sim (2.67564\cdots) \tilde{a}^n$$

である。ただし，$\tilde{a}=2.3091385933\cdots=(0.4330619231\cdots)^{-1}$ である。$\tilde{A}(n)$ の正確な母関数の定式化はつい最近発見された[31-33]が，とても複雑なのでここに述べることはできない。Bender（ベンダー）[30]は，さらに凸ポリオミノの期待される形を解析した。遠くから見ると，多くの凸ポリオミノは，水平（と垂直）の厚さが大ざっぱに $2.37597\cdots$ に等しく，垂直から 45 度傾いた棒のように見える，ことを見つけた。この種のさらなる結果は，[34-36]にある。

凸ポリオミノに対する増大定数 $\tilde{a}$ は，**平行四辺形ポリオミノ** (parallelogram polyominoes)，すなわち左と右の境界がともに北東（右上）の方向に登ってゆくようなポリオミノに対する増大定数 a' と同じである。その値は

$$A'(1)=1, \quad A'(2)=2, \quad A'(3)=4, \quad A'(4)=9, \quad A'(5)=20,$$
$$A'(6)=46, \quad A'(7)=105, \quad \cdots$$

である。これらは簡単な母関数 $f(q)$ という長所をもつ。$(q)_0=1$, $(q)_n=\prod_{j=1}^{n}(1-q^j)$ とする。このとき，$f(q)$ はベッセル関数の q-アナログ (q-analogs) の比 $J_1(q)/J_0(q)$ である。

$$J_0(q) = 1 + \sum_{n=1}^{\infty} \frac{(-1)^n q^{\binom{n+1}{2}}}{(q)_n(q)_n}, \quad J_1(q) = -\sum_{n=1}^{\infty} \frac{(-1)^n q^{\binom{n+1}{2}}}{(q)_{n-1}(q)_n}$$

により，$a'=\tilde{a}$ を得るが，異なる乗法的定数 (multiplicative constant) $0.29745\cdots$ を

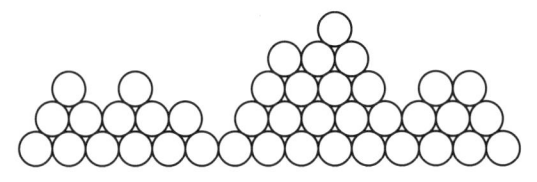

図 5.19 n-泉の例.

得る.

われわれがなんとか要約できる問題よりも，この種の数え上げの問題は数多くある！Glasser, Privman & Svrakic（グラッサー，プリブマン&スブラキッチ）[37]とOdlyzko & Wilf（オドリズコ&ヴィルフ）[38-40]によって独立に研究された，もう1つの例がある. n-泉 (n-fountain)（図 5.19）は，垂直な壁に対して三角格子配列における n 個の硬貨の連結した，自己支持積み重ね配列として，最もよく図示されている．

底辺の行は空隙をもつことはできないが，上の行は空隙をもつことができる．1つ上の行にある各硬貨は，下の行にある2つの隣接硬貨に接触しなければならない．n-泉の個数を $B(n)$ とする．$B(n)$ の母関数は，Ramanujan（ラマヌジャン）の連分数を含む美しい恒等式である．

$$1+\sum_{n=1}^{\infty}B(n)x^n = 1+x+x^2+2x^3+3x^4+5x^5+9x^6+15x^7+26x^8+45x^9+\cdots$$

$$= \frac{1|}{|1} - \frac{x|}{|1} - \frac{x^2|}{|1} - \frac{x^3|}{|1} - \frac{x^4|}{|1} - \frac{x^5|}{|1} - \cdots$$

次の増大評価が得られる．

$$\lim_{n\to\infty} B(n)^{1/n} = \beta = 1.7356628245\cdots = (0.5761487691\cdots)^{-1}$$

$$B(n) = (0.3123633245\cdots)\beta^n + O\left(\left(\frac{5}{3}\right)^n\right)$$

他の関連する数え上げの問題については[41]を参照せよ．

[1] S. W. Golomb, *Polyominoes*, 2nd ed., Princeton Univ. Press, 1994; MR 95k:00006.
[2] A. J. Guttmann, Planar polygons: Regular, convex, almost convex, staircase and rowconvex, *Computer-Aided Statistical Physics*, Proc. 1991 Taipei symp., ed. C.-K. Hu, Amer. Inst. Phys., 1992, pp. 12-43.
[3] M. P. Delest, Polyominoes and animals: Some recent results, *J. Math. Chem.* 8 (1991) 3-18; MR 93e:05024.
[4] C. Godsil, M. Grötschel, and D. J. A. Welsh, Combinatorics in statistical physics, *Handbook of Combinatorics*, v. II, ed. R. Graham, M. Grötschel, and L. Lovász, MIT Press, 1995, pp. 1925-1954; MR 96h:05001.
[5] D. A. Klarner, Polyominoes, *Handbook of Discrete and Computational Geometry*, ed. J. E. Goodman and J. O'Rourke, CRC Press, 1997, pp. 225-240; MR 2000j:52001.
[6] D. Eppstein, Polyominoes and Other Animals (Geometry Junkyard).
[7] N. J. A. Sloane, On-Line Encyclopedia of Integer Sequences, A001168, A001169, A001207, A001420, A005169, A006958, and A067675.
[8] D. H. Redelmeier, Counting polyominoes: Yet another attack, *Discrete Math.* 36 (1981) 191-203;

MR 84g:05049.
- [9] A. R. Conway and A. J. Guttmann, On two-dimensional percolation, *J. Phys. A* 28 (1995) 891-904.
- [10] T. Oliveira e Silva, Enumeration of animals on the (44) regular tiling of the Euclidean plane, unpublished note (1999).
- [11] I. Jensen and A. J. Guttmann, Statistics of lattice animals (polyominoes) and polygons, *J. Phys. A* 33 (2000) L257-L263; MR 2001e:82025.
- [12] I. Jensen, Enumerations of lattice animals and trees, *J. Stat. Phys.* 102 (2001) 865-881; MR 2002b:82026.
- [13] D. E. Knuth, The number of 47-ominoes, unpublished note (2001).
- [14] D. A. Klarner, Cell growth problems, *Canad. J. Math.* 19 (1967) 851-863; MR 35 #5339.
- [15] D. A. Klarner, My life among the polyominoes, *Nieuw Arch. Wisk.* 29 (1981) 156-177; also in *The Mathematical Gardner*, ed. D. A. Klarner,Wadsworth, 1981, pp. 243-262; MR 83e:05038.
- [16] M. Eden, A two-dimensional growth process, *Proc. Fourth Berkeley Symp. Math. Stat. Probab.*, v. 4, ed. J. Neyman, Univ. of Calif. Press, 1961, pp. 223-239; MR 24 # B2493.
- [17] D. A. Klarner and R. L. Rivest, A procedure for improving the upper bound for the number of n-ominoes, *Canad. J. Math.* 25 (1973) 585-602; MR 48 #1943.
- [18] B. M. I. Rands and D. J. A. Welsh, Animals, trees and renewal sequences, *IMA J. Appl. Math.* 27 (1981) 1-17; corrigendum 28 (1982) 107; MR 82j:05049 and MR 83c:05041.
- [19] S. G. Whittington and C. E. Soteros, Lattice animals: Rigorous results and wild guesses, *Disorder in Physical Systems:A Volume in Honour of J. M. Hammersley*, ed. G. R. Grimmett and D. J. A. Welsh, Oxford Univ. Press, 1990, pp. 323-335; MR 91m:82061.
- [20] A. J. Guttmann, On the number of lattice animals embeddable in the square lattice, *J. Phys. A* 15 (1982) 1987-1990.
- [21] D. A. Klarner, S. Shende, and W. Satterfield, Lower-bounding techniques for lattice animals, unpublished work (2000).
- [22] W. F. Lunnon, Counting hexagonal and triangular polyominoes, *Graph Theory and Computing*, ed. R. C. Read, Academic Press, 1972, pp. 87-100; MR 49 #2439.
- [23] G. Pólya, On the number of certain lattice polygons, J. Combin. Theory 6 (1969) 102-105; also in *Collected Papers*, v. 4, ed. G.-C. Rota, MIT Press, 1984, pp. 441-444, 630; MR 38 #4329.
- [24] H. N. V. Temperley, Combinatorial problems suggested by the statistical mechanics of domains and of rubber-like molecules, *Phys. Rev.* 103 (1956) 1-16; MR 17,1168f.
- [25] D. A. Klarner, Some results concerning polyominoes, *Fibonacci Quart.* 3 (1965) 9-20; MR 32 #4028.
- [26] M. P. Delest, Generating functions for column-convex polyominoes, *J. Combin. Theory Ser. A* 48 (1988) 12-31; MR 89e:05013.
- [27] V. Domocos, A combinatorial method for the enumeration of column-convex polyominoes, *Discrete Math.* 152 (1996) 115-123; MR 97c:05005.
- [28] D. Hickerson, Counting horizontally convex polyominoes, *J. Integer Seq.* 2 (1999) 99.1.8; MR 2000k:05023.
- [29] D. A. Klarner and R. L. Rivest, Asymptotic bounds for the number of convex n-ominoes, *Discrete Math.* 8 (1974) 31-40; MR 49 #91.
- [30] E. A. Bender, Convex n-ominoes, *Discrete Math.* 8 (1974) 219-226; MR 49 #67.
- [31] M. Bousquet-Mélou and J.-M. Fédou, The generating function of convex polyominoes: The resolution of a q-differential system, *Discrete Math.* 137 (1995) 53-75; MR 95m:05009.
- [32] M. Bousquet-Mélou and A. Rechnitzer, Lattice animals and heaps of dimers, *Discrete Math.*, to appear.
- [33] K. Y. Lin, Exact solution of the convex polygon perimeter and area generating function, *J. Phys. A* 24 (1991) 2411-2417; MR 92f:82023.
- [34] G. Louchard, Probabilistic analysis of some (un)directed animals, *Theoret. Comput. Sci.* 159 (1996) 65-79; MR 97b:68164.

[35] G. Louchard, Probabilistic analysis of column-convex and directed diagonally-convex animals, *Random Structures Algorithms* 11 (1997) 151-178; MR 99c:05047.
[36] G. Louchard, Probabilistic analysis of column-convex and directed diagonally-convex animals. II: Trajectories and shapes, *Random Structures Algorithms* 15 (1999) 1-23; MR 2000e: 05042.
[37] M. L. Glasser, V. Privman, and N. M. Svrakic, Temperley's triangular lattice compact cluster model: Exact solution in terms of the q series, *J. Phys. A* 20 (1987) L1275-L1280; MR 89b:82107.
[38] R. K. Guy, The strong law of small numbers, *Amer. Math. Monthly* 95 (1988) 697-712; MR 90c: 11002.
[39] A. M. Odlyzko and H. S. Wilf, The editor's corner: n coins in a fountain, *Amer. Math. Monthly* 95 (1988) 840-843.
[40] A. M. Odlyzko, Asymptotic enumeration methods, *Handbook of Combinatorics*, v. I, ed. R. Graham, M. Grötschel, and L. Lovász, MIT Press, 1995, pp. 1063-1229; MR 97b:05012.
[41] V. Privman and N. M. Svrakic, Difference equations in statistical mechanics. I: Cluster statistics models, *J. Stat. Phys.* 51 (1988) 1091-1110; MR 89i:82079.

5.20 最長部分数列の定数

5.20.1 増加部分数列

記号 $1, 2, \cdots, N$ のランダム置換を π とする. π の**増加部分数列** (increasing subsequence) とは, $1 \leq j_1 < j_2 < \cdots < j_k \leq N$ と $\pi(j_1) < \pi(j_2) < \cdots < \pi(j_k)$ をともに満たす, 数列 $(\pi(j_1), \pi(j_2), \cdots, \pi(j_k))$ のことである. π の最も長い増加部分数列の長さを L_N と定義する. たとえば, 置換 $\pi = (2,7,4,1,6,3,9,5,8)$ は, 最長の部分数列 $(2,4,6,9)$ および $(1,3,5,8)$ をもつから, $L_9 = 4$ である. $N \to \infty$ のとき, L_N の確率分布 (たとえば, 平均と分散) について, 何がいえるだろうか？

この問題の研究に, 研究者は殺到している[1-4]. Vershik & Kerov (ヴェルシック＆ケロフ)[5] と Logan & Shepp (ローガン＆シェップ)[6] は, [7-10] の先行研究を基礎にして,

$$\lim_{N \to \infty} N^{-1/2} \mathrm{E}(L_N) = 2$$

を証明した. Odlyzko & Rains (オドリズコ＆レインズ)[11]は, 1993年に, 極限

$$\lim_{N \to \infty} N^{-1/3} \mathrm{Var}(L_N) = c_0, \quad \lim_{N \to \infty} N^{-1/6}(\mathrm{E}(L_N) - 2\sqrt{N}) = c_1$$

がともに存在し, 0 でない有限値であると予想した. 数値的な近似値は, モンテカルロシミュレーションによって計算された. (数理物理からのモデルを使った) 解析を試験的に用いて, Baik, Deift & Johansson (バイク, デイフト＆ヨハンソン)[12] は, [11] で予想されたものを裏付け,

$$c_0 = 0.81318 \cdots (\text{すなわち} \sqrt{c_0} = 0.90177 \cdots), \quad c_1 = -1.77109 \cdots$$

を得た. これらの定数は, Painlevé (パンルヴェ) II 型方程式の解により正確に定義さ

れる．（ちなみに，パンルヴェ III 型は【5.22】に現れ，パンルヴェ V 型は【2.15.3】に現れる．）偏差は，ランダム行列とランダム置換との間の関係に影響する[13,14]．より正確には，ガウス・ユニタリー集合（Gaussian unitary ensemble, GUE）の確率法則に従って生成されるランダムエルミート行列の最大固有値を特徴づける，ある種の確率分布関数 $F(x)$ を，Tracy & Widom（トレーシー&ウィドム）[15-17]は導いた．L_N の極限分布は Tracy & Widom の $F(x)$ であることを，Baik, Deift & Johansson が証明した．したがって，[16]で示されたモーメントにより定数 c_0 と c_1 の推定値が得られる．

より詳細な結果を示す前に，一般化する．π の 2-**増加部分数列**（2-increasing subsequence）とは，π の 2 つの素な増加部分数列の和集合である．π の最も長い 2-増加部分数列の長さから L_N を減じた値を $\tilde{L}_N$ と定義する．たとえば，置換 $(2,4,7,9,5,1,3,6,8)$ は，最長の増加部分数列 $(2,4,5,6,8)$ および最長の 2-増加部分数列 $(2,4,7,9) \cup (1,3,6,8)$ をもつ．したがって，$\tilde{L}_9 = 8-5 = 3$ である．前のように

$$\lim_{N \to \infty} N^{-1/3} \mathrm{Var}(\tilde{L}_N) = \tilde{c}_0, \quad \lim_{N \to \infty} N^{-1/6}(\mathrm{E}(\tilde{L}_N) - 2\sqrt{N}) = \tilde{c}_1$$

がともに存在し，その値が

$$\tilde{c}_0 = 0.5405\cdots, \quad \tilde{c}_1 = -3.6754\cdots$$

であることが証明できる[18]．

対応する分布関数 $\tilde{F}(x)$ は，GUE の下でランダムエルミート行列の 2 番目に大きい固有値を特徴づける．このような証明は，任意の $m>2$ に対して，m-増加数列に拡張でき，また[19-21]にあるランダムヤング図形（Young tableaux）からの行の長さの結合分布に拡張される．

約束した詳細を述べる[12,18]．$0 < t \leq 1$ を固定する．パンルヴェ II 型微分方程式

$$q_t''(x) = 2q_t(x)^3 + xq_t(x), \quad q_t(x) \sim \frac{1}{2}\left(\frac{t}{\pi}\right)^{1/2} x^{-1/4} \exp\left(-\frac{2}{3}x^{3/2}\right) \quad (x \to \infty)$$

の解を $q_t(x)$ とし，$\Phi(x,t)$ を

$$\Phi(x,t) = \exp\left[-\int_x^\infty (y-x)q_t(y)^2 dy\right]$$

と定義する．トレーシー-ウィドム関数は

$$F(x) = \Phi(x,1), \quad \tilde{F}(x) = \Phi(x,1) - \frac{\partial \Phi}{\partial t}(x,t)\bigg|_{t=1}$$

である．したがって，

$$c_0 = \int_{-\infty}^\infty x^2 F'(x)\,dx - \left(\int_{-\infty}^\infty xF'(x)\,dx\right)^2, \quad c_1 = \int_{-\infty}^\infty xF'(x)\,dx$$

$$\tilde{c}_0 = \int_{-\infty}^\infty x^2 \tilde{F}'(x)\,dx - \left(\int_{-\infty}^\infty x\tilde{F}'(x)\,dx\right)^2, \quad \tilde{c}_1 = \int_{-\infty}^\infty x\tilde{F}'(x)\,dx$$

は，望んでいた式である．$c_0, c_1, \tilde{c}_0, \tilde{c}_1$ の値は，[16]の図 2 の説明に現れていることに注意せよ．だから，これらを Odlyzko-Rains-Tracy-Widom 定数とよぶように主張してよかろう．

この成果がとくに好奇心をそそるのは[1,22]，これが普遍的なカードゲームであるソ

リティア（うまくいった解析は未だに完成されていない）に関連があり，また素数論における未解決なリーマン予想【1.6】にもたぶん関連があることである．他の応用については，[23,24]を参照せよ．

5.20.2 共通部分数列

アルファベット $\{0,1,\cdots,k-1\}$ から取られる値 a_i, b_j を項とする，長さ n のランダム数列をそれぞれ a, b とする．数列 c が a と b の**共通部分数列**（common subsequences）とは，c が a, b の共通の部分数列であるときをいう．すなわち，c が a から0個以上の項 a_i を取り除いて得られ，かつ，c が b から0個以上の項 b_j を取り除いて得られることを意味する．a と b の最長の共通部分数列の長さを $\lambda_{n,k}$ とする．たとえば，数列 $a=(1,0,0,2,3,2,1,1,0,2)$，$b=(0,1,1,1,3,3,3,0,2,1)$ は，最長共通部分数列 $c=(0,1,1,0,2)$ をもつから，$\lambda_{10,3}=5$ である．k の関数として，$n\to\infty$ のとき，$\lambda_{n,k}$ の平均について何がいえるだろうか？

$\mathrm{E}(\lambda_{n,k})$ は n に関して優加法性（superadditive），すなわち $\mathrm{E}(\lambda_{m,k})+\mathrm{E}(\lambda_{n,k}) \leq \mathrm{E}(\lambda_{m+n,k})$，をもつことが証明される．したがって，Fekete（フェケテ）の定理[25,26]によって，極限値

$$\gamma_k = \lim_{n\to\infty} \frac{\mathrm{E}(\lambda_{n,k})}{n} = \sup_n \frac{\mathrm{E}(\lambda_{n,k})}{n}$$

が存在する．Chvátal & Sankoff（ホバタル&サンコフ）[27-30]からはじまり，多くの研究者[31-37]は γ_k を研究した．表5.5に，現在までに知られている γ_k の最良の数値的推定値[37]および，γ_k に対する厳密な下界と上界を示す．

すべての k に対して，$1 \leq \gamma_k\sqrt{k} \leq e$ であることが知られ[27,31]，$\lim_{k\to\infty} \gamma_k\sqrt{k}=2$ であると予想されている[38]．n の1次式で増大すると予想されている $\mathrm{Var}(\lambda_{n,k})$ [39,41,42]のみならず，極限の比の収束速度

$$\gamma_k n - O(\sqrt{n\ln(n)}) \leq \mathrm{E}(\lambda_{n,k}) \leq \gamma_k n$$

にも興味がある[39-41]．

数列 c が a と b の**共通優数列**（common supersequence）とは，a および b がともに c の部分数列であるという意味で，c が a と b 両方の優数列であるときをいう．a, b の最も短い共通部分数列の長さ $\Lambda_{n,k}$ は，

$$\lim_{n\to\infty} \frac{\mathrm{E}(\Lambda_{n,k})}{n} = 2 - \gamma_k$$

を満たすことが示される[34,35,44]．

表5.5 比 γ_k の推定値

k	下界	推定値	上界
2	0.77391	0.8118	0.83763
3	0.63376	0.7172	0.76581
4	0.55282	0.6537	0.70824
5	0.50952	0.6069	0.66443

このような魅力のある双対性は,しかしながら,2個より多いランダム数列の集合から,最長の部分数列/最短の優数列を探すときには,失敗する.

[1] D. Aldous and P. Diaconis, Longest increasing subsequences: From patience sorting to the Baik-Deift-Johansson theorem, *Bull. Amer. Math. Soc.* 36 (1999) 413-432; MR 2000g:60013.
[2] P. Deift, Integrable systems and combinatorial theory, *Notices Amer. Math. Soc.* 47 (2000) 631-640; MR 2001g:05012.
[3] C. A. Tracy and H. Widom, Universality of the distribution functions of random matrix theory, *Integrable Systems: From Classical to Quantum*, Proc. 1999 Montréal conf., ed. J. Harnad, G. Sabidussi, and P. Winternitz, Amer. Math. Soc, 2000, pp. 251-264; mathph/9909001; MR 2002f:15036.
[4] R. P. Stanley, Recent progress in algebraic combinatorics, presentation at *Mathematical Challenges for the 21st Century* conf., Los Angeles, 2000; *Bull. Amer. Math. Soc.*, to appear; math.CO/0010218.
[5] A. M. Vershik and S. V. Kerov, Asymptotic behavior of the Plancherel measure of the symmetric group and the limit form of Young tableaux (in Russian), *Dokl. Akad. Nauk SSSR* 233 (1977) 1024-1027; Engl. transl. in *Soviet Math. Dokl.* 233 (1977) 527-531; MR 58 #562.
[6] B. F. Logan and L. A. Shepp, A variational problem for random Young tableaux, *Adv. Math.* 26 (1977) 206-222; MR 98e:05108.
[7] S. M. Ulam, Monte Carlo calculations in problems of mathematical physics, *Modern Mathematics for the Engineer: Second Series*, ed. E. F. Beckenbach, McGraw-Hill, 1961, pp. 261-281; MR 23 #B2202.
[8] R. M. Baer and P. Brock, Natural sorting over permutation spaces, *Math. Comp.* 22 (1968) 385-410; MR 37 #3800.
[9] J. M. Hammersley, A few seedlings of research, *Proc. Sixth Berkeley Symp. Math. Stat. Probab.*, v. 1, ed. L. M. Le Cam, J. Neyman, and E. L. Scott, Univ. of Calif. Press, 1972, pp. 345-394; MR 53 #9457.
[10] J. F. C. Kingman, Subadditive ergodic theory, *Annals of Probab.* 1 (1973) 883-909; MR 50 #8663.
[11] A. M. Odlyzko and E. M. Rains, On longest increasing subsequences in random permutations, *Analysis, Geometry, Number Theory: The Mathematics of Leon Ehrenpreis*, Proc. 1998 Temple Univ. conf., ed. E. L. Grinberg, S. Berhanu, M. Knopp, G. Mendoza, and E. T. Quinto, Amer. Math. Soc., 2000, pp. 439-451; MR 2001d:05003.
[12] J. Baik, P. Deift, and K. Johansson, On the distribution of the length of the longest increasing subsequence of random permutations, *J. Amer. Math. Soc.* 12 (1999) 1119-1178; math.CO/9810105; MR 2000e:05006.
[13] C. A.Tracy and H.Widom, Random unitary matrices, permutations and Painlevé, *Commun. Math. Phys.* 207 (1999) 665-685; math.CO/9811154; MR 2001h:15019.
[14] E. M. Rains, Increasing subsequences and the classical groups, *Elec. J. Combin.* 5 (1998) R12; MR 98k:05146.
[15] C. A. Tracy and H. Widom, Level-spacing distributions and the Airy kernel, *Phys. Lett. B* 305 (1993) 115-118; hep-th/9210074; MR 94f:82046.
[16] C. A. Tracy and H. Widom, Level-spacing distributions and the Airy kernel, *Commun. Math. Phys.* 159 (1994) 151-174; hep-th/9211141; MR 95e:82003.
[17] C. A. Tracy and H. Widom, The distribution of the largest eigenvalue in the Gaussian ensembles: $\beta=1, 2, 4$, *Calogero-Moser-Sutherland Models*, Proc. 1997 Montréal conf., ed. J. F. van Diejen and L. Vinet, Springer-Verlag, 2000, pp. 461-472; solv-int/9707001; MR 2002g:82021.
[18] J. Baik, P. Deift, and K. Johansson, On the distribution of the length of the second row of a Young diagram under Plancherel measure, *Geom. Func. Anal.* 10 (2000) 702-731; addendum 10 (2000) 1606-1607; math.CO/9901118; MR 2001m:05258.
[19] A. Okounkov, Random matrices and random permutations, *Int. Math. Res. Notices* (2000) 1043-

1095; math.CO/9903176; MR 2002c:15045.
- [20] A. Borodin, A. Okounkov, and G. Olshanski, Asymptotics of Plancherel measures for symmetric groups. *J. Amer. Math. Soc.* 13 (2000) 481-515; math.CO/9905032; MR 2001g:05103
- [21] K. Johansson, Discrete orthogonal polynomial ensembles and the Plancherel measure, *Annals of Math.* 153 (2001) 259-296; math.CO/9906120; MR 2002g:05188.
- [22] D. Mackenzie, From solitaire, a clue to the world of prime numbers, *Science* 282 (1998) 1631-1632; 283 (1999) 794-795.
- [23] I. M. Johnstone, On the distribution of the largest principal component, *Annals of Statist.* 29 (2001) 295-327; Stanford Univ. Dept. of Statistics TR 2000-27.
- [24] C. A. Tracy and H. Widom, A distribution function arising in computational biology, *MathPhys Odyssey 2001: Integrable Models and Beyond*, Proc. 2001 Okayama/Kyoto conf., ed. M. Kashiwara and T. Miwa, Birkhäuser, 2002, pp. 467-474; math.CO/0011146.
- [25] M. Fekete, Über die Verteilung der Wurzeln bei gewissen algebraischen Gleichungen mit ganzzahligen Koeffizienten, *Math. Z.* 17 (1923) 228-249.
- [26] G. Pólya and G. Szegö, *Problems and Theorems in Analysis*, v. 1, Springer-Verlag, 1972, ex. 98, pp. 23, 198; MR 81e:00002.
- [27] V. Chvátal and D. Sankoff, Longest common subsequences of two random sequences, *J. Appl. Probab.* 12 (1975) 306-315; MR 53 # 9324.
- [28] V. Chvátal and D. Sankoff, An upper-bound technique for lengths of common subsequences, *Time Warps, String Edits, and Macromolecules: The Theory and Practice of Sequence Comparison*, ed. D. Sankoff and J. B. Kruskal, Addison-Wesley, 1983, pp. 353-357; MR 85h:68073.
- [29] R. Durrett, *Probability: Theory and Examples*, Wadsworth, 1991, pp. 361-373; MR 91m:60002.
- [30] I. Simon, Sequence comparison: Some theory and some practice, *Electronic Dictionaries and Automata in Computational Linguistics*, Proc. 1987 St. Pierre d'Oléron conf., ed. M. Gross and D. Perrin, Lect. Notes in Comp. Sci. 377, Springer-Verlag, 1987, pp. 79-92.
- [31] J. G. Deken, Some limit results for longest common subsequences, *Discrete Math.* 26 (1979) 17-31; MR 80e:68100.
- [32] J. G. Deken, Probabilistic behavior of longest-common-subsequence length, *Time Warps, String Edits, and Macromolecules*, ed.D. Sankoff and J.B. Kruskal, Addison-Wesley, 1983, pp. 359-362; MR 85h:68073.
- [33] V. Dancik and M. S. Paterson, Upper bounds for the expected length of a longest common subsequence of two binary sequences, *Random Structures Algorithms* 6 (1995) 449-458; also in *Proc. 1994 Symp. on Theoretical Aspects of Computer Science (STACS)*, Caen, ed. P. Enjalbert, E.W. Mayr, and K.W.Wagner, Lect. Notes in Comp. Sci. 775, Springer-Verlag, 1994, pp. 669-678; MR 95g:68040 and MR 96h:05016.
- [34] V. Dancik, *Expected Length of Longest Common Subsequences*, Ph.D. thesis, Univ. of Warwick, 1994.
- [35] M. Paterson and V. Dancík, Longest common subsequences, *Mathematical Foundations of Computer Science (MFCS)*, Proc. 1994 Kosice conf., ed. I. Privara, B. Rovan, and P. Ruzicka, Lect. Notes in Comp. Sci. 841, Springer.Verlag, Berlin, 1994, pp. 127-142; MR 95k:68180.
- [36] V. Dancík, Upper bounds for the expected length of longest common subsequences, unpublished note (1996); abstract in *Bull. EATCS* 54 (1994) 248.
- [37] R. A. Baeza-Yates, R. Gavaldà, G. Navarro, and R. Scheihing, Bounding the expected length of longest common subsequences and forests, *Theory Comput. Sys.* 32 (1999) 435-452; also in *Proc. Third South American Workshop on String Processing (WSP)*, Recife, ed. R. Baeza-Yates, N. Ziviani, and K. Guimar.aes, Carleton Univ. Press, 1996, pp. 1-15; MR 2000k:68076.
- [38] D. Sankoff and S. Mainville, Common subsequences and monotone subsequences, *Time Warps, String Edits, and Macromolecules*, ed. D. Sankoff and J. B. Kruskal, Addison-Wesley, 1983, pp. 363-365; MR 85h:68073.
- [39] K. S. Alexander, The rate of convergence of the mean length of the longest common subsequence, *Annals of Appl. Probab.* 4 (1994) 1074-1082; MR 95k:60020.

[40] W. T. Rhee, On rates of convergence for common subsequences and first passage time, *Annals of Appl. Probab.* 5 (1995) 44-48; MR 96e:60023.
[41] J. Boutet de Monvel, Extensive simulations for longest common subsequences: Finite size scaling, a cavity solution, and configuration space properties, *Europ. Phys. J. B 7* (1999) 293-308; cond-mat/9809280.
[42] M. S.Waterman, Estimating statistical significance of sequence alignments, *Philos. Trans. Royal Soc. London Ser. B* 344 (1994) 383-390.
[43] V. Dancik, Common subsequences and supersequences and their expected length, *Combin. Probab. Comput.* 7 (1998) 365-373; also in *Sixth Combinatorial Pattern Matching Symp. (CPM)*, Proc. 1995 Espoo conf., ed. Z. Galil and E. Ukkonen, Lect. Notes in Comp. Sci. 937, Springer-Verlag, 1995, pp. 55-63; MR 98e:68100 and MR 99m:05007.
[44] T. Jiang and M. Li, On the approximation of shortest common supersequences and longest common subsequences, *SIAM J. Comput.* 24 (1995) 1122-1139; also in *Proc. 1994 Int. Colloq. on Automata, Languages and Programming (ICALP)*, Jerusalem, ed. S. Abiteboul and E. Shamir, Lect. Notes in Comp. Sci. 820, Springer-Verlag, 1994, pp. 191-202; MR 97a:68081.

5.21 k-充足可能の定数

$x_1, x_2, \cdots, x_n$ をブール変数とする。x_j と $\neg x_j$ (否定) の両方は同時に選択できないという制限のもとで、集合 $\{x_1, \neg x_1, x_2, \neg x_2, \cdots, x_n, \neg x_n\}$ からランダムに k 個の要素を選ぶ。これらの k-**文字** (literal) は**節** (clause) を決定し、節は文字の選言 ($\vee$, すなわち「または」(inclusive or)) である。

この選択過程を m 回行う。m 個の独立な節は、節の連言 ($\wedge$, すなわち「かつ」) からなる**式** (formula) を決定する。特別な場合 $n=5, k=3, m=4$ における実例の式は、

$$[x_1 \vee (\neg x_5) \vee (\neg x_2)] \wedge [(\neg x_3) \vee x_2 \vee (\neg x_1)] \wedge [x_5 \vee x_2 \vee x_4] \wedge [x_4 \vee (\neg x_3) \vee x_1]$$

である。式が**充足可能** (satisfiable) とは、式が真 (すなわち、値1をもつ) となるような x_i への0か1の割り付けが存在する、ときをいう。与えられた大きな式に対してこのような割り付けを発見するための、または式が**充足不可能** (unsatisfiable) であることを証明するための、効果的なアルゴリズムの設計は、理論的なコンピュータ科学において重要な話題である[1-3]。

k-充足 (k-satisfiability) 問題、略して k-SAT は、$k=2$ と $k \geq 3$ に対して、異なるふるまいをする。$k=2$ のときは、問題は線型時間アルゴリズムによって解くことができる。だが、$k \geq 3$ のとき、問題は NP-完全である。

浸透理論【5.18】からの考え方を含む別の特徴付けがある。極限比 $m/n \to r$ であるように $m \to \infty$, $n \to \infty$ とすると、実験的証拠により、ランダム k-SAT 問題はパラメータ r の臨界値 $r_c(k)$ で相転移が起こることを示している。$r < r_c$ に対して、ランダムな式は $m, n \to \infty$ のとき確率 $\to 1$ で充足可能である。$r > r_c$ に対して、ランダムな式はさらに、ほとんど確実に (almost surely) 充足不可能である。境界から離れると、

k-SAT 問題を解くことは比較的やさしい．計算の困難さは，閾値 $r=r_c$ 自身において最大となる．この考察は，巡回セールスマン問題【8.5】および他の組み合わせ論的な悪夢の問題を解くためのアルゴリズムの改良の助けに，究極的にはなるだろう．

2-SAT の場合には，$r_c(2)=1$ であることが証明されている[4-6]．統計力学の観点から，2-SAT の厳密な理解は，[7]で成し遂げられた．

$k \geq 3$，k-SAT の場合には，それと比べてわずかなことしか証明されていない．すべての k に対して，正しい不等式がある[4]．

$$\frac{3}{8}\frac{2^k}{k} \leq r_c(k) \leq \ln(2) \cdot \ln\left(\frac{2^k}{2^k-1}\right)^{-1} \sim \ln(2) \cdot 2^k$$

多くの研究者は，3-SAT 閾値

$$3.26 \leq r_c(3) \leq 4.506$$

に関する，ぴったりした上界[8-16]および下界[17-20]を定めるために貢献した．大規模な計算[21-23]により，推定値 $r_c(3)=4.25\cdots$ を得る．大きな k に対する推定値[1]は，$r_c(4)=9.7\cdots, r_c(5)=20.9\cdots, r_c(6)=43.2\cdots$ である．しかし，これらは改良できる．2-SAT とは異なって，$k \geq 3$ に対して $r_c(k)$ が存在するという証明は未だにない．しかし，Friedgut（フリージット）[24]はこの方向で，重要な一歩を踏みだした．ランダムグラフのある種の性質に対応する強相転移は，彼の論文で本質的な役割をする．$r_c(k)$ が $O(2^k/k)$ と $O(2^k)$ の間の範囲を振動する可能性は完全に除かれていないが，これは意外であろう．

ランダムグラフに対する閾値現象の類似の事実を述べる．極限比 $m/n \to r$ を満たしながら $m \to \infty, n \to \infty$ とするとき，n 個の頂点と m 個の辺をもつランダムグラフ G において，G は，$s < s_c(k)$ に対して確率 $\to 1$ で k-彩色可能（colorable）であり，$s > s_c(k)$ に対して確率 $\to 1$ で k-彩色可能でない．前のように，$s_c(k)$ の存在は，$k \geq 3$ のときには予想にすぎない．しかし，範囲[25-33]

$$1.923 \leq s_c(3) \leq 2.495, \quad 2.879 \leq s_c(4) \leq 4.587$$
$$3.974 \leq s_c(5) \leq 6.948, \quad 5.190 \leq s_c(6) \leq 9.539$$

が成り立ち，$s_c(3)$ の推定値[34]は $s_c(3)=2.3$ である．

さらに，$(\pm 1, \pm 1, \pm 1, \cdots, \pm 1)$ の形をしたベクトルからなる離散 n 次元立方体（discrete n-cube）Q を考える．任意の $v \in Q$ によって生成される**半立方体**（half cube）H_v は，v との内積が負であるような，すべての $w \in Q$ の集合から生成される．ベクトル $u \in H_v$ のとき，H_v は u を**覆う**（cover），というのは自然である．$v_1, v_2, \cdots, v_m$ を Q からランダムに選んだものとする．極限比 $m/n \to t$ を満たしながら $m \to \infty$，$n \to \infty$ とするとき，$t > t_c$ に対して確率 $\to 1$ で $\bigcup_{k=1}^{m} H_{v_k}$ は Q のすべてを覆う．しかし，$t < t_c$ に対して確率 $\to 1$ ではそのようにはならない．t_c の存在は[35]で予想されたが，証明は知られていない．範囲[36,37]は

$$0.005 \leq t_c \leq 0.9963 = 1 - 0.0037$$

であり，t_c の推定値は $t_c=0.82$ である[38,39]．この問題を研究する動機は，2値ニューラルネットワークで生じる．

2-SATと3-SATの両方を含む，おもしろい変形がある．実数 $p, 0 \leq p \leq 1$ を固定する．ランダムに m 個の句 (clause) を選ぶとき，3-句を確率 p で，2-句を確率 $1-p$ で選ぶ．これは $(2+p)$-SAT として知られていて，2-SAT から 3-SAT へ移るとき，複雑性の始まりを理解するために役に立つ [3, 40-42]．明らかに，このモデルの臨界値は，すべての p に対して

$$r_c(2+p) \leq \min\left\{\frac{1}{1-p}, \frac{1}{p}r_c(3)\right\}$$

を満たす．さらに，[43]より，$p \leq 2/5$ のとき，確率 → 1 のもとで，ランダム $(2+p)$-SAT 式が充足可能であるための必要十分条件は，その 2-SAT 部分式が充足可能である．これは注目に値する結果，「2-句を60%と3-句を40%含むランダム混合は，2-SAT のようにふるまう！」である．臨界閾値の予想 $p_c=2/5$ の証拠は，[44]にある．さらに[45]を参照せよ．

句を作るときの「または (inclusive or)」を「排他的論理和 (exclusive or)」に置きかえることを含んだ別の変形がある．k-SAT, $k \geq 3$ の対比として，XOR-SAT 問題は，多項式時間で解け，かつ，充足可能から充足不可能への推移は，完全に理解されている[46].

[1] S. Kirkpatrick and B. Selman, Critical behavior in the satisfiability of random Boolean expressions, *Science* 264 (1994) 1297-1301; MR 96e:68063.
[2] B. Hayes, Can't get no satisfaction, *Amer. Scientist*, v. 85 (1997) n. 2, 108-112.
[3] R. Monasson, R. Zecchina, S. Kirkpatrick, B. Selman, and L. Troyansky, Determining computational complexity from characteristic phase transitions, *Nature* 400 (1999) 133-137; MR 2000f:68055.
[4] V. Chvátal and B. Reed, Mick gets some (the odds are on his side), *Proc. 33rd Symp. on Foundations of Computer Science (FOCS)*, Pittsburgh, IEEE, 1992, pp. 620-627.
[5] A. Goerdt, A threshold for unsatisfiability, *J. Comput. Sys. Sci.* 53 (1996) 469-486; also in *Mathematical Foundations of Computer Science (MFCS)*, Proc. 1992 Prague conf., ed. I. M. Havel and V. Koubek, Lect. Notes in Comp. Sci. 629, Springer-Verlag, 1992, pp. 264-274; MR 94j:03011 and MR 98i:03012.
[6] W. Fernandez de la Vega, On random 2-SAT, unpublished note (1992).
[7] B. Bollobás, C. Borgs, J. T. Chayes, J. H. Kim, and D. B.Wilson, The scaling window of the 2-SAT transition, *Random Structures Algorithms* 18 (2001) 201-256; MR 2002a:68052.
[8] J. Franco and M. Paull, Probabilistic analysis of the Davis-Putnam procedure for solving the satisfiability problem, *Discrete Appl. Math.* 5 (1983) 77-87; correction 17 (1987) 295-299; MR 84e:68038 and MR 88e:68050.
[9] A. Z. Broder, A. M. Frieze, and E. Upfal, On the satisfiability and maximum satisfiability of random 3-CNF formulas, *Proc. 4th ACM-SIAM Symp. on Discrete Algorithms (SODA)*, Austin, ACM, 1993, pp. 322-330; MR 94b:03023.
[10] A. El Maftouhi and W. Fernandez de la Vega, On random 3-SAT, *Combin. Probab. Comput.* 4 (1995) 189-195; MR 96f:03007.
[11] A. Kamath, R. Motwani, K. Palem, and P. Spirakis, Tail bounds for occupancy and the satisfiability threshold conjecture, *Random Structures Algórithms* 7 (1995) 59-80; MR 97b:68091.
[12] L. M. Kirousis, E. Kranakis, D. Krizanc, and Y. C. Stamatiou, Approximating the unsatisfiability threshold of random formulas, *Random Structures Algorithms* 12 (1998) 253-269; MR 2000c:68069.

[13] O. Dubois and Y. Boufkhad, A general upper bound for the satisfiability threshold of random r-SAT formulae, *J. Algorithms* 24 (1997) 395-420; MR 98e:68103.
[14] S. Janson, Y. C. Stamatiou, and M. Vamvakari, Bounding the unsatisfiability threshold of random 3-SAT, *Random Structures Algorithms* 17 (2000) 103-116; erratum 18 (2001) 99-100; MR 2001c:68065 and MR 2001m:68064.
[15] A. C. Kaporis, L. M. Kirousis, Y. C. Stamatiou, M. Vamvakari, and M. Zito, Coupon collectors, q-binomial coefficients and the unsatisfiability threshold, *Seventh Italian Conf. on Theoretical Computer Science (ICTCS)*, Proc. 2001 Torino conf., ed. A. Restivo, S. Ronchi Della Rocca, and L. Roversi, Lect. Notes in Comp. Sci. 2202, Springer-Verlag, 2001, pp. 328-338.
[16] O. Dubois, Y. Boufkhad, and J. Mandler, Typical random 3-SAT formulae and the satisfiability threshold, *Proc. 11th ACM-SIAM Symp. on Discrete Algorithms (SODA)*, San Francisco, ACM, 2000, pp. 126-127.
[17] M.-T. Chao and J. Franco, Probabilistic analysis of two heuristics for the 3-satisfiability problem, *SIAM J. Comput.* 15 (1986) 1106-1118; MR 88b:68079.
[18] A. Frieze and S. Suen, Analysis of two simple heuristics on a random instance of k-SAT, *J. Algorithms* 20 (1996) 312-355; MR 97c:68062.
[19] D. Achlioptas, Setting 2 variables at a time yields a new lower bound for random 3-SAT, *Proc. 32nd ACMSymp. on Theory of Computing (STOC)*, Portland, ACM, 2000, pp. 28-37.
[20] D. Achlioptas and G. B Sorkin, Optimal myopic algorithms for random 3-SAT, *Proc. 41st Symp. on Foundations of Computer Science (FOCS)*, Redondo Beach, IEEE, 2000, pp. 590-600.
[21] T. Larrabee and Y. Tsuji, Evidence for a satisfiability threshold for random 3CNF formulas, presentation at *AAAI Spring Symp. on AI and NP-Hard Problems*, Palo Alto, 1993; Univ. of Calif. at Santa Cruz Tech. Report UCSC-CRL-92-42.
[22] B. Selman, D. G. Mitchell, and H. J. Levesque, Generating hard satisfiability problems, *Artificial Intellig.* 81 (1996) 17-29; Hard and easy distributions of SAT problems, *Proc. Tenth Nat. Conf. on Artificial Intelligence*, San Jose, AAAI Press, 1992, pp. 459-465; MR 98b:03013.
[23] J. M. Crawford and L. D. Auton, Experimental results on the crossover point in random 3-SAT, *Artificial Intellig.* 81 (1996) 31-57; also in *Proc. Eleventh Nat. Conf. on Artificial Intelligence*, Washington DC, AAAI Press, 1993, pp. 21-27; MR 97d:03055.
[24] E. Friedgut, Sharp thresholds of graph properties, and the k-SAT problem (appendix by J. Bourgain), *J. Amer. Math. Soc.* 12 (1999) 1017-1054; MR 2000a:05183.
[25] T. Luczak, Size and connectivity of the k-core of a random graph, *Discrete Math.* 91 (1991) 61-68; MR 92m:05171.
[26] V. Chvátal, Almost all graphs with $1.44n$ edges are 3-colorable, *Random Structures Algorithms* 2 (1991) 11-28; MR 92c:05056.
[27] M. Molloy and B. Reed, A critical point for random graphs with a given degree sequence, *Random Structures Algorithms* 6 (1995) 161-179; MR 97a:05191.
[28] B. Pittel, J. Spencer, and N. Wormald, Sudden emergence of a giant k-core in a random graph, *J. Combin. Theory Ser. B* 67 (1996) 111-151; MR 97e:05176.
[29] M. Molloy, A gap between the appearances of a k-core and a $(k+1)$-chromatic graph, *Random Structures Algorithms* 8 (1996) 159-160.
[30] D. Achlioptas and M. Molloy, The analysis of a list-coloring algorithm on a random graph, *Proc. 38th Symp. on Foundations of Computer Science (FOCS)*, Miami Beach, IEEE, 1997, pp. 204-212.
[31] P. E. Dunne and M. Zito, An improved upper bound on the non-3-colourability threshold, *Inform. Process. Lett.* 65 (1998) 17-23; MR 98i:05073.
[32] D. Achlioptas and M. Molloy, Almost all graphs with $2.522n$ edges are not 3-colorable, *Elec. J. Combin.* 6 (1999) R29; MR 2000e:05140.
[33] A. C. Kaporis, L. M. Kirousis, and Y. C. Stamatiou, A note on the non-colorability threshold of a random graph, *Elec. J. Combin.* 7 (2000) R29; MR 2001e:05116.
[34] T. Hogg and C. P. Williams, The hardest constraint problems: A double phase transition,

Artificial Intellig. 69 (1994) 359-377.
[35] E. Gardner, Maximum storage capacity in neural networks, *Europhys. Lett.* 4 (1987) 481-485.
[36] J. H. Kim and J. R. Roche, Covering cubes by random half cubes, with applications to binary neural networks, *J. Comput. Sys. Sci.* 56 (1998) 223-252; MR 2000g:68129.
[37] M. Talagrand, Intersecting random half cubes, *Random Structures Algorithms* 15 (1999) 436-449; MR 2000i:60011.
[38] W. Krauth and M. Opper, Critical storage capacity of the $J=\pm 1$ neural network, *J. Phys. A* 22 (1989) L519-L523.
[39] W. Krauth and M. Mézard, Storage capacity of memory networks with binary couplings, *J. Physique* 50 (1989) 3057-3066.
[40] R. Monasson and R. Zecchina, The entropy of the k-satisfiability problem, *Phys. Rev. Lett.* 76 (1996) 3881-3885; cond-mat/9603014; MR 97a:82053.
[41] R. Monasson and R. Zecchina, Statistical mechanics of the random k-satisfiability model, *Phys. Rev. E* 56 (1997) 1357-1370; MR 98g:82022.
[42] R. Monasson, R. Zecchina, S. Kirkpatrick, B. Selman, and L. Troyansky, $(2+p)$-SAT: Relation of typical-case complexity to the nature of the phase transition, *Random Structures Algorithms* 15 (1999) 414-435; cond-mat/9910080; MR 2000i:68076.
[43] D. Achlioptas, L. M. Kirousis, E. Kranakis, and D. Krizanc, Rigorous results for $(2+p)$-SAT, *Theoret. Comput. Sci.* 265 (2001) 109-129; MR 2002g:68047.
[44] G. Biroli, R. Monasson, and M.Weigt, A variational description of the ground state structure in random satisfiability problems, *Europ. Phys. J. B* 14 (2000) 551-568; cond-mat/9907343.
[45] I. Gent and T. Walsh, The SAT phase transition, *Proc. 11th European Conf. on Artificial Intelligence (ECAI)*, Amsterdam, Wiley, 1994, pp. 105-109.
[46] C. Creignou and H. Daudé, Satisfiability threshold for random XOR-CNF formulas, *Discrete Appl. Math.* 96-97 (1999) 41-53; MR 2000i:68072.

5.22 レンツ-イジング定数

イジングモデル (Ising model) は，相転移の物理学に関したものである．たとえば，熱せられるとき，磁石の磁力が失われ，ある有限の臨界温度より高い温度で，完全に磁性が失われる性質に関したものである．この小論ではこの話題をほとんど紹介できない．強正方形 (hard square)【5.12】と浸透クラスター【5.18】の場合と違って，簡潔に完全な問題を述べることは不可能である．パラメータ T が増加するとき，それらの結合分布が特異点 (singularity) を経るような，1 と −1 からなる大きな配列に関心がある．結合分布の定義および特徴づけは，念入りにできるが，我々の取り扱いは組合せ論的であり，級数展開に焦点を当てる．全体の背景については[1-10]を参照せよ．

正則 d 次元立方格子で $N=n^d$ 個の位置をもつものを L とする．たとえば，2 次元では，L は $N=n^2$ の $n \times n$ 正方格子 (square lattice) である．境界の影響を取り除くために，例外なくすべての位置が，最近接点を $2d$ 個もつように，d 次元トーラスを形成するために，L の周囲を合わせる．この約束は，大きな N に対しては無視できる誤差である．

5.22.1 低温級数展開

L の N 個の位置は黒または白に色づけされていると仮定する．L の dN 個の辺は 3 種に分類される．黒-黒，黒-白，白-白．これらの相対的数について，共通に何がいえるだろうか？ このような彩色のすべての可能性において，ちょうど p 個の黒の位置とちょうど q 個の黒-白の辺（図 5.20）が存在するような彩色の個数を $A(p,q)$ とする．

このとき，充分大きな N に対して[11-14]，

$$A(0,0) = 1 \qquad \text{(すべて白)}$$
$$A(1,2d) = N \qquad \text{(黒 1 個)}$$
$$A(2,4d-2) = dN \qquad \text{(黒 2 個，隣接)}$$
$$A(2,4d) = \frac{1}{2}(N-2d-1)N \qquad \text{(黒 2 個，隣接していない)}$$
$$A(3,6d-4) = (2d-1)dN \qquad \text{(黒 3 個，隣接)}$$

が成り立つ．この数列の性質は，2 変数母関数

$$a(x,y) = \sum_{p,q} A(p,q) x^p y^q$$

によって研究できる．形式的ベキ級数

$$\alpha(x,y) = \lim_{N \to \infty} \frac{1}{N} \ln(a(x,y))$$
$$= xy^{2d} + dx^2 y^{4d-2} - \frac{2d+1}{2} x^2 y^{4d} + (2d-1) dx^3 y^{6d-4} + \cdots$$

は，N の 1 次の係数をまとめるだけで得られる．後者は，ときに，係数がすべて整数の級数[15]

$$\exp(\alpha(x,y)) = 1 + xy^{2d} + dx^2 y^{4d-2} - dx^2 y^{4d} + (2d-1) dx^3 y^{6d-4} + \cdots$$

のように書かれる．これは，物理学者が**位置ごとのイジング自由エネルギー**（Ising free energy per site）に対する**低温級数展開**（low-temperature series）とよんでいるものである．文字 x, y は形式的な変数ではなく，温度と磁場に関係している．級数 $\alpha(x,y)$ は単なる数学的な構造体ではなく，物理実験で測定できる性質をもつ熱力学的

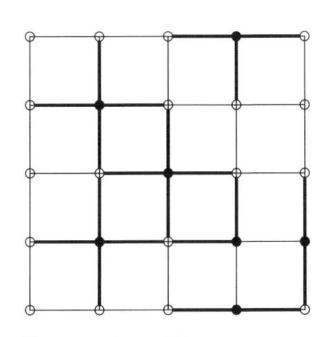

図 5.20　$d=2$，$N=25$，$p=7$，$q=21$ の彩色例（周囲は無視した）．

関数である[16]. $x=1$ という特別な場合は，**零磁場の場合**（zero magnetic field case）として知られている．簡単のために，$\alpha(y)=\alpha(1,y)$ と書く．

$d=2$ のとき [11,17],
$$\exp(\alpha(y))=1+y^4+2y^6+5y^8+14y^{10}+44y^{12}+152y^{14}+566y^{16}+\cdots$$
が成り立つ．Onsager（オンサガー）[18-23]は驚くべき閉じた式を発見した．
$$\alpha(y)=\frac{1}{2}\int_0^1\int_0^1\ln\left[(1+y^2)^2-2y(1-y^2)(\cos(2\pi u)+\cos(2\pi v))\right]dudv$$
である．これにより，任意の次数[24]の級数の係数やさらにそれ以上の対象を計算できる．

$d=3$ のとき [11,25-30],
$$\exp(\alpha(y))=1+y^6+3y^{10}-3y^{12}+15y^{14}-30y^{16}+101y^{18}-261y^{20}+\cdots$$
が成り立つ．この級数のいかなる閉じた式も，発見されていない．必要とする計算は，$d=2$ の計算よりもずっと多くの手間を要する．

5.22.2 高温級数展開

関連する高温級数は，見たところ無関係な組合せ論の問題から生じる．L の空でない「部分グラフ」は，連結でかつ少なくとも1つの辺を含むと仮定する．いくつかの部分グラフが，2つの性質

・L の各辺は，多くとも1つだけ使われ，かつ
・L の各位置は「偶数」回（0回も可能）使われる

をもつように，L に描かれると仮定する．

L のこのような構成を，**偶多角形描画**（even polygonal drawing）とよぶ（図5.21）．偶多角形描画は，連結である必要はなく，単純，閉集合でかつ，共通の辺をもたない多角形（edge-disjoint polygon）の和集合である．

ちょうど r 個の辺をもつ偶多角形描画の個数を $B(r)$ とする．このとき，十分大き

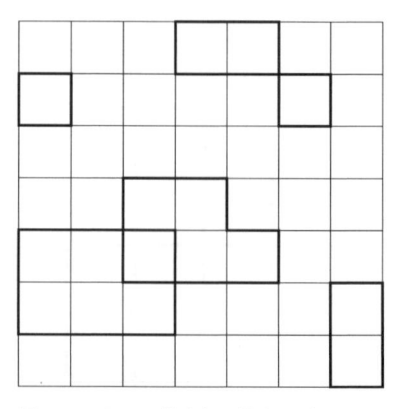

図 5.21　$d=2$ の偶多角形描画．閉部分グラフまたはユークリッド部分グラフともいう．

な N に対して [4,11,31],

$$B(4) = \frac{1}{2}d(d-1)N \qquad \text{(正方形)}$$

$$B(6) = \frac{1}{3}d(d-1)(8d-13)N \qquad \text{(2つの正方形，隣接している)}$$

$$B(8) = \frac{1}{8}d(d-1)(d(d-1)N + 216d^2 - 848d + 850)N \qquad \text{(多くの可能性)}$$

が成り立つ．一方，$d \geq 3$ に対して，描画はからみあうことができ，結び目となりうる [32]．だから，大きな r に対して，$B(r)$ を計算することは，とても複雑である！他方，$d=2$ に対しては明らかに，つねに $B(q) = \sum_p A(p,q)$ である．前のように，(1変数) 母関数

$$b(z) = 1 + \sum_r B(r) z^r$$

を定義すると，形式的ベキ級数

$$\beta(z) = \lim_{n \to \infty} \frac{1}{N} \ln(b(z))$$

$$= \frac{1}{2}d(d-1)z^4 + \frac{1}{3}d(d-1)(8d-13)z^6 + \frac{1}{4}d(d-1)(108d^2 - 424d + 425)z^8$$

$$+ \frac{2}{15}d(d-1)(2976d^3 - 19814d^2 + 44956d - 34419)z^{10} + \cdots$$

は，**イジング自由エネルギー** (Ising free energy) に対する**高温零場級数展開** (high-temperature zero-field series) とよばれている．$d=3$ のとき [11,25,29,33-36]

$$\exp(\beta(z)) = 1 + 3z^4 + 22z^6 + 192z^8 + 2046z^{10} + 24853z^{12} + 329334z^{14} + \cdots$$

が成り立つ．しかし，ここでも，この級数の係数の知識はかぎられている．

5.22.3 強磁性体モデルにおける相転移

イジング (Ising) モデルに関する，主な未解決問題は次の2つ [4,31,37]
- $d=2$ のとき，$a(x,y)$ の閉じた式を見つけること
- $d=3$ のとき，$\beta(z)$ の閉じた式を見つけること

である．なぜ，これらがとても重要なのだろうか．いま，基本的な物理および，さらに，前記の組合せ問題との関係を議論する．

一定の絶対温度 T で，ある外部の磁場に，鉄の棒を置く．磁場は，棒の中にある量の磁化 (magnetization) を誘発する．外部磁場を，ゆっくりなくすとき，低い温度 T において棒はその内部磁化をいくらか保持するが，高い温度 T において棒の内部磁化は完全に消えることを経験的に観測している．

一意の**臨界温度** (critical temperature) T_c は，**キューリー点** (Curie point) ともよばれ，磁力の性質が変化する温度のことである．イジングモデルは，ミクロ的な視点からこの物理現象を説明するための単純な方法である．

格子 L の各位置で，位置 i が「上向き (up)」であるとき「スピン変数」$\sigma_i = 1$ とし，

位置 i が「下向き (down)」であるとき $\sigma_i = -1$ と定義する．これは，**スピン 1/2 モデル**（spin-1/2 model）として知られている．**分配関数**（partition function）

$$Z(T) = \sum_{\sigma} \exp\left[\frac{1}{\kappa T}\left(\sum_{(i,j)} \xi \sigma_i \sigma_j + \sum_k \eta \sigma_k\right)\right]$$

を調べる．ただし，ξ は，最近接スピン間の結合 (coupling)（または相互作用 (interaction)）定数である．$\eta \geq 0$ は外部磁場の強さの定数であり，$\kappa > 0$ はボルツマン (Boltzmann) 定数である．

関数 $Z(T)$ は，物理システムの熱力学的特徴のすべてをとらえ，状態確率を計算する場合，一種の「分母」のようにふるまう．最初の和はベクトル $\sigma = (\sigma_1, \sigma_2, \cdots, \sigma_N)$ がとりうるすべての可能な 2^N の値にわたる和であり，2 番目の和は格子（位置 i と j が異なり，隣接する）のすべての辺にわたる和である．したがって，$\xi > 0$ と仮定する．これは**強磁性の場合**に対応する．いくぶん異なる理論は，反強磁性の場合（$\xi < 0$）に現れる．

前に議論した組合せ問題に，Z はどのように関連づけられるか？ スピン 1 を白色に，スピン -1 を黒色に対応させる．このとき，

$$\sum_{(i,j)} \sigma_i \sigma_j = (dN-q)\cdot 1 + q\cdot(-1) = dN - 2q$$

$$\sum_k \sigma_k = (N-p)\cdot 1 + p\cdot(-1) = N - 2p$$

である．したがって，

$$Z = x^{-N/2} y^{-dN/2} a(x, y)$$

である．ただし，

$$x = \exp\left(-\frac{2\eta}{\kappa T}\right), \quad y = \exp\left(-\frac{2\xi}{\kappa T}\right)$$

である．温度 T が低いと x と y は小さくなるから，$a(x, y)$ に対する表現である低温級数展開は正しい．（$T = \infty$ は，格子サイトの彩色が等しい確率でつけられる場合に対応する．これは正確に，前に述べた組合せ論的問題である．小さい値 p, q の状態を強調するために，範囲 $0 < T < \infty$ は不均一な重み (unequal weighting) に対応する．点 $T = 0$ は，すべてのスピンが整列した理想的な場合に対応する．熱はシステム内に無秩序をもたらす．）

高温の場合に，Z を

$$Z = \left(\frac{4}{(1-z^2)^d(1-w^2)}\right)^{N/2} \frac{1}{2^N} \sum_{\sigma}\left(\prod_{(i,j)}(1+\sigma_i\sigma_j z)\cdot \prod_k (1+\sigma_k w)\right)$$

と書き直す．ただし，

$$z = \tanh\left(\frac{\xi}{\kappa T}\right), \quad w = \tanh\left(\frac{\eta}{\kappa T}\right)$$

である．零場シナリオ (zero-field scenario)（$\eta = 0$）では，この式は簡単に

$$Z = \left(\frac{4}{(1-z^2)^d}\right)^{N/2} b(z)$$

となる.大きな T は小さい Z を与えるから,用語は再び意味をもつ.

5.22.4 臨界温度

いくつかのおもしろい定数に注意を向ける.低温級数 $\alpha(y)=\sum_{k=0}^{\infty}\alpha_k y^k$ の複素数平面での収束半径 y_c は,[29] により

$$y_c = \lim_{k\to\infty}|\alpha_{2k}|^{-1/2k} = \begin{cases} \sqrt{2}-1=0.4142135623\cdots & (d=2 \text{ のとき}) \\ \sqrt{0.2853\cdots}=0.5341\cdots & (d=3 \text{ のとき}) \end{cases}$$

である.したがって,$d=2$ のとき,強磁性体の臨界温度 T_c は

$$K_c = \frac{\xi}{\kappa T_c} = \frac{1}{2}\ln\left(\frac{1}{y_c}\right) = \frac{1}{2}\ln(\sqrt{2}+1) = 0.4406867935\cdots$$

を満たす.2次元の結果は,Kramers & Wannier(クラマース&ワンニア)[38] および Onsager[18] による有名な研究成果である.$d=3$ に対して,$y^2=-0.2853\cdots$ における特異点は物理的でない.したがって,強磁性体と関連はない.$y^2=0.412048\cdots$ における2番目の特異点は,望んでいたものであるが,直接計算することは難しい [29,39].ここで臨界温度を精密に得るために,代わりに高温級数 $\beta(z)=\sum_{k=0}^{\infty}\beta_k z^k$ を調べ,

$$z_c = \lim_{k\to\infty}\beta_{2k}^{-1/2k} = 0.218094\cdots, \quad K_c = \frac{1}{2}\ln\left(\frac{1+z_c}{1-z_c}\right) = 0.221654\cdots$$

を計算する.この推定値を導く級数およびモンテカルロ解析の莫大な文献がある [40-53].([54] における z_c に対して予想された正確な式は,間違いであるようだ [55].)$d>3$ に対して,次の推定値が知られている [56-65].

$$z_c = \begin{cases} 0.14855\cdots & (d=4 \text{ のとき}) \\ 0.1134\cdots & (d=5 \text{ のとき}) \\ 0.0920\cdots & (d=6 \text{ のとき}) \\ 0.0775\cdots & (d=7 \text{ のとき}) \end{cases}, \quad K_c = \begin{cases} 0.14966\cdots & (d=4 \text{ のとき}) \\ 0.1139\cdots & (d=5 \text{ のとき}) \\ 0.0923\cdots & (d=6 \text{ のとき}) \\ 0.0777\cdots & (d=7 \text{ のとき}) \end{cases}$$

関連する臨界指数 γ は,まもなく議論する.

5.22.5 磁化率

別の組み合わせ論的問題がある.いくつかの部分グラフが,性質
- L の各辺は多くとも1回用いられる
- L のすべての位置は,2つを除き,偶数である,かつ
- 残りの2つの位置は奇数であり,かつ同じ(連結な)部分グラフにある

を満たす L 上に描かれていると仮定する.

この構成を**奇多角形描画**(odd polygonal drawing)という(図5.22).奇多角形描画は,1つの偶多角形描画と2つの奇数の位置が結合している(向き付けされていない)自己回避歩行【5.10】との,辺が共通しない(edge-disjoint)和集合である.

ちょうど r 個の辺が存在する,奇多角形描画の個数の2倍を $C(r)$ とする.このとき,充分大きな N に対して [12,66],

$C(1)=2dN$ \hfill (SAW)

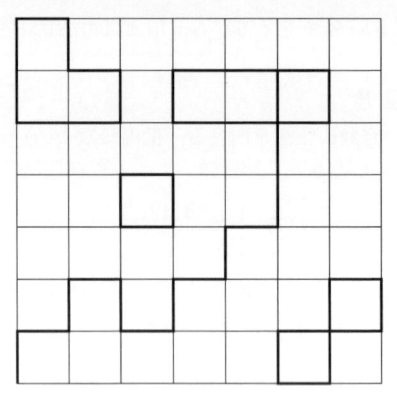

図5.22 $d=2$ の奇多角形描画.

$$C(2) = 2d(2d-1)N \qquad \text{(SAW)}$$
$$C(3) = 2d(2d-1)^2 N \qquad \text{(SAW)}$$
$$C(4) = 2d(2d(2d-1)^3 - 2d(2d-2))N \qquad \text{(SAW)}$$
$$C(5) = d^2(d-1)N^2 + 2d(16d^4 - 32d^3 + 16d^2 + 4d - 3)N \qquad \text{(正方形あるいは SAW)}$$

が成り立つ．前のように，母関数と形式的ベキ級数

$$c(z) = N + \sum_r C(r) z^r, \quad \chi(z) = \lim_{n \to \infty} \frac{1}{N} \ln(c(z)) = \sum_{k=0}^{\infty} \chi_k z^k$$

を定義する．これは，物理学者が**位置ごとのイジング磁化率**（Ising magnetic susceptibility per site）に対する**高温零場級数展開**（high-temperature zero-field series）とよんでいるものである．$\chi(z)$ の収束半径 z_c は，$d>1$ のときの $\beta(z)$ の収束半径に等しい．たとえば，$d=3$ のとき，級数[67-73]

$$\chi(z) = 1 + 6z + 30z^2 + 150z^3 + 726z^4 + 3510z^5 + 16710z^6 + \cdots$$

の解析は，（いくつか利用できる級数のうちで最もよいふるまいをする）臨界パラメータの推定値を得るために好まれる方法である．さらに，極限

$$\lim_{k \to \infty} \frac{\chi_k}{z_c^{-k} k^{\gamma-1}}$$

は存在し，かつ次元に従属するある正の定数 γ に対して，0 でない．例として，$d=2$ のとき，級数[67,74,75]

$$\chi(z) = 1 + 4z + 12z^2 + 36z^3 + 100z^4 + 276z^5 + 740z^6 + 1972z^7 + 5172z^8 + \cdots$$

を取り巻く数値的証拠は，**臨界磁化率指数**（critical susceptibility exponent）γ が 7/4 であり，かつ γ は（格子の選び方に独立であるという意味で）「普遍的（universal）」であることを示唆している．$d \geq 3$ のとき，γ に対する類似の厳密な式は，1つとして妥当なものには思えない．$d=3$ に対する一致した意見は，$\gamma = 1.238 \cdots$ である [40, 44, 46, 49-52, 71, 73]．

最後に，イジングモデル[76]と $\chi(z)$ の関係を明確にする．

$$\lim_{n\to\infty}\frac{1}{N}\ln(Z(z,w)) = \ln(2) - \frac{d}{2}\ln(1-z^2) - \frac{1}{2}\ln(1-w^2) + \beta(z)$$
$$+ \frac{1}{2}(\chi(z)-1)w^2 + O(w^4)$$

である．ただし，大きな O は z に依存する．したがって，$\chi(z)$ は w に関して2階導関数を評価する場合に生ずる．とくに $\chi(z)$ は（すぐに定義される）P の分散を計算する場合に生じる．

5.22.6 Q モーメントと P モーメント

温度混合へうまく一般化した，ランダム彩色問題（random coloring problem）に戻る．便宜上

$$Q = d - \frac{2}{N}q = \frac{1}{N}\sum_{(i,j)}\sigma_i\sigma_j, \quad P = 1 - \frac{2}{N}p = \frac{1}{N}\sum_k \sigma_k$$

とする．これ以降は $d=2$ と仮定する．Q の漸近分布を調べるために，$F(z)$ を

$$F(z) = \lim_{n\to\infty}\frac{1}{N}\ln(Z(z))$$

と定義する．このとき，明らかに，$\ln(Z)$ を項別微分することによって，

$$\lim_{n\to\infty}\mathrm{E}(Q) = (\kappa T)\frac{dF}{d\xi}, \quad \lim_{n\to\infty}N\mathrm{Var}(Q) = (\kappa T)^2\frac{d^2F}{d\xi^2}$$

を得る．両方のモーメントに対する厳密な式は，オンサガーの公式

$$F(z) = \ln\left(\frac{2}{1-z^2}\right) + \frac{1}{2}\int_0^1\int_0^1 \ln\left[(1+z^2)^2 - 2z(1-z^2)(\cos(2\pi u) + \cos(2\pi v))\right]du\,dv$$

を使って可能であるが，2つの特別な温度だけの結果を与える．$T=\infty$ の場合，等しい重みを割り当てた状態に対して，$\mathrm{E}(Q)\to 0$ かつ $N\mathrm{Var}(Q)\to 2$ であり，[77]では推論を確認している．$T=T_c$ の場合，$F(z)$ とその1階導関数は，ともに正しく定義される[11]から，特異性はかなり難解なことに注意せよ．

$$F(z_c) = \frac{\ln(2)}{2} + \frac{2G}{\pi} = 0.9296953983\cdots = \frac{1}{2}(\ln(2) + 1.1662436161\cdots)$$

$$\lim_{n\to\infty}\mathrm{E}(Q) = \sqrt{2}$$

である．ただし，G はカタランの定数【1.7】である．しかしながら，F の2階導関数は，$z=z_c$ の近くで非有界である．実際[5]，

$$\lim_{n\to\infty}N\mathrm{Var}(Q) \approx -\frac{8}{\pi}\left(\ln\left|\frac{T}{T_c}-1\right| + g\right)$$

である．ただし，g は定数であり，

$$g = 1 + \frac{\pi}{4} + \ln\left(\frac{\sqrt{2}}{4}\ln(\sqrt{2}+1)\right) = 0.6194036984\cdots$$

である．これは，物理学者が**イジング比熱**（Ising specific heat）の**対数発散**（logarithmic divergence）とよぶものに関係がある（図 5.23）．

余談として，三角形および六角形平面格子に対応する $F(z_c)$ の値は，それぞれ[11]，

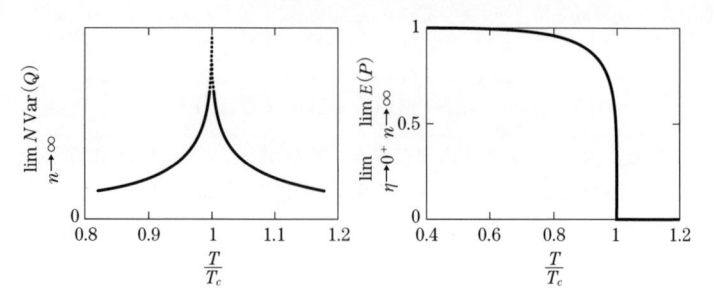

図 5.23 イジング比熱と自発磁化グラフ.

$$\ln(2) + \frac{\ln(3)}{4} + \frac{H}{2} = 0.8795853862\cdots$$

$$\frac{3\ln(2)}{4} + \frac{\ln(3)}{2} + \frac{H}{4} = 1.0250590965\cdots$$

である.両方の結果は,新しい定数を特徴づける[78,79].

$$H = \frac{5\sqrt{3}}{6\pi}\psi'\left(\frac{1}{3}\right) - \frac{5\sqrt{3}}{9}\pi - \ln(6) = \frac{\sqrt{3}}{6\pi}\psi'\left(\frac{1}{6}\right) - \frac{\sqrt{3}}{3}\pi - \ln(6) = -0.1764297331\cdots$$

ただし,$\psi'(x)$ はトリガンマ (trigamma) 関数 (ディガンマ (digamma) 関数 $\psi(x)$ の導関数) である【1.5.4】.他のところでの H の出現については,[80-82]を参照せよ.
式

$$\ln(2) + \ln(3) + H$$
$$= \frac{1}{4\pi^2}\int_{-\pi}^{\pi}\int_{-\pi}^{\pi}\ln[6 - 2\cos(\theta) - 2\cos(\phi) - 2\cos(\theta+\phi)]d\theta d\phi$$
$$= \frac{3\sqrt{3}}{\pi}\left(1 - \frac{1}{5^2} + \frac{1}{7^2} - \frac{1}{11^2} + \frac{1}{13^2} - \frac{1}{17^2} + \cdots\right) = 1.6153297360\cdots$$

は,【3.10】と【5.23】における類似した結果にうまく相当していることに注意せよ.

より難しい解析により,P に対応する2つのモーメントを計算することができる.さらに,磁化率および臨界指数の意味をよりはっきりと見ることができる.

$$F(z, w) = \lim_{n \to \infty}\frac{1}{N}\ln(Z(z, w))$$

とする.このとき,明らかに,前のように

$$\lim_{\eta \to 0^+}\lim_{n \to \infty}\mathrm{E}(P) = (\kappa T)\frac{\partial F}{\partial \eta}\bigg|_{\eta=0}, \quad \lim_{\eta \to 0^+}\lim_{n \to \infty}N\mathrm{Var}(P) = (\kappa T)^2\frac{\partial^2 F}{\partial \eta^2}\bigg|_{\eta=0}$$

である.もちろん,$w \neq 0$ のとき,正確な $F(z,w)$ を知らない.しかしながら,$w = 0$ での導関数には,すべての z に対して妥当である簡単な式が成り立つ.

$$\lim_{\eta \to 0^+}\lim_{n \to \infty}\mathrm{E}(P) = \begin{cases}\left[1 - \sinh\left(\frac{2\xi}{\kappa T}\right)^{-4}\right]^{1/8} & (T < T_c \text{ のとき}) \\ 0 & (T > T_c \text{ のとき})\end{cases}$$

$$= \begin{cases} (1+y^2)^{1/4}(1-6y^2+y^4)^{1/8}(1-y^2)^{-1/2} & (T<T_c \text{ のとき}) \\ 0 & (T>T_c \text{ のとき}) \end{cases}$$

この式は Onsager と Yang（ヤング）[83-85]による．厳密な正当化は[86-88]にある．特別な温度 $T=\infty$ に対して，p は二項分布 $B(N,1/2)$ に従うから，$\mathrm{E}(P) \to 0$，$N\mathrm{Var}(P) \to 1$ が成り立つ．臨界点では，さらに $\mathrm{E}(P) \to 0$ である．しかし，2階導関数はたいへんおもしろい複雑なふるまいをする．

$$\lim_{\eta \to 0^+} \lim_{n \to \infty} N\mathrm{Var}(P)$$
$$= \chi(z) \approx c_0^+ t^{-7/4} + c_1^+ t^{-3/4} + d_0 + c_2^+ t^{1/4} + e_0 t \ln(t) + d_1 t + c_3^+ t^{5/4}$$

が成り立つ．ただし，$0 < t = 1-T_c/T$，$c_0^+ = 0.9625817323\cdots$，$d_0 = -0.1041332451\cdots$，$e_0 = 0.0403255003\cdots$，$d_1 = -0.14869\cdots$ であり，

$$c_1^+ = \frac{\sqrt{2}}{8} K_c c_0^+, \quad c_2^+ = \frac{151}{192} K_c^2 c_0^+, \quad c_3^+ = \frac{615\sqrt{2}}{512} K_c^3 c_0^+$$

である．

Wu, McCoy, Tracy & Barouch（ウー，マッコイ，トレーシー&バロック）[89-99]は，（次項で説明する）Painlevé（パンルヴェ）III型微分方程式の解によって，これらの級数の係数に対する厳密な式を決定した．係数の異なる数値は，反強磁性体の場合に対してのみならず，$T < T_c$ に対しても応用される[100, 101]．たとえば，$t < 0$ のとき，対応する先頭の係数は $c_0^- = 0.0255369745\cdots$ である．磁化率 $\chi(z)$ の研究は，この小論で述べた他の熱力学的関数よりもかなり多くのものに含まれる．思考の厳密な流れには，依然として問題点がある[102]．さらに，最近の発展[103, 104]により，$\chi(z)$ の完全な漸近構造が，今では大部分決定された．

5.22.7 パンルヴェIII型方程式

Painlevé（パンルヴェ）III型微分方程式は[105]

$$\frac{f''(x)}{f(x)} = \left(\frac{f'(x)}{f(x)}\right)^2 - \frac{1}{x}\frac{f'(x)}{f(x)} + f(x)^2 - \frac{1}{f(x)^2}$$

であり，境界条件

$$f(x) \sim 1 - \frac{e^{-2x}}{\sqrt{\pi x}} \quad (x \to \infty), \quad f(x) \sim x(2\ln(2) - \gamma - \ln(x)) \quad (x \to +0)$$

を満たす解を $f(x)$ とする．ただし，γ はオイラー定数【1.5】である．

$$g(x) = \left[\frac{xf'(x)}{2f(x)} + \frac{x^2}{4f(x)^2}((1-f(x)^2)^2 - f'(x)^2)\right]\ln(x)$$

と定義する．このとき，c_0^+ と c_0^- の厳密な式は

$$c_0^+ = 2^{5/8}\pi\ln(\sqrt{2}+1)^{-7/4}\int_0^\infty y(1-f(y))\exp\left[\int_y^\infty x\ln(x)(1-f(x)^2)\,dx - g(y)\right]dy$$

$$c_0^- = 2^{5/8}\pi\ln(\sqrt{2}+1)^{-7/4}\int_0^\infty y\left\{(1+f(y))\exp\left[\int_y^\infty x\ln(x)(1-f(x)^2)\,dx - g(y)\right] - 2\right\}dy$$

である．パンルヴェII型は，最長増加部分数列問題（longest increasing subsequence

problem)【5.20】の議論で現れた．パンルヴェV型は，GUE予想【2.15.3】に関連して現れる．

これらの結果をわずかばかり変形したものがある．任意の定数 $c>0$ に対して，

$$h(x) = -\ln\left(f\left(\frac{x}{c}\right)\right)$$

と定義する．このとき，関数 $h(x)$ は，双曲正弦Gordon（ゴードン）微分方程式として知られている次の方程式を満たす．

$$h''(x) + \frac{1}{x}h'(x) = \frac{2}{c^2}\sinh(2h(x))$$

$$h(x) \sim \sqrt{\frac{c}{\pi x}}\exp\left(-\frac{2x}{c}\right) \quad (x \to \infty)$$

最後に，美しい式を述べる．

$$\int_0^\infty x\ln(x)\,(1-f(x)^2)\,dx = \frac{1}{4} + \frac{7}{12}\ln(2) - 3\ln(A)$$

である．ただし，A はGlaisher（グレイシャー）定数【2.15】である．ひょっとすると，c_0^+ と c_0^- は，いつの日にか，A にも関係づけられるかもしれない．

[1] G. F. Newell and E. W. Montroll, On the theory of the Ising model of ferromagnetism, *Rev. Mod. Phys.* 25 (1953) 353-389; MR 15,88b.
[2] M. E. Fisher, The theory of cooperative phenomena, *Rep. Prog. Phys.* 30 (1967) 615-730.
[3] S. G. Brush, History of the Lenz-Ising model, *Rev. Mod. Phys.* 39 (1967) 883-893.
[4] C. Thompson, *Mathematical Statistical Mechanics*, Princeton Univ. Press, 1972; MR 80h:82001.
[5] B. M. McCoy and T. T. Wu, *The Two-Dimensional Ising Model*, Harvard Univ. Press, 1973.
[6] R. J. Baxter, *Exactly Solved Models in Statistical Mechanics*, Academic Press, 1982; MR 90b:82001.
[7] G. A. Baker, *Quantitative Theory of Critical Phenomena*, Academic Press, 1990; MR 92d:82039.
[8] C. Domb, *The Critical Point: A Historical Introduction to the Modern Theory of Critical Phenomena*, Taylor and Francis, 1996.
[9] G. M. Bell and D. A. Lavis, *Statistical Mechanics of Lattice Systems*, v. 1-2, Springer-Verlag, 1999.
[10] D. S. Gaunt and A. J. Guttmann, Asymptotic analysis of coefficients, *Phase Transitions and Critical Phenomena*, v. 3, ed. C. Domb and M. S. Green, Academic Press, 1974, pp. 181-243; A. J. Guttmann, Asymptotic analysis of power-series expansions, *Phase Transitions and Critical Phenomena*, v. 13, ed. C. Domb and J. L. Lebowitz, Academic Press, 1989, pp. 1-234; MR 50 # 6393 and MR 94i:82003.
[11] C. Domb, On the theory of cooperative phenomena in crystals, *Adv. Phys.* 9 (1960) 149-361.
[12] T. Oguchi, Statistics of the three-dimensional ferromagnet. II and III, *J. Phys. Soc. Japan* 6 (1951) 27-35; MR 15,188h.
[13] A. J. Wakefield, Statistics of the simple cubic lattice, *Proc. Cambridge Philos. Soc.* 47 (1951) 419-429, 719-810; MR 13,308a and MR 13,417c.
[14] A. Rosengren and B. Lindström, A combinatorial series expansion for the Ising model, *Europ. J. Combin.* 8 (1987) 317-323; MR 89c:82062.
[15] R. J. Baxter and I. G. Enting, Series expansions for corner transfer matrices: The square lattice Ising model, *J. Stat. Phys.* 21 (1979) 103-123.
[16] C. Domb, Some statistical problems connected with crystal lattices, *J. Royal Statist. Soc. B* 26 (1964) 367-397.

[17] N. J. A. Sloane, On-Line Encyclopedia of Integer Sequences, A001393, A002891, A001408, and A002916.
[18] L. Onsager, Crystal physics. I: A two-dimensional model with an order-disorder transition, *Phys. Rev.* 65 (1944) 117-149.
[19] M. Kac and J. C. Ward, A combinatorial solution of the two-dimensional Ising model, *Phys. Rev.* 88 (1952) 1332-1337.
[20] S. Sherman, Combinatorial aspects of the Ising model for ferromagnetism. I: A conjecture of Feynman on paths and graphs, *J. Math. Phys.* 1 (1960) 202-217, addendum 4 (1963) 1213-1214; MR 22 #10273 and MR 27 #6560.
[21] N. V. Vdovichenko, A calculation of the partition function for a plane dipole lattice (in Russian), *Z. Èksper. Teoret. Fiz.* 47 (1965) 715-719; Engl. transl. in *Soviet Physics JETP* 20 (1965) 477-488; MR 31 #5622.
[22] M. L. Glasser, Exact partition function for the two-dimensional Ising model, *Amer. J. Phys.* 38 (1970) 1033-1036.
[23] F. Harary, A graphical exposition of the Ising problem, *J. Austral. Math. Soc.* 12 (1971) 365-377; MR 45 #5032.
[24] P. D. Beale, Exact distribution of energies in the two-dimensional Ising model, *Phys. Rev. Lett.* 76 (1996) 78-81.
[25] N. J. A. Sloane, On-Line Encyclopedia of Integer Sequences, A002890, A029872, A029873, and A029874.
[26] M. Creutz, State counting and low-temperature series, *Phys. Rev. B* 43 (1991) 10659-10662.
[27] G. Bhanot, M. Creutz, and J. Lacki, Low temperature expansion for the 3-d Ising model, *Phys. Rev. Lett.* 69 (1992) 1841-1844; hep-lat/9206020.
[28] G. Bhanot, M. Creutz, I. Horvath, J. Lacki, and J. Weckel, Series expansions without diagrams, *Phys. Rev. E* 49 (1994) 2445-2453; hep-lat/9303002.
[29] A. J. Guttmann and I. G. Enting, Series studies of the Potts model. I: The simple cubic Ising model, *J. Phys. A* 26 (1993) 807-821; MR 94a:82009.
[30] C. Vohwinkel, Yet another way to obtain low temperature expansions for discrete spin systems, *Phys. Lett. B* 301 (1993) 208-212.
[31] B. A. Cipra, An introduction to the Ising model, *Amer. Math. Monthly* 94 (1987) 937-959; MR 89g:82001.
[32] M. E. Fisher, The nature of critical points, *Statistical Physics, Weak Interactions, Field Theory-Lectures in Theoretical Physics*, v. 7C, ed. W. E. Brittin, Univ. of Colorado Press, 1964, pp. 1-159.
[33] G. S. Rushbrooke and J. Eve, High-temperature Ising partition function and related noncrossing polygons for the simple cubic lattice, *J. Math. Phys.* 3 (1962) 185-189; MR 25 #1907.
[34] S. McKenzie, Derivation of high temperature series expansions: Ising model, *Phase Transitions-Cargèse 1980*, ed. M. Lévy, J.-C. Le Guillou, and J. Zinn-Justin, Plenum Press, 1982, pp. 247-270.
[35] G. Bhanot, M. Creutz, U. Glässner, and K. Schilling, Specific heat exponent for the 3-d Ising model from a 24-th order high temperature series, *Phys. Rev. B* 49 (1994) 12909-12914; Nucl. Phys. B (Proc. Suppl.) 42 (1995) 758-760; hep-lat/9312048.
[36] A. J. Guttmann and I. G. Enting, The high-temperature specific heat exponent of the 3D Ising model, *J. Phys. A* 27 (1994) 8007-8010.
[37] F. Harary and E. M. Palmer, *Graphical Enumeration*, Academic Press, 1973, pp. 11-16, 113-117, 236-237; MR 50 #9682.
[38] H. A. Kramers and G. H. Wannier, Statistics of the two-dimensional ferromagnet. I and II, *Phys. Rev.* 60 (1941) 252-276; MR 3,63i and MR 3,64a.
[39] C. Domb and A. J. Guttmann, Low-temperature series for the Ising model, *J. Phys. C* 3 (1970) 1652-1660.
[40] J. Zinn-Justin, Analysis of Ising model critical exponents from high temperature series

expansions, *J. Physique* 40 (1979) 969-975.
[41] J. Adler, Critical temperatures of the $d=3$, $s=1/2$ Ising model: The effect of confluent corrections to scaling, *J. Phys. A* 16 (1983) 3585-3599.
[42] A. J. Liu and M. E. Fisher, The three-dimensional Ising model revisited numerically, *Physica A* 156 (1989) 35-76; MR 90c:82059.
[43] A. M. Ferrenberg and D. P. Landau, Critical behavior of the three-dimensional Ising model: A high-resolution Monte Carlo study, *Phys. Rev. B* 44 (1991) 5081-5091.
[44] B. G. Nickel, Confluent singularities in 3D continuum ϕ^4 theory: Resolving critical point discrepancies, *Physica A* 177 (1991) 189-196.
[45] N. Ito and M. Suzuki, Monte Carlo study of the spontaneous magnetization of the three-dimensional Ising model, *J. Phys. Soc. Japan* 60 (1991) 1978-1987.
[46] Y. Kinosita, N. Kawahima, and M. Suzuki, Coherent-anomaly analysis of series expansions and its application to the Ising model, *J. Phys. Soc. Japan* 61 (1992) 3887-3901; MR 93j:82037.
[47] C. F. Baillie, R. Gupta, K. A. Hawick, and G. S. Pawley, Monte Carlo renomalization-group study of the three-dimensional Ising model, *Phys. Rev. B* 45 (1992) 10438-10453.
[48] H.W. J. Blöte and G. Kamieniarz, Finite-size calculations on the three-dimensional Ising model, *Physica A* 196 (1993) 455-460.
[49] D. P. Landau, Computer simulation studies of critical phenomena, *Physica A* 205 (1994) 41-64.
[50] M. Kolesik and M. Suzuki, Accurate estimates of 3D Ising critical exponents using the coherent-anomaly method, *Physica A* 215 (1995) 138-151.
[51] H. W. J. Blöte, E. Luijten, and J. R. Heringa, Ising universality in three dimensions: A Monte Carlo study, *J. Phys. Math. A* 28 (1995) 6289-6313.
[52] R. Gupta and P. Tamayo, Critical exponents of the 3D Ising model, *Int. J. Mod. Phys. C* 7 (1996) 305-319.
[53] A. L. Talapov and H.W. J. Blöte, The magnetization of the 3D Ising model, *J. Phys. A* 29 (1996) 5727-5733.
[54] A. Rosengren, On the combinatorial solution of the Ising model, *J. Phys. A* 19 (1986) 1709-1714; MR 87k:82149.
[55] M. E. Fisher, On the critical polynomial of the simple cubic Ising model, *J. Phys. A.* 28 (1995) 6323-6333; corrigenda 29 (1996) 1145; MR 97b:82018a and MR 97b:82018b.
[56] N. J. A. Sloane, On-Line Encyclopedia of Integer Sequences, A010556, A010579, A010580, A030008, A030044, A030045, A030046, A030047, A030048, and A030049.
[57] M. E. Fisher and D. S. Gaunt, Ising model and self-avoiding walks on hypercubical lattices and high density expansions, *Phys. Rev.* 133 (1964) A224-A239.
[58] M. A. Moore, Critical behavior of the four-dimensional Ising ferromagnet and the breakdown of scaling, *Phys. Rev. B* 1 (1970) 2238-2240.
[59] D. S. Gaunt, M. F. Sykes, and S. McKenzie, Susceptibility and fourth-field derivative of the spin-1/2 Ising model for $T>T_c$ and $d=4$, *J. Phys. A* 12 (1979) 871-877.
[60] M. F. Sykes, Derivation of low-temperature expansions for Ising model. X: The four-dimensional simple hypercubic lattice, *J. Phys. A* 12 (1979) 879-892.
[61] C. Vohwinkel and P. Weisz, Low-temperature expansion in the $d=4$ Ising model, *Nucl. Phys. B* 374 (1992) 647-666; MR 93a:82015.
[62] A. J. Guttmann, Correction to scaling exponents and critical properties of the n-vector model with dimensionality >4, *J. Phys. A* 14 (1981) 233-239.
[63] A. B. Harris and Y. Meir, Recursive enumeration of clusters in general dimension on hyper-cubic lattices, *Phys. Rev. A* 36 (1987) 1840-1848; MR 88m:82037.
[64] C. Münkel, D. W. Heermann, J. Adler, M. Gofman, and D. Stauffer, The dynamical critical exponent of the two-, three- and five-dimensional kinetic Ising model, *Physica A* 193 (1993) 540-552.
[65] M. Gofman, J. Adler, A. Aharony, A. B. Harris, and D. Stauffer, Series and Monte Carlo study of high-dimensional Ising models, *J. Stat. Phys.* 71 (1993) 1221-1230.

[66] M. F. Sykes, Some counting theorems in the theory of the Ising model and the excluded volume problem, *J. Math. Phys*. 2 (1961) 52-59; MR 22 #8749.

[67] N. J. A. Sloane, On-Line Encyclopedia of Integer Sequences, A002906, A002913, A002926, and A002927.

[68] D. S. Gaunt and M. F. Sykes, The critical exponent γ for the three-dimensional Ising model, *J. Phys. A* 12 (1979) L25-L28.

[69] D. S. Gaunt, High temperature series analysis for the three-dimensional Ising model: A review of some recent work, *Phase Transitions-Cargèse 1980*, ed. M. Lévy, J.-C. Le Guillou, and J. Zinn-Justin, Plenum Press, 1982, pp. 217-246.

[70] A. J. Guttmann, The high temperature susceptibility and spin-spin correlation function of the three-dimensional Ising model, *J. Phys. A* 20 (1987) 1855-1863; MR 88h:82075.

[71] P. Butera and M. Comi, N-vector spin models on the simple-cubic and the body-centered-cubic lattices: A study of the critical behavior of the susceptibility and of the correlation length by high-temperature series extended to order 21, *Phys. Rev. B* 56 (1997) 8212-8240; hep-lat/9703018.

[72] M. Campostrini, Linked-cluster expansion of the Ising model, *J. Stat. Phys*. 103 (2001) 369-394; cond-mat/0005130.

[73] P. Butera and M. Comi, Extension to order β^{23} of the high-temperature expansions for the spin-1/2 Ising model on the simple-cubic and the body-centered-cubic lattices, *Phys. Rev. B* 62 (2000) 14837-14843; hep-lat/0006009.

[74] M. F. Sykes, D. S. Gaunt, P. D. Roberts, and J. A. Wyles, High temperature series for the susceptibility of the Ising model. I: Two dimensional lattices, *J. Phys. A* 5 (1972) 624-639.

[75] B. G. Nickel, On the singularity structure of the 2D Ising model, *J. Phys. A* 32 (1999) 3889-3906; addendum 33 (2000) 1693-1711; MR 2000d:82013 and MR 2001a: 82022.

[76] C. Domb, Ising model, *Phase Transitions and Critical Phenomena*, v. 3, ed. C. Domb and M. S. Green, Academic Press, 1974, pp. 357-484; MR 50 #6393.

[77] P. A. P. Moran, Random associations on a lattice, *Proc. Cambridge Philos. Soc*. 34 (1947) 321-328; 45 (1949) 488; MR 8,592b.

[78] R. M. F. Houtappel, Order-disorder in hexagonal lattices, *Physica* 16 (1950) 425-455; MR 12, 576j.

[79] V. S. Adamchik, Exact formulas for some Ising-related constants, unpublished note (1997).

[80] R. Burton and R. Pemantle, Local characteristics, entropy and limit theorems for spanning trees and domino tilings via transfer-impedances, *Annals of Probab*. 21 (1993) 1329-1371; MR 94m:60019.

[81] F. Y. Wu, Number of spanning trees on a lattice, *J. Phys. A* 10 (1977) L113-L115; MR 58 #8974.

[82] R. Shrock and F. Y. Wu, Spanning trees on graphs and lattices in d dimensions, *J. Phys. A* 33 (2000) 3881-3902; MR 2001b:05111.

[83] L. Onsager, Statistical hydrodynamics, *Nuovo Cimento Suppl*. 6 (1949) 279-287.

[84] C. N. Yang, The spontaneous magnetization of a two-dimensional Ising model, *Phys. Rev.* 85 (1952) 808-817; MR 14,522e.

[85] E. W. Montroll and R. B. Potts, Correlations and spontaneous magnetization of the two-dimensional Ising model, *J. Math. Phys*. 4 (1963) 308-319; MR 26 #5913.

[86] G. Benettin, G. Gallavotti, G. Jona-Lasinio, and A. L. Stella, On the Onsager-Yang value of the spontaneous magnetization, *Commun. Math. Phys*. 30 (1973) 45-54.

[87] D. B. Abraham and A. Martin-Löf, The transfer matrix for a pure phase in the two-dimensional Ising model, *Commun. Math. Phys*. 32 (1973) 245-268; MR 49 #6851.

[88] G. Gallavotti, *Statistical Mechanics: A Short Treatise*, Springer-Verlag, 1999.

[89] M. E. Fisher, The susceptibility of the plane Ising model, *Physica* 25 (1959) 521-524.

[90] E. Barouch, B. M. McCoy, and T. T. Wu, Zero-field susceptibility of the two-dimensional Ising model near T_c, *Phys. Rev. Lett*. 31 (1973) 1409-1411.

[91] C. A. Tracy and B. M. McCoy, Neutron scattering and the correlation functions of the Ising

model near T_c, *Phys. Rev. Lett.* 31 (1973) 1500-1504.
- [92] T. T. Wu, B. M. McCoy, C. A. Tracy, and E. Barouch, Spin-spin correlation functions for the two-dimensional Ising model: Exact theory in the scaling region, *Phys. Rev. B* 13 (1976) 316-374.
- [93] B. M. McCoy, C. A. Tracy, and T. T. Wu, Painlevé functions of the third kind, *J. Math. Phys.* 18 (1977) 1058-1092; MR 57 #12993.
- [94] C. A. Tracy, Painlevé transcendents and scaling functions of the two-dimensional Ising model, *Nonlinear Equations in Physics and Mathematics*, ed. A. O. Barut, Reidel, 1978, pp. 221-237; MR 84k:35001.
- [95] X.-P. Kong, H. Au-Yang, and J. H. H. Perk, New results for the susceptibility of the two-dimensional Ising model at criticality, *Phys. Lett. A* 116 (1986) 54-56.
- [96] X.-P. Kong, *Wave-Vector Dependent Susceptibility of the Two-Dimensional Ising Model*, Ph. D. thesis, State Univ. of New York at Stony Brook, 1987.
- [97] S. Gartenhaus and W. S. McCullough, Higher order corrections for the quadratic Ising lattice susceptibility at criticality, *Phys. Rev. B* 38 (1988) 11688-11703; *Phys. Lett. A* 127 (1988) 315-318; MR 89b:82105.
- [98] C. A. Tracy, Asymptotics of a tau-function arising in the two-dimensional Ising model, *Commun. Math. Phys.* 142 (1991) 298-311; MR 93c:82014.
- [99] O. Babelon and D. Bernard, From form factors to correlation functions; the Ising model, *Phys. Lett. B* 288 (1992) 113-120; MR 94e:82015.
- [100] X.-P. Kong, H. Au-Yang, and J. H. H. Perk, Comment on a paper by Yamadi and Suzuki, *Prog. Theor. Phys.* 77 (1987) 514-516.
- [101] S. S. C. Burnett and S. Gartenhaus, Zero-field susceptibility of an antiferromagnetic square Ising lattice, *Phys. Rev. B* 47 (1993) 7944-7956.
- [102] J. Palmer and C. A. Tracy, Two-dimensional Ising correlations: Convergence of the scaling limit, *Adv. Appl. Math* 2 (1981) 329-388; MR 84m:82024.
- [103] W. P. Orrick, B. G. Nickel, A. J. Guttmann, and J. H. H. Perk, Critical behaviour of the two-dimensional Ising susceptibility, *Phys. Rev. Lett.* 86 (2001) 4120-4123.
- [104] W. P. Orrick, B. G. Nickel, A. J. Guttmann, and J. H. H. Perk, The susceptibility of the square lattice Ising model: New developments, *J. Stat. Phys.* 102 (2001) 795-841; MR 2002e:82013.
- [105] C. A. Tracy and H. Widom, Universality of the distribution functions of random matrix theory, *Integrable Systems: From Classical to Quantum*, Proc. 1999 Montréal conf., ed. J. Harnad, G. Sabidussi, and P. Winternitz, Amer. Math. Soc, 2000, pp. 251-264; mathph/9909001; MR 2002f:15036.

5.23 モノマー・ダイマー定数

L をグラフとする【5.6】. **ダイマー** (dimer) とは, L の 2 つの隣接頂点とそれらを結ぶ(向き付けられていない)枝 (bond) からなる. **ダイマー配置** (dimer arrangement) とは, 共通部分をもたない L のダイマーの集まりである. 被覆されていない頂点は, **モノマー** (monomer) とよばれる. だから, ダイマー配置は, **モノマー・ダイマー被覆** (monomer-dimer covering) としても知られている. このような被覆だけを, 次の節のはじめに手短に議論する.

ダイマー被覆（dimer covering）は，ダイマー配置の和集合がLのすべての頂点を含む配置のことである．この小論の残りの部分は，ダイマー被覆およびそれと密接に関連したタイル張り（充填）の話題である．

5.23.1　2次元ドミノによる充填

$n \times n$正方格子Lの，異なるモノマー・ダイマー被覆の個数をa_nとし，$N=n^2$とする．このとき，$a_1=1$, $a_2=7$, $a_3=131$, $a_4=10012$[1,2]であり，漸近的に[3-6]

$$A = \lim_{n \to \infty} a_n^{1/N} = 1.940215351\cdots = (3.764435608\cdots)^{1/2}$$

が成り立つ．定数Aの厳密な式は知られていない．Aを推定するBaxter（バクスター）の方法は，角転送行列変分法（corner transfer matrix variational approach）に基づいている．さらに，これは【5.12】で活躍した．モノマー・ダイマー問題を議論する物理学者にとり自然な方法は，各ダイマーに活性度（activity）zを関連付けることである．だから，Aは$z=1$に対応する．ダイマー/頂点（1頂点あたりのダイマーの個数）の平均値ρは，$z=0$のとき0であり，$z=\infty$のとき1/2である．$z=1$のとき，$\rho=0.3190615546\cdots$であり，再び，閉じた式はない[3]．他のモデルとは違い（【5.12】，【5.18】，【5.22】を参照せよ），モノマー・ダイマーシステムは相転移をしない[7]．

a_nを計算することは，Lでの（必ずしも完全でない）**マッチング**（matching）の個数を数えること，すなわちLにおける辺の独立な集合を数えることと同値である．これは，ある種の2進接続行列（2進隣接行列，binary incident matrices）のパーマネント（permanents）[†]を計算する難しい問題に関連している[8-14]．Kenyon, Randall & Sinclair（ケニョン，ランドール&シンクレア）[15]は，ρが与えられたと仮定して，Lのモノマー・ダイマー被覆の個数を計算するための，ランダム化した多項式時間近似アルゴリズムを与えた．

これ以降，ゼロモノマー密度の場合，すなわち$z=\infty$に注意を向ける．Lの異なるダイマー被覆の個数をb_nとする．このとき，nが奇数のとき$b_n=0$であり，nが偶数のとき

$$b_n = 2^{N/2} \prod_{j=1}^{n/2} \prod_{k=1}^{n/2} \left(\cos^2 \frac{j\pi}{n+1} + \cos^2 \frac{k\pi}{n+1} \right)$$

が成り立つ．この厳密な式は，Kastelyn（カステリン）[16]とFisher & Temperley（フィシャー&テンペルレイ）[17,18]による．さらに，

$$\lim_{\substack{n \to \infty \\ n\,\text{偶数}}} \frac{1}{N} \ln(b_n) = \frac{1}{16\pi^2} \int_{-\pi}^{\pi} \int_{-\pi}^{\pi} \ln[4 + 2\cos(\theta) + 2\cos(\phi)] \, d\theta d\phi = \frac{G}{\pi} = 0.2915609040\cdots$$

である．すなわち，

$$B = \lim_{\substack{n \to \infty \\ n\,\text{偶数}}} b_n^{1/N} = \exp\left(\frac{G}{\pi}\right) = 1.3385151519\cdots = (1.7916228120\cdots)^{1/2}$$

[†]訳注：正方行列について行列式の古典的定義に対して，$n!$個の項を符号をつけずに総和した値．行列式の場合と違って効率的な計算法は知られていない．

である．ただし，Gはカタランの定数【1.7】である．これは，グラフ理論の用語で，正方格子上の**完全マッチング**（perfect matching）を数え上げる問題に対する注目すべき解である．さらに，次の問題にも答えるものである．2×1または1×2**ドミノ**（domino）で$n \times n$チェス盤を充填する方法の個数は何個であるのか？　より詳しい結果については，[19-26]を参照せよ．定数B^2は【3.10】でδとよばれており，【1.8】にも現れる．式$4G/\pi$は【5.22】に，$G/(\pi \ln(2))$は【5.6】に，$8G/\pi^2$は【7.7】に現れる．

正方格子をトーラスを形作るように包むと，個数b_nはいくぶん異なるものになる．しかし，極限定数Bは同じままである[16,27]．代わりに，アステカダイヤモンド（Aztec diamond）[28]のような形をしたチェス盤を仮定すると，関連した定数は$B=2^{1/4}=1.189\cdots<1.338\cdots=e^{G/\pi}$である．したがって，たとえ正方形チェス盤が，ダイヤモンドチェス盤よりもわずかに少ない面積をもつとしても，正方形チェス盤はより多くのドミノ充填をもつ[29]．格子の境界効果は，したがって，自明ではないことがわかる．

5.23.2　菱形とバイボーン

巻きつけられた六角形（蜂の巣）格子（hexagonal (honeycomb) lattice with wraparound）上のダイマーに対する$\exp(2G/\pi)$と類似なものは[30-32]，

$$C^2 = \lim_{n \to \infty} c_n^{2/N} = \exp\left(\frac{1}{8\pi^2}\int_{-\pi}^{\pi}\int_{-\pi}^{\pi}\ln[3+2\cos(\theta)+2\cos(\phi)+2\cos(\theta+\phi)]d\theta d\phi\right)$$
$$= 1.3813564445\cdots$$

である．この定数は【3.10】ではβとよばれ，別の式によっても表現できる．それは周期的な境界条件を満たす三角形マス目をもつチェス盤上にある，菱形充填（lozenge tilings）を特徴づける．[33-38]をも参照せよ．

トーラスのような巻きつけがないとき，数列[39]

$$c_n = \prod_{j=1}^{n}\prod_{k=1}^{n}\frac{n+j+k-1}{j+k-1}$$

が現れ，異なる増大定数$3\sqrt{3}/4$が当てはまる．六角格子（hexagonal grid）が，n, n, n（すなわち，考えられる最も単純な境界条件）の辺をもつ中心対称（center-symmetric）であると仮定する．さらに，この数列は$n \times n \times n$の箱の中に含まれる平面分割数を数え上げる[40,41]．

巻きつけられた三角格子（triangular lattice with wraparound）上のダイマーに対する，対応する類似な量は[30,42,43]

$$D^2 = \lim_{n \to \infty} d_n^{2/N} = \exp\left(\frac{1}{8\pi^2}\int_{-\pi}^{\pi}\int_{-\pi}^{\pi}\ln[6+2\cos(\theta)+2\cos(\phi)+2\cos(\theta+\phi)]d\theta d\phi\right)$$
$$= 2.3565273533\cdots$$

である．式$4\ln(D)$は，【5.22】で述べた定数$\ln(6)+H$に密接した類似をもつ．さらに，周期的な境界条件を満たす六角形のマス目をもつチェス盤上のバイボーン（bibone，2つの単位を合わせたドミノの牌）による充填を特徴づけている．巻きつけ

5.23.3 3次元ドミノによる充填

$n \times n \times n$ の立方体格子 L において，異なるダイマー被覆の個数を h_n とし，$N = n^3$ とする．このとき，n が奇数の場合は $h_n = 0$ であり，$h_2 = 9$, $h_4 = 5051532105$ [46,47] である．固体物理化学における重要な未解決問題は

$$\lim_{\substack{n \to \infty \\ n \text{ 偶数}}} h_n^{1/N} = \exp(\lambda)$$

または，同値な

$$\lambda = \lim_{\substack{n \to \infty \\ n \text{ 偶数}}} \frac{1}{N} \ln(h_n)$$

の推定値である．Hammersley（ハマースレイ）[48]は，λ が存在し，$\lambda \geq 0.29156$ であることを証明した．下界は，Fisher[49]によって，0.30187 に改良された．Hammersley[50,51]は 0.418347 に，Priezzhev（プリジェフ）[52,53]は 0.419989 に改良した．[54]の結果報告で，ある種の 2 進行列のパーマネントに対する下界の Schrijver & Valiant（シュライベル&ヴァリアント）による予想は，$\lambda \geq 0.44007584$ であると Minc（ミンク）は指摘した．Schrijver[55]はこの予想を証明した．これはこれまでの最良の結果である．

Fowler & Rushbrooke（フローラ&ラッシュブルック）[56]は，（λ の存在を仮定して）60 年以上前に λ に対する上界 0.54931 を得た．上界は，Minc によって[8,57,58]，0.5482709 に改良された．Ciucu（シウク）[59]は 0.463107 に，Lundow（ルンドウ）[60]は 0.457547 に改良した．

Nagle（ナグル）[30]，Gaunt（ガント）[31]，Beichl & Sullivan（ビーチル&サリヴァン）[61]らによる一連の厳密でない数値的な推定値は，結果的に $\lambda = 0.4466\cdots$ となっている．a_n から，小さい偶数 n に対して h_n を計算することは難しいので，行列パーマネント近似スキーム（matrix permanent approximation schemes）だけが，唯一の希望である．この分野は危険なくらい難しい．[62,63]にある，h_n を予想した厳密な漸近式は間違っている．

関連する話題は，n 次元単位立方体の 2^n 個の頂点を，$\{0,1\}$ から選ぶ n 個の組すべてから作られる単位立方体のダイマー被覆の個数 k_n である[47,64]．個数 $k_6 = 16332454526976$ は，Lundow[46]と Weidemann（ヴェイドマン）[65]により，独立に計算された．この場合，k_n の漸近的ふるまいは，かなり正確に知られている[44,65,66]．

$$\lim_{n \to \infty} \frac{1}{n} k_n^{2^{1-n}} = \frac{1}{e} = 0.3678794411\cdots$$

である．ただし，e は自然対数の底である【1.3】．

[1] N. J. A. Sloane, On-Line Encyclopedia of Integer Sequences, A004003, A006125, A008793,

A028420, and A039907.
[2] J. J. Henry, Monomer-dimer counts, $1 \le n \le 21$, unpublished note (1997).
[3] R. J. Baxter, Dimers on a rectangular lattice, *J. Math. Phys.* 9 (1968) 650-654.
[4] L. K. Runnels, Exact finite method of lattice statistics. III: Dimers on the square lattice, *J. Math. Phys.* 11 (1970) 842-850.
[5] M. Heise, Upper and lower bounds for the partition function of lattice models, *Physica A* 157 (1989) 983-999; MR 91c:82014.
[6] F. Cazals, Monomer-dimer tilings, *Studies in Automatic Combinatorics*, v. 2, Algorithms Project, INRIA, 1997.
[7] O. J. Heilmann and E. H. Lieb, Theory of monomer-dimer systems, *Commun. Math. Phys.* 25 (1972) 190-232; MR 45 #6337.
[8] H. Minc, *Permanents*, Addisison-Wesley, 1978; MR 80d:15009.
[9] R. A. Brualdi and H. J. Ryser, *Combinatorial Matrix Theory*, Cambridge Univ. Press, 1991; MR 93a:05087.
[10] J. M. Hammersley, A. Feuerverger, A. Izenman, and K. Makani, Negative finding for the three-dimensional dimer problem, *J. Math. Phys.* 10 (1969) 443-446.
[11] M. Jerrum, Two-dimensional monomer-dimer systems are computationally intractable, *J. Stat. Phys.* 48 (1987) 121-134; erratum 59 (1990) 1087-1088; MR 89d:82008 and MR 91h:82002.
[12] M. Luby, A survey of approximation algorithms for the permanent, *Sequences: Combinatorics, Compression, Security and Transmission*, Proc. 1988 Naples workshop, ed. R. M. Capocelli, Springer-Verlag, 1990, pp. 75-91; MR 90m:68059.
[13] M. Jerrum and A. Sinclair, Approximating the permanent, *SIAM J. Comput.* 18 (1989) 1149-1178; MR 91a:05075.
[14] A. Frieze and M. Jerrum, An analysis of a Monte-Carlo algorithm for estimating the permanent, *Combinatorica* 15 (1995) 67-83; MR 96g:68052.
[15] C. Kenyon, D. Randall, and A. Sinclair, Approximating the number of monomer-dimer coverings of a lattice, *J. Stat. Phys.* 83 (1996) 637-659; MR 97g:82003.
[16] P.W. Kasteleyn, The statistics of dimers on a lattice. I: The number of dimer arrangements on a quadratic lattice, *Physica* 27 (1961) 1209-1225.
[17] M. E. Fisher, Statistical mechanics of dimers on a plane lattice, *Phys. Rev.* 124 (1961) 1664-1672; MR 24 #B2437.
[18] H. N. V. Temperley and M. E. Fisher, Dimer problem in statistical mechanics . An exact result, *Philos. Mag.* 6 (1961) 1061-1063; MR 24 #B2436.
[19] E. W. Montroll, Lattice statistics, *Applied Combinatorial Mathematics*, ed. E. F. Beckenbach, Wiley, 1964, pp. 96-143; MR 30 #4687.
[20] P.W. Kasteleyn, Graph theory and crystal physics, *Graph Theory and Theoretical Physics*, ed. F. Harary, Academic Press, 1967, pp. 43-110; MR 40 #6903.
[21] J. K. Percus, *Combinatorial Methods*, Springer-Verlag, 1971; MR 49 #10555.
[22] L. Lovász and M. D. Plummer, *Matching Theory*, North-Holland, 1986; MR 88b:90087.
[23] P. John, H. Sachs, and H. Zernitz, Counting perfect matchings in polyminoes with an application to the dimer problem, *Zastos. Mat.* 19 (1987) 465-477; MR 89e: 05158.
[24] H. Sachs and H. Zernitz, Remark on the dimer problem, *Discrete Appl. Math.* 51 (1994) 171-179; MR 97e:05067.
[25] D. M. Cvetkovic, M. Doob, and H. Sachs, *Spectra of Graphs: Theory and Applications*, 3rd ed., Johann Ambrosius Barth Verlag, 1995, pp. 245-251; MR 96b:05108.
[26] D. A. Lavis and G. M. Bell, *Statistical Mechanics of Lattice Systems*, v. 2, Springer-Verlag, 1999; MR 2001g:82002.
[27] J. Propp, Dimers and dominoes, unpublished note (1992).
[28] N. Elkies, G. Kuperberg, M. Larsen, and J. Propp, Alternating-sign matrices and domino tilings. I, *J. Algebraic Combin.* 1 (1992) 111-132; MR 94f:52035.
[29] H. Cohn, R. Kenyon, and J. Propp, A variational principle for domino tilings, *J. Amer. Math.*

Soc. 14 (2001) 297-346; math.CO/0008220.
[30] J. F. Nagle, New series-expansion method for the dimer problem, *Phys. Rev.* 152 (1966) 190-197.
[31] D. S. Gaunt, Exact series-expansion study of the monomer-dimer problem, *Phys. Rev.* 179 (1969) 174-186.
[32] A. J. Phares and F. J. Wunderlich, Thermodynamics and molecular freedom of dimers on plane honeycomb and Kagomé lattices, *Nuovo Cimento B* 101 (1988) 653-686; MR 89m:82067.
[33] G. H. Wannier, Antiferromagnetism: The triangular Ising net, *Phys. Rev.* 79 (1950) 357-364; MR 12,576e.
[34] P. W. Kasteleyn, Dimer statistics and phase transitions, *J. Math. Phys.* 4 (1963) 287-293; MR 27 #3394.
[35] F. Y.Wu, Remarks on the modified potassium dihydrogen phosphate model of a ferroelectric, *Phys. Rev.* 168 (1968) 539-543.
[36] J. F. Nagle, Critical points for dimer models with 3/2-order transitions, *Phys. Rev. Lett.* 34 (1975) 1150-1153.
[37] H.W. J. Blöte and H. J. Hilhorst, Roughening transitions and the zero-temperature triangular Ising antiferromagnet, *J. Phys. A* 15 (1982) L631-L637; MR 83k:82046.
[38] C. Richard,M.Höffe, J. Hermisson, and M. Baake, Random tilings: Concepts and examples, *J. Phys. A* 31 (1998) 6385-6408; MR 99g:82033.
[39] V. Elser, Solution of the dimer problem on a hexagonal lattice with boundary, *J. Phys. A* 17 (1984) 1509-1513.
[40] G. David and C. Tomei, The problem of the calissons, *Amer. Math. Monthly* 96 (1989) 429-431; MR 90c:51024.
[41] P. A. MacMahon, *Combinatory Analysis*, Chelsea, 1960, pp. 182-183, 239-242; MR 25 #5003.
[42] A. J. Phares and F. J. Wunderlich, Thermodynamics and molecular freedom of dimers on plane triangular lattices, *J. Math. Phys.* 27 (1986) 1099-1109; MR 87c:82071.
[43] R. Kenyon, The planar dimer model with boundary: A survey, *Directions in Mathematical Quasicrystals*, ed. M. Baake and R. V. Moody, CRM Monogr. 13, Amer. Math. Soc., 2000, pp. 307-328; MR 2002e:82011.
[44] J. Propp, Enumeration of matchings: Problems and progress, *New Perspectives in Algebraic Combinatorics*, Proc. 1996-1997 MSRI Berkeley Program on Combinatorics, ed. L. J. Billera, A. Björner, C. Greene, R. E. Simion, and R. P. Stanley, Cambridge Univ. Press, 1999, pp. 255-291; math.CO/9904150; MR 2001c:05008.
[45] H. Sachs, A contribution to problem 18 in "Twenty open problems in enumeration of matchings," unpublished note (1997).
[46] P. H. Lundow, Computation of matching polynomials and the number of 1-factors in polygraphs, research report 12-1996, Umeå Universitet.
[47] N. J. A. Sloane, On-Line Encyclopedia of Integer Sequences, A005271, A028446, A033535, and A045310.
[48] J. M. Hammersley, Existence theorems and Monte Carlo methods for the monomer-dimer problem, *Research Papers in Statistics: Festschrift for J. Neyman*, ed. F. N. David, Wiley, 1966, pp. 125-146; MR 35 #2595.
[49] J. A. Bondy and D. J. A.Welsh, A note on the monomer-dimer problem, *Proc. Cambridge Philos. Soc.* 62 (1966) 503-505; MR 34 #3958.
[50] J. M. Hammersley, An improved lower bound for the multidimensional dimer problem, *Proc. Cambridge Philos. Soc.* 64 (1968) 455-463; MR 38 #5639.
[51] J. M. Hammersley and V. V. Menon, A lower bound for the monomer-dimer problem, *J. Inst. Math. Appl.* 6 (1970) 341-364; MR 44 #3886.
[52] V. B. Priezzhev, The statistics of dimers on a three-dimensional lattice. I: An exactly solved model, *J. Stat. Phys.* 26 (1981) 817-828; MR 84h:82059a.
[53] V. B. Priezzhev, The statistics of dimers on a three-dimensional lattice. II: An improved lower bound, *J. Stat. Phys.* 26 (1981) 829-837; MR 84h:82059b.

[54] A. Schrijver and W. G. Valiant, On lower bounds for permanents, *Proc. Konink. Nederl. Akad. Wetensch. Ser. A* 83 (1980) 425-427; *Indag. Math.* 42 (1980) 425-427; MR 82a:15004.
[55] A. Schrijver, Counting 1-factors in regular bipartite graphs, *J. Combin. Theory B* 72 (1998) 122-135; MR 99b:05117.
[56] R. H. Fowler and G. S. Rushbrooke, An attempt to extend the statistical theory of perfect solutions, *Trans. Faraday Soc.* 33 (1937) 1272-1294.
[57] H. Minc, An upper bound for the multidimensional dimer problem, *Math. Proc. Cambridge Philos. Soc.* 83 (1978) 461-462; MR 58 # 289.
[58] H. Minc, An asymptotic solution of the multidimensional dimer problem, *Linear Multilinear Algebra* 8 (1980) 235-239; MR 81e:82063.
[59] M. Ciucu, An improved upper bound for the 3-dimensional dimer problem, *Duke Math. J.* 94 (1998) 1-11; MR 99f:05026.
[60] P. H. Lundow, Compression of transfer matrices, *Discrete Math.* 231 (2001) 321-329; MR 2002a:05200.
[61] I. Beichl and F. Sullivan, Approximating the permanent via importance sampling with application to the dimer covering problem, *J. Comput. Phys.* 149 (1999) 128-147; MR 99m:82021.
[62] A. J. Phares and F. J. Wunderlich, Thermodynamics of dimers on a rectangular $L \times M \times N$ lattice, *J. Math. Phys.* 26 (1985) 2491-2499; comment by J. C. Wheeler, 28 (1987) 2739-2740; MR 86m:82050 and MR 88m:82038.
[63] H. Narumi, H. Kita, and H. Hosoya, Expressions for the perfect matching numbers of cubic $l \times m \times n$ lattices and their asymptotic values, *J. Math. Chem.* 20 (1996) 67-77; MR 98c:82030.
[64] N. Graham and F. Harary, The number of perfect matchings in a hypercube, *Appl. Math. Lett.* 1 (1988) 45-48.
[65] L. H. Clark, J. C. George, and T. D. Porter, On the number of 1-factors in the n-cube, *Proc. 28th Southeastern Conf. on Combinatorics, Graph Theory and Computing*, Boca Raton, 1997, Congr. Numer. 127, Utilitas Math., 1997, pp. 67-69; MR 98i:05127.
[66] H. Sachs, Problem 298, How many perfect matchings does the graph of the n-cube have?, *Discrete Math.* 191 (1998) 251.

5.24 リーブの正方氷定数

巻きつけた $n \times n$ 平面正方格子を L とし, $N = n^2$ とする. L の**向き** (orientation) とは, L の各辺へ方向 (または矢) を割り付けたものである. 各頂点でちょうど2つの内側と2つの外側へ指し示す辺があるような, L の向きの個数 f_n は何個あるのか? このような向きは, **アイスルール** (ice rule, 氷の規則) に従うといい, また**オイラーの向き** (Eulerian orientations) ともいう. 数列 $\{f_n\}$ は, $f_1 = 4, f_2 = 18, f_3 = 148, f_4 = 2970$ からはじまる [1,2]. 難解な解析により, Lieb (リーブ) [3-5]は,

$$\lim_{n \to \infty} f_n^{1/N} = \left(\frac{4}{3}\right)^{3/2} = \sqrt{\frac{64}{27}} = 1.5396007178\cdots$$

であることを証明した. この定数は, **正方氷剰余エントロピー** (residual entropy for square ice) として知られている. 基礎にある物理の簡単な議論は【5.24.3】にある. さらに, 各頂点で, 6つの矢の配置の可能性[6-9]があるので, モデルは**6頂点モデル**

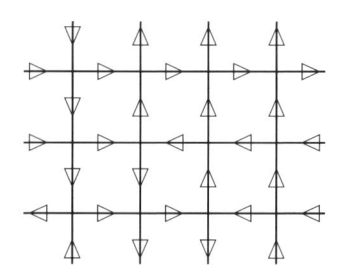

図5.24 氷の規則を満たす矢の平面配置例.

(six-vertex model) ともよばれる. 図5.24 を参照せよ.

いくつかの関連する結果に取りかかる. 各頂点で, 偶数個の辺が内側を指し示し, 偶数個の辺が外側を指し示すような, L の向きの個数を $\tilde{f}_n$ とする. 明らかに, $\tilde{f}_n \geq f_n$ である. このモデルは **8頂点モデル** (eight-vertex model) とよばれる. しかしながら, この場合には解析は必ずしもひどく難解なものではない. 基本的な線型代数学によって $\tilde{f}_n = 2^{N+1}$ が成り立つ. **16頂点モデル** (sixteen-vertex model) (矢について何の制約もない) の対応する式は, 明らかに 2^{2N} である.

代わりに, N 個の頂点をもつ平面三角格子 L に焦点をあてる. 各頂点で, ちょうど3つの内側に向かう辺と3つの外側へ向かう辺が存在するような, L の向きの個数 g_n は何個であるのか? (表現「オイラーの向き」がここで当てはまる. しかし, これは「氷の規則」ではない.) Baxter[10]は, **20頂点モデル** (twenty-vertex model) が

$$\lim_{n \to \infty} g_n^{1/N} = \sqrt{\frac{27}{4}} = 2.5980762113\cdots$$

を満たすことを証明した. f_n と g_n を計算する問題は, L 上において, 法3でいたるところゼロフローでない個数を数える問題と同じである[9,11,12]. Mihail & Winkler (ミハイル&ウィンクラー)[13]は関連する計算複雑性問題 (computational complexity issues) を研究した.

有名な交代符号行列予想 (alternating sign matrix conjecture)[1,14-16]のいくつかの解のうちの1つは, 正方氷の規則モデル (square ice model) と密接に関係がある. この業績は, 組合せ論と統計物理の共通性を (もう一度) 明白にするには十分である.

5.24.1 塗り分け

次の節【5.25】を予期するたいへんおもしろい話題がある. どの2つの隣接頂点も同じ色に彩色されていないように, 正方格子の頂点を3色で塗り分けて彩色する方法の数を u_n とする. Lenard[5]は, $u_n = 3f_n$ であることを指摘した. いいかえると, 正方形の地図を3色で彩色する個数は, 正方氷配置の個数の3倍である. 後に一般化することを心に留めておいて, しばらくの間 u_n に戻る.

正方格子を三角格子 L に置きかえて, 整数 $q \geq 4$ を固定する. どの2つの隣接頂点も

同じ色に彩色されていないように，L の格子の頂点を q 色で彩色する方法の数を v_n とする．Baxter[17,18] は次のことを示した．パラメータ $-1<x<0$ が $q>4$ に対して，$q=2-x-x^{-1}$ によって定義されているとする．このとき，

$$\lim_{n\to\infty} v_n^{1/N} = -\frac{1}{x}\prod_{j=1}^{\infty}\frac{(1-x^{6j-3})(1-x^{6j-2})^2(1-x^{6j-1})}{(1-x^{6j-5})(1-x^{6j-4})(1-x^{6j})(1-x^{6j+1})}$$

が成り立つ．とくに，$q\to 4+0$ とすると（実数 q で式は意味をもつことに注意），

$$C^2 = \lim_{n\to\infty} v_n^{1/N} = \prod_{j=1}^{\infty}\frac{(3j-1)^2}{(3j-2)(3j)} = \frac{3}{4\pi^2}\Gamma\left(\frac{1}{3}\right)^3 = 1.4609984862\cdots = (1.2087177032\cdots)^2$$

である．これを，三角格子の**バクスターの 4 彩色定数**（Baxter's 4-coloring constant）とよぶ．

N 個の頂点をもつ正方格子および六角形（蜂の巣）格子を，q 色で彩色（q-coloring）する方法の数を，それぞれ u_n, w_n と定義する．対応する極限値の解析的な式は，得られていない．しかし，ある種の級数展開の数値的な評価により，表 5.6 [19-26] のリストが得られる．この表でただ 1 つ知られている量は，左上の隅にある Lieb（リーブ）定数のみである．染色多項式（chromatic polynomial）に関連する議論は，【5.25】を参照せよ．

5.24.2 折り曲げ

「正方対角折り曲げ（square-diagonal folding）」問題は，次の塗り分け問題（coloring problem）に翻訳される．次ページの 2 つのタイル，タイル 1 またはタイル 2 のどちらかで正方格子の表面を覆う．

N 個の正方形からなる格子に対して，そのような被覆（covering）は 2^N 個存在する．いま，もとの格子の各頂点を取り囲むことにより，隣にある 4 個のタイルから作られる正方形「ループ」が存在する．もとの格子の巻きつけを仮定して，純粋に白線からなるループの個数 K_w と純粋に黒線からなるループの個数 K_b を数える．図 5.25 の被覆の実例において，K_w と K_b はともに 0 である．

正方対角格子（square-diagonal lattice）の**折り曲げエントロピー**（entropy of folding）を

$$s = \lim_{n\to\infty}\frac{1}{4N}\ln\left(\sum_{\text{被覆}} 2^{K_w+K_b}\right)$$

と定義する．ただし，和は 2^N 個の充填配置すべてにわたる．（このエントロピーは，タ

表 5.6 $u_n^{1/N}$ と $w_n^{1/N}$ の極限値

q	$\lim_{n\to\infty} u_n^{1/N}$	$\lim_{n\to\infty} w_n^{1/N}$
3	$1.5396\cdots$	$1.6600\cdots$
4	$2.3360\cdots$	$2.6034\cdots$
5	$3.2504\cdots$	$3.5795\cdots$
6	$4.2001\cdots$	$4.5651\cdots$
7	$5.1667\cdots$	$5.5553\cdots$

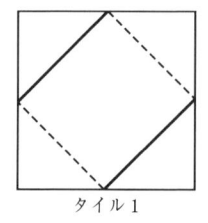

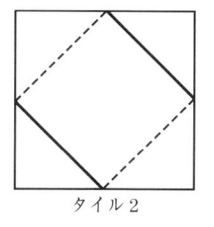

タイル1：正方形を囲む連続した4辺の中央を交互に黒と白の線分で結ぶ；西から北への線分は黒，北から東への線は白，東から南への線分は黒，南から西への線分は白．
タイル2：白と黒を反対に定める；西から北への線分は白，北から東への線は黒，東から南への線分は白，南から西への線分は黒．
(太実線が黒，破線が白を表す)

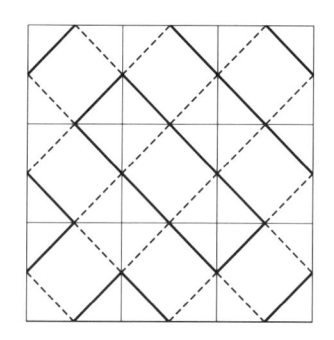

図5.25 2種類のタイルによる格子の被覆例．

イルごとよりも三角形ごとであり，これは付加因子 1/4 を説明する．)

明らかな s の下界は，

$$s \geq \lim_{n\to\infty} \frac{1}{4N}\ln(2^N+2^N) = \lim_{n\to\infty}\frac{N+1}{4N}\ln(2) = \frac{1}{4}\ln(2) = 0.1732\cdots$$

である．これは，チェス盤のように交互にタイルの配置を変えることにより得られる．このような可能性は（すべてのタイル1をタイル2に，すべてのタイル2をタイル1に単純に取りかえることによって）2通りある．より精密な議論[22,23]によって，$s=0.2299\cdots$ を得る．

対応する**三角格子**（triangular lattice）の**折り曲げエントロピー**は，Baxter[17,18]により，$\ln(C)=0.1895600483\cdots$ である．すでに述べたように，より簡単な彩色による解釈がある．

5.24.3 氷の結晶内の原子配列

正方氷（square ice）とは，固体相における水（H_2O）の2次元における理想化である．酸素原子 O を正方格子の頂点として表し，水素原子 H を外側に向かう辺として解

釈する．しかしながら，現実問題として，温度と圧力に依存して，何種類もの3次元氷が存在する[24,25]．「普通の六方晶系の氷 (ordinary hexagonal ice)」Ice-Ih および「等軸晶系の氷 (cubic ice)」Ice-Ic に対する，剰余エントロピー (residual entropy) は，[3,26-30]

$$1.5067 < W < 1.5070$$

を満たし，Nagle の評価誤差の範囲内にある．これらの複雑な3次元格子は，数学者が精神を集中させがちな単純なモデルと同じではない．

実際には，氷の規則 (ice rule) (2つの矢が外側を向き，2つの矢が内側を向き，2つが向きをもたない) をもつか，またはオイラーの向き (3つの矢が外側を向き，3つの矢が内側を向く) のどちらかをもつ，通例の $n \times n \times n$ 立方格子に対する W の値を調べることはおもしろい．誰もこれを成し遂げた人はいないと思われる．

[1] N. J. A. Sloane, On-Line Encyclopedia of Integer Sequences, A005130, A050204, and A054759.
[2] J. J. Henry, Ice configuration counts, $1 \leq n \leq 13$, unpublished note (1998).
[3] E. H. Lieb and F. Y. Wu, Two-dimensional ferroelectric models, *Phase Transitions and Critical Phenomena*, v. 1, ed. C. Domb and M. S. Green, Academic Press, 1972, pp. 331-490; MR 50 # 6391.
[4] E. H. Lieb, Exact solution of the problem of the entropy of two-dimensional ice, *Phys. Rev. Lett.* 18 (1967) 692-694.
[5] E. H. Lieb, Residual entropy of square ice, *Phys. Rev.* 162 (1967) 162-172.
[6] J. K. Percus, *Combinatorial Methods*, Springer-Verlag, 1971; MR 49 #10555.
[7] R. J. Baxter, *Exactly Solved Models in Statistical Mechanics*, Academic Press, 1982; MR 90b:82001.
[8] G. M. Bell and D. A. Lavis, *Statistical Mechanics of Lattice Systems*, v. 1-2, Springer-Verlag, 1999.
[9] C. Godsil, M. Grötschel, and D. J. A. Welsh, Combinatorics in Statistical Physics, *Handbook of Combinatorics*, v. II, ed. R. Graham, M. Grötschel, and L. Lovász, MIT Press, 1995, pp. 1925-1954; MR 96h:05001.
[10] R. J. Baxter, F model on a triangular lattice, *J. Math. Phys.* 10 (1969) 1211-1216.
[11] D. H. Younger, Integer flows, *J. Graph Theory* 7 (1983) 349-357; MR 85d:05223.
[12] C.-Q. Zhang, *Integer Flows and Cycle Covers of Graphs*, Dekker, 1997; MR 98a:05002.
[13] M. Mihail and P.Winkler, On the number of Eulerian orientations of a graph, *Algorithmica* 16 (1996) 402-414; also in *Proc. 3rd ACM-SIAM Symp. on Discrete Algorithms (SODA)*, Orlando, ACM, 1992, pp. 138-145; MR 93f:05059 and MR 97i:68096.
[14] D. Bressoud and J. Propp, How the Alternating Sign Matrix conjecture was solved, *Notices Amer. Math. Soc.* 46 (1999) 637-646; corrections 1353-1354, 1419; MR 2000k:15002.
[15] D. Bressoud, *Proofs and Confirmations: The Story of the Alternating Sign Matrix Conjecture*, Math. Assoc. Amer. and Cambridge Univ. Press, 1999; MR 2000i:15002.
[16] A. Lascoux, Square-ice enumeration, *Sémin. Lothar. Combin.* 42 (1999) B42p; MR 2000f:05086.
[17] R. J. Baxter, Colorings of a hexagonal lattice, *J. Math. Phys.* 11 (1970) 784-789; MR 42 #1457.
[18] R. J. Baxter, q colourings of the triangular lattice, *J. Phys. A* 19 (1986) 2821-2839; MR 87k:82123.
[19] J. F. Nagle, A new subgraph expansion for obtaining coloring polynomials for graphs, *J. Combin. Theory* 10 (1971) 42-59; MR 44 #2663.
[20] A. V. Bakaev and V. I. Kabanovich, Series expansions for the q-colour problem on the square and cubic lattices, *J. Phys. A.* 27 (1994) 6731-6739; MR 95i:82014.
[21] R. Shrock and S.-H. Tsai, Asymptotic limits and zeros of chromatic polynomials and ground state entropy of Potts antiferromagnets, *Phys. Rev. E* 55 (1997) 5165; cond-mat/9612249.
[22] P. Di Francesco, Folding the square-diagonal lattice, *Nucl. Phys. B* 525 (1998) 507-548; cond-

mat/9803051; MR 99k:82033.
- [23] P. Di Francesco, Folding and coloring problems in mathematics and physics, *Bull. Amer. Math. Soc.* 37 (2000) 251-307; MR 2001g:82004.
- [24] L. Pauling, *The Nature of the Chemical Bond*, 3rd ed., Cornell Univ. Press, 1960.
- [25] *Advances in Physics* 7 (1958) 171-297 (entire issue devoted to the physics of water and ice).
- [26] H. Takahasi, Zur Slaterschen Theorie der Umwandlung von KH2PO4 (Teil 1), *Proc. Physico-Math. Soc. Japan* 23 (1941) 1069-1079.
- [27] E. A. DiMarzio and F. H. Stillinger, Residual entropy of ice, *J. Chem. Phys.* 40 (1964) 1577-1581.
- [28] J. F. Nagle, Lattice statistics of hydrogen bonded crystals. I: The residual entropy of ice, *J. Math. Phys.* 7 (1966) 1484-1491.
- [29] J. F. Nagle, Ferroelectric models, *Phase Transitions and Critical Phenomena*, v. 3, ed. C. Domb and M. S. Green, Academic Press, 1974, pp. 653-666; MR 50 # 6393.
- [30] L. K. Runnels, Ice, *Sci. Amer.*, v. 215 (1966) n. 6, 118-126.

5.25 タット-ベラハ定数

n個の頂点v_j【5.6】をもつグラフをGとし，λを正の整数とする．Gのλ彩色（λ-coloring）とは，隣接する頂点が異なる色で彩色されていなければならないという性質をもつ関数$\{v_1, v_2, \cdots, v_n\} \to \{1, 2, \cdots, \lambda\}$のことである．$G$の$\lambda$彩色の個数を$P(\lambda)$と定義する．このとき，$P(\lambda)$は次数$n$の多項式であり，$G$の**染色多項式**（chromatic polynomial）または**クロミアル**（chromial）とよばれる．たとえば，Gが三角形（各頂点が結ばれた3つの頂点）のとき，$P(\lambda) = \lambda(\lambda-1)(\lambda-2)$である．染色多項式は，Birkhoff & Lewis（バーコフ&ルイス）[1]によって，最初に広範に研究された．入門的な資料は[2-6]を参照せよ．

グラフGが**平面的**（planar）とは，共通の頂点を除いて，どの2つの辺も交差しないように平面にGを描くことができるときをいう．平面地図の有名な4色問題は，次のようにいい直すことができる．Gが平面グラフならば，$P(4) > 0$である．定理のいくつかのいいかえのうち，Kauffmann（カウフマン）の組合せ論的3次元ベクトル交差積（cross product）の結果は[7-9]にある．

他の実数での$P(\lambda)$のふるまいについて，何がいえるだろうか？ 明らかに，$P(0) = 0$であり，Gが連結のとき$P(1) = 0$であり，$\lambda < 0$または$0 < \lambda < 1$のとき$P(\lambda) \neq 0$である．さらに，φが黄金比【1.2】であるとき，$P(\varphi + 1) \neq 0$である．φに関するもっと多くのことは，まもなく述べる．

連結な平面グラフGは，（立体射影のもとでの）単連結領域（面）からなる2次元球面の細分を定める．これらの各領域が，Gのちょうど3つの辺で作られる単純閉曲線で囲まれるとき，Gを**球面三角分割**（spherical triangulation）とよぶ．今後は，この条件はつねに満たされていると仮定する．

明らかに，任意の球面三角分割Gに対して$P(2) = 0$である．典型的なGの実験研

究は，$1<\lambda<2$ に対して $P(\lambda) \neq 0$ であるが，区間 $2<\lambda<3$ において1点のみで0となる予想を示唆している．Tutte（タット）[10,11]は，
$$0<|P(\varphi+1)|\leq \varphi^{5-n}$$
を証明した．したがって，$\varphi+1$ 自身は $P(\lambda)$ の零点ではないけれども，充分大きな n に対して，$\varphi+1$ は，$P(\lambda)$ の値がいくらでも0に近くすることができる．この理由により，定数 $\varphi+1$ を**黄金根**（golden root）とよぶ．

$P(3)>0$ であるための必要十分条件は，G がオイラー型（Eulerian），すなわち各頂点に接続している辺の数が偶数である[5]ことが知られている．したがって，オイラー型でない三角分割に対しては，$P(3)=0$ である．

Tutte[12-14]は，注目すべき等式を後に証明した．
$$P(\varphi+2)=(\varphi+2)\varphi^{3n-10}(P(\varphi+1))^2$$
これから $P(\varphi+2)>0$ を得る．$\varphi+2=\sqrt{5}\varphi=3.6180339887\cdots$ であることに注意せよ．前に述べたように，$P(4)>0$ で，$\lambda \geq 5$ に対して $P(\lambda)>0$ である[1]．（$2.618\cdots$ の後の）零点の次の集積点が，どのあたりに存在するかについての可能性を問題とすることは自然である．

ここでは，厳密な理論は役に立たない．したがって，数値的証拠だけで十分である[15-18]．以下では，球面三角分割の族 $\{G_k\}$ を固定する．n_k を G_k の位数で，$k\to\infty$ のとき $n_k\to\infty$ であるものとする．典型的に，グラフ G_k は，各 k に対して，G_{k-1} から帰納的に構成される．しかし，これは本質的でない．実験結果から，染色多項式の零点の次の一群は，点
$$\psi=2+2\cos\left(\frac{2\pi}{7}\right)=4\cos\left(\frac{\pi}{7}\right)^2=3.2469796037\cdots$$
すなわち3次方程式 $\psi^3-5\psi^2+6\psi-1=0$ の解の周りに集まっていることが明らかになっている．黄金根 $\varphi+1$ と類似なものとして，定数 ψ は**白銀根**（silver root）とよばれている．

また，零点はある他の点，$>\psi$ だが ≤ 4，の周りに集まっている．Beraha（ベラハ）[19]は，可能な集積点において，$\{G_k\}$ の選び方に無関係なあるパターンを得た．任意の $\{G_k\}$ に対して，染色多項式の零点 z_k が実数 x の周りに集積しているならば，ある $N\geq 1$ に対して，$x=B_N$ であることを Beraha は予想した．ただし，
$$B_N=2+2\cos\left(\frac{2\pi}{N}\right)=4\cos\left(\frac{\pi}{N}\right)^2$$
である．いいかえると，極限値 x は，ある加算無限集合の外の値をとることはできない．Tutte-Beraha（**タット-ベラハ**）定数 B_N は，すでに議論したすべての根を含むことに注意せよ．
$$B_2=0, \quad B_3=1, \quad B_4=2, \quad B_5=\varphi+1$$
$$B_6=3, \quad B_7=\psi, \quad B_{10}=\varphi+2, \quad \lim_{N\to\infty}B_N=4$$
である．集積点として B_5, B_7, B_{10} をもつことを証明することができるように，特別な族 $\{G_k\}$ が構成された[20-23]．ベラハの予想の驚くべきものは，その一般性にある．

G_k の構成とは無関係に成り立つ，という予想である．

Beraha & Kahane（ベラハ&カハネ）はさらに，集積点として $B_1=4$ をもつ族 $\{G_k\}$ を作った．このことは，つねに $P(4)>0$ であることは知られているが，すべての k で $P_k(z_k)=0$ で $\lim_{k\to\infty} z_k=4$ であるから，驚くべきことである．したがって，4色問題は，真であるけれども，もう少しで偽である[24]．

さらに，格子の上をわたり $P(\lambda)$ を評価することは，λ状態零温度反強磁性体の Potts（ポッツ）モデル（λ-state zero-temperature antiferromagnetic Potts model）を解くことと同値であるから，タット-ベラハ定数は数理物理学に現れる[25-28]．[27]での Beraha 予想の発見的な説明は，洞察力があるが，厳密な証明ではない[8]．塗り分けと氷モデル（ice model）との関連する議論は【5.24】を参照せよ．$\cos(\pi/7)$ を含む他の式については，【2.23】と【8.2】で述べられている．

[1] G. D. Birkhoff andD. C. Lewis, Chromatic polynomials, *Trans. Amer. Math. Soc.* 60 (1946) 355-451; MR 8,284f.
[2] T. L. Saaty and P. C. Kainen, *The Four-Color Problem: Assaults and Conquest*, McGraw-Hill, 1977; MR 87k:05084.
[3] R. C. Read and W. T. Tutte, Chromatic polynomials, *Selected Topics in Graph Theory*. 3, ed. L. W. Beineke and R. J. Wilson, Academic Press, 1988, pp. 15-42; MR 93h:05003.
[4] W. T. Tutte, Chromials, *Hypergraph Seminar*, Proc. 1972 Ohio State Univ. conf., ed. C. Berge and D. Ray-Chaudhuri, Lect. Notes in Math. 411, Springer-Verlag, 1974, pp. 243-266; MR 51 # 5370.
[5] W. T. Tutte, The graph of the chromial of a graph, *Combinatorial Mathematics III*, Proc. 1974 Univ. of Queensland conf., ed. A. A. Street and W. D. Wallis, Lect. Notes in Math. 452, Springer-Verlag, 1975, pp. 55-61; MR 51 # 10152.
[6] W. T. Tutte, Chromials, *Studies in Graph Theory. Part II*, ed. D. R. Fulkerson, Math. Assoc. Amer., 1975, pp. 361-377; MR 53 # 10637.
[7] L. H. Kauffman, Map coloring and the vector cross product, *J. Combin. Theory Ser. B* 48 (1990) 145-154; MR 91b:05078.
[8] L. H. Kauffman and H. Saluer, An algebraic approach to the planar coloring problem, *Commun. Math. Phys.* 152 (1993) 565-590; MR 94f:05056.
[9] R. Thomas, An update on the four-color theorem, *Notices Amer. Math. Soc.* 45 (1998) 848-859; MR 99g:05082.
[10] G. Berman and W. T. Tutte, The golden root of a chromatic polynomial, *J. Combin. Theory* 6 (1969) 301-302; MR 39 # 98.
[11] W. T. Tutte, On chromatic polynomials and the golden ratio, *J. Combin. Theory* 9 (1970) 289-296; MR 42 # 7557.
[12] W. T. Tutte, More about chromatic polynomials and the golden ratio, *Combinatorial Structures and Their Applications*, Proc. 1969 Univ. of Calgary conf., ed. R. K. Guy, H. Hanani, N. Sauer, and J. Schönheim, Gordon and Breach, 1970, pp. 439-453; MR 41 # 8299.
[13] W. T. Tutte, The golden ratio in the theory of chromatic polynomials, *Annals New York Acad. Sci.* 175 (1970) 391-402; MR 42 # 130.
[14] D. W. Hall, On golden identities for constrained chromials, *J. Combin. Theory Ser. B* 11 (1971) 287-298; MR 44 # 6510.
[15] D.W. Hall andD. C. Lewis, Coloring six-rings, *Trans. Amer. Math. Soc.* 64 (1948) 184-191; MR 10,136g.
[16] D. W. Hall, J. W. Siry, and B. R. Vanderslice, The chromatic polynomial of the truncated icosahedron, *Proc. Amer. Math. Soc.* 16 (1965) 620-628; MR 31 # 3361.

[17] D. W. Hall, Coloring seven-circuits, *Graphs and Combinatorics*, Proc. 1973 George Washington Univ. conf., ed. R. A. Bari and F. Harary, Lect. Notes in Math. 406, Springer-Verlag, 1974, pp. 273-290; MR 51 #5366.
[18] W. T. Tutte, Chromatic sums for rooted planar triangulations. V: Special equations, *Canad. J. Math.* 26 (1974) 893-907; MR 50 #167.
[19] S. Beraha, *Infinite Non-trivial Families of Maps and Chromials*, Ph.D. thesis, Johns Hopkins Univ., 1974.
[20] S. Beraha, J. Kahane, and R. Reid, B_7 and B_{10} are limit points of chromatic zeroes, *Notices Amer. Math. Soc.* 20 (1973) A-5.
[21] S. Beraha, J. Kahane, and N. J.Weiss, Limits of zeroes of recursively defined polynomials, *Proc. Nat. Acad. Sci. USA* 72 (1975) 4209; MR 52 #5946.
[22] S. Beraha, J. Kahane, and N. J. Weiss, Limits of zeros of recursively defined families of polynomials, *Studies in Foundations and Combinatorics*, ed. G.-C. Rota, Academic Press, 1978, pp. 213-232; MR 80c:30005.
[23] S. Beraha, J. Kahane, and N. J.Weiss, Limits of chromatic zeros of some families of maps, *J. Combin. Theory Ser. B* 28 (1980) 52-65; MR 81f:05076.
[24] S. Beraha and J. Kahane, Is the four-color conjecture almost false?, *J. Combin. Theory Ser. B* 27 (1979) 1-12; MR 80j:05061.
[25] R. J. Baxter, *Exactly Solved Models in Statistical Mechanics*, Academic Press, 1982; MR 90b: 82001.
[26] R. J. Baxter, Chromatic polynomials of large triangular lattices, *J. Phys. A.* 20 (1987) 5241-5261; MR 89a:82033.
[27] H. Saleur, Zeroes of chromatic polynomials: A new approach to Beraha conjecture using quantum groups, *Commun. Math. Phys.* 132 (1990) 657-679; MR 91k:17014.
[28] H. Saleur, The antiferromagnetic Potts model in two dimensions: Berker Kadanoff phases, antiferromagnetic transition and the role of Beraha numbers, *Nucl. Phys. B* 360 (1991) 219-263; MR 92j:82016.

6 関数の反復に関する定数

6.1 ガウスのレムニスケート定数

Gauss (ガウス) は 1799 年に, $n \geq 1$ に対して

$$a_0 = 1, \quad b_0 = \sqrt{2}, \quad a_n = \frac{a_{n-1} + b_{n-1}}{2}, \quad b_n = \sqrt{a_{n-1} b_{n-1}}$$

のように定義される数列 a_n, b_n は

$$\frac{1}{M} = \lim_{n \to \infty} \frac{1}{a_n} = \lim_{n \to \infty} \frac{1}{b_n} = \frac{2}{\pi} \int_0^1 \frac{dx}{1-x^4} = 0.8346268416\cdots = \frac{1}{1.1981402347\cdots}$$

という極限値 M をもつことを証明した.

この漸化数列式は, いわゆる**算術幾何平均アルゴリズム** (AGM) とよばれているものに基づいている. Gauss は, この定数が特別のものであり, 幾何学との興味深い関係があることを同時に指摘している. $r^2 = \cos(2\theta)$ で定義されるレムニスケート (lemniscate) 曲線の全長 $2L$ は,

$$L = \int_0^\pi \frac{d\theta}{1+\sin(\theta)^2} = 2 \int_0^1 \frac{dx}{1-x^4} = 2.6220575542\cdots$$

で与えられ, したがって $L = \frac{\pi}{M}$ なのである.

この**レムニスケート定数** L は, π が円に対して果たすのと同様の働きをレムニスケートに対してするもので, AGM アルゴリズムはこの L を計算する際に 2 乗収束する (効率のよい) 方法を与えるものである [1-5].

L の他の表し方として

$$L = \sqrt{2} K\left(\frac{1}{\sqrt{2}}\right) = \frac{1}{2\sqrt{2}} \Gamma\left(\frac{1}{4}\right)^2 = \frac{\pi}{\sqrt{2}} \exp\left(\frac{1}{2}\left[\gamma - \frac{\beta'(1)}{\beta(1)}\right]\right)$$

がある．ここで，K は第1種の完全楕円積分【1.4.6】であり，$\Gamma(x)$ は Euler のガンマ関数【1.5.4】で，γ はオイラー-マスケローニの定数，$\beta(x)$ は Dirichlet のベータ関数【1.7】である．【2.10】でも述べたように，これは明らかに多くのアイデアが出会う場所である．

もっと速く収束する2個の級数があり[4,6]，それは

$$\frac{1}{M}=\left[\sum_{n=-\infty}^{\infty}(-1)^n e^{-\pi n^2}\right]=2^{5/4}e^{-\pi/3}\left[\sum_{n=-\infty}^{\infty}(-1)^n e^{-2\pi(3n+1)n}\right]^2$$

である．第3の式としては中央2項係数を含むものがある【1.5.4】．

[7,8]では何人かの人々が，$M/\sqrt{2}=0.8472130848\cdots$ をいわゆる「普遍的 (ubiquitous) 定数」とし，$L/\sqrt{2}=1.8540746773\cdots$ の値も[9]で与えられている．定積分

$$\int_0^1 \frac{dx}{\sqrt{1-x^4}}=\frac{L}{2}=1.3110287771\cdots, \quad \int_0^1 \frac{x^2 dx}{\sqrt{1-x^4}}=\frac{M}{2}=0.59990701173\cdots$$

はしばしば，それぞれ第1種および第2種のレムニスケート定数とよばれている[4,10,11]．

Gauss はまさしく，彼の極限値結果と，他の類するものとが来たるべき未来への発火点となることを期待したのである．**楕円モジュラー関数**の広大な分野は Abel, Jacobi, Cayley, Klein, Fricke などの名前と関連しているが，Gauss の洞察[1,4]から始まったということができる．この理論が1900年頃までは明確化されなかったとはいえ，近年すばらしいルネッサンスを迎えたのである．このルネッサンスへの2つの貢献要素は，1つにはラマヌジャンの諸結果の広範囲にわたる見直しであり，他は AGM 式の反復法に基づく π を計算する速いアルゴリズムである．

定数 L が超越数であることは1937年に Schneider（シュナイダー）によって証明された[12]．さてここで，もう少し複雑なことを考えてみよう．

0 ではないすべての Gauss の整数をわたる無限積

$$\sigma(z)=z\prod_{\omega\neq 0}\left(1-\frac{z}{\omega}\right)\exp\left(\frac{z}{\omega}+\frac{z^2}{2\omega^2}\right)$$

は Weierstrass（**ワイエルシュトラス**）**のシグマ関数**とよばれる[13,14]．次の等式が成り立つ[15-17]．

$$\sigma\left(\frac{1}{2}\right)=2^{5/4}\pi^{1/2}e^{\pi/8}\Gamma\left(\frac{1}{4}\right)^{-2}=2^{-1/4}e^{\pi/8}L^{-1}=0.4749493799\cdots$$

そしてこれが超越数であることが Nesterenko（ネステレンコ）によって1996年に示された．したがって，$\sigma(1/2)$ 中の要素 $\exp(\pi/8)$ を処理するのに十分な進歩がおよそ60年ぶりになされたのである．この種のものが他にもある【1.5.4】．

Gauss の整数 ω のかわりに，次のような点の格子を考えよう†．

$$\left\{\tilde{\omega}=j\cdot\left(\frac{1}{2}-i\frac{\sqrt{3}}{2}\right)+k\cdot\left(\frac{1}{2}+i\frac{\sqrt{3}}{2}\right); \ -\infty<j,k<\infty \text{は整数}\right\}$$

そして，$\tilde{\sigma}(z)$ を，すべての $\tilde{\omega}\neq 0$ について上のように定義しよう．

---

†訳注：この格子点は Eisenstein（アイゼンシュタイン）の整数とよばれる．

この関数は【6.1.1】でもいくらか必要になる．

Erdös, Herzog & Piranian（エルデーシュ，ヘルツォーク＆ピラニアン）[18]による成果を出発点として，Borwein（ボーウィン）[19]は1つの興味ある問題を研究した．いま $p(z)$ を，次数 n のモニックな[最高次の係数が1の]多項式とする．そして，複素数平面上 $|p(z)|=1$ で定義される曲線を考える．はたしてこの曲線の全周長は，$p(z)=z^n-1$ に対する値以下であろうか？ 特に $n=2$ の場合，これはレムニスケート $r^2=2\cos(2\theta)$ となり，その周長は $2\sqrt{2}L$ となる．この問題に対する最近の結果については[20]を参照されたい．

積分

$$\int_0^1 \sqrt{1-x^4}dx = \frac{L}{3} = 0.8740191847\cdots$$

は，Friedlander & Iwaniec（フリードランダー＆イワニエツ）による最近の数論的研究の仕事として，Landau-Ramanujan 定数の議論の中で現れている【2.3】．また，幾何学的確率論の立場から，M は正方形内の N 個のランダムにとった点の凸包の周長の期待値に関する式の中にも現れる【8.1】．

6.1.1 ワイエルシュトラスのペー関数

前節で定義した $\sigma(z)$ と $\tilde{\sigma}(z)$ について，

$$\wp(z) = -\frac{d^2}{dz^2}\ln(\sigma(z)), \quad \tilde{\wp}(z) = -\frac{d^2}{dz^2}\ln(\tilde{\sigma}(z))$$

とする．【1.4.6】における Jacobi の楕円関数と同様に，$\wp(z)$ と $\tilde{\wp}(z)$ は2重周期有理型関数である．$\wp(x)$ の実半周期は $L/\sqrt{2}=1.8540746773\cdots$ であるのに対して，$\tilde{\wp}(x)$ のそれは

$$\frac{\sqrt[3]{2}}{\sqrt[4]{3}}K\left(\frac{\sqrt{2-\sqrt{3}}}{2}\right) = \frac{1}{4\pi}\Gamma\left(\frac{1}{3}\right)^3 = 1.5299540370\cdots$$

である[1, 9, 21]．さらに，すべての $0<x\leq\tilde{r}$, $0<y\leq\tilde{r}$ に対して，

$$x = \int_{\wp(x)}^\infty \frac{dt}{\sqrt{(4t^2-1)t}}, \quad y = \int_{\tilde{\wp}(y)}^\infty \frac{dt}{\sqrt{4t^3-1}}$$

が成り立ち，これらはなぜ $\wp(z)$ や $\tilde{\wp}(z)$ が楕円曲線論において重要であるかを示唆している[22]．**ワイエルシュトラスのペー関数**は，実は2つの助変数をもつ関数族の1つであり，ここに示した2個の例をその中に含んでいるのである．

[1] J. M. Borwein and P. B. Borwein, *Pi and the AGM: A Study in Analytic Number Theory and Computational Complexity*, Wiley, 1987, pp. 1-15, 27-32; MR 99h：11147.

[2] D. A. Cox, The arithmetic-geometric mean of Gauss, *Enseign. Math.* 30 (1984) 275-330; MR 86a：01027.

[3] G. Miel, Of calculations past and present: The Archimedean algorithm, *Amer. Math. Monthly* 90 (1983) 17-35; MR 85a：01006.

[4] J. Todd, The lemniscate constants, *Commun. ACM* 18 (1975) 14-19, 462; MR 51 # 11935.

[5] D. Shanks, The second-order term in the asymptotic expansion of $B(x)$, *Math. Comp.* 18 (1964)

75-86; MR 28 #2391.
- [6] D. H. Lehmer, The lemniscate constant, *Math. Tables Other Aids Comput.* 3 (1948-49) 550-551.
- [7] J. Spanier and K. B. Oldham, *An Atlas of Functions*, Hemisphere, 1987.
- [8] M. Schroeder, How probable is Fermat's last theorem?, *Math. Intellig.* 16 (1994) 19-20; MR 95e：11040.
- [9] M. Abramowitz and I. A. Stegun, *Handbook of Mathematical Functions*, Dover, 1972, pp. 652, 658; MR 94b：00012.
- [10] R. W. Gosper, A calculus of series rearrangements, *Algorithms and Complexity: New Directions and Recent Results*, Proc. 1976 Carnegie-Mellon conf., ed. J. F.Traub, Academic Press, 1976, pp. 121-151; MR 56 #9899.
- [11] S. Lewanowicz and S. Paszkowski, An analytic method for convergence acceleration of certain hypergeometric series, *Math. Comp.* 64 (1995) 691-713; MR 95h：33006.
- [12] C. L. Siegel, *Transcendental Numbers*, Princeton Univ. Press, 1949, pp. 95-100; MR 11,330c.
- [13] S. Lang, *Complex Analysis*, 3rd ed., Springer-Verlag, 1993; MR 99i：30001.
- [14] K. Chandrasekharan, *Elliptic Functions*, Springer-Verlag, 1985; MR 87e：11058.
- [15] M. Waldschmidt, Nombres transcendants et fonctions sigma de Weierstrass, *C. R. Math. Rep. Acad. Sci.* Canada 1 (1978/79) 111-114; MR 80f：10044.
- [16] F. Le Lionnais, *Les Nombres Remarquables*, Hermann, 1983.
- [17] S. Plouffe, Weierstrass constant (Plouffe's Tables).
- [18] P. Erdös, F. Herzog, and G. Piranian, Metric properties of polynomials, *J. d'Analyse Math.* 6 (1958) 125-148; MR 21 #123.
- [19] P. Borwein, The arc length of the lemniscate $|p(z)|=1$, *Proc. Amer. Math. Soc.* 123 (1995) 797-799; MR 95d：31001.
- [20] A. Eremenko and W. Hayman, On the length of lemniscates, *Michigan Math. J.* 46 (1999) 409-415; MR 2000k：30001.
- [21] T. H. Southard, Approximation and table of the Weierstrass $\wp$ function in the equianharmonic case for real argument, *Math. Tables Aids Comput.* 11 (1957) 99-100; MR 19,182c.
- [22] J. H. Silverman, *The Arithmetic of Elliptic Curves*, Springer-Verlag, 1986, pp. 150-159; MR 87g：11070.

6.2　オイラー-ゴンパーツ定数

正則連分数

$$c_1 = 0 + \frac{1|}{|1} + \frac{1|}{|2} + \frac{1|}{|3} + \frac{1|}{|4} + \frac{1|}{|5} + \cdots$$

は収束する（したがって，これは調和級数とその点で異なる）．その極限値は

$$\frac{I_1(2)}{I_0(2)} = c_1 = 0.6977746579\cdots$$

であり[1-3]，ここで $I_0(x)$，$I_1(x)$ は変形ベッセル関数である【3.6】．この公式を用いて，Siegel（ジーゲル）[4,5]は c_1 が超越数であることを証明した．

　さて，もし c_1 における分子と分母を入れかえたらどうなるであろうか？　[6,7]において，

$$C_1 = 0 + \frac{1|}{|1} + \frac{1|}{|1} + \frac{2|}{|1} + \frac{3|}{|1} + \frac{4|}{|1} + \frac{5|}{|1} + \cdots = \sqrt{\frac{\pi e}{2}} \operatorname{erfc}\left(\frac{1}{2}\right)$$

$$= \int_1^\infty \exp\left[\frac{1}{2}(1-x^2)\right] dx = \sqrt{\frac{\pi e}{2}} - \widetilde{C}_1 = 0.6556795424\cdots$$

であることが示されている．ここで erfc は補誤差関数【4.6】であり，

$$\widetilde{C} = \sum_{n=1}^\infty \frac{1}{1 \cdot 3 \cdot 5 \cdots (2n-1)} = \sqrt{\frac{\pi e}{2}} \operatorname{erf} \frac{1}{\sqrt{2}} = 1.4106861346\cdots$$

である．もし C_1 の各分子を2度反復させて

$$C_2 = 0 + \frac{1|}{|1} + \frac{1|}{|1} + \frac{1|}{|1} + \frac{2|}{|1} + \frac{2|}{|1} + \frac{3|}{|1} + \frac{3|}{|1} + \frac{4|}{|1} + \frac{4|}{|1} + \frac{5|}{|1} + \frac{5|}{|1} + \cdots$$

としたらどうなるであろうか．この場合は

$$C_2 = -\operatorname{Ei}(-1) = \int_1^\infty \frac{\exp(1-x)}{x} dx = 0.5963473623\cdots$$

となる[6,8]．ここで Ei は指数積分である【6.2.1】．Euler-Gopertz (**オイラー-ゴンパーツ**) **定数** C_2 についての他の結果はすぐ後に再出する．

c_1 の各分子について上と同様な反復をするとどのような明確な結果が出るかはわかっていないが，数値的には $c_2 = 0.5851972651\cdots$ である．

Euler[9-11]は，

$$0 + \frac{1|}{|1} + \frac{1^2|}{|1} + \frac{2^2|}{|1} + \frac{3^2|}{|1} + \frac{4^2|}{|1} + \frac{5^2|}{|1} + \cdots = \ln(2) = 0.6931471805\cdots$$

であることを見出し，Ramanujan[12,13]は

$$0 + \frac{1^2|}{|1} + \frac{1^2|}{|1} + \frac{2^2|}{|1} + \frac{2^2|}{|1} + \frac{3^2|}{|1} + \frac{3^2|}{|1} + \frac{4^2|}{|1} + \frac{4^2|}{|1} + \frac{5^2|}{|1} + \frac{5^2|}{|1} + \cdots$$

$$= 4\int_1^\infty \frac{x \exp(-\sqrt{5}x)}{\cosh(x)} dx = 0.5683000031\cdots$$

であることを示した．しかしながら，ここでもまた，分子と分母を入れかえた場合や，指数が3以上の場合についての正確な結果は得られていない．

6.2.1 指数積分

γ をオイラー-マスケローニの定数とする【1.5】．**指数積分** $\operatorname{Ei}(x)$ は次のように定義される．

$$\operatorname{Ei}(x) = \gamma + \ln|x| + \sum_{k=1}^\infty \frac{x^k}{k \cdot k!} + \begin{cases} \lim_{\varepsilon \to 0^+}\left(\int_{-\infty}^{-\varepsilon} \frac{e^t}{t} dt + \int_\varepsilon^x \frac{e^t}{t}\right) & (x>0 \text{ のとき}) \\ \int_{-\infty}^x \frac{e^t}{t} dt & (x<0 \text{ のとき}) \end{cases}$$

すなわち，$\operatorname{Ei}(x)$ は広義積分に関する Cauchy の主値である．$\operatorname{Ei}(x)$ の応用例として，Raabe（ラーベ）積分[14-16]を計算する際の次の式がある．

$$A = \int_0^\infty \frac{\sin(x)}{1+x^2} dx = \frac{1}{2}(e^{-1}\operatorname{Ei}(1) - e\operatorname{Ei}(-1))$$

$$B = \int_0^\infty \frac{x\cos(x)}{1+x^2} dx = -\frac{1}{2}(e^{-1}\operatorname{Ei}(1) + e\operatorname{Ei}(-1))$$

6.2.2 対数積分

$0<x\neq 1$ に対して,**対数積分**を $\mathrm{Li}(x)=\mathrm{Ei}(\ln(x))$ と定義する.すると,$\mathrm{Li}(\mu)=0$ となるただ1つの数 $\mu>1$ が存在する.Ramanujan と Soldner（ソルドナー）[17-22] は,μ を計算して $\mu=1.4513692348\cdots$ を得た.たとえば[23],

$$\mathrm{Li}(2)=\lim_{\varepsilon\to 0+}\left(\int_0^{1-\varepsilon}\frac{dt}{\ln(t)}+\int_{1+\varepsilon}^2\frac{dt}{\ln(t)}\right)=\int_\mu^2\frac{dt}{\ln(t)}=1.0451637801\cdots$$

である.

有名な素数定理【2.1】は,通常 $\mathrm{Li}(x)$ や $\mathrm{li}(x)=\mathrm{Li}(x)-\mathrm{Li}(2)$ を用いて表される.これらはともに,$x\to\infty$ のとき $O(x/\ln x)$ なので,両者の違い $\mathrm{Li}(2)$ は,解析数論の人たちには（漸近的な意味で）重要ではないとみなされている.

6.2.3 発散級数

交代発散級数
$$0!-1!+2!-3!+-\cdots$$
については一体どのような意味づけができるだろうか.Euler は形式的に次の結果を与えた[24-28].

$$\sum_{n=0}^\infty (-1)^n n!=\sum_{n=0}^\infty\left((-1)^n\int_0^\infty x^n e^{-x}dx\right)=\int_0^\infty\frac{e^{-x}}{1+x}dx=C_2$$

この級数の偶数項と奇数項は別々に計算されていて[29-31],

$$\sum_{n=0}^\infty(2n)!=A=0.6467611227\cdots,\quad \sum_{n=0}^\infty(2n+1)!=-B=0.0504137604\cdots$$

となる.ここで A と B は先に【6.2.1】で定義された定積分である.また,同じ拡張の考えで[32,33]

$$\sum_{n=1}^\infty(-1)^{n+1}(2n+1)!!=1\cdot 3-1\cdot 3\cdot 5+1\cdot 3\cdot 5\cdot 7-1\cdot 3\cdot 5\cdot 7\cdot 9+\cdots=C_1$$

が知られている.

6.2.4 生存解析

Le Lionnais（ル・リオンネ）は C_2 のことをゴンパーツ定数とよんでいる.その意味する所を説明することは興味深い.個人の一生の時間 X を,累積分布関数 $F(x)=\mathrm{P}(X\leq x)$ をランダム変数とし,確率密度関数 $f(x)=F'(x)$ をもつものとしよう.すると時間 x まで生きている人が最大限時間 t まで生きる確率は

$$\mathrm{P}(X-x\leq t\,|\,X>x)=\frac{\mathrm{P}(x<X\leq x+t)}{\mathrm{P}(X>x)}=\frac{F(x+t)-F(x)}{1-F(x)}$$

である.これは保険統計学において,**死亡率力**とか,**危険関数**[35,36]として知られているものと関係する.$X>x$ のとき,$X-x$ の条件つき期待値はしたがって,

$$\mathrm{E}(X-x\,|\,X>x)=\int_0^\infty \frac{t\cdot f(x+t)}{1-F(x)}\,dt$$

で与えられる．

有名なゴンパーツ分布[37]

$$F(x)=1-\exp\left[\frac{b}{a}(1-e^{ax})\right] \qquad (x>0,\ a>0,\ b>0)$$

を考えよう．そして f のモード（最頻値）を $x=m$ とすると，これは $f'(m)=0$ となる唯一の点である．すると，すべての a と b に対して

$$\mathrm{E}(X-m\,|\,X>m)=\frac{C_2}{a}$$

となることが容易に示される[38]．これは本来のオイラー定数の不思議な現れ方である．

同様に，$\Phi(x)=\mathrm{erf}(x/\sqrt{2})$, $\varphi(x)=\Phi'(x)$ とする．すなわち X が半正規（折りたたみ）分布に従うならば，変曲点である $x=1$ において

$$\mathrm{E}(X-1\,|\,X>1)=\sqrt{\frac{\pi}{2}}\left(\frac{1}{C_1}-1\right),\quad \frac{1-\Phi(1)}{\varphi(1)}=C_1$$

が成り立つ．

最後に，次のような2個の関連する連分数展開がある[6,10,39-41]．

$$\widetilde{C}_1=0+\frac{1|}{|1}-\frac{1|}{|3}+\frac{2|}{|5}-\frac{3|}{|7}+\frac{4|}{|9}-+\cdots$$

$$C_2=0+\frac{1|}{|2}-\frac{1^2|}{|4}-\frac{2^2|}{|6}-\frac{3^2|}{|8}-\frac{4^2|}{|10}-\cdots$$

ここで $(1-C_2)/e=0.1484955067\cdots$ は両側一般化フィボナッチ数列[42]と関連していることを注意する．オイラー-ゴンパーツ定数は【5.6.2】においても，生長木に関係して現れる．

[1] D. H. Lehmer, Continued fractions containing arithmetic progressions, *Scripta Math.* 29 (1973) 17-24; MR 48 # 5979.

[2] S. Rabinowitz, Asymptotic estimates for convergents of a continued fraction, *Amer. Math. Monthly* 97 (1990) 157.

[3] N. Robbins, A note regarding continued fractions, *Fibonacci Quart.* 33 (1995) 311-312; MR 96d: 11010.

[4] C. L. Siegel, Über einige Anwendungen diophantischer Approximationen, *Abh. Preuss. Akad. Wiss., Phys.-Math. Klasse* (1929) n. 1, 1-70; also in *Gesammelte Abhandlungen*, v. 1, ed. K. Chandrasekharan and H. Maass, Springer-Verlag, 1966, pp. 209-266.

[5] C. L. Siegel, *Transcendental Numbers*, Chelsea 1965, pp. 59, 71-72; MR 11, 330c.

[6] H. S. Wall, *Analytic Theory of Continued Fractions*, Van Nostrand, 1948, pp. 356-358, 367; MR 10,32d

[7] B. C. Berndt, Y.-S. Choi, and S.-Y. Kang, The problems submitted by Ramanujan to the Journal of the Indian Mathematical Society, *Continued Fractions: From Analytic Number Theory to Constructive Approximation*, Proc. 1998 Univ. of Missouri conf., ed.B. C. Berndt and F. Gesztesy, Amer. Math. Soc., 1999, pp. 15-56; MR 2000i:11003.

[8] L. Lorentzen and H. Waadeland, *Continued Fractions with Applications*, North-Holland 1992,

pp. 576-577; MR 93g:30007.
- [9] L. Euler, *Introduction to Analysis of the Infinite. Book I*, 1748, transl. J.D. Blanton, Springer-Verlag, 1988, pp. 311-312; MR 89g:01067.
- [10] B. C. Berndt, *Ramanujan's Notebooks: Part II*, Springer-Verlag, 1989, pp. 147-149, 167-172; MR 90b:01039.
- [11] B. C. Berndt, *Ramanujan's Notebooks: Part V*, Springer-Verlag, 1998, p. 56;MR99f:11024.
- [12] G. H. Hardy, Srinivasa Ramanujan obituary notice, *Proc. London Math. Soc.* 19 (1921) xl.lviii.
- [13] C. T. Preece, Theorems stated by Ramanujan. X, *J. London Math. Soc.* 6 (1931) 22-32.
- [14] A. Erdélyi, W. Magnus, F. Oberhettinger, and F. G. Tricomi, *Higher Transcendental Functions*, v. 2, McGraw-Hill, 1953, pp. 143-147; MR 15, 419.
- [15] I. S. Gradshteyn and I. M. Ryzhik, *Table of Integrals, Series and Products*, 6th ed., Academic Press, 2000, pp. 417-418; MR 2001c:00002.
- [16] M. Abramowitz and I. A. Stegun, *Handbook of Mathematical Functions*, Dover, 1972, pp. 78, 230 -233, 302; MR 94b:00012.
- [17] G. H. Hardy, P. V. Seshu Aiyar, and B. M.Wilson (eds.), Further extracts from Ramanujan's letters, *Collected Papers of Srinivasa Ramanujan*, Cambridge Univ. Press, 1927, pp. 349-355.
- [18] G. H. Hardy, *Ramanujan: Twelve Lectures on Subjects Suggested by His Life and Work*, Chelsea, 1959, p. 45; MR 21 # 4881.
- [19] N. Nielsen, *Theorie des Integrallogarithmus und verwandter Transzendenten*, Chelsea, 1965, p. 88; MR 32 # 2622.
- [20] B. C. Berndt and R. J. Evans, Some elegant approximations and asymptotic formulas for Ramanujan, *J. Comput. Appl. Math.* 37 (1991) 35-41; MR 93a:41055.
- [21] B. C. Berndt, *Ramanujan's Notebooks: Part IV*, Springer-Verlag, 1994, pp. 123-124; MR 95e: 11028.
- [22] P. Sebah, 75500 digits of the Ramanujan-Soldner constant via fast quartic Newton iteration, unpublished note (2001).
- [23] A. E. Ingham, *The Distribution of Prime Numbers*, Cambridge Univ. Press, 1932, p. 3.
- [24] G. N. Watson, Theorems stated by Ramanujan. VIII: Theorems on divergent series, *J. London Math. Soc.* 4 (1929) 82-86.
- [25] T. J. I'A. Bromwich, *An Introduction to the Theory of Infinite Series*, 2nd ed., MacMillan, 1942, pp. 323-324, 336.
- [26] G. H. Hardy, *Divergent Series*, Oxford Univ. Press, 1949, pp. 26-29; MR 11,25a.
- [27] E. J. Barbeau, Euler subdues a very obstreperous series, *Amer. Math. Monthly* 86 (1979) 356-372; MR 80i:01007.
- [28] B. C. Berndt, *Ramanujan's Notebooks: Part I*, Springer-Verlag, 1985, pp. 101-103, 143-145; MR 86c:01062.
- [29] J. Keane, Subseries of the alternating factorial series, unpublished note (2001).
- [30] R. B. Dingle, Asymptotic expansions and converging factors. I: General theory and basic converging factors, *Proc. Royal Soc. London. Ser.* A 244 (1958) 456-475; MR 21 # 2145.
- [31] R. B. Dingle, *Asymptotic Expansions: Their Derivation and Interpretation*, Academic Press, 1973, pp. 447-448; MR 58 # 17673.
- [32] L. Euler, De seriebus divergentibus, 1754, *Opera Omnia Ser. I*, v. 14, Lipsiae, 1911, pp. 585-617.
- [33] E. J. Barbeau and P. J. Leah, Euler's 1760 paper on divergent series, *Historia Math.* 3 (1976) 141-160; 5 (1978) 332; MR 58 # 21162a-b.
- [34] F. Le Lionnais, *Les Nombres Remarquables*, Hermann 1983.
- [35] R. C. Elandt-Johnson and N. L. Johnson, *Survival Models and Data Analysis*, Wiley, 1980; MR 81j:62195.
- [36] M. L. Garg, B. Raja Rao, and C. K. Redmond, Maximum-likelihood estimation of the parameters of the Gompertz survival function, *J. Royal Statist. Soc. Ser. C Appl. Statist.* 19 (1970) 152-159; MR 42 # 2581.
- [37] B. Gompertz, On the nature of the function expressive of the law of human mortality, and on

the new mode of determining the value of life contingencies, *Philos. Trans. Royal Soc. London* 115 (1825) 513-585.
[38] J. H. Pollard, Expectation of life and Gompertz's distribution, unpublished note (2001).
[39] L. R. Shenton, Inequalities for the normal integral including a new continued fraction, *Biometrika* 41 (1954) 177-189; MR 15,884e.
[40] N. L. Johnson and S. Kotz, *Distributions in Statistics: Continuous Univariate Distributions*, v. 2. Houghton Mifflin, 1970; MR 42 # 5364.
[41] T. J. Stieltjes, Recherches sur les fractions continues, *Annales Faculté Sciences Toulouse* 8 (1894) J1-J122; 9 (1895) A1-A47; also in *Oeuvres Complètes*, t. 2, ed. W. Kapteyn and J. C. Kluyver, Noordhoff, 1918, pp. 402-566; Engl. transl. in *Collected Papers*, v. 2, ed. G. van Dijk, Springer-Verlag, 1993, pp. 406-570, 609-745; MR 95g:01033.
[42] P. C. Fishburn, A. M. Odlyzko, and F. S. Roberts, Two-sided generalized Fibonacci sequences, *Fibonacci Quart.* 27 (1989) 352-361; MR 90k:11019.

6.3 ケプラー-ボウカンプ定数

半径1の円 C_1 を描き，それに内接する正三角形を描く．そしてその三角形に内接する円を C_2 とし，C_2 に正方形を内接させる．円 C_3 はその正方形に内接するもので，それに正五角形が内接するものとする．この過程を無限にくりかえし，各過程ごとに正多角形の辺の数を1つずつ増やしていく．極限状態の円 C_∞ の半径は，[1-3]によれば

$$\rho = \prod_{j=3}^{\infty} \cos\left(\frac{\pi}{j}\right) = 0.1149420448\cdots = (8.77000366252\cdots)^{-1}$$

である．この構成は Kepler（ケプラー）[4,5]によるもので，彼は一時，木星と土星の太陽を回る軌道が，うまく選ばれて描かれた円 C_1 と C_2 により，正三角形内に外接したり内接したりする円たちによって近似されるのではないかと信じていた．正三角形が最初の正多角形なので彼は火星の軌道が C_3 に対応し，地球の軌道は C_4 に対応するもの等々と考えた（この形式はしかしながら，わずか6個の惑星しか知られていなかったという事実を説明することはできなかった．ケプラーはさらに，2次元の正多角形を3次元のちょうど5個の正多面体に置きかえて，天文学上のデータと以前のよりもよく合う結果を得た）．

上に述べた過程において，「内接」とある所をすべて「外接」に置きかえて同じ過程を考えてみよう．すると極限の半径は新しいものではなく ρ^{-1} となる[6]．同様に，無限積

$$\sigma = \prod_{j=2}^{\infty} \frac{j}{\pi} \sin\left(\frac{\pi}{j}\right) = 0.3287096916\cdots = \frac{2}{\pi}(0.5163359762\cdots)$$

を考えると，これは ρ との明確な関係がない．対照的に，無限積

$$\prod_{j=3}^{\infty} \left(1 - \sin\left(\frac{\pi}{j}\right)\right)$$

は0に発散する．

Bouwkamp（ボウカンプ）はたぶん，もっと速く収束する式を開発した最初の数学者らしい[7,8]．それは，

$$\rho = \frac{2}{\pi}\prod_{m=1}^{\infty}\prod_{n=1}^{\infty}\left(1-\frac{1}{m^2\left(n+\frac{1}{2}\right)^2}\right) = \frac{2}{\pi}\exp\left[-\sum_{k=1}^{\infty}\frac{\zeta(2k)2^{2k}(\lambda(2k)-1)}{k}\right],$$

$$\sigma = \prod_{m=1}^{\infty}\prod_{n=1}^{\infty}\left(1-\frac{1}{m^2 n^2}\right) = \exp\left[-\sum_{k=1}^{\infty}\frac{\zeta(2k)(\zeta(2k)-1)}{k}\right]$$

である．ここで $\zeta(x)$ は【1.6】で，$\lambda(x)$ は【1.7】でそれぞれ定義されている．

関数

$$f(x) = \prod_{j=1}^{\infty}\cos\left(\frac{x}{j}\right), \quad \lim_{x\to\pi}\frac{f(x)}{x-\pi} = \frac{\rho}{2}$$

については最近，次の結果が知られた[9]．

$$\int_0^{\infty} f(x)\,dx = 0.78533805572\cdots < \frac{\pi}{4} = 0.7853981633\cdots$$

関数

$$g(x) = \prod_{j=1}^{\infty}\frac{j}{x}\sin\left(\frac{x}{j}\right), \quad \lim_{x\to\pi}\frac{g(x)}{x-\pi} = -\frac{\sigma}{\pi}$$

についても同様に分析されている．$f(x)$，$g(x)$ と数論の約数問題との間の興味ある関係については[10-12]を参照せよ．

[1] R. W. Hamming, *Numerical Methods for Scientists and Engineers*, 2nd ed., McGraw-Hill, 1973, pp. 193-194.
[2] T. Curnow, Falling down a polygonal well, *Math. Spectrum* 26 (1993/94) 110-118.
[3] R. S. Pinkham, Mathematics and modern technology, *Amer. Math. Monthly* 103 (1996) 539-545.
[4] J. Kepler, *Mysterium Cosmographicum*, transl. by A. M. Duncan, Abaris Books, 1981.
[5] O. Lodge, Johann Kepler, *The World of Mathematics*, v. 1, ed. J. R. Newman, Simon and Schuster, 1956, pp. 220-234.
[6] E. Kasner and J. Newman, *Mathematics and the Imagination*, Simon and Schuster, 1940, pp. 310-312.
[7] C. J. Bouwkamp, An infinite product, *Proc. Konink. Nederl. Akad. Wetensch. Ser. A* 68 (1965) 40-46; *Indag. Math.* 27 (1965) 40-46; MR 30 #5468.
[8] P. T. Wahl and S. Robins, Evaluating Kepler's concentric circles product, unpublished note (1993).
[9] D. Borwein and J. M. Borwein, Some remarkable properties of sinc and related integrals, *Ramanujan J.* 5 (2001) 73-89; CECM preprint 99:142.
[10] G. af Hällström, Zwei Beispiele ganzer Funktionen mit algebraischen Höchstindex einer Stellensorte, *Math. Z.* 47 (1941) 161-174; MR 4,7a.
[11] D. J. Newman and N. J. Fine, A trigonometric limit, *Amer. Math. Monthly* 63 (1956) 128-129.
[12] N. A. Bowen and A. J. Macintyre, Entire functions related to the Dirichlet divisor problem, *Entire Functions and Related Parts of Analysis*, ed. J. Korevaar, S. S. Chern, L. Ehrenpreis, W. H. J. Fuchs, and L. A. Rubel, Proc. Symp. Pure Math., 11, Amer. Math. Soc., pp. 66-78; MR 38 #3434.

6.4 グロスマンの定数

Grossman (グロスマン)[1]は，次のような非線型漸化式による実数列を定義した．
$$a_0=1, \quad a_1=y, \quad a_{n+2}=\frac{a_n}{1+a_{n+1}} \quad (n \geq 0)$$
確実な数値的証拠により，彼はこの数列がただ1つの値 $y=\eta$ に対して収束すると予想し，その値を $\eta=0.737338033\cdots$ と計算した．

Janssen & Tjaden (ヤンセン&チャーデン)[2]は，このグロスマン予想の証明に成功した．Nyerges (ニェルゲス)[3]はさらに進んで，与えられた**任意の**出発点 $a_0=x \geq 0$ に対して，$y=F(x)$ が成立する値の存在と単独性とを示した．これは次のような関数等式
$$x=(1+F(x))F(F(x)) \quad (F:[0,\infty) \to [0,\infty) \text{ は連続関数})$$
を導き，グロスマン定数は $\eta=F(1)$ という特別な値である．この他には，よく知られた定数や関数を用いてこの η を容易に表す方法は知られていない．

Ewing & Foias (ユーイング&フォイアス)[4]は，次のような漸化式
$$b_1=x>0, \quad b_{n+1}=\left(1+\frac{1}{b_n}\right)^n \quad (n \geq 1)$$
を考え，$b_n \to \infty$ となるただ1つの値 $x=\xi$ が存在することを示した．この場合，Ross (ロス)[4]による計算で，$\xi=1.1874523511\cdots$ であることがわかっている．ここでもまた，ξ の簡単な表現や η との関連は得られていない．

【3.5】と【6.10】において，グロスマン定数を思い起こす他の定数について述べる．

[1] J. W. Grossman, Problem 86-2, *Math. Intellig.* 8 (1986) 31.
[2] A. J. E. M. Janssen and D. L. A. Tjaden, Solution to Problem 86-2, *Math. Intellig.* 9 (1987) 40-43.
[3] G. Nyerges, The solution of the functional equation $x=(1+F(x))F^2(x)$, unpublished note (2000).
[4] J. Ewing and C. Foias, An interesting serendiptous real number, *Finite Versus Infinite: Contributions to an Eternal Dilemma*, ed. C. S. Calude and G. Paun, Springer-Verlag, 2000, pp. 119-126; MR 2001k:11267.

6.5 プルーフェの定数

一見すると驚くような次式

から出発しよう．ここで

$$a_n = \sin(2^n) = \begin{cases} \sin(1) & (n=0 \text{ のとき}) \\ 2a_0\sqrt{1-a_0^2} & (n=1 \text{ のとき}) \\ 2a_{n-1}(1-2a_{n-2}^2) & (n\geq 2 \text{ のとき}) \end{cases}$$

であり，$x<0$ のとき $\rho(x)=1$，$x\geq 0$ のとき $\rho(x)=0$ とする．

$$\sum_{n=0}^{\infty} \frac{\rho(a_n)}{2^{n+1}} = \frac{1}{2\pi}$$

言葉でいえば，$1/(2\pi)$ の2進展開は，2次数列[2次の漸化式で定まる数列] $\{a_n\}$ の符号だけで完全に決定されるということである．その容易な証明法は，正弦関数と余弦関数の2倍角公式を用いるものである．そういうと $1/(2\pi)$ の2進展開を計算する早道が見つかったと思うかもしれないが，それは誤りといえよう．その理由は，$\sin(1)$ の値を高度に正確に計算することがまず必要であるが，その計算は $1/(2\pi)$ の計算よりも容易ではないからである．

余弦関数に対する2倍角公式は，さらに簡単な1次数列

$$b_n = \cos(2^n) = \begin{cases} \cos(1) & (n=0 \text{ のとき}) \\ 2b_{n-1}^2 - 1 & (n\geq 1 \text{ のとき}) \end{cases}$$

を与えるが，しかしその和

$$K = \sum_{n=0}^{\infty} \frac{\rho(b_n)}{2^{n+1}} = 0.4756260767\cdots$$

にはまとまった表現が得られていないようである．（この問題は後に再び取り上げよう．）

正接関数に対する2倍角公式は，しかしながら1次数列を生じ

$$c_n = \tan(2^n) = \begin{cases} \tan(1) & (n=0 \text{ のとき}) \\ \dfrac{2c_{n-1}}{1-c_{n-1}^2} & (n\geq 1 \text{ のとき}) \end{cases}$$

となり，その和に対するまとまった式として

$$\sum_{n=0}^{\infty} \frac{\rho(c_n)}{2^{n+1}} = \frac{1}{\pi}$$

が前記と同様に容易に証明できる．ここでもまた，$\tan(1)$ の値を計算することは $1/\pi$ の値を計算するよりも容易ではない．

ここまで，正弦関数と正接関数に対して，ある無理数を入れると，はっきりしている無理数が生ずることを見てきた．Plouffe（プルーフェ）[1-3]は，この過程がいくらかでも調整できはしないかと考えた．彼はまずこれら3つの数列を**有理数値**，たとえば $1/2$ から始めて，なおかつ既知の無理数の2進展開が得られはしないかと問題にしたのである．次のように定義する．

6.5 プルーフェの定数

$$a_n = \sin\left(2^n \arcsin\left(\frac{1}{2}\right)\right) = \begin{cases} \dfrac{1}{2} & (n=0 \text{ のとき}) \\ \dfrac{\sqrt{3}}{2} & (n=1 \text{ のとき}) \\ 2a_{n-1}(1-2a_{n-1}^2) & (n \geq 2 \text{ のとき}) \end{cases}$$

$$\beta_n = \cos\left(2^n \arccos\left(\frac{1}{2}\right)\right) = \begin{cases} \dfrac{1}{2} & (n=0 \text{ のとき}) \\ 2\beta_{n-1}^2 - 1 & (n \geq 1 \text{ のとき}) \end{cases}$$

$$\gamma_n = \tan\left(2^n \arctan\left(\frac{1}{2}\right)\right) = \begin{cases} \dfrac{1}{2} & (n=0 \text{ のとき}) \\ \dfrac{2\gamma_{n-1}}{1-\gamma_{n-1}^2} & (n \geq 1 \text{ のとき}) \end{cases}$$

すると最初の 2 つの和は

$$\sum_{n=0}^{\infty} \frac{\rho(a_n)}{2^{n+1}} = \frac{1}{12}, \quad \sum_{n=0}^{\infty} \frac{\rho(\beta_n)}{2^{n+1}} = \frac{1}{2}$$

と有理数となるが，3 番目の和は

$$C = \sum_{n=0}^{\infty} \frac{\rho(\gamma_n)}{2^{n+1}} = 0.1475836176 \cdots$$

となり，ずっと不思議である．Plouffe は，数値計算を行って，

$$C = \frac{1}{\pi} \arctan\left(\frac{1}{2}\right)$$

ではないかとしたが，その正確な証明は未解決であった．

Borwein & Girgensohn（ボーウィン＆ギルゲンゾーン）[4] は C に対する Plouffe の式を証明すると同時に，それ以上の結果を得た．彼らは与えられた任意の実数 x に対して，

$$\xi_n = \tan(2^n \arctan(x)) = \begin{cases} x & (n=0 \text{ のとき}) \\ \dfrac{2\xi_{n-1}}{1-\xi_{n-1}^2} & (n \geq 1, \ |\xi_{n-1}| \neq 1 \text{ のとき}) \\ -\infty & (n \geq 1, \ |\xi_{n-1}| = 1 \text{ のとき}) \end{cases}$$

とすると，

$$\sum_{n=0}^{\infty} \frac{\rho(\xi_n)}{2^{n+1}} = \begin{cases} \dfrac{\arctan(x)}{\pi} & (x \geq 0 \text{ のとき}) \\ 1 + \dfrac{\arctan(x)}{\pi} & (x < 0 \text{ のとき}) \end{cases}$$

を証明した．これを**プルーフェの反復級数**（recursion）とよぶ．

しかしこれは彼らの論文のほんの一面であった．上記の和を $f(x)$ とすると，次の関数等式を満たすことが決定的である．

$$2f(x) = f\left(\frac{2x}{1-x^2}\right) \quad (x \geq 0 \text{ のとき})$$

$$2f(x)-1=f\left(\frac{2x}{1-x^2}\right) \quad (x<0 \text{ のとき})$$

きわめて一般的な関数等式が他の興味ある漸化式や2進展開を生み出す．ここではそれらの結果をまとめることはしないが，プルーフェの反復級数がその理論のもっとも簡単な場合のようであることを注意しておく．他の例として，対数関数，双曲線関数，第1種の楕円積分などと関連する例も[4]に述べられている．

よく知られた Lehmer（レーマー）[5]の定理は C が無理数であることを示している．実は C は超越数である[6]．

Chowdhury（チョウドゥリ）[7]は，前に定義された定数 K が，2進法において $1/(2\pi)$ と $1/\pi$ とのビットごとの排他的論理和として表されることを最近述べた．すなわち，

$$0.00101000101111100110\cdots$$
$$\oplus\, 0.01010001011111001100\cdots$$
$$=0.01111001100000101010\cdots$$

であり，「排他的論理和」は桁上りをしない，2を法とする加法と同値である．

$1/(2\pi)$ は $1/\pi$ のビット列の桁ずらしにすぎないので，定数 K はまことに興味深い！　さらに一般的に，$1\leq x\leq 1$ のとき $\arccos(x)/(2\pi)$ と $\arccos(x)/\pi$ のビットごとの排他的論理和は $\sum_{n=0}^{\infty}\rho(\eta_n)2^{-n-1}$ となる．ここで

$$\eta_n=\cos(2^n\arccos(x))=\begin{cases} x & (x=0 \text{ のとき}) \\ 2\eta_{n-1}^2-1 & (n\geq 1 \text{ のとき}) \end{cases}$$

である．これは非常によく研究されている課題である．数列 $\{1-2\eta_n\}$ はカオスになるロジスティック写像 $y\mapsto 4y(1-y)$ の反復と同値であり，これは【1.9】において初期値 $1-2x$ をもつものとして定義されている．不幸にして，この見方は，K をこれ以上に詳しく追及する助けにならない．

[1] S. Plouffe, Why build the Inverse Symbolic Calculator?, unpublished note (1995).
[2] S. Plouffe, The computation of certain numbers using a ruler and compass, *J. Integer Seq.* 1 (1998) 98.1.3; MR 2000c:11211.
[3] N. J. A. Sloane, On-Line Encyclopedia of Integer Sequences, A004715, A004716, and A004717.
[4] J. M. Borwein and R. Girgensohn, Addition theorems and binary expansions, *Canad. J. Math.* 47 (1995) 262-273; MR 96i:39037.
[5] I. Niven, *Irrational Numbers*, Math. Assoc. Amer., 1956, pp. 36-41; MR 18,195c.
[6] B. H. Margolius, Plouffe's constant is transcendental, unpublished note (2002).
[7] M. Chowdhury, A formula for 0.4756260767..., unpublished note (2001).

6.6　レーマーの定数

任意の無理数 x は，次のような無限連分数展開をただ1つもつ．

6.6 レーマーの定数

$$x = a_0 + \frac{1|}{|a_1} + \frac{1|}{|a_2} + \frac{1|}{|a_3} + \cdots$$

ここで a_k は $k \geq 1$ のときは正の整数で，a_0 は整数である[1]．逆に，上のような連分数展開は収束する．黄金比

$$\frac{1+\sqrt{5}}{2} = 1 + \frac{1|}{|1} + \frac{1|}{|1} + \frac{1|}{|1} + \cdots$$

は，その収束の速度が最も遅い場合であることが知られている．Lehmer（レーマー）[2,3]は連分数の興味ある類似を発見した．任意の正の無理数 x は，次のような**無限余接表現**をただ1つ有する．

$$x = \cot\left(\sum_{k=0}^{\infty}(-1)^k \mathrm{arccot}(b_k)\right)$$

ここで各 b_k は $k \geq 0$ のとき非負の整数で，$k \geq 1$ のとき

$$b_k \geq b_{k-1}^2 + b_{k-1} + 1$$

である．逆に，上記の形の級数はつねに収束する．レーマーの定数 ξ は，黄金比の場合に相当するものとして，

$$\xi = \cot(\mathrm{arccot}(0) - \mathrm{arccot}(1) + \mathrm{arccot}(3) - \mathrm{arccot}(13) + - \cdots$$
$$+ (-1)^k c_k + \cdots) = 0.5926327182\cdots$$

で与えられ，その収束の速度が最も遅いものである．

ここで，k 番目の逆余接関数の要素は2次の漸化式をもつ数列[4]によって，

$$c_0 = 0, \quad c_k = c_{k-1}^2 + c_{k-1} + 1 \quad (k \geq 1)$$

で定義され，それ自身が興味ある研究課題である．Lehmer は ξ が次数 <4 の代数的数ではないことを示した．Roth の定理【2.22】と合わせると，それは1938年当時のLehmer は知るよしもなかったことだが，彼の推論から ξ は超越数であることが示される[5]．

一体何が Lehmer にこのような連続余接表現を考えさせることになったのであろうか．彼は次のような簡単な2変数関数の反復

$$f(x,y) = x+y, \quad g(x,y) = x + \frac{1}{y}, \quad h(x,y) = \frac{xy+1}{y-x} = \cot(\mathrm{arccot}(x) - \mathrm{arccot}(y))$$

を考え，これらが次のような等式を与えることに注意した．

$$f(x_1, f(x_2, f(x_3, \cdots))) = \sum_{j=1}^{\infty} x_j$$

$$g(x_1, g(x_2, g(x_3, \cdots))) = x_1 + \frac{1|}{|x_2} + \frac{1|}{|x_3} + \cdots$$

$$h(x_1, h(x_2, h(x_3, \cdots))) = \cot\left(\sum_{j=1}^{\infty}(-1)^{j+1} \mathrm{arccot}(x_j)\right)$$

最初の2つの結果，すなわち無限和と無限連分数展開は数学のいたる所に現れるものである．Lehmer の結果と，考えうる他の結果は，近い将来に応用を見出すであろう．

[1] G. H. Hardy and E. M. Wright, *An Introduction to the Theory of Numbers*, 5th ed., Oxford Univ. Press, 1985; MR 81i:10002.
[2] D. H. Lehmer, A cotangent analogue of continued fractions, *Duke Math. J.* 4 (1938) 323-340.
[3] J. Shallit, Predictable regular continued cotangent expansions, *J. Res. Nat. Bur. Standards B*. 80 (1976) 285-290; MR 55 #2734.
[4] N. J. A. Sloane, On-Line Encyclopedia of Integer Sequences, A002065.
[5] P. Borwein, Lehmer's constant is transcendental, unpublished note (1999).

6.7 カーエンの定数

ここに，あまり知られていない**自己生成連分数**の1つの例がある．

$$\frac{0}{1}=0, \quad \frac{1}{1}=0+\frac{1}{1}$$

から出発し，$q_0=1$, $q_1=1$ と左辺の分母を定義して

$$\frac{p_2}{q_2}=0+\frac{1}{1+\frac{1}{q_0}}=0+\frac{1\rfloor}{\lfloor 1}+\frac{1\rfloor}{\lfloor q_0}$$

と続けて $\gcd(p_2,q_2)=1$ とすると，$q_2=2$ となる．（これ以後，分数 p/q と書いたときには，簡単のため，p,q は最小のものとする．）操作を続けて

$$\frac{p_3}{q_3}=0+\frac{1\rfloor}{\lfloor 1}+\frac{1\rfloor}{\lfloor q_0}+\frac{1\rfloor}{\lfloor q_1}$$

とすると $q_3=3$ となる．さらに続けて，

$$\frac{p_4}{q_4}=0+\frac{1\rfloor}{\lfloor 1}+\frac{1\rfloor}{\lfloor q_0}+\frac{1\rfloor}{\lfloor q_1}+\frac{1\rfloor}{\lfloor q_2}$$

とすると $q_4=8$ となる．この過程の各々において，n 番目の部分分母 q_n は q_{n-2} までの部分商の有限連分数として定義される．これを無限に続けると，q の数列

$$1, 1, 2, 3, 8, 27, 224, 6075, 1361024, 8268226875, 112532552156 81024, \cdots$$

は $q_{n+2}=q_n(q_{n+1}+1)$ を満たすことがわかり，その連分数の極限値は次のような交代無限級数の和と一致する．

$$\lim_{n\to\infty}\frac{p_n}{q_n}=\sum_{j=0}^{\infty}\frac{(-1)^j}{q_j q_{j+1}}=0.6294650204\cdots$$

この定数はたぶん Davidson & Shallit（ダヴィソン&シャリット）[1]において最初に論じられたもので，彼らは上の極限値が超越数であることを証明している．

さてさらに一般的に，先に進もう．$w_0, w_1, w_2, \cdots$ を正の整数の無限列とする．

$$\frac{0}{1}=0, \quad \frac{1}{w_0}=0+\frac{1}{w_0}$$

から $q_0=1$, $q_1=w_0$ と定義する．

$$\frac{p_2}{q_2} = 0 + \frac{1}{|w_0|} + \frac{1}{|w_1 q_0|}$$

から，$q_2 = q_0(w_1 q_1 + 1)$ を得る．

$$\frac{p_3}{q_3} = 0 + \frac{1}{|w_0|} + \frac{1}{|w_1 q_0|} + \frac{1}{|w_2 q_1|}$$

から，$q_3 = q_1(w_2 q_2 + 1)$ を得る．これを無限に続けると，q の数列は $q_{n+2} = q_n(w_{n+1}q_{n+1} + 1)$ を満たし，その連分数の極限値は次のような級数と一致する．

$$\xi(w) = \lim_{n \to \infty} \frac{p_n}{q_n} = \sum_{j=0}^{\infty} \frac{(-1)^j}{q_j q_{j+1}}$$

この数 $\xi(w)$ は，w の選び方に関係なく常に超越数であることが証明されている[1]．

k を正の整数としよう．上記の特別な場合として，$w_0 = 1$，$w_{j+1} = q_j^{k-1}$ をすべての $j \geq 0$ に対して定義する．すると q の数列は $q_{n+2} = q_n(q_n^{k-1}q_{n+1} + 1)$ を満たし，この場合の極限値 $\xi(w)$ は，

$$0 + \frac{1}{|1|} + \frac{1}{|q_0^k|} + \frac{1}{|q_1^k|} + \frac{1}{|q_2^k|} + \cdots = \xi_k = \sum_{j=0}^{\infty} \frac{(-1)^j}{q_j q_{j+1}}$$

である．Davidson-Shallit の定数は $k=1$ の場合に生じる．$k=2$ の場合はしばしば $s_n = q_n q_{n+1} + 1$ と書きかえられ，したがって

$$\xi_2 = c = \sum_{j=0}^{\infty} \frac{(-1)^j}{s_j - 1} = 0.6434105462\cdots$$

を得る．ここで，$s_0 = 2$，$s_{n+1} = s_n^2 - s_n + 1$ となり，これを Sylvester（**シルヴェスターの数列**）とよぶ．この数列は【6.10】でふたたび論じられる．Cahen（カーエン）[2]は定数 c の研究を最初に行った人であり，関連する文献は[3-6]にある．1930 年代，Mahler（マーラー）はすべての超越数を S, T, U の 3 族に分類した．その分類の仕方は，問題とする点で計算したとき，有界な次数と高さをもつ多項式が，どのくらい小さくなり得るか，ということで決められる．Töpfner（テプナー）[7]は，c が族 S に属することを示すことに成功した．$k \geq 3$ の場合は研究されておらず，ただ次の数値だけが得られている．

$\xi_3 = 0.6539007091\cdots$，$\xi_4 = 0.6600049346\cdots$，$\xi_5 = 0.6632657345\cdots$

カーエンの定数 c のいくつかの変形は指摘しておく価値がある．数

$$c' = \sum_{j=0}^{\infty} \frac{(-1)^j}{s_j}$$

は $2c = c' + 1$ を満たし，したがって c' は超越数であるが，他方

$$\sum_{j=0}^{\infty} \frac{1}{s_j} = 1$$

である．

$$\sum_{j=0}^{\infty} \frac{1}{s_j - 1} = 1.6910302067\cdots$$

については一体何がいえるだろうか．最後に，他のどんな種類の自己生成連分数が歴史

的文献中に現れていたであろうか？

[1] J. L. Davison and J. O. Shallit, Continued fractions for some alternating series, *Monatsh. Math.* 111 (1991) 119-126; MR 92f:11094.
[2] M. E. Cahen, Note sur un développement des quantites numériques, qui présente quelque analogie avec celui des fractions continues, *Nouvelles Annales de Mathematiques* 10 (1891) 508-514.
[3] E. Ya. Remez, On series with alternating signs which may be connected with two algorithms of M. V. Ostrogadskii for the approximation of irrational numbers (in Russian), *Uspekhi Mat. Nauk*, v. 6 (1951) n. 5, 33-42; MR 13, 444d.
[4] P.-G. Becker, Algebraic independence of the values of certain series by Mahler's method, *Monatsh. Math.* 114 (1992) 183-198; MR 94b:11066.
[5] T. Töpfer, On the transcendence and algebraic independence of certain continued fractions, *Monatsh. Math.* 117 (1994) 255-262; MR 95e:11079.
[6] C. Baxa, Fast growing sequences of partial denominators, *Acta Math. Inform. Univ. Ostraviensis* 2 (1994) 81-84; MR 95k:11091.
[7] T. Töpfer, Algebraic independence of the values of generalized Mahler functions, *Acta Arith.* 70 (1995) 161-181; MR 96a:11070.

6.8 プルーエ-トゥエ-モース定数

Prouhet-Thue-Morse（プルーエ-トゥエ-モース）の数列
$$\{t_n\}=\{0,1,1,0,1,0,0,1,1,0,\cdots\}$$
には，いくつかの同値な定義がある[1]．

- $t_0=0$, $t_{2n}=t_n$, $t_{2n+1}=1-t_n, \cdots$ $(n\geq 0)$
- t_n は，2 を基底とする n の 2 進数表示における，数字 1 の個数【2.16】
- $(-1)^{t_n}$ は，無限関 $\prod_{k=0}^{\infty}(1-x^{2^k})$ を展開したときの x^n の係数
- $\{0,0,1,0,0,1,1-t_0,1-t_1,1-t_2,1-t_3,\cdots\}$ は辞書的順序で最小な無重複無限語【5.17】

次の定数から始める．定数
$$\tau=\sum_{n=0}^{\infty}\frac{t_n}{2^{n+1}}=0.4124540336\cdots=\frac{1}{2}(0.8249080672\cdots)$$
はしばしば**パリティー** (parity) **定数**とよばれるもので，超越数であることが知られている[2-6]．もっと「人工的」でない式として，次の無限関積[7,8]
$$\prod_{k=0}^{\infty}\left(1-\frac{1}{2^{2^k}}\right)=2(1-2\tau)$$
と連分数
$$2-\frac{1|}{|4}-\frac{3|}{|16}-\frac{15|}{|256}-\frac{255|}{|65536}-\frac{65535|}{|4294967296}-\cdots=\frac{\tau}{3\tau-1}$$

とがある．連分数の形は，2^{2^n} と $2^{2^n}-1$ によって生成されるものである．

6.8.1 確率的数え上げ

Woods & Robbins（ウッズ&ロビンス）[9]は，

$$\prod_{m=0}^{\infty}\left(\frac{2m+1}{2m+2}\right)^{(-1)^{t_m}}=\frac{1}{\sqrt{2}}$$

を示した．Shallit（シャリット）[10]はさらにこれを一般化し，3を基数とする場合を述べている．他の一般化が[11-14]にあり，そこには次のものがある．

$$\prod_{m=0}^{\infty}\left(\frac{(2m+1)^2}{(m+1)(4m+1)}\right)^{(-1)^{u_m}}=\frac{1}{\sqrt{2}}$$

ここで，u_m は Golay-Rudin-Shapiro（ゴーレイ-ルーディン-シャピロ）数列であり，m の2進展開において（重複する可能性もこめて）11 の個数を2を法として数えた値を表す．

ここで，n 個のコインにまつわる問題がある．各 $1 \leq k \leq n$ に対して，コインを独立に投げたとき，k 個目のコインに表が出るのに要する回数-1 を x_k としよう．R_n をすべての k に対して x_k とは異なる最小の非負整数とすれば，明らかに $0 \leq R_n \leq n$ である．Flajolet & Martin（フラジョレット&マーティン）[15]は，

$$\mathrm{E}(R_n)=\frac{1}{\ln(2)}\ln(\psi n)+\delta(n)+o(1)$$

であることを示した．ここに，

$$\psi=\frac{e^{\gamma}}{\sqrt{2}}\prod_{m=1}^{\infty}\left(\frac{2m+1}{2m}\right)^{(-1)^{t_m}}=0.7735162909\cdots$$

であり，γ はオイラー-マスケローニの定数【1.5】で，$\delta(n)$ は「無視しうる」大きさの周期関数（$|\delta(n)|<10^{-5}$）で，【5.14】で述べられた型のものである．分散に対するさらに複雑な表現 $\mathrm{Var}(R_n)\sim 1.257\cdots+\varepsilon(n)$ は[15-17]で述べられている．その証明は関数

$$F(z)=\sum_{k=1}^{\infty}\frac{(-1)^{t_k}}{k^z},\quad \mathrm{Re}(z)>1$$

の，全複素数平面への解析接続に関係している．これはデータマイニングのための**確率論的数え上げアルゴリズム**を評価するのに有用であり，数列 $\{t_n\}$ がどのように続いてゆくのか興味深い．Plouffe[18]は次のような無限積を示した．

$$\prod_{m=1}^{\infty}\left(\frac{m}{m+1}\right)^{(-1)^{t_{m-1}}}=0.8116869215\cdots,$$

$$\prod_{m=1}^{\infty}\left(\frac{2m}{2m+1}\right)^{(-1)^{t_{m-1}}}=0.8711570464\cdots,$$

$$\prod_{m=1}^{\infty}\left(\frac{2m}{2m+1}\right)^{(-1)^{t_m}}=1.6281601297\cdots,$$

$$\prod_{m=1}^{\infty}\left(\frac{m}{m+1}\right)^{(-1)^{t_m}}=2.3025661371\cdots$$

これらは Flajolet によるものである．3番目はむろん $2^{-1/2}e^\gamma\psi^{-1}$ である．もっと身近な定数によるこれらの有限表現は得られていない．この状況が，Woods-Robbins の公式や他のすべてをもっと注目させるものである！

6.8.2 非整数の基

q を，$1<q\leq2$ である実数として固定する．q-展開を次のように定義する．

$$\sum_{n=1}^{\infty}\varepsilon_n q^{-n}=1$$

ここで各 n について $\varepsilon_n=0$ あるいは $\varepsilon_n=1$ である．

いわゆる貪欲アルゴリズムは，q-展開が存在することを示す．もし $q=2$ ならば，すべての n について $\varepsilon_n=1$ となり，これがただ1つの2-展開である．$1<q<2$ である他の q で，ただ1つの q-展開をもつものが存在するであろうか？

直感的に，その答えは否であろうと期待される．実際，黄金比【1.2】に対して $1<q<\varphi$ を固定すると，非可算的に無限に多くの q-展開が存在する．また $q=\varphi$ の場合は可算的に無限に多くの q-展開が存在する[19-21]．

しかしながら，$\varphi<q<2$ の場合は上の直感は正しくない．例外的な q の値のなす測度0の非可算部分集合があり，その q に対してただ1つの q-展開が存在する．さらに，その例外部分集合は最小の要素をもち，その性格は正確に決定されている[22]．この特別な q の値は，次の方程式のただ1つの解である．

$$\prod_{k=0}^{\infty}\left(1-\frac{1}{q^{2^k}}\right)=\left(1-\frac{1}{q}\right)^{-1}-2$$

したがって，$q=1.7872316501\cdots$ となる．それに対応する q-展開は，すべての $n\geq1$ に対して $\varepsilon_n=t_n$ であり，予期せぬ Prouhet-Thue-Morse の数列の出現である．また，この Komornik-Loreti（コモルニク-ロレティ）定数 q が超越数であることが Allouche & Cosnard（アルーチェ＆コスナード）[23]によって証明されている．

6.8.3 外　　角

τ と Myrberg（ミルベルグ）定数 $c_\infty=1.4011551890\cdots$ との間には，フラクタル幾何学【1.9】における関係がある．Mandelbrot 集合 M【6.10】が充電されているとする．したがってそれは，**等ポテンシャル曲線**（これは M に内接する）と**場の軌跡**（電気力線，これは等ポテンシャル曲線に直交する）とを決定する．遠くから見れば，M は点電荷に似ていて，場の軌跡は $r\to\infty$ のとき $r\exp(2xi\theta)$ の形の半直線に近づく．$1-cx^2$ の分岐点 c_k に対応する**外角** θ_k は【1.9】に与えられたとおり，点

$$c_2=\frac{5}{4}=1.25, \quad c_3=1.3680\cdots, \quad c_4=1.3940\cdots$$

に対してそれぞれ（2進法で）

$$\theta_2=0.\overline{01}=\frac{1}{3}, \quad \theta_3=0.\overline{0110}=\frac{2}{5}, \quad \theta_4=0.\overline{01101001}=\frac{7}{17}$$

であり，極限値は $\theta_\infty = \tau$ である．不幸にして，その詳細はあまりに複雑で，これ以上は述べられない[24-26]．

6.8.4 フィボナッチ語

他の「自己生成」定数は，いわゆる**うさぎ定数**とよばれるもので，無限2項フィボナッチ語へと導く反復ビット変換 $0 \mapsto 1$, $1 \mapsto 10$ によって定義される[27-32]．(Thue-Morse 語に対する類似の変換は $0 \mapsto 01$, $1 \mapsto 10$ である．) もっと簡単な定義は，

$$\rho = \sum_{k=1}^{\infty} \frac{1}{2^{\lfloor k\varphi \rfloor}} = 0.7098034428\cdots$$

であり，ここで φ は黄金比である【1.2】．次の事実が知られている[33-37]．

$$\rho = 0 + \frac{1|}{|2^0} + \frac{1|}{|2^1} + \frac{1|}{|2^1} + \frac{1|}{|2^2} + \frac{1|}{|2^3} + \frac{1|}{|2^5} + \frac{1|}{|2^8} + \cdots$$

ここで各指数は古典的なフィボナッチ (Fibonacci) 数列そのものであり，したがって ρ は超越数である．

6.8.5 折り紙

細長い1枚の紙を2つに折って，右側が左側の上になるようにしよう[38]．この操作を繰り返すと，折り目の列ができてゆき，それを開くと谷 (1) か山 (0) が現れる．**折り紙数列** $\{s_n\} = \{1,1,0,1,1,0,0,1,1,1,\cdots\}$ は $s_{4n-3} = 1$, $s_{4n-1} = 0$, $s_{2n} = s_n$ ($n \geq 1$) として定義される．あるいはまた語変換 $w \mapsto w 1 \tilde{w}$ で表される．ここに $\tilde{w}$ は w の中の 0 を 1 に置きかえ，1 を 0 に置きかえて w の鏡像 (左右反転) をしたもの，とも定義できる．次の事実が証明できる．

$$\sigma = \sum_{k=1}^{\infty} \frac{s_n}{2^n} = 0.8507361882\cdots = \sum_{k=0}^{\infty} \frac{1}{2^{2^k}} \left(1 - \frac{1}{2^{2^{k+2}}}\right)^{-1}$$

これから σ の超越性が導かれる[5,39]．

[1] J.-P. Allouche and J. Shallit, The ubiquitous Prouhet-Thue-Morse sequence, *Sequences and Their Applications (SETA)*, Proc. 1998 Singapore conf., ed. C. Ding, T. Helleseth, and H. Niederreiter, Springer-Verlag, 1999, pp. 1-16; MR 2002e:11025.

[2] K. Mahler, Arithmetische Eigenschaften der Lösungen einer Klasse von Funktionalgleichungen, *Math. Annalen* 101 (1929) 342-366; corrigendum 103 (1930) 532.

[3] A. Cobham, *A Proof of Transcendence Based on Functional Equations*, IBM Tech. Report RC-2041, 1968.

[4] M. Dekking, Transcendance du nombre de Thue-Morse, *C. R. Acad. Sci. Paris Sér. A-B* 285 (1977) A157-A160; MR 56 # 15571.

[5] J. H. Loxton and A. J. van der Poorten, Arithmetic properties of the solutions of a class of functional equations, *J. Reine Angew. Math.* 330 (1982) 159-172; MR 83i: 10046.

[6] J. Shallit, Number theory and formal languages, *Emerging Applications of Number Theory*, Proc. 1996 Minneapolis conf., ed. D. A. Hejhal, J. Friedman, M. C. Gutzwiller, and A. M. Odlyzko, Springer-Verlag, 1999, pp. 547-570; MR 2000d:68123.

[7] M. Beeler, R. W. Gosper, and R. Schroeppel, Parity number, HAKMEM, MIT AI Memo 239, item 122.

[8] R. W. Gosper, Some identities for your amusement, *Ramanujan Revisited*, Proc. 1987 Urbana conf., ed. G. E. Andrews, R. A. Askey, B. C. Berndt, K. G. Ramanathan, and R. A. Rankin, Academic Press, 1987, pp. 607-609; MR 89d:05016.

[9] D. R.Woods and D. Robbins, Solution to problem E2692, *Amer. Math. Monthly* 86 (1979) 394-395.

[10] J. O. Shallit, On infinite products associated with sums of digits, *J. Number Theory* 21 (1985) 128-134; MR 86m:11007.

[11] J.-P. Allouche and H. Cohen, Dirichlet series and curious infinite products, *Bull. London Math. Soc.* 17 (1985) 531-538; MR 87b:11085.

[12] J.-P. Allouche, H. Cohen, M. Mendès France, and J. O. Shallit, De nouveaux curieux produits infinis, *Acta Arith.* 49 (1987) 141-153; MR 89f:11022.

[13] J.-P. Allouche and J. O. Shallit, Infinite products associated with counting blocks in binary strings, *J. London Math. Soc.* 39 (1989) 193-204; MR 90g:11013.

[14] J.-P. Allouche, P. Hajnal, and J. O. Shallit, Analysis of an infinite product algorithm, *SIAM J. Discrete Math.* 2 (1989) 1-15; MR 90f:68073.

[15] P. Flajolet and G. N. Martin, Probabilistic counting algorithms for data base applications, *J. Comput. Sys. Sci.* 31 (1985) 182-209; MR 87h:68023.

[16] P. Kirschenhofer and H. Prodinger, On the analysis of probabilistic counting, *Number-Theoretic Analysis: Vienna 1988-89*, ed. E. Hlawka and R. F. Tichy, Lect. Notes in Math. 1452, Springer-Verlag, 1990, pp. 117-120; MR 92a:11087.

[17] P. Kirschenhofer, H. Prodinger, and W. Szpankowski, How to count quickly and accurately, *Proc. 1992 Int. Colloq. on Automata, Languages and Programming (ICALP)*, Vienna, ed. W. Kuich, Lect. Notes in Comp. Sci. 623, Springer-Verlag, 1992, pp. 211-222; MR 94j:68099.

[18] S. Plouffe, Flajolet constants (Plouffe's Tables).

[19] P. Erdös, I. Joó, and V. Komornik, Characterization of the unique expansions $1=\sum_{i=1}^{\infty} q^{-n_i}$ and related problems, *Bull. Soc. Math. France* 118 (1990) 377-390; MR 91j:11006.

[20] P. Erdös and I. Joó, On the expansion $1=\sum q^{-n_i}$, *Period. Math. Hungar.* 23 (1991) 27-30; corrigendum 25 (1992) 113; MR 92i:11030 and MR 93k:11017.

[21] P. Erdös, M. Horváth, and I. Joó, On the uniqueness of the expansions $1=\sum q^{-n_i}$, *Acta Math. Hungar.* 58 (1991) 333-342; MR 93e:11012.

[22] V.Komornik and P. Loreti, Unique developments in non-integer bases, *Amer. Math. Monthly* 105 (1998) 636-639; MR 99k:11017.

[23] J.-P. Allouche and M. Cosnard, The Komornik-Loreti constant is transcendental, *Amer. Math. Monthly* 107 (2000) 448-449.

[24] H.-O. Peitgen and P. H. Richter, *The Beauty of Fractals. Images of Complex Dynamical Systems*, Springer-Verlag, 1986; MR 88e:00019.

[25] A. Douady, Algorithms for computing angles in the Mandelbrot set, *Chaotic Dynamics and Fractals*, Proc. 1985 Atlanta conf., ed. M. F. Barnsley and S. G. Demko, Academic Press, 1986, pp. 155-168; MR 88a:58142.

[26] S. Bullett and P. Sentenac, Ordered orbits of the shift, square roots, and the devil's staircase, *Math. Proc. Cambridge Philos. Soc.* 115 (1994) 451-481; MR 95j: 58043.

[27] M. R. Schroeder, *Number Theory in Science and Communication: With Applications in Cryptography, Physics, Digital Information, Computing and Self-Similarity*, 2nd ed., Springer-Verlag, 1986; MR 99c:11165.

[28] M. Gardner, *Penrose Tiles and Trapdoor Ciphers*, W. H. Freeman, 1989, pp. 21-22; MR 97m: 00004.

[29] M. Schroeder, *Fractals, Chaos, Power Laws: Minutes from an Infinite Paradise*, W. H. Freeman, 1991, pp. 53-57, 306-310; MR 92m:00018.

[30] J. C. Lagarias, Number theory and dynamical systems, *The Unreasonable Effectiveness of Number Theory*, Proc. 1991 Orono conf., ed. S. A. Burr, Amer. Math. Soc., 1992, pp. 35-72; MR 93m:11143.

[31] T. Gramss, Entropy of the symbolic sequence for critical circle maps, *Phys. Rev. E* 50 (1994)

2616-2620; MR 96m:58072.
- [32] J. Grytczuk, Infinite self-similar words, *Discrete Math*. 161 (1996) 133-141; MR 98a:68154.
- [33] P. E. Böhmer, Über die Transzendenz gewisser dyadischer Brüche, *Math. Annalen* 96 (1927) 367-377, 735.
- [34] D. E. Knuth, Transcendental numbers based on the Fibonacci sequence, *Fibonacci Quart*. 2 (1964) 43-44.
- [35] J. L. Davison, A series and its associated continued fraction, *Proc. Amer. Math. Soc*. 63 (1977) 29-32; MR 55 # 2788.
- [36] W. W. Adams and J. L. Davison, A remarkable class of continued fractions, *Proc. Amer. Math. Soc*. 65 (1977) 194-198; MR 56 # 270.
- [37] D. Bowman, A new generalization of Davison's theorem, *Fibonacci Quart*. 26 (1988) 40-45; MR 89c:11016.
- [38] M. Dekking, M. Mendès France, and A. van der Poorten, Folds, *Math. Intellig*. 4 (1982) 130-138, 173-181, 190-195; MR 84f:10016.
- [39] J. H. Loxton, A method of Mahler in transcendence theory and some of its applications, *Bull. Austral. Math. Soc*. 29 (1984) 127-136; MR 85i:11059.

6.9 ミンコフスキー-バウアーの定数

関数 ? : $[0,1] \to [0,1]$ を次のように定義する.

$$?\left(0+\frac{1}{\lfloor a}+\frac{1}{\lfloor b}+\frac{1}{\lfloor c}+\frac{1}{\lfloor d}+\cdots\right) = 0.\underbrace{00\cdots0}_{a-1\text{個}}\underbrace{11\cdots1}_{b\text{個}}\underbrace{00\cdots0}_{c\text{個}}\underbrace{11\cdots1}_{d\text{個}}00\cdots^\dagger$$

ここで入力は正則連分数であり,出力は2進小数で表される[1-3].これが Minkowski (ミンコフスキー) の ? 関数である (図 6.1).それは連続で狭義単調増加であるが,フラクタル的なものである.実はこれは,その微分がほとんどいたる所 (ルベーグ測度 0 のある集合を除いて) で 0 であるという意味で**特異関数**である.特別な値として,次のものがある.

$$?\left(\frac{-1+\sqrt{5}}{2}\right)=\frac{2}{3},\quad ?(-1+\sqrt{2})=\frac{2}{5},\quad ?\left(\frac{-1+\sqrt{3}}{2}\right)=\frac{2}{7}$$

Bower (バウアー) [4,5] は,0, 1/2, 1 以外の ? の固定点について考えた.中心点のまわりに対称的に少なくとも 2 個の固定点があるように見える.ではちょうど 2 個だけなのか.彼は小さい方の値が (10 進小数で) $0.4203723394\cdots$ であることを計算した.この定数はまとまった形をもつであろうか.それは代数的数なのだろうか.Farey 分数を用いての ? の定義も可能である.

人工的な定数の問題を論じたついでに,Champernowne (シャンパーノウン) 定数

---
†訳注:連分数が有限項で終わるときは,最後のビットの 0, 1 と反対のビットを以下無限に並べる.

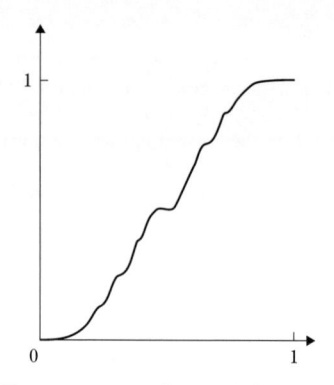

図 6.1 ミンコフスキーの？関数のグラフ

[6]

$$C = 0.12345678910111213141516171819202122232425\cdots$$

に注意しよう．これは小数部分にすべての正の整数の数字を次々に並べて作られている．Copeland-Erdös (コープランド-エルデーシュ) 数[7]は，

$$0.235711131719232931374143475359616771 7379\cdots$$

というもので，小数部分にすべての素数の数字を上と同様に並べて作られたものである．これら2つの数はともに無理数であることが知られている．最近の証明については[8-10]を見よ．Mahler[11]は C が超越数であることを最初に証明した．彼の証明は，比較的に「短い」有理数（たとえば，10/81 や 60499999499/490050000000）が C の大変によい近似値であるという考察から導かれるものである．この考察は一方で，C の正規連分数展開中にきわめて大きな分母が存在することを意味する（たとえば，1709番目の部分分母は $\approx 10^{4911098}$ である．これは Sofroniou & Spaletta（ソフロニウ&スパレッタ）[12]による．

またさらに，Trott（トロット）の定数 E を次のような数（明らかにただ1つ）と定義する．それは E の小数部分の列 $\{\varepsilon_k\}$ がその部分分数の分母の列と同じで[12]，すべての k について

$$E = 0.\varepsilon_1\varepsilon_2\varepsilon_3\varepsilon_4\cdots = 0 + \frac{1|}{|\varepsilon_1} + \frac{1|}{|\varepsilon_2} + \frac{1|}{|\varepsilon_3} + \frac{1|}{|\varepsilon_4} + \cdots \quad (0 \leq \varepsilon_k \leq 9)$$

となるものである．この和の値は $E = 0.1084101512\cdots$ である．はたして E は超越数であろうか．E に対する他の表現は可能であろうか．

[1] R. Salem, On some singular monotonic functions which are strictly increasing, *Trans. Amer. Math. Soc.* 53 (1943) 427-439; MR 4, 217b.
[2] J. H. Conway, *On Numbers and Games*, Academic Press, 1976, pp. 82-86; MR 56 # 8365.
[3] P. Viader and J. Paradis, A new light on Minkowski's ?(x) function, *J. Number Theory* 73 (1998) 212-227; MR 2000a:11104.
[4] N. J. A. Sloane, On-Line Encyclopedia of Integer Sequences, A048817, A048818, A048819,

A048820, A048821, and A048822.
[5] C. Bower, Fixed points of Minkowski's ?(x) function, unpublished note (1999).
[6] D. G. Champernowne, The construction of decimals normal in the scale of ten, *J. London Math. Soc.* 8 (1933) 254-260.
[7] A. H. Copeland and P. Erdös, Note on normal numbers, *Bull. Amer. Math. Soc.* 52 (1946) 857-860; MR 8,194b.
[8] N. Hegyvári, On some irrational decimal fractions, *Amer. Math. Monthly* 100 (1993) 779-780; MR 94e:11080.
[9] A. M. Mercer, A note on some irrational decimal fractions, *Amer. Math. Monthly* 101 (1994) 567-568; MR 95d:11088.
[10] P. Martinez, Some new irrational decimal fractions, *Amer. Math. Monthly* 108 (2001) 250-253; MR 2002b:11096.
[11] K. Mahler, Arithmetische Eigenschaften einer Klasse von Dezimalbrüchen, *Konink. Akad. Wetensch. Proc. Sci. Sect.* 40 (1937) 421-428; Zbl. 17, 56.
[12] N. J. A. Sloane, On-Line Encyclopedia of Integer Sequences, A030167, A030168, A033307, A033308, A039662, and A039663.

6.10 2次数列の定数

1次数列にはFibonacci（フィボナッチ）数列があり，それは【1.2】で論じられている．2次数列は，1次数列よりもはるかにわかっていないし，はるかに不思議である．その最も簡単な例は

$$a_0=2, \quad a_n=a_{n-1}^2 \quad (n\geq 1)$$

であり，解 $a_n=2^{2^n}$ をもつ．さらに手強い他の例は，たかだか n の高さをもつ強2分木【5.6】のすべての個数である．

$$b_0=1, \quad b_n=b_{n-1}^2+1 \quad (n\geq 1)$$

（図6.2）．Aho & Sloane（エイホ&スローン）[1,2]は，この2次の数列が同様に2重指数解 $b_n=\lfloor \beta^{2^n} \rfloor$ をもつことを示したが，β は正確には与えられていない．実は

$$\beta=\exp\left[\sum_{j=0}^{\infty} 2^{-j-1}\ln(1+b_j^{-2})\right]=1.5028368010\cdots$$

である．もし β に $\{b_n\}$ と独立な表し方が見出せるならば，それは非常に驚くべきものであろう．

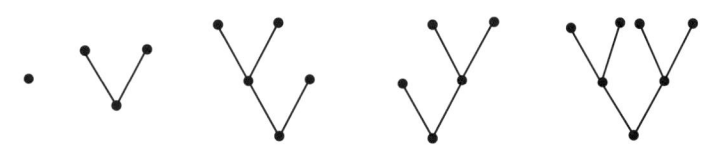

図 6.2 高さ2以下の強2分木は5種ある．

他の例は，数1の真に下からの最良近似

$$C_n = \sum_{j=0}^{n} \frac{1}{c_j}$$

であり，ここで$1 < c_1 < c_2 < c_3 \cdots < c_n$ はすべて整数とする．それは次の2次の数列

$$c_1 = 2, \quad c_n = c_{n-1}^2 - c_{n-1} + 1 \quad (n \geq 2)$$

で与えられ【6.7】，Sylvester（シルヴェスター）の数列として知られているものである．さらに，等式

$$C_n = 1 - \frac{1}{c_{n+1} - 1}$$

は，C_nが貪欲アルゴリズムで形成されることを示している，すなわち最大の実現可能な単位分数を次々と項とし選ぶことである．Aho & Sloane は $c_n = \lfloor \chi^{2^n} + 1/2 \rfloor$ を示した．ここで

$$\chi = \frac{\sqrt{6}}{2} \exp\left[\sum_{j=0}^{\infty} 2^{-j-1} \ln(1 + (2c_j - 1)^{-2})\right] = 1.2640847353\cdots$$

である．ここでもまた，χ に対する独立な表現は非常に驚くべきものであろう．このような2重指数関数は，すでに【2.13】，【5.7】，【5.16】などにおいても現れている．

有名な他の例は Lucas（リュカ）の反復数列[14-21]

$$u_n = u_{n-1}^2 - 2$$

であり，これは$|u_0| > 2$のとき Mersenne 素数との関係[その判定に利用される]によって非常によく研究されている．ここで

$$u_n = \left(\frac{1}{2}u_0 + \frac{1}{2}\sqrt{u_0^2 - 4}\right)^{2^n} + \left(\frac{1}{2}u_0 - \frac{1}{2}\sqrt{u_0^2 - 4}\right)^{2^n}$$

であり，この種の話ではいつでも発散が起こる．$|u_0| < 2$ のときは，もっと込みいったものになり，力学系の研究者たちにとって興味深い．漸化式

$$0 \leq x_0 \leq 1, \quad x_n = ax_{n-1}(1 - x_{n-1}) \quad (n \geq 1 \text{について，} 0 \leq a \leq 4)$$

について循環構造と周期倍増を含む議論は【1.9】にある．

他の有名な例は Lehmer（レーマー）の反復数列

$$\nu_0 = 1, \quad \nu_n = \nu_{n-1}^2 + \nu_{n-1} + 1 \quad (n \geq 1)$$

であり，もっとも遅く収束する連余接列の係数を生み出す【6.6】．

2次の数列は，木関連の題材に関係して他の方法で現れる【5.6】．Golton-Watson（ゴルトン-ワトソン）の分岐過程に関係する消滅確率においては

$$y_0 = 0, \quad y_n = (1-p) + py_{n-1}^2 \quad (n \geq 1, \ 0 < p < 1)$$

であり，その非同型な二分木の近似においては

$$w_0 = 2, \quad w_n = w_{n-1}^2 + 2 \quad (n \geq 1)$$

である．

1-加法数列の研究においては，次の3項2次漸化式

$$t_n = 2(t_{n-2}(t_{n-2} + 1) + t_{n-4k-3}(t_{n-4k-3} + 1) + t_{n-8k-4}(t_{n-8k-4} + 1)) \mod 3$$

に初期データ $(t_1, t_2, \cdots, t_{8k+3}, t_{8k+4}) = (0, 0, \cdots, 0, 1)$ をもつものが本質的である[22]．

そしてそれはStolarsky-Harborth（ストラルスキ-ハルボルス）の定数と関係している【2.16】.

しかし最も有名な2次数列は
$$s_0=0, \quad s_n=s_{n-1}^2-\mu \quad (n\geq 1)$$
であり，ここでμは任意の複素数でよい．Mandelbrot（マンデルブロ）**集合** M は，$s_n \not\to \infty$ であるようなすべてのμの集合と定義される（図6.3）．Mの境界はHausdorff次元が2のフラクタルであり【8.20】，それは無限の長さをもつ[23]．しかしながら，Mの面積は厳密に1.506302と1.561303との間にあり，発見的に1.50659177と評価されている．詳しくは[24-29]を見よ．いまだに誰もMの面積についての正確な式をあえて予想しようともしていない．

Davison & Shallit[30]は，次のような2次漸化式を研究した[30].
$$q_0=q_1=1, \quad q_{n+2}=q_n(q_{n+1}+1) \quad (n\geq 0)$$
そしてこの数列が
$$q_n=\lfloor \xi^{\varphi^n}\eta^{(1-\varphi)^n} \rfloor$$
であることを示した．ここでφは黄金比【1.2】であり，
$$\xi=1.3503061\cdots, \quad \eta=1.4298155\cdots$$
である．他のこのような漸化式[31]は，
$$r_0=0, \quad r_1=1, \quad r_{n+2}=r_{n+1}+r_n^2 \quad (n\geq 0)$$
であり，これは
$$r_{2n}\sim(1.436\cdots)^{\sqrt{2}^{2n}}, \quad r_{2n+1}\sim(1.451\cdots)^{\sqrt{2}^{2n+1}}$$

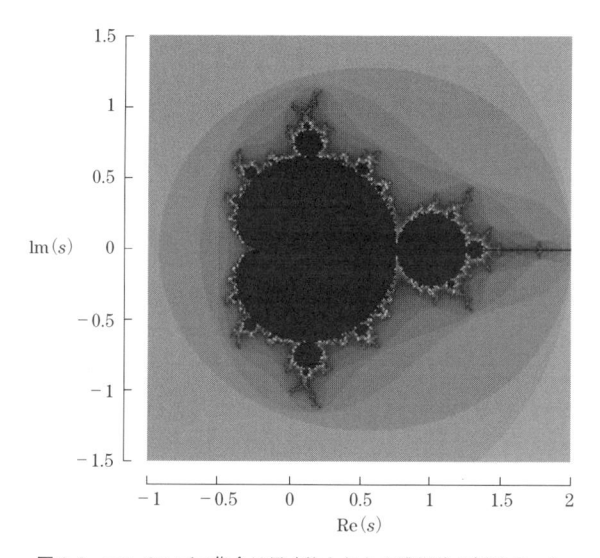

図 6.3 マンデルブロ集合は黒く塗られた心臓形状の領域で，上に示した長方形に完全に含まれている．実軸との交わりは区間$[-1/4, 2]$である．

を満たす．ここで添え数の奇偶性が増大度に関係しているのが面白い．

Greenfield & Nussboum（グリーンフィールド＆ヌスバウム）[32]は，正の実数の両方向無限数列 $\{z_n : n=\cdots, -2, -1, 0, 1, 2, \cdots\}$ で，すべての n について漸化式
$$z_0=1, \quad z_n=z_{n+1}+z_{n-2}^2$$
を満たすものの可能性を考えた．これは $z_1=1.5078747554\cdots$ に対してのみ起こることが知られている．

Stein & Everett（スタイン＆エヴェレット）[33]と Wright（ライト）[34]は，数列
$$d_1=1, \quad d_{n+1}=(n+\delta)\sum_{k=1}^{n} d_k d_{n-k+1} \quad (n\geq 1)$$
を色々な δ の値について研究した．$\delta=0$ と $\delta=-1/3$ のときそれぞれ
$$d_n \sim \frac{1}{e}\prod_{j=2}^{n}(2j-1), \quad d_n \sim (0.35129898\cdots)\prod_{j=2}^{n}(2j-2)$$
となることを得た．ここで，e は自然対数の底である【1.3】．どちらの場合にも組み合せ論的な説明がある．

Lenstra（レンストラ）[12]と Zagier（ザギエ）[35]は，Göbel（ゲーベル）の数列，
$$f_0=1, \quad f_n=\frac{1}{n}\left(1+\sum_{k=0}^{n-1}f_k^2\right) \quad (n\geq 1)$$
を研究し，その最初の非整数項が $f_{43}>10^{178485291567}$ であることを示した．さらに，次の式が成り立つことも示した．
$$f_n \sim (1.0478314475\cdots)^{2^n}(n+2-n^{-1}+4n^{-2}-21n^{-3}+137n^{-4}+\cdots)$$

Somos（ソモス）[36]は，関連する数列
$$g_0=1, \quad g_n=ng_{n-1}^2 \quad (n\geq 1)$$
を考え，
$$g_n \sim \gamma^{2^n}(n+2-n^{-1}+4n^{-2}-21n^{-3}+137n^{-4}-+\cdots)^{-1}$$
であることを発見した．ここで，γ は無限ベキ根展開
$$\gamma=1.6616879496\cdots=\sqrt{1\cdot\sqrt{2\cdot\sqrt{3\cdot\sqrt{4\cdots}}}}=\prod_{j=1}^{\infty} j^{2^{-j}}$$
をもつ．他の Somos 定数 $\lambda=0.3995246670\cdots$ は次のようにして生じる．$x<\lambda$ ならば数列
$$h_0=0, \quad h_1=x, \quad h_n=h_{n-1}(1+h_{n-1}-h_{n-2}) \quad (n\geq 2)$$
は1より小さい極限値に収束し，$x>\lambda$ ならばこの数列は無限大に発散する．これはグロスマンの定数【6.4】に似ている．

[1] A. V. Aho and N. J. A. Sloane, Some doubly exponential sequences, *Fibonacci Quart*. 11 (1973) 429-437; MR 49 # 209.
[2] D. H. Greene and D. E. Knuth, *Mathematics for the Analysis of Algorithms*, Birkhäuser, 1982, pp. 31-34; MR 92c:68067.
[3] O. D. Kellogg, On a Diophantine problem, *Amer. Math. Monthly* 28 (1921) 300-303.
[4] T. Takenouchi, On an indeterminate equation, *Proc. Physico-Math. Soc. Japan* 3 (1921) 78-92.

[5] D. R. Curtiss, On Kellogg's Diophantine problem, *Amer. Math. Monthly* 29 (1922) 380-387.
[6] H. E. Salzer, The approximation of numbers as sums of reciprocals, *Amer. Math. Monthly* 54 (1947) 135-142; 55 (1948) 350-356; MR 8,534e and MR 10,18c.
[7] S. W. Golomb, On certain nonlinear recurring sequences, *Amer. Math. Monthly* 70 (1963) 403-405; MR 26 # 6112.
[8] S.W. Golomb, On the sum of the reciprocals of the Fermat numbers and related irrationalities, *Canad. J. Math.* 15 (1963) 475-478; MR 27 # 105.
[9] J. N. Franklin and S. W. Golomb, A function-theoretic approach to the study of nonlinear recurring sequences, *Pacific J. Math.* 56 (1975) 455-468; MR 51 # 10212.
[10] P. Erdös and R. Graham, *Old and New Problems and Results in Combinatorial Number Theory*, Enseignement Math. Monogr. 28, 1980, pp. 30-32, 41; MR 82j:10001.
[11] R. L. Graham, D. E. Knuth, and O. Patashnik, *Concrete Mathematics*, 2nd ed., Addison-Wesley, 1994, pp. 109, 147, 518; MR 97d:68003.
[12] R. K. Guy, *Unsolved Problems in Number Theory*, 2nd ed., Springer-Verlag, 1994; sect. D11, E15; MR 96e:11002.
[13] L. Brenton and R. R. Bruner, On recursive solutions of a unit fraction equation, *J. Austral. Math. Soc.* 57 (1994) 341-356; MR 95i:11024.
[14] T. W. Chaundy and E. Phillips, The convergence of sequences defined by quadratic recurrence-formulae, *Quart. J. Math.* 7 (1936) 74-80.
[15] G. H. Hardy and E. M. Wright, *An Introduction to the Theory of Numbers*, 5th ed., Oxford Univ. Press, 1985, pp. 148-149, 223-225; MR 81i:10002.
[16] F. Lazebnik and Y. Pilipenko, Convergence of $a_{n+1}=a_n^2-2$, *Amer. Math. Monthly* 94 (1987) 789-793.
[17] R. André-Jeannin, H. Kappus, and I. Sadoveanu, A radical limit, *Fibonacci Quart.* 30 (1992) 369-371.
[18] F. S. Pirvanescu, G. Williams, and H.-J. Seiffert, A dynamical system recursion, *Math. Mag.* 66 (1993) 127-129.
[19] J. L. King, Y. L.Wong, R. J. Chapman, and C. Cooper, A mean limit, *Amer. Math. Monthly* 102 (1995) 556-557.
[20] P. Flajolet, J.-C. Raoult, and J. Vuillemin, The number of registers required for evaluating arithmetic expressions, *Theoret. Comput. Sci.* 9 (1979) 99-125; MR 80e:68101.
[21] R. Sedgewick and P. Flajolet, *Introduction to the Analysis of Algorithms*, Addison-Wesley, 1996, p. 55.
[22] J. Cassaigne and S. R. Finch, A class of 1-additive sequences and quadratic recurrences, *Experim. Math.* 4 (1995) 49-60; MR 96g:11007.
[23] M. Shishikura, The Hausdorff dimension of the boundary of the Mandelbrot set and Julia sets, *Annals of Math.* 147 (1998) 225-267; MR 2000f:37056.
[24] B. Branner, The Mandelbrot set, *Chaos and Fractals:The Mathematics Behind the Computer Graphics*, ed. R. L. Devaney and L. Keen, Proc. Symp. Appl. Math. 39, Amer. Math. Soc., 1989, pp. 75-106; MR 91a:58130.
[25] J. H. Ewing and G. Schober, The area of the Mandelbrot set, *Numer. Math.* 61 (1992) 59-72; MR 93e:30006.
[26] R. Munafo, The area of the Mandelbrot set via pixel counting, unpublished note (1997).
[27] J. R. Hill, Fractals and the grand Internet parallel processing project, *Fractal Horizons:The Future Use of Fractals*, ed. C. Pickover, St. Martins Press, 1996, pp. 299-324.
[28] Y. Fisher and J. Hill, Bounding the area of the Mandelbrot set, unpublished note (1997).
[29] J. R. Hill, Area of Mandelbrot set components and clusters, unpublished note (1998).
[30] J. L. Davison and J. O. Shallit, Continued fractions for some alternating series, *Monatsh. Math.* 111 (1991) 119-126; MR 92f:11094.
[31] W. Duke, S. J. Greenfield, and E. R. Speer, Properties of a quadratic Fibonacci recurrence, *J. Integer Seq.* 1 (1998) 98.1.8; MR 99k:11024.

[32] S. J. Greenfield and R. D. Nussbaum, Dynamics of a quadratic map in two complex variables, *J. Differ. Eq.* 169 (2001) 57-141; MR 2002b:39013.
[33] P. R. Stein and C. J. Everett, On a quadratic recurrence rule of Faltung type, *J. Combin. Inform. Sys. Sci.* 3 (1978) 1-10; MR 58 # 10704.
[34] E. M. Wright, A quadratic recurrence of Faltung type, *Math. Proc. Cambridge Philos. Soc.* 88 (1980) 193-197; corrigenda 92 (1982) 379; MR 81i:10016 and MR 83k:10024.
[35] D. Zagier, Problems posed at the St. Andrews Colloquium, day 5, problem 3 (1996).
[36] M. Somos, Several constants related to quadratic recurrences, unpublished note (1999).

6.11 反復指数の定数

$y>0$ が与えられたとき，一体どんな $x>0$ が $y=x^x$ を満たすのか．その答は思った以上に複雑である．たとえば，

- $x=3$ は $x^x=27$ の唯一の解であり，
- $x=2$ は $x^x=4$ の唯一の解であるが，
- $x=1/2$ と $x=1/4$ はともに $x^x=2^{-1/2}$ の解であり，それ以外の解はない．

もっと一般的には[1-3]，

- $x=(\cdots\log_{1/y}\log_{1/y}(1/e))^{-1}$ は，$y\geq e^e=15.154$ のとき $x^x=y$ の唯一の解である．
- $x=y^{1/y^{1/y\cdots}}$ は $1\leq y\leq e^e$ のとき，$x^x=y$ の唯一の解である．
- しかし $0.692\cdots=e^{-1/e}\leq y<1$ のときは，$x=y^{1/y^{1/y\cdots}}$ と $x=(\cdots\log_{1/y}\log_{1/y}e)^{-1}$ の双方が $x^x=y$ の解であり，それ以外の解はない．

これは，**反復指数数列** $\xi^{\xi^{\xi^{\cdots}}}$ が $0.065\cdots=e^{-e}\leq\xi\leq e^{1/e}=1.444\cdots$ のとき収束し，それ以外の範囲の $\xi>0$ に対しては発散するという事実の部分的な結果である．これと同じ型のものに**超指数数列**と**指数塔**などがある．

y の関数として x を表す他の表現に $\exp(W(\ln(y)))$ がある．ここで，Lambert（ランベルト）の W 関数[3,4]は

$$W(\eta)=\begin{cases}-\ln\left(\cdots\log_{e^{-\eta}}\log_{e^{-\eta}}\dfrac{1}{e}\right) & (\eta\geq e=2.718\cdots\text{のとき})\\ \eta(e^{-\eta})^{(e^{-\eta})\cdots} & (-0.367=-e^{-1}\leq\eta\leq e\text{のとき})\end{cases}$$

であり，$W(\eta)=\exp(W(\eta))=\eta$ を満たす．特に，$W(\ln(27))=\ln(3)$，$W(\ln(4))=\ln(2)$，$W(-\ln(2)/2)=-\ln(2)$ である．本書の残りのいたる所でこの Lambert 関数に言及するであろう．

方程式 $x^2=2^x$ を考えると，これには $x=2$, $x=4$ という2つの解の他に第3番目の実数解がある．それは

$$x=-1\cdot 2^{-(1/2)\cdot 2^{-(1/2)\cdot 2^{\cdots}}}=-\frac{2}{\ln(2)}W\frac{\ln(2)}{2}=-0.7666646959\cdots$$

であり，超越数であることが知られている[5]．$W(-\ln(2)/2)$ が初等的であるのに，

$W(\ln(2)/2)$ はそうではないという事実は興味深い．

次に方程式 $x+e^x=0$ を考えよう．これはただ1つの実数解，
$$x=-1\cdot e^{-1\cdot e^{-1\cdot e^{\cdots}}}=-W(1)=-0.5671432904\cdots=-\ln(1.7632228343\cdots)$$
をもつ[†]．他の例も同様に考えられる．

調和数列の類似である超指数列
$$H_n=\left(\frac{1}{2}\right)^{(1/3)^{\cdots(1/n)}}$$
は，偶数部分指数の列と奇数部分指数の列とがそれぞれ異なる極限値に収束するという意味で発散する[6-8]．すなわち，
$$\lim_{n\to\infty}H_{2n}=0.6583655992\cdots<0.6903471261\cdots=\lim_{n\to\infty}H_{2n+1}$$
である．これらの定数を表す他の表現は知られていない．

i を虚数単位としよう．すると，無限多価の値をもつ表現 i^i はすべて実数である．
$$i^i=\exp\left(-\frac{\pi}{2}(4n+1)\right)$$
であるので，$n=0$ のときは $i^i=\exp(-\pi/2)=0.2078795764\cdots$ となる．もしも対数関数の主値である $n=0$ のときにのみ注目すれば，その反復指数は
$$\frac{2}{\pi}iW\left(-\frac{\pi}{2}i\right)=0.4382829367\cdots+(0.3605924718\cdots)i$$
に収束することが証明できる[9-13]．

ここに，次のような2つの著しい結果がある[14-16]．
$$\int_0^1 x^x dx=\sum_{n=1}^{\infty}\frac{(-1)^{n+1}}{n^n}=0.7834305107\cdots, \quad \int_0^1\frac{1}{x^x}dx=\sum_{n=1}^{\infty}\frac{1}{n^n}=1.2912859970\cdots$$
これらは，マクローリン級数展開の項別積分によって容易に証明できる．

次のような級数の計算はさらに困難である[17]．
$$\lim_{N\to\infty}\sum_{n=1}^{2N}(-1)^n n^{1/n}=\sum_{k=1}^{\infty}\left((2k)^{1/2k}-(2k-1)^{1/(2k-1)}\right)=\sum_{m=1}^{\infty}(-1)^m(m^{1/m}-1)$$
$$=1+\lim_{N\to\infty}\sum_{n=1}^{2N+1}(-1)^n n^{1/n}=0.1878596424\cdots$$

これはゆっくりと収束する．その級数が【2.15】で得られた表現とある意味で似てはいるものの，何の正確な式も知られていない．Cesàro（チェザロ）の和と，Cohen-Villegas-Zagier（コーエン-ヴィレガス-ザギエ）[18]の加速法は，その和を計算するのに使われる2つの技法である．

はるか昔，Poisson（ポアソン）[19]は次のような注目すべき等式を得た．
$$-\frac{\pi}{2}W(-x)=\int_0^{\pi}\frac{\sin\left(\frac{3}{2}\theta\right)-xe^{\cos(\theta)}\sin\left(\frac{5}{2}\theta-\sin(\theta)\right)}{1-2xe^{\cos(\theta)}\cos(\theta-\sin(\theta))+x^2 e^{2\cos(\theta)}}\sin\left(\frac{\theta}{2}\right)d\theta$$
これは $|x|<e^{-1}$ のときに成り立つ．彼の方法がいつの日にか，「コンパクト」な定積分

[†]訳注：複素数解は無限個ある．

を用いて他の超越方程式（たとえばKeplerの方程式【4.8】）の解へと導くのではないだろうか．

6.11.1 指数的再帰数列

2次再帰数列に比べて，指数的再帰数列についてはあまり言うことがない．その最も簡単な例は，
$$c_0=0, \quad c_n=2^{c_{n-1}} \quad (n\geq 1)$$
である[20]．もし∅が空集合を表すとすると，∅のベキ集合 $P(\emptyset)$ の濃度（カーディナル数）は $c_1=1$ であり，$P(P(\emptyset))$ の濃度は $c_2=2$, $P(P(P(\emptyset)))$ の濃度は $c_3=4$, などである．$\{c_n\}$ の増大度はAckermann関数のそれにも似て，いかなる指数関数の増大度よりも大きい．

$\{c_n\}$ の他の例は次のようなものである．根付き恒等木は，根を固定する自己同型写像が恒等写像のみであるような根付き木のことである【5.6】．整数 $h>0$ を固定しよう．高さ h の恒等木は根と，高さ $h-1$ 個の（すべて異なる）恒等木の空でない集合と，高さ $<h-1$ 個の（すべて異なる）恒等木の（空であってもよい）集合とから成り立つ（図6.4）．したがって，そのようなすべての恒等木の数は，繰り返しを許さないので，
$$(2^{c_h-c_{h-1}}-1)2^{c_{h-1}}=c_{h+1}-c_h$$
となる．これらは，集合論で**ランク付け集合**（ranked set）とよばれるものと同値である．

この変形の1つとして
$$\gamma_0=0.1490279983\cdots, \quad \gamma_n=2^{\gamma_{n-1}} \quad (n\geq 1)$$
が組み合せゲーム論[21,22]に現れる．n 日目における**不偏ミゼールゲーム**[†]の個数は

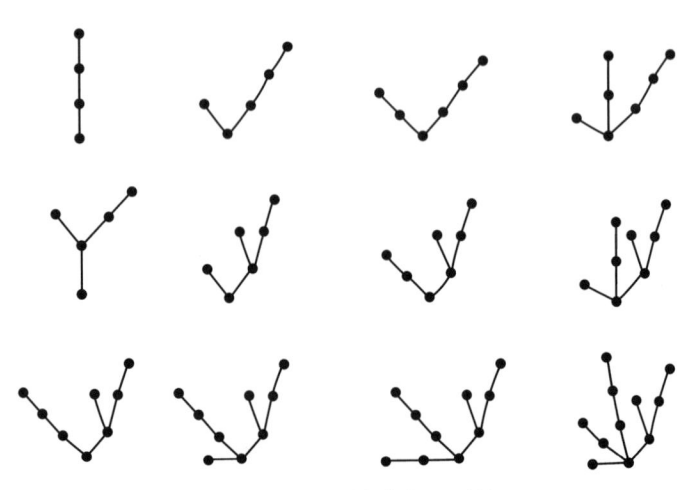

図6.4 高さ3の根付き恒等木は12個ある．

[†]訳注：最終着手者が敗けとなるゲーム．出力 $O(t)$ は t の次に可能な局面を表す．

$g_m=\lceil\gamma_n\rceil$ であり，そのようなすべてのゲームは，特別な条件を満たす根付き恒等木 t とみなすことができる．t に隣接する木の（相異なる）部分恒等木の集合を $S(t)$ としよう．t の**出力** $O(t)$ は，ある $s \in S(t)$ について $O(s)=P$ となるか，または t が単独の頂点であるとき N であり，それ以外のとき $O(t)=P$ である．木 t が**可逆的**であるとは，ある木 u に対して $S(u)$ が $S(t)$ の真部分集合となるか，あるいは $v \in S(t)-S(u)$ であるとき $u \in S(v)$ となることをいう．さらに，もし u が単独の頂点ならば，ある $w \in S(t)$ について $O(w)=P$ となる．最後に木 t が**正準的** (canonical) とは，t が可逆的でなく，かつおのおのの $s \in S(t)$ が正準的であるときと定義する．高さ $\leq n$ の正準的な木の個数 g_n は $0 \leq n \leq 5$ のとき $1, 2, 3, 5, 22, 4171780$ である．g_6 の正しい値が $\approx 2^{4171780}$ であることが[23-25]に述べられている．Conway[21]は，すべての n について有効な定数 γ_0 の存在証明を構成することは困難ではないと主張している．

[1] R. A. Knoebel, Exponentials reiterated, *Amer. Math. Monthly* 88 (1981) 235-252; MR 82e:26004.
[2] J. M. de Villiers and P. N. Robinson, The interval of convergence and limiting functions of a hyperpower sequence, *Amer. Math. Monthly* 93 (1986) 13-23; MR 87d:26005.
[3] Y. Cho and K. Park, Inverse functions of $y=x^{1/x}$, *Amer. Math. Monthly* 108 (2001) 963-967.
[4] R. M. Corless, G. H. Gonnet, D. E. G. Hare, D. J. Jeffrey, and D. E. Knuth, On the Lambert W function, *Adv. Comput. Math.* 5 (1996) 329-359; MR 98j:33015.
[5] P. Pollack, The intersection of $y=x^2$ and $y=2^x$, unpublished note (1998).
[6] D. F. Barrow, Infinite exponentials, *Amer. Math. Monthly* 43 (1936) 150-160.
[7] M. Creutz and R. M. Sternheimer, On the convergence of iterated exponentiation, *Fibonacci Quart.* 18 (1980) 341-347; 19 (1981) 326-335; 20 (1982) 7-12; MR 82b:26006, MR 83b:26006, and MR 83i:26002.
[8] L. Baxter, Divergence of an iterated exponential, unpublished note (1996).
[9] W. J. Thron, Convergence of infinite exponentials with complex elements, *Proc. Amer. Math. Soc.* 8 (1957) 1040-1043; MR 20 # 2552.
[10] D. L. Shell, On the convergence of infinite exponentials, *Proc. Amer. Math. Soc.* 13 (1962) 678-681; MR 25 # 5307.
[11] A. J. Macintyre, Convergence of $i^{i^{\cdot^{\cdot^{\cdot}}}}$, *Proc. Amer. Math. Soc.* 17 (1966) 67; MR 32 # 5855.
[12] I. N. Baker and P. J. Rippon, A note on complex iteration, *Amer. Math. Monthly* 92 (1985) 501-504; MR 86m:30024.
[13] I. N. Baker and P. J. Rippon, Towers of exponentials and other composite maps, *Complex Variables Theory Appl.* 12 (1989) 181-200; MR 91b:30068.
[14] Putnam Competition Problem A-4, *Amer. Math. Monthly* 77 (1970) 723, 725.
[15] M. S. Klamkin, R. A. Groeneveld, and C. D. Olds, A comparison of integrals, *Amer. Math. Monthly* 77 (1970) 1114; 78 (1971) 675-676.
[16] B. C. Berndt, *Ramanujan's Notebooks: Part IV*, Springer-Verlag, 1994, p. 308; MR 95e:11028.
[17] M. R. Burns, An alternating series involving n^{th} roots, unpublished note (1999).
[18] H. Cohen, F. Rodriguez Villegas, and D. Zagier, Convergence acceleration of alternating series, *Experim. Math.* 9 (2000) 3-12; MR 2001m:11222.
[19] S. D. Poisson, Suite du memoire sur les integrales definies, *Journal de l'École Royale Polytechnique*, t. 12, c. 19 (1823) 404-509.
[20] R. C. Buck, Mathematical induction and recursive definitions, *Amer. Math. Monthly* 70 (1963) 128-135.
[21] J. H. Conway, *On Numbers and Games*, Academic Press, 1976, pp. 139-141;MR56 # 8365.
[22] S. Plouffe, γ_0 of Conway (Plouffe's Tables).
[23] C. Thompson, Count of day 6 misère-inequivalent impartial games, unpublished note (1999).

[24] D. Hoey and S. Huddleston, Confirmation of C. Thompson's corrected g_6, unpublished note (1999).

[25] N. J. A. Sloane, On-Line Encyclopedia of Integer Sequences, A014221, A038081, A038093, A047995, and A048828.

6.12 コンウェイの定数

たとえば，数字の列 13 を考えよう．これを「1つの1，1つの3」と読んで並べ，そしてその**続きの列** 1113 を「3つの1，1つの3」と読み，このようにして，次の列 3113 を得る．これを続けると，次のような数字並びの列が得られる[1]．

$$132113,$$
$$1113122113,$$
$$311311222113,$$
$$13211321322113,$$
$$1113122113121113222113,$$
$$31131122211311123113322113,$$
$$132113213221133112132123222113,$$
$$11131221131211132221232112111312111213322113,$$
$$3113112221131112311332111213122112311311123112112113222113$$

このように，ここには最初の 12 個の列（$k=1$ から $k=12$ まで）を与えた．数 1, 2, 3 のみが各列に現れることが証明でき，したがってこの過程は無限に続けられる．k 番目の列の長さについて何がいえるであろうか．その増大度は指数的であるように見え，一見した所ではそれをさらに詳しく特徴づけるのは不可能な程に困難であると予想するであろう．Conway（コンウェイ）[2-5]はその予想に反して，その大きさが $C\lambda^k$ に漸近的に等しいことを証明した．ここで，$\lambda = 1.3035772690\cdots = (0.7671198507\cdots)^{-1}$ であり，それは次の多項式

$$\begin{aligned}
& x^{71} - x^{69} - 2x^{68} - x^{67} + 2x^{66} + 2x^{65} + x^{64} - x^{63} - x^{62} - x^{61} - x^{60} \\
& - x^{59} + 2x^{58} + 5x^{57} + 3x^{56} - 2x^{55} - 10x^{54} - 3x^{53} - 2x^{52} + 6x^{51} + 6x^{50} \\
& + x^{49} + 9x^{48} - 3x^{47} - 7x^{46} - 8x^{45} - 8x^{44} + 10x^{43} + 6x^{42} + 8x^{41} - 5x^{40} \\
& - 12x^{39} + 7x^{38} - 7x^{37} + 7x^{36} + x^{35} - 3x^{34} + 10x^{33} + x^{32} - 6x^{31} - 2x^{30} \\
& - 10x^{29} - 3x^{28} + 2x^{27} + 9x^{26} - 3x^{25} + 14x^{24} - 8x^{23} - 7x^{21} + 9x^{20} \\
& + 3x^{19} - 4x^{18} - 10x^{17} - 7x^{16} + 12x^{15} + 7x^{14} + 2x^{13} - 12x^{12} - 4x^{11} \\
& - 2x^{10} + 5x^9 + x^7 - 7x^6 + 7x^5 - 4x^4 + 12x^3 - 6x^2 + 3x - 6
\end{aligned}$$

の最大の零点である．

この多項式と λ については Atkin（アトキン）によって最初に計算された．Vardi（ヴァルディ）[6]は誤植（x^{35} の符号が[4]では落ちている）を指摘している．

6.12 コンウェイの定数

さらに，その同じ定数 λ はすべてのそのような数列の大きさの程度に出発列に関係なく適用される．ただし2つの単純な例外がある．上では列 13 から出発した．その際の定数 λ は，空な初期列と列 22 とを除いて，「普遍的」に適用可能である．この驚くべき事実は，**宇宙論的定理**として知られているものの1つの系であり，その定理の証明はつい最近なされた[7]ものである．Ekhad & Zeilberger（エッカド&ザイルバーガー）の力まかせの方法は，定理を証明する際にソフトウェアを使用した著明な実例である．

さらに多くのことがいえる．時としてある列はその縮小列が決して互いに干渉しない2つの列 L と R の連接となる．列 LR は $L.R$ のように分割され，LR はその**合成**とよばれる．自明でない分割をもたない列は**要素**あるいは**原子**とよばれる．そして，92個の特別な原子（水素，ヘリウム，…，ウラニウムの名がつけられた）が存在することがわかった．1,2（そしてあるいは）3からなる「どんな」列も結局これらの要素（元素）の合成（化合物）に退化する．加えて，かなりの要素が初期列とは無関係に，固定された正の極限値をもつ．したがって，どの100万個の原子にも平均約91790個の水素（最も普遍的）があり，一方約27個がヒ素（最も普遍的でない）である．

Conway による，元素の周期表[3,4]は，上掲の列 13 の展開を描いているが，長い三進列でなく，元素の観点からの進化をも示している．たとえば，$k=1$ から $k=6$ までのときその列は元素 Pa, Th, Ac, Ra, Fr, Rn が要素であるが，$k=7$ のときは最初の化合物は 13211321322113 を生じ，Ho.At のように書きかえられる．なぜならば，Ho は 1321132 であり，At は 1322113 だからである．他の例 $k=91$ の場合は，ヘリウムが化合物 Hf.Pa.H.Ca.Li を生じさせる．というのも，H が 22 だからである．

さらに進んで 11 から始めよう．すると 21 を得，1211 を得て

$$111221,$$
$$312211,$$
$$13112221,$$
$$1113213211 = 1132 \cdot 13211 = \text{Hf.Sn}$$

となる．もし 12 から出発すると，最初の列は元素そのもので，12=Ca である．しかし一方 32 あるいは 23 から出発すると，

$$1312, \quad 1213$$
$$11131112 = 1113.1112 = \text{Th.K}, \quad 11121113 = 1112.1113 = \text{K.Th}$$

を得る．またもっと一般的な場合の列で，数 1,2,3 以外の数字を含むものもある．たとえば 14 か 55 から出発すると，比較的多数について定理がまだ使えるが，ちょうど2つの元素を加えてみよう（プルトニウムとネプツニウムの同位元素）．すると

$$\text{Pu}_4 = 312211322212211211123222114,$$
$$\text{Np}_4 = 13112221133211322112112113322114,$$
$$\text{Pu}_5 = 312211322212211211123222115,$$
$$\text{Np}_5 = 13112221133211322112211133222115$$

であり，その比較的多数は 0 に収束する．このことは，数 6,7,8,9,… 等をもつ列につ

いても同様である．

最後にコンウェイの定数 λ に戻ろう．λ は，92×92 移行行列 M の（ただ 1 つの）最大の（絶対値について）固有値であり，行列の (i,j) 番目の要素は，元素 i の 1 つの原子の崩壊から生じるような元素 j の原子番号である．同様に，比較的多数は M の注意深い固有値解析から生まれる．Conway の 71 番目の次数多項式は Galois（ガロア）群 S_{71} を有するので，λ はべキ根を用いて表すことができない[8]．同様に[9-12]をも参照せよ．

[1] N. J. A. Sloane, On-Line Encyclopedia of Integer Sequences, A001140, A001141, A001143, A001145, A001151, A001154, A001155, A005150, A005341 A006751, and A006715.
[2] M. Mowbray, R. Pennington, and E. Welbourne, Prelude to Professor Conway's article, *Eureka* 46 (1986) 4.
[3] J. H. Conway, The weird and wonderful chemistry of audioactive decay, *Eureka* 46 (1986) 5-16.
[4] J. H. Conway, The weird and wonderful chemistry of audioactive decay, *Open Problems in Communication and Computation*, ed. T. M. Cover and B. Gopinath, Springer-Verlag, 1987, pp. 173-188.
[5] J. H. Conway and R. K. Guy, *The Book of Numbers*, Springer-Verlag, 1996, pp. 208-209; MR 98g:00004.
[6] I. Vardi, *Computational Recreations in Mathematica*, Addison-Wesley, 1991; MR 93e:00002.
[7] S. B. Ekhad and D. Zeilberger, Proof of Conway's lost cosmological theorem, *Elec. Res. Announce. Amer. Math. Soc.* 3 (1997) 78-82; MR 98i:05010.
[8] J. H. Conway, Comments on the "Look and Say sequence," unpublished note (1997).
[9] X. Gourdon and B. Salvy, Effective asymptotics of linear recurrences with rational coefficients, *Discrete Math.* 153 (1996) 145-163; MR 97c:11013.
[10] M. Hilgemeier, One metaphor fits all: A fractal voyage with Conway's audioactive decay, *Fractal Horizons:The Future Use of Fractals*, ed. C. Pickover, St. Martins Press, 1996, pp. 137-162.
[11] J. Sauerberg and L. Shu, The long and the short on counting sequences, *Amer. Math. Monthly* 104 (1997) 306-317; MR 98i:11013.
[12] N. J. A. Sloane, My favorite integer sequences, *Sequences and Their Applications (SETA)*, Proc. 1998 Singapore conf., ed. C. Ding, T. Helleseth, and H. Niederreiter, Springer-Verlag, 1999, pp. 103-130; math.CO/0207175.

7 複素解析に関する定数

7.1 ブロック-ランダウ定数

　原点を中心とする単位円板 D 上で定義された複素解析関数 f で $f(0)=0$, $f'(0)=1$ を満たすようなもの全体の集合を F と書く．

　F の各元 f に対して，f が 1 対 1 となるような D の部分領域 S が存在し，$f(S)$ が半径 r の円板を含むとき，この r の上限を $b(f)$ とする．Bloch（ブロック）[1-7] は，$b(f)$ が少なくとも 1/12 であることを示した．**ブロックの定数** B は，$\inf\{b(f):f\in F\}$ により定義される．B の正確な値は知られていないが，次の不等式が Ahlfors-Grunsky（アールフォルス-グルンスキー）[8] や Heins（ハインズ）[9] により得られた．すなわち

$$0.433 < \frac{\sqrt{3}}{4} < B \leq \frac{1}{\sqrt{1+\sqrt{3}}} \frac{\Gamma\left(\frac{1}{3}\right)\Gamma\left(\frac{11}{12}\right)}{\Gamma\left(\frac{1}{4}\right)} = 0.4718616534\cdots$$

Ahlfors-Grunsky はさらに，B は上の不等式の上界に等しいであろうと予想した．

　関連のある定数が次のように定義されている．F の各元 f に対して，$f(D)$ が半径 r の円板を含むとき，この r の上限を $l(r)$ とおく．Landau（**ランダウ**）**の定数** L [3,5,7,10] は，$\inf\{l(f):f\in F\}$ により定義される．L が B 以上であることは明らかである．ブロックの定数 B と同様にランダウの定数 L の正確な値は知られていない．Robinson（ロビンソン）[11] と Rademacher（ラーデマッヘル）[12] により独立に，次の不等式

$$0.5 = \frac{1}{2} < L \leq \frac{\Gamma\left(\frac{1}{3}\right)\Gamma\left(\frac{5}{6}\right)}{\Gamma\left(\frac{1}{6}\right)} = 0.5432589653\cdots$$

が示された. Rademacher は, L がこの不等式の上界に等しいだろうとの予想もした.

ブロックの定数 B, ランダウの定数 L についての予想は両方とも今日まで証明されてはいない[13-17]. ガンマ関数の値の比【1.5.4】というこれらの予想の厳密な表現形式は, 魅力的である.

Bonk (ボンク) [18]は, 1990年にブロックの定数 B の下界が $\sqrt{3}/4 + 10^{-14}$ であることを示した. Minda (ミンダ) [19]は, これを定数 B の評価に対するこの半世紀間で初めての数値的な改良であるとした. Chen & Gauthier (チェン&ゴウティエ) [20,21]は, Bonk の方法を改良して 10^{-14} を 2×10^{-4} に置きかえた. また Yanagihara (柳原宏) [22]は, L の下界を $1/2 + 10^{-335}$ に改良した.

F の元のうち 1 対 1 の関数となるものからなる部分集合を G と書く. このような関数は, **単葉**であるといわれる. G では, ブロックの定数とランダウの定数が一致することは明らかである. **単葉ブロック-ランダウ定数** K を $K = \inf\{l(f) : f \in G\}$ と定義する. K に関する最も新しい評価は, $0.57088 < K < 0.6564155$ [23-28]である. K の厳密な表現は, いまだ誰も予想さえもしていない.

ここで述べたことには, いろいろな拡張が考えられている. たとえば領域 D を円板ではなく円環にする[29]とか, 関数 f を 1 変数ではなく多変数にする[30,31]などである. これらを議論すると, わき道にそれてしまうことになる.

MacGregor (マクレガー) [32]は, $f(D)$ の別の幾何学的な性質に関する興味深い問題をいくつか提起した. f を F の元とするとき, $\sup\{|f(z) - f(w)| : z, w \in D\}$ で定義された $f(D)$ の**直径**は少なくとも 2 であることが示される. 証明は[33]を見てほしい. 他にどんなことが示されるのか? f を G の元とするとき, $f(D)$ と単位円板とが交わった部分の面積を $a(f)$ とする. Goodman, Jenkins & Reich (グッドマン, ジェンキンス&ライヒ) の研究により, $0.62\pi < A = \inf\{a(f) : f \in G\} < 0.7728\pi$ が出る. この定数 A の正確な値は何なのか? また Strohhäcker (シュトロヘッカー) [17]は, G の元 f が与えられたとき端点の一方を原点とし長さが 0.73 より長い線分が $f(D)$ に存在することを示した. 0.73 を改良するような最大な数は何か? ひょっとすると, この問題は Hayman-Wu (ヘイマン-ウー) 定数【7.5】として知られているものと関連があるかもしれない. 他の関連事項については,【8.19】も見てほしい.

G の各元 f に対して $f(D)$ が原点を中心とする半径 r の円板を含むとき, この r の上限を $m(f)$ とする. ここでの (原点を中心とする円板という) 仮定に注意しよう. Koebe (ケーベ) 定数 M [38,39]を $\inf\{m(f) : f \in G\}$ と定義する. Koebe は, M が存在することを示し, Bieberbach (ビーベルバッハ) [41]が $M = 1/4$ というケーベの予想をうちたてた. 極値関数は, 写像

$$f(z)=\frac{z}{(1-z)^2}$$

とその回転だけからなる．集合 F に対して M に対応した定数は 0 以外に存在しないことがわかる．任意に大きな整数 n に対して，$f(z)=(\exp(nz)-1)/n$ は，F に属するが，指数関数は決して 0 にはならないので，$f(z)$ は，$-1/n$ という値はとらない．よって原点中心のどんな円板も，十分に大きな n をとれば $f(D)$ には含まれない．

[1] A. Bloch, Les théorèmes M. Valiron sur les fonctions entières et la théorie de l'uniformisation, Annales Faculté Sciences Toulouse 17 (1925) 1-22.
[2] M. Heins, Selected Topics in the Classical Theory of Functions of a Complex Variable, Holt, Rinehart, and Winston, 1962, pp. 45-47, 83-86; MR 29 #217.
[3] E. Hille, Analytic Function Theory, v. 2, Ginn, 1962, pp. 382-389; MR 34 #1490.
[4] L. V. Ahlfors, Conformal Invariants: Topics in Geometric Function Theory, McGraw-Hill, 1973, pp. 14-15; MR 50 #10211.
[5] J. B. Conway, Functions of One Complex Variable, 2nd ed., Springer-Verlag, 1978, pp. 292-296; MR 80c:30003.
[6] W. K. Hayman, Multivalent Functions, 2nd ed., Cambridge Univ. Press, 1994, pp. 136-140; MR 96f:30003.
[7] R. Remmert, Classical Topics in Complex Function Theory, Springer-Verlag, 1998, pp. 225-242; MR 98g:30002.
[8] L. V. Ahlfors and H. Grunsky, Über die Blochsche Konstante, Math. Z. 42 (1937) 671-673.
[9] M. Heins, On a class of conformal metrics, Nagoya Math. J. 21 (1962) 1-60;MR26 #1451.
[10] E. Landau, Über die Blochsche Konstante und zwei verwandte Weltkonstanten, Math. Z. 30 (1929) 608-634.
[11] L. V. Ahlfors, An extension of Schwarz's lemma, Trans. Amer. Math. Soc. 43 (1938) 359-364 (footnote, p. 364).
[12] H. Rademacher, On the Bloch-Landau constant, Amer. J. Math. 65 (1943) 387-390; MR 4,270d.
[13] G. M. Goluzin, Geometric Theory of Functions of a Complex Variable (in Russian), 2nd ed., Izdat. Akad. Nauk SSSR, 1966; Engl. transl. in Amer. Math. Soc. Transl., v. 26, 1969; MR 36 #2793 and MR 40 #308.
[14] C. D. Minda, Bloch constants, J. d'Analyse Math. 41 (1982) 54-84; MR 85e:30013.
[15] A. Baernstein and J. P. Vinson, Local minimality results related to the Bloch and Landau constants, Quasiconformal Mappings and Analysis: A Collection of Papers Honoring F.W. Gehring, ed. P. L. Duren, J. M. Heinonen, B. G. Osgood, and B. P. Palka, Springer-Verlag, 1998, pp. 55-89; MR 99m:30036.
[16] A. Baernstein, Landau's constant, and extremal problems involving discrete subsets of C, Linear and Complex Analysis Problem Book 3, Part II, ed. V. P. Havin and N. K Nikolski, Lect. Notes in Math. 1543, Springer-Verlag, 1994, pp. 404-407.
[17] A. Baernstein, A minimum problem for heat kernels of flat tori, Extremal Riemann Surfaces, ed. J. R. Quine and P. Sarnak, Contemp. Math. 201, Amer. Math. Soc., 1998, pp. 227-243; MR 98a:58153.
[18] M. Bonk, On Bloch's constant, Proc. Amer. Math. Soc. 110 (1990) 889-894;MR91c:30011.
[19] D. Minda, The Bloch and Marden constants, Computational Methods and Function Theory, Proc. 1989 Valparáiso conf., ed. St. Ruscheweyh, E. B. Saff, L. C. Salinas, and R. S. Varga, Lect. Notes in Math. 1435, Springer-Verlag, 1990, pp. 131-142; MR 91k: 30060.
[20] H. Chen and P. M. Gauthier, On Bloch's constant, J. d'Analyse Math. 69 (1996) 275-291; MR 97j: 30002.
[21] C. Xiong, Lower bound of Bloch's constant, Nanjing Daxue Xuebao Shuxue Bannian Kan 15 (1998) 174-179; MR 2000c:30011.

[22] H. Yanagihara, On the locally univalent Bloch constant, *J. d'Analyse Math.* 65 (1995) 1-17; MR 96e:30041.
[23] R. E. Goodman, On the Bloch-Landau constant for schlicht functions, *Bull. Amer. Math. Soc.* 51 (1945) 234-239; MR 6,262a.
[24] E. Reich, On a Bloch-Landau constant, *Proc. Amer. Math. Soc.* 7 (1956) 75-76; MR 17,1066g.
[25] J. A. Jenkins, On the schlicht Bloch constant, *J. Math. Mech.* 10 (1961) 729-734; MR 23 # A1804.
[26] S. Y. Zhang, On the schlicht Bloch constant, *Beijing Daxue Xuebao* 25 (1989) 537-540 (in Chinese); MR 91a:30014.
[27] E. Beller and J. A. Hummel, On the univalent Bloch constant, *Complex Variables Theory Appl.* 4 (1985) 243-252; MR 86i:30005.
[28] J. A. Jenkins, On the schlicht Bloch constant. II, *Indiana Math. J.* 47 (1998) 1059-1063; MR 99m:30071.
[29] P.-S. Chiang and A. J. Macintyre, Upper bounds for a Bloch constant, *Proc. Amer. Math. Soc.* 17 (1966) 26-31; MR 32 # 5883.
[30] H. Chen and P. M. Gauthier, Bloch constants in several variables, *Trans. Amer. Math. Soc.* 353 (2001) 1371-1386; MR 2001m:32035.
[31] H. Chen, P. M. Gauthier, and W. Hengartner, Bloch constants for planar harmonic mappings, *Proc. Amer. Math.* Soc. 128 (2000) 3231-3240; MR 2001b:30027.
[32] T. H. MacGregor, Geometric problems in complex analysis, *Amer. Math. Monthly* 79 (1972) 447-468; MR 45 # 7034.
[33] G. Pólya and G. Szegö, *Problems and Theorems in Analysis*, v. 1, Springer-Verlag, 1972, ex. 239; MR 81e:00002.
[34] A. W. Goodman, Note on regions omitted by univalent functions, *Bull. Amer. Math. Soc.* 55 (1949) 363-369; MR 10,601f.
[35] J. A. Jenkins, On values omitted by univalent functions, *Amer. J. Math.* 75 (1953) 406-408; MR 14,967a.
[36] A. W. Goodman and E. Reich, On regions omitted by univalent functions. II, *Canad. J. Math.* 7 (1955) 83-88; MR 16,579e.
[37] E. Strohhäcker, Beiträge zur Theorie der schlichten Funktionen, *Math. Z.* 37 (1933) 356-380.
[38] P. L. Duren, *Univalent Functions*, Springer-Verlag, 1983, pp. 26-31; MR 85j:30034.
[39] J. B. Conway, *Functions of One Complex Variable II*, Springer-Verlag, 1995, pp. 61-67; MR 96i:30001.
[40] P. Koebe, Über die Uniformisierung beliebiger analytischer Kurven, *Nachr. Königlichen Ges. Wiss. Göttingen, Math.-Phys. Klasse* (1907) 191-210.
[41] L. Bieberbach, Über die Koeffizienten derjenigen Potenzreihen welche eine schlichte Abbildungen des Einheitskreises vermitteln, *Sitzungsber. Preuss. Akad. Wiss.* (1916) 940-955.

7.2 マサー-グラメイン定数

$f(z)$ を任意の正の整数 n に対して値 $f(n)$ が整数となるような整関数とする．どのような条件があれば f が多項式になるといえるのか？ Pólya (ポリヤ) [1]は，$M_r = \sup_{|z| \leq r} |f(z)|$ とするとき

$$\limsup_{r \to \infty} \frac{\ln(M_r)}{r} < \ln(2) = 0.6931471805\cdots$$

7.2 マサー-グラメイン定数

ならば，f が多項式になることを証明した．さらに，$f(z)=2^z$ という特別な場合を考えると，$\ln(2)$ が，上記の結果が成立する最大の定数（すなわち最良定数）であることを示した[2-4]．

ここでより難しいが関連した問題を示す．それは，実部，虚部ともに整数である複素数すべてよりなる集合，すなわちガウス整数環を考慮したものである．$f(z)$ を任意のガウス整数 n に対して $f(n)$ がガウス整数となるような整関数とする．（上と同様に）どのような条件があれば，f が多項式であるといえるのか？ Gel'fond（ゲルフォント）[5]は，Fukasawa（深沢）の研究[6]をもとにして，

$$\limsup_{r\to\infty}\frac{\ln(M_r)}{r^2}<\alpha$$

ならば f が多項式となるような正の定数 α が存在することを証明した．驚くほどのことではないが，f が多項式であるためには，（分母に r ではなく r^2 を含む）より強い極限に関する条件が必要である．最良の定数 α については後で論ずることにする．α を見つけようとした試みのなかから生じた別の定数 δ について注目する．

Masser（マサー）[7]は，α が $\pi/(2e)=0.5778636748\cdots$ より大きくはなりえないことを証明し，α は $\pi/(2e)$ に等しいと考えた．彼は，また次のより弱い結果も証明した．それは，次の不等式が成立すれば，f は多項式になるというものである．すなわち

$$\limsup_{r\to\infty}\frac{\ln(M_r)}{r^2}<\alpha_0=\frac{1}{2}\exp\left(-\delta+\frac{4c}{\pi}\right)$$

ここで

$$c=\gamma\beta(1)+\beta'(1)=\frac{\pi}{4}(-\ln(2)+2\ln(\pi)+2\gamma-2\ln(L))=0.6462454398\cdots$$

であり，γ は Euler-Mascheroni（オイラー-マスケローニ）の定数【1.5】，$\beta(x)$ は Dirichlet（ディリクレ）のベータ関数【1.7】，L は Gauss（ガウス）のレムニスケート定数【6.1】であり，δ はこの後すぐに定義されるものである．これに類似した式がこの本の Landau-Ramanujan（ランダウ-ラマヌジャン）定数【2.3】や Sierpinski（シェルピンスキー）の定数 S【2.10】に関する部分に現れる．実は，$c=\pi S/4$ である．

δ をオイラー-マスケローニの定数の2次元への自然な一般化として定義する．すなわち

$$\delta=\lim_{n\to\infty}\left(\sum_{k=2}^{n}\frac{1}{\pi r_k^2}-\ln(n)\right)$$

と定義する．ここで r_k は，ある複素数 z を中心とし半径 r の閉円板が少なくとも k 個の異なるガウス整数を含むときの r の最小値とする．

δ の計算は非常に難しい．Gramain & Weber（グラメイン&ウェーバー）[8]は，$1.811447299<\delta<1.897327117$ を示した．これより $0.170739<\alpha_0<0.1860446$ が出る．α_0 は Gel'fond の（級数補間法として知られる）方法により与えうる最大の定数とわかる．明らかに α_0 は予想された最良の定数 $\pi/(2e)$ からはかけ離れているが，α_0 が $1/(2e)=0.1839397205\cdots$ に近いということは興味深い．Gramain[9,10]は，$\alpha_0=1/(2e)$ であると予想した．もしそうなら $\delta=1+4c/\pi=1.8228252496\cdots$ が導かれるが，これが正しいかどうかは，わかっていない．

Masser-Gramain（マサー-グラメイン）定数 δ を，たとえば小数第4位まで求めるにはどのように計算したらよいのだろうか？ r_k に関する公式は知られていない．そこで Gramain & Weber は，定義に基づいて，大きな k に対する r_k の評価を行う以外に方法がなかった．たとえば[7]で $r_2=1/2$, $r_3=r_4=1/\sqrt{2}$ が示され，[9]で

$$\frac{\sqrt{\pi(k-1)+4}-2}{\pi} < r_k < \sqrt{\frac{k-1}{\pi}}$$

が示された．上界はかなりよいが，下界については，δ の正確な評価を得るためには，改良する必要がある．$k \geq 6$ に対しては，

$$\frac{\sqrt{\pi(k-6)+2}-\sqrt{2}}{\pi} \leq r_k$$

とわかるが，必要とするさらなる改良[9,10]については，複雑すぎるのでここで述べることはできない．[8]によると δ を小数第4位まで計算しようとすると 5×10^{13} までの k に対して r_k を計算しなければならないとのことである．r_k を計算するアルゴリズムがもっと効率的なものになること，r_k に対する不等式が改良されること，δ を計算する別の手順を見つけること，あるいは計算機のハードウェアの大進歩などがなくては，δ の値は未知のままになってしまうだろう．

　オイラー定数のまったく異なる n 次元格子和への一般化が【1.10.1】で議論されている．

　最後に残っていた問題を解決しよう．Gramain[9,10]は Gruman（グルマン）[11]の研究をもとにして最良定数 a は $\pi/(2e)$ であるという Masser の予想を証明した．しかしながらこのことは，δ すなわち a_0 の値を知ることに対して何の光明も投げかけない．

[1] G. Pólya, Über ganzwertige ganze Funktionen, *Rend. Circ. Mat. Palmero* 40 (1915) 1-16; also in *Collected Papers*, v. 1, ed. R. P. Boas, MIT Press, 1974, pp. 1-16, 771-773.
[2] R. C. Buck, Integral valued entire functions, *Duke Math. J.* 15 (1948) 879-891; MR 10,693c.
[3] R. P. Boas, *Entire Functions*, Academic Press, 1954; MR 16,914f.
[4] L. A. Rubel, *Entire and Meromorphic Functions*, Springer-Verlag, 1996; MR 97c:30001.
[5] A. Gel'fond, Sur les propriétés arithmétiques des fonctions entières, *Tôhoku Math. J.* 30 (1929) 280-285.
[6] S. Fukasawa, Über ganzwertige ganze Funktionen, *Tôhoku Math.* J. 27 (1926) 41-52.
[7] D. W. Masser, Sur les fonctions entières à valeurs entières, *C. R. Acad. Sci. Paris Sér. A-B* 291 (1980) 1-4; MR 81i:10048.
[8] F. Gramain and M.Weber, Computing an arithmetic constant related to the ring of Gaussian integers, *Math. Comp.* 44 (1985) 241-245; corrigendum 48 (1987) 854; MR 87a:11028 and MR 88a:11034.
[9] F. Gramain, Sur le théorème de Fukasawa-Gel'fond-Gruman-Masser, *Séminaire Delange-Pisot-Poitou, Théorie des Nombres, 1980-1981*, Birkhäuser, 1982, pp. 67-86; MR 85d:11071.
[10] F. Gramain, Sur le théorème de Fukasawa-Gel'fond, *Invent. Math.* 63 (1981) 495-506;MR 83g:30028.
[11] L. Gruman, Propriétés arithmétiques des fonctions entières, *Bull. Soc. Math. France* 108 (1980) 421-440; MR 82g:10072.

7.3 ホイッタカー-ゴンチャロフ定数

f を f 自身とその導関数 $f^{(n)}$ ($n=1,2,3,\cdots$) がそれぞれ単位円板に少なくとも1つの零点 z_n をもつような整関数とする．どのような条件があれば，f が恒等的に 0 であると結論づけることができるのか？ $M_r = \sup_{|z| \leq r} |f(z)|$ とするとき

$$\limsup_{r \to \infty} \frac{\ln(M_r)}{r} < \ln(2)$$

ならば $f=0$ であることを示すのは難しくない[1-5]．この上界は最良のものではない．

$$\limsup_{r \to \infty} \frac{\ln(M_r)}{r} < W$$

ならば $f=0$ となるような最大の数 W を Whittaker（**ホイッタカー**）**の定数**と定義する．すると前の結果と $f(z) = \sin(z) + \cos(z)$ という例から $0.693\cdots \leq W \leq 0.785\cdots = \pi/4$ とわかる．あるいはまた，任意に選んだ z に対して等式

$$\limsup_{n \to \infty} |f^{(n)}(z)|^{1/n} = \limsup_{r \to \infty} \frac{\ln(M_r)}{r}$$

も使う．この式によれば，$f^{(n)}$ の漸近的な局所挙動は，最大絶対値関数 M の大域的な性質によって決まってしまう．

点列 $\{z_n\}$ の挙動あるいは f のもちうる単葉性に関連する条件とを使うほか，マクローリン級数の係数を用いる W の別のいくつかの定式化もある．これらについてここで述べるのはやめ，Goncharov（**ゴンチャロフ**）**の定数** G が上記のようなところから生じ[6,7]，後に Buckholtz（バックホルツ）[8,9]により $W=G$ であることが示されたことのみを述べておく．ゴンチャロフ多項式として知られていることに関連のある定式化については【7.3.1】で議論する．

W に関する最もよく知られた精密な評価は Macintyre（マッキンタイヤ）[10-12]による

$$0.7259\cdots < W < 0.7378\cdots$$

であり，これは Pólya, Boas, Levinson（ポリヤ，ボアス，レビンソン）の初期の研究に基づいてうちたてられたものである．上界は，関数微分方程式

$$\frac{d}{dz} \varphi(z, q) = \varphi(qz, q)$$

の整関数の解，すなわち

$$\varphi(z, q) = \sum_{n=0}^{\infty} \frac{1}{n!} q^{n(n-1)/2} z^n \qquad (|q| \leq 1)$$

の研究から生じる．より正確にいうと，W は，すべての q を考えたとき $\varphi(z, q)$ の零点の絶対値の最小値より大きくない．W の下界の方は，異なる方法によって生じる．

数値に関する発見的な方法により Varga & Wang（ヴァルガ&ワン）[12,13]は，W

=2/e であるという Boas の予想が正しくないことを示すことになる $0.7360 < W$ を推論した．Waldvogel（ヴァルドフォゲル）[16]は，さらに計算することにより $0.73775075 < W$ と推論したが，この研究に対する厳密な理論的な裏づけはいまだ終了していないことを強調しておきたい．しかしながら Macintyre の上界を精密化[12,13,16]することにより $W < 0.7377507574\cdots$ が得られる．それゆえ Varga & Waldvogel は，W は，この上界に等しいであろうと予想している．どんなに多倍長の浮動小数点数計算をしても，このようなきちんとした等式を証明するのには十分ではないであろう．

W のいくつかの一般化が[4,17-22]において定義された．Oskolkov（オスコルコフ）[23]は，W にいくらでも近い下界を計算する新しい方法を得たと主張した．

関連した話題を述べよう．ベキ級数を微分すること

$$\sum_{n=0}^{\infty} a_n z^n \to \sum_{n=1}^{\infty} n\, a_n z^{n-1}$$

とベキ級数を「ずらす」こと（すなわち正規化された剰余をつくること）

$$\sum_{n=0}^{\infty} a_n z^n \to \sum_{n=1}^{\infty} a_n z^{n-1}$$

は，幾分似た操作である．W，ゴンチャロフ多項式，微分に関する前述の理論は「ずらし」という操作に対しても同じように成り立つ．話を簡単にし多様性をもたせるために，別の観点から考えよう．

f を解析関数としそのマクローリン級数

$$f(z) = \sum_{k=0}^{\infty} a_k z^k$$

が，ちょうど 1 に等しい収束半径をもつものとする．

$$S_n(z, f) = \sum_{k=0}^{n} a_k z^k, \quad n = 1, 2, 3, \cdots$$

を f の n 番目の部分和とし，この多項式 S_n の零点の最大絶対値を $\rho_n(f)$ とする．

$$\rho(f) = \liminf_{n \to \infty} \rho_n(f)$$

とおき，**ベキ級数定数 P** を

$$P = \sup_f \rho(f)$$

と定義する．Porter（ポーター）[24]と Kakeya（掛谷宗一）[25,26]は，$P \leq 2$ を示した．Clunie & Erdös（クルニー&エルデーシュ）[25]は，$\sqrt{2} < P < 2$ であることを示した．Backholtz（バックホルツ）[27]は，これを $1.7 < P < 1.862$ に改良し，Frank（フランク）は，これを $1.7818 < P < 1.82$ に改良した．$1/P$ の評価に関する研究が Pommiez（ポミエ）[28-30]により独立になされた．ホイッタカー定数 W がゴンチャロフ多項式を用いて定式化されるのと同じように，ベキ級数定数 P はいわゆる剰余多項式【7.3.2】を用いて定式化される．

この場合は，関数微分方程式ではなく，ずらし操作に関する関数方程式を考える．改めて解

$$\psi(z,q) = \sum_{n=0}^{\infty} q^{n(n-1)/2} z^n \qquad (|q| \leq 1)$$

の零点を研究すると，下界 $P \geq 1.7818046151\cdots$ が得られる．Waldvogel[16]は，この下界が実は P の真の値であろうと予想した．これは前述したことに似ている．しかしながらこれを解析することは，もっと複雑である．

3番目の定数は[16]で研究されたものであるが，ある Padé（パデ）近似に関連したものである．関連した論文としては[31-33]がある．

7.3.1 ゴンチャロフ多項式

ホイッタカー-ゴンチャロフ定数 W に対する限界は理論的には**ゴンチャロフ多項式**[7]，すなわち $n \geq 1$ に対して

$$G_0(z)=1, \quad G_n(z,z_0,z_1,\cdots,z_{n-1}) = \int_{z_0}^{z} \int_{z_1}^{t_1} \cdots \int_{z_{n-2}}^{t_{n-2}} \int_{z_{n-1}}^{t_{n-1}} 1 \, dt_n \, dt_{n-1} \cdots dt_2 \, dt_1$$

を経由して決定しうる．同値な再帰的な定義は，

$$G_n(z,z_0,z_1,\cdots,z_{n-1}) = \frac{z^n}{n!} - \sum_{k=0}^{n-1} \frac{z_k^{n-k}}{(n-k)!} G_k(z,z_0,z_1,\cdots,z_{k-1})$$

である．Evgrafov（エフグラフォフ）[34]は，

$$\left(\limsup_{n \to \infty} g_n^{1/n} \right)^{-1} = W$$

であることを示した．ここで

$$g_n = \max_{\substack{|z_k|=1 \\ 0 \leq k \leq n-1}} |G_n(0,z_0,z_1,\cdots,z_{n-1})|$$

である．Buckholtz[35]は，さらに

$$\left(\frac{2}{5}\right)^{1/n} g_n^{-1/n} < W \leq g_n^{-1/n}$$

であり，それゆえ上極限を極限に置きかえうることを示した．残念なことに，これらの公式を用いたときの収束の速さが W の正確な値を得るためには遅すぎる[12]．別の方法を用いなければならない．

7.3.2 剰余多項式

ベキ級数定数 P の下界は理論的には**剰余多項式**[30,36,37]，すなわち $n \geq 1$ に対して

$$B_0(z)=1, \quad B_n(z,z_0,z_1,\cdots,z_{n-1}) = z^n - \sum_{k=0}^{n-1} z_k^{n-k} B_k(z,z_0,z_1,\cdots,z_{k-1})$$

を経由して決定しうる．Buckholtz[36]は，

$$\lim_{n \to \infty} b_n^{1/n} = P$$

であることを証明した．ここで

$$b_n = \max_{\substack{|z_k|=1 \\ 0 \leq k \leq n-1}} |B_n(0,z_0,z_1,\cdots,z_{n-1})|$$

である．残念なことに，ゴンチャロフ多項式のときと同様，これらの公式を用いたとき

の収束の速さは P の正確な値を得るためには遅すぎる.

[1] S. Kakeya, An extension of power series, *Proc. Physico-Math. Soc. Japan* 14 (1932) 125. 138.
[2] S. Takenaka, On the expansion of integral transcendental functions in generalized Taylor's series, *Proc. Physico-Math. Soc. Japan* 14 (1932) 529-542.
[3] J. M. Whittaker, *Interpolatory Function Theory*, Cambridge Univ. Press, 1935, pp. 44-45.
[4] R. P. Boas, Entire functions of exponential type, *Bull. Amer. Math. Soc.* 48 (1942) 839-849; MR 4,136c.
[5] R. P. Boas, *Entire Functions*, Academic Press, 1954, pp. 8-13, 172-177; MR 16,914f.
[6] R. P. Boas, An upper bound for the Gontcharoff constant, *Duke Math. J.* 15 (1948) 953-954; MR 10,443c.
[7] W. Goncharov, Recherches sur les dérivées successives des fonctions analytiques, *Annales Sci. École Norm. Sup.* 47 (1930) 1-78.
[8] J. D. Buckholtz, Successive derivatives of analytic functions, *Indian J. Math.* 13 (1971) 83-88; MR 48 #11501.
[9] J. D. Buckholtz and J. L. Frank, Whittaker constants. II, *J. Approx. Theory* 10 (1974) 112-122; MR 50 #2490.
[10] S. S. Macintyre, An upper bound for the Whittaker constant W, *J. London Math. Soc.* 22 (1947) 305-311; MR 10,27a.
[11] S. S. Macintyre, On the zeros of successive derivatives of integral functions, *Trans. Amer. Math. Soc.* 67 (1949) 241-251; MR 11,340b.
[12] R. S. Varga, *Topics in Polynomial and Rational Interpolation and Approximation*, Les Presses de l'Université de Montréal, 1982, pp. 120-129; MR 83h:30041.
[13] R. S. Varga and P. S. Wang, Improved bounds for the Whittaker constant using MACSYMA, *Proc. 1979 MACSYMA Users Conf.*, Washington DC, pp. 121-125.
[14] R. P. Boas, Functions of exponential type. II, *Duke Math. J.* 11 (1944) 17-22; MR 5,175h.
[15] R. P. Boas, Functions of exponential type. IV, *Duke Math. J.* 11 (1944) 799; MR 6,123a.
[16] J. Waldvogel, Zero-free disks in families of analytic functions, *Approximation Theory, Tampa*, ed. E. B. Saff, Lect. Notes in Math. 1287, Springer-Verlag, 1987, pp. 209-228; MR 89g:30012.
[17] R. P. Boas, Expansions of analytic functions, *Trans. Amer. Math. Soc.* 48 (1940) 467-487; MR 2,80e.
[18] V. Ganapathy Iyer, A property of the zeros of the successive derivatives of integral functions, *J. Indian Math. Soc.* 2 (1937) 289-294.
[19] H. S. Wilf, Whittaker's constant for lacunary entire functions, *Proc. Amer. Math. Soc.* 14 (1963) 238-242; MR 26 #2611.
[20] J. L. Frank and J. K. Shaw, Univalence of odd derivatives of even entire functions, *J. Reine Angew. Math.* 277 (1975) 1-4; MR 52 #11039.
[21] J. L. Frank and J. K. Shaw, On analytic functions with gaps, *J. London Math. Soc.* 6 (1973) 577-582; MR 54 #13041.
[22] J. L. Frank and J. K. Shaw, On analytic functions with gaps. II, *Acta Math. Acad. Sci. Hungar.* 27 (1976) 37-42; MR 54 #13042.
[23] V. A. Oskolkov, On a lower bound on the Whittaker constant, *Current Problems in Function Theory*, Proc. 1985 Teberda conf., ed. Yu. F. Korobeinik, Rostov. Gos. Univ., 1987, pp. 34-39, 177; MR 91b:30084.
[24] M. B. Porter, On the polynomial convergents of a power series, *Annals of Math.* 8 (1906-1907) 189-192.
[25] J. Clunie and P. Erdös, On the partial sums of power series, *Proc. Royal Irish Acad. Sect. A* 65 (1967) 113-123; MR 36 #5314.
[26] W. K. Hayman, *Research Problems in Function Theory*, Athlone Press, Univ. of London, 1967, problem 7-7; MR 36 #359.

[27] J. D. Buckholtz, Zeros of partial sums of power series, *Michigan Math. J.* 15 (1968) 481-484; MR 38 # 3409.
[28] M. Pommiez, Sur les restes successifs des séries de Taylor, *Annales Faculté Sciences Toulouse* 24 (1960) 77-165; MR 27 # 2614.
[29] M. Pommiez, Sur les différences divisées successives et les restes des séries de Newton généralisées, *Annales Faculté Sciences Toulouse* 28 (1964) 101-110; MR 33 # 7559b.
[30] J. D. Buckholtz and J. L. Frank, Whittaker constants, *Proc. London Math. Soc.* 23 (1971) 348-370; MR 45 # 5358.
[31] G. W. Crofts and J. K. Shaw, Successive remainders of the Newton series, *Trans. Amer. Math. Soc.* 181 (1973) 369-383; MR 47 # 8825.
[32] D. S. Lubinsky and E. B. Saff, Convergence of Padé approximants of partial theta functions and the Rogers-Szegö polynomials, *Constr. Approx.* 3 (1987) 331-361; MR 88m:30092.
[33] G. Szegö, Ein Beitrag zur Theorie der Thetafunktionen, *Sitzungsber. Preuss. Akad. Wiss.* (1926) 242-252; also in *Collected Papers*, v. 1, ed. R. Askey, Birkhäuser, 1982, pp. 795-805.
[34] M. A. Evgrafov, *The Abel-Goncarov Interpolation Problem* (in Russian), Gosudarstv. Izdat. Tehn.-Teor. Lit., 1954; MR 16,1104a.
[35] J. D. Buckholtz, The Whittaker constant and successive derivatives of entire functions, *J. Approx. Theory* 3 (1970) 194-212; MR 41 # 5618.
[36] J. D. Buckholtz, Zeros of partial sums of power series. II, *Michigan Math. J.* 17 (1970) 5-14; MR 41 # 3718.
[37] J. D. Buckholtz and J. K. Shaw, Zeros of partial sums and remainders of power series, *Trans. Amer. Math. Soc.* 166 (1972) 269-284; MR 45 # 8810.

7.4 ジョン定数

X, Y を実バナッハ空間とし (たとえば X, Y を有限次元のユークリッド空間としてもよい), D を X の開部分集合とする. m, M を $0 < m \leq M < \infty$ を満たす2つの数とする. 写像 $f : D \to Y$ が, 連続な開写像で, 局所的に1対1であり, さらに任意の $x \in D$ に対して

$$m \leq \liminf_{y \to x} \frac{|f(y)-f(x)|}{|y-x|}, \quad \limsup_{y \to x} \frac{|f(y)-f(x)|}{|y-x|} \leq M$$

を満たすとき, f は (m, M) **等長写像**であると定義する.

この定義の最後の部分はどういう意味だろうか? f が領域 D を変形させると考えると, f は, D での線素の長さを m と M との間に制限された因子分だけ変えるような変形をする. このような写像 f は, **擬等長写像**あるいは**両側リプシッツ写像**としても知られている.

John (ジョン) [1-3] は, $m = M$ とすると, すべての $x, y \in D$ に対して f は, $|f(y) - f(x)|/|y - x| = m$ を満たさなければならないので, f は m だけスケールを変える剛体運動であることを証明した. とくに f は (大域的に) D 上1対1である.

この結果に留意すると, $M/m < \mu$ ならば, D のすべての (m, M) 等長写像が1対1

になるという性質をもつ最大値 $\mu=\mu(D)$ は何かと考えるのは当然である．今後 D は，X での開球と仮定しておく．Gevirtz（ゲビルツ）[4]は，$\mu\geq r=1.114305\cdots$ を証明した．ここで r は，方程式

$$r=\frac{r+\sqrt{25r^2-8r}}{2r(3r-1)}$$

の唯一の実数解である．数値的な最良な下界は知られていない．μ に対する上界については，この節の終りの部分で少し述べる．

X がさらにヒルベルト空間（よって X での角度を測ることができる）のときは，この付け加えた構造により限界を改良することができる．Gevirtz[4,5]は，John の結果を拡張し，$\mu\geq\sqrt{2}=1.414213\cdots$ であることを示した．X も Y も「両方とも」ヒルベルト空間のとき，Gevirtz[4]は，John の[3]を精密化し $\mu\geq\sqrt{1+\sqrt{2}}=1.553773\cdots$ を示した．そして[5]で実は，$\mu\geq s=1.65743\cdots$ であることを示した．ここで s は関数

$$s=s(t)=\frac{\pi+2\sqrt{1+t^2}}{1+\frac{\pi}{2}+t}$$

の $t>0$ での最小値である．これらの下界についての証明は，John によってなされた擬等長写像に関する基本原理を用いたかなり複雑なものである．しかし，このような方法は，数値的に精密な値を求めるために充分な力をもっているとはいいがたい．

John[6]は，1変数複素解析関数によってもたらされた写像という特別な場合を考察した．すなわち，彼は，z 平面での単位円板 D 上で定義された解析関数 f で任意の点 $z\in D$ に対して $m\leq|f'(z)|\leq M$ を満たすものを考えた．前と同様，$M/m\leq\gamma$ のとき f が（円板で）単葉となるような最大値 γ は何か？　この値 γ を D に対する**ジョン定数**という．これは前述のものの特別な場合なので γ は μ より大きいと考えてよい．

Avhadiev & Aksentev（アブハディエフとアクセンテフ）[7]，John[6]，Yamashita（山下慎二）[8]そして Gevirtz[9,10]を含む数人の研究者たちが γ を決定しようと研究した．最もよく知られた評価[6,9]は，

$$4.810477\cdots\leq\exp\left(\frac{1}{2}\pi\right)\leq\gamma\leq\exp(\lambda\pi)=7.1879033516\cdots$$

であり，ここでの $\lambda=0.6278342677\cdots$ は，次の超越方程式

$$\frac{\pi}{\exp(2\pi\lambda)-1}=\sum_{k=1}^{\infty}\frac{k}{k^2+\lambda^2}\exp\left(-\frac{k\pi}{2\lambda}\right)$$

を満たす．Gevirtz[10]は，実は $\gamma=\exp(\lambda\pi)$ であろうと予想し，どうしてこの等式が成立しなければならないのかという有力な証拠を示した．正確な証明は知られていない．

再び f を円板 D を変形させる写像として思い描くと，有用な物理的な解釈が現れてくる．D が，m と M の間の因子に関する無限小伸縮には抵抗がないが，この限界を越えた場合は無限の抵抗を示すような仮想的物質からつくられているとするとき，D を D 自身に接触させたまま曲げるには，比 M/m は，どれぐらいの大きさでなくてはな

らないのか？　解析関数 f に対しては，その答えは，$7.1879033516\cdots$ になるようである．

ジョン定数は，複素数平面上の単位円板以外の領域 D に対しても定義することができる．このような道具立て[10,11]で始めたいろいろな研究により Gevirtz の予想が正しいという証拠が得られる．

最後に，はじめに述べた，より一般的な条件にもどろう．$X=Y$ とし X を1次元実数直線とすると，区間での実数値局所同相写像は，（その単調性により）必ず大域的同相写像になるので，$\mu=\infty$ となる．$X=Y$ で X の次元が少なくとも2とすると μ に関して上界が存在する．たとえば，X をヒルベルト空間とすると $2\geq\mu\geq 1.65743\cdots$ となる．これは John[3]による単純な2次元の例からの派生である．X がバナッハ空間とだけ仮定したときにいえることは，$64\geq\mu\geq 1.114305\cdots$ である．この証明は[5]にある．

ここでの説明は，部分的には Julian Gevirtz（ジュリアン・ゲビルツ）からの手紙に基づいている．彼は，またその中で Fritz John（フリッツ・ジョン）との長い個人的な交流について言及している．このような理由により，この文章を John を回想したささやかな賛辞として捧げる．

[1] F. John, Rotation and strain, *Commun. Pure. Appl. Math.* 14 (1961) 391.413; also in *Collected Papers*, v. 2, ed. J. Moser, Birkhäuser, 1985, pp. 643-665, 703; MR 25 #1672.
[2] F. John, On quasi-isometric mappings. I, *Commun. Pure. Appl. Math.* 21 (1968) 77-110; note 25 (1972) 497; also in *Collected Papers*, v. 2, ed. J. Moser, Birkhäuser, 1985, pp. 568-602, 636; MR 36 #5716 and MR 46 #715.
[3] F. John, On quasi-isometric mappings. II, *Commun. Pure. Appl. Math.* 22 (1969) 265-278; also in *Collected Papers*, v. 2, ed. J. Moser, Birkhäuser, 1985, pp. 603-616, 637.
[4] J. Gevirtz, Injectivity of quasi-isometric mappings of balls, *Proc. Amer. Math. Soc.* 85 (1982) 345-349; MR 84h:47072.
[5] J. Gevirtz, Upper and lower bounds in injectivity criteria for quasi-isometries, unpublished note (1997).
[6] F. John, A criterion for univalency brought up to date, *Commun. Pure. Appl. Math.* 29 (1976) 293-295; also in Collected Papers, v. 2, ed. J. Moser, Birkhäuser, 1985, pp. 633-635, 637; MR 54 #10592.
[7] F. G. Avhadiev and L. A. Aksentev, Sufficient conditions for univalence of analytic functions (in Russian), *Dokl. Akad. Nauk SSSR* 198 (1971) 743-746; Engl. transl. in *Soviet Math. Dokl.* 12 (1971) 859-863; MR 44 #2916.
[8] S. Yamashita, On the John constant, *Math. Z.* 161 (1978) 185-188; MR 58 #22516.
[9] J. Gevirtz, An upper bound for the John constant, *Proc. Amer. Math. Soc.* 83 (1981) 476-478; MR 82j:30026.
[10] J. Gevirtz, On extremal functions for John constants, *J. London Math. Soc.* 39 (1989) 285-298; MR 90h:30049.
[11] J. Gevirtz, The theory of sharp first-order univalence criteria, *J. d'Analyse Math.* 64 (1994) 173-202; MR 96d:30017.
[12] S. Hildebrandt, Remarks on the life and work of Fritz John, *Commun. Pure Appl. Math.* 51 (1998) 971-989; MR 99i:01033.

7.5 ヘイマン定数

7.5.1 ヘイマン-キェルベリ

f を超越整関数とする．すなわち，f は全複素数平面上で解析的であるが多項式ではないとする．各 $r>0$ に対して原点を中心とし半径 r の円周上での f の最大絶対値を，

$$M(r) = \max_{|z|=r} |f(z)|$$

とおく．関数

$$a(r) = \frac{d^2}{d\ln(r)^2}\ln(M(r)) = \left(r\frac{d}{dr}\right)^2 \ln(M(r))$$

を考える．これは，存在しかつ孤立点を除いて連続である．Hadamard（アダマール）の三円定理[1]により $a(r) \geq 0$ である．$a(r)$ について他にどんなことがいえるのか？ Hayman（ヘイマン）[2]は，すべての f に対して

$$\limsup_{r\to\infty} a(r) \geq A$$

となる定数 $A>0.18$ が存在することを証明した．彼は，$A=1/4$ と予想したが，Kjellberg（キェルベリ）[3]により $0.24<A<0.25$ が示され，この予想は正しくないことが証明された．Kjellberg は，Richardson（リチャードソン）が，$A<0.245$ を示したようだと述べている．より正確な計算機を用いた A の評価は，いまだに知られていない．

7.5.2 ヘイマン-コレンブラム

p を $p \geq 1$ である実数とする．$c(p)$ を次を満たすような 1 未満の最大の実数とする．すなわち，f, g を単位円板上で解析的な関数とし，$c(p)<|z|<1$ を満たすすべての z に対して $|f(z)| \leq |g(z)|$ ならば，

$$\int_{|z|\leq 1} |f(z)|^p dx\, dy \leq \int_{|z|\leq 1} |g(z)|^p dx\, dy \quad (\text{ここで } z = x+iy)$$

となる．

Hayman（ヘイマン）[4]は，$c(2)$ が存在し $0.04 = 1/25 \leq c(2) \leq 1/\sqrt{2} = 0.7071\cdots$ を証明し，Korenblum（コレンブラム）の予想を確かめた．（より正確にいうと，Korenblum は，$c(2)$ が存在することを予想し，条件付でその上界を示した．）重要な拡張として Hinkkanen（ヒンカネン）[6]は，$c(p)$ が存在することと，$0.15724 \leq c(p)$ であることを示し，$p \to \infty$ のとき $c(p) \to 1$ となるかどうかを問題にした．$c(p)$ については言うまでもなく $c(2)$ の正確な値についての予想もまったくなされていない[†]．

[†]訳注：C. Wang, "Refining the constant in a maximum principle for the Bergman Space" *Proc. Amer. Math. Soc.* (132) (2004) no. 3, 853-855 によると $c(2)<0.69472$ と改良された．

7.5.3 ヘイマン-スチュワート

f を有理型関数とする．すなわち，f は（孤立した）極を除いて全複素数平面で解析的である．f は，2つの整関数の商であることが証明できる．いつものように f をリーマン球面 S への写像とみなす．これは，f が極をもつところで ∞ をとると考えることができるからである．任意の $r>0$ と任意の点 $a\in S$ に対して f の a-点の個数関数を

$n(r,a)=$ 円板 $|z|\leq r$ での方程式 $f(z)=a$ の解の（重複度をこめた）個数

として定義する．ここで2つの関連した量を定義する．それらは，

$$n(r)=\max_{a\in S} n(r,a)$$

と

$$A(r)=\underset{a\in S}{\mathrm{mean}}\, n(r,a)=\frac{1}{\pi}\int_S n(r,a)\,da=\frac{1}{\pi}\int_{|z|\leq r}\frac{|f'(z)|^2}{(1+|f(z)|^2)^2}\,dx\,dy$$

である．ここで $z=x+iy$ である．$r\to\infty$ のときこれらの量を比較するのは自然である．f が有理関数（2つの多項式の商）の場合を除いて $A(r)\to\infty$，$n(r)\to\infty$ となるので，これら自体には興味がない．

最大値はつねに平均値を越えているので，すべての r に対して $n(r)\geq A(r)$ であることは明らかである．$\limsup_{r\to\infty} n(r)/A(r)=\infty$ となるような有理型関数を構成することができる．それゆえ，われわれの注意を比

$$H(f)=\liminf_{r\to\infty}\frac{n(r)}{A(r)}$$

に向ける．Hayman & Stewart（ヘイマン&スチュワート）[7-9]は，すべての f に対して $1\leq H(f)\leq e$ であることを証明した．$H(f)>1$ となる有理型関数の最初の例が Toppila（トッピラ）[10]によって構成された．実際には，彼の例では $H(f)$ は少なくとも 80/79 であるというものであった．しかし Miles（マイルズ）[11]は，すべての f に対して $H(f)$ が $e-10^{-28}$ より大きくないことを証明した．それゆえ定数 $h=\sup_f H(f)$（ここで上限は，すべての非定数有理型関数 f をわたるものとする）とすると $80/79\leq h\leq e-10^{-28}$ が成立する．

さて興味深い変種がある．S の各有限部分集合 T に対して

$$n_T(r)=\max_{a\in T} n(r,a)$$

と定義する．固定した T に対して明らかに $n_T(r)<n(r)$ である．Gary（ギャリー）[12]は，すべての f に対して

$$\liminf_{r\to\infty}\frac{n_T(r)}{A(r)}\leq 2.65$$

であることを証明した．これは Miles のより精巧な結果とは好対照をなす．これらの実数の評価をより精密にすることを近い将来に期待することができるのであろうか．

Alexandre Eremenko（アレクサンドル・エレメンコ）は，手紙の中で「ヘイマンの定数は，ある種の複雑な極値問題（極値は有理型関数の族にわたってとるものとする）の解としてすべて定義される．これらの極値問題のどれも，都合のよい対称性をもつ解

をもたないようである．それゆえ，これらの定数に対する数値的な評価を見つける以上のことは期待できまい．こういう型の別の定数が単葉 Bloch-Landau（ブロック-ランダウ）定数【7.1】である．これとは対照的に普通のブロック-ランダウ定数は（多分）異なる性質をもっている．（もし予想された値が正しいのならば）それらは，何らかの美しい対称性をもつ極値的配置と関連がある．Carleson & Jones（カールソン＆ジョーンズ）は，Clunie-Pommerenke（クルニー-ポメレンケ）定数 β が $1/4$ であるという予想【7.6】から β がここでの第 2 の類であると信じている．もちろん偶然に $\beta=1/4$ ということは起こりえない．ある隠れた対称性がその原因であるにちがいない．」と書いている．

7.5.4 ヘイマン-ウー

Hayman & Wu（ヘイマン＆ウー）[13]は，$f(z)$ を単位開円板で単葉であり，L を平面上の任意の直線とするとき逆像 $f^{-1}(L)$ が $|f^{-1}(L)| \leq C$ を満たすような定数 C が存在することを証明した．Øyma（エィマ）[14,15]は，C の最小可能値が $\pi^2 \leq C < 4\pi$ を満たすことを証明し，さらに C はここでの下界に等しくなると予想した．

[1] J. B. Conway, *Functions of One Complex Variable*, 2nd ed., Springer-Verlag, 1978, p. 137; MR 80c：30003.
[2] W. K. Hayman, Note on Hadamard's convexity theorem, *Entire Functions and Related Parts of Analysis*, ed. J. Korevaar, S. S. Chern, L. Ehrenpreis, W. H. J. Fuchs, and L. A. Rubel, Proc. Symp. Pure Math. 11, Amer. Math. Soc., 1968, pp. 210-213; MR 40 # 5858 and MR 41 errata/addenda, p. 1965.
[3] B. Kjellberg, The convexity theorem of Hadamard-Hayman, *Proc. Symp. Math.*, Royal Institute of Technology, Stockholm, 1973, pp. 87-114.
[4] W. K. Hayman, On a conjecture of Korenblum, *Analysis* 19 (1999) 195-205; MR 2000e:30041.
[5] B. Korenblum, A maximum principle for the Bergman space, *Publicacions Matèmatiques* 35 (1991) 479-486; MR 93j:30018.
[6] A. Hinkkanen, On a maximum principle in Bergman space, *J. d'Analyse Math.* 79 (1999) 335-344; MR 2000m:30033.
[7] W. K. Hayman and F. M. Stewart, Real inequalities with applications to function theory, *Proc. Cambridge Philos. Soc.* 50 (1954) 250-260; MR 15,857g.
[8] W. K. Hayman, *Meromorphic Functions*, Oxford Univ. Press, 1964, pp. 1-20; MR 29 # 1337.
[9] W. K. Hayman, *Research Problems in Function Theory*, Athlone Press, Univ. of London, 1967, problem 1-16; MR 36 # 359.
[10] S. Toppila, On the counting function for the a-values of a meromorphic function, *Annales Acad. Sci. Fenn. Ser. A I Math.* 2 (1976) 565-572; MR 58 # 22563.
[11] J. Miles, On a theorem of Hayman and Stewart, *Complex Variables* 37 (1998) 425-455; MR 2000a: 30062.
[12] J. D. Gary, *On the Supremum of the Counting Function for the a-Values of a Meromorphic Function*, Ph.D. thesis, Univ. of Illinois, 1984.
[13] W. K. Hayman and J. M. G. Wu, Level sets of univalent functions, *Comment. Math. Helv.* 56 (1981) 366-403; MR 83b:30008.
[14] K. Oyma, Harmonic measure and conformal length. *Proc. Amer. Math. Soc.* 115 (1992) 687-689; MR 92i:30007.
[15] K. Øyma, The Hayman-Wu constant, *Proc. Amer. Math. Soc.* 119 (1993) 337-338; MR 93k:30031.

7.6 リトルウッド-クルニー-ポメレンケ定数

7.6.1 アルファ

$p(z)$ を次数 n の多項式とする．リーマン球面の中への写像として，p が z に関してどのように変化するかを測るという意味で $|p'(z)|/(1+|p(z)|^2)$ と表される式を $p(z)$ の**球面導関数**という[1].

$$P(p) = \iint_{|z|\leq 1} \frac{|p'(z)|}{1+|p(z)|^2} \, dx \, dy$$

と定義する．ここで $z=x+iy$ である．この二重積分は単位円板上の $p(z)$ の平均球面導関数に比例する．最大値

$$F(n) = \sup\{P(p) : P は次数 n の多項式\}$$

と上極限

$$\alpha = \limsup_{n\to\infty} \frac{\ln(F(n))}{\ln(n)}$$

を考える．Littlewood（リトルウッド）[2]は $F(n)$ は有限であり，$F(n) \leq \pi\sqrt{n}$ すなわち $\alpha \leq 1/2$ であることを証明した．彼は $\alpha < 1/2$ を予想した．Eremenko & Sodin（エレメンコ&ソドン）[3,4]は，$n\to\infty$ のとき $F(n) = o(\sqrt{n})$ であることを証明した．すぐその後，Lewis & Wu（ルイス&ウー）[5]は，$\alpha \leq 1/2 - 2^{-264}$ を証明した．それゆえ Littlewood の予想は確かめられた．しかし，Eremenko は，$\alpha > 0$ を示し，Baker & Stallard（ベーカー&スタラード）[7]は，これを $\alpha \geq 1.11 \times 10^{-5}$ に改良した．

（多項式とは対照的に）有理関数の場合，α に対応する値は $1/2$ である[2,8,9]．Littlewood[2]は，球面導関数を含まない α の別な定義もいくつか与えた．ここで与えた α の定義は Eremenko[10]によってなされたものである．

7.6.2 ベータとガンマ

平面開領域上で定義された複素解析関数 f が 1 対 1，すなわち $f(z) = f(w)$ であることと $z = w$ であることが同値であるとき，f は**単葉**であるという．

$$D = \{z : |z| < 1\} \quad (円板), \quad E = \{z : |z| > 1\} \quad (開円環領域)$$

$$S = \left\{f(z) = z + \sum_{n=2}^{\infty} c_n z^n \text{ となる } D \text{ での単葉関数 } f\right\},$$

$$S_1 = \left\{f(z) = \sum_{n=1}^{\infty} a_n z^n, \sup_{z\in D}|f(z)| \leq 1 \text{ となる } D \text{ での有界単葉関数 } f\right\},$$

$$S_2 = \left\{f(z) = z + \sum_{n=0}^{\infty} b_n z^{-n} \text{ となる } E \text{ での単葉関数 } f\right\}$$

とする．族 S に対して de Branges（ド・ブランジュ）[11,12]は，$|c_n|\leq n$ であることを証明し，Bieberbach（ビーベルバッハ）の有名な予想[13]を解決した．この不等式は最良のものである．S_1, S_2 に対する同様の最良の不等式は知られていない．S_1 と S_2 に対する係数の減少していく割合についての評価には密接な関係があることがわかる．

$$A_n = \sup_{f \in S_1} |a_n|, \quad B_n = \sup_{f \in S_2} |b_n|,$$

$$-\gamma_1 = \lim_{n \to \infty} \frac{\ln(A_n)}{\ln(n)}, \quad -\gamma_2 = \lim_{n \to \infty} \frac{\ln(B_n)}{\ln(n)}$$

とする．$k=1,2$ に対して比較的簡単な不等式 $1/2 \leq \gamma_k \leq 1$ がわかる．Littlewood による初期の研究[14]をもとにし，Clunie & Pommerenke（クルニー＆ポメレンケ）[15-18]は，

$$0.509 < \gamma_k < 0.83$$

であることを示し，Carleson & Jones（カールソン＆ジョーンズ）[19]は，上界を $\gamma_k < 0.76 \cdots$ に改良した[†]．

別のより幾何学的な定式化をしよう．$\varepsilon > 0$，$f \in S_k$ に対して写像 f での円周 $|z| = \exp((-1)^k \varepsilon)$ の像の弧長を考える．

$$L_\varepsilon = \sup_{f \in S_1} |\{f(z) : |z| = \exp(-\varepsilon)\}|, \quad M_\varepsilon = \sup_{f \in S_2} |\{f(z) : |z| = \exp(\varepsilon)\}|,$$

$$-\beta_1 = \lim_{\varepsilon \to 0^+} \frac{\ln(L_\varepsilon)}{\ln(\varepsilon)}, \quad -\beta_2 = \lim_{\varepsilon \to 0^+} \frac{\ln(M_\varepsilon)}{\ln(\varepsilon)}$$

とする．Carleson & Jones の議論により，$0.509 < \gamma_1 = \gamma_2 = 1 - \beta_1 = 1 - \beta_2 < 0.76\cdots$ が示される（実際には，彼らはこれ以上のことを証明した）．ベキ級数の係数と円の像の弧長との間の $\beta + \gamma = 1$ という関係は，もっと早い段階での論文で予想されていたようであるが，Carleson & Jones が，初めてこれを明示的にそして正確に証明した．

Eremenko[10]は，グリーン関数の等高線の弧長を用いてこれらの定数の3番目の定式化を与えた．

7.6.3 予想される関係

Carleson & Jones[19]は，数値実験をもとにして $\gamma = 3/4$（よって $\beta = 1/4$）であることを予想した．これは幾分疑問視されているが，このことを示す信頼すべき方法もまだない．

Eremenko[6,10]は，$\alpha = \beta$ であろうと予想し，さらに α あるいは β がわからなくてもこれを証明する（あるいは予想が成り立たないと反証する）ことは可能であることにも言及した．$\alpha = \beta$ かどうかという問題は実際に α, β 自身の値を求めるよりもたぶんやさしい．

1つの関連のない問題でしめくくる．すべての $f \in S$ に対して

$$\int_{|z| \leq 1} |f'(z)|^\lambda dx\, dy < \infty$$

[†] 訳注：最後の結果は厳密ではなかった．現在の最良の限界は $0.54 < \gamma_k < 0.83$ である．

となるような実数 λ の集合を考える. Brennan（ブレナン）[20-22] は，この積分があ
る $\delta>0$ に対して $-1-\delta<\lambda<2/3$ を満たす λ に対しては有限であり，$\lambda=2/3$ あるいは
$\lambda=-2$ のときは無限になることを証明した. 彼は，この積分が $-2<\lambda<2/3$（すなわち
$\delta=1$ ととってよい）とき有限であろうと予想した. δ の最良値は未知のままである.

[1] C. A. Berenstein and R. Gay, *Complex Variables: An Introduction*, Springer-Verlag, 1991, pp. 134-136; MR 92f:30001.
[2] J. E. Littlewood, Some conjectural inequalities with applications to the theory of integral functions, *J. London Math. Soc.* 27 (1952) 387-393; also in *Collected Papers*, v. 2, Oxford Univ. Press, 1982, pp. 1296-1303; MR 14,154f.
[3] A. E. Eremenko and M. L. Sodin, Hypothesis of Littlewood and distribution of values of entire functions, *Funct. Anal. Appl.* 20 (1986) 60-62; MR 87f:30070.
[4] A. E. Eremenko and M. L. Sodin, A proof of the conditional Littlewood theorem on the distribution of the values of entire functions, *Math. USSR-Izv.* 30 (1988) 395-402; MR 88m:30073.
[5] J. L. Lewis and J.-M.Wu, On conjectures of Arakelyan and Littlewood, *J. d'Analyse Math.* 50 (1988) 259-283; MR 89i:30038.
[6] A. E. Eremenko, Lower estimate in Littlewood's conjecture on the mean spherical derivative of a polynomial and iteration theory, *Proc. Amer. Math. Soc.* 112 (1991) 713-715; MR 92k:30008.
[7] I.N. Baker and G. M. Stallard, Error estimates in a calculation of Ruelle, *Complex Variables Theory Appl.* 29 (1996) 141-159; MR 97b:30037.
[8] Y. M. Chen and M. C. Liu, On Littlewood's conjectural inequalities, *J. London Math. Soc.* 1 (1969) 385-397; MR 43 # 2192.
[9] W. K. Hayman, On a conjecture of Littlewood, *J. d'Analyse Math.* 36 (1979) 75-95; MR 81j:30042.
[10] A. E. Eremenko, Some constants coming from the work of Littlewood, unpublished note (1999).
[11] L. de Branges, A proof of the Bieberbach conjecture, *Acta Math.* 154 (1985) 137-152; MR 86h: 30026.
[12] A. Z. Grinshpan, The Bieberbach conjecture and Milin's functionals, *Amer. Math. Monthly* 106 (1999) 203-214; MR 2000b:30027.
[13] L. Bieberbach, Über die Koeffizienten derjenigen Potenzreihen welche eine schlichte Abbildungen des Einheitskreises vermitteln, *Sitzungsber. Preuss. Akad. Wiss.* (1916) 940-955.
[14] J. E. Littlewood, On the coefficients of schlicht functions, *Quart. J. Math.* 9 (1938) 14-20.
[15] Ch. Pommerenke, On the coefficients of univalent functions, *J. London Math. Soc.* 42 (1967) 471-474; MR 36 # 5329.
[16] J. Clunie and Ch. Pommerenke, On the coefficients of univalent functions, *Michigan Math.* J. 14 (1967) 71-78; MR 34 # 7786.
[17] Ch. Pommerenke, *Univalent Functions*, Vandenhoeck and Ruprecht, 1975, pp. 125-137; MR 58 # 22526.
[18] P. L. Duren, *Univalent Functions*, Springer-Verlag, 1983, pp. 1-40, 234-243; MR 85j:30034.
[19] L. Carleson and P. W. Jones, On coefficient problems for univalent functions and conformal dimension, *Duke Math. J.* 66 (1992) 169-206; MR 93c:30022.
[20] J. E. Brennan, The integrability of the derivative in conformal mapping, *J. London Math. Soc.* 18 (1978) 261-272; MR 80b:30009.
[21] Ch. Pommerenke, On the integral means of the derivative of a univalent function. II, *Bull. London Math. Soc.* 17 (1985) 565-570; MR 87f:30050b.
[22] L. Carleson and N. G. Makarov, Some results connected with Brennan's conjecture, *Ark. Mat.* 32 (1994) 33-62; MR 95g:30030.

7.7 リース-コルモゴロフ定数

$F(z) = f(z) + i\tilde{f}(z)$ を単位閉円板上で定義された解析関数でその虚部が $\tilde{f}(0) = 0$ という性質をもつものとする。p-Hardy（ハーディ）ノルム[1,2]を

$$\|f\|_p = \left(\frac{1}{2\pi}\int_0^{2\pi}|f(e^{i\theta})|^p d\theta\right)^{1/p} \qquad (0 < p < \infty)$$

と定義する。共役な関数 $f, \tilde{f}$ の（このノルムでの）相対的な大きさについてどんなことがいえるのだろうか？ Riesz（リース）[3]は，

$$\|\tilde{f}\|_p \leq C_p \cdot \|f\|_p \qquad (1 < p < \infty)$$

を証明し，Pichorides（ピコライズ）[4]と Cole（コール）[5]は，この不等式の最良の定数は

$$C_p = \begin{cases} \tan\left(\dfrac{\pi}{2p}\right) & (1 < p \leq 2) \\ \cot\left(\dfrac{\pi}{2p}\right) & (2 < p < \infty) \end{cases}$$

である[6-8]ことを示した。$p=1$ のときは，$\|f\|_1 < \infty$ であるが $\|\tilde{f}\|_1 = \infty$ となるような関数 F が存在する。よってこの場合には，修正した「相対的な大きさ」というものが必要になる。

S を単位円周の可測な部分集合とするとき，$|S|$ を 2π で割った S のルベーグ測度とする。$t \geq 0$ に対して

$$S_t(f) = \{z : |f(z)| \geq t \text{ かつ } |z| = 1\}$$

と定義する。Kolmogorov（コルモゴロフ）[9]は**弱型 1-1 不等式**

$$|S_t(\tilde{f})| \leq C_1 \cdot \frac{1}{t} \cdot \|f\|_1$$

がすべての $t > 0$ に対して成立することを証明し，Davis（デービス）[10]が，最良定数は，

$$C_1 = \frac{\pi^2}{8G} = 1.3468852519\cdots = (0.7424537454\cdots)^{-1}$$

であることを示した。ここで G はカタランの定数【1.7】である。コルモゴロフの定理の系として

$$\|\tilde{f}\|_p \leq C_p \cdot \|f\|_1 \qquad (0 < p < 1)$$

である。Davis[11,12]は，ここでの最良定数は，

$$C_p = \left(\frac{1}{2\pi}\int_0^{2\pi}|\operatorname{cosec}(\theta)|^p d\theta\right)^{1/p} = \left(\frac{1}{\sqrt{\pi}}\frac{\Gamma\left(\dfrac{1-p}{2}\right)}{\Gamma\left(\dfrac{2-p}{2}\right)}\right)^{1/p}$$

であることを示した．ここで $\Gamma(x)$ はガンマ関数【1.5.4】である．F と f の相対的大きさに関する問題があるが，それについては議論しない．[13-17] も参照してほしい．

[1] A. G. Zygmund, *Trigonometric Series*, v. 1, 2nd ed., Cambridge Univ. Press, 1959, pp. 131, 252, 377; MR 89c:42001.
[2] P. L. Duren, *Theory of H_p Spaces*, Academic Press, 1970, pp. 53-58; MR 42 #3552.
[3] M. Riesz, Sur les fonctions conjuguées, *Math. Z.* 27 (1927) 218-244.
[4] S. K. Pichorides, On the best values of the constants in the theorems of M. Riesz, Zygmund and Kolmogorov, *Studia Math.* 44 (1972) 165-179 (errata insert); MR 47 #702.
[5] T. W. Gamelin, *Uniform Algebras and Jensen Measures*, Cambridge Univ. Press, 1978, pp. 107-145; MR 81a:46058.
[6] I. E. Verbickii, Estimate of the norm of a function in a Hardy space in terms of the norms of its real and imaginary parts (in Russian), *Mat. Issled.* 54 (1980) 16-20, 164-165; MR 81k:30046.
[7] M. Esśen, A superharmonic proof of the M. Riesz conjugate function theorem, *Ark. Mat.* 22 (1984) 241-249; MR 86c:30068.
[8] M. Esśen, D. Shea, and C. Stanton, Best constant inequalities for conjugate functions, *J. Comput. Appl. Math.* 105 (1999) 257-264; MR 2000k:42008.
[9] A.N.Kolmogorov, Sur les fonctions harmoniques conjuguées et les séries de Fourier, *Fund. Math.* 7 (1925) 24-29; Engl. transl. in *Selected Works*, v. 1., ed. V. M. Tikhomirov, Kluwer, 1991, pp. 35-40; MR 93d:01096.
[10] B. Davis, On the weak type (1, 1) inequality for conjugate functions, *Proc. Amer. Math.* Soc. 44 (1974) 307-311; MR 50 #879.
[11] B. Davis, On Kolmogorov's inequalities $\|\tilde{f}\|_p \leq C_p \cdot \|f\|_1$, $0 < p < 1$, *Trans. Amer. Math. Soc.* 222 (1976) 179-192; MR 54 #10967.
[12] A. Baernstein, Some sharp inequalities for conjugate functions, *Indiana Univ. Math. J.* 27 (1978) 833-852; MR 80g:30022.
[13] C. Bennett, A best constant for Zygmund's conjugate function inequality, *Proc. Amer. Math. Soc.* 56 (1976) 256-260; MR 53 #6214.
[14] S. P. Grushevskii, A generalization of the Kolmogorov inequality, *Soobshch. Akad. Nauk Gruzin.* SSR 103 (1981) 33-36; MR 84h:42020.
[15] C. Choi, A weak-type inequality for differentially subordinate harmonic functions, *Trans. Amer. Math. Soc.* 350 (1998) 2687-2696; MR 99e:31006.
[16] A. B. J. Kuijlaars, Best constants in one-sided weak-type inequalities, *Methods Appl. Anal.* 5 (1998) 95-108; MR 99h:42014.
[17] R. Bañuelos and G. Wang, Davis's inequality for orthogonal martingales under differential subordination, *Michigan Math. J.* 47 (2000) 109-124; MR 2001g:60100.

7.8　グレッチュ環状領域定数

R を平面領域とする．すなわち R は複素数平面 $\mathbb{C}$ での連結開部分集合である．2つの領域 R_1, R_2 に対して1対1で上への解析写像 $f: R_1 \to R_2$ が存在するとき R_1 と R_2 とは**等角同値**である．明らかにこれは同値関係になる．有名なリーマンの写像定理から次の結果がわかる．

- 単連結領域に関しては，ちょうど2つの同値類が存在する．一方は $\mathbb{C}$ のみからなるもので，他方は単位円板（そして単位円板に等角同値なたくさんの領域）を含むものである．
- 二重連結領域に関しては，非可算無限の同値類が存在する．そしてそれぞれの同値類は，ある一意的に決まった実数 $r>1$ に対する円環領域 $A(1,r)=\{z:1<|z|<r\}$（そして $A(1,r)$ に等角同値なたくさんの領域）を含む．

とくに2つの円環領域 $A(s,t)$ と $A(u,v)$ が等角同値になるのは $t/s=v/u$ のときであり，このときに限る．すなわち外側の円の半径と内側の円の半径の比は等角不変量である [1,2]．

しばらく話題を変えよう．n 次元ユークリッド空間での**環状領域** R とは，その補集合が2つの成分 C_0, C_1 からなり，C_0 は有界で C_1 は非有界となる領域のこととする．B_0, B_1 を R の境界成分とする．R の**等角容量**（conformal capacity）を

$$\mathrm{cap}(R) = \inf_\varphi \int_R |\nabla \varphi|^n dx$$

と定義する．ここで下限は，B_0 上で 0，B_1 上で 1 となるような R 上のすべての実連続可微分関数をわたるものとする．R の**モジュラス**（modulus）を

$$\mathrm{mod}(R) = \left(\frac{\sigma}{\mathrm{cap}(R)}\right)^{1/(n-1)}$$

で定義する．ここで $\sigma = n\pi^{n/2}\Gamma(1+n/2)^{-1}$ は n 次元空間での半径 1 をもつ超球面の表面積である．n 次元超球環状領域 $A(s,t)$ に対しては，

$$\mathrm{mod}(A(s,t)) = \ln\left(\frac{t}{s}\right)$$

が知られている [3-8]．よって $n=2$ のとき環状領域のモジュラスは等角不変量である．$n \geq 3$ に対しては，もはやリーマンの写像定理が適用できないので，これをうまく幾何学的に解釈することはできない．n 次元単位超球は，別の超球あるいは半空間とのみ等角同値である．それにもかかわらずモジュラスは，別のところ（たとえば，擬等角写像に付随した歪曲定理において）で重要である．

$G(n, a)$ を n 次元 Grötzsch（**グレッチュ**）**環状領域**とする．すなわち，その補集合の成分が

$$C_0 = \{(x, 0, 0, \cdots, 0) : 0 \leq x \leq a\} \quad (\text{ここで } 0 < a < 1)$$
$$C_1 = \left\{(x_1, x_2, \cdots, x_n) : \sum_{i=1}^n x_i^2 \geq 1\right\}$$

となるような環状領域のことである．いいかえると $G(n, a)$ は，n 次元単位超球に，0 から a へ半径方向のベクトルに沿って切れ目を入れた領域である．次の極限が存在し，有限であることが知られている [7-15]．すなわち

$$\ln(\lambda_n) = \lim_{a \to 0^+} (\mathrm{mod}(G(n,a)) + \ln(a))$$

であり，これは，$\mathrm{mod}(G(n,a))$ が，a が 0 に減少していくにつれて対数的に大きくなることを示している．$n=2$ という特別な場合では，

7.8 グレッチュ環状領域定数

表 7.1 パラメータ λ_n の評価

n	下界	λ_n に対する最良の評価	上界
3	9.341	9.37 ± 0.02	9.9002
4	21.85	22.6 ± 0.2	26.046

$$\mathrm{mod}(G(2,a)) = \frac{\pi}{2} \frac{K(\sqrt{1-a^2})}{K(a)}$$

であり[4,13,14], これより $\lambda_2 = 4$ が出る. K は第 1 種完全楕円積分である. 同様な形のものが[4,5]にある. また興味深い漸近的な結果[9]

$$\lim_{n \to \infty} \lambda_n^{1/n} = e$$

もある. ここで e は自然対数の底【1.3】である.

λ_3 や λ_4 に対するこのような厳密な表現は知られていない. λ_n に対する精密な上界, 下界, そして最もよく知られている数値的な評価を表 7.1[12,15]に示す. $3 \leq n \leq 22$ のときの $\lambda_n \exp(-n)$ に対する評価の表が[14]にあり, これらは, 次の単純な不等式

$$2\exp(0.76(n-1)) \leq \lambda_n \leq 2\exp(n-1)$$

に沿ったものである. 最後に $n=2$ の場合に戻る. $A(1,r)$ を $G(2,a(r))$ 上に写像する等角写像 f に対する公式は何であろうか？ ここで $a(r)$ は以下で定義される切れ目の長さである. この写像は, Jacobi (ヤコビ) の楕円正弦関数 sn【1.4.6】と関連してくる. この方面の研究には, より高い超越性をもつ関数がしばしば登場する. $n \geq 3$ に対する適切な一般化の発見が待たれている.

7.8.1 $a(r)$ に対する公式

円環領域 $A(1,r)$ とグレッチュ環状領域 $G(2,a)$ は

$$\ln(r) = \mathrm{mod}(A(1,r)) = \mathrm{mod}(G(2,a)) = \frac{\pi}{2} \frac{K(\sqrt{1-a^2})}{K(a)}$$

を満たすとき, そしてそのときに限り等角同値になる. a を r の関数として解きたい. $a(r)$ は, 無限積として次のように書ける[14,16]. すなわち

$$a(r) = \frac{2b(r)}{1+b(r)^2}, \quad \text{ここで } b(r) = \frac{2}{r} \prod_{j=1}^{\infty} \left(\frac{1+r^{-8j}}{1+r^{-8j+4}} \right)^2$$

その補集合成分として

$$D_0 = \{(x,0,0,\cdots,0) : -b \leq x \leq b\} \quad (\text{ここで } 0 < b < 1)$$
$$D_1 = \left\{ (x_1, x_2, \cdots, x_n) : \sum_{i=1}^{n} x_i^2 \geq 1 \right\}$$

をもつ環状領域 $H(n,b)$ を考える. すなわち $H(n,b)$ は, n 次元単位超球に $-b$ から原点, b から原点という対称的な切れ目を入れた領域である. すると $H(2,b(r))$, $A(1,r)$ そして $G(2,a(r))$ は互いに等角同値になる. ひょっとすると $\mathrm{mod}(G(n,a))$ に対する結果と同様の結果が $\mathrm{mod}(H(n,b))$ に対してもあるのかもしれない. [17,18] も参照してほしい.

[1] J. B. Conway, *Functions of One Complex Variable*, 2nd ed., Springer-Verlag, 1978, pp. 160-164; MR 80c:30003.
[2] P. K. Kythe, *Computational Conformal Mapping*, Birkhäuser, 1998, pp. 295-303; MR 99k:65027.
[3] F. W. Gehring, Symmetrization of rings in space, *Trans. Amer. Math. Soc.* 101 (1961) 499-519; MR 24 # A2677.
[4] O. Lehto and K. I. Virtanen, *Quasiconformal Mappings in the Plane*, Springer-Verlag, 1973, pp. 30-33, 59-62; MR 49 # 9202.
[5] G. D. Anderson, Symmetrization and extremal rings in space, *Annales Acad. Sci. Fenn. Ser. A I* 438 (1969) 1-24; MR 41 # 465.
[6] G. D. Anderson, Extremal rings in n-space for fixed and varying n, *Annales Acad. Sci. Fenn. Ser. A I* 575 (1974) 1-21; MR 53 # 799.
[7] G. D. Anderson, Limit theorems and estimates for extremal rings of high dimension, *Romanian-Finnish Seminar on Complex Analysis*, Proc. 1976 Bucharest conf., ed. C. A. Cazacu, A. Cornea, M. Jurchescu, and I. Suciu, Lect. Notes in Math. 743, Springer-Verlag, 1979, pp. 10-34; MR 81c: 30042.
[8] P. Caraman, *n-Dimensional Quasiconformal (QCF) Mappings*, Editura Academiei/Abacus Press, 1974, pp. 51-52, 232; MR 50 # 10249.
[9] G. D. Anderson, Dependence on dimension of a constant related to the Grötzsch ring, *Proc. Amer. Math. Soc.* 61 (1976) 77-80; MR 56 # 603.
[10] G.D. Anderson and M. K.Vamanamurthy, Estimates for the asymptotic order of a Grötzsch ring constant, *Tôhoku Math. J.* 34 (1982) 133-139; MR 83i:30012.
[11] G. D. Anderson and M. K. Vamanamurthy, Asymptotic estimates for moduli of extremal rings, *Tôhoku Math. J.* 37 (1985) 533-540; MR 87d:30020.
[12] G. D. Anderson and J. S. Frame, Numerical estimates for a Grötzsch ring constant, *Constr. Approx.* 4 (1988) 223-242; MR 89i:30017.
[13] M. Vuorinen, *Conformal Geometry and Quasiregular Mappings*, Lect. Notes in Math. 1319, Springer-Verlag, 1988, pp. 65-67, 88-89; MR 89k:30021.
[14] G. D. Anderson, M. K. Vamanamurthy, and M. Vuorinen, *Conformal Invariants, Inequalities, and Quasiconformal Maps*, Wiley, 1997, pp. 119-120, 149-151, 166-169, 247-264; MR 98h:30033.
[15] K. Samuelsson and M. Vuorinen, Computation of capacity in 3D by means of a posteriori estimates for adaptive FEM, Royal Institute of Technology preprint TRITA-NA-9508 (1995).
[16] Z. Nehari, *Conformal Mapping*, Dover, 1975, pp. 280-296, 333-336; MR 51 # 13206.
[17] R. Wegmann, An iterative method for the conformal mapping of doubly connected regions, *J. Comput. Appl. Math.* 14 (1986) 79-98; *Numerical Conformal Mapping*, ed. L. N. Trefethen, North-Holland, 1986, 79-98; MR 87e:30008.
[18] N. Papamichael, I. E. Pritsker, E. B. Saff, and N. S. Stylianopoulos, Approximation of conformal mappings of annular regions, *Numer. Math.* 76 (1997) 489-513;MR98e:65016.

8 幾何学に関する定数

8.1 幾何確率定数

幾何確率[1]の広範な話題をほんの手短に述べるつもりだが，ここで2,3の問題を紹介するには充分である．

n 次元単位立方体から1点がランダムに選ばれたとする．この点と立方体の中心とのユークリッド距離の期待値 $\delta(n)$ は，次のようなきちんとした形をもっている[2-7]．すなわち

$$\delta(1) = \frac{1}{4}, \quad \delta(2) = \frac{1}{6}(\sqrt{2} + \ln(1+\sqrt{2})) = 0.3825978582\cdots$$

$$\delta(3) = \frac{1}{48}(6\sqrt{3} + 12\ln(2+\sqrt{3}) - \pi) = 0.4802959782\cdots$$

である．(すべての n に対して) 次の不等式，漸近値がある．

$$\frac{1}{4}n^{1/2} \leq \delta(n) \leq \frac{1}{2}\left(\frac{n}{3}\right)^{1/2}, \quad \delta(n) \sim \frac{1}{2}\left(\frac{n}{3}\right)^{1/2}$$

(とくに $\delta(n)$ は有界ではない．) $\delta(4) = 0.5609498093\cdots$，$\delta(5) = 0.6312033175\cdots$ をきちんとした形で表現することが可能であろうか？ ついでながら，$2\delta(n)$ は，この点から n 立方体の任意な頂点への平均距離になる．n 次元単位球に対して同様の問題を考えると[8-14]，そのユークリッド距離の期待値は $n/(n+1)$ となる (これは明らかに有界である)．

n 次元単位立方体から2点が独立かつ一様に選ばれたとする．これら2点間のユークリッド距離の期待値 $\Delta(n)$ は，

$$\varDelta(1)=\frac{1}{3}, \quad \varDelta(2)=\frac{1}{15}(\sqrt{2}+2+5\ln(1+\sqrt{2}))=0.5214054331\cdots$$

$$\varDelta(3)=\frac{1}{105}(4+17\sqrt{2}-6\sqrt{3}+21\ln(1+\sqrt{2})+42\ln(2+\sqrt{3})-7\pi)=0.6617071822\cdots$$

となり，対応する評価式，漸近値はそれぞれ

$$\frac{1}{3}n^{1/2}\leq\varDelta(n)\leq\left(\frac{n}{6}\right)^{1/2}, \quad \varDelta(n)\sim\left(\frac{n}{6}\right)^{1/2}$$

となる．$\varDelta(4)=0.7776656535\cdots$ や $\varDelta(5)=0.8785309152\cdots$ を，きちんとした形で表現することができるか？　この問題を n 次元単位球にしたときについては，もっと多くのことが知られている．このときの平均距離は，ガンマ関数の値の比となり，$n\to\infty$ のとき $\sqrt{2}$ に収束する．n が大きくなるとき，n 次元球に対する $\varDelta(n)$ は有限であるが，n 次元立方体のときの $\varDelta(n)$ は無限になることは，とても興味深い．さらに n 次元球での点どうしを隔てている距離の分散は0に近づく．よって大きな n に対して，無作為に選んだ2点の距離は，2つの直交する半径の端の点どうしの距離にほとんど常に等しくなる[1]．

3次元単位立方体の無作為に選んだ2点間のユークリッド距離の逆数の期待値は，

$$2\left(\frac{\sqrt{2}+1-2\sqrt{3}}{5}-\frac{\pi}{3}-\ln[(\sqrt{2}-1)(2-\sqrt{3})]\right)=1.8823126444\cdots$$

であり[15,16]，これは明らかに一般化が可能である．

(2点ではなく) 3点が n 次元単位立方体から無作為に選ばれたとする．これらの3点が鈍角三角形をなす確率 $\Pi(n)$ は何になるか？　Langford (ランフォード)[17,18]が

$$\Pi(2)=\frac{97}{150}+\frac{\pi}{40}=0.7252064830\cdots$$

であることを証明したが，$n>2$ での $\Pi(n)$ に対する同様な計算は，なされていない．この問題の n 次元球版については，より多くのことが知られている[19,20]．n が十分大きいときには，n 次元球での無作為に選んだ三角形は鋭角三角形になりがちである．なぜなら n 次元球の体積の大部分がその表面近くにあるからである[21]．実際このように無作為に選んだ三角形は，だいたい正三角形になる傾向があるので鈍角三角形になる確率は小さい．関連した議論については[22-28]を参照してほしい．

今度は n 次元単位立方体から無作為に N 個の点 $p_1, p_2, \cdots, p_N$ を選ぶことにする．$p_1, p_2, \cdots, p_N$ の凸胞を C と書くことにする．すなわち，

$$C=\left\{\sum_{j=1}^{N}\lambda_j p_j : \text{すべての } j \text{ に対して } \lambda_j\geq 0 \text{ であり } \sum_{j=1}^{N}\lambda_j=1 \text{ をみたす}\right\}$$

は，$p_1, p_2, \cdots, p_N$ を含むすべての凸集合の交わりである．このとき Rényi & Sulanke (レーニイ&スランケ)[29-39]によれば

・C の n 次元体積の期待値 $\mathrm{E}(V_n(N))$
・C の $(n-1)$ 次元表面積の期待値 $\mathrm{E}(S_n(N))$
・C の (多角形的) 境界上の頂点の個数の期待値 $\mathrm{E}(P_n(N))$

は，それぞれ

$$\lim_{N\to\infty}\frac{N}{\ln(N)}(1-\mathrm{E}(V_2(N)))=\frac{8}{3}$$

$$\lim_{N\to\infty}\sqrt{N}(4-\mathrm{E}(S_2(N)))=2\sqrt{\pi}M=4.2472965459\cdots$$

$$\lim_{N\to\infty}\mathrm{E}(P_2(N))-\frac{8}{3}\ln(N)=\frac{8}{3}(\gamma-\ln(2))=-0.3091507084\cdots$$

を満たす．ここで γ はオイラー-マスケローニの定数【1.5】であり，M はガウスのレムニスケート定数【6.1】である．Affentranger & Wieacker（アフェントランジャ&ウィッカー）[40,41] は，$n\geq 3$ のときの $V_n(N)$ と $P_n(N)$ の漸近値を得た．Cabo & Groeneboom（カボ&グロエンブーム）[42-45]は，

$$I=\sqrt{\frac{\pi}{8}}\left[2-\int_1^\infty(\sqrt{1+s^2}-s)s^{-3/2}ds\right]=\frac{\sqrt{\pi}}{2}M=1.0618241364\cdots$$

$$J=2-4\int_1^\infty(\sqrt{1+s^2}-s)\varphi(s-1)\,ds+\frac{4}{5}\int_1^\infty(\sqrt{1+s^2}-s)^2 s^{-2}ds$$

$$+\frac{1}{4}\int_1^\infty\int_1^t(\sqrt{1+s^2}-s)(\sqrt{1+t^2}-t)\psi\left(\frac{t}{s}-1\right)s^{-3}ds\,dt$$

$$+\frac{1}{8}\int_1^\infty\int_1^\infty(\sqrt{1+s^2}-s)(\sqrt{1+t^2}-t)\psi(st-1)\,ds\,dt$$

$$=1.37575\cdots$$

そして

$$\varphi(s)=\frac{1}{2(s+1)^2}-\frac{1}{4s(s+1)}+\frac{1}{4s}\frac{\arctan(\sqrt{s})}{\sqrt{s}}$$

$$\psi(s)=\frac{15}{s^3}+\frac{1}{s^2}-\left(\frac{15}{s^3}+\frac{6}{s^2}-\frac{1}{s}\right)\frac{\arctan(\sqrt{s})}{\sqrt{s}}$$

とするとき

$$\lim_{N\to\infty}N\,\mathrm{Var}(S_2(N))=4(J-I^2)=0.9932\cdots$$

であることを示した．この結果の高次元版については何もわかっていない．

今度は，正方形内に無作為に引かれた N 本の「直線」があるとする[46,47]．これらの直線は正方形を領域に分けるが，この領域の個数の平均は[48,49]により

$$\frac{N(N-1)\pi}{16}+N+1$$

と与えられる．ここにも幾何学での Archimedes（アルキメデス）の定数 π が出現するのは興味深い．立方体を N 個の無作為に選んだ「平面」で分割するときの領域の数の平均は，

$$\frac{(2N+23)N(N-1)\pi}{324}+N+1$$

である．これらの高次元版はどうなるのか？　起こりうる領域の数の最大値に関する事柄が[50-52]にある．

最後に（実際には幾何確率に関して生じたものではないが）異なる型の問題を述べる．

ここで問題になるのはその存在性である.面積 C の任意な平面可測集合が,面積がちょうど 1 のある三角形の三頂点を必ず含むという性質を満たす正の数 C は存在するか? Erdös (エルデーシュ) [53,54] は, C は $4\pi/(3\sqrt{3})$ ぐらい小さくなるのではと考えたが,今までのところ C が有限かどうか決定することにも何の進歩もない.ユークリッド平面で面積 1 をもつ凸領域は,面積がたかだか 2 である三角形に内接しうるかどうかという関連のある問題がかなり以前に解決された[55,56]. 3 次元のときどうかということは未解決のままである[57].

[1] L. A. Santaló, *Integral Geometry and Geometric Probability*, Addison-Wesley, 1976, pp. 21-26, 49-58, 212-213, 294-295; MR 55 #6340.
[2] B. Ghosh, Random distances within a rectangle and between two rectangles, *Bull. Calcutta Math. Soc.* 43 (1951) 17-24; MR 13,475a.
[3] R. S. Anderssen, R. P. Brent, D. J. Daley, and P. A. P. Moran, Concerning $\int_0^1 \cdots \int_0^1 (x_1^2 + \cdots + x_k^2)^{1/2} dx_1 \ldots dx_k$ and a Taylor series method, *SIAM J. Appl. Math.* 30 (1976) 22-30; MR 52 #15773.
[4] H. J. Oser and D. J. Daley, An average distance, *SIAM Rev.* 18 (1976) 497-500.
[5] D. P. Robbins and T. S. Bolis, Average distance between two points in a box, *Amer. Math. Monthly* 85 (1978) 277-278.
[6] R. E. Pfiefer, A. L. Holshouser, L. R. King, and B. G. Klein, Minimum average distance between points in a rectangle, *Amer. Math. Monthly* 96 (1989) 64-65.
[7] S. R. Dunbar, The average distance between points in geometric figures, *College Math. J.* 28 (1997) 187-197; MR 98a:52007.
[8] R. D. Lord, The distribution of distance in a hypersphere, *Annals of Math. Statist.* 25 (1954) 794-798; MR 16,377d.
[9] J. M. Hammersley, The distribution of distance in a hypersphere, *Annals of Math. Statist.* 21 (1950) 447-452; MR 12,268e.
[10] M. G. Kendall and P. A. P. Moran, *Geometrical Probability*, Hafner, 1963, pp. 41-42, 53-55; MR 30 #4275.
[11] V. S. Alagar, The distribution of the distance between random points, *J. Appl. Probab.* 13 (1976) 558-566; MR 54 #6225.
[12] R. Pinkham and R. Holzsager, A random distance, *Amer. Math. Monthly* 106 (1999) 171-172.
[13] B. Eisenberg and R. Sullivan, Crofton's differential equation, *Amer. Math. Monthly* 107 (2000) 129-139; MR 2001g:60021.
[14] K. Brown, Distances in bounded regions; Distributions of distances (MathPages).
[15] H. Essén and A. Nordmark, Some results on the electrostatic energy of ionic crystals, *Canad. J. Chem.* 74 (1996) 885-891.
[16] Z. F. Seidov and P. I. Skvirsky, Gravitational potential and energy of homogeneous rectangular parallelepiped, astro-ph/0002496.
[17] E. Langford, The probability that a random triangle is obtuse, *Biometrika* 56 (1969) 689-690.
[18] E. Langford, A problem in geometrical probability, *Math Mag.* 43 (1970) 237-244; MR 45 #7774.
[19] G. R. Hall, Acute triangles in the n-ball, *J. Appl. Probab.* 19 (1982) 712-715; MR 83h:60016.
[20] C. Buchta, A note on the volume of a random polytope in a tetrahedron, *Illinois J. Math.* 30 (1986) 653-659; MR 87m:60038.
[21] B. Eisenberg and R. Sullivan, Random triangles in n dimensions, *Amer. Math. Monthly* 103 (1996) 308-318; MR 96m:60025.
[22] V. S. Alagar, On the distribution of a random triangle, *J. Appl. Probab.* 14 (1977) 284-297; MR 55 #9203.
[23] D. G. Kendall, Exact distributions for shapes of random triangles in convex sets, *Adv. Appl. Probab.* 17 (1985) 308-329; MR 86m:60033.

[24] D. G.Kendall and H.-L. Le, Exact shape-densities for random triangles in convex polygons, *Adv. Appl. Probab.* Suppl. (1986) 59-72; MR 88h:60022.
[25] R. K. Guy, There are three times as many obtuse-angled triangles as there are acute-angled ones, *Math. Mag.* 66 (1993) 175-178.
[26] S. Portnoy, A Lewis Carroll pillow problem: Probability of an obtuse triangle, *Statist. Sci.* 9 (1994) 279-284; MR 95h:60003.
[27] D. E. Dobbs and J. L. Zarestky, Acute random triangles, *Nieuw Arch. Wisk.* 15 (1997) 141-162.
[28] Z. F. Seidov, Random triangle in square: Geometrical approach, math.GM/0002134.
[29] A. Rényi and R. Sulanke, Über die konvexe Hülle von n zufällig gewählten Punkten. I, *Z. Wahrsch. Verw. Gebiete* 2 (1963) 75-84; II, 3 (1964) 138-147; also in *Selected Papers of Alfréd Rényi*, v. 3, Akadémiai Kiadó, 1976, pp. 143-152 and 242-251; MR 27 #6190 and MR 29 #6392.
[30] B. Efron, The convex hull of a random set of points, *Biometrika* 52 (1965) 331-343; MR 34 #6820.
[31] H. Raynaud, Sur l'enveloppe convexe des nuages de points aléatoires dans $\mathbb{R}^n$. I, *J. Appl. Probab.* 7 (1970) 35-48; MR 41 #2736.
[32] H. Carnal, Die konvexe Hülle von n rotationssymmetrisch verteilten Punkten, *Z. Wahrsch. Verw. Gebiete* 15 (1970) 168-176; MR 44 #3367.
[33] H. Ziezold, Über die Eckenanzahl zufälliger konvexer Polygone, *Izv. Akad. Nauk Armjan. SSR Ser. Mat.* 5 (1970) 296-312; MR 44 #4800.
[34] C. Buchta, Stochastische Approximation konvexer Polygone, *Z. Wahrsch. Verw. Gebiete* 67 (1984) 283-304; MR 85k:60021.
[35] R. A. Dwyer, Convex hulls of samples from spherically symmetric distributions, *Discrete Appl. Math.* 31 (1991) 113-132; MR 92i:60020.
[36] D. J. Aldous, B. Fristedt, P. S. Griffin, and W. E. Pruitt, The number of extreme points in the convex hull of a random sample, *J. Appl. Probab.* 28 (1991) 287-304; MR 92j:60022.
[37] W.Weil and J. A.Wieacker, Stochastic geometry, *Handbook of Convex Geometry*, v. B, ed. P. M. Gruber and J. M. Wills, North-Holland, 1993, pp. 1391-1431.
[38] R. Schneider, Discrete aspects of stochastic geometry, *Handbook of Discrete and Computational Geometry*, ed. J. E. Goodman and J. O'Rourke, CRC Press, 1997, pp. 167-184.
[39] A. M. Mathai, *An Introduction to Geometrical Probability: Distributional Aspects with Applications*, Gordon and Breach, 1999, pp. 364-383; MR 2001a:60014.
[40] F. Affentranger and J. A. Wieacker, On the convex hull of uniform random points in a simple d-polytope, *Discrete Comput. Geom.* 6 (1991) 291-305; MR 92c:52004.
[41] B. F. vanWel, The convex hull of a uniform sample from the interior of a simple d-polytope, *J. Appl. Probab.* 26 (1989) 259-273; MR 91a:60040.
[42] P. Groeneboom, Limit theorems for convex hulls, *Probab. Theory Relat. Fields* 79 (1988) 327-368; MR 89j:60024.
[43] A. J. Cabo and P. Groeneboom, Limit theorems for functionals of convex hulls, *Probab. Theory Relat. Fields* 100 (1994) 31-55; MR 95g:60017.
[44] I. Hueter, Limit theorems for the convex hull of random points in higher dimensions, *Trans. Amer. Math. Soc.* 351 (1999) 4337-4363; MR 2000a:52008.
[45] J. Keane, Convex hull integrals and the "ubiquitous constant," unpublished note (2000).
[46] S. Goudsmit, Random distribution of lines in a plane, *Rev. Mod. Phys.* 17 (1945) 321-322.
[47] H. Solomon, *Geometric Probability*, SIAM, 1978, pp. 39-44; MR 58 #7777.
[48] L. A. Santaló,Valor medio del numero de regiones en que un cuerpo del espacio es dividido por n planos arbitrarios, *Rev. Unión Mat. Argentina* 10 (1945) 101-108.
[49] L. A. Santaló, Sobre la distribucion de planos en el espacio, *Rev. Unión Mat. Argentina* 13 (1948) 120-124.
[50] A. M. Yaglom and I. M. Yaglom, *Challenging Mathematical Problems with Elementary Solutions*, v. 1, Dover, 1987, pp. 13, 102-106; MR 88m:00012a.
[51] D. J. Price, Some unusual series occurring in n-dimensional geometry, *Math. Gazette* 30 (1946) 149-150.

[52] H. Dörrie, *100 Great Problems of Elementary Mathematics: Their History and Solution*, Dover, 1965, pp. 283-285; MR 84b:00001.
[53] P. Erdös, Some combinatorial, geometric and set theoretic problems in measure theory, *Measure Theory*, Proc. 1983 Oberwolfach conf., ed. D. Kölzow and D. Maharam-Stone, Lect. Notes in Math. 1089, Springer-Verlag, 1984, pp. 321-327; MR 85m:28002.
[54] H. T. Croft, K. J. Falconer, and R. K. Guy, *Unsolved Problems in Geometry*, Springer-Verlag, 1991, sect. G13; MR 95k:52001.
[55] W. Gross, Über affine Geometrie. XIII: Eine Minimumeigenschaft der Ellipse und des Ellipsoids, *Berichte über die Verhandlungen der Königlich Sächsischen Gesellschaft der Wissenschaften zu Leipzig, Math.-Phys. Klasse* 70 (1918) 38-54.
[56] H. G. Eggleston, *Problems in Euclidean Space: Applications of Convexity*, Pergamon, 1957, pp. 149-160; MR 23 # A3228.
[57] E. Calabi, Circumscribing tetrahedron of least volume, unpublished note (1999).

8.2 円による被覆定数

単位区間 $[0,1]$ を 1 以下の等しい長さをもつ N 個の部分区間で完全に覆うという問題は簡単である．長さ $1/N$ の部分区間で充填（タイル貼り）をすればよい．それらの境界点においてのみ重なりが生じる．

単位円板 D を 1 以下の等しい半径をもつ N 個の円板で完全に覆うという問題は，もっと難しい．ここでは，重なりの部分が本質的であり，これがこの問題を解くのを難しくしている．$r(N)$ をこのような被覆が可能な最小の半径とする．D が被覆されれば，特にその境界 C（単位円）も覆われていなくてはならない．単位円の長さ $2\pi/N$ の弧を覆うためには，少なくとも $\sin(\pi/N)$ の半径をもつ円板が必要である．ゆえに $r(N) \geq \sin(\pi/N)$ とわかる．実際には $N=2,3,4$ のときは，等号が成立する（表 8.1）．$N=7$ のときも直接に示すことができる．C に内接する正六角形の一辺の長さは 1 なので，C を覆うには半径 $1/2$ の円板が少なくとも 6 個必要である．半径 $1/2$ の 7 番目の円板は，D の残った中心部分を十分覆うことができる．

$N=5$ のときが，自明でない最初の場合になる．Neville（ネビル）[1,2]が，初めて公刊された解答を与えた（図 8.1）．しかし最後の段階で $r(5)$ の正しくない値を示してしまった．[3]の初期の版では，この誤りをそのままくりかえした．正しくは $r(5)=$

表8.1 単位円板を被覆する N 個の小円板の最小共通半径 $r(N)$

N	1	2	3	4	5	6
$r(N)$	1	1	$\frac{\sqrt{3}}{2}=0.866025\cdots$	$\frac{\sqrt{2}}{2}=0.707106\cdots$	$0.609382\cdots$	$0.555905\cdots$

N	7	8	9	10	11	12
$r(N)$	$\frac{1}{2}=0.5$	$0.445041\cdots$	$0.414213\cdots$	$0.394930\cdots$	$0.380006\cdots$	$0.361103\cdots$

8.2 円による被覆定数

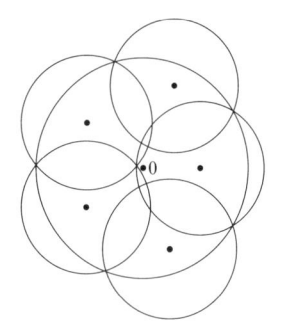

図 8.1 5つの円に対するネビルの極小配置．3つの円板は0の近くを通るが2つは通らないので，これは対称な図ではない．

$0.6093828640\cdots$ であり，これは $\cos(\theta+\varphi/2)$ の値として得られる．ここで θ と φ は，次の4つの未知数をもつ4つの非線形方程式系

$$2\sin(\theta)-\sin\left(\theta+\frac{1}{2}\varphi+\psi\right)-\sin\left(\psi-\theta-\frac{1}{2}\varphi\right)=0$$

$$2\sin(\varphi)-\sin\left(\theta+\frac{1}{2}\varphi+\chi\right)-\sin\left(\chi-\theta-\frac{1}{2}\varphi\right)=0$$

$$2\sin(\theta)+\sin(\chi+\theta)-\sin(\chi-\theta)-\sin(\psi+\varphi)-\sin(\psi-\varphi)-2\sin(\psi-2\theta)=0$$

$$\cos(2\psi-\chi+\varphi)-\cos(2\psi+\chi-\varphi)-2\cos(2\chi)+\cos(2\psi+\chi-2\theta)+\cos(2\psi-\chi-2\theta)=0$$

の解である．

別の特徴づけが Bezdek（ベズデク）[4-6]によって与えられた．すなわち $r(5)^{-1}$ は，

$a(y)=80y^2+64y,\quad b(y)=416y^3+384y^2+64y$

$c(y)=848y^4+928y^3+352y^2+32y$

$d(y)=768y^5+992y^4+736y^3+288y^2+96y$

$e(y)=256y^6+384y^5+592y^4+480y^3+336y^2+96y+16$

$f(y)=128y^5+192y^4+256y^3+160y^2+96y+32,\quad g(y)=64y^2+64y+16$

とし，$\sqrt{2}<x<2y+1,\ -1<y<1$ という条件を満たすすべての y に対して最大となるような多項式

$$a(y)x^6-b(y)x^5+c(y)x^4-d(y)x^3+e(y)x^2-f(y)x+g(y)$$

の最大の実の零点である．

Neville[2]は，$r(5)$ が代数的数であることを知っていた．なぜなら彼は自分の方程式系について「これらの特別な方程式は，明らかに代数的であり角度 $\theta/2, \varphi/4, \psi/2, \chi/2$ の正接に関して有理的でさえあるので，$\cos(\theta+\varphi/2)$ に対する代数方程式が見つけ出せる…」と記しているからである．Mellisen（メリセン）[7]と Zimmermann（ツィンマーマン）[8]は，$r(5)$ の最小多項式

$$1296x^8+2112x^7-3480x^6+1360x^5+1665x^4-1776x^3+22x^2-800x+625$$

を独立に得た．しかし彼らが，これを見つけた最初の人びとではなかったようである．

Zahn（ザーン）[9]は，コンピュータ実験により $N=6$ と $8≦N≦10$ のときの $r(N)$ を計算した．Bezdek[10]は，[5-7]で報告されているように数値的に $r(6)=0.559052114\cdots$ を得た．たぶん彼は，$r(5)$ と同様に $r(6)$ の多項式的な最適化の特徴づけを見出したようである．Nagy（ナジィ）[11]と Krotozynski（クロトジンスキー）[12]は，$8≦N≦10$ のとき

$$r(N)=\left(1+2\cos\left(\frac{2\pi}{N-1}\right)\right)^{-1}=\begin{cases} 0.4450418679\cdots & (N=8 \text{のとき}) \\ \sqrt{2}-1=0.4142135623\cdots & (N=9 \text{のとき}) \\ 0.3949308436\cdots & (N=10 \text{のとき}) \end{cases}$$

と予想し，Fejes Tóth（フェイエトス）[13]が $r(8)$ と $r(9)$ の公式を証明するのに成功した．$r(10)$ の公式に対する根拠が Melissen[7]により示された．彼は，この話題についてのみごとな総合報告を書いた．ごく最近 Faugère & Zimmermann（フォウゲール&ツィンマーマン）[14]は，$r(6)$ の最小多項式が

$$7841367x^{18}-3344997x^{16}+62607492x^{14}-63156942x^{12}+41451480x^{10}$$
$$-19376280x^{8}+5156603x^{6}-746832x^{4}+54016x^{2}+3072$$

であることを発見した．$N≧10$ であるすべての $r(N)$ については，未解決である．Melissen & Schuur（メリセン&シューア）[7]により，$r(11)<(1+2\cos(\pi/5))^{-1}$ であり，

$$r(12)=\frac{1}{3}(1+(1+3\sqrt{57})^{1/3}-8(1+3\sqrt{57})^{-1/3})=0.3611030805\cdots$$

という予想もあることを述べておく．

Kerschner（カーシュナ）[15]と Verblunsky（ヴェルブランスキー）[16]による興味深い「逆」の結果がいくつかある．たとえば D を覆うのに必要な半径 ε の円板の最小個数を $N(\varepsilon)$ とすると，D の面積の，小さな円板の面積の総和に対する比の極限

$$\lim_{\varepsilon\to 0^+}\frac{\pi}{(\pi\varepsilon^2)N(\varepsilon)}=\lim_{\varepsilon\to 0^+}\frac{1}{\varepsilon^2 N(\varepsilon)}=\frac{3\sqrt{3}}{2\pi}=0.8269933431\cdots$$

は，被覆の「漸近的効率」を測るものと考えることができる．単位円板を正方形にかえると，より正確なことがいえる．

関連した問題を示そう．単位区間は，長さ $1/2, 1/4, 1/8, 1/16, 1/32, \cdots$ の区間により自然な方法で被覆できる．その上この共通の比 $1/2$ は，より小さいものにすることはできない．この結果の2次元版はどうなるのか？ Eppstein（エプステイン）[17]は，$s=0.77$ のとき，D は半径 s^k ($k=1,2,3,\cdots$) の円板によって被覆しうることと，$s=0.765$ のときは明らかに被覆できないことを示した．$s≦0.77$ を満たす最小の s の正確な値を見つけられればすばらしい．

[7,18]において N 個のより小さな等しい半径の円板で単位正方形を被覆する問題が紹介されている．単位円板[1,7,19-22]あるいは単位正方形[1,7,23-28]における「充填」円板という双対的問題に心が引かれるが，このことについては少しだけ述べる．単位正方形内の任意な N 個の点の間の最短距離のうちの最大なものを $t(N)$ とおく（表8.2を参照）．$t(10)=0.4212795439\cdots$ の計算が最近までの最も大きな障害であった．

8.2 円による被覆定数

表8.2 単位正方形に中心をもつ N 個の円板が正方形を充填するときの最大の共通直径 $t(N)$

N	2	3	4	5
$t(N)$	$\sqrt{2}=1.414213\cdots$	$\sqrt{6}-\sqrt{2}=1.035276\cdots$	1	$\dfrac{\sqrt{2}}{2}=0.707106\cdots$

N	6	7	8	9
$t(N)$	$\dfrac{\sqrt{13}}{6}=0.600925\cdots$	$2(2-\sqrt{3})=0.535898\cdots$	$\dfrac{\sqrt{6}-\sqrt{2}}{2}=0.517638\cdots$	$\dfrac{1}{2}=0.5$

Schlüter（シュルター）の予想[29, 30]が正しいことが証明され[31]，そして $t(10)$ に対する最小多項式は，

$1180129x^{18} - 11436428x^{17} + 98015844x^{16} - 462103584x^{15} + 1145811528x^{14} - 1398966480x^{13}$
$+ 227573920x^{12} + 1526909568x^{11} - 1038261808x^{10} - 2960321792x^{9} + 7803109440x^{8}$
$- 9722063488x^{7} + 7918461504x^{6} - 4564076288x^{5} + 1899131648x^{4} - 563649536x^{3}$
$+ 114038784x^{2} - 14172160x + 819200$

である．余談であるが，ちょうど2つの円のからむ2つの初等的な問題を述べよう．

それぞれ半径1の2つの円の重なりを想像しよう．内側の重なり合った部分の領域の面積 A が外側にある2つの三日月型の面積の和に等しいとすると，明らかに $A=2\pi/3$ である．このとき2つの円の中心間の距離 $2u$ はいくらか？ $u=0.2649320846\cdots$ であり，これは，方程式

$$u\sqrt{1-u^2} + \arcsin(u) = \frac{\pi}{6}$$

の区間 $[0,1]$ でのただ1つの解であることが証明できる．u は超越的か？ これは「Mrs. Miniver（ミニバー夫人）の問題」とよばれている[32, 33]．

2番目の問題は「放牧山羊の問題」とよばれている[34, 35]．山羊が半径1の円形放牧地の境界の杭につながれている．この山羊がこの放牧地の草のちょうど半分を食べることができるような綱の長さは何になるか？ 綱の長さ v は，

$$v\sqrt{4-v^2} - 2(v^2-2)\arccos\left(\frac{v}{2}\right) = \pi$$

を満たすので $v=1.1587284730\cdots$ である．v は超越的か？ v と u は代数的に独立か？

[1] H. T. Croft, K. J. Falconer, and R. K. Guy, *Unsolved Problems in Geometry*, Springer-Verlag, 1991, sect. D1, D2, D3; MR 95k:52001.
[2] E. H. Neville, Solutions of numerical functional equations, *Proc. London Math. Soc.* 14 (1915) 308-326.
[3] W. W. Rouse Ball, *Mathematical Recreations and Essays*, 11th ed., MacMillan, 1939, pp. 97-99; MR 88m:00013.
[4] J. Molnár, Über eine elementargeometrische Extremalaufgabe, *Mat. Fiz. Lapok* 49 (1942) 249-253; MR 8,218j.
[5] K. Bezdek, Über einige Kreisüberdeckungen, *Beiträge Algebra Geom.* 14 (1983) 7-13; MR 85a:

52012.
[6] K. Bezdek, Über einige optimale Konfigurationen von Kreisen, *Annales Univ. Sci. Budapest Eötvös Sect.* Math. 27 (1984) 143-151; MR 87f:52020.
[7] J. B. M. Melissen, *Packing and Coverings with Circles*, Ph.D. thesis, Universiteit Utrecht, 1997.
[8] P. Zimmermann, Computation of the minimal polynomial of Neville's five disc constant, unpublished note (1997).
[9] C. T. Zahn, Black box maximization of circular coverage, *J. Res. Nat. Bur. Standards B* 66 (1962) 181-216; MR 29 # 1583.
[10] K. Bezdek, *Körök optimális fedélsei*, Ph.D. thesis, Eötvös Loránd Univ., Budapest, 1979.
[11] D. Nagy, *Fedések és alkalmazásaik*, M.Sc. thesis, Eötvös Loránd Univ., Budapest, 1974.
[12] S. Krotoszynski, Covering a disk with smaller disks, *Studia Sci. Math. Hungar.* 28 (1993) 277-283; MR 95c:52035.
[13] G. Fejes Tóth, Thinnest covering of a circle by eight or nine congruent circles, unpublished note (1996).
[14] J.-C. Faugère and P. Zimmermann, The minimal polynomial of Bezdek's constant, unpublished note (1998).
[15] R. Kershner, The number of circles covering a set, *Amer. J. Math.* 61 (1939) 665-671; MR 1,8b.
[16] S. Verblunsky, On the least number of unit circles which can cover a square, *J. London Math. Soc.* 24 (1949) 164-170; MR 11,455g.
[17] D. Eppstein, Covering and Packing (Geometry Junkyard).
[18] J. B. M. Melissen and P. C. Schuur, Improved coverings of a square with six and eight equal circles, *Elec. J. Combin.* 3 (1996) R32; MR 97h:52027.
[19] S. Kravitz, Packing cylinders into cylindrical containers, *Math. Mag.* 40 (1967) 65-71.
[20] U. Pirl, Der Mindestabstand von n in der Einheitskreisscheibe gelegenen Punkten, *Math. Nachr.* 40 (1969) 111-124; MR 40 # 6379.
[21] H. Melissen, Densest packings of eleven congruent circles in a circle, *Geom. Dedicata* 50 (1994) 15-25; MR 95e:52032.
[22] R. L. Graham, B. D. Lubachevsky, K. J. Nurmela, and P. R. J. Östergård, Dense packings of congruent circles in a circle, *Discrete Math.* 181 (1998) 139-154; MR 99b: 52040.
[23] J. Schaer, The densest packing of 9 circles in a square, *Canad. Math. Bull.* 8 (1965) 273-277; MR 31 # 6164.
[24] M. Goldberg, The packing of equal circles in a square, *Math. Mag.* 43 (1970) 24-30.
[25] G.Wengerodt, Die dichteste Packung von 16 Kreisen in einem Quadrat, *Beiträge Algebra Geom.* 16 (1983) 173-190; 25 (1987) 25-46; MR 85j:52024 and MR 88g:52014.
[26] K. J. Nurmela, Packing up to 50 equal circles in a square, *Discrete Comput. Geom.* 18 (1997) 111-120; MR 98e:52020.
[27] K. J. Nurmela and P. R. J. Östergård, More optimal packings of equal circles in a square, *Discrete Comput. Geom.* 22 (1999) 439-457; MR 2000h:05052.
[28] K. J. Nurmela, P. R. J. Östergård, and R. aus dem Spring, Asymptotic behavior of optimal circle packings in a square, *Canad. Math. Bull.* 42 (1999) 380-385; MR 2000g: 52012.
[29] K. Schlüter, Kreispackung in Quadraten, *Elem. Math.* 34 (1979) 12-14; MR 80c:52014.
[30] M. Mollard and C. Payan, Some progress in the packing of equal circles in a square, *Discrete Math.* 84 (1990) 303-307; MR 92d:52043.
[31] R. Peikert, D.Würtz, M. Monagan, and C. de Groot, Packing circles in a square: A review and new results, *System Modelling and Optimization*, Proc 1991 Zurich conf., ed. P. Kall, Lect. Notes in Control and Inform. Sci. 180, Springer-Verlag, 1992, pp. 45-54.
[32] J. Struther, *Mrs. Miniver*, Harcourt Brace, 1940, ch. 12, pp. 94-95.
[33] L. A. Graham, *Ingenious Mathematical Problems and Methods*, Dover, 1959, pp. 6, 64-66; MR 22 # 1.
[34] M. Fraser, A tale of two goats, *Math. Mag.* 55 (1982) 221-227.
[35] M. Fraser, The grazing goat in n dimensions, *College Math. J.* 15 (1984) 126-134; MR 85k:00001.

8.3 普遍被覆定数

　平面上で直径1のすべての集合からなる族を U とする．平面領域 R が U のすべての集合の合同なコピーを含みうるとき，R は U に対する**移動被覆**（あるいは，**普遍被覆**）という．すなわち，直径1の集合は，平行移動と回転移動を適当に行うことにより R により被覆される[1-6]．

　S を平面上の特別な領域の族とする（たとえば，すべての円板の族）．S の元の中に，U に対する移動被覆であり，できるだけ小さな面積をもつものが存在するだろうか？もしこの答が肯定的なときは，そのような要素の面積を $A(S)$ と定義する．

　例として，すべての円板からなる族を考える [3, 4] と

$$A(円板の族)=\frac{\pi}{3}=1.0471975511\cdots$$

であり，これは半径が $1/\sqrt{3}$ の円の面積になる．同じような推論により一辺が1の長さの正方形で十分であることもわかる．すなわち

$$A(正方形の族)=1$$

となる．正六角形領域の族に対してはさらによい値

$$A(正六角形の族)=\frac{\sqrt{3}}{2}=0.8660254037\cdots$$

が得られる．

　さて今度は，平面凸領域すべてよりなる族 C を考えよう．Lebesgue（ルベーグ）[7] は，$\mu=A(C)$ の値，すなわち直径1の集合すべてを被覆することができる凸毛布の最小面積はどうなるかを考えた．Pál（パル）[7]，Sprague（スプラーグ）[8] と Hansen（ハンセン）[9, 10] による

$$0.8257117836\cdots=\frac{\pi}{8}+\frac{\sqrt{3}}{4}\leq\mu\leq\frac{\sqrt{3}}{2}-2\varepsilon_P-\varepsilon_S-\varepsilon_H\leq 0.84413770$$

が最もよく知られた限界である．下界は，直径1の正三角形とその重心を中心とする直径1の円との合併集合の凸胞の面積である．上界の評価は，Pál の考えた正六角形被覆の角を落としていくことによって改良を加えたものであり

$$\varepsilon_P=\frac{7\sqrt{3}}{12}-1\sim 10^{-2}, \quad \varepsilon_S\sim 10^{-3}$$

である．Sprague の上界の評価に関する Hansen の2つの改良は小さく，$\sim 10^{-19}$ と $\sim 10^{-11}$ である．より劇的な改良が[11]にあり，それは，0.8441 を 0.8430 にするというものであるが，これは予想されただけである．Hansen の研究について興味深い点の1つに，彼がコンピュータシミュレーションを用いたことがあげられる．たとえば[10]でシミュレーションによりある種の形を除外した．Hansen が1981年の自分の予想を

撤回したかどうかはっきりしない．Klee & Wagon（クレー & ワゴン）は[5]に，「この問題の進展は，過去においても，なかなか骨の折れるものであったが，これからはよりいっそう進展は遅くなるかもしれない」と書いている．

非凸の被覆に関しては，Duff（ダフ）[12,13]が $0.84413570\cdots$ という面積をもつ領域を構成した．これは凸領域に関して知られていたどの例よりも小さい値である．非凸性が事態を改善しうることは驚くべきことではない．【8.4】や【8.17】での関連した議論についても参照のこと．

これらの問題を変形したものが多数存在している．「被覆」の意味を（今までのように移動，すなわち平行移動と回転移動ではなく）平行移動のみに制限したときの様々な結果が【8.3.1】で与えられる．また，（被覆の）面積ではなく周の長さや平均幅を最小にするという問題を考えることもできる[14,16]．異なる意味での最小性（たとえば，真部分集合は被覆にならないような被覆）が，n 次元の場合について Eggleston（エグルストン）[17]によって研究された．

μ の上界の評価に関しては，食い違った報告がされている．Meschkowski（メシコフスキ）[2]と Hansen[9]は，Sprague の評価は 0.844144 であるといい，一方 Duff [12]と Klee & Wagon[5]は 0.84413770 であるといっている．この食い違いについては何の説明もされていない．

最後に，この話題についての初期の論文では，これを「ベシコヴィッチ（Besicovitch）の問題」[18-20]と間違って引用していることを指摘しておきたい．

8.3.1 平行移動による被覆

平面領域 R が U に対する**平行移動被覆**（あるいは**強普遍被覆**）であるというのは，直径が 1 である集合が，適当な平行移動により R によって覆われるようにできるときをいう[5,14,21]．回転移動は認めない．前に用いたのと同様の記号を使うと，円板が，明らかに回転対称であることにより $\tilde{A}$(円板の族)$=\pi/3$ とわかる．また $\tilde{A}$(正方形の族)$=1$ も成立する．しかし Pál の正六角形は，平行移動被覆「ではない」[21,22]．

では，$\tilde{A}$(正六角形の族) の値は，何になるのか？

すべての平面凸領域の族を C とすると，

$$\tilde{A}(C) = \frac{\pi}{6} + 2\sqrt{3} - 3 = 0.9877003907\cdots$$

であるという予想がある[5,15,16]．この値は，図 8.2 の切項単位正方形の面積である．$\tilde{A}(C)$ の厳密な評価は文献には現れていないようである．Watts（ワッツ）の正方形の穴をあける工具との奇妙な関係については[23]を参照のこと．Reulaux（ルーロー）三角形に関連した他の定数を【8.10】で示す．

8.3 普遍被覆定数

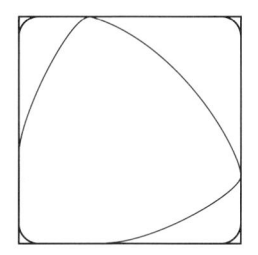

図 8.2 内接するルーロー三角形を完全に正方形内にあるように回転し，4つの角の集合を接触しないように取り除くことにより得られた切頂単位正方形．

[1] H. T. Croft, K. J. Falconer, and R. K. Guy, *Unsolved Problems in Geometry*, Springer-Verlag, 1991, sect. D15; MR 95k:52001.
[2] H. Meschkowski, *Unsolved and Unsolvable Problems in Geometry*, Oliver and Boyd, 1966; MR 35 # 7206.
[3] S. R. Lay, *Convex Sets and Their Applications*, Wiley, 1972; MR 93g:52001.
[4] I. M. Yaglom and V. G. Boltyanskii, *Convex Figures*, Holt, Rinehart and Winston, 1961, pp. 18, 31-33, 122-125, 140-156; MR 14,197d.
[5] V. Klee and S. Wagon, *Old and New Unsolved Problems in Plane Geometry and Number Theory*, Math. Assoc. Amer., 1991, pp. 23, 92-93; MR 92k:00014.
[6] C. S. Ogilvy, *Excursions in Geometry*, Dover, 1969, pp. 142-144.
[7] J. Pál, Über ein elementares Variationsproblem, *Mathematisk-Fysiske Meddelelser, Kongelige Danske Videnskabernes Selskabs*, v. 3 (1920) n. 2, 1-35.
[8] R. Sprague, Über ein elementares Variationsproblem, *Mathematisk Tidsskrift B* (1936) 96-99.
[9] H. C. Hansen, A small universal cover of figures of unit diameter, *Geom. Dedicata* 4 (1975) 165-172; MR 53 # 1423.
[10] H. C. Hansen, Small universal covers for sets of unit diameter, *Geom. Dedicata* 42 (1992) 205-213; MR 93c:52018.
[11] H. C. Hansen, Towards the minimal universal cover, *Normat* 29 (1981) 115-119, 148; MR 83g:52011.
[12] G. F. D. Duff, A smaller universal cover for sets of unit diameter, *C. R. Math. Rep. Acad. Sci. Canada* 2 (1980) 37-42; MR 81k:52025.
[13] M. D. Kovalev, A minimal Lebesgue covering exists (in Russian), *Mat. Zametki* 40 (1986) 401-406, 430; Engl. transl. in Math. Notes 40 (1986) 736-739; MR 88h:52008.
[14] K. Bezdek and R. Connelly, Covering curves by translates of a convex set, *Amer. Math. Monthly* 96 (1989) 789-806; MR 90k:52020.
[15] K. Bezdek and R. Connelly, Minimal translation covers for sets of diameter 1, *Period. Math. Hungar.* 34 (1997) 23-27; MR 98k:52042.
[16] K. Bezdek and R. Connelly, The minimum mean width translation cover for sets of diameter one, *Beiträge Algebra Geom.* 39 (1998) 473-479; MR 99i:52021.
[17] H. G. Eggleston, Minimal universal covers in E^n, *Israel J. Math.* 1 (1963) 149-155; MR 28 # 4432.
[18] W.W. Rouse Ball and H. S. M. Coxeter, *Mathematical Recreations and Essays*, MacMillan, 1944, p. 99; MR 88m:00013.
[19] A. Edwards, Professor Besicovitch's minimal problem: A challenge, *Eureka* 20 (1957) 26-27.
[20] H. S. M. Coxeter, Lebesgue's minimal problem, *Eureka* 21 (1958) 13.
[21] G. D. Chakerian, Intersection and covering properties of convex sets, *Amer. Math. Monthly* 76 (1969) 753-766; MR 40 # 3433.

[22] D. Chakerian and D. Logothetti, Minimal regular polygons serving as universal covers in $\mathbb{R}^2$, *Geom. Dedicata* 26 (1988) 281-297; MR 89f:52035.
[23] M. Gardner, Curves of constant width, one of which makes it possible to drill square holes, *Sci. Amer.*, v. 208 (1963) n. 2, 148-156 and v. 208 (1963) n. 3, 154; also in *Mathematics: An Introduction to Its Spirit and Use*, W. H. Freeman, 1979, pp. 107-111 and 238-239.

8.4 モザーのミミズ定数

平面上の長さ1をもつ連続で求長可能な弧のことを**ミミズ**（worm）とよぶことにする．すべてのミミズからなる族を W とする．平面領域 R が W のすべてのミミズに合同なコピーを含むとき，R は W に対する**移動被覆**（あるいは**普遍被覆**）であるとよぶ．すなわち，単位長さの弧はすべて適当に平行移動や回転移動をほどこすと，R によって覆うことができる[1,2]．

平面上の特別な領域の族を S とする（たとえば円板の族などである）．S の元で W に対する移動被覆となるもののうち最小面積をもつものが存在するのだろうか？　もしそれが存在するときには，そのときの面積を $A(S)$ と書くことにする．例としてすべての円板からなる族を考えると[3]

$$A(\text{円板の族}) = \frac{\pi}{4} = 0.7853981633\cdots$$

となり，これは直径1の円の面積である．対角線の長さが1の正方形領域でも十分であることを証明する[4,5]のは幾分難しい．$A(\text{正方形の族}) = 1/2 = 0.5$ である．長方形領域の族というより大きな族に対しては，[4,5]で

$$A(\text{長方形の族}) = \beta\sqrt{1-\beta^2} = 0.3943847688\cdots$$

が示された．この値は，$\beta, \sqrt{1-\beta^2}$ という長さの辺をもつ長方形の面積であり，この β は長さ1の最大幅曲線に関して得られた値である【8.4.1】．半円領域の族に対しては，もっとよくわかっている[6]．Meir（メイア）が証明したように

$$A(\text{半円の族}) = \frac{\pi}{8} = 0.3926690816\cdots$$

である．おもしろいことに，正三角形の族については，依然として謎のままである．Besicovitch（ベシコヴィッチ）[7]は，

$$A(\text{正三角形の族}) \geq \frac{7\sqrt{3}}{27} = 0.4490502094\cdots$$

を証明した．右辺の値は，一辺が $2\sqrt{21}/9$ の正三角形の面積であり，彼はここで等号が成立するようだと思っていた．A に対するきちんとした表現に関する予想が Knox（ノックス）[8]によって見出された．この予想に対する反例は，もしあるとすると，そのミミズの2つの端点を結ぶ線分が，ミミズと3番目（あるいは，それ以上）の点で交わる

という意味で**ジグザグ**でなければならない[9].

さて，すべての凸平面領域からなる族 C を考えよう．Moser（**モザー**）**のミミズ定数** μ は $A(C)$ の値で定義される．すなわち，すべてのミミズを被覆することができる最小凸毛布の面積である．最もよく知られた評価は，Schaer & Wetzel（シャエア&ウェツェル）[5,6]と Poole, Garriets, Norwood & Laidacker（プール，ゲリエツ，ノーウッド&ライダッカー）[10-12]によって見つけられたように

$$0.2194626846\cdots = \frac{\beta}{2} \leq \mu \leq 0.27524\cdots$$

である．この上界は，ある菱形の2つの隣り合う辺の一部を円弧によって置きかえたものの面積である．最近，この上界を改良しようとする試みがなされたが成功していない[13,14]．多数の予想のうち[6,11,12]にある1つを述べる．それは半径1，中心角度 $\pi/6$ の扇形は，すべてのミミズを被覆するというものである．これが正しいとすると，μ に関する上界を $\pi/12 = 0.261799\cdots$ に減少させることになる．

これらの問題に関するたくさんの変種がある．ミミズが「閉じている」，すなわち始点と終点とが一致すると仮定したときの結果は【8.4.2】で示す．「被覆」の意味を（今までみてきたような移動すなわち平行移動，回転移動ではなく）平行移動のみに制限する場合でのいろいろな結果が【8.4.3】で示される．被覆の面積ではなく周の長さを最小にすることもできる[15]．また，たとえばミミズは必ず被覆の境界に近いかどうかというような被覆の効率を考えることもできる．このような問題を考えていくと自然に Bellman（ベルマン）の「森で道に迷う」問題に行きつく[16-19]．

次のような関連した問題がある[20]．それは，「任意のミミズは面積が $1/4$ のある長方形の毛布で被覆できること，そしてこの $1/4$ が最良の値であることを証明せよ」というものである．「ミミズを与えたとき，これを被覆する S の元を見つけよ」という問題は，「すべてのミミズを被覆できるような S の元を見つけよ」という前述の問題に類似しているが，同じような注目を集めてはこなかった．もう1つの問題とは，次のようなものである．「任意にミミズが与えられたとき，その最小凸被覆のうちでの最大の面積は $1/(2\pi) = 0.159154\cdots$ であることを示せ」．この最大値は，円弧の長さが1であるような半円のときに生じる．この結果の3次元版はどうなるのか？

凸であるという条件をはずすと興味深いことがいろいろと起こる[1,2]．Hansen は，（証明なしで）面積が $0.246\cdots$ である非凸普遍被覆について述べている．この値は，凸普遍被覆での知られた最良の値よりも小さい．しかし彼の主張は確かめられていない．非凸の場合の証明可能な上界は，$0.26044\cdots$ である[23]．Davis（デービス）[24]は，平面上でのすべての多角形的弧（折れ線）の族に対して平行移動被覆となる非凸集合で測度が0となるものを構成した．これは掛谷-ベシコヴィッチ問題【8.17】と密接な関係がある．ところが Marstrand（マーストランド）[25,26]は，すべての求長可能な弧よりなる族に対する移動被覆は，必ず正の測度をもつことを証明した．

8.4.1 単位長さの最大幅曲線

W に含まれるミミズに合同なコピーをすべて含むような無限に長い平面帯領域の最小幅は何か？ これは次のようにいっても同値である．ミミズ w を1つ固定し，この w と $0 \leq \theta \leq \pi$ に対してミミズ w を含む帯領域を決める x 軸と θ の角度をなす平行線の間の距離を $d(w, \theta)$ とする．すべての θ にわたる $d(w, \theta)$ の最小値を w の幅と定義する．当面の問題は「最大の幅をもつミミズは何か？」になる．

これの答えは，**幅広ミミズ**あるいは**カリパス（測径器）**であり，Zalgaller（ザルガラ）により最初に見つけられた[17,27,28]．図 8.3 を参照のこと．

この曲線は，次のようにきちんと与えられた幅をもっている．すなわち

$$\beta = \sup_w \min_\theta d(w, \theta) = \frac{1}{2}\left(\frac{\pi}{2} - \varphi - 2\psi + \tan(\varphi) + \tan(\psi)\right)^{-1}$$
$$= 0.4389253692\cdots = (2.2782916414\cdots)^{-1}$$

ここで角度 φ, ψ は

$$\varphi = \arcsin\left[\frac{1}{6} + \frac{4}{3}\sin\left(\frac{1}{3}\arcsin\left(\frac{17}{64}\right)\right)\right], \quad \psi = \arctan\left(\frac{1}{2}\sec(\varphi)\right)$$

によって定義されたものである．

このことより普遍長方形被覆は（カリパスを覆うために）両辺とも β 以上の長さであり，（長さ 1 の線分を覆うために）対角線の長さが 1 以上でなくてはならないことがすぐにわかる．$\beta \times \sqrt{1-\beta^2}$ の長方形が普遍被覆であることを証明するには，実はさらなる作業が要求される．

Zalgaller[29]も，この問題の 3 次元版を考え，3 次元空間での長さ 1 をもつ最大幅曲線は幅 $1/3.921545\cdots = 0.255001$ をもつと予想した．

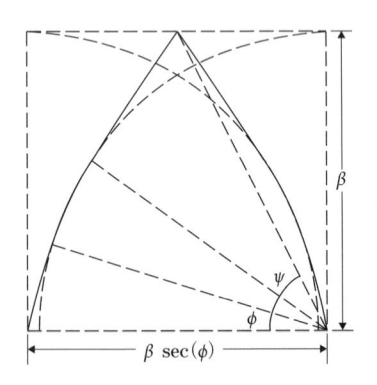

図 8.3 4 つの接線部分と 2 つの円弧よりなるカリパス．図はかなり精密に描かれている．

8.4.2 閉じたミミズ

平面上の単位長さの連続で求長可能な閉曲線(始点と終点とが一致する)を**閉じたミミズ**とよぶ.前のように最小面積をもつ移動被覆に興味がある.このようにより制限が加わった場合には,

$$A'(\text{円の族}) = \frac{\pi}{16} = 0.1963495408\cdots$$

とわかる.これは,直径が $1/2$ の円の面積である[3,30,31].また,

$$A'(\text{正方形の族}) = \frac{1}{8} = 0.125$$

となるが,これは対角線の長さが $1/2$ の正方形の面積である[5,13].そして

$$A'(\text{長方形の族}) = \frac{\sqrt{\pi^2 - 4}}{2\pi^2} = 0.122737657\cdots$$

とわかる.これは辺の長さが $1/\pi$,$\sqrt{\pi^2-4}/(2\pi)$ の長方形の面積である[5,31].また

$$A'(\text{三角形の族}) = \frac{3\sqrt{3}}{4\pi^2} = 0.1316200785\cdots$$

となり,これは一辺の長さが $\sqrt{3}/\pi$ の正三角形の面積である[32,33].

閉じたミミズを被覆する三角形についての方が,任意のミミズ(弧)を被覆する三角形のときよりもずっと多くのことがわかっているのは不思議なことである.さてここに関連した結果がある[34].周の長さが2の三角形を被覆することができる最小の正三角形の一辺の長さは1ではなく,次のようにして決まる,$s = 2/y = 1.0028514266\cdots$ である.ここでの y は,三角関数(を含む関数)

$$f(x) = \sqrt{3}\left(1 + \sin\frac{x}{2}\right)\sec\left(\frac{\pi}{6} - x\right)$$

の区間 $[0, \pi/6]$ での最小値である.定数 s は,より発展的な問題に関連して[35]でも出現する.

ここでの場合モザーのミミズ定数に類似したことについて,すなわち,すべての閉じたミミズを被覆することができる凸毛布の最小面積 $\mu' = A'(C)$ については,どんなことがわかるのか? Schaer & Wetzel[5]と Chakerian & Klamkin(チャケリアン&クラムキン)[31]は,円周の長さが1の円と,この円の中心を中点とする長さ $1/2$ の線分の凸胞の面積に等しい下界を提案した.すなわち

$$\mu' \geq \frac{1}{4\pi^2}\left(\sqrt{\pi^2 - 4} + \pi - 2\arccos\left(\frac{2}{\pi}\right)\right) = 0.0963296165\cdots$$

である.さらに最近になって Füredi & Wetzel(フェレディ&ウェツェル)[35]は,改良した不等式 $0.09666 \leq \mu' \leq 0.11754$ を得た.ここでの上界は,1つの小さな角を取り去った(以前言及した)最良の長方形の面積から生じる.

Schaer によるもので[31]にある関連の問題をとりあげる.それは「閉じたミミズが $1/\pi^2$ の面積をもつ長方形毛布によって被覆しうること,そして $1/\pi^2$ が最良の値であることを証明せよ」というものである.「ミミズが与えられたとき,これを被覆する S の

元を見つけよ」という問題は，前述の「すべてのミミズを被覆する S の元を見つけよ」という問題と類似している．

8.4.3 平行移動による被覆

W のミミズが適当な平行移動をすることにより平面領域 R で被覆されるとき，R は W に対する**平行移動被覆**（あるいは，**強普遍被覆**）であるという．回転移動は許されない．ミミズは 2 種類あるので，これらを別々に調べる．任意のミミズ（弧）に対して，すべての凸平面領域よりなる族 C を考えよう．この場合には，次のような Pál [36] による完全な解答がある．それは

$$\tilde{\mu} = \tilde{A}(C) = \frac{\sqrt{3}}{3} = 0.5773502692\cdots$$

であり，これは高さが 1 の正三角形の面積である．この場合は，多分すべての中で最も単純である．

閉じたミミズに対しては，明らかに円板が回転について対称であることより

$$\tilde{A}(\text{円の族}) = \frac{\pi}{16} = 0.1963495408\cdots$$

とわかり [3,30,31]，また

$$\tilde{A}(\text{三角形の族}) = \frac{\sqrt{3}}{9} = 0.1924500897\cdots$$

とわかる．これは一辺の長さが 2/3 の正三角形 [32,33] の面積である．そして

$$\tilde{A}(\text{長方形の族}) = \frac{1}{4} = 0.25$$

となり，これは一辺の長さが 1/2 の正方形の面積である．また凸領域の場合は，

$$0.1554479088\cdots \leq \tilde{\mu} = \tilde{A}(C) \leq 0.16526\cdots$$

とわかる．これは，Wetzel [6] と Bezdek & Connelly（ベズデク＆コネリー）[15] によるものである．

[1] H. T. Croft, K. J. Falconer, and R. K. Guy, *Unsolved Problems in Geometry*, Springer-Verlag, 1991, sect. D18, G6; MR 95k:52001.
[2] V. Klee and S. Wagon, *Old and New Unsolved Problems in Plane Geometry and Number Theory*, Math. Assoc. Amer., 1991, pp. 23, 92-93; MR 92k:00014.
[3] J. E. Wetzel, Covering balls for curves of constant length, *Enseign. Math.* 17 (1971) 275-277; MR 48 #12315.
[4] J. P. Jones and J. Schaer, The worm problem, Univ. of Calgary research paper 100, 1970.
[5] J. Schaer and J. E. Wetzel, Boxes for curves of constant length, *Israel J. Math.* 12 (1972) 257-265; MR 47 #5726.
[6] J. E. Wetzel, Sectorial covers for curves of constant length, *Canad. Math. Bull.* 16 (1973) 367-375; MR 50 #14451.
[7] A. S. Besicovitch, On arcs that cannot be covered by an open equilateral triangle of side 1, *Math. Gazette* 49 (1965) 286-288; MR 32 #6320.
[8] S. Knox, Calculating the length of Besicovitch's short arc, unpublished note (1994).
[9] J. E. Wetzel, On Besicovitch's problem, unpublished note (2000).

[10] G. Poole and J. Gerriets, Minimum covers for arcs of constant length, *Bull. Amer. Math. Soc.* 79 (1973) 462-463; MR 47 #4150.
[11] J. Gerriets and G. Poole, Convex regions which cover arcs of constant length, *Amer. Math. Monthly* 81 (1974) 36-41; MR 48 #12310.
[12] R. Norwood, G. Poole, and M. Laidacker, The worm problem of Leo Moser, *Discrete Comput. Geom.* 7 (1992) 153-162; MR 92j:52014.
[13] D. Reynolds, The worm's patchwork quilt, unpublished note (1994).
[14] I. Stewart, Mother Worm's blanket, *Sci. Amer.*, v. 274 (1996) n. 1, 98-99; errata, v. 274 (1996) n. 6, 103.
[15] K. Bezdek and R. Connelly, Covering curves by translates of a convex set, *Amer. Math. Monthly* 96 (1989) 789-806; MR 90k:52020.
[16] R. Bellman, Minimization problem, *Bull. Amer. Math. Soc.* 62 (1956) 270.
[17] V. A. Zalgaller, How to get out of the woods? On a problem of Bellman (in Russian), *Matematicheskoe Prosveshchenie* 6 (1961) 191-195.
[18] H. Joris, Le chasseur perdu dans la forêt, *Elem. Math.* 35 (1980) 1-14; MR 81d:52001.
[19] S. R. Finch and J. E. Wetzel, Lost in a forest, unpublished note (2002).
[20] Putnam Competition Problem B-4, *Amer. Math. Monthly* 77 (1970) 723, 727.
[21] P. A. P. Moran, On a problem of S. Ulam, *J. London Math. Soc.* 21 (1946) 175-179; MR 8,597n.
[22] H. C. Hansen, Ormeproblemet, *Normat* 40 (1992) 119-123, 143; MR 93h:52025.
[23] G. Poole and R. Norwood, An improved upper bound to Leo Moser's worm problem, *Discrete Comput. Geom.*, to appear.
[24] R. O. Davies, Some remarks on the Kakeya problem, *Proc. Cambridge Philos. Soc.* 69 (1971) 417-421; MR 42 #7869.
[25] J. M. Marstrand, Packing smooth curves in $\mathbb{R}^q$, *Mathematica* 26 (1979) 1-12; MR 81d:52009.
[26] J. M. Marstrand, Packing planes in $\mathbb{R}^q$, *Mathematica* 26 (1979) 180-183; MR 81f:2800.
[27] J. Schaer, The broadest curve of length 1, Univ. of Calgary research paper 52, 1968.
[28] A. Adhikari and J. Pitman, The shortest planar arc of width 1, *Amer. Math. Monthly* 96 (1989) 309-327; MR 90d:52016.
[29] V. A. Zalgaller, The problem of the shortest space curve of unit width (in Russian), *Mat. Fiz. Anal. Geom.* 1 (1994) 454-461; MR 98i:52012.
[30] J. C. C. Nitsche, The smallest sphere containing a rectifiable curve, *Amer. Math. Monthly* 78 (1971) 881-882; MR 45 #480.
[31] G. D. Chakerian and M. S. Klamkin, Minimal covers for closed curves, *Math. Mag.* 46 (1973) 55-61; MR 47 #2496.
[32] J. E.Wetzel, Triangular covers for closed curves of constant length, *Elem. Math.* 25 (1970) 78-81; MR 42 #960.
[33] J. E. Wetzel, On Moser's problem of accommodating closed curves in triangles, *Elem. Math.* 27 (1972) 35-36; MR 45 #4282.
[34] J. E. Wetzel, The smallest equilateral cover for triangles of perimeter two, *Math. Mag.* 70 (1997) 125-130.
[35] Z. Füredi and J. E.Wetzel, The smallest convex cover for triangles of perimeter two, *Geom. Dedicata* 81 (2000) 285-293; MR 2001c:52001.
[36] J. Pál, Ein minimumprobleme für ovale, *Math. Annalen* 83 (1921) 311-319.

8.5 巡回セールスマン定数

d 次元単位立方体内の n 個の異なる点を考える．各点をちょうど1回ずつ通る $(n-1)!/2$ 通りの**巡回**（閉路）を考えるとき，最も短い距離 $L_d(n)$ は何か？

最短巡回距離 $L_d(n)$ を決定することは，**巡回セールスマン問題** (travelling salesman problem, TSP) として知られている．これは，オペレーションズリサーチ，アルゴリズム開発，複雑系理論といった分野を席捲する最もよく知られた組合せ論的最適問題の1つである．これは多項式時間では計算不可能な問題，すなわちNP困難な問題なので，この問題を解決するのは難しい．

それにもかかわらずいくつかの興味深い漸近値に出会う．立方体でのすべての最適な巡回に対して

$$\limsup_{n\to\infty} \frac{L_d(n)}{n^{(d-1)/d}} \leq \alpha_d, \quad \alpha_d' = \frac{\alpha_d}{\sqrt{d}}$$

となるような最小の定数 α_d が存在する．また極限は，立方体での「ほとんどすべての」最適な巡回に対して，すなわち巡回の無視しうる（測度0の）部分集合においてのみ成立しないという意味において，

$$\lim_{n\to\infty} \frac{L_d(n)}{n^{(d-1)/d}} = \beta_d, \quad \beta_d' = \frac{\beta_d}{\sqrt{d}}$$

となるような別の定数 β_d が存在する．これらの定数は，最初，Bearwood, Halton & Hammersley (ベアウッド，ハルトン&ハマースレイ)[1,2] によって研究された．厳密な上界，下界については，表8.3にあげた[3-9]．

[10-14]によると

$$\lim_{d\to\infty} \beta_d' = \frac{1}{\sqrt{2\pi e}} = 0.2419707245\cdots$$

$$\frac{1}{\sqrt{2\pi e}} \leq \lim_{d\to\infty} \alpha_d' \leq \frac{2(3-\sqrt{3})\theta}{\sqrt{2\pi e}} = 0.40509\cdots$$

と，わかっている．ここで

$$\frac{1}{2} \leq \theta = \lim_{d\to\infty} \theta_d^{1/d} \leq 0.66019$$

表 8.3 巡回セールスマン定数 α_d' と β_d' に関する上界，下界

d	β_d'の下界	β_d'の上界	α_d'の下界	α_d'の上界
2	0.44194	0.6508	0.75983	0.98398
3	0.37313	0.61772	0.64805	0.90422
4	0.34207	0.55696	0.5946	0.8364

であり，θ_d は d 次元空間での最良の球面充填密度である【8.7】．たとえいつの日にか θ_d に対する上界 $2^{-0.59905d+o(d)}$ が，正しいと信じられている[15]$2^{-d+o(d)}$ に改良されたとしても，$\lim_{d\to\infty}\alpha_d'$ に対する上界は，0.30681 になるだけであろう．この極限値をきちんと評価するためには新しい見方が要請されるであろう[13]．

Johnson, McGeoch & Rothberg（ジョンソン，マックゲオク&ロスバーグ）[16]や Percus & Martin（パーカス&マーチン）[17,18]による

$$\beta_2 = 0.7124\cdots, \quad \beta_3 = 0.6979\cdots, \quad \beta_4 = 0.7234\cdots$$

というあまり厳密ではない β_d の数値的な推定がある．β_2 の初期の推定値とあまりよく一致しないのは，コンピュータによる異なる実験方法に付随する有限の大きさの効果に関連があるらしい．Applegate, Cook & Rohe（アップルゲート，クック&ロー）[19]による β_2 の最近の別の評価では，$0.714\cdots$ である．これは，（フラクタル空間充填曲線に基づく）β_2 に対する Norman & Moscato（ノーマン&モスカト）[20]の予想した表現

$$\beta_2 = \frac{4(1+2\sqrt{2})\sqrt{51}}{153} = 0.7147827007\cdots$$

が正統化されることを示唆している．これは，確かに TSP シミュレーションの際の乱数生成に関する質の問題を再検討する必要性を示している．

n 個の点が単位正方形内に独立かつ一様に分布しているとき（最適とは必ずしもかぎらない）「ランダムな」巡回の長さ $\Lambda_2(n)$ は

$$\lim_{n\to\infty}\frac{\mathrm{E}(\Lambda_2(n))}{n} = \frac{1}{15}(\sqrt{2}+2+5\ln(1+\sqrt{2})) = 0.521405433\cdots$$

を満たす[21]．ここで E は，すべての巡回にわたる平均と，すべての点集合にわたる平均の両方を表しているものとする．この $0.5214\cdots$ というきちんとした表現は，Ghosh（ゴーシュ）[22]によるもので，【8.1】でさらに議論した．$\mathrm{E}(\Lambda_2(n))$ は n のオーダーで増加するが，$L_2(n)$ は典型的に $\sqrt{n}$ のオーダーで増加することに注意しよう．

$\lim_{d\to\infty}\beta_d'$ のより正確な値が[18]に予想されており，それは

$$\beta_d = \sqrt{\frac{d}{2\pi e}}(\pi d)^{1/2d}\left[1 + \frac{2-\ln(2)-2\gamma}{d} + O\left(\frac{1}{d^2}\right)\right]$$

である．ここで γ は，オイラー-マスケローニの定数【1.5】を表している．この公式の根拠は，すぐ次に述べる話題の特別な場合のランダムリンク TSP として知られている．

8.5.1 ランダムリンク TSP

K_n を n 個の頂点をもつ完全グラフとする．すなわち異なる頂点のすべての組が 1 つの辺を決定する．ここでは，d 次元空間にあるというような設定はない．よってどんな距離も取り除かれている．各辺に独立に**長さ**と呼ばれる一様 [0,1] 確率変数を割りあてる．この長さは，必ずしも三角不等式を満たさないので普通の意味での距離ではないことに注意しよう．各頂点をちょうど 1 回ずつ通る $(n-1)!/2$ 通りのすべての巡回の中で長さの和が最小値 $L(n)$ をもつような最短路を決めることができる．そして

$$\lim_{n\to\infty} L(n) = \beta \quad \text{（確率 1 で）}$$

と定義する．Krauth & Mézard（クラウス&メザー）[23]はあまり厳密ではないが，谷下り法を用いてβの解析的表現を得た．それは

$$\beta = \frac{1}{2}\int_{-\infty}^{\infty} f(x)(1+f(x))\exp(-f(x))dx = 2.0415\cdots = 2(1.0208\cdots)$$

であり，ここで$f(x)$は積分方程式

$$f(x) = \int_{-x}^{\infty}(1+f(y))\exp(-f(y))dy$$

の解である．実際にはこれは，d次元ユークリッド的TSPをd-パラメータ化したランダムリンクの族により近似するという（$d=1$に対応する場合での）1つの筋書にすぎない．

8.5.2 最小全域木

d次元単位立方体内に，異なるn個の点を考えるというおなじみの設定に戻ろう．これらの点の集合をVと書く．**最小全域木**（minimal spanning tree，MST）とは，（普通のユークリッドの意味での辺の長さの和とした）最小な長さ$L_d(n)$となるような頂点集合Vをもつ連結グラフ【5.6】のことである．

$$\lim_{n\to\infty}\frac{L_d(n)}{n^{(d-1)/d}} = \beta_d \quad （確率1で）$$

と定義する．数値的評価[25,26]や理論的な結果[11]の中に

$$\beta_2 = 0.6331\cdots, \quad \beta_3 = 0.6232\cdots, \quad \beta_d \sim \sqrt{\frac{d}{2\pi e}} \quad (d\to\infty \text{ のとき})$$

が含まれている．β_dに対する完全な（しかし複雑な）表現が存在することは驚きである[27,28]．$d=2$の場合に対してのみ公式を与える．$x_0=0$とする．点$x_j(0\leq j\leq i-1)$を中心とし半径$1/2$の円板$D_j(0\leq j\leq i-1)$が連結集合をなすようにとった平面上のすべての点$\{x_1, x_2, \cdots, x_{i-1}\}$の集合を$\Delta_i$と書く．$g_i(x_1, x_2, \cdots, x_{i-1})$を$\bigcup_{j=0}^{i-1}D_j$の面積と定義すると，

$$\beta_2 = \frac{1}{2} + \frac{1}{2}\sum_{i=2}^{\infty}\frac{\Gamma\left(i-\frac{1}{2}\right)}{i!}\int_{\Delta_i}g_i(x_1, x_2, \cdots x_{i-1})^{-i+1/2}dx_1 dx_2\cdots dx_{i-1}$$

となる．この級数の初めの5項を用いると精密な下界$\beta_2 \geq 0.600822$ [27]が得られる．

最小全域木が与えられたとき，$L_d(n)$以外の特徴づけについて考えることができる．たとえば，辺の長さの平方和$\tilde{L}_d(n)$を考え

$$\lim_{n\to\infty}\frac{\tilde{L}_d(n)}{n^{(d-1)/d}} = \tilde{\beta}_d \quad （確率1で）$$

と定義する．$\tilde{\beta}_d$の存在性は，Aldous & Steele（アルダス&スティール）[29,30]により証明された．数値的評価には$\tilde{\beta}_2 = 0.4769\cdots$（しばしばBland（**ブランド**）の定数とよばれる），$\tilde{\beta}_3 = 0.4194\cdots$が含まれる[26]．前と同じように$\tilde{\beta}_d$に対するきちんとした表現も見出すことができ[27]，これにより，厳密な下界$\tilde{\beta}_d \geq 0.401\cdots$がわかる．

最適な巡回が与えられたとき，TSPに対する辺の長さの平方和パラメータ$\tilde{L}_d(n)$

も興味深い．$\tilde{\beta}_d$ の存在証明は未知であるが，ある $c_d > 0$ に対して $n \to \infty$ のとき

$$\frac{\tilde{L}_d(n)}{n^{(d-1)/d}} > c_d \ln(n)$$

となる特別な点の配置が構成できる[31-33]．このことより $\tilde{a}_d$ は，明らかに「存在しない」．他にも変種は豊富である．最適な巡回を計算する際，$L_d(n)$ ではなく $\tilde{L}_d(n)$ を最小にすると，(べき乗の重みづけのため) 異なる路が決定されることがしばしば起こり，最悪な場合の定数[34-37]

$$\limsup_{n\to\infty} \frac{\tilde{L}_d(n)}{n^{(d-2)/d}} \leqq \hat{a}_d$$

は，$d=2$ のとき 4 である．Yukich (ユキチ)[38,39]は，対応する平均の場合の定数 $\tilde{\beta}_d$ も存在することを証明したが，$\tilde{\beta}_2$ の値は未知である．

独立な一様[0,1]確率的な辺の長さを付けた n 頂点完全グラフ K_n に対して，長さの和 $L(n)$ をもつ MST を考える．Frieze (フリーゼ)[40-42]は，確率的に

$$\lim_{n\to\infty} L(n) = \zeta(3) = 1.2020569031\cdots$$

であることを示した．ここで $\zeta(3)$ は Apéry (アペリ) の定数【1.6】である．なんと美しい結果だろう！ Janson (ジャンソン)[43]は，$\sqrt{n}(L(n) - \zeta(3))$ が，漸近的に分散 σ^2 が

$$\sigma^2 = \frac{\pi^4}{45} - 2 \sum_{i=0}^{\infty} \sum_{j=1}^{\infty} \sum_{k=1}^{\infty} \frac{(i+k-1)! k^k (i+j)^{i-2} j}{i! k! (i+j+k)^{i+k+2}} = 1.6857\cdots$$

となる正規分布 $(0, \sigma^2)$ であることを示したが，この定数を簡単なものにすることは不可能なようである．$\zeta(3)$ に関連して生じる別の事項が[44]にある．

8.5.3 最小マッチング

再び d 次元単位立方体内での n 個の異なる点を考える．今回は n が偶数という仮定を付け加える．n 個の各点が，ちょうど 1 つの辺と出会うような $n/2$ 個の辺からなる (不連結な) グラフを**マッチング**という．(普通のユークリッドの意味での辺の長さの和とした) 最小の長さ $L_d(n)$ となるマッチングを**最小マッチング** (minimum matching, MM) という．β_d は前のように定義する．平面での β_2 は，しばしば Papadimitrious (パパディミトリアス) の定数[45,46]とよばれる．数値的評価[47-54]や理論的な結果[11]には，

$$\beta_2 = 0.3104\cdots, \quad \beta_3 = 0.3172\cdots, \quad \beta_d \sim \frac{1}{2}\sqrt{\frac{d}{2\pi e}} \quad (d \to \infty \text{ のとき})$$

が含まれている．対応する最悪の場合の定数 α_2 は，$0.537 \leqq \alpha_2 \leqq 0.707$[47]を満たす．

独立な一様[0,1]確率的な辺の長さをもつ n 頂点の完全グラフ K_n に対して，長さの和 $L(n)$ をもつ MM を考える．Mézard & Parisi (メザー&パリシ)[55-57]は，レプリカ法 (と前の $f(x)$ のときより単純な積分方程式) を用いて $\beta = \pi^2/12 = 0.8224670334\cdots$ を求めた．そして Aldous (アルダス)[58]は，精密な証明を見出した．実験的証明が [53,54]に現れた．前と同じように，これは d 次元ユークリッド的 MM 問題を，d-パ

ラメータ化したランダムリンクの族により近似するという ($d=1$ に対応する場合の) 1
つの筋書にすぎない.

[1] J. Beardwood, J. H. Halton, and J. M. Hammersley, The shortest path through many points, *Proc. Cambridge Philos. Soc.* 55 (1959) 299-327; MR 22 #202.
[2] R. M. Karp and J. M. Steele, Probabilistic analysis of heuristics, *The Traveling Salesman Problem: A Guided Tour of Combinatorial Optimization*, ed. E. L. Lawler, J. K. Lenstra, A. H. G. Rinnoov Kan, and D. B. Shmoys, Wiley, 1985, pp. 181-205; MR 87f: 90057.
[3] J. M. Steele, *Probability Theory and Combinatorial Optimization*, SIAM, 1997, pp. 30-51; MR 99d: 60002.
[4] L. Fejes Tóth, Über einen geometrischen Satz, *Math. Z.* 46 (1940) 83-85; MR 1,263g.
[5] S. Verblunsky, On the shortest path through a number of points, *Proc. Amer. Math. Soc.* 2 (1951) 904-913; MR 13,577c.
[6] L. Few, The shortest path and the shortest road through n points, *Mathematika* 2 (1955) 141-144; MR 17,1235f.
[7] H. J. Karloff, How long can a Euclidean traveling salesman tour be?, *SIAM J. Discrete Math.* 2 (1989) 91-99; MR 89m:90063.
[8] S. Moran, On the length of optimal TSP circuits in sets of bounded diameter, *J. Combin. Theory Ser. B* 37 (1984) 113-141; MR 86i:05087.
[9] L. A. Goddyn, Quantizers and the worst case Euclidean traveling salesman problem, *J. Combin. Theory Ser. B* 50 (1990) 65-81; MR 91k:90121.
[10] W. D. Smith, *Studies in Computational Geometry Motivated by Mesh Generation*, Ph.D. thesis, Princeton Univ., 1988.
[11] D. Bertsimas and G. Van Ryzin, An asymptotic determination of the minimum spanning tree and minimum matching constants in geometrical probability, *Oper. Res. Lett.* 9 (1990) 223-231; MR 92b:90172.
[12] W. T. Rhee, On the travelling salesperson problem in many dimensions, *Random Structures Algorithms* 3 (1992) 227-233; MR 93d:90053.
[13] J. M. Steele and T. L. Snyder, Worst-case growth rates of some classical problems of combinatorial optimization, *SIAM J. Comput.* 18 (1989) 278-287; MR 90k:90131.
[14] G. A. Kabatyanskii and V. I. Levenshtein, Bounds for packings on a sphere and in space (in Russian), *Problemy Peredachi Informatsii* 14 (1978) 3-25; Engl. transl. in *Problems Information Transmission* 14 (1978) 1-17.
[15] P. M. Gruber, Geometry of numbers, *Handbook of Convex Geometry*, v. B, ed. P. M. Gruber and J. M. Wills, North-Holland, 1993, pp. 739-763; MR 94k:11074.
[16] D. S. Johnson, L. A. McGeoch, and E. E. Rothberg, Asymptotic experimental analysis for the Held-Karp traveling salesman bound, *Proc. 7th ACM-SIAM Symp. on Discrete Algorithms (SODA)*, Atlanta, ACM, 1996, pp. 341-350; MR 96j:68007.
[17] A. G. Percus and O. C. Martin, Finite size and dimensional dependence in the Euclidean traveling salesman problem, *Phys. Rev. Lett.* 76 (1996) 1188-1191; MR 96k: 90057.
[18] N. J. Cerf, J. Boutet de Monvel, O. Bohigas, O.C. Martin, and A. G. Percus, The random link approximation for the Euclidean traveling salesman problem, *J. Physique I* 7 (1997) 117-136; cond-mat/9607080.
[19] D. Applegate, W. Cook, and A. Rohe, Chained Lin-Kernighan for large traveling salesman problems, *INFORMS J. Comput.*, to appear; Forschungsinstitut für Diskrete Mathematik report 99887, Universität Bonn.
[20] M. G. Norman and P. Moscato, The Euclidean traveling salesman problem and a space-filling curve, *Chaos Solitons Fractals* 6 (1995) 389-397.
[21] E. Bonomi and J.-L. Lutton, The n-city travelling salesman problem: Statistical mechanics and the Metropolis algorithm, *SIAM Rev.* 26 (1984) 551-568; MR 86e:90041.

[22] B. Ghosh, Random distances within a rectangle and between two rectangles, *Bull. Calcutta Math. Soc.* 43 (1951) 17-24; MR 13,475a.
[23] W. Krauth and M. Mézard, The cavity method and the travelling salesman problem, *Europhys. Lett.* 8 (1989) 213-218.
[24] A. G. Percus and O. C. Martin, The stochastic traveling salesman problem: Finite size scaling and the cavity prediction, *J. Stat. Phys.* 94 (1999) 739-758; MR 2000b:82027.
[25] F. D. K. Roberts, Random minimal trees, *Biometrika* 55 (1968) 255-258; MR 37 #6210.
[26] M. Cortina-Borja and T. Robinson, Estimating the asymptotic constants of the total length of Euclidean minimal spanning trees with power-weighted edges, *Statist. Probab. Lett.* 47 (2000) 125-128; MR 2000k:90060.
[27] F. Avram and D. Bertsimas, The minimum spanning tree constant in geometrical probability and under the independent model: A unified approach, *Annals of Appl. Probab.* 2 (1992) 113-130; MR 93b:60027.
[28] P. Jaillet, Cube versus torus models and the Euclidean minimum spanning tree constant, *Annals of Appl. Probab.* 3 (1993) 582-592; MR 94d:60018.
[29] J. M. Steele, Growth rates of Euclidean minimal spanning trees with power weighted edges, *Annals of Probab.* 16 (1988) 1767-1787; MR 89j:60049.
[30] D. Aldous and J. M. Steele, Asymptotics for Euclidean minimal spanning trees on random points, *Probab. Theory Relat. Fields* 92 (1992) 247-258; MR 93c:60007.
[31] J. Gao and J. M. Steele, Sums of squares of edge lengths and spacefilling curve heuristics for the traveling salesman problem, *SIAM J. Discrete Math.* 7 (1994) 314-324;MR95a:90083.
[32] T. L. Snyder and J. M. Steele, A priori bounds on the Euclidean traveling salesman, *SIAM J. Comput.* 24 (1995) 665-671; MR 96d:90109.
[33] M. Bern and D. Eppstein, Worst case bounds for subadditive geometric graphs, *Proc. 9th ACM Symp. on Computational Geometry (SCG)*, San Diego, ACM, 1993, pp. 183-188.
[34] D. J. Newman, A Problem Seminar, Springer-Verlag, 1982, pp. 9-10, 74-75; MR 84d:00004.
[35] B. Bollobás and A. Meir, A travelling salesman problem in the k-dimensional unit cube, *Oper. Res. Lett.* 11 (1992) 19-21; MR 93c:90046.
[36] J. E. Yukich, Worst case asymptotics for some classical optimization problems, *Combinatorica* 16 (1996) 575-586; MR 97i:90061.
[37] S. Lee, Worst case asymptotics of power-weighted Euclidean functionals, *Discrete Math.*, to appear.
[38] J. E. Yukich, Asymptotics for the Euclidean TSP with power weighted edges, *Probab. Theory Relat. Fields* 102 (1995) 203-220; MR 96d:60022.
[39] J. E.Yukich, *Probability Theory of Classical Euclidean Optimization Problems*, Lect. Notes in Math. 1675, Springer-Verlag, 1998; MR 2000d:60018.
[40] A. M. Frieze, On the value of a random minimum spanning tree problem, *Discrete Appl. Math.* 10 (1985) 47-56; MR 86d:05103.
[41] J. M. Steele, On Frieze's ζ (3) limit for lengths of minimal spanning trees, *Discrete Appl. Math.* 18 (1987) 99-103; MR 88i:05063.
[42] B. Bollobás, *Random Graphs*, Academic Press, 1985, pp. 141-144; MR 87f:05152.
[43] S. Janson, The minimal spanning tree in a complete graph and a functional limit theorem for trees in a random graph, *Random Structures Algorithms* 7 (1995) 337-355;MR97d:05244.
[44] M. D. Penrose, Random minimal spanning tree and percolation on the N-cube, *Random Structures Algorithms* 12 (1998) 63-82; MR 99g:60024.
[45] C. H. Papadimitriou, The probabilistic analysis of matching heuristics, *Proc. 15th Allerton Conf. on Communication, Control, and Computing*, Univ. of Illinois, 1978, pp. 368-378.
[46] J. M. Steele, Subadditive Euclidean functionals and nonlinear growth in geometric probability, *Annals of Probab.* 9 (1981) 365-376; MR 82j:60049.
[47] K. J. Supowit, E. M. Reingold, and D. A. Plaisted, The travelling salesman problem and minimum matching in the unit square, *SIAM J. Comput.* 12 (1983) 144-156; MR 84f:90032b.

[48] M. Iri, K. Murota, and S. Matsui, Heuristics for planar minimum-weight perfect matchings, *Networks* 13 (1983) 67-92; MR 84e:90029.
[49] M.Weber and Th. M. Liebling, Euclidean matching problems and the Metropolis algorithm, *Z. Oper. Res. Ser. A-B* 30 (1986) A85-A110; MR 87i:90174.
[50] P. Grassberger and H. Freund, An efficient heuristic algorithm for minimum matching, *Z. Oper. Res.* 34 (1990) 239-253; MR 91e:90088.
[51] D. P. Williamson and M. X. Goemans, Computational experience with an approximation algorithm on large-scale Euclidean matching instances, *INFORMS J. Comput.* 8 (1996) 29-40; MR 95b:68018.
[52] W. Cook and A. Rohe, Computing minimum-weight perfect matchings, *INFORMS J. Comput.* 11 (1999) 138-148.
[53] J. H. Boutet de Monvel and O. C. Martin, Mean field and corrections for the Euclidean minimum matching problem, *Phys. Rev. Lett.* 79 (1997) 167-170; cond-mat/9701182.
[54] J. Houdayer, J. H. Boutet de Monvel, and O. C. Martin, Comparing mean field and Euclidean matching problems, *Europ. Phys. J. B* 6 (1998) 383-393; cond-mat/9803195; MR 99i:82023.
[55] M. Mézard and G. Parisi, On the solution of the random link matching problems, *J. Physique* 48 (1987) 1451-1459.
[56] M. Mézard and G. Parisi, The Euclidean matching problem, *J. Physique* 49 (1988) 2019-2025; MR 90e:90048.
[57] R. Brunetti, W. Krauth, M. Mézard, and G. Parisi, Extensive numerical simulations of weighted matchings: Total length and distribution of links in the optimal solution, *Europhys. Lett.* 14 (1991) 295-301.
[58] D. J. Aldous, The $\zeta(2)$ limit in the random assignment problem, *Random Structures Algorithms* 18 (2001) 381-418; MR 2002f:60015.

8.6 シュタイナー木定数

d 次元空間で n 個の点の集合を P とする．シュタイナー極小木と最小架橋木の定義を与える．

- P を連結する最短の連結グラフ【5,6】を P の**シュタイナー極小木**(Steiner minimal tree，SMT) という．
- P を連結する頂点集合 P だけによる最短の連結グラフを**最小架橋木**（minimum spanning tree，MST）という．

n 個の辺をもつ平面正 n 角形の n 個の頂点を P_n と書く．図 8.4 と図 8.5 は，頂点間を結ぶ辺のみを許す MST と，他方木を最適化にするために新たな頂点を付け加えることも認める SMT（こちらは頂点の位置が無限にあることになり，より計算が難しくなる）の例である．$|G|$ をグラフ G の辺の長さの和を表すものとすると，明らかに

$$\frac{|\mathrm{SMT}(P_3)|}{|\mathrm{MST}(P_3)|}=\frac{\sqrt{3}}{2}=0.866\cdots, \quad \frac{|\mathrm{SMT}(P_4)|}{|\mathrm{MST}(P_4)|}=\frac{1+\sqrt{3}}{3}=0.910\cdots$$

である．ついでにいうと，$\mathrm{SMT}(P_5)$ も同じように三辺が 120° で交わるように三項点を付け加えたもので構成される．（付け加えた点は Torricelli（**トリチェリ**）点あるい

図 8.4 P_3 の SMT と MST.

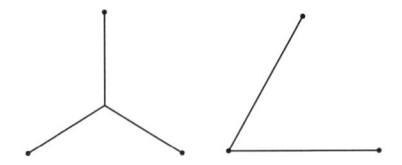

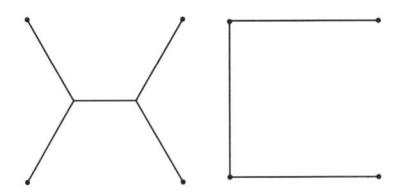

図 8.5 P_4 の SMT と MST.

は Steiner (**シュタイナー**) 点とよばれている), しかし $n \geqq 6$ の場合は, SMT(P_n)＝MST(P_n) である. これは,「プラトー (Plateau) 問題」[1] の最小面積解と同様に石けん膜実験によって確かめることができる.

　任意の集合 P に対して MST(P) を決定するのは比較的やさしい. それゆえ

$$\rho_d = \inf_{n,P} \frac{|\text{SMT}(P)|}{|\text{MST}(P)|}$$

の値に興味がわく. ここでの下限は, d 次元空間の n 点集合および正の整数 n のすべてにわたるものとする. Steiner (**シュタイナー**) 比 ρ_d は, MST の全長が, シュタイナー点を付け加えていくにつれてどのくらい減少しうるかということを示している. この下限に達する点集合は,「最大限に近道をもつ」ものと見なしうる [2-5].

　Du & Hwang (ドウ&フワン) [6] は Gilbert & Pollak (ギルバート&ポラック) の予想

$$\rho_2 = \frac{\sqrt{3}}{2} = 0.8660254037\cdots$$

を証明し, Smith & Smith (スミス&スミス) [8,9] は,

$$\rho_3 \leqq s_3 = \frac{3\sqrt{3} + \sqrt{7}}{10} = 0.7841903733\cdots$$

であることを 3-**ソーセージ** (その点は, 円螺線に沿って等間隔に配置されている [10-13]) とよばれる集合 P を用いて証明した. 彼等は, $\rho_3 = s_3$ であるという広範な直観的な証拠を与えたが, 厳密な証明は知られていない. ρ_3 の最良下界, 実際には任意の ρ_d に対する最良下界は [14,15]

$$\rho_d \geqq \frac{2 + x - \sqrt{x^2 + x + 1}}{\sqrt{3}} = 0.6158277481\cdots$$

であり, ここで x は

$$128x^6+456x^5+783x^4+764x^3+408x^2+108x-28=0$$

のただ1つの正の解である.

上界についてもう少し詳しく議論しよう. $d=2$ のときの正三角形, $d=3$ のときの正四面体の自然な一般化として d-**単体**（正単体）を定義する. Chung & Gilbert（チュン＆ギルバート）[16] は, この場合でのシュタイナー比 r_d に関する評価を計算した. そして

$$\limsup_{d \to \infty} r_d \le \frac{\sqrt{3}}{4-\sqrt{2}} = 0.6698352124\cdots$$

を示した. Smith[15,17]は, この式の「上極限」を「極限」に, 不等号は実際には等号におきかえられると予想した. $d \ge 3$ のときは, $r_d > \rho_d$ とわかっており, そしてたとえば,

$$r_3 = \frac{1+\sqrt{6}}{3\sqrt{2}} = 0.813053\cdots, \quad r_4 = \frac{\sqrt{3}+\sqrt{5}+2\sqrt{6}}{8\sqrt{2}} = 0.783748\cdots$$

である.

類似した高次元 d-**ソーセージ**に対応するシュタイナー比 s_d も $d \ge 3$ のとき $s_d < r_d$ であることも知られている. s_d は d の関数として真に減少関数である. しかし $\lim_{d \to \infty} s_d$ の数値, および $d \ge 3$ に対して $\rho_d = s_d$ になるかどうかもわかっていない. Du & Smith（デュ＆スミス）[15]は, d が小さいときはたぶん $\rho_d = s_d$ となるが, $d \ge 15$ のときは, これは成立しないと考えた. たとえば $s_4 = 0.7439856178\cdots$ は, 最小多項式[18]

$$900s^8 - 1863s^6 + 2950s^4 - 1511s^2 + 164$$

をもつことがわかり, $s_5 = 0.7218101748\cdots$ についても同様のことがわかるのもそんなに遠くないと思われる. ([8.5]でのMSTについて考えたのと同じように) 別の観点もある. P の n 個の点がすべて単位正方形内にあるという制限をつけると, Chung & Graham（チュン＆グラハム）[19-21]によって発見されたように, すべての n に対して

$$0.930\sqrt{n}+c < \left(\frac{3}{4}\right)^{1/4}\sqrt{n}+c \le |\text{SMT}(P)| < 0.995\sqrt{n}+C$$

を満たす定数 c と C が存在する. 上の制限を n 個の点が d 次元単位立方体内にあるというふうに変えると, $n \to \infty$ のとき, d を十分大きくとると

$$|\text{SMT}(P)| \le \sqrt{\frac{d}{2\pi e}} n^{1-1/d}$$

となる[2,22]. 両方の漸近的な結果が改良されるのを見たいものである.

[1] R. Courant and H. Robbins, *What Is Mathematics?*, Oxford Univ. Press, 1941; MR 80i:00001.
[2] F. K. Hwang, D. S. Richards, and P. Winter, *The Steiner Tree Problem*, North-Holland, 1992; MR 94a:05051.
[3] A. O. Ivanov and A. A. Tuzhilin, *Minimal Networks: The Steiner Problem and Its Generalizations*, CRC Press, 1994; MR 95h:05050.
[4] D.-Z. Du and F. K. Hwang, The state of art on Steiner ratio problems, *Computing in Euclidean Geometry*, 2nd ed., ed. D.-Z. Du and F. K. Hwang, World Scientific, 1995, pp. 163-191; MR 94i:68304.

[5] D. Cieslik, *Steiner Minimal Trees*, Kluwer, 1998; MR 99i:05062.
[6] D.-Z. Du and F. K. Hwang, A proof of the Gilbert-Pollak conjecture on the Steiner ratio, *Algorithmica* 7 (1992) 121-135; MR 92m:05060.
[7] E. N. Gilbert and H. O. Pollak, Steiner minimal trees, *SIAM J. Appl. Math.* 16 (1968) 1-29; MR 36 # 6317.
[8] W. D. Smith and J. M. Smith, On the Steiner ratio in 3-space, *J. Combin. Theory Ser. A* 69 (1995) 301-332; MR 95k:05053.
[9] J. Keane, Simplifying the 3-sausage constant s_3, unpublished note (1999).
[10] A. H. Boerdijk, Some remarks concerning close-packing of equal spheres, *Philips Res. Rep.* 7 (1952) 303-313; MR 14,310b.
[11] H. S. M. Coxeter, *Introduction to Geometry*, 2nd ed., Wiley, 1969, p. 412; MR 90a:51001.
[12] R. Buckminster Fuller, *Synergetics*, MacMillan, 1975, pp. 520-524.
[13] H. S. M. Coxeter, The simplicial helix and the equation $\tan n\theta = n \tan \theta$, *Canad. Math. Bull.* 28 (1985) 385-393; MR 87b:51018.
[14] D.-Z. Du, On Steiner ratio conjectures, *Annals Oper. Res.* 33 (1991) 437-451; MR 92m:90053.
[15] D.-Z. Du and W. D. Smith, Disproofs of generalized Gilbert-Pollak conjecture on the Steiner ratio in three or more dimensions, *J. Combin. Theory Ser. A* 74 (1996) 115-130; MR 97h:05044.
[16] F. R. K. Chung and E. N. Gilbert, Steiner trees for the regular simplex, *Bull. Inst. Math. Acad. Sinica* 4 (1976) 313-325; MR 58 # 21738.
[17] W. D. Smith, How to find Steiner minimal trees in Euclidean d-space, *Algorithmica* 7 (1992) 137-177; MR 93a:68064.
[18] J. Keane and G. Niklasch, Evaluating the 4-sausage constant s_4, unpublished note (1999).
[19] F. R. K. Chung and R. L. Graham, On Steiner trees for bounded point sets, *Geom. Dedicata* 11 (1981) 353-361; MR 82j:05072.
[20] H. T. Croft, K. J. Falconer, and R. K. Guy, *Unsolved Problems in Geometry*, Springer-Verlag, 1991, sect. F15; MR 95k:52001.
[21] T. L. Snyder, Worst-case minimum rectilinear Steiner trees in all dimensions, *Discrete Comput. Geom.* 8 (1992) 73-92; MR 93c:90067.
[22] W. D. Smith, *Studies in Computational Geometry Motivated by Mesh Generation*, Ph.D. thesis, Princeton Univ., 1988.

8.7 エルミートの定数

n 次元空間において等しい大きさで重なり合うことのない球による最も密な(格子あるいは非格子)充填(packing)は何か[1,2]? $n=1$ のとき,この問題は,等しい長さの線分による直線の充填(タイル貼り)に対応している.よって最大密度 Δ_n は明らかに $\Delta_1=1$ である.$n=2$ のときは,平面での円による六角格子充填を考えることにより $\Delta_2=\pi/\sqrt{2}=0.9068996821\cdots$ とわかる.これは Thue(トゥエ)[3,4]により初めて証明された.続いて Fejes Tóth(フェイエトス)[5,6]や Segre & Mahler(セグレ&マーラー)[7]によって証明が見出された.$n=3$ のときは,3次元の球による面心立方充填により $\Delta_3=\pi/\sqrt{18}=0.7404804896\cdots$ とわかる.これは,Hales(ヘイルズ)[8-10]によって初めて証明されるまでは,Keler(ケプラー)予想という名でよく知られていた.

表 8.4 エルミートの定数 δ_n, γ_n^n

n	δ_n	γ_n^n
1	1	1
2	$\dfrac{\pi}{2\sqrt{3}}=0.9068996821\cdots$	$\dfrac{4}{3}$
3	$\dfrac{\pi}{3\sqrt{2}}=0.7404804896\cdots$	2
4	$\dfrac{\pi^2}{16}=0.6168502750\cdots$	4
5	$\dfrac{\pi^2}{15\sqrt{2}}=0.4652576133\cdots$	8
6	$\dfrac{\pi^3}{48\sqrt{3}}=0.3729475455\cdots$	$\dfrac{64}{3}$
7	$\dfrac{\pi^3}{105}=0.2952978731\cdots$	64
8	$\dfrac{\pi^4}{384}=0.2536695079\cdots$	256

では $n≧4$ のとき, Δ_n はどうなるのか? 4次元での非格子充填は, 格子充填よりよい結果を導き出せるのだろうか?

格子充填に限れば, $n≦8$ のときの最大密度 δ_n がわかっている. n 次元の単位球の体積を $\omega_n=\pi^{n/2}\Gamma(n/2+1)^{-1}$ とし, 位数 n の Hermite (エルミート) の定数を

$$\gamma_n = 4\left(\frac{\delta_n}{\omega_n}\right)^{2/n}$$

とする. 表 8.4 に, 小さい n に対してわかっている結果をまとめる. また十分大きな n に対しては,

$$-1 \leq \frac{\log_2(\delta_n)}{n} \leq \frac{\log_2(\Delta_n)}{n} \leq -0.59905\cdots, \quad \frac{1}{2\pi e} \leq \frac{\gamma_n}{n} \leq \frac{1.74338\cdots}{2\pi e}$$

であることを証明できる. $c=-0.59905$ や $4^{1+c}=1.74338\cdots$ を用いた評価式は, 複雑である. これは, Kabatyanskii & Levenshtein (カバチャンスキー&レベンシュテイン) [1,13,14] によるものである. $c=-1$ と信じられており [15]. これを用いると $n\to\infty$ のとき $\gamma_n/n \to 1/(2\pi e)$ となるが, 極限が存在するかどうかさえわかっていない [11]. すべての n に対して γ_n^n は必ず有理数になるのか? エルミート定数 γ_n は, 2次形式の研究や符号理論においても重要である.

[1] J. H. Conway and N. J. A. Sloane, *Sphere Packings, Lattices and Groups*, Springer-Verlag, 1988, pp. 1-21; MR 2000b:11077.
[2] H. T. Croft, K. J. Falconer, and R. K. Guy, *Unsolved Problems in Geometry*, Springer-Verlag, 1991, sect. D10, D11; MR 95k:52001.
[3] C. A. Rogers, *Packing and Covering*, Cambridge Univ. Press, 1964, pp. 1-11, 80-85; MR 30 # 2405.
[4] A. Thue, Über die dichteste Zusammenstellung von kongruenten Kreisen in einer Ebene, *Videnskapsselskapets Skrifter I, Matematisk-Naturvidenskapelig Klasse, Kristiania*, n. 1, Dybwad, 1910, pp. 1-9; also in *Selected Mathematical Papers*, ed. T. Nagell, A. Selberg, S. Selberg, and K. Thalberg, Universitetsforlaget, 1977, pp. 257-263; MR 57 # 46.
[5] L. Fejes Tóth, Über einen geometrischen Satz, *Math. Z.* 46 (1940) 83-85; MR 1,263g.

[6] L. Fejes Tóth, *Lagerungen in der Ebene, auf der Kugel und im Raum*, Springer-Verlag, 1953, pp. 58-61; MR 15,248b.
[7] B. Segre and K. Mahler, On the densest packing of circles, *Amer. Math. Monthly* 51 (1944) 261-270; MR 6,16c.
[8] T. C. Hales, Cannonballs and honeycombs, *Notices Amer. Math. Soc.* 47 (2000) 440-449; MR 2000m:52027.
[9] T. C. Hales, An overview of the Kepler conjecture, math.MG/9811071.
[10] N. J. A. Sloane, Kepler confirmed, *Nature* 395 (1998) 435-436.
[11] P. M. Gruber and C. G. Lekkerkerker, *Geometry of Numbers*, North-Holland, 1987, pp. 385-392, 409-410; MR 88j:11034.
[12] J. C. Lagarias, Point lattices, *Handbook of Combinatorics*, v. I, ed. R. Graham, M. Grötschel, and L. Lovász, MIT Press, 1995, pp. 919-966; MR 96m:11051.
[13] G. A. Kabatyanskii and V. I. Levenshtein, Bounds for packings on a sphere and in space (in Russian), *Problemy Peredachi Informatsii* 14 (1978) 3-25; Engl. transl. in *Problems Information Transmission* 14 (1978) 1-17.
[14] W. D. Smith, *Studies in Computational Geometry Motivated by Mesh Generation*, Ph.D. thesis, Princeton Univ., 1988.
[15] P. M. Gruber, Geometry of numbers, *Handbook of Convex Geometry*, v. B, ed. P. M. Gruber and J. M. Wills, North-Holland, 1993, pp. 739-763; MR 94k:11074.

8.8 タムスの定数

$S=\{(u,v,w):u^2+v^2+w^2=1\}$ で3次元空間での単位球面を，そして $|p-q|$ で2点 p,q の間のユークリッド距離を表す．N を2以上の整数とし，a を実数とする．S 上の有限集合 $\omega_N=\{x_1,x_2,\cdots,x_N\}$ に付随する a-**エネルギー**とは，

$$\varepsilon(a,\omega_N)=\begin{cases}\sum_{i<j}|x_i-x_j|^a & (a\neq 0 \text{ のとき})\\ \sum_{i<j}\ln\left(\dfrac{1}{|x_i-x_j|}\right) & (a=0 \text{ のとき})\end{cases}$$

である．S 上の N 個の点に対する**極値エネルギー** $\varepsilon(a,N)$ を

$$E(a,N)=\begin{cases}\min_{\omega_N\subseteq S}\varepsilon(a,\omega_N) & (a\leq 0 \text{ のとき})\\ \max_{\omega_N\subseteq S}\varepsilon(a,\omega_N) & (a>0 \text{ のとき})\end{cases}$$

と定義する．$E(a,N)$ の値や最小あるい最大エネルギーとなる ω_N の代表的配置などはとても興味深い．応用として，符号理論，静電気学，結晶学，植物学，幾何学，計算の複雑さの理論を含んでいる．ここではいくつかの結果だけを注意しよう．1-エネルギーを最大にすることは，すべての点の組の間の平均距離を最大にすることと同じである [1-5]．

$$\lim_{N\to\infty}\frac{1}{N^2}E(1,N)=\frac{2}{3}$$

が証明できる．そして
$$\lim_{N\to\infty}\frac{E(1,N)-\frac{2}{3}N^2}{N^{1/2}}=\lambda$$
であることがわかる．ここで精密な不等式 $-2.5066282746\cdots=-\sqrt{2\pi}\leqq\lambda<0$ と推定値 $\lambda=-0.40096\cdots$ がある[6,7]．

$E(-1,N)$ を決定することは，(粒子がクーロン (Coulomb) ポテンシャルに従って互いに反発していると仮定するとき) 球面上で平衡状態になるように等しい点電荷を配置することに対応する．これは Thomson (トムソン) の電子問題として知られている．そして最適な点の配置は Fekete (フェケテ) 点とよばれる[8-13]．
$$\lim_{N\to\infty}\frac{1}{N^2}E(-1,N)=\frac{1}{2}$$
が証明できる．Wagner (ワグナー)[14,15]の研究に基づき Kuijlaars & Saff (クァイラース&サフ)[16,17]は
$$\lim_{N\to\infty}\frac{E(-1,N)-\frac{1}{2}N^2}{N^{3/2}}=\sqrt{3}\left(\frac{\sqrt{3}}{8\pi}\right)^{1/2}\zeta\left(\frac{1}{2}\right)\left(\zeta\left(\frac{1}{2},\frac{1}{3}\right)-\zeta\left(\frac{1}{2},\frac{2}{3}\right)\right)$$
$$=-0.5530512933\cdots$$
であると予想した．ここで $\zeta(s)$ は普通の Riemann のゼータ関数で
$$\zeta(s,a)=\sum_{\substack{k=0\\k+a\neq 0}}^{\infty}\frac{1}{(k+a)^s}$$
は，(解析接続した) Hurwitz (フルヴィツ) のゼータ関数である．この予想が正しいという注目に値するような理論的かつ実験的証拠がある．

0-エネルギーを最小にすることは，距離の積 $\prod_{i<j}|x_i-x_j|$ を最大にすることと同値であり，
$$\lim_{N\to\infty}\frac{E(0,N)-\left(-\frac{1}{4}\ln\left(\frac{4}{e}\right)N^2-\frac{1}{4}N\ln(N)\right)}{N}=\mu$$
が知られている．ここで $-0.1127687700\cdots\leqq\mu\leqq-0.0234972918\cdots$ という精密な不等式と $\mu=-0.026422\cdots$ という推定値がある[6,7,18,19]．

$a\to-\infty$ のとき a-エネルギーは最小の距離を含む項にますます影響を受けるようになる．すなわち
$$\lim_{a\to-\infty}\varepsilon(a,\omega_N)^{1/a}=\min_{i<j}|x_i-x_j|$$
となる．それゆえ，最小エネルギー問題は，
$$d_N=\max_{\omega_N}\min_{i<j}|x_i-x_j|$$
を計算することに帰し，Tammes (タムス) の花粉に関する1930年の問題[8,21-28]に対する解答になる．同値な問題として「(重なり合うことなく) S 上につめこみうる N 個の合同な円の直径の最大は何か？」がある．$N\to\infty$ のとき

$$d_N = \left(\frac{8\pi}{3}\right)^{1/2} N^{-1/2} + O(N^{-2/3})$$

とわかっている．誤差項のより精密な評価は，明らかにまだなされていない．Fejes Tóth（フェイエトス）[26,29,30]と van der Waerden（ファン・デル・ヴェルデン）[26,31]により限界

$$2\left[\frac{\sqrt{3}}{2\pi}N + 3\left(\frac{N}{4\pi}\right)^{2/3} + 3\left(\frac{N}{4\pi}\right)^{1/3}\right]^{-1/2} \leq d_N \leq \left[4 - \csc\left(\frac{\pi}{6}\frac{N}{N-2}\right)^2\right]^{1/2}$$

が示された．関連する問題として S を「被覆」しうる N 個の合同な円の直径の最小値を求めるもの[32]や，単位「円板」上の N 個の平衡をなす点電荷の配置を求めるものがある[33]．

[1] L. Fejes Tóth, On the sum of distances determined by a pointset, *Acta Math. Acad. Sci. Hungar.* 7 (1956) 397-401; MR 21 # 5937.
[2] R. Alexander, On the sum of distances between n points on a sphere, *Acta Math. Acad. Sci. Hungar.* 23 (1972) 443-448; MR 47 # 957.
[3] K. B. Stolarsky, Sum of distances between n points on a sphere. II, *Proc. Amer. Math. Soc.* 41 (1973) 575-582; MR 48 # 12314.
[4] J. Berman and K. Hanes, Optimizing the arrangement of points on the unit sphere, *Math. Comp.* 31 (1977) 1006-1008; MR 57 # 17502.
[5] J. Beck, Sums of distances between points on a sphere—An application of the theory of irregularities of distribution to discrete geometry, *Mathematica* 31 (1984) 33-31; MR 86d:52004.
[6] Y. M. Zhou, *Arrangements of Points on the Sphere*, Ph.D. thesis, Univ. of South Florida, 1995.
[7] E. A. Rakhmanov, E. B. Saff, and Y. M. Zhou, Minimal discrete energy on the sphere, *Math. Res. Lett.* 1 (1994) 647-662; MR 96e:78011.
[8] L. L. Whyte, Unique arrangements of points on a sphere, *Amer. Math. Monthly* 59 (1952) 606-611; MR 14,310c.
[9] T. Erber and G. M. Hockney, Equilibrium configurations of N equal charges on a sphere, *J. Phys. A* 24 (1991) L1369-L1377.
[10] L. Glasser and A. G. Every, Energies and spacings of point charges on a sphere, *J. Phys. A* 25 (1992) 2473-2482.
[11] K. S. Brown, Min-energy configurations of electrons on a sphere (MathPages).
[12] R. H. Hardin, N. J. A. Sloane, and W. D. Smith, Minimal energy arrangements of points on a sphere (AT&T Labs Research).
[13] T. Erber and G. M. Hockney, Complex systems: Equilibrium configurations of N equal charges on a sphere ($2 \leq N \leq 112$), *Advances in Chemical Physics*, v. 98, ed. I. Prigogine and S. A. Rice, Wiley, 1997, pp. 495-594; MR 98e:78002.
[14] G. Wagner, On means of distances on the surface of a sphere (lower bounds), *Pacific J. Math.* 144 (1990) 389-398; MR 91e:52014.
[15] G. Wagner, On means of distances on the surface of a sphere (upper bounds), *Pacific J. Math.* 154 (1992) 381-396; MR 93b:52007.
[16] A. B. J. Kuijlaars and E. B. Saff, Asymptotics for minimal discrete energy on the sphere, *Trans. Amer. Math. Soc.* 350 (1998) 523-538; MR 98e:11092.
[17] E. B. Saff and A. B. J. Kuijlaars, Distributing many points on a sphere, *Math. Intellig.* 19 (1997) 5-11; MR 98h:70011.
[18] G. Wagner, On the product of distances to a point set on a sphere, *J. Austral. Math. Soc. A* (1989) 466-482; MR 90j:11080.
[19] B. Bergersen, D. Boal, and P. Palffy-Muhoray, Equilibrium configurations of particles on a

sphere: The case of logarithmic interactions, *J. Phys. A* 27 (1994) 2579-2568.
[20] S. Smale, Mathematical problems for the next century, *Math. Intellig.*, v. 20 (1998) n. 2, 7-15; MR 99h:01033.
[21] J. H. Conway and N. J. A. Sloane, *Sphere Packings, Lattices and Groups*, Springer-Verlag, 1988; MR 2000b:11077.
[22] H. T. Croft, K. J. Falconer, and R. K. Guy, *Unsolved Problems in Geometry*, Springer-Verlag, 1991, sect. D7; MR 95k:52001.
[23] B.W. Clare and D. L. Kepert, The closest packing of equal circles on a sphere, *Proc. Royal Soc. London A* 405 (1986) 329-344; MR 87i:52024.
[24] T. Tarnai and Zs. Gáspár, Arrangement of 23 points on a sphere (on a conjecture of R. M. Robinson), *Proc. Royal Soc. London A* 433 (1991) 257-267; MR 92m:52043.
[25] D. A. Kottwitz, The densest packing of equal circles on a sphere, *Acta Cryst.* A47 (1991) 158-165; errata A47 (1991) 851; MR 92m:52042.
[26] J. B. M. Melissen, *Packing and Coverings with Circles*, Ph.D. thesis, 1997, Universiteit Utrecht.
[27] K. J. Nurmela, Constructing spherical codes by global optimization methods, research report A32, Helsinki Univ. of Technol.
[28] R. H. Hardin, N. J. A. Sloane, and W. D. Smith, Spherical codes (packings) (AT&T Labs Research).
[29] L. Fejes Tóth, On the densest packing of spherical caps, *Amer. Math. Monthly* 56 (1949) 330-331; MR 10,731b and MR 11,870 errata/addenda.
[30] L. Fejes Tóth, *Regular Figures*, MacMillan, 1964; MR 29 # 2705.
[31] B. L. van der Waerden, Punkte auf der Kugel. Drei Zusätze, *Math. Annalen* 125 (1952) 213-222; corrigendum 152 (1963) 94; MR 14,401c.
[32] R. H. Hardin, N. J. A. Sloane, and W. D. Smith, Spherical coverings (AT&T Labs Research).
[33] K. J. Nurmela, Minimum-energy point charge configurations on a circular disk, *J. Phys. A* 31 (1998) 1035-1047.

8.9 双曲的体積定数

まず，ある数え上げの問題を述べる．n を正の整数とする．n 次元ユークリッド空間内で一般の位置にある $n+1$ 個の点の凸胞を **n 単体**という．たとえば，1 単体は線分であり，2 単体は三角形（とその内部）であり，3 単体は四面体（とその内部）である．

n 立方体が下記のような条件の下で，異なる内部をもつ有限個の n 単体に分割されるとき**三角形分割される**（あるいは，より正確には，**面対面頂点三角形分割される**）という．

・任意の n 単体の頂点は，立方体の頂点でもある．
・任意の 2 つの n 単体の交わりは，その 2 つの単体の各々の面である．

n 立方体を三角形分割するために必要な n 単体の最小数を**単体度** $f(n)$ と定義する（図 8.6）．膨大な量の計算により表 8.5 の $f(n)$ の値や表 8.6 の $f(n)$ の上界，下界が導かれる[1-7]．1 つの未解決問題は，すべて n に対して有効な $f(n)$ のきちんとした下界を決定することである．これを行う 1 つの試みを手短に述べる．

8.9 双曲的体積定数

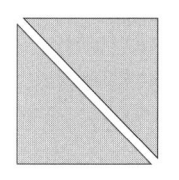

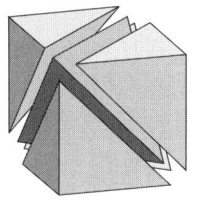

図 8.6 n 立方体の三角形分割. $f(2)=2$, $f(3)=5$.

表 8.5 単位度の値

n	1	2	3	4	5	6	7
$f(n)$	1	2	5	16	67	308	1493

表 8.6 単体度 $f(n)$ の評価

n	8	9	10
下界	5522	26593	131269
上界	13136	105341	928780

n 次元単位球面に内接する正 n 単体を**標準 n 単体** S_n とよぶ. (たとえば S_2 は, 面積が $3\sqrt{3}/4$ の正三角形である.) 各辺が $2/\sqrt{n}$ の長さをもち, 原点を中心とする n 立方体を**標準 n 立方体** C_n という. 明らかに

$$S_n \text{の体積} = \frac{\sqrt{n+1}}{n!}\left(1+\frac{1}{n}\right)^{n/2}, \quad C_n \text{の体積} = \left(\frac{4}{n}\right)^{n/2}$$

である. $f(n)$ を最小にしようとする最もよく知られた試みには, 次の積分

$$\xi_n = \text{理想双曲的 } n \text{ 立方体の体積} = \int_{C_n}\left(1-\sum_{k=1}^{n}x_k^2\right)^{-(n+1)/2}dx_1dx_2\cdots dx_n$$

$$\eta_n = \text{正則理想双曲的 } n \text{ 単体の体積} = \int_{S_n}\left(1-\sum_{k=1}^{n}x_k^2\right)^{-(n+1)/2}dx_1dx_2\cdots dx_n$$

が含まれている. さらに精密には, Smith (スミス) [8] や Marshall (マーシャル) により独立に示されたように

$$f(n) \geqq \frac{\xi_n}{\eta_n} \geqq \frac{1}{2}6^{n/2}(n+1)^{-(n+1)/2}n!$$

である. これは, かなり改良する余地が残っている——それは $f(n)$ と表にある限界との差がとても大きいからである[†]——しかし定数 ξ_n, η_n が現れるのは興味深い.

$\eta_2 = \pi$, $\eta_3 = \pi\ln(\beta) = 1.0149416064\cdots$ と示しうる. ここで β は【3.10】で定義された量である. そして [9-11] で

[†] 訳注: たとえばこの式では $f(8) \geqq \dfrac{1}{2}\cdot 6^{8/2}\cdot 9^{-9/2}\cdot 8! = \dfrac{1}{2}\cdot 6^4\cdot 3^{-9}\cdot 8! = 1327.\cdots$ である.

$$\eta_4 = \frac{10\pi}{3}\arcsin\left(\frac{1}{3}\right) - \frac{\pi^2}{3} = 0.2688956601\cdots, \quad \eta_5 = 0.05756\cdots$$

が示された．また $\xi_2 = 2\pi, \xi_3 = 5\eta_3 = 5.0747080320\cdots, \xi_4 = 3.92259368\cdots$ そして $\xi_5 = 2.75861972\cdots$ である[11,12]．漸近的には[8,9,12] $n \to \infty$ のとき

$$\eta_n \sim e\frac{\sqrt{n}}{n!}, \quad \xi_n \sim 2\sqrt{\pi}\frac{c^n}{\Gamma\left(\frac{n+1}{2}\right)}$$

ここで e は自然対数の底【1.3】であり，$c = 1.0820884492\cdots$ は，次の Dawson（ドウソン）の積分[13,14]の最大値の 2 倍である．すなわち

$$D(x) = \exp(-x^2)\int_0^x \exp(t^2)\,dt, \quad \frac{c}{2} = 0.5410442246\cdots = \frac{1.2615225101\cdots}{\sqrt{2e}}$$

これは，$x = 0.9241388730\cdots = 1/c$ のとき一意的に生じる．

この詳しい漸近的な情報にもかかわらず，ある定数 $\gamma > 0$ に対して $f(n) \geq \gamma^n n!$ が成立するかどうかは未知のままである[15]．

[1] R. K. Guy, A couple of cubic conundrums, *Amer. Math. Monthly* 91 (1984) 624-629.
[2] P. S. Mara, Triangulations for the cube, *J. Combin. Theory Ser. A* 20 (1976) 170-177; MR 53 # 10624.
[3] J. F. Sallee, A triangulation of the n-cube, *Discrete Math.* 40 (1982) 81-86;MR84d:05065b.
[4] J. F. Sallee, A note on minimal triangulation of the n-cube, *Discrete Appl. Math.* 4 (1982) 211-215; MR 84g:52019.
[5] J. F. Sallee, The middle-cut triangulations of the n-cube, *SIAM J. Algebraic Discrete Methods* 5 (1984) 407-419; MR 86c:05054.
[6] R. B. Hughes and M. R. Anderson, Simplexity of the cube, *Discrete Math.* 158 (1996) 99-150; MR 97g:90083.
[7] D. Eppstein, How many tetrahedra? (Geometry Junkyard).
[8] W. D. Smith, A lower bound for the simplexity of the N-cube via hyperbolic volumes, *Europ. J. Combin.* 21 (2000) 131-137; MR 2001c:52004.
[9] U. Haagerup and H. J. Munkholm, Simplices of maximal volume in hyperbolic n-space, *Acta Math.* 147 (1981) 1-11; MR 82j:53116.
[10] J. J. Seidel, On the volume of a hyperbolic simplex, *Studia Sci. Math. Hungar.* 21 (1986) 243-249; MR 88k:51040.
[11] W. D. Smith, *Studies in Computational Geometry Motivated by Mesh Generation*, Ph.D. thesis, Princeton Univ., 1988.
[12] J. Keane, Volumes of ideal hyperbolic cubes, unpublished note (1996).
[13] M. Abramowitz and I. A. Stegun, *Handbook of Mathematical Functions*, Dover, 1972, p. 298; MR 94b:00012.
[14] J. Spanier and K. B. Oldham, *An Atlas of Functions*, Hemisphere, 1987, pp. 405-409.
[15] J. A. De Loera, F. Santos, and F. Takeuchi, Extremal properties for dissections of convex 3-polytopes, *SIAM J. Discrete Math.* 14 (2001) 143-161; MR 2002g:52014.

8.10 ルーロー三角形定数

平面での，定幅1の集合のうちでReuleaux（ルーロー）三角形（図8.7参照）は，最小の面積をもち[1-11]，最も不均斉なものである[12-15]．この定理の内容で鍵になっているいくつかの言葉を注意深く調べてみよう．するといくつかの関連する定数を導くことができる．

コンパクトな凸集合$C \subseteq \mathbb{R}^2$が，すべての直線上へのCの正射影が同じ長さwとなるとき，**定幅**wをもつという．さらに一般に$d>2$の場合での$C \subseteq \mathbb{R}^d$に対する要求される条件は，「Cをはさむ2つの平行な$(d-1)$次元平面が常に同じ距離wだけ離れている」となる[†]．話を簡単にするため$w=1$とする．定理の最初の部分は，$C \subseteq \mathbb{R}^2$の面積$\mu(C)$が

$$\mu(C) \geq \frac{\pi - \sqrt{3}}{2} = 0.7047709230\cdots$$

を満たすということである．$C \subseteq \mathbb{R}^3$の体積$\mu(C)$は

$$\mu(C) \geq \left(\frac{2}{3} - \frac{\sqrt{3}}{4}\arccos\left(\frac{1}{3}\right)\right)\pi = 0.4198600459\cdots$$

を満たすと信じられている．これはReuleaux（ルーロー）三角形に類似のMeisser（メイサー）の四面体に対応している[1,16]．今までのところ最もよく知られた下界は$(3\sqrt{6}-7)\pi/3 = 0.364916225\cdots$ である．それゆえ，改良の余地がかなりある[8,11]．

不均斉性を定義するのは，もっと難しい．それは，主に互いに両立しえない概念があるからである．対称性に関するちょうど2つの測度に焦点をおく．それらはKovner-

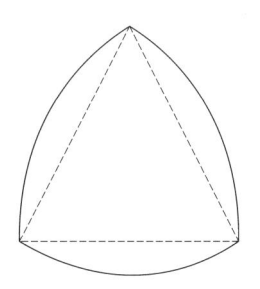

図8.7 ルーロー三角形（実線）は，正三角形（点線）の頂点と，この頂点の1つを中心にし，他の2項点を端点にもつ円弧より構成されている．

[†]原注：以前から使われているという理由で【8.4.1】では"width"ではなく"breadth"という語を用いた．

Besicovitch（コフナー-ベシコヴィッチ）（内）測度と Estermann（エスターマン）（外）測度とよばれるもので，それぞれ[14]

$$\sigma(C) = \frac{\mu(A)}{\mu(C)}, \quad \tau(C) = \frac{\mu(C)}{\mu(B)}$$

である．ここで A は C の最大の凸で中心対称[†]な部分集合であり，B は最小の凸で中心対称な優集合である．定理の2番目の部分は，$C \subseteq \mathbb{R}^2$ [8,12] に対して

$$\sigma(C) \geq \frac{6\arccos\left(\frac{5+\sqrt{33}}{12}\right)+\sqrt{3}-\sqrt{11}}{\pi - \sqrt{3}} = 0.8403426028\cdots$$
$$= 1 - 0.1596573971\cdots$$

$$\tau(C) \geq \frac{\pi - \sqrt{3}}{\sqrt{3}} = 0.8137993642\cdots = 1 - 0.1862006357\cdots$$

である．対応する優集合 B は，最小になるルーロー三角形 C のまわりに外接する正六角形である．部分集合 A は，C をその中心に関して鏡映を行い，できた新しい部分集合を C' とよぶことにし，$C \cap C'$ を作ることによって得られる円弧六角形である．この評価の高次元版は，知られていない．

もう1つ結果を示す．整数正方格子のどの頂点も含まず，最大の定幅 w をもつような集合 $C \subseteq \mathbb{R}^2$ は何か？ 答は，1つの対称軸が，2つの平行な格子辺の間にあるような向きにしたルーロー三角形である．その幅 $w = 1.5449417003\cdots$ は，最小多項式[9]

$$4x^6 - 12x^5 + x^4 + 22x^3 - 14x^2 - 4x + 4$$

をもつ．ルーロー三角形が，平面凸平行移動【8.3.1】，最大平面的出会い定数【8.21】や Bloch-Landau（ブロック-ランダウ）定数の厳密な値【7.1】に関する予想にも出現することを述べておく．

[1] T. Bonnesen and W. Fenchel, *Theory of Convex Bodies*, BCS Associates, 1987, pp. 135-149; MR 49 #9736.
[2] W. Blaschke, Konvexe Bereiche gegebener konstanter Breite und kleinsten Inhalts, *Math. Annalen* 76 (1915) 81-93.
[3] M. Fujiwara, Analytic proof of Blaschke's theorem on the curve of constant breadth with minimum area, *Proc. Imperial Akad. Japan* 3 (1927) 307-309; 7 (1931) 300-302.
[4] A. E. Mayer, Der Inhalt der Gleichdicke, *Math. Annalen* 110 (1934-35) 97-127.
[5] H. G. Eggleston, A proof of Blaschke's theorem on the Reuleaux triangle, *Quart. J. Math.* 3 (1952) 296-297; MR 14,496a.
[6] H. G. Eggleston, *Convexity*, Cambridge Univ. Press, 1958, pp. 122-131; MR 23 #A2123.
[7] A. S. Besicovitch, Minimum area of a set of constant width, *Convexity*, ed. V. L. Klee, Proc. Symp. Pure Math. 7, Amer. Math. Soc., 1963, pp. 13-14; MR 27 #1878.
[8] G. D. Chakerian, Sets of constant width, *Pacific J. Math.* 19 (1966) 13-21; MR 34 #4986.
[9] G. T. Sallee, The maximal set of constant width in a lattice, *Pacific J. Math.* 28 (1969) 669-674; MR 39 #2069.
[10] M. Gardner, Curves of constant width, one of which makes it possible to drill square holes, *Sci.*

[†]訳注：凸集合が**中心対称**であるとは，ある点（中心）が存在し，これが，この点を通るすべての弦を2等分するときをいう．

Amer., v. 208 (1963) n. 2, 148-156 and v. 208 (1963) n. 3, 154; also in *Mathematics: An Introduction to Its Spirit and Use*, W. H. Freeman, 1979, pp. 107-111 and 238-239.
[11] G. D. Chakerian and H. Groemer, Convex bodies of constant width, *Convexity and Its Applications*, ed. P. M. Gruber and J. M.Wills, Birkhäuser, 1983, pp. 49-96;MR85f:52001.
[12] A. S. Besicovitch, Measure of asymmetry of convex curves. II: Curves of constant width, *J. London Math. Soc.* 26 (1951) 81-93; MR 12,850g.
[13] H. G. Eggleston, Measure of asymmetry of convex curves of constant width and restricted radii of curvature, *Quart. J. Math.* 3 (1952) 63-72; MR 13,768d.
[14] B. Grünbaum, Measures of symmetry for convex sets, *Convexity*, ed. V. L. Klee, Proc. Symp. Pure Math. 7, Amer. Math. Soc., 1963, pp. 233-270; MR 27 #6187.
[15] H. Groemer and L. J. Wallen, A measure of asymmetry for domains of constant width, *Beiträge Algebra Geom.* 42 (2001) 517-521.
[16] I. M. Yaglom and V. G. Boltyanskii, *Convex Figures*, Holt, Rinehart and Winston, 1961, pp. 70-82, 242-264; MR 14,197d.

8.11 光探知定数

単位円 C に対する**光探知機**とは，C を横切るすべての直線（すなわち**光線**）を遮断する点集合のことをいう．明らかに C 自身は，C に対する光探知機になる．しかしこれは効率がよくない．この要請された条件を満たすもっと短い長さの曲線が存在する [1,2]．続きを述べる前に，「曲線」が何を意味するのかを説明する必要がある．

路とは，平面上のある区間の連続像であり，**弧**とは，自己交差しない路のことである．これが「曲線」という語の意味だとすると，完全な解答がある．Joris（ジョリス）[3] や Faba, Mycielski & Pederson（フェバー，ミシェルスキー&ペデルセン）[4,5] は，弓形弧（図 8.8）が，単位円と交わるすべての直線と交わる最短路であることを証明した．

「曲線」という概念を弱くすると，長さを相当短くすることができる．n 個の（連結でないこともありうる）弧の和を n **弧**という．Makai（マカイ）は，最小で，長さを

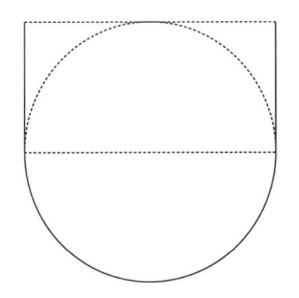

図 8.8　長さ $\pi+2 = 5.1415926535\cdots$ の弓形弧．

実際に求めることができる2弧を見つけた．これは，Thurston（サーストン）[1]により**弓と矢**の図とよばれた（図8.9）．Faber & Mycielska（ファベル&ミシェルスキー）[5]は，これを改良し，最小で，長さも求められる3弧を見つけた（図8.10）．これらの例はDay（デイ）[8]により再発見された．2弧の場合は，連立方程式

$$2\cos(\theta_1) - \sin\left(\frac{\theta_2}{2}\right) = 0, \quad \tan\left(\frac{\theta_1}{2}\right)\cos\left(\frac{\theta_2}{2}\right) + \sin\left(\frac{\theta_2}{2}\right)\left(\sec\left(\frac{\theta_2}{2}\right)^2 + 1\right) = 2, \quad \theta_3 = \theta_1$$

の解が角度

$$\theta_1 = \theta_3 = 1.2865112676\cdots \approx 73.71°, \quad \theta_2 = 1.1910478286\cdots \approx 68.24°$$

を与え，これは，2弧の長さに関する上界

$$L_2 \leq 2\pi - 2\theta_1 - \theta_2 + 2\tan\left(\frac{\theta_1}{2}\right) + \sec\left(\frac{\theta_2}{2}\right) - \cos\left(\frac{\theta_2}{2}\right) + \tan\left(\frac{\theta_1}{2}\right)\sin\left(\frac{\theta_2}{2}\right) = 4.8189264563\cdots$$

を導く．同様な方程式が3弧の長さに対する上界 $L_3 \leq 4.799891547\cdots$ を導く．しかし4弧や5弧に対応する結果については何も知られていない．**光探知定数**を

$$L = \inf_{n \geq 1} L_n \geq \pi$$

と定義する．ここで下界はCroft（クロフト）[9]やThurston（サーストン）[1]によるものである．数列 $\{L_n\}$ は真に減少すると思う人もいるが，$n \geq 4$ である n 弧で，3弧

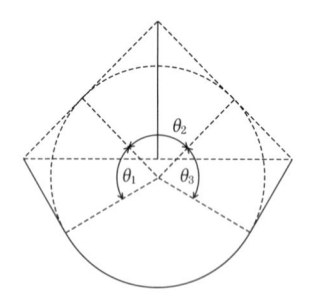

図8.9　2弧の長さは $4.8189264563\cdots$，ここで $\theta_1 = 1.2865\cdots$，$\theta_2 = 1.1910\cdots$，$\theta_3 = 1.2865\cdots$．「弓と矢」の名称は至極もっともである．

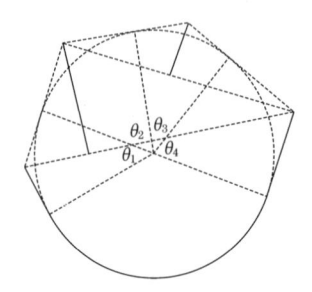

図8.10　3弧の長さは $4.799891547\cdots$，ここで $\theta_1 = 0.96\cdots$，$\theta_2 = 1.04\cdots$，$\theta_3 = 0.7\cdots$，$\theta_4 = 1.2\cdots$．

を改良することはできないと思う人もいる．

L は，しばしば**溝掘り人の定数**ともよばれる．どちらの方向へかはわからないが，地中にまっすぐなケーブルが埋められており，与えられた地点から単位長さ以内の所をそのケーブルが通っていることだけがわかっているとする．任意の $\varepsilon > 0$ に対して，全長が $L + \varepsilon$ の溝（多くの場合高度に不連結）を掘ることにより，このケーブルの位置を確実につきとめることができるような方策が存在する．関連する方策には，深いジャングルで道に迷った探険家や深い霧の海で泳いでいる人が，自分が（ジャングルや海の）まっすぐな境界から単位距離以内のところにいることがわかっているときの脱出路が含まれる[5]．これらは，「森での遭難」問題として知られているものの特別な場合である．

「路」の異なる一般化も可能である．区間の連続像の代わりに平面上の任意の連結開集合を考える．普通の長さの代わりに 1 次元 Hausdorff（ハウスドルフ）測度を考える．Eggleston（エグルストン）[9,13]は，このように拡張された曲線の族に対しても C に対する最適探知機は長さ $\pi + 2$ の弓形弧であると結論づけた．不思議なことに単位円を正三角形や正方形に変えると，最適な知られている連結探知機は，いくつかの枝をもつ「木」のような形で，頂点のシュタイナー生成木（Steiner span）とよばれるものになる【8.6】．正方形に対しては，円のときと同様に，連結性をはずした場合にはもっとよい結果をえる[5, 14-19]．単位正方形に対して予想された最適光探知機は，（図 8.11 に示したように）2 つの成分をもち，長さは $(2+\sqrt{3})/\sqrt{2} = 2.6389584337\cdots = 4(0.6597396084\cdots)$ である．

Eppstein（エプステイン）[1]は，凸多角形の最小の「不透明の森」を計算するためのアルゴリズムの設計との興味深い関連を指摘した[20-22]．光探知機の別の変種が[23-25]に現れている．

Zalgaller（ザルガラ）[26]は，最初の問題を次のように再定式化した．すなわち，平面上の単位開円板の外にある曲線で，この曲線に沿って動くとき，単位円 C のすべての点を見ることができるような曲線は何か？ そしてこれの 3 次元版を研究した．3 次元の単位開球の外にある曲線でこの曲線に沿って動くとき，単位球面 S のすべての点を見ることができるような最短曲線は何か？ 精密な検討を加えたものではないが，

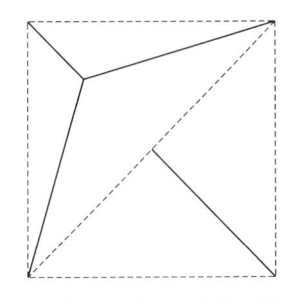

図 8.11 予想された最短の不透明正方形壁の長さは 2.6389584337… である．

Zalgaller は,長さ 9.56778… の「視察路」を得た.

[1] D. Eppstein, Building a better beam detector? (Geometry Junkyard).
[2] H. T. Croft, K. J. Falconer, and R. K. Guy, *Unsolved Problems in Geometry*, Springer-Verlag, 1991, sect. A30; MR 95k:52001.
[3] H. Joris, Le chasseur perdu dans la forêt, *Elem. Math.* 35 (1980) 1-14; MR 81d:52001.
[4] V. Faber, J. Mycielski, and P. Pedersen, On the shortest curve which meets all the lines which meet a circle, *Annales Polonici Math.* 44 (1984) 249-266; MR 87b:52023.
[5] V. Faber and J. Mycielski, The shortest curve that meets all the lines that meet a convex body, *Amer. Math. Monthly* 93 (1986) 796-801; MR 87m:52017.
[6] L. Fejes Tóth, Remarks on the dual to Tarski's plank problem (in Hungarian), *Mat. Lapok* 25 (1974) 13-20 (1977); MR 56 #1204.
[7] E. Makai, On a dual of Tarski's plank problem, *Diskrete Geometrie*, 2 Kolloq., Inst. Math. Univ. Salzburg, 1980, pp. 127-132; Zbl. 459/52005.
[8] I. Stewart, The great drain robbery, *Sci. Amer.*, v. 273 (1995) n. 3, 206-207; supplement, v. 273 (1995) n. 6, 106 and v. 274 (1996) n. 2, 125.
[9] H. T. Croft, Curves intersecting certain sets of great circles on the sphere, *J. London Math. Soc.* 1 (1969) 461-469; MR 40 #865.
[10] R. Bellman, Minimization problem, *Bull. Amer. Math. Soc.* 62 (1956) 270.
[11] J. R. Isbell, An optimal search pattern, *Naval Res. Logist. Quart.* 4 (1957) 357-359; MR 19,820a.
[12] S. R. Finch and J. E. Wetzel, Lost in a forest, unpublished note (2002).
[13] H. G. Eggleston, The maximal inradius of the convex cover of a plane connected set of given length, *Proc. London Math. Soc.* 45 (1982) 456-478; MR 84e:52004.
[14] R. E. D. Jones, Opaque sets of degree α, *Amer. Math. Monthly* 71 (1964) 535-537; MR 29 #2189.
[15] R. Honsberger, *Mathematical Morsels*, Math. Assoc. Amer., 1978, pp. 22-25; MR 58 #9950.
[16] K. A. Brakke, The opaque cube problem, *Amer. Math. Monthly* 99 (1992) 866-871; MR 94b:51025.
[17] B. Kawohl, The opaque square and the opaque cube, *General Inequalities* 7, ed. C. Bandle, W. N. Everitt, L. Losonczi, and W.Walter, Birkhäuser, 1997, pp. 339-346; MR 98f:52006.
[18] B. Kawohl, Symmetry or not?, *Math. Intellig.*, v. 20 (1998) n. 2, 16-22; MR 99e:00010.
[19] B. Kawohl, Some nonconvex shape optimization problems, *Optimal Shape Design*, Proc. 1998 Tróia conf., ed. A. Cellina and A. Ornelas, Lect. Notes in Math. 1740, Springer-Verlag, 2000, pp. 7-46.
[20] V. Akman, An algorithm for determining an opaque minimal forest of a convex polygon, *Inform. Process. Lett.* 24 (1987) 193-198; MR 88d:68094.
[21] P. Dublish, An $O(n^3)$ algorithm for finding the minimal opaque forest of a convex polygon, *Inform. Process. Lett.* 29 (1988) 275-276; MR 90a:68079.
[22] T. Shermer, A counterexample to the algorithms for determining opaque minimal forests, *Inform. Process. Lett.* 40 (1991) 41-42; MR 93e:68134.
[23] I. Bárány and Z. Füredi, Covering all secants of a square, *Intuitive Geometry*, Proc. 1985 Siófok conf., ed. K. Böröczky and G. Fejes Tóth, Colloq. Math. Soc. János Bolyai 48, North-Holland, 1987, pp. 19-27; MR 88h:52020.
[24] W. Kern and A. Wanka, On a problem about covering lines by squares, *Discrete Comput. Geom.* 5 (1990) 77-82; MR 90i:52016.
[25] T. J. Richardson and L. Shepp, The 'point' goalie problem, unpublished note (1997).
[26] V. A. Zalgaller, Shortest inspection lines for a sphere (in Russian), unpublished note (1992).

8.12 ソファー移動定数

単位幅の回廊の直角の角を回って動かすことのできる最長のはしご L の長さは何か？ はしごは，まっすぐで曲がらないものと仮定し，角を曲がるときもこの回廊の中に完全に入っていなければならないとしておく（ここでは，つねに 2 次元の場合に制限して考えることにする．図 8.12）．この問題の答は簡単である．L は点 c で交わる最短の線分 ab と同じ長さになる．これは明らかに $2\sqrt{2}$ という長さである[1]．

別の問題もある．W を角を回って動かしうる連結な折れ曲がらない針金とするとき，W の直径は，どれほどまで大きくできるか？ 連続可微分な曲線の**直径**とは，この曲線上の点 x, y の距離 $|x-y|$ の最大値として定義する．もしまったく曲がらない場合，これははしごの問題に帰着する．一般の場合，直径の最大は $2(1+\sqrt{2})$ になる（図 8.13）．最良の曲線 W は，点 c で交わり，一意的に決まる四半円弧 $\widehat{pq}$ である[2]．

もう少し難しい問題もある．角を回って動かすことができるソファー S の最大の面積は何か[3-5]？ S が平面連結領域であることのみ仮定する．Hammersley（ハマースレイ）[6]が，最大面積は少なくとも $\pi/2 + 2/\pi = 2.2074\cdots$ であること（図 8.14）を示したが，直観に反して，この領域は最適なものではない．

Gerver（ガーバー）（そして独立に Logan（ローガン））は，それまで調べられていたどれよりも大きな面積をもち複雑な境界をもつある種のソファーを構成した[7,8]（図 8.15）．さらに彼のソファーは次の条件を満たすソファーの族 Σ での「証明可能な」最適なものである．

- S が角を回るとき，90° 回転する．
- S が回転を始めたとき，初めは 2 点が壁に接触する．そして 4 点が接触し，（S が 45° 回転するとき）接触する点は，3 点になる．そして再び 4 点が接触し，S が回

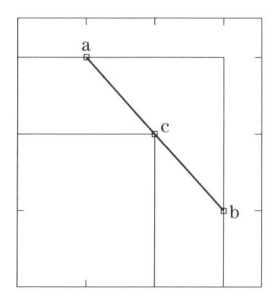

図 8.12 これは回廊の角を回って通ることのできる最長のはしごである．

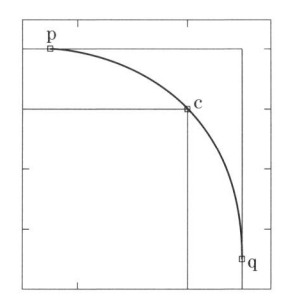

図 8.13 これは回廊の角を回って通ることのできる最大直径の針金である．

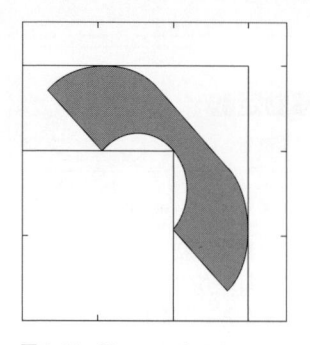

図 8.14 Hammersley のソファーは，半径 $2/\pi$ の半円を取り除いた $1\times 4/\pi$ 長方形と 1 の長さの辺の各々上の四半円とからなる．

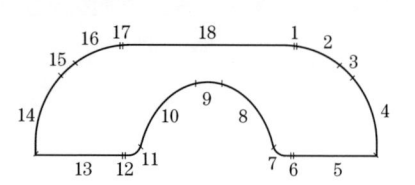

図 8.15 Gerver が予想した最適ソファーの境界は，18 個の分かれた辺からなる．

転を終わるとき，再び 2 点が壁に接触する．
もしこれより大きなソファーが見つかれば，とても驚異である．なぜなら，それは，Σ には属しえないはずだからである．

Gerver のソファーの面積は，何になるか？ この問題に答えるためには，まず連立する 4 つの方程式

$A(\cos(\theta)-\cos(\varphi))-2B\sin(\varphi)+(\theta-\varphi-1)\cos(\theta)-\sin(\theta)+\cos(\varphi)+\sin(\varphi)=0$
$A(3\sin(\theta)+\sin(\varphi))-2B\cos(\varphi)+3(\theta-\varphi-1)\sin(\theta)+3\cos(\theta)-\sin(\varphi)+\cos(\varphi)=0$

$$A\cos(\varphi)-\left(\sin(\varphi)+\frac{1}{2}-\frac{1}{2}\cos(\varphi)+B\sin(\varphi)\right)=0$$

$$\left(A+\frac{\pi}{2}-\varphi-\theta\right)-\left(B-\frac{1}{2}(\theta-\varphi)(1+A)-\frac{1}{4}(\theta-\varphi)^2\right)=0$$

より定数 A, B, φ, θ を計算し，$A=0.0944265608\cdots$, $B=1.3992037273\cdots$, $\varphi=0.039177647\cdots$, そして $\theta=0.6813015093\cdots$ を得る．次に，

$$r(\alpha)=\begin{cases}\dfrac{1}{2} & (0\leq\alpha<\varphi \text{ のとき})\\ \dfrac{1}{2}(1+A+\alpha-\varphi) & (\varphi\leq\alpha<\theta \text{ のとき})\\ A+\alpha-\varphi & \left(\theta\leq\alpha<\dfrac{\pi}{2}-\theta \text{ のとき}\right)\\ B-\dfrac{1}{2}\left(\dfrac{\pi}{2}-\alpha-\varphi\right)(1+A)-\dfrac{1}{4}\left(\dfrac{\pi}{2}-\alpha-\varphi\right)^2 & \left(\dfrac{\pi}{2}-\theta\leq\alpha<\dfrac{\pi}{2}-\varphi \text{ のとき}\right)\end{cases}$$

$$s(\alpha)=1-r(\alpha)$$

$$u(\alpha) = \begin{cases} B - \frac{1}{2}(\alpha-\varphi)(1+A) - \frac{1}{4}(\alpha-\varphi)^2 & (\varphi \leq \alpha < \theta \text{ のとき}) \\ A + \frac{\pi}{2} - \varphi - \alpha & \left(\theta \leq \alpha < \frac{\pi}{4} \text{ のとき}\right) \end{cases}$$

とし，u' は u の導関数とする．3つの関数 y_1, y_2, y_3 を

$$y_1(\alpha) = 1 - \int_0^\alpha r(t)\sin(t)\,dt, \quad y_2(\alpha) = 1 - \int_0^\alpha s(t)\sin(t)\,dt,$$
$$y_3(\alpha) = y_2(\alpha) - u(\alpha)\sin(\alpha)$$

により定義する．このとき，最適ソファーの面積は，$2.2195316688\cdots$ である．すなわち

$$2\int_0^{\pi/2-\varphi} y_1(\alpha)\,r(\alpha)\cos(\alpha)\,d\alpha + 2\int_0^\theta y_2(\alpha)\,s(\alpha)\cos(\alpha)\,d\alpha$$
$$+ 2\int_\varphi^{\pi/4} y_3(\alpha)(u(\alpha)\sin(\alpha) - u'(\alpha)\cos(\alpha) - s(\alpha)\cos(\alpha))\,d\alpha$$

である．この3つの積分はそれぞれ外側の境界の凸部分より下の面積，内側の境界の凸部分より上の面積，そして内側の境界（ここで回廊の角がソファーにこすれる）の凹部より上の面積を表している．

Sommers（ゾマーズ）[9]は，S が凸であるという条件を付け加えてこの問題を考え，数値的に最適面積が $1.644703\cdots$ 以上であると決定した．S が長方形のときは，たとえ回廊の角が直角ではなく，2つの通路が異なる幅のときでも，もっとよく知られている [10]．

この話題は，ロボット工学での動作計画に関連があり，特にピアノ運搬者問題[11]として知られている．n 次元空間での開部分集合と，U の2つのコンパクト部分集合 C_0, C_1 を与える．ここで C_1 は連続運動により C_0 から得られたものとする．このとき C_0 を U 内に完全に入れたまま C_1 へ動かすことができるか[12-15]？

[1] G. P. Vennebush, Move that sofa!, *Math. Teacher*, v. 95 (2002) n. 2, 92-97.
[2] P. E. Manne and S. R. Finch, A solution to the bent wire problem, *Amer. Math. Monthly* 109 (2002) 750-752.
[3] L. Moser, Moving furniture through a hallway, Problem 66-11, *SIAM Rev.* 8 (1966) 381-382.
[4] N. R. Wagner, The sofa problem, *Amer. Math. Monthly* 83 (1976) 188-189; MR 53 # 1422.
[5] H. T. Croft, K. J. Falconer, and R. K. Guy, *Unsolved Problems in Geometry*, Springer-Verlag, 1991, sect. G5; MR 95k:5200.
[6] J. M. Hammersley, On the emfeeblement of mathematical skills by "Modern Mathematics" and by similar soft intellectual trash in schools and universities, *Bull. Inst. Math. Appl.* 4 (1968) 66-85.
[7] J. L. Gerver, On moving a sofa around a corner, *Geom. Dedicata* 42 (1992) 267-283; MR 93d: 51040.
[8] I. Stewart, *Another Fine Math You've Got Me Into...*, W. H. Freeman, 1992, pp. 255-269; MR 93i:00003.
[9] J. A. Sommers, The convex sofa problem, unpublished note (2001).
[10] G. Eriksson, H. Eriksson, and K. Eriksson, Moving a food trolley around a corner, *Theoret. Comput. Sci.* 191 (1998) 193-203; MR 98k:68164.
[11] J. H. Davenport, A "piano movers" problem, *SIGSAM Bull.*, v. 20 (1986) n. 1, 15-17.

[12] B. Buchberger, G. E. Collins, and B. Kutzler, Algebraic methods in geometry, *Annual Review of Computer Science*, v. 3, ed. J. F. Traub, B. J. Grosz, B. W. Lampson, and N. J. Nilsson, Annual Reviews Inc., 1988, pp. 85-119; MR 91g:68156.

[13] D. Leven and M. Sharir, An efficient and simple motion planning algorithm for a ladder moving in two-dimensional space amidst polygonal barriers, *J. Algorithms* 8 (1987) 192-215; MR 88h:68035.

[14] E. B. Feinberg and C. H. Papadimitriou, Finding feasible points for a two-point body, *J. Algorithms* 10 (1989) 109-119; MR 90b:68099.

[15] M.-F. Roy, Géométrie algébrique réelle et robotique: La complexité du déménagement des pianos, *Gazette Math.* (Paris) 51 (1992) 75-96; MR 93f:93090.

8.13 カラビの三角形定数

T で正三角形を表すものとする．T 内にはめこまれる3つの合同な最大の正方形が明らかに存在する（図8.16）．この性質をもつ正三角形ではない三角形が存在するか？初めは，答は「否」だと思われるであろう．たとえば，直角三角形 U には，常に U 内にはめこまれるただ一つの最大な正方形，つまり U の垂直な2つの辺に沿った辺をもつ正方形があることがわかるからである．

Calabi（カラビ）は，この問題を研究し予想に反する答を見つけた[1,2]．3つの合同な最大正方形をもつ正三角形ではない三角形が一意的に存在する（図8.17）．それは二等辺三角形で，AB を底辺とし $AC=BC$ とすると，比

$$\frac{AB}{AC}=2\cos(\alpha)=1.5513875245\cdots$$

は，最小多項式 $2x^3-2x^2-3x+2$ をもつ代数的数である．また頂点 A の角度は

$$\alpha=0.6829826991\cdots \text{（ラジアン）} \sim 39.13°$$

と与えられる．さらに，関連した研究が Wetzel（ウェツェル）[3,4]によってなされた．1つの未解決問題を述べる．それは「この結果の3次元四面体版は何か？」である．

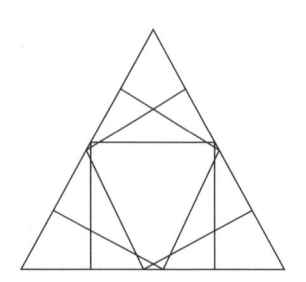

図8.16 最大の大きさの3つの（異なる）内接正方形をもつ正三角形．

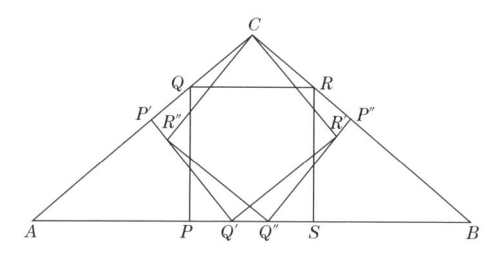

図 8.17 最大の大きさをもつ 3 つの内接正方形をもつ正三角形ではない三角形.

[1] J. H. Conway and R. K. Guy, *The Book of Numbers*, Springer-Verlag, 1996, p. 206; MR 98g:00004.
[2] E. Calabi, Outline of proof regarding squares wedged in triangle, unpublished note (1997).
[3] J. E. Wetzel, Squares in triangles, *Math. Gazette* 86 (2002) 28-34.
[4] J. E. Wetzel, Rectangles in triangles, submitted (2001).

8.14 デヴィッチの4次元立方体定数

単位立方体にどれくらいの大きさの正方形が内接しうるか？ これは Rupert（ルパート）王子の問題として知られている．さらに一般に，$m<n$ のとき単位 n 立方体にどれくらいの大きさの m 次元立方体が内接しうるか？

$f(m,n)$ を最適 m 立方体の辺の長さとする．すべての n に対して明らかに $f(1,n)=\sqrt{n}$ である．図 8.18 は，

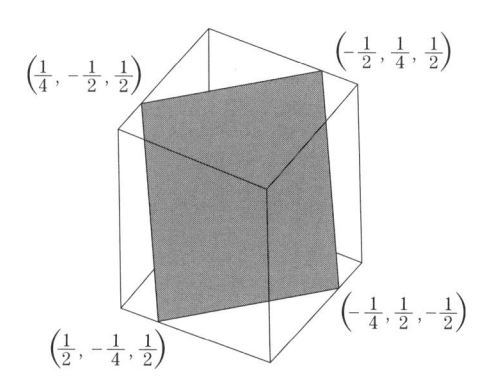

図 8.18 最大の内接正方形をもち（±1/2, ±1/2, ±1/2）を頂点とする 3 次元立方体.

$$f(2,3) = \frac{3}{4}\sqrt{2} = 1.0606601717\cdots$$

であることを示唆している，そしてこの結果は，古くから正しいことが知られている[1-4].

Devicci（デヴィッチ）[5]は，

$$f(m,n) = \sqrt{\frac{n}{m}} \quad (m\text{ が }n\text{ の約数のとき}), \quad f(2,n) = \begin{cases} \sqrt{\dfrac{n}{2}} & (n\text{ が偶数のとき}) \\ \sqrt{\dfrac{4n-3}{8}} & (n\text{ が奇数のとき}) \end{cases}$$

を証明した．手のこんだ議論により[5],

$$f(3,4) = 1.0074347569\cdots$$

とわかり，これは最小多項式 $4x^8 - 28x^6 - 7x^4 + 16x^2 + 16$ をもつ．実際に $f(3,4)$ を根号を用いて解くことができる．テッセラクト（tesseract）[†]という名称は，4次元立方体を表すものとしてしばしば用いられる[7]ので，$f(3,4)$ をここでは**デヴィッチの4次元立方体（テッセラクト）定数**とよぶ．Gardner（ガードナー）[8,9]によるとこの結果を数値的に予想した人々の中に Baer（ベーア），Bosch（ボッシュ），そして de Josselin de Jong（ド・ジョスラン・ド・ジョン）がいる．

Huber（ヒューバー）[10]は，$f(m,n)$ のより完全な値を決定した．たとえば

$$f(3,5) = \sqrt{11 - 4\sqrt{6}} = 1.0963763171\cdots$$

である．$f(m,n)$ は，つねに代数的数であることがわかっている[6]．対応する最小多項式の次数を，m,n の妥当なものと思われるような関数として表しうるか？

立方体で（正方形の代わりに）辺の比の決まった最大の長方形に対する同じ問題については，正方形の場合に比べると最近まで軽視されてきた[11].

---

[1] H. T. Croft, K. J. Falconer, and R. K. Guy, *Unsolved Problems in Geometry*, Springer-Verlag, 1991, sect. B4; MR 95k:52001.
[2] D. J. E. Schrek, Prince Rupert's problem and its extension by Pieter Nieuwland, *Scripta Math.* 16 (1950) 73-80, 261-267.
[3] D. Wells, *The Penguin Dictionary of Curious and Interesting Numbers*, Penguin, 1986, p. 33.
[4] J. G. Mauldon and R. J. Chapman, A variant of Prince Rupert's problem, *Amer. Math. Monthly* 102 (1995) 465-467.
[5] K. R. DeVicci, Largest m-cube in an n-cube, unpublished manuscript (1996).
[6] R. K. Guy and R. J. Nowakowski, Monthly unsolved problems, 1969-1997, *Amer. Math. Monthly* 104 (1997) 967-968.
[7] M. L'Engle, *A Wrinkle in Time*, Dell, 1962.
[8] M. Gardner, Hypercubes, *Sci. Amer.*, v. 215 (1966) n. 5, 138-145 and v. 215 (1966) n. 6, 131-132.
[9] M. Gardner, *The Colossal Book of Mathematics*, W. W. Norton, 2001, pp. 162-174.
[10] G. Huber, Cubing the square: A progress report on the Rupert problems, unpublished note (1999).
[11] R. P. Jerrard and J. E. Wetzel, Prince Rupert's rectangles, submitted (2002).

[†] 訳注：4方向に広がった図形の意味．四方体という仮の訳語がある．

8.15 グラハムの六角形定数

P を平面上の凸 n 角形とする．P の直径を 1 と仮定する．このことは，P の任意の 2 頂点間の最大距離が 1 であることと同値である．P によって囲まれる最大の面積は何か？　明らかに $F_3 = \sqrt{3}/4 = 0.4330127018\cdots$ であり，これは一辺の長さが 1 の正三角形のときにのみ成立する．さらに一般に，すべての n に対して

$$\frac{n}{8}\sin\left(\frac{2\pi}{n}\right) \le F_n \le \frac{n}{2}\cos\left(\frac{\pi}{n}\right)\tan\left(\frac{\pi}{2n}\right)$$

がわかる．Reinhardt（ラインハルト）[1] は，すべての奇数 n に対して右側の不等式は等式になること，そしてこの等号は，直径 1 の正 n 角形に対してのみ成立することを証明した．単純に，左の不等式は，偶数の n に対しては等式となり，同様な一意性が成立するとの期待をもつだろう．

$n=4$ のとき，左側の不等式は等式になる．しかしまったく別の観点から，偶数 n についての状況は，意外なものであることがわかる．直径が 1 の正方形のときだけでなく，直径が 1 である四辺形の無限個の族に対しても $F_4 = 1/2$ が成立する．よって $n=4$ のときには一意性は成り立たない．おもしろいことに，$n=6$ のときには一意性が成り立つ．$n = 8, 10, 12, \cdots$ に対して一意性が成立するかどうかは，わかっていない．

$n=6$ のときに焦点をしぼろう．直径が 1 の正六角形の面積は

$$\left.\frac{n}{8}\sin\frac{2\pi}{n}\right|_{n=6} = \frac{3\sqrt{3}}{8} = 0.6495190528\cdots$$

である．Graham（グラハム）[2-5] は，これは最適では「ない」という驚くべき結果を証明した．彼は，直径が 1 の六角形でその面積が $F_6 = 0.6749814429\cdots$ となるものを構成した．この F_6 の値は最小多項式

$$4096x^{10} + 8192x^9 - 3008x^8 - 30848x^7 + 21056x^6 + 146496x^5 - 221360x^4 + 1232x^3$$
$$+ 144464x^2 - 78488x + 11993$$

をもつ代数的数である（図 8.19）．

直径が 1 の八角形（$n=8$）のときの最大面積については何がいえるか？　Briggs（ブリグス），Prieto（プリエト），Vanderbei（ヴァンダベイ），Wright（ライト），Gay（ゲイ）などの人々は，$F_8 = 0.726868\cdots$ を数値的な大域的最適化手法を経て得た．さらに最近，Audet（オーデット）ら [6] が，2 次プログラミング手法を用いて最適八角形の形状についての Graham の予想を証明した．対応する最小多項式については未だわかっていない．

十角形（$n=10$），十二角形（$n=12$）について，きちんとした結果は知られていないが，数値的推定値は，それぞれ $F_{10} = 0.749137\cdots$，$F_{12} = 0.760729\cdots$ である．周の長さの方が面積よりも最大化しやすい [7]．高次元の場合は，あまりよく知られていない．

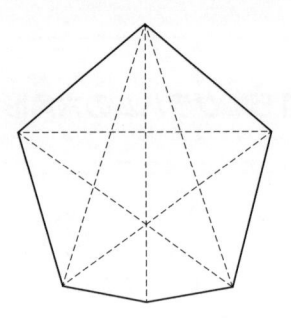

図 8.19 グラハムの六角形は,直径の数値が 1 の(最大の面積をもつという意味での)最適六角形である.

$d+2$ 個の頂点をもつ d 次元凸多面体の最大体積は既知であるが[8], $d+2$ 個より多くの頂点をもつ場合は未解決のままであることが確かである.

[1] K. Reinhardt, Extremale Polygone gegebenen Durchmessers, *Jahresbericht Deutsch. Math.-Verein.* 31 (1922) 251-270.
[2] R. L. Graham, The largest small hexagon, *J. Combin. Theory Ser. A* 18 (1975) 165-170; MR 50 # 12803.
[3] J. H. Conway and R. K. Guy, *The Book of Numbers*, Springer-Verlag, 1996, pp. 206-207; MR 98g: 00004.
[4] R. K. Guy and J. L. Selfridge, Optimal coverings of the square, *Infinite and Finite Sets*, Proc. 1973 Keszthely conf., ed. A. Hajnal, R. Rado, and V. T. Sós, Colloq. Math. Soc. János Bolyai 10, North-Holland, 1975, pp. 745-799; MR 51 # 13873.
[5] H. T. Croft, K. J. Falconer, and R. K. Guy, *Unsolved Problems in Geometry*, Springer-Verlag, 1991, sect. B6; MR 95k:52001.
[6] C. Audet, P. Hansen, F. Messine, and J. Xiong, The largest small octagon, *J. Combin. Theory Ser. A* 98 (2002) 46-59.
[7] N. K. Tamvakis, On the perimeter and the area of the convex polygons of a given diameter, *Bull. Soc. Math. Grèce* 28 A (1987) 115-132; MR 89g:52008.
[8] B. Kind and P. Kleinschmidt, On the maximal volume of convex bodies with few vertices, *J. Combin. Theory Ser. A* 21 (1976) 124-128; MR 53 # 11500.

8.16 ハイルブロンの三角形定数

第 n Heilbronn(ハイルブロン)の三角形定数とは,次が成り立つようなすべての数 H の下限 H_n のことである[1].すなわち単位正方形にどのように n 個の点の配置が与えられたときも,これらの点からとった任意の 3 点からなる三角形の最小の面積が H

8.16 ハイルブロンの三角形定数

以下になる.

Goldberg (ゴールドバーグ)[2]は, $H_3=H_4=1/2=0.5$ を含む最初のいくつかのハイルブロン定数を導き, いくつかの予想もした. Yang, Shang & Zeng (ヤン, シャン&ゼン)[3,4]は, $H_5=\sqrt{3}/9=0.1924500897\cdots$ を示し, Goldberg の予想の 1 つが正しくないことを示した. しかし, 彼らは Goldberg の $H_6=1/8=0.125$ という予想は正しいことを示した. 図 8.20, 図 8.21 を参照. また, $H_7 \geq 0.0838590090\cdots$ が知られている. この下界は, 最小多項式 $152x^3+12x^2-14x+1$ をもつ. また

$$H_8 \geq \frac{\sqrt{13}-1}{36}=0.0723764243\cdots, \quad H_9 \geq \frac{9\sqrt{65}-55}{320}=0.0548759991\cdots$$

である. Comellas & Yebra (コメラス&イェブラ)[5]が, これらの下界はたぶん最適な値のようだと表明したが, 今のところ証明は, (まだ) できていないことを認めた.

H_n の漸近値についてはどんなことがいえるのか? Heilbronn は, 1950 年に $n \to \infty$ のとき $H_n=O(n^{-2})$ であると予想した. Roth (ロス), Schmidt (シュミット) らは, 任意の $\varepsilon>0$ に対して十分大きなすべての n に対して[6,7]

$$H_n=O(n^{-8/7+\varepsilon})$$

を示すことによって, 上記の予想の証明に向かって進んだ. しかしながら Komlós, Pintz & Szemerédi (コムロス, ピンツ&セメレディ)[8]は, 十分大きな n に対して

$$\frac{c\ln(n)}{n^2} \leq H_n$$

[8]が成り立つような定数 $c>0$ の存在を示すことにより, Heilbronn の予想が正しくないことを証明した. 彼らの証明はまったく構成的ではない. 下界に関する最近の別証明[9]は, 各 n に対してすべての三角形の面積が $c\ln(n)/n^2$ 以上となる n 個の点の配置を見つけるための多項式時間アルゴリズムを与える. 上界に関しては, ベキ指数の 8/7 を 2 に置きかえることができるか? これは難しい問題であり, すぐに完全な解答ができると期待する人は誰もいない.

Jiang, Li & Vitányi (ジャン, リー&ヴィターニ)[10,11]は, 単位正方形に n 個の

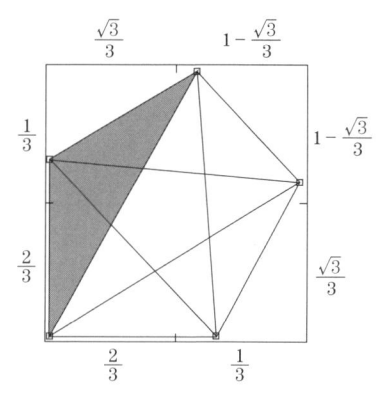

図 8.20　$n=5$ に対する最良の点配置. 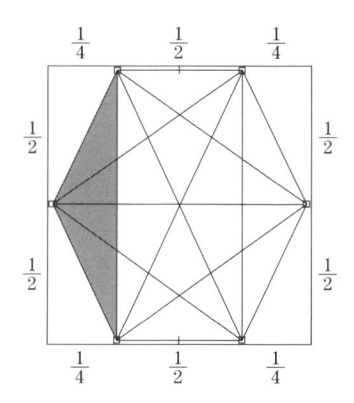　　図 8.21　$n=6$ に対する最良の点配置.

一様に分布した点が与えられたときの（最悪の場合ではなく）平均の場合の筋書を解析し，$0<a<b$ である定数 a,b に対して最小の三角形の面積の期待値が an^{-3} と bn^{-3} の間にあることを見つけた．Heilbronn の問題の高次元版の研究は，[12,13]でなされた．

H_n の定義における単位正方形を単位面積の正三角形に置きかえると $\tilde{H}_3=1$, $\tilde{H}_4=1/3$, $\tilde{H}_5=3-2\sqrt{2}$, $\tilde{H}_6=1/8$ である[14]．実際には，領域を等辺のものに限る必要はない．なぜなら $\tilde{H}_n$ は，考える単位面積三角形の形状には無関係だからである[6]．さらに，前に議論した漸近値が，平面上のコンパクト凸領域での n 個の点の場合にも（定数因子を付け加えるぐらいで）そのまま適用できる．

ここで漠然と関連のありそうな問題を述べる．単位正方形が m 個の連結成分に分解されているとする．d をこの m 個の集合の直径の最大値とする．d の最小値は何か[15-19]？ たとえば，$m=3$ とすると $d=\sqrt{65}/8=1.0077822185\cdots$ である．

もう1つの問題は，Dirichlet-Voronoi（ディリクレ-ボロノイ）細胞や別の幾何学的近接問題を思い起こさせる．k 個の点を単位正方形内にどのように配置すれば，この k 個の点から正方形への最短距離の平均距離を最小にできるか[20-21]？ $k\to\infty$ のとき，k 個の点は正六角形格子の頂点に近づいていく．たくさんの変種がある．最後に，単位正方形内の l 個の円板充塡の問題が，この正方形内の l 個の点の間の最小距離のうちの最大値を決める問題と同値であることを述べておく．

[1] R. K. Guy, *Unsolved Problems in Number Theory*, 2nd ed., Springer-Verlag, 1994, sect. F4; MR 96e:11002.
[2] M. Goldberg, Maximizing the smallest triangle made by N points in a square, *Math. Mag.* 45 (1972) 135-144; MR 45 # 5875.
[3] L.Yang, J. Z. Zhang, and Z. B. Zeng, Heilbronn problem for five points, Int. Centre Theoret. Physics preprint IC/91/252 (1991).
[4] L. Yang, J. Z. Zhang, and Z. B. Zeng, A conjecture on the first several Heilbronn numbers and a computation (in Chinese), *Chinese Annals Math. Ser. A* 13 (1992) 503-515; MR 93i:51045.
[5] F. Comellas and J. L. A. Yebra, New lower bounds for Heilbronn numbers, *Elec. J. Combin.* 9 (2002) R6.
[6] K. F. Roth, Developments in Heilbronn's triangle problem, *Adv. Math.* 22 (1976) 364-385; MR 55 # 2771.
[7] J. Komlós, J. Pintz, and E. Szemerédi, On Heilbronn's triangle problem, *J. London Math. Soc.* 24 (1981) 385-396; MR 82m:10051.
[8] J. Komlós, J. Pintz, and E. Szemerédi, A lower bound for Heilbronn's problem, *J. London Math. Soc.* 25 (1982) 13-24; MR 83i:10042.
[9] C. Bertram-Kretzberg, T. Hofmeister, and H. Lefmann, An algorithm for Heilbronn's problem, *SIAM J. Comput.* 30 (2000) 383-390; also in *Proc. 3rd Computing and Combinatorics Conf. (COCOON)*, Shanghai, 1997, ed. T. Jiang and D. T. Lee, Lect. Notes in Comp. Sci. 1276, Springer-Verlag, 1997, pp. 23-31; MR 2001b:05200.
[10] T. Jiang, M. Li, and P. Vitányi, The expected size of Heilbronn's triangles, *Proc. 14th IEEE Conf. on Computational Complexity (CCC)*, Atlanta, 1999, IEEE, pp. 105-113; math.CO/9902043.
[11] T. Jiang, M. Li, and P. Vitányi, The average-case area of Heilbronn-type triangles, *Random Structures Algorithms* 20 (2002) 206-219.
[12] G. Barequet, A lower bound for Heilbronn's triangle problem in d dimensions, *SIAM J. Discrete Math.* 14 (2001) 230-236; also in *Proc. 10th ACM-SIAM Symp. on Discrete Algorithms*

(SODA), Baltimore, ACM, 1999, pp. 76-81; MR 2002h:05159.
[13] G. Barequet, A duality between small-face problems in arrangements of lines and Heilbronn-type problems, *Discrete Math.* 237 (2001) 1-12; MR 2002c:52032.
[14] L. Yang, J. Z. Zhang, and Z. B. Zeng, On the Heilbronn numbers of triangular regions (in Chinese), *Acta Math. Sinica* 37 (1994) 678-689; MR 95k:52014.
[15] J. Lipman, Covering a square, *Amer. Math. Monthly* 65 (1958) 775.
[16] C. S. Ogilvy, *Tomorrow's Math: Unsolved Problems for the Amateur*, Oxford Univ. Press, 1962, pp. 20-21, 146-147.
[17] R. Honsberger, On sets of points in the plane, *Two Year College Math. J.* 11 (1980) 116-117.
[18] C. H. Jepsen, Coloring points in the unit square, *College Math. J.* 17 (1986) 231-237.
[19] R. K. Guy and J. L. Selfridge, Optimal coverings of the square, *Infinite and Finite Sets*, Proc. 1973 Keszthely conf., ed. A. Hajnal, R. Rado, and V. T. Sós, Colloq. Math. Soc. János Bolyai 10, North-Holland, 1975, pp. 745-799; MR 51 #13873.
[20] L. Fejes Tóth, The isepiphan problem for n-hedra, *Amer. J. Math.* 70 (1948) 174-180; MR 9,460f.
[21] L. Fejes Tóth, *Lagerungen in der Ebene auf der Kugel und im Raum*, Springer-Verlag, 1972; MR 50 #5603.
[22] B. Bollobás and N. Stern, The optimal structure of market areas, *J. Econom. Theory* 4 (1972) 174-179; MR 57 #8996.
[23] C. H. Papadimitriou, Worst-case and probabilistic analysis of a geometric location problem, *SIAM J. Comput.* 10 (1981) 542-557; MR 82h:90039.
[24] P. M. Gruber, A short analytic proof of Fejes Tóth's theorem on sums of moments, *Aequationes Math.* 58 (1999) 291-295; MR 2000j:52012.
[25] P. M. Gruber, Optimal configurations of finite sets in Riemannian 2-manifolds, *Geom. Dedicata* 84 (2001) 271-320; MR 2002f:52017.
[26] F. Morgan and R. Bolton, Hexagonal economic regions solve the location problem, *Amer. Math. Monthly* 109 (2002) 165-172.

8.17 掛谷-ベシコヴィッチ定数

平面での領域 R が**掛谷領域**であるとは，R 内で単位長さの線分を**逆転**させることができる，すなわち180°回転させた線分を元の位置に R からはみ出ることなくうまく連続的に動かすことができるときをいう．Kakeya（掛谷宗一）[1]は，このような領域 R の最小面積を問題にした[†]．

$$K = \inf_{R \text{ 掛谷}} \text{area}(R)$$

とする．ここで下限は，すべての掛谷集合をわたるものとする．Besicovitch（ベシコヴィッチ）[2,3]は，$K=0$ であるという驚くべき結果を証明した．これは，単位長さの線分が，いくらでも小さい面積の領域の中で逆転できることを示している．彼の証明では，非有界な（すなわち，大きな直径をもち）とても多重な連結領域（すなわち，たく

[†]訳注：掛谷問題の起源については，舟を旋回させるための水面とか，狭い室内で長い棒（たとえば槍）をふり回すためのスペースとかの諸説がある．掛谷自身は最小領域をアストロイドあるいはデルトイドだと予想したが，意外にも面積の下限は 0 という結果が得られた．

さんの穴をもつもの）を用いた．人々は，このように複雑な領域で考える必要があるのか，そしてRにさらなる制限を加えるとどうなるかと考えた．

van Alphen（ファン・アルペン）[8]は，任意の $\varepsilon>0$ に対して R を $2+\varepsilon$ の半径の円内にあると制限すると $K=0$ となることを証明した．よって有界性は，問題の核心ではない．後に，Cunningham（カニンガム）[9]は，R が単連結（すなわち穴をもたない）で半径1の円内にあるときでさえ $K=0$ となることを証明した．よって穴があいていないということも問題の論点ではない．これらは，非常に複雑な結果であり，幾何学以外でもこれらの重要性についての説明が[10-13]に見うけられる．

異なる制限は，異なる結果を生じさせる．

$$K_c = \inf_{\substack{R:凸 \\ 掛谷}} \text{area}(R)$$

とおく．（ここで R が凸とは，任意の2点 $P,Q \in R$ に対して線分 $PQ \subseteq R$ という意味である．）また

$$K_s = \inf_{\substack{R:星型 \\ 掛谷}} \text{area}(R)$$

とおく．（ここで R が星型とは，適当な点 $O \in R$ が存在し，任意の点 $P \in R$ に対して線分 $OP \subseteq R$ という意味である）．Pál[14]は，

$$K_c = \frac{\sqrt{3}}{3} = 0.5773502691\cdots$$

を証明した．この値は，高さ1の正三角形に対応している．

それとは対照的に，Bloom, Schoenberg & Cunningham（ブルーム，シェーンベルグ＆カニンガム）[6,9,15]は，

$$0.0290888208\cdots = \frac{\pi}{108} \le K_s \le \frac{5-2\sqrt{2}}{24}\pi = 0.2842582246\cdots = (0.0904822031\cdots)\pi$$

を証明し，Schoenberg は，さらに K_s が，この上界に等しいだろうと予想した．これがまだ未解決のままであるのは確かである．

[1] S. Kakeya, Some problems on maxima and minima regarding ovals, *Science Reports, Tôhoku Imperial Univ.* 6 (1917) 71-88.
[2] A. S. Besicovitch, On Kakeya's problem and a similar one, *Math. Z.* 27 (1928) 312-320.
[3] A. S. Besicovitch, The Kakeya problem, *Amer. Math. Monthly* 70 (1963) 697-706; MR 28 # 502.
[4] H. T. Croft, K. J. Falconer, and R. K. Guy, *Unsolved Problems in Geometry*, Springer-Verlag, 1991, sect. G6; MR 95k:52001.
[5] H. Meschkowski, *Unsolved and Unsolvable Problems in Geometry*, Oliver and Boyd, 1966, pp. 103-109; MR 35 # 7206.
[6] I. J. Schoenberg, On the Kakeya-Besicovitch problem, *Mathematical Time Exposures*, Math. Assoc. Amer., 1982, pp. 168-184; MR 85b:00001.
[7] K. J. Falconer, *The Geometry of Fractal Sets*, Cambridge Univ. Press, 1985, pp. 95-109; MR 88d:28001.
[8] H. J. van Alphen, Generalization of a theorem of Besicovitsch (in Dutch), *Mathematica, Zutphen B* 10 (1942) 144-157; MR 7,320b.
[9] F. Cunningham, The Kakeya problem for simply connected and for star-shaped sets, *Amer.*

Math. Monthly 78 (1971) 114-129; MR 43 # 1044.
[10] T. Wolff, Recent work connected with the Kakeya problem, *Prospects in Mathematics*, Proc. 1996 Princeton conf., ed. H. Rossi, Amer. Math. Soc., 1999, pp. 129-162; MR 2000d:42010.
[11] J. Bourgain, Harmonic analysis and combinatorics: How much may they contribute to each other?, *Mathematics: Frontiers and Perspectives*, ed. V. Arnold, M. Atiyah, P. Lax, and B. Mazur, Amer. Math. Soc., 2000, pp. 13-32; MR 2001c:42009.
[12] T. Tao, From rotating needles to stability of waves: Emerging connections between combinatorics, analysis, and PDE, *Notices Amer. Math. Soc.* 48 (2001) 294-303; amplification 566; commentary 678; MR 2002b:42021.
[13] N. Katz and T. Tao, Recent progress on the Kakeya conjecture, *Proc. 6th Conf. on Harmonic Analysis and Partial Differential Equations*, El Escorial, *Publicacions Matèmatiques* Extra Volume (2002) 161-179; math.CA/0010069.
[14] J. Pál, Ein minimumprobleme für ovale, *Math. Annalen* 83 (1921) 311-319.
[15] F. Cunningham and I. J. Schoenberg, On the Kakeya constant, *Canad. J. Math.* 17 (1965) 946-956; MR 31 # 6163.

8.18 直線的交点定数

G をグラフとする【5.6】．**直線的図**とは，G から平面内への写像で，頂点を異なる点へ，辺を線分に写すという性質をもつものをいう．このような G の直線的図の中で，平面での辺の交点の最小数 $\bar{\nu}(G)$ を決定せよ．この $\bar{\nu}(G)$ を G の**直線的交点数**とよぶ [1-4]．

n 個の頂点，$\binom{n}{2}$ 個の辺をもつ完全グラフ K_n に対して $\bar{\nu}(K_n)$ の既知の値，評価を表 8.7, 8.8 に示す [5-8]（$n=17$ まで既知）．

漸近的に，

$$0.311507 < \rho = \lim_{n \to \infty} \frac{\bar{\nu}(K_n)}{\binom{n}{4}} = \sup_n \frac{\bar{\nu}(K_n)}{\binom{n}{4}} \leq \frac{6467}{16848} < 0.383844$$

とわかる [8,9]．ρ に対する正確な値は未知である．

幾何確率論から Sylvester（シルベスター）[10] による関連があまりなさそうな問題をとりあげる．R を平面上の有限面積の開凸集合とする．R で独立かつ一様に配置された点から無作為に，4点を選ぶ．確率1で（4点のうちの）どの3点も同一直線上にはない．よってこの4点のつくる凸胞は，三角形（他の3点の凸胞内に残りの1点がある）か四辺形かのいずれかである．$q(R)$ は，この凸胞が四辺形である確率を表すものとする．Sylvester は，平面でのすべての凸集合 R に関する $q(R)$ の最小値，最大値

表 8.7 $\bar{\nu}(K_n)$ の値

n	4	5	6	7	8	9	10	11	12
$\bar{\nu}(K_n)$	0	1	3	9	19	36	62	102	153

表 8.8 $\bar{\nu}(K_n)$ の評価

n	13	14	15
上界	229	324	447
下界	221	310	423

を問題にした．
Blaschke（ブラシュケ）[11,12]は，$q(R)$ の最大値は，

$$1-\frac{35}{12\pi^2}=0.7044798810\cdots$$

であり，R が楕円のときこの値に達し，最小値は2/3で R が三角形のときこの値をとることを証明した．詳細および関連する問題については[13-21]を参照のこと．

R に課した条件をゆるめると，どのような対応した結果が成立するか？ R を平面上の有限面積の開集合とする（すなわち，凸性はもはや課さない）．$q(R)$ は前と同様に定義する．すると明らかに $q(R)=1$ である．なぜなら R としてごく薄い円環領域をとればよいからであり，この R で無作為に選ばれた4点は，ほとんど確実に四辺形を形づくるからである．

$q(R)$ の下限は，調べるのがもっと難しい．Scheinerman & Wilf（シェイナーマン&ウィルフ）[22,23]は，

$$\inf_R q(R)=\rho$$

であるという驚くべき事実を証明した．これゆえ，2つの関連がなさそうに見えた定数の間に関係がついた．ρ の価値が高まったので，多分いつかこの値を計算しなければという関心をひくものと思われる．

直線的図を議論してきたが，これとは対照的に，**普通図**は，曲がった辺も許容し，**普通交点数** $\nu(G)$ を導く．Gay（ゲイ）[1]は

$$\nu(K_n)=\frac{1}{4}\left\lfloor\frac{n}{2}\right\rfloor\left\lfloor\frac{n-1}{2}\right\rfloor\left\lfloor\frac{n-2}{2}\right\rfloor\left\lfloor\frac{n-3}{2}\right\rfloor$$

と予想した．$n\leq 12$ に対しては，これは確かめられている[24]．$\bar{\nu}(K_n)$ に対しては，同様な予想された公式は知られていない．十分大きな n に対して $\bar{\nu}(K_n)>\nu(K_n)$ であると信じられている[25]．

グラフの「厚さ」に関連した概念がいくつかある．定義や文献については[25,26]を参照のこと．ρ のような多数の基本的な定数が確かに幾何確率論（古い文献では，かつて「積分幾何学」とよばれていた）に存在しているが，計算するのは極めて難しい．

[1] P. Erdös and R. K. Guy, Crossing number problems, *Amer. Math. Monthly* 80 (1973) 52-57; MR 52 # 2894.
[2] A. T. White and L.W. Beineke, Topological graph theory, *Selected Topics in Graph Theory*, ed. L. W. Beineke and R. J. Wilson, Academic Press, 1978, pp. 15-49; MR 81e:05059.
[3] M. Gardner, *Knotted Doughnuts and Other Mathematical Entertainments*,W. H. Freeman, 1986, pp. 133-144; MR 87g:00007.
[4] N. J. A. Sloane, On-Line Encyclopedia of Integer Sequences, A000241 and A014540.
[5] D. Singer, The rectilinear crossing number of certain graphs, unpublished manuscript (1971).
[6] W. D. Smith, *Studies in Computational Geometry Motivated by Mesh Generation*, Ph.D. thesis, Princeton Univ., 1988.
[7] A. Brodsky, S. Durocher, and E. Gethner, The rectilinear crossing number of K_{10} is 62, *Elec. J. Combin.* 8 (2001) R23; cs.DM/0009023; MR 2002f:05058.
[8] O. Aichholzer, F. Aurenhammer, and H. Krasser, Progress on rectilinear crossing numbers,

Institut für Grundlagen der Informationsverarbeitung report, Technische Universität Graz (2002).
[9] A. Brodsky, S. Durocher, and E. Gethner, Toward the rectilinear crossing number of K_n: New drawings, upper bounds, and asymptotics, *Discrete Math.* to appear; cs.DM/0009028.
[10] J. J. Sylvester, On a special class of questions on the theory of probabilities, *Birmingham British Assoc. Report* 35 (1865) 8-9; also in *Collected Mathematical Papers*, v. 2, ed. H. F. Baker, Cambridge Univ. Press, 1904, pp. 480-481.
[11] W. J. E. Blaschke, Über affine Geometrie. XI: Lösung des Vierpunktproblems von Sylvester aus der Theorie der geometrischen Wahrscheinlichkeiten, *Berichte über die Verhandlungen der Königlich Sächsischen Gesellschaft der Wissenschaften zu Leipzig, Math.-Phys. Klasse* 69 (1917) 436-453.
[12] W. J. E. Blaschke, *Vorlesungen über Differentialgeometrie. II: Affine Differentialgeometrie*, Springer-Verlag, 1923.
[13] M. G. Kendall and P. A. P. Moran, *Geometrical Probability*, Hafner, 1963, pp. 42-46; MR 30 # 4275.
[14] V. Klee, What is the expected volume of a simplex whose vertices are chosen at random from a given convex body?, *Amer. Math. Monthly* 76 (1969) 286-288.
[15] W. J. Reeds, Random points in a simplex, *Pacific J. Math.* 54 (1974) 183-198; MR 51 # 8959.
[16] L. A. Santaló, *Integral Geometry and Geometric Probability*, Addison-Wesley, 1976, pp. 63-65; MR 55 # 6340.
[17] H. Solomon, *Geometric Probability*, SIAM, 1978, pp. 101-125; MR 58 # 7777.
[18] R. E. Pfiefer, The historical development of J. J. Sylvester's problem, *Math. Mag.* 62 (1989) 309-317.
[19] H. T. Croft, K. J. Falconer, and R. K. Guy, *Unsolved Problems in Geometry*, Springer-Verlag, 1991, sect. B5; MR 95k:52001.
[20] N. Peyerimhoff, Areas and intersections in convex domains, *Amer. Math. Monthly* 104 (1997) 697-704; MR 98g:52009.
[21] A. M. Mathai, *An Introduction to Geometrical Probability: Distributional Aspects with Applications*, Gordon and Breach, 1999, pp. 159-171; MR 2001a:60014.
[22] E. R. Scheinerman and H. S. Wilf, The rectilinear crossing number of a complete graph and Sylvester's "Four Point Problem" of geometric probability, *Amer. Math. Monthly* 101 (1994) 939-943; MR 95k:52006.
[23] H. S. Wilf, Some crossing numbers, and some unsolved problems, *Combinatorics, Geometry and Probability: A Tribute to Paul Erdös*, Proc. 1993 Trinity College conf., ed. B. Bollobás and A. Thomason, Cambridge Univ. Press, 1997; MR 98d:00027.
[24] D. Applegate, W. Cook, S. Dash, and N. Dean, Ordinary crossing numbers for K_{11} and K_{12}, unpublished note (2001).
[25] C. Thomassen, Embeddings and minors, *Handbook of Combinatorics*, v. I, ed. R. Graham, M. Grötschel, and L. Lovász, MIT Press, 1995, pp. 301-349; MR 97a:05075.
[26] M. B. Dillencourt, D. Eppstein, and D. S. Hirschberg, Geometric thickness of complete graphs, *Graph Drawing*, Proc. 1998 Montréal conf., ed. S. H. Whitesides, Lect. Notes in Comp. Sci. 1547, Springer-Verlag, 1998, pp. 102-110; math.CO/9910185; MR 2000g:68118.

8.19 外半径-内半径定数

平面上のコンパクト凸集合 K を含む最小の円板の半径を K の**外半径** $R(K)$ といい，K に含まれる最大の円板の半径を**内半径** $r(K)$ という．よく知られた集合に対する外半径，内半径の公式は[1-3]に見出される．R や r にまつわる興味深い定数が，集合族に関するいろいろな幾何学的最適問題に登場する．多数ありうる中から3つの例をあげよう．

幅1のコンパクト凸集合 F 内にあるすべての三角形を考える（F の**幅**とは，F から任意の直線への正射影の長さのうちの最小値のことである）．いくつかの特別な集合 F に含まれるようなすべての三角形について最大内半径 $a(F)=\max_{\Delta} r(\Delta)$ を調べよう．

・ F_4 を幅1の（すなわち一辺の長さが1の）正方形とすると[4,5]
$$a(F_4)=\frac{-1+\sqrt{5}}{4}=0.3090169943\cdots$$
である．

・ F_5 を幅1の（すなわち一辺の長さが $2\cot(2\pi/5)$ の）正五角形とすると
$$a(F_5)=0.2440155280\cdots$$
となり，これは最小多項式[6,7]
$$5x^9-170x^8+436x^7-205x^6-96x^5+440x^4-120x^3+64x^2-80x+16$$
をもつ．

・ F_6 を幅1の（すなわち一辺の長さが $1/\sqrt{3}$ の）正六角形とすると
$$a(F_6)=\frac{1}{4}=0.25$$
である．

$a(F_5)<\min\{a(F_4),a(F_6)\}$ に注意する．実際，
$$0.166<\frac{1}{6}\leq\inf_{F} a(F)\leq a(F_5)$$
とわかっている．ここで下限は，すべての F にわたってとる．実際には，この下限は，（上の式の）上界に等しいのではあるまいか？ これは未解決の問題である．

以下のために少し記号を用意する．S は頂点（$\pm 1,\pm 1$）をもつ正方形を表すものとし，$h_1,h_2,\cdots,h_8$ を（反時計回りに番号をつけた）S の辺の半分ずつとする．$(0,0)$ を通る横軸に垂直でない直線 L が与えられたとき，L^+ で右半平面での半直線を，L^- で左半平面での半直線を表すものとする．L^+ が h_i と交わり，L^- は h_j と交わるものとしておく．ここで $i\equiv j\pmod 4$ である．M^+ を $(0,0)$ を通り h_k（$k\neq i$ かつ $k\neq j$）と交わる3番目の半直線とする．このとき，M^+ は，L と**適切に異なっている**という．最後に，Z は平面での標準整数格子，すなわち $(1,0)$，$(0,1)$ という基底をもつ格子を表す

ものとする.

　その内部に原点を含み,これ以外の格子点をまったく含まないようなすべてのコンパクト凸集合 G を考える.（【2.23】の言葉使いでは,G は Z **容認的**である）.さらに,G の外接円の中心を原点とし,この外接円 C と $(0,0)$ を通る任意の直線 L に対して,$G \cap M^+ \cap C \neq \emptyset$ となる適切に異なる半直線が存在しないときは,$G \cap L^+ \cap C \neq \emptyset$ であり,かつ $G \cap L^- \cap C \neq \emptyset$ であるということはありえないと仮定する（いいかえれば,G が,どこか L 以外の方向で突出していないときは,L の両方向に同時に S の外側に突出することはない.）このとき[9]

$$\sup_G R(G) = 1.6847127097\cdots$$

であり,これは最小多項式 $5x^6 - 15x^4 + 3x^2 - 2$ をもつ.最大外半径をもつ集合は図 8.22 に示した（二等辺三角形ではない）三角形である.L^+, L^-, M に関する技巧的な条件をつけないと,上限は無限大になってしまう.（原点を通り,他の格子点を通らない幅 ε,長さ $1/\varepsilon$ の薄い板を思い浮かべればよい.）

　重心（すなわち重力の中心）をもつコンパクト凸集合 K の外接円の中心に関連する結果がある.$b(K)$ は,外接円の中心と重心との距離を $R(K)$ で割った商を表すものとする.円板や正三角形を考えれば明らかに $\inf_K b(K) = 0$ である.

$$\sup_K b(K) = \frac{2}{3} x = 0.4278733971\cdots$$

[10]が知られており,ここで x は,超越的方程式

$$x^2 + 2\sqrt{1-x^2} = 2x(x + \arccos(x)) \quad (-1 \leq x \leq 1)$$

の一意の解である.この場合の極値集合は,平行な辺の片方が円弧に置きかえられたある種の対称な台形である.

　内半径は,幾何学からは遠く隔ったある種の問題,例えば,Bloch-Landau（ブロック-ランダウ）定数【7.1】や薄膜の振動の固有値解析などの定式化[11-13]に関連がある.

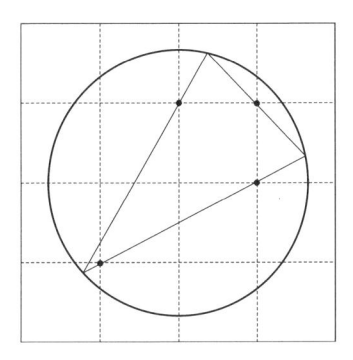

図 8.22 この Z-許容的な三角形 T は,最大外半径 $R(T) = 1.6847127097\ldots$ をもつ.

[1] H. S. M. Coxeter, *Regular Polytopes*, Methuen, 1948, pp. 2-3, 20-23; MR 51 #6554.
[2] H. S. M. Coxeter, *Introduction to Geometry*, Wiley, 1964, pp. 10-16; MR 90a:51001.
[3] M. Berger, *Geometry I*, Springer-Verlag, 1987, pp. 280-287; MR 95g:51001.
[4] L. Funar and A. Bondesen, Circles, triangles, squares and the Golden mean, *Amer. Math. Monthly* 96 (1989) 945-946.
[5] F. F. Abi-Khuzam and R. Barbara, A sharp inequality and the inradius conjecture, *Math. Inequal. Appl.* 4 (2001) 323-326; MR 2002a:51021.
[6] R. Stong, A pentagonal maximum problem, *Amer. Math. Monthly* 104 (1997) 169-171.
[7] W. Li and Y. Cheng, Solution to the regular pentagon problem, unpublished note (1991).
[8] L. Funar and C. A. Rogers, Inradii and width, *Amer. Math. Monthly* 97 (1990) 858.
[9] P. W. Awyong and P. R. Scott, On the maximal circumradius of a planar convex set containing one lattice point, *Bull. Austral. Math. Soc.* 52 (1995) 137-151; MR 96g:52030.
[10] P. R. Scott, Centre of gravity and circumcentre of a convex body in the plane, *Quart. J. Math.* 40 (1989) 111-117; MR 89m:52017.
[11] R. Bañuelos and T. Carroll, Brownian motion and the fundamental frequency of a drum, *Duke Math. J.* 75 (1994) 575-602; addendum 82 (1996) 227; MR 96m:31003 and MR 97f:31004.
[12] R. Bañuelos, T. Carroll, and E. Housworth, Inradius and integral means for Green's functions and conformal mappings, *Proc. Amer. Math. Soc.* 126 (1998) 577-585; MR 98g:30016.
[13] D. Betsakos, Harmonic measure on simply connected domains of fixed inradius, *Ark. Mat.* 36 (1998) 275-306; MR 2000a:30048.

8.20 アポロニウス充填定数

図8.23の2つの図形を考える．左側のは，大きな円周とその中の3つの円板から出発している．右側のは，曲線三角形型境界と1つの円板から出発している．両方の充填は，繰り返し操作の1段階前で覆われないで残っている場所にできるだけ大きな円板 D_i をはめ込むことによって得られる．すべての新しい円板は，それが接触する既存のすべての円板と接する．こうしてできた図形は，三方対称（3方回転について対称）である．

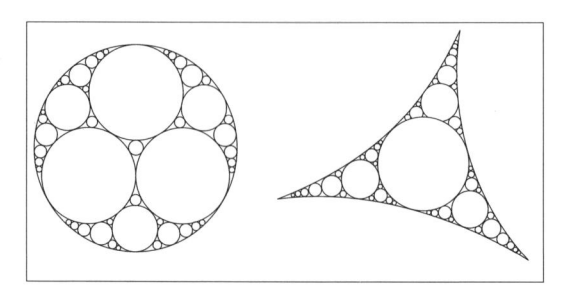

図8.23 最初の円と最初の曲線三角形について描かれたアポロニウス充填.

8.20 アポロニウス充填定数

表 8.9 充填定数 ε の評価

推定値	出典
1.306951	Melzak [2]
1.3058	Boyd (Mandelbrot[11]によって報告された)
1.305636	Boyd[12]
1.305684	Manna & Herrmann[13]
1.305686729	Thomas & Dhar[14]
1.305688	McMullen[15]

充填の**残留集合** E, すなわち円板により覆われない点集合については, どんなことがいえるのか? 集合 E は, ルベーグ測度 0 であることが示される. 1 つの重要な量は**充填指数** ε で, これは,

$$\sum_{i=1}^{\infty} |D_i|^e < \infty$$

となる [1,2] ような e の下限として定義される. ここで $|D|$ は D の直径を表している. もう 1 つの重要な量は, Hausdorff (**ハウスドルフ**) 次元 $\dim(E)$ で, これは, [1,3]

$$\sup_{\delta>0} \inf_{\substack{可算 \delta\text{-被覆} \\ U_i}} \sum_{i=1}^{\infty} |U_i|^s = \begin{cases} \infty & (0 \leq s < \dim(E) \text{ のとき}) \\ 0 & (s > \dim(E) \text{ のとき}) \end{cases}$$

によって一意的に定義される. ここで δ-**被覆** U_i とは, $E \subseteq \cup_{i=1}^{\infty} U_i$ で各 U_i は開集合であり, すべての i に対して $0 < |U_i| \leq \delta$ であることを意味している. Larman (ラーマン)[4] や Boyd (ボイド)[5-7] が示したように

$$\varepsilon = \dim(E)$$

となる. さらに Boyd[8-10] や他の人たちによる研究により精密な限界

$$1.300197 < \varepsilon < 1.314534$$

が導かれる. また, いろいろな文献から数値的推定値が得られる (表 8.9 を参照).

円板によるあらゆる別の充填のやり方を考えたとしても $\dim(E)$ は最小になるか? Boyd[6] は, これは難しい問題であると答えた. これを解決するための進展があったかどうかは, 不明である. 【2.16】を参照してほしい. そこでは Sierpinski (シェルピンスキー) ガスケット, すなわち, 上向きの三角形の, 下向きの正三角形による充填に言及している. (このシェルピンスキーガスケットについては, $\dim(E) = \ln(3)/\ln(2)$ であると完全にわかっている.) この話題は, 最近, 数論研究者たちの関心を引いている [16].

[1] K. J. Falconer, *The Geometry of Fractal Sets*, Cambridge Univ. Press, 1985, pp. 7-9, 14-15, 125-131; MR 88d:28001.

[2] Z. A. Melzak, On the solid-packing constant for circles, *Math. Comp.* 23 (1969) 169-172; MR 39 #6179.

[3] K. E. Hirst, The Apollonian packing of circles, *J. London Math. Soc.* 42 (1967) 281-291; MR 35 #876.

[4] D. G. Larman, On the exponent of convergence of a packing of spheres, *Mathematika* 13 (1966)

57-59; MR 34 #1928.
[5] D. W. Boyd, Osculatory packing by spheres, *Canad. Math. Bull.* 13 (1970) 59-64; MR 41 #4387.
[6] D.W. Boyd, The residual set dimension of the Apollonian packing, *Mathematika* 20 (1973) 170-174; MR 58 #12732.
[7] C. Tricot, A new proof for the residual set dimension of the Apollonian packing, *Math. Proc. Cambridge Philos. Soc.* 96 (1984) 413-423; MR 85j:52022.
[8] D.W. Boyd, Lower bounds for the disk packing constant, *Math. Comp.* 24 (1970) 697-704; MR 43 #3924.
[9] D. W. Boyd, The disk-packing constant, *Aequationes Math.* 7 (1971) 182-193; MR 46 #2557.
[10] D. W. Boyd, Improved bounds for the disk-packing constant, *Aequationes Math.* 9 (1973) 99-106; MR 47 #5728.
[11] B. B. Mandelbrot, *The Fractal Geometry of Nature*, W. H. Freeman, 1983, pp. 169-172, 360; MR 84h:00021.
[12] D. W. Boyd, The sequence of radii of the Apollonian packing, *Math. Comp.* 39 (1982) 249-254; MR 83i:52013.
[13] S. S. Manna and H. J. Herrmann, Precise determination of the fractal dimensions of Apollonian packing and space-filling bearings, *J. Phys. A* 24 (1991) L481-L490; MR 92c:52027.
[14] P. B. Thomas and D. Dhar, The Hausdorff dimension of the Apollonian packing of circles, *J. Phys. A* 27 (1994) 2257-2268; MR 95f:28010.
[15] C. T. McMullen, Hausdorff dimension and conformal dynamics. III: Computation of dimension, *Amer. J. Math.* 120 (1998) 691-721; MR 2000d:37055.
[16] R. L. Graham, J. C. Lagarias, C. L. Mallows, A. R.Wilks, and C. H. Yan, Apollonian circle packings: Number theory, *J. Number Theory*, to appear; math.NT/0009113.

8.21 出会い定数

E で, d 次元ユークリッド空間のコンパクトな連結部分集合を表す.Gross(グロス)[1]と Stadje(スタディエ)[2]は,独立に次の定理を証明した.すなわち(必ずしも異なっている必要はない)任意の点 $x_1, x_2, \cdots, x_n \in E$ に対して

$$\frac{1}{n}\sum_{i=1}^{n}|x_i - y| = a(E)$$

となる $y \in E$ が存在するような一意的な実数 $a(E)$ が存在する.いいかえると,$x_1, x_2, \cdots, x_n \in E$ からの平均距離が,$a(E)$ であるような $y \in E$ が存在する.この定数 $a(E)$ は,任意の正の整数 n について,そしてすべての n 個の点集合に対して有効であろう.さらに,別の定数では成り立たないであろう.これはまったくの驚きである.

たとえば,E が凸とすると,$a(E)$ は,E の外半径になる.そこで非凸集合に焦点をあてよう.C が直径1の円のとき $a(E)=2/\pi=0.6366197723\cdots$ である[3,4].$\varDelta$ を底辺の長さが 2 で,周の長さが $2\lambda+2$ の二等辺三角形とすると[5],

$$a(\Delta) = \begin{cases} \dfrac{\lambda^2+2\lambda-\sqrt{\lambda^2-1}-2\sqrt{(\lambda-\sqrt{\lambda^2-1})\lambda(\lambda+1)}}{\lambda^2+3\lambda-1-\lambda\sqrt{\lambda^2-1}-2\sqrt{(\lambda-\sqrt{\lambda^2-1})\lambda(\lambda+1)}} & (\sqrt{2}\leq\lambda\leq\xi \text{ に対して}) \\ \dfrac{\lambda^2+1}{2\lambda} & (\lambda\geq\xi \text{ に対して}) \end{cases}$$

となる．ここで $\xi=2.3212850380\cdots$ は，最小多項式 $2x^5-4x^4-5x^2+4x-1$ をもつ．E が任意の楕円や鋭角三角形のときでさえも，$a(E)$ に対するきちんとした表現は，まだ誰にも見つけられていない．

$a(E)$ の2つの別の定義は次のようなものである．すなわち

$$a(E) = \sup_{n\geq 1} \sup_{x_1,x_2,\cdots,x_n\in E} \min_{y\in E} \frac{1}{n}\sum_{i=1}^{n}|x_i-y| = \inf_{n\geq 1}\inf_{x_1,x_2,\cdots,x_n\in E}\max_{y\in E}\frac{1}{n}\sum_{i=1}^{n}|x_i-y|$$

であり，これがゲーム理論のミニマックス定理と関連があるのは明らかである[1,3,6]．

E の**出会い定数** $r(E)$ を，正規化された比

$$r(E) = \frac{a(E)}{\text{diam}(E)} \qquad (\text{ここで diam}(E)=\max_{u,v\in E}|u-v|)$$

として定義する．これを意味のあるものにするために，（直径が0にならないように）E は1点 p ではありえず，また（連結性をもつために）有限集合ではありえない．これらの制限の下で Gross と Stadje は，$1/2\leq r(E)<1$ であることを証明した．d 次元ユークリッド空間のすべての集合 E を考えたとき $r(E)$ の最大値 R_d は何か？ 明らかに $R_1=1/2$ である．$d=2$ のときは，ルーローの三角形 T が答えを与えるように思われる【8.10】．Nickolas & Yost (ニコラス&ヨスト)[7] そして Wolf (ヴォルフ) は，

$$\max\left\{\frac{2}{3}, r(T)\right\} \leq R_2 \leq \frac{1}{2}+\frac{\pi}{16}<0.69634955$$

という不等式を厳密に示した．また $r(T)$ に対する最もよく知られた数値的推定値は，$0.6675277360\cdots$[9] である．$r(T)$ のきちんとした形の表現は，まだ発見されていない．$R_2=r(T)$ という予想は，もっと注目する価値がある！

$d>2$ のときは[7]，限界

$$\frac{d}{d+1}\leq R_d \leq \frac{\Gamma\left(\dfrac{d}{2}\right)^2 2^{d-2}\sqrt{2d}}{\Gamma\left(d-\dfrac{1}{2}\right)\sqrt{\pi(d+1)}} < \sqrt{\frac{d}{d+1}}$$

が知られている．ここで $\Gamma(x)$ はガンマ関数【1.5.4】である．この不等式は $d=2$ のときよりも精密ではない．知られている限りでは，まだ誰も出会い定数を最大にする高次元の形状を推定してみようという試みを行っていない．

2番目の関連のある予想は $R_2=S_2$ である，ここで[8-11]

$$S_d = \sup_{n\geq 1}\sup_{\substack{x_1,x_2,\cdots,x_n \\ |x_i-x_j|\leq 1}}\frac{1}{n^2}\sum_{i=1}^{n}\sum_{j=1}^{n}|x_i-x_j|$$

である．いいかえると，S_d は，d 次元空間で，どの2点 x_i, x_j も1以上には離れていないような点 $x_1, x_2, \cdots, x_n$ を組にしたときの距離の平均である．

さらに $a(E)\leq b(E)$ とわかる，ここで[8]

$$b(E) = \sup_{n \geq 1} \sup_{x_1, x_2, \cdots, x_n \in E} \frac{1}{n^2} \sum_{i=1}^{n} \sum_{j=1}^{n} |x_i - x_j|$$

である．$b(E)$ の研究を一般化により始める．すなわち，和を積分に，点質量 x_i を確率密度に置きかえポテンシャル論を応用する [12-16]．3番目の予想は，$a(T) = b(T)$ である [9]．E が2次元球面という特別な場合は【8.8】で議論した．

前述の結果も一般化できる．すなわち，E は任意のコンパクトで連結な距離空間としてよい．Stadje[2] は，E は，$x, y \in E$ に対して実数値連続で対称な関数 $f(x, y)$ (一種の「弱距離」) をもつコンパクトで連結なハウスドルフ空間でありさえすればよいことを証明した．

最後に，E を長軸の長さが2，短軸の長さが1の楕円とする．数値的に $a(E) = 2.1080540666\cdots$ [9] と知られている．前に述べたように $a(E)$ の精密な公式は存在していないが，それにもかかわらず，将来この定数の性質をもっとよく理解することが重要だと思われる．

[1] O. Gross, The rendezvous value of metric space, *Advances in Game Theory*, ed. M. Dresher, L. S. Shapley, and A. W. Tucker, Princeton Univ. Press, 1964, pp. 49-53; MR 28 # 5841.
[2] W. Stadje, A property of compact connected spaces, *Arch. Math. (Basel)* 36 (1981) 275-280; MR 83e:54028.
[3] J. Cleary, S. A. Morris, and D. Yost, Numerical geometry—Numbers for shapes, *Amer. Math. Monthly* 93 (1986) 260-275; MR 87h:51043.
[4] H. T. Croft, K. J. Falconer, and R. K. Guy, *Unsolved Problems in Geometry*, Springer-Verlag, 1991, sect. G1; MR 95k:52001.
[5] D. K. Kulshestha, T. W. Sag, and L. Yang, Average distance constants for polygons in spaces with nonpositive curvature, *Bull. Austral. Math. Soc.* 42 (1990) 323-333; MR 92b:51025.
[6] C. Thomassen, The rendezvous number of a symmetric matrix and a compact connected metric space, *Amer. Math. Monthly* 107 (2000) 163-166; MR 2001a:54040.
[7] P. Nickolas and D. Yost, The average distance property for subsets of Euclidean space, *Arch. Math. (Basel)* 50 (1988) 380-384; MR 89d:51026.
[8] R. Wolf, Averaging distances in real quasihypermetric Banach spaces of finite dimension, *Israel J. Math.* 110 (1999) 125-151; MR 2001f:46012.
[9] G. Larcher, W. C. Schmid, and R. Wolf, On the approximation of certain mass distributions appearing in distance geometry, *Acta Math. Hungar.* 87 (2000) 295-316; MR2001f:65013.
[10] L. Fejes Tóth, Über eine Punktverteilung auf der Kugel, *Acta Math. Acad. Sci. Hungar.* 10 (1959) 13-19; MR 21 # 4393.
[11] H. S. Witsenhausen, On the maximum of the sum of squared distances under a diameter constraint, *Amer. Math. Monthly* 81 (1974) 1100-1101; MR 51 # 6594.
[12] L. Fejes Tóth, On the sum of distances determined by a pointset, *Acta Math. Acad. Sci. Hungar.* 7 (1956) 397-401; MR 21 # 5937.
[13] G. Björck, Distributions of positive mass, which maximize a certain generalized energy integral, *Ark. Mat.* 3 (1956) 255-269; MR 17,1198b.
[14] R. Alexander and K. B. Stolarsky, Extremal problems of distance geometry related to energy integrals, *Trans. Amer. Math. Soc.* 193 (1974) 1-31; MR 50 # 3121.
[15] R. Alexander, Generalized sums of distances, *Pacific J. Math.* 56 (1975) 297-304; MR 58 # 24005.
[16] G. D. Chakerian and M. S. Klamkin, Inequalities for sums of distances, *Amer. Math. Monthly* 80 (1973) 1009-1017; MR 48 # 9558.

補　　遺

　以下の結果はとても美しいので見すごしたくはない．ガウス整数 $a+bi$（ここで a, b は整数，$i^2=-1$）は，単数 $\{\pm1, \pm i\}$ をもつ一意分解整域をなす．2 つのガウス整数が無作為に選ばれたとする．それらが互いに素である確率は，次々に大きな円板をとり極限をとると

$$\frac{6}{\pi^2 G}=0.6637008046\cdots$$

となる[1,2]．

　ここで，G はカタランの定数【1.7】である．これは，2 つの普通の整数が互いに素である【1.4】という対応する確率 $(6/\pi^2)$ よりもいくぶん大きい．

　同様に Eisenstein-Jacobi（アイゼンシュタイン-ヤコビ）整数 $a+b\omega$（ここで a, b は整数，$\omega=(-1+i\sqrt{3})/2$ は，単数 $\{\pm1, \pm\omega, \pm\omega^2\}$ をもつ一意分解整域をなす．このような 2 つの数が無作為に選ばれたとき，それらが互いに素である確率は，次々に大きな円板をとり極限をとると

$$\frac{6}{\pi^2 H}=0.7780944891\cdots$$

である[1,3]．ここで

$$H=\frac{4\pi}{3\sqrt{3}}\ln(\beta)=\sum_{k=0}^{\infty}\left(\frac{1}{(3k+1)^2}-\frac{1}{(3k+2)^2}\right)=0.7813024128\cdots$$

であり，β は【3.10】で広範囲に議論される．

　また，定数 $6/(\pi^2 G)$, $6/(\pi^2 H)$ は，それぞれ，無作為にとったガウス整数が無平方である確率，および無作為にとったアイゼンシュタイン-ヤコビ整数が無平方である確率である．【2.5】でのように「無関心性（carefree）」という概念に関連があるが，対応する定数は未だ知られていない．

　ところで，【2.5】の終わりで予想された k (>3) 個の数が互いに素である確率に関する結果が正しいことが証明された[4]．

　そしてこの本が印刷にまわっている時点では，【2.13】の終わりに与えられた素数下極限問題が解決されたか（それに近いか）どうかは不明である．

[1] G. E. Collins and J. R. Johnson, The probability of relative primality of Gaussian integers, *Proc. 1988 Int. Symp. Symbolic and Algebraic Computation (ISSAC)*, Rome, ed. P. Gianni, Lect. Notes in Comp. Sci. 358, Springer-Verlag, 1989, pp. 252-258; MR 90 m:11165.

[2] E. Pegg, The neglected Gaussian integers (MathPuzzle).

[3] E. Kowalski, Coprimality and squarefreeness within quadratic fields, unpublished note (2003).
[4] J.-Y. Cai and E. Bach, On testing for zero polynomials by a set of points with bounded precision, *Theoret. Comput. Sci.* 296 (2003) 15-25.
[5] D. Goldston and C. Yildirim, Small gaps between primes, submitted (2003).

付録　定数一覧

0	零，ド・ブルイン-ニューマン定数の予想値【2.32】
0.0001111582⋯	スティルチェス定数に関連【2.21】
0.00016⋯	カメロンの無和集合定数の1つ【2.25】
0.0002206747⋯	6番目のデュボア・レイモン定数【3.12】
0.0005367882⋯	ヘンズリー定数【2.18】
0.0007933238⋯	5番目のスティルチェス定数【2.21】
0.0013176411⋯	ヒース-ブラウン-モロズ定数，アルティンの定数に関連【2.4】
0.0017486751⋯	λ_7，ガウス-クズミン-ヴィルジング定数に関連【2.17】
0.0019977469⋯	ソボレフ等周定数に関連【3.6】
0.0020538344⋯	3番目のスティルチェス定数【2.21】
0.0023253700⋯	4番目のスティルチェス定数【2.21】
0.0031816877⋯	メルザクの定数，ソボレフ等周定数に関連【3.6】
0.0044939231⋯	ゴロム-ディックマン定数に関連【5.4】
0.0047177775⋯	$-\lambda_6$，ガウス-クズミン-ヴィルジング定数に関連【2.17】
0.0052407047⋯	4番目のデュボア・レイモン定数【3.12】
0.0063564559⋯	スティルチェス定数に関連【2.21】
0.0072973525⋯	微細構造定数，ファイゲンバウム-クーレ-トレサー定数に関連【1.9】
0.0095819302⋯	ソボレフ等周定数に関連【3.6】
0.0096903631⋯	2番目のスティルチェス定数の反数【2.21】
0.0102781647⋯	p_3，ヴァレーの定数に関連【2.19】
0.0125537906⋯	ゴロム-ディックマン定数に関連【5.4】
0.0128437903⋯	λ_5，ガウス-クズミン-ヴィルジング定数に関連【2.17】
0.0173271405⋯	b_3，デュボア・レイモン定数に関連【3.12】
0.0176255⋯	浸透集積密度定数に関連【5.18】
0.0177881056⋯	$-41/32+3\sqrt{3}/4$，浸透集積密度に関連【5.18】
0.0183156388⋯	e^{-4}，レーニィの駐車定数の1つ【5.3】

$0.0186202233\cdots$	ポリヤの酔歩定数の1つ【5.9】
$0.0219875218\cdots$	ゴウチャマンの定数, シャピロ-ドリンフェルド定数に関連【3.1】
$0.0230957089\cdots$	アペリの定数【1.6】, スティルチェス定数【2.21】, ド・ブルイン-ニューマン定数【2.32】に関連
$0.0231686908\cdots$	ヘンズリー定数【2.18】
$0.0255369745\cdots$	c^-_0, レンツ-イジング定数に関連【5.22】
$0.0261074464\cdots$	4番目のマシューズ定数【2.4】
$0.026422\cdots$	タムスの定数に関連【8.8】
$0.0275981\cdots$	$\kappa_S(p_c)$, 浸透集積密度定数に関連【5.18】
$0.0282517642\cdots$	3番目のデュボア・レイモン定数【3.12】
$0.0333810598\cdots$	ファイゲンバウム-クーレ-トレサー定数の1つ【1.9】
$0.0354961590\cdots$	$-\lambda_4$, ガウス-クズミン-ヴィルジング定数に関連【2.17】
$0.0355762113\cdots$	$-41/16+3\sqrt{3}/2$, 浸透集積密度に関連【5.18】
$0.0369078300\cdots$	ゴロム-ディックマン定数に関連【5.4】
$0.0370072165\cdots$	ゴロム-ディックマン定数に関連【5.4】
$0.0381563991\cdots$	レーニィの駐車定数の1つ【5.3】
$0.0403255003\cdots$	e_0, レンツ-イジング定数に関連【5.22】
$0.0461543172\cdots$	スティルチェス定数に関連【2.21】
$0.0482392690\cdots$	$m_{3,4}$, マイセル-メルテンス定数に関連【2.2】
$0.0484808014\cdots$	p_2, ヴァレーの定数に関連【2.19】
$0.0494698522\cdots$	ゴロム-ディックマン定数に関連【5.4】
$0.0504137604\cdots$	オイラー-ゴンペルツ定数に関連【6.2】
$0.0548759991\cdots$	H_9の予想値, ハイルブロンの三角形定数【8.16】
$0.05756\cdots$	双曲的体積定数に関連【8.9】
$0.0585498315\cdots$	$1/(2\pi e)$, エルミートの定数に関連【8.7】
$0.0605742294\cdots$	オイラーのφ定数の1つ【2.7】
$0.0608216553\cdots$	3番目のマシューズ定数【2.4】
$0.0648447153\cdots$	ポリヤの酔歩定数の1つ【5.9】
$0.0653514259\cdots$	ノートンの定数【2.18】
$0.0653725925\cdots$	スティルチェス定数に関連【2.21】
$0.065770\cdots$	$\kappa_S(1/2)$, 浸透集積密度定数に関連【5.18】
$0.0657838882\cdots$	ギップス-ウィルブラハム定数に関連【4.1】
$0.0659880358\cdots$	e^{-e}, 反復指数定数の1つ【6.11】
$0.0723764243\cdots$	H_8の予想値, ハイルブロンの三角形定数【8.16】
$0.0728158454\cdots$	1番目のスティルチェス定数の反数【2.21】
$0.0729126499\cdots$	ポリヤの酔歩定数の1つ【5.9】
$0.0757395140\cdots$	ヴァレーの定数に関連【2.19】
$0.0773853773\cdots$	ヴァレーの定数に関連【2.19】
$0.0810614667\cdots$	スティルチェス定数に関連【2.21】
$0.0838590090\cdots$	H_7の予想値, ハイルブロンの三角形定数【8.16】
$0.0858449341\cdots$	ポリヤの酔歩定数の1つ【5.9】
$0.0883160988\cdots$	ゴロム-ディックマン定数に関連【5.4】

付録 定 数 一 覧

$0.0894898722\cdots$	$G/\pi-1/2$, ギップス-ウィルブラハム定数に関連【4.1】
$0.0904822031\cdots$	K_s/π の予想値, 掛谷-ベシコヴィッチ定数に関連【8.17】
$0.0923457352\cdots$	ド・ブルイン-ニューマン定数に関連【2.32】
$0.0931878229\cdots$	$\exp(-\pi^2/(6\ln(2)))$, ヒンチン-レヴィ定数に関連【1.8】
$0.0946198928\cdots$	マイセル-メルテンス定数に関連【2.2】
$0.0948154165\cdots$	オッターの木数え上げ定数に関連【5.6】
$0.097\cdots$	10 進自己数密度定数【2.24】
$0.0980762113\cdots$	$(3\sqrt{3}-5)/2$, 浸透集積密度に関連【5.18】
$0.1008845092\cdots$	λ_3, ガウス-クズミン-ヴィルジング定数に関連【2.17】
$0.1013211836\cdots$	$1/\pi^2$, ソボレフ等周定数に関連【3.6】
$0.1041332451\cdots$	$-d_0$, レンツ-イジング定数に関連【5.22】
$0.1047154956\cdots$	ポリヤの酔歩定数の1つ【5.9】
$0.1076539192\cdots$	「1/9」定数【4.5】
$0.1084101512\cdots$	トロトの定数, ミンコフスキー-バウアーの定数に関連【6.9】
$0.1118442752\cdots$	浸透集積密度定数に関連【5.18】
$0.1149420448\cdots$	ケプラー-ボウカンプ定数【6.3】
$0.1227367657\cdots$	モザーのミミズ定数に関連【8.4】
0.125	$1/8$, モザーのミミズ定数【8.4】, ハイルブロンの三角形定数【8.16】に関連
$0.1316200785\cdots$	モザーのミミズ定数に関連【8.4】
$0.1351786098\cdots$	ポリヤの酔歩定数の1つ【5.9】
$0.14026\cdots$	自己回避歩行定数の1つ【5.10】
$0.1433737142\cdots$	ハフナー-サルナック-マッカレー定数に関連【2.5】
$0.1473494003\cdots$	2 番目のマシューズ定数, アルティンの定数に関連【2.4】
$0.1475836176\cdots$	$\arctan(1/2)/\pi$, プルーフェの定数【6.5】
$0.1484955067\cdots$	オイラー-ゴンペルツ定数に関連【6.2】
$0.14855\cdots$	4 D 臨界点, レンツ-イジング定数に関連【5.22】
$0.14869\cdots$	$-d_1$, レンツ-イジング定数に関連【5.22】
$0.1490279983\cdots$	コンウェイの不偏ミゼールゲーム定数【6.11】
$0.14966\cdots$	4 D 逆臨界温度, レンツ-イジングに関連【5.22】
$0.1544313298\cdots$	$2\gamma-1$, オイラー-マスケローニの定数に関連【1.5】
$0.1596573971\cdots$	ルーロー三角形定数に関連【8.10】
$0.1598689037\cdots$	スティルチェス定数に関連【2.21】
$0.1599\cdots$	自己回避歩行定数の1つ【5.10】
$0.1624329213\cdots$	強正方形エントロピー定数に関連【5.12】
$0.164\cdots$	浸透集積密度定数に関連【5.18】
$0.1709096198\cdots$	ゴロム-ディックマン定数に関連【5.4】
$0.1715004931\cdots$	δ_0, ホール-モンゴメリー定数【2.33】
$0.1715728753\cdots$	$3-2\sqrt{2}$, H_5 の値, ハイルブロンの三角形定数に関連【8.16】
$0.1724297877\cdots$	ポリヤの酔歩定数の1つ【5.9】
$0.1729150690\cdots$	ガウス-クズミン-ヴィルジング定数に関連【2.17】
$0.1763470368\cdots$	レーニィの駐車定数の1つ【5.3】

$0.1764297331\cdots$	$-G_2$, レンツ-イジング定数に関連【5.22】
$0.1770995223\cdots$	オッターの木数え上げ定数に関連【5.6】
$0.1789797444\cdots$	$2G/\pi-1$, ギップス-ウィルブラハム定数に関連【4.1】
$0.1807171047\cdots$	ザギエの定数, フレイマンの定数に関連【2.31】
$0.1824878875\cdots$	シャピロ-ドリンフェルド定数に関連【3.1】
$0.183\cdots$	浸透集積密度定数に関連【5.18】
$0.1839397205\cdots$	$1/(2e)$, マサー-グラメイン定数に関連【7.2】
$0.1862006357\cdots$	ルーロー三角形定数に関連【8.10】
$0.1866142973\cdots$	C_6, ハーディ-リトルウッド定数の1つ【2.1】
$0.186985\cdots$	レーニィの駐車定数の1つ【5.3】
$0.1878596424\cdots$	反復指数定数に関連【6.11】
$0.1895600483\cdots$	折り畳み三角形エントロピー, リーブの正方形氷定数に関連【5.24】
$0.1924225474\cdots$	オッターの木数え上げ定数に関連【5.6】
$0.1924500897\cdots$	$\sqrt{3}/9$, H_5の値, ハイルブロンの三角形定数【8.16】
$0.1932016732\cdots$	ポリヤの酔歩定数の1つ【5.9】
$0.1963495408\cdots$	モザーのミミズ定数に関連【8.4】
$0.1945280495\cdots$	2番目のデュボア・レイモン定数【3.12】
$0.1994588183\cdots$	ヴァレー定数【2.19】
$0.1996805161\cdots$	弱無三重集合定数の予想値【2.26】
$0.2007557220\cdots$	マイセル-メルテンス定数に関連【2.2】
$0.2076389205\cdots$	ド・ブルイン-ニューマン定数に関連【2.32】
$0.2078795764\cdots$	$i^i=\exp(-\pi/2)$, 反復指数定数に関連【6.11】
$0.209\cdots$	4進自己数密度定数【2.24】
$0.2095808742\cdots$	ゴロム-ディックマン定数に関連【5.4】
$0.21\cdots$	カメロンの無和集合定数の1つ【2.25】
$0.2173242870\cdots$	ロックスの定数, ポーター-ヘンズリー定数に関連【2.18】
$0.218094\cdots$	3D臨界点, レンツ-イジング定数に関連【5.22】
$0.2183801414\cdots$	ポリヤの酔歩定数の1つ【5.9】
$0.2192505830\cdots$	グレイシャー-キンケリン定数に関連【2.15】
$0.221654\cdots$	3D逆臨界温度, レンツ-イジング定数に関連【5.22】
$0.2221510651\cdots$	オッターの木数え上げ定数に関連【5.6】
$0.2265708154\cdots$	強正方形エントロピー定数に関連【5.12】
$0.2299\cdots$	折り畳み正方対角エントロピー, リーブの正方形氷定数に関連【5.24】
$0.2351252848\cdots$	コンウェイ-ガイ定数, エルデーシュの和違い集合定数に関連【2.28】
$0.2387401436\cdots$	オッターの木数え上げ定数に関連【5.6】
$0.24\cdots$	ヘイマン定数の1つ【7.5】
$0.2419707245\cdots$	$1/\sqrt{2\pi e}$, ソボレフの等周定数【3.6】, 巡回セールスマン定数【8.5】
$0.2424079763\cdots$	強正方形エントロピー定数に関連【5.12】

付 録 定 数 一 覧　　　　　　553

$0.2440155280\cdots$	外半径，内半径定数の1つ【8.19】
$0.247\cdots$	過剰数密度定数【2.11】
0.25	1/4，ケーベの定数，ブロック-ランダウ定数に関連【7.1】
$0.2503634293\cdots$	オッターの木数え上げ定数に関連【5.6】
$0.2526602590\cdots$	2進自己数密度定数【2.24】
$0.2536695079\cdots$	δ_8，エルミートの定数に関連【8.7】
$0.2545055235\cdots$	カルマーの合成定数に関連【5.5】
$0.255001\cdots$	モザーのミミズ定数に関連【8.4】
$0.2614972128\cdots$	M，マイセル-メルテンス定数の1つ【2.2】
$0.2649320846\cdots$	ミニバー夫人の定数，円による被覆定数に関連【8.2】
$0.2665042887\cdots$	オッターの木数え上げ定数に関連【5.6】
$0.2677868402\cdots$	偽造不可能語定数，無パターン語に関連【5.17】
$0.2688956601\cdots$	双曲的体積定数に関連【8.9】
$0.2696063519\cdots$	マイセル-メルテンス定数に関連【2.2】
$0.2697318462\cdots$	ペル-スティーヴンハーゲン定数の1つ【2.8】
$0.272190\cdots$	カサイニュ-フィンチ定数，ストラルスキ-ハルボルス定数に関連【2.16】
$0.2731707223\cdots$	ハフナー-サルナック-マッカレー定数に関連【2.5】
$0.2746\cdots$	タラニコフ定数，無パターン語に関連【5.17】
$0.2749334633\cdots$	マイセル-メルテンス定数に関連【2.2】
$0.2763932022\cdots$	$(5-\sqrt{5})/10$，強正方形エントロピー定数に関連【5.12】
$0.2801694990\cdots$	ベルンシュテインの定数【4.4】
$0.28136\cdots$	ペル-スティーヴンハーゲン定数の1つ【2.8】
$0.2842582246\cdots$	掛谷-ベシコヴィッチ定数の予想値【8.17】
$0.2853\cdots$	レンツ-イジング定数に関連【5.22】
$0.2857142857\cdots$	2/7，2番目のディオファンタス近似の予想値【2.23】
$0.2867420562\cdots$	$-m_{1,4}$，マイセル-メルテンス定数に関連【2.2】
$0.2867474284\cdots$	強無関心定数，ハフナー-サルナック-マッカレー定数に関連【2.5】
$0.2887880950\cdots$	Q，デジタル検索木定数【5.14】，レンジェル【5.7】に関連
$0.2898681336\cdots$	$p_1=\pi^2/3-3$，ヴァレーの定数に関連【2.19】
$0.29\cdots$	ポリヤの酔歩定数の1つ【5.9】
$0.2915609040\cdots$	G/π，2Dダイマー定数【5.23】
$0.2952978731\cdots$	δ_7，エルミートの定数に関連【8.7】
$0.29745\cdots$	クラーナーのポリオミノ定数に関連【5.18】
$0.2974615529\cdots$	ピタゴラス三角形定数の1つ【5.2】
$0.2979521902\cdots$	ポリヤの酔歩定数の1つ【5.9】
$0.2993882877\cdots$	オッターの木数え上げ定数に関連【5.6】
$0.3036552633\cdots$	カルマーの合成定数に関連【5.5】
$0.3036630028\cdots$	ガウス-クズミン-ヴィルジング定数【2.17】
$0.3042184090\cdots$	オッターの木数え上げ定数に関連【5.6】
$0.3061875165\cdots$	オッターの木数え上げ定数に関連【5.6】

$0.3074948787\cdots$	$C_4=2E/27$, ハーディ-リトルウッド定数の1つ【2.1】
$0.3084437795\cdots$	ジグムンドの定数, ヤング-フェイエール-ジャクソンに関連【3.14】
$0.3090169943\cdots$	外半径, 内半径定数の1つ【8.19】
$0.3091507084\cdots$	$(8/3)(\ln(2)-\gamma)$, 幾何確率定数の1つ【8.1】
$0.3104\cdots$	パパディミトリアスの定数, 巡回セールスマン定数に関連【8.5】
$0.3110788667\cdots$	ゾロタリョーフ-シュア定数【3.9】
$0.312\cdots$	τ, 浸透集積密度定数に関連【5.18】
$0.3123633245\cdots$	クラーナーの格子動物定数に関連【5.18】
$0.3148702313\cdots$	ポリヤの酔歩定数の1つ【5.9】
$0.3157184521\cdots$	$\gamma-M$, マイセル-メルテンス定数に関連【2.2】
$0.3166841737\cdots$	アトキンソン-ネグロ-サントロ定数【2.28】
$0.3172\cdots$	巡回セールスマン定数に関連【8.5】
$0.3181736521\cdots$	カルマーの合成定数【5.5】
$0.3187590609\cdots$	ソボレフ等周定数に関連【3.6】
$0.3187766258\cdots$	オッターの木数え上げ定数に関連【5.6】
$0.3190615546\cdots$	モノマー-ダイマー定数に関連【5.23】
$0.3230659472\cdots$	$\ln(\beta)$, クネーザー-マーラー多項式定数に関連【3.10】
$0.3271293669\cdots$	ランダウ-ラマヌジャン定数に関連【2.3】
$0.3287096916\cdots$	ケプラー-ボウカンプ定数に関連【6.3】
$0.3289868133\cdots$	$\pi^2/30$, ハフナー-サルナック-マッカレー定数に関連【2.5】
$0.3332427219\cdots$	強六角形エントロピー定数, 強正方形に関連【5.12】
$0.3333333333\cdots$	$1/3$, レーニィの駐車定数に関連【5.3】
$0.3349813253\cdots$	マイセル-メルテンス定数に関連【2.2】
$0.3383218568\cdots$	オッターの木数え上げ定数に関連【5.6】
$0.3405373295\cdots$	ポリヤの酔歩定数の1つ【5.9】
$0.3472963553\cdots$	$2\sin(\pi/18)$, 浸透集積密度定数に関連【5.18】
$0.35129898\cdots$	2次再帰定数の1つ【6.10】
$0.3522211004\cdots$	ハフナー-サルナック-マッカレー定数に関連【2.5】
$0.3529622229\cdots$	オッターの木数え上げ定数に関連【5.6】
$0.3532363719\cdots$	ハフナー-サルナック-マッカレー定数【2.5】
$0.3551817423\cdots$	オッターの木数え上げ定数に関連【5.6】
$0.359072\cdots$	浸透集積密度定数に関連【5.18】
$0.3605924718\cdots$	$\mathrm{Im}(i^{i^i})$, 反復指数定数に関連【6.11】
$0.3607140971\cdots$	オッターの木数え上げ定数に関連【5.6】
$0.3611030805\cdots$	$r(12)$ 予想値, 円による被覆定数に関連【8.2】
$0.3625364234\cdots$	オッターの木数え上げ定数の1つ【5.6】
$0.364132\cdots$	レーニィの駐車定数の1つ【5.3】
$0.3678794411\cdots$	$1/e$, 自然対数の底の逆数【1.3】, 反復指数【6.11】
$0.368\cdots$	強正方形エントロピー定数に関連【5.12】
$0.3694375103\cdots$	C_7, ハーディ-リトルウッド定数の1つ【2.1】
$0.3720486812\cdots$	デジタル検索木定数に関連【5.14】

$0.3728971438\cdots$	極値定数の1つ【5.16】
$0.3729475455\cdots$	δ_6, エルミートの定数に関連【8.7】
$0.3733646177\cdots$	二分検索木定数の1つ【5.13】
$0.3739558136\cdots$	アルティンの定数【2.4】
$0.3790522777\cdots$	自己回避歩行定数の1つ【5.10】
$0.380006\cdots$	$r(11)$ 予想値, 円による被覆定数に関連【8.2】
$0.3825978582\cdots$	幾何確率定数の1つ【8.1】
$0.3919177761\cdots$	極値定数の1つ【5.16】
$0.3926990816\cdots$	$\pi/8$, モザーのミミズ定数に関連【8.4】
$0.3943847688\cdots$	モザーのミミズ定数に関連【8.4】
$0.3949308436\cdots$	$r(10)$ 予想値, 円による被覆定数に関連【8.2】
$0.3972130965\cdots$	オッターの木数え上げ定数に関連【5.6】
$0.3995246670\cdots$	2次再帰定数の1つ【6.10】
$0.3995352805\cdots$	a^{-1}, ファイゲンバウム-クーレ-トレサー定数の1つ【1.9】
$0.40096\cdots$	タムスの定数に関連【8.8】
$0.402\cdots$	浸透集積密度定数に関連【5.18】
$0.4026975036\cdots$	オッターの木数え上げ定数に関連【5.6】
$0.4074951009\cdots$	強正方形エントロピー定数【5.12】
$0.4080301397\cdots$	$(2-e^{-1})/4$, レーニィの駐車定数の1つ【5.3】
$0.4097321837\cdots$	ベリー-エッセン定数の予想値【4.7】
$0.4098748850\cdots$	C_5, ハーディ-リトルウッド定数の1つ【2.1】
$0.412048\cdots$	レンツ-イジング定数に関連【5.22】
$0.4124540336\cdots$	プルーエ-トゥエ-モース定数【6.8】
$0.4127732370\cdots$	カルマーの合成定数に関連【5.5】
$0.4142135623\cdots$	$\sqrt{2}-1$, 円による被覆定数【8.2】, レンツ-イジング定数【5.22】に関連
$0.4159271089\cdots$	極値定数の1つ【5.16】
$0.4194\cdots$	巡回セールスマン定数の1つ【8.5】
$0.4198600459\cdots$	ルーロー三角形定数に関連【8.10】
$0.4203723394\cdots$	ミンコフスキー-バウアー定数【6.9】
$0.4207263771\cdots$	整数型チェビシェフ定数の予想値【4.9】
$0.4212795439\cdots$	シュルターの定数 $t(10)$, 円による被覆定数に関連【8.2】
$0.4213829566\cdots$	$(6\ln(2))/\pi^2$, レヴィの定数【1.8】
$0.4217993614\cdots$	ピゾー-ヴィジャヤラハヴァン-サーレム定数に関連【2.30】
$0.422\cdots$	強正方形エントロピー定数に関連【5.12】
$0.4227843351\cdots$	$1-\gamma$, オイラー-マスケローニの定数【1.5】, スティルチェス定数【2.21】に関連
$0.4278733971\cdots$	外半径, 内半径定数に関連【8.19】
$0.4281657248\cdots$	オイラー-マスケローニの定数に関連【1.5】
$0.4282495056\cdots$	無関心定数, ハフナー-サルナック-マッカレー定数に関連【2.5】
$0.4302966531\cdots$	ヤング-フェイエール-ジャクソン定数に関連【3.14】
$0.4323323583\cdots$	$(1-e^{-2})/2$, レーニィの駐車定数の1つ【5.3】

$0.4330619231\cdots$	クラーナーのポリオミノ定数に関連【5.19】
$0.434\cdots$	ハーディ-リトルウッド定数に関連【2.1】
$0.4381562356\cdots$	オッターの木数え上げ定数に関連【5.6】
$0.4382829367\cdots$	$\mathrm{Re}(i^{i^{i^{\cdots}}})$, 反復指数定数に関連【6.11】
$0.4389253692\cdots$	モザーのミミズ定数に関連【8.4】
$0.43961\cdots$	自己回避歩行定数の1つ【5.10】
$0.4399240125\cdots$	オッターの木数え上げ定数に関連【5.6】
$0.4406867935\cdots$	$\ln(\sqrt{2}+1)/2$, レンツ-イジング定数に関連【5.22】
$0.4428767697\cdots$	オッターの木数え上げ定数に関連【5.6】
$0.4450418679\cdots$	$r(8)$, 円による被覆定数に関連【8.2】
$0.4466\cdots$	3Dダイマー定数【5.23】
$0.4472135955\cdots$	$1/\sqrt{5}$, 1番目のディオファンタス近似定数【2.23】
$0.4490502094\cdots$	モザーのミミズ定数に関連【8.4】
$0.4522474200\cdots$	マイセル-メルテンス定数の1つ【2.2】
$0.4545121805\cdots$	アラディ-グリンステッド定数に関連【2.9】
$0.4567332095\cdots$	オッターの木数え上げ定数に関連【5.6】
$0.461543\cdots$	スティルチェス定数に関連【2.21】
$0.4645922709\cdots$	ランダウ-ラマヌジャン定数に関連【2.3】
$0.4652576133\cdots$	δ_5, エルミートの定数に関連【8.7】
$0.4656386467\cdots$	オッターの木数え上げ定数に関連【5.6】
$0.4702505696\cdots$	2・(コンウェイ-ガイ定数), エルデーシュの和違い集合定数に関連【2.28】
$0.4718616534\cdots$	ブロックの定数の予想値【7.1】
$0.4749493799\cdots$	ワイエルシュトラス定数, ガウスのレムニスケート定数に関連【6.1】
$0.4756260767\cdots$	プルーフェの定数に関連【6.5】
$0.4769\cdots$	ブランドの定数, 巡回セールスマン定数に関連【8.5】
$0.4802959782\cdots$	幾何確率定数の1つ【8.1】
$0.4834983471\cdots$	ゴロム-ディックマン定数に関連【5.4】
$0.4865198884\cdots$	ランダウ-ラマヌジャン定数に関連【2.3】
$0.4876227781\cdots$	ガウスの双子素数定数, ハーディ-リトルウッド定数【2.1】に関連
$0.4906940504\cdots$	ランダウ-ラマヌジャン定数に関連【2.3】
$0.4945668172\cdots$	シャピロ-ドリンフェルド定数【3.1】
$0.4956001805\cdots$	$1-\gamma_0-\gamma_1$, スティルチェス定数に関連【2.21】
0.5	$1/2$, 浸透集積密度【5.18】, ランダウ-ラマヌジャン定数【2.3】に関連
$0.5163359762\cdots$	ケプラー-ボウカンプ定数に関連【6.3】
$0.5178759064\cdots$	オッターの木数え上げ定数に関連【5.6】
$0.5212516264\cdots$	ルベーグ定数に関連【4.2】
$0.5214054331\cdots$	ゴーシュの定数, 幾的確率【8.1】, 巡回セールスマン定数【8.5】に関連
$0.5235987755\cdots$	$\pi/6$, アルキメデスの定数【1.4】, マデラングの定数【1.10】

$0.531280\cdots$	ガウス-クズミン-ヴィルジング定数に関連【2.17】
$0.5313399499\cdots$	ピタゴラス三角形定数の1つ【5.2】
$0.5341\cdots$	レンツ-イジング定数に関連【5.22】
$0.5349496061\cdots$	オッターの木数え上げ定数の1つ【5.6】
$0.5351070126\cdots$	アルティンの定数に関連【2.4】
$0.5392381750\cdots$	ポリヤの酔歩定数の1つ【5.9】
$0.5396454911\cdots$	$\zeta(1/2)+2$, オイラー-マスケロニの定数に関連【1.5】
$0.5405\cdots$	最長部分数列定数の1つ【5.20】
$0.5410442246\cdots$	双曲的体積定数に関連【8.9】
$0.5432589653\cdots$	ランダウの定数の予想値【7.1】
$0.5530512933\cdots$	クァイラス-サフ定数,タムスの定数に関連【8.8】
$0.5559052114\cdots$	ベズデクの定数 $r(6)$, 円による被覆定数に関連【8.2】
$0.5598656169\cdots$	アラディ-グリンステッド定数に関連【2.9】
$0.5609498093\cdots$	幾何確率定数の1つ【8.1】
$0.5614594835\cdots$	$e^{-\gamma}$, オイラーの定数【1.5】, 約数関数【2.7】, ゴロム-ディックマン定数【5.4】に関連
$0.562009\cdots$	レーニィの駐車定数に関連【5.3】
$0.5671432904\cdots$	$W(1)$, $xe^x=1$ の解,反復指数定数に関連【6.11】
$0.5682854937\cdots$	過剰数密度定数に関連【2.11】
$0.5683000031\cdots$	オイラー-ゴンパーツ定数に関連【6.2】
$0.5697515829\cdots$	弱無関心定数,ハフナー-サルナック-マッカレー定数に関連【2.5】
$0.5731677401\cdots$	$(3/4) \cdot$(ランダウ-ラマヌジャン定数)【2.3】
$0.57339\cdots$	ペル-スティーヴンハーゲン定数の1つ【2.8】
$0.5743623733\cdots$	オッターの木数え上げ定数に関連【5.6】
$0.5759599688\cdots$	ステファンの定数,アルティン定数に関連【2.4】
$0.5761487691\cdots$	クラーナーのポリオミノ定数に関連【5.18】
$0.5767761224\cdots$	ランダウ-ラマヌジャン定数に関連【2.3】
$0.5772156649\cdots$	オイラー-マスケロニの定数, γ【1.5】, さらにスティルチェス定数【2.21】に関連
$0.5773502691\cdots$	$1/\sqrt{3}$, 掛谷-ベシコヴィッチ定数に関連【8.17】
$0.5778636748\cdots$	$\pi/(2e)$, マサー-グラメイン定数に関連【7.2】
$0.5784167628\cdots$	$(8/7)\cos(2\pi/7)\cos(\pi/7)^2$, ディオファンタス近似定数に関連【2.23】
$0.5801642239\cdots$	最適停止定数の1つ【5.15】
$0.5805775582\cdots$	ペル定数【2.8】
$0.5817480456\cdots$	マデラングの定数に関連【1.10】
$0.5819486593\cdots$	ランダウ-ラマヌジャン定数に関連【2.3】
$0.5825971579\cdots$	ρ, ポリヤの酔歩定数の1つ【5.9】
$0.5831218080\cdots$	$2G/\pi$, 2Dダイマー定数【5.23】, さらにクネーザー-マーラー定数【3.10】に関連
$0.5851972651\cdots$	オイラー-ゴンパーツ定数に関連【6.2】

$0.5877\cdots$	自己回避歩行定数の1つ【5.10】
$0.5878911617\cdots$	強正方形エントロピー定数に関連【5.12】
$0.59\cdots$	最適停止定数の1つ【5.15】
$0.5926327182\cdots$	レーマーの定数【6.6】
$0.5927460\cdots$	p_c, 浸透集積密度定数に関連【5.18】
$0.5947539639\cdots$	オッターの木数え上げ定数に関連【5.6】
$0.5963473623\cdots$	オイラー-ゴンパーツ定数【6.2】
$0.5990701173\cdots$	$M/2$, ガウスのレムニスケート定数に関連【6.1】
$0.6069\cdots$	最長部分数列定数の1つ【5.20】
$0.6079271018\cdots$	$6/\pi^2$, アルキメデスの定数【1.4】, ハフナー-サルナック-マッカレー定数【2.5】に関連
$0.6083817178\cdots$	オイラーの φ 定数の1つ【2.7】
$0.6093828640\cdots$	ネヴィルの定数 $r(5)$, 円による被覆定数に関連【8.2】
$0.6134752692\cdots$	強無三重集合定数【2.26】
$0.6168502750\cdots$	δ_4, エルミートの定数に関連【8.7】
$0.6168878482\cdots$	オッターの木数え上げ定数に関連【5.6】
$0.6180339887\cdots$	$\varphi-1$, 黄金比に関連【1.2】
$0.6194036984\cdots$	レンツ-イジング定数の1つ【5.22】
$0.6223065745\cdots$	バックハウスの定数, カルマーの定数に関連【5.5】
$0.6231198963\cdots$	オッターの木数え上げ定数に関連【5.6】
$0.6232\cdots$	巡回セールスマン定数の1つ【8.5】
$0.6243299885\cdots$	ゴロム-ディックマン定数【5.4】
$0.6257358072\cdots$	グレイシャー-キンケリン定数に関連【2.15】
$0.6278342677\cdots$	ジョン定数に関連【7.4】
$0.6294650204\cdots$	ダヴィソン-シャリット定数 ξ_1, カーエンの定数に関連【6.7】
$0.6312033175\cdots$	幾何確率定数の1つ【8.1】
$0.6321205588\cdots$	$1-1/e$, 自然対数の底に関連【1.3】
$0.6331\cdots$	巡回セールスマン定数の1つ【8.5】
$0.6333683473\cdots$	$2\cdot$(アトキンソン-ネグロ-サントロ定数), エルデーシュの和違い集合定数に関連【2.28】
$0.6351663546\cdots$	$C_3=2D/9$, ハーディ-リトルウッド定数の1つ【2.1】
$0.6366197723\cdots$	$2/\pi$, アルキメデスの定数【1.4】, 出会い定数に関連【8.21】
$0.6389094054\cdots$	ランダウ-ラマヌジャン定数に関連【2.3】
$0.6419448385\cdots$	マイセル-メルテンス定数に関連【2.2】
$0.6434105462\cdots$	カーエンの定数 ξ_2【6.7】
$0.6462454398\cdots$	マサー-グラメイン定数に関連【7.2】
$0.6467611227\cdots$	オイラー-ゴンパーツ定数に関連【6.2】
$0.6537\cdots$	最長部分数列定数の1つ【5.20】
$0.6539007091\cdots$	ξ_3, カーエンの定数に関連【6.7】
$0.6556795424\cdots$	オイラー-ゴンパーツ定数に関連【6.2】
$0.6563186958\cdots$	オッターの木数え上げ定数に関連【5.6】
$0.6569990137\cdots$	$-\delta_0$, ホール-モンゴメリー定数【2.33】

付録 定 数 一 覧

$0.6583655992\cdots$	反復指数定数の1つ【6.11】
$0.6594626704\cdots$	ポリヤの酔歩定数の1つ【5.9】
$0.6597396084\cdots$	$(2+\sqrt{3})/(4\sqrt{2})$, 光探知定数に関連【8.11】
$0.6600049346\cdots$	ξ_4, カーエンの定数に関連【6.7】
$0.6601618158\cdots$	双子素数定数, ハーディ-リトルウッド定数に関連【2.1】
$0.6613170494\cdots$	フェラー-トーニー定数, アルティンの定数に関連【2.4】
$0.6617071822\cdots$	幾何確率定数の1つ【8.1】
$0.6627434193\cdots$	ラプラス極限定数【4.8】
$0.6632657345\cdots$	ξ_5, カーエンの定数に関連【6.7】
$0.6672538227\cdots$	フェラーの硬貨投げ定数に関連【5.11】
$0.6675277360\cdots$	出会い定数に関連【8.21】
$0.6697409699\cdots$	シャンクスの定数, ハーディ-リトルウッド定数に関連【2.1】
$0.67\cdots$	エルデーシュ-レーベンソールド定数【2.27】
$0.6709083078\cdots$	マデラングの定数に関連【1.10】
$0.6749814429\cdots$	グラハムの六角形定数【8.15】
$0.676339\cdots$	浸透集積密度定数の1つ【5.18】
$0.6774017761\cdots$	カルマーの定数に関連【5.5】
$0.6821555671\cdots$	オッターの木数え上げ定数に関連【5.6】
$0.6829826991\cdots$	カラビの三角形定数に関連【8.13】
$0.6844472720\cdots$	オッターの木数え上げ定数に関連【5.6】
$0.6864067314\cdots$	C_{quad}, ハーディ-リトルウッド定数の1つ【2.1】
$0.6867778344\cdots$	カルマーの定数に関連【5.5】
$0.6903471261\cdots$	反復指数定数の1つ【6.11】
$0.6922006276\cdots$	$e^{-1/e}$, 反復指数定数の1つ【6.11】
$0.6931471805\cdots$	$\ln(2)$, 自然対数の底に関連【1.3】
$0.6962\cdots$	浸透集積密度定数の1つ【5.18】
$0.6975013584\cdots$	2番目のパパラルディ定数, アルティンの定数に関連【2.4】
$0.6977746579\cdots$	$I_1(2)/I_0(2)$, オイラー-ゴンペルツ定数に関連【6.2】
$0.6979\cdots$	巡回セールスマン定数の1つ【8.5】
$0.6995388700\cdots$	オッターの木数え上げ定数の1つ【5.6】
$0.70258\cdots$	アンブリー-トレフェセン定数, 黄金比に関連【1.2】
$0.7041699604\cdots$	フランセン-ロビンソン定数に関連【4.6】
$0.7044798810\cdots$	$1-35/(12\pi^2)$, 直線的交点定数に関連【8.18】
$0.7047534517\cdots$	ランダウ-ラマヌジャン定数に関連【2.3】
$0.7047709230\cdots$	$(\pi-\sqrt{3})/2$, ルーロー三角形定数に関連【8.10】
$0.7059712461\cdots$	ポーター-ヘンズリー定数に関連【2.18】
$0.708\cdots$	ポリヤの酔歩定数の1つ【5.9】
$0.7098034428\cdots$	うさぎ定数, プルーエ-トゥエ-モース定数に関連【6.8】
$0.7124\cdots$	巡回セールスマン定数の1つ【8.5】
$0.7147827007\cdots$	予想値, 巡回セールスマン定数の1つ【8.5】
$0.7172\cdots$	最長部分数列定数の1つ【5.20】
$0.7213475204\cdots$	$1/(2\ln(2))$, レンジェルの定数【5.7】, フェラーの硬貨投げ定数

	【5.11】に関連
0.7218106748…	5Dシュタイナー比，シュタイナー木定数に関連【8.6】
0.7234…	巡回セールスマン定数の1つ【8.5】
0.7235565167…	ポリヤの酔歩定数の1つ【5.9】
0.7236067977…	$(1+1/\sqrt{5})/2$，ディオファンタス近似定数に関連【2.23】
0.7252064830…	$97/150+\pi/40$，ランフォードの定数，幾何確率に関連【8.1】
0.7266432468…	ファン・デル・コルプトの定数に関連【3.15】
0.726868…	グラハムの六角形定数に関連【8.15】
0.7322131597…	偽造不可能語定数，無パターン語に関連【5.17】
0.7326498193…	ランダウ–ラマヌジャン定数に関連【2.3】
0.7373383033…	グロスマンの定数【6.4】
0.7377507574…	ホイッタカー–ゴンチャロフ定数の予想値【7.3】
0.7404804896…	$\pi/\sqrt{18}$，最密球充塡，エルミートの定数に関連【8.7】
0.7424537454…	リース-コルモゴロフ定数の1つ【7.7】
0.7439711933…	サルナックの定数，アルティン定数に関連【2.4】
0.7439856178…	4Dシュタイナー比，シュタイナー木定数に関連【8.6】
0.7475979202…	レーニィの駐車定数の1つ【5.3】
0.749137…	グラハムの六角形定数に関連【8.15】
0.7493060013…	クネーザー–マーラー多項式定数に関連【3.10】
0.75	$3/4$，自己回避歩行定数の1つ【5.10】
0.7520107423…	アーベル群の数え上げ定数の1つ【5.1】
0.7578230112…	フラジョレット-オドリズコ定数，ゴロム–ディックマン定数に関連【5.4】
0.760729…	グラハムの六角形定数に関連【8.15】
0.7608657675…	$(1/2)\cdot$（ベイトマン–ステムラー定数），ハーディ–リトルウッド定数に関連【2.1】
0.7642236535…	ランダウ–ラマヌジャン定数【2.3】
0.7647848097…	マイセル–メルテンス定数に関連【2.2】
0.7656250596…	リューヴィル–ロス定数に関連【2.22】
0.7666646959…	反復指数定数に関連【6.11】
0.7669444905…	ニーヴェン定数に関連【2.6】
0.7671198507…	コンウェイの定数【6.12】
0.77100…	自己回避歩行定数の1つ【5.10】
0.7711255236…	ガウス–クズミン–ヴィルジング定数に関連【2.17】
0.7735162909…	フラジョレット–マーティン定数，プルーエ–トゥエ–モース定数に関連【6.8】
0.7759021363…	ベンダーの定数，レンジェルの定数に関連【5.7】
0.7776656535…	幾何確率定数の1つ【8.1】
0.7824816009…	ゴロム–ディックマン定数に関連【5.4】
0.7834305107…	反復指数定数の1つ【6.11】
0.7841903733…	3Dシュタイナー比，シュタイナー木定数に関連【8.6】
0.7853805572…	ケプラー–ボウカンプ定数に関連【6.3】

付録 定数一覧　　　　　　　　　561

$0.7853981633\cdots$	$\pi/4$, ケプラー‐ボウカンプ定数【6.3】, モザーのミミズ定数【8.4】に関連
$0.7885305659\cdots$	ヒンチンの定数のルロスによる近似【1.8】
$0.79\cdots$	最適停止定数の1つ【5.15】
$0.7916031835\cdots$	オッターの木数え上げ定数の1つ【5.6】
$0.7922082381\cdots$	ラルの定数, ハーディ‐リトルウッド定数に関連【2.1】
$0.8003194838\cdots$	予想値, 弱無三重集合定数【2.26】
$0.8008134543\cdots$	ベンダーの定数, レンジェルの定数に関連【5.7】
$0.8019254372\cdots$	オイラー‐マスケロニの定数に関連【1.5】
$0.8043522628\cdots$	最適停止定数の1つ【5.15】
$0.8086525183\cdots$	ソロモンの駐車定数, レーニィの駐車に関連【5.3】
$0.8093940205\cdots$	アラディ‐グリンステッド定数【2.9】
$0.8116869215\cdots$	フラジョレットの定数の1つ, トゥエ‐モースに関連【6.8】
$0.8118\cdots$	最長部分数列定数の1つ【5.20】
$0.8125565590\cdots$	ストラルスキー‐ハルボルス定数【2.16】
$0.8128252421\cdots$	ヤング‐フェイエール‐ジャクソン定数に関連【3.14】
$0.81318\cdots$	c_0, 最長部分数列定数の1つ【5.20】
$0.8137993642\cdots$	ルーロー三角形定数に関連【8.10】
$0.8175121124\cdots$	シャピロ‐ドリンフェルド定数に関連【3.1】
$0.82\cdots$	k‐充足可能定数に関連【5.21】
$0.822\cdots$	ポリヤの酔歩定数の1つ【5.9】
$0.8224670334\cdots$	$\pi^2/12$, 巡回セールスマン定数に関連【8.5】
$0.8247830309\cdots$	$(\sqrt{5}-1)/\sqrt{2}$, トゥランのベキ和定数の1つ【3.16】
$0.8249080672\cdots$	$2\cdot$(ブルーエ‐トゥエ‐モース定数)【6.8】
$0.8269933431\cdots$	$3\sqrt{3}/(2\pi)$, 円による被覆定数に関連【8.2】
$0.8319073725\cdots$	$1/\zeta(3)$, アペリの定数に関連【1.6】
$0.8324290656\cdots$	ローザーの定数, ハーディ‐リトルウッド定数に関連【2.1】
$0.8346268416\cdots$	$1/M$, ガウスのレムニスケート定数に関連【6.1】
$0.8351076361\cdots$	ホール‐モンゴメリー定数に関連【2.33】
$0.8371132125\cdots$	A'_3, ブルンの定数に関連【2.14】
$0.8403426028\cdots$	ルーロー三角形定数に関連【8.10】
$0.8427659133\cdots$	$(12\ln(2))/\pi^2$, レヴィの定数【1.8】
$0.8472130848\cdots$	$M/\sqrt{2}$, ユビキタス（普遍）定数, ガウスのレムニスケート定数に関連【6.1】
$0.8507361882\cdots$	折り紙定数, プルーエ‐トゥエ‐モースに関連【6.8】
$0.8561089817\cdots$	ランダウ‐ラマヌジャン定数に関連【2.3】
$0.8565404448\cdots$	3番目のパパラルディ定数, アルティン定数に関連【2.4】
$0.8621470373\cdots$	ガウス‐クズミン‐ヴィルジング定数に関連【2.17】
$0.8636049963\cdots$	ストラルスキー‐ハルボルス定数に関連【2.16】
$0.8657725922\cdots$	整数型チェビシェフ定数の予想値【4.9】
$0.8660254037\cdots$	$\sqrt{3}/2$, 2Dシュタイナー比【8.6】, 普遍被覆定数【8.3】に関連
$0.8689277682\cdots$	ランダウ‐ラマヌジャン定数に関連【2.3】

$0.8705112052\cdots$	オッターの木数え上げ定数に関連【5.6】
$0.8705883800\cdots$	A_4, ブルンの定数に関連【2.14】
$0.8711570464\cdots$	フラジョレットの定数の1つ, トゥエ-モースに関連【6.8】
$0.8728875581\cdots$	ランダウ-ラマヌジャン定数に関連【2.3】
$0.8740191847\cdots$	$L/3$, ランダウ-ラマヌジャン【2.3】, ガウスのレムニスケート【6.1】に関連
$0.8740320488\cdots$	トゥランのベキ和定数の1つ【3.16】
$0.8744643684\cdots$	ニーヴェン定数に関連【2.6】
$0.8785309152\cdots$	幾何確率定数の1つ【8.1】
$0.8795853862\cdots$	レンツ-イジング定数に関連【5.22】
$0.8815138397\cdots$	平均類数, アルティンの定数に関連【2.4】
$0.8856031944\cdots$	$\Gamma(x)$ の最小値, オイラー-マスケローニの定数に関連【1.5.4】
$0.8905362089\cdots$	$e^\gamma/2$, ハーディ-リトルウッド定数に関連【2.1】
$0.8928945714\cdots$	ニーヴェン定数に関連【2.6】
$0.8948412245\cdots$	ランダウ-ラマヌジャン定数に関連【2.3】
$0.90177\cdots$	$\sqrt{c_0}$, 最長部分数列定数の1つ【5.20】
$0.90682\cdots$	レーニィの駐車定数の1つ【5.3】
$0.9068996821\cdots$	$\pi/\sqrt{12}$, 最密充填, エルミートの定数に関連【8.7】
$0.9089085575\cdots$	「1/9」定数に関連【4.5】
$0.91556671\cdots$	レーニィの駐車定数の1つ【5.3】
$0.9159655941\cdots$	カタランの定数, G【1.7】
$0.9241388730\cdots$	双曲的体積定数に関連【8.9】
$0.9285187329\cdots$	ガウス-クズミン-ヴィルジング定数に関連【2.17】
$0.9296953983\cdots$	$\ln(2)/2+2G/\pi$, レンツ-イジング定数に関連【5.22】
$0.9312651841\cdots$	4番目のパパラルディ定数, アルティン定数に関連【2.4】
$0.9375482543\cdots$	$-\zeta'(2)$, ポーターの定数に関連【2.18】
$0.9468064072\cdots$	ランダウ-ラマヌジャン定数に関連【2.3】
$0.9625228267\cdots$	ルベーグ定数に関連【4.2】
$0.9625817323\cdots$	$c^+{}_0$, レンツ-イジング定数に関連【5.22】
$0.9730397768\cdots$	ランダウ-ラマヌジャン定数に関連【2.3】
$0.9780124781\cdots$	エルバートの定数, シャピロ-ドリンフェルドに関連【3.1】
$0.9795555269\cdots$	3番目のベンダースキー定数, グレイシャー-キンケリンに関連【2.15】
$0.9848712825\cdots$	レーニィの駐車定数の1つ【5.3】
$0.9852475810\cdots$	ランダウ-ラマヌジャン定数に関連【2.3】
$0.9877003907\cdots$	普遍被覆定数に関連【8.3】
$0.9878490568\cdots$	$\ln$(ヒンチンの定数)【1.8】
$0.9891336344\cdots$	$2\cdot$(シャピロ-ドリンフェルド定数)【3.1】
$0.9894312738\cdots$	ルベーグ定数に関連【4.2】
$0.9920479745\cdots$	4番目のベンダースキー定数, グレイシャー-キンケリンに関連【2.15】
$0.9932\cdots$	幾何確率定数に関連【8.1】

1	リンニクの定数の予想値，ベイカーの定数【2.12】
1.0028514266…	モザーのミミズ定数に関連【8.4】
1.0031782279…	一般化スターリング定数，スティルチェス定数に関連【2.21】
1.0074347569…	デヴィッチの4次元立方体定数【8.14】
1.0077822185…	$\sqrt{65}/8$，ハイルブロンの三角形定数に関連【8.16】
1.0096803872…	5番目のベンダースキー定数，グレイシャー-キンケリンに関連【2.15】
1.0149416064…	$\pi\ln(\beta)$【8.9】，ギーゼキングの定数，クネーザー-マーラーに関連【3.10】
1.0174087975…	h_3，オイラー-マスケロニの定数に関連【1.5.4】
1.0185012157…	ポーター-ヘンズリー定数に関連【2.18】
1.0208…	巡回セールスマン定数の1つ【8.5】
1.0250590965…	レンツ-イジング定数に関連【5.22】
1.0306408341…	$\pi^2/(6\ln(2)\ln(10))$，レヴィ定数【1.8】
1.0309167521…	2番目のベンダースキー定数，グレイシャー-キンケリンに関連【2.15】
1.0346538818…	マイセル-メルテンス定数の1つ【2.2】
1.0451637801…	$\text{Li}(2)$，オイラー-ゴンペルツ定数に関連【6.2】
1.0471975511…	$\pi/3$，普遍被覆定数に関連【8.3】
1.0478314475…	2次再帰定数の1つ【6.10】
1.0544399448…	ランダウ-ラマヌジャン定数に関連【2.3】
1.0547001962…	自己回避歩行定数の1つ【5.10】
1.0606601717…	デヴィッチの4次元立方体定数に関連【8.14】
1.0662758532…	ルベーグ定数に関連【4.2】
1.0693411205…	ポリヤの酔歩定数の1つ【5.9】
1.0786470120…	ポリヤの酔歩定数の1つ【5.9】
1.0786902162…	ソボレフ等周定数に関連【3.6】
1.0820884492…	双曲的体積定数に関連【8.9】
1.0873780254…	フェラーの硬貨投げ定数の1つ【5.11】
1.0892214740…	ヴァレーの定数に関連【2.19】
1.0894898722…	$1/2+G/\pi$，ウィルブラハム-ギッブス定数に関連【4.1】
1.0939063155…	ポリヤの酔歩定数の1つ【5.9】
1.0963763171…	デヴィッチの4次元立方体定数に関連【8.14】
1.0978510391…	A_3，ブルンの定数に関連【2.14】
1.0986419643…	パリの定数，黄金比に関連【1.2】
1.0986858055…	レンジェルの定数【5.7】
1.1009181908…	デジタル検索木定数に関連【5.14】
1.1038396536…	ガウス-クズミン-ヴィルジング定数に関連【2.17】
1.1061028674…	ポリヤの酔歩定数の1つ【5.9】
1.1064957714…	コプソン-ド・ブルイン定数の1つ【3.5】
1.1128357889…	$(4L)/(3\pi)$，ランダウ-ラマヌジャン定数に関連【2.3】
1.1169633732…	ポリヤの酔歩定数の1つ【5.9】

1.1178641511…	ゴー-シュムツ定数，ゴロム-ディックマンに関連【5.4】
1.1180339887…	$\sqrt{5}/2$, シュタイニッツ定数の1つ【3.13】
1.128057…	浸透集積密度定数の1つ【5.18】
1.1289822228…	オッターの木数え上げ定数に関連【5.6】
1.13198824…	ヴィスワナスの定数，黄金比に関連【1.2】
1.1365599187…	オッターの木数え上げ定数に関連【5.6】
1.1373387363…	デジタル検索木定数の1つ【5.14】
1.1481508398…	ポーターの定数に関連【2.18】
1.1504807723…	ゴールドバッハ-ヴィノグラドフ定数，ハーディ-リトルウッド定数に関連【2.1】
1.1530805616…	ランダウ-ラマヌジャン定数に関連【2.3】
1.1563081248…	ポリヤの酔歩定数の1つ【5.9】
1.1574198038…	オッターの木数え上げ定数に関連【5.6】
1.1575…	自己回避歩行定数の1つ【5.10】
1.1587284730…	放牧山羊定数，円による被覆定数に関連【8.2】
1.159…	自己回避歩行定数の1つ【5.10】
1.1662436161…	$4G/\pi$, レンツ-イジング定数に関連【5.22】
1.1762808182…	サレム定数【2.30】
1.177043…	自己回避歩行定数の1つ【5.10】
1.1789797444…	$2G/\pi$, ウィルブラハム-ギッブス定数に関連【4.1】
1.1803405990…	h_1, オイラー-マスケロニの定数に関連【1.5.4】
1.1865691104…	$\pi^2/(12\ln(2))$, レヴィ定数【1.8】
1.1874523511…	フォイアズの定数，グロスマンの定数に関連【6.4】
1.1981402347…	M, ガウスのレムニスケート定数【6.1】
1.1996786402…	ラプラス極限定数に関連【4.8】
1.2013035599…	ローザーの定数，ハーディ-リトルウッド定数に関連【2.1】
1.2020569031…	$\zeta(3)$, アペリの定数【1.6】
1.205…	自己回避歩行定数の1つ【5.10】
1.2087177032…	バクスターの定数，リーブの正方形氷定数に関連【5.24】
1.2160045618…	オッターの木数え上げ定数の1つ【5.6】
1.21667…	自己回避歩行定数の1つ【5.10】
1.2241663491…	オッターの木数え上げ定数の1つ【5.6】
1.2267420107…	フィボナッチ数の積定数，黄金比に関連【1.2】
1.2368398446…	フェラーの硬貨投げ定数の1つ【5.11】
1.238…	レンツ-イジング定数に関連【5.22】
1.2394671218…	ポリヤの酔歩定数の1つ【5.9】
1.257…	プルーエ-トゥエ-モース定数に関連【6.8】
1.2577468869…	アラディ-グリンステッド【2.9】，ヒンチン-レヴィ【1.8】に関連
1.2599210498…	$\sqrt[3]{2}$, ピタゴラスの定数に関連【1.1】
1.2610704868…	二分検索木定数に関連【5.13】
1.2615225101…	双曲的体積定数に関連【8.9】
1.2640847353…	2次再帰定数の1つ【6.10】

$1.2672063606\cdots$	μ_6, 極値定数の1つ【5.16】
$1.272\cdots$	自己回避歩行定数の1つ【5.10】
$1.275\cdots$	自己回避歩行定数の1つ【5.10】
$1.2824271291\cdots$	グレイシャー-キンケリン定数【2.15】
$1.2885745539\cdots$	ファイゲンバウム-クーレ-トレサー定数に関連【1.9】
$1.2910603681\cdots$	ヴァレーの定数に関連【2.19】
$1.2912859970\cdots$	反復指数定数の1つ【6.11】
$1.2923041571\cdots$	ランダウ-ラマヌジャン定数に関連【2.3】
$1.2940\cdots$	自己回避歩行定数の1つ【5.10】
$1.2985395575\cdots$	ベイトマンの A 定数, ハーディ-リトルウッド定数に関連【2.1】
$1.302\cdots$	無二重語定数【5.17】
$1.3035772690\cdots$	コンウェイの定数【6.12】
$1.30568\cdots$	アポロニウス充填定数【8.20】
$1.3063778838\cdots$	ミルズの定数【2.13】
$1.3110287771\cdots$	レムニスケートの四半弧長 $L/2$, ガウスの定数【6.1】
$1.3135070786\cdots$	K_{-3}, ヒンチン定数に関連【1.8】
$1.3203236316\cdots$	$2C_{\text{twin}}$, ハーディ-リトルウッド定数の1つ【2.1】
$1.3247179572\cdots$	黄金比【1.2】, ピゾー-ヴィジャヤラハヴァン定数【2.30】に関連
$1.3325822757\cdots$	マイセル-メルテンス定数【2.2】, φ 定数【2.7】に関連
$1.3385151519\cdots$	$\exp(G/\pi)$, 2Dダイマー定数【5.23】
$1.3426439511\cdots$	強正方形エントロピー定数に関連【5.12】
1.34375	$43/32$, 自己回避歩行定数の1つ【5.10】
$1.3468852519\cdots$	リース-コルモゴロフ定数の1つ【7.7】
$1.3505061\cdots$	2次再帰定数の1つ【6.10】
$1.3511315744\cdots$	ヴァレーの定数に関連【2.19】
$1.3521783756\cdots$	μ_7, 極値定数の1つ【5.16】
$1.3531302722\cdots$	最適停止定数に関連【5.15】
$1.3694514039\cdots$	シャリットの定数, シャピロ-ドリンフェルド定数に関連【3.1】
$1.3728134628\cdots$	$2C_{\text{quad}}$, ハーディ-リトルウッド定数の1つ【2.1】
$1.3750649947\cdots$	マイセル-メルテンス定数の1つ【2.2】
$1.37575\cdots$	幾何確率定数に関連【8.1】
$1.3813564445\cdots$	β, クネーザー-マーラー多項式定数に関連【3.10】
$1.3905439387\cdots$	ベイトマンの B 定数, ハーディ-リトルウッドに関連【2.1】
$1.3932039296\cdots$	ポリヤの酔歩定数の1つ【5.9】
$1.3954859724\cdots$	強六角形エントロピー定数, 強正方形に関連【5.12】
$1.3994333287\cdots$	カルマーの合成定数に関連【5.5】
$1.4011551890\cdots$	ミルベリの定数, ファイゲンバウム-クーレ-トレサー定数に関連【1.9】
$1.4045759346\cdots$	複素グロタンディーク定数の予想値【3.11】
$1.4092203477\cdots$	ストラルスキー-ハルボルス定数に関連【2.16】
$1.4106861346\cdots$	オイラー-ゴンペーツ定数に関連【6.2】
$1.4142135623\cdots$	$\sqrt{2}$, ピタゴラスの定数【1.1】

1.4236003060…	μ_8, 極値定数の1つ【5.16】
1.4298155…	2次再帰定数の1つ【6.10】
1.4359911241…	$1/3+2\sqrt{3}/\pi$, 1番目のルベーグ定数【4.2】
1.4426950408…	$\ln(2)^{-1}$, ポーター-ヘンズリー定数に関連【2.18】
1.4446678610…	$e^{1/e}$, 反復指数定数の1つ【6.11】
1.4503403284…	K_{-2}, ヒンチン定数に関連【1.8】
1.4513692348…	ラマヌジャン-ソルドナー定数, オイラー-ゴンパーツに関連【6.2】
1.4560749485…	バックハウスの定数,【5.5】に関連
1.457…	無三重語定数【5.17】
1.4603545088…	$-\zeta(1/2)$, アペリの定数に関連【1.6】
1.4609984862…	バクスターの定数, リーブの正方形氷に関連【5.24】
1.4616321449…	$\Gamma(x)$ を最小にする x, オイラー-マスケローニの定数に関連【1.5.4】
1.4655712318…	ムーアの定数, 黄金比に関連【1.2】
1.4670780794…	ポーターの定数【2.18】
1.4677424503…	ファイゲンバウム-クーレ-トレサー定数の1つ【1.9】
1.4681911223…	アラディ-グリンステッド定数に関連【2.9】
1.4741726868…	オッターの木数え上げ定数の1つ【5.6】
1.4762287836…	カルマーの合成定数に関連【5.5】
1.4767…	自己回避歩行定数の1つ【5.10】
1.4879506635…	$-\zeta(2/3)/\zeta(2)$, ニーヴェン定数に関連【2.6】
1.4880785456…	オッターの木数え上げ定数の1つ【5.6】
1.5028368010…	2次再帰定数の1つ【6.10】
1.5030480824…	強正方形エントロピー定数【5.12】
1.50659177…	マンデルブロ集合の面積, 2次再帰定数【6.10】
1.50685…	ナグルの定数, リーブの正方形氷定数に関連【5.24】
1.5078747554…	グリーンフィールド-ナスバウム定数, 2次再帰定数【6.10】
1.5163860591…	ポリヤの酔歩定数の1つ【5.9】
1.5217315350…	ベイトマン-ステムラー定数, ハーディ-リトルウッド定数【2.1】
1.5299540370…	ガウスのレムニスケート定数に関連【6.1】
1.5353705088…	デジタル検索木定数に関連【5.14】
1.5396007178…	$(4/3)^{3/2}$, リーブの正方形氷定数【5.24】
1.5422197217…	平面六角形格子に対するマデラング定数【1.10】
1.5449417003…	ルーロー三角形定数に関連【8.10】
1.5464407087…	強正方形エントロピー定数に関連【5.12】
1.5513875245…	カラビの三角形定数【8.13】
1.5557712501…	ファイゲンバウム-クーレ-トレサー定数の1つ【1.9】
1.5707963267…	$\pi/2$, アルキメデスの定数に関連【1.4】
1.5849625007…	$\ln(3)/\ln(2)$, ストラルスキー-ハルボルス定数に関連【2.16】
1.5868266790…	ファイゲンバウム-クーレ-トレサー定数に関連【1.9】
1.6066951524…	デジタル検索木定数の1つ【5.14】

1.6153297360…	レンツ-イジング定数に関連【5.22】
1.6155426267…	2D NaCl マデラング定数の反数【1.10】
1.6180339887…	黄金比，φ【1.2】
1.6222705028…	オドリズコ-ヴィルフ定数，ミルズの定数に関連【2.13】
1.6281601297…	フラジョレット-マーティン定数，プルーエ-トゥエ-モースに関連【6.8】
1.6366163233…	エルデーシュ-レーベンソールド定数に関連【2.27】
1.6421884352…	2番目のルベーグ定数【4.2】
1.644703…	移動ソファー定数に関連【8.12】
1.6449340668…	$\pi^2/6$，アペリ定数【1.6】，ハフナー-サルナック-マッカレー定数【2.5】に関連
1.6467602581…	デジタル検索木定数に関連【5.14】
1.6600…	リーブの正方形氷定数に関連【5.24】
1.6616879496…	2次再帰定数の1つ【6.10】
1.6813675244…	オッターの木数え上げ定数の1つ【5.6】
1.6824415102…	ベイトマン-グロスワルド c_{23} 定数，ニーヴェン定数に関連【2.6】
1.6847127097…	外半径，内半径定数の1つ【8.19】
1.6857…	巡回セールスマン定数に関連【8.5】
1.6903029714…	ファイゲンバウム-クーレ-トレサー定数の1つ【1.9】
1.6910302067…	カーエンの定数に関連【6.7】
1.6964441175…	ヒンチン-レヴィ定数に関連【1.8】
1.7052111401…	ニーヴェン定数【2.6】
1.7091579853…	$\pi^2/(12\ln(\varphi))$，ヒンチン-レヴィ定数に関連【1.8】
1.7286472389…	カルマーの合成定数【5.5】
1.7356628245…	クラーナーのポリオミノ定数に関連【5.19】
1.7374623212…	ゴロム-ディックマン定数に関連【5.4】
1.7410611252…	オッターの木数え上げ定数の1つ【5.6】
1.7454056624…	K_{-1}，ヒンチンの定数に関連【1.8】
1.7475645946…	3D NaCl マデラング定数の反数【1.10】
1.75	7/4，レンツ-イジング定数に関連【5.22】
1.7548776662…	フェラーの硬貨投げ定数の1つ【5.11】
1.7555101394…	オッターの木数え上げ定数の1つ【5.6】
1.756…	自己回避歩行定数の1つ【5.10】
1.7579327566…	無限入れ子根号，黄金比に関連【1.2】
1.7587436279…	アラディ-グリンステッド定数に関連【2.9】
1.7632228343…	反復指数定数に関連【6.11】
1.76799378…	最適停止定数に関連【5.15】
1.77109…	$-c_1$，最長部分数列定数の1つ【5.20】
1.7724538509…	$\sqrt{\pi}$，オイラーの定数【1.5.4】，カールソン-レヴィン定数【3.2】に関連
1.7783228615…	3番目のルベーグ定数【4.2】
1.7810724179…	e^γ，オイラーの定数【1.5】，エルデーシュ-レーベンソールド定数

	【2.27】に関連
1.7818046151⋯	ベキ級数定数の予想値【7.3】
1.7822139781⋯	実グロタンディーク定数の予想値【3.11】
1.7872316501⋯	コモルニク-ロレティ定数, プルーエ-トゥエ-モース定数に関連【6.8】
1.7916228120⋯	$\exp(2G/\pi)$, 2Dダイマー定数【5.23】
1.7941471875⋯	カルマーの合成定数に関連【5.5】
1.8173540210⋯	自己回避歩行定数の1つ【5.10】
1.8228252496⋯	マサー-グラメイン定数の予想値【7.2】
1.8356842740⋯	マイセル-メルテンス定数に関連【2.2】
1.8392867552⋯	トリボナッチ数列と黄金比とプラスティック定数に関連【1.2】
1.8393990840⋯	4D NaCl マデラング定数の反数【1.10】
1.8442049806⋯	ランダウ-コルモゴロフ定数の1つ【3.3】
1.8477590650⋯	$\sqrt{2}+\sqrt{2}$, 自己回避歩行定数の予想値【5.10】
1.8519370519⋯	ウィルブラハム-ギッブス定数【4.1】
1.8540746773⋯	$L/\sqrt{2}$, ガウスのレムニスケート定数に関連【6.1】
1.8823126444⋯	幾何確率定数の1つ【8.1】
1.9021605831⋯	ブルンの定数【2.14】
1.9081456268⋯	β^2, クネーザー-マーラー多項式定数に関連【3.10】
1.9093378156⋯	5D NaCl マデラング定数の反数【1.10】
1.9126258077⋯	オッターの木数え上げ定数の1つ【5.6】
1.9276909638⋯	ファイゲンバウム-クーレ-トレサー定数の1つ【1.9】
1.9287800⋯	ライトの定数, ミルズの定数に関連【2.13】
1.940215351⋯	2Dモノマー-ダイマー定数【5.23】
1.9435964368⋯	オイラーのφ定数の1つ【2.7】
1.9484547890⋯	c_4, クネーザー-マーラー多項式定数に関連【3.10】
1.9504911124⋯	4・(ガウス整数の双子素数定数), ハーディ-リトルウッドに関連【2.1】
1.9655570390⋯	6D NaCl マデラング定数の反数【1.10】
1.9670449011⋯	c_5, クネーザー-マーラー多項式定数に関連【3.10】
1.9771268308⋯	c_6, クネーザー-マーラー多項式定数に関連【3.10】
1.9954559575⋯	フランセン-ロビンソン定数に関連【4.6】
2	高速行列乗法定数の予想値【2.29】
2.006⋯	エルデーシュ-レーベンソールド定数に関連【2.27】
2.0124059897⋯	7D NaCl マデラング定数の反数【1.10】
2.0287578381⋯	デュボア・レイモン定数に関連【3.12】
2.0415⋯	巡回セールスマン定数に関連【8.5】
2.0462774528⋯	レヴィ定数のルロスによる近似【1.8】
2.05003⋯	ホイットニー-ミクリン拡張定数の1つ【3.8】
2.0524668272⋯	8D NaCl マデラング定数の反数【1.10】
2.0531987328⋯	自己回避歩行定数に関連【5.10】
2.0780869212⋯	$\ln(\varphi)^{-1}$, ポーター-ヘンズリー定数に関連【2.18】

$2.1080540666\cdots$	出会い定数に関連【8.21】
$2.1102339661\cdots$	ブラウン-ワン定数,ヤング-フェイエール-ジャクソン定数から【3.14】
$2.158\cdots$	ミアン-チョウラ定数,エルデーシュの逆数和に関連【2.20】
$2.1732543125\cdots$	$\zeta(3/2)/\zeta(3)$, ニーヴェン定数に関連【2.6】
$2.1760161352\cdots$	クネーザー-マーラー多項式定数に関連【3.10】
$2.1894619856\cdots$	オッターの木数え上げ定数の1つ【5.6】
$2.1918374031\cdots$	オッターの木数え上げ定数の1つ【5.6】
$2.2001610580\cdots$	ヒンチン定数のルロスによる近似【1.8】
$2.2038565964\cdots$	オイラーの φ 定数の1つ【2.7】
$2.2195316688\cdots$	移動ソファー定数【8.12】
$2.2247514809\cdots$	ロビンソンの C 定数,ヒンチンの定数に関連【1.8】
$2.2394331040\cdots$	竹内-プレルバーグ定数【5.8】
$2.2665345077\cdots$	フランセン-ロビンソン定数に関連【4.6】
$2.2782916414\cdots$	モザーのミミズ定数の1つ【8.4】
$2.2948565916\cdots$	アーベル群の数え上げ定数の1つ【5.1】
$2.3\cdots$	$s_c(3)$ の推定値,k-充足可能定数に関連【5.21】
$2.3025661371\cdots$	フラジョレットの定数の1つ,トゥエ-モースに関連【6.8】
$2.3038421962\cdots$	ロビンソンの A 定数,ヒンチン定数に関連【1.8】
$2.3091385933\cdots$	クラーナーのポリオミノ定数に関連【5.19】
$2.3136987039\cdots$	マデラングの定数に関連【1.10】
$2.3212850380\cdots$	出会い定数に関連【8.21】
$2.3360\cdots$	リーブの正方形氷定数に関連【5.24】
$2.3507\cdots$	ランダウ-コルモゴロフ定数の1つ【3.3】
$2.3565273533\cdots$	モノマー-ダイマー定数の1つ【5.23】
$2.3731382208\cdots$	$\pi^2/(6\ln(2))$, レヴィの定数【1.8】
$2.37597\cdots$	クラーナーのポリオミノ定数に関連【5.19】
$2.3768417063\cdots$	整数型チェビシェフ定数の予想値【4.9】
$2.3979455861\cdots$	デュボア・レイモン定数に関連【3.12】
$2.4048255576\cdots$	$J_0(x)$ の最初の零点,ソボレフ等周定数に関連【3.6】
$2.4149010237\cdots$	ゴロム-ディックマン定数に関連【5.4】
$2.4413238136\cdots$	ルベーグ定数に関連【4.2】
$2.4725480752\cdots$	ソボレフ等周定数に関連【3.6】
$2.4832535361\cdots$	オッターの木数え上げ定数の1つ【5.6】
$2.4996161129\cdots$	アーベル群の数え上げ定数の1つ【5.1】
$2.5029078750\cdots$	α, ファイゲンバウム-クーレ-トレサー定数の1つ【1.9】
$2.5066282746\cdots$	$\sqrt{2\pi}$, スターリングの定数,アルキメデスの定数【1.4】,グレイシャー-キンケリン定数【2.15】に関連
$2.5175403550\cdots$	オッターの木数え上げ定数の1つ【5.6】
$2.5193561520\cdots$	マデラングの定数に関連【1.10】
$2.5695443449\cdots$	$e^{\gamma}/\ln(2)$, オイラー-マスケローニの定数に関連【1.5】
$2.5849817595\cdots$	シェルピンスキーの定数【2.10】

$2.5980762113\cdots$	$\sqrt{27/4}$，リーブの正方形氷定数に関連【5.24】
$2.6034\cdots$	リーブの正方形氷定数に関連【5.24】
$2.6180339887\cdots$	黄金比 $\varphi+1$，タット-ベラハ定数【5.25】，ガウス-クズミン-ヴィルジング定数【2.17】に関連
$2.6220575542\cdots$	レムニスケートの半弧長 L，ガウスの定数【6.1】
$2.6381585303\cdots$	2D自己回避歩行定数の推定値【5.10】
$2.6389584337\cdots$	$(2+\sqrt{3})/\sqrt{2}$，光探知定数に関連【8.11】
$2.67564\cdots$	クラーナーのポリオミノ定数に関連【5.19】
$2.6789385347\cdots$	$\Gamma(1/3)$，オイラー-マスケローニの定数に関連【1.5.4】
$2.6789638796\cdots$	4・(シャンクスの定数)，ハーディ-リトルウッド定数に関連【2.1】
$2.6811281472\cdots$	オッターの木数え上げ定数の1つ【5.6】
$2.6854520010\cdots$	ヒンチンの定数【1.8】
$2.7182818284\cdots$	自然対数の底，e【1.3】
$2.72062\cdots$	自己回避歩行定数の1つ【5.10】
$2.7494879027\cdots$	オッターの木数え上げ定数の1つ【5.6】
$2.75861972\cdots$	双曲的体積定数に関連【8.9】
$2.7865848321\cdots$	フランセン-ロビンソン定数に関連【4.6】
$2.8077702420\cdots$	フランセン-ロビンソン定数【4.6】
$2.8154600332\cdots$	オッターの木数え上げ定数の1つ【5.6】
$2.8264199970\cdots$	村田の定数，アルティン定数【2.4】，φ 関数【2.7】に関連
$2.8336106558\cdots$	ファイゲンバウム-クーレ-トレサー定数の1つ【1.9】
$2.8372974794\cdots$	マデラングの定数に関連【1.10】
$2.8582485957\cdots$	ハーディ-リトルウッド定数の1つ【2.1】
$2.9409823408\cdots$	$c_o/\sqrt{2\pi}$，レンジェル定数に関連【5.7】
$2.9409900447\cdots$	$c_e/\sqrt{2\pi}$，レンジェル定数に関連【5.7】
$2.9557652856\cdots$	オッターの木数え上げ定数の1つ【5.6】
$2.9904703993\cdots$	ゴーシュムツ定数，ゴロム-ディックマン定数に関連【5.4】
3	タット-ベラハ定数に関連【5.25】
$3.0079\cdots$	エルデーシュの逆数和定数に関連【2.20】
$3.01\cdots$	エルデーシュの逆数和定数に関連【2.20】
$3.1415926535\cdots$	アルキメデスの定数，π（円周率）【1.4】
$3.1477551485\cdots$	平方剰余定数，マイセル-メルテンス定数に関連【2.2】
$3.1704593421\cdots$	オイラーの φ 定数の1つ【2.7】
$3.1962206165\cdots$	「板」定数，ソボレフ等周定数に関連【3.6】
$3.2099123007\cdots$	$\exp(4G/\pi)$，2Dダイマー定数【5.23】，またクネーザー-マーラー定数【3.10】
$3.2469796037\cdots$	白銀比，タット-ベラハ定数の1つ【5.25】
$3.2504\cdots$	リーブの正方形氷定数に関連【5.24】
$3.2659724710\cdots$	オッターの木数え上げ定数の1つ【5.6】
$3.2758229187\cdots$	$\exp(\pi^2)/(12\ln(2))$，レヴィ定数【1.8】
$3.2871120555\cdots$	オッターの木数え上げ定数の1つ【5.6】

$3.2907434386\cdots$	オッターの木数え上げ定数の1つ【5.6】
$3.3038421963\cdots$	ロビンソンの B 定数,ヒンチン定数に関連【1.8】
$3.33437\cdots$	バンビーの定数,フレイマンの定数に関連【2.31】
$3.3412669407\cdots$	オッターの木数え上げ定数に関連【5.6】
$3.3598856662\cdots$	デジタル検索木定数に関連【5.14】
$3.3643175781\cdots$	ファン・デル・コルプトの定数【3.15】
$3.4070691656\cdots$	マガタの定数,カルマーの合成定数に関連【5.5】
$3.4201328816\cdots$	自己回避歩行定数に関連【5.10】
$3.4493588902\cdots$	ロビンソンの D 定数,ヒンチンの定数に関連【1.8】
$3.4627466194\cdots$	Q^{-1},デジタル検索木定数【5.14】,レンジェル【5.7】に関連
$3.501838\cdots$	自己回避歩行定数に関連【5.10】
$3.5070480758\cdots$	フェラーの硬貨投げ定数に関連【5.11】
$3.5795\cdots$	リーブの正方形氷定数に関連【5.24】
$3.6096567319\cdots$	ρ_2 の予想値,ディオファンタス近似【2.23】
$3.6180339887\cdots$	$\varphi+2$,タット-ベラハ定数の1つ【5.25】
$3.6256099082\cdots$	$\Gamma(1/4)$,オイラー-マスケロニ定数に関連【1.5.4】
$3.63600703\cdots$	ファイゲンバウム-クーレ-トレサー定数の1つ【1.9】
$3.6746439660\cdots$	平方剰余定数,マイセル-メルテンスに関連【2.2】
$3.6754\cdots$	最長部分数列定数の1つ【5.20】
$3.7038741039\cdots$	$2\cdot($ウィルブラハム-ギッブス定数$)$【4.1】
$3.764435608\cdots$	2Dモノマー-ダイマー定数【5.23】
$3.7962\cdots$	z_c,強正方形エントロピー定数に関連【5.12】
$3.8264199970\cdots$	村田の定数$+1$,アルティン定数【2.4】,φ 関数【2.7】に関連
$3.8695192413\cdots$	最適停止定数に関連【5.15】
$3.9002649200\cdots$	マデラングの定数に関連【1.10】
$3.921545\cdots$	モザーのミミズ定数に関連【8.4】
$3.92259368\cdots$	双曲的体積定数に関連【8.9】
4	タット-ベラハ【5.25】,2Dグレッチュ環状領域定数【7.8】
$4.0180767046\cdots$	ファイゲンバウム-クーレ-トレサー定数の1つ【1.9】
$4.062570\cdots$	クラナーのポリオミノ定数【5.19】
$4.121326\cdots$	ファイゲンバウム-クーレ-トレサー定数の1つ【1.9】
$4.1327313541\cdots$	$\sqrt{2\pi e}$,ソボレフの等周定数【3.6】,巡回セールスマン定数【8.5】
$4.1507951\cdots$	自己回避歩行定数の1つ【5.10】
$4.1511808632\cdots$	ハーディ-リトルウッド定数の1つ【2.1】
$4.2001\cdots$	リーブの正方形氷定数に関連【5.24】
$4.2472965459\cdots$	幾何確率定数の1つ【8.1】
$4.25\cdots$	$r_c(3)$ の推定値,k-充足可能定数に関連【5.21】
$4.3076923076\cdots$	$56/13$,3D球体に対するコーン定数【3.7】
$4.3110704070\cdots$	二分検索木定数の1つ【5.13】
$4.5278295661\cdots$	フレイマンの定数【2.31】
$4.5651\cdots$	リーブの正方形氷定数に関連【5.24】
$4.5678018826\cdots$	ガスパーの定数,ヤング-フェイエール-ジャクソンに関連【3.14】

4.5860790989…	ポリヤの酔歩定数の1つ【5.9】
4.5908437119…	$\Gamma(1/5)$, オイラー-マスケロニの定数に関連【1.5.4】
4.6592661225…	ベイトマン-グロスワルド c_{03} 定数, ニーヴェン定数に関連【2.6】
4.6692016091…	δ, ファイゲンバウム-クーレ-トレサー定数の1つ【1.9】
4.68404…	3D自己回避歩行定数の推定値【5.10】
4.7300407448…	棒の定数, ソボレフの等周定数に関連【3.6】
4.799891547…	光探知定数の3弧近似【8.11】
4.8189264563…	光探知定数の2弧近似【8.11】
4.8426…	自己回避歩行定数の1つ【5.10】
4.9264…	アルティン定数に関連【2.4】
5.0747080320…	双曲的体積定数に関連【8.9】
5.1387801326…	ソボレフ等周定数に関連【3.6】
5.1667…	リーブの正方形氷定数に関連【5.24】
5.2441151086…	レムニスケート弧長 $2L$, ガウスの定数【6.1】
5.2569464048…	オイラー-マスケロニの定数に関連【1.5.4】
5.4545172445…	ヒンチン-レヴィ定数に関連【1.8】
5.5243079702…	ヒンチン-レヴィ定数に関連【1.8】
5.5553…	リーブの正方形氷定数に関連【5.24】
5.5663160017…	$\Gamma(1/6)$, オイラー-マスケロニの定数に関連【1.5.4】
5.6465426162…	オッターの木数え上げ定数の1つ【5.6】
5.6493764966…	整数型チェビシェフ定数の予想値【4.9】
5.7831859629…	ソボレフ等周定数に関連【3.6】
5.8726188208…	ベイトマン-グロスワルド- c_{13} 定数, ニーヴェン定数に関連【2.6】
5.9087…	アルティン定数に関連【2.4】
5.9679687038…	ファイゲンバウム-クーレ-トレサー定数の1つ【1.9】
6.0…	カメロンの無和集合定数の1つ【2.25】
6.2831853071…	2π, アルキメデスの定数に関連【1.4】
6.3800420942…	オッターの木数え上げ定数の1つ【5.6】
6.77404…	4D自己回避歩行定数の推定値【5.10】
6.7992251609…	ファイゲンバウム-クーレ-トレサー定数の1つ【1.9】
6.8…	カメロンの無和集合定数の1つ【2.25】
7.1879033516…	ジョン定数の予想値【7.4】
7.2569464048…	オイラー-マスケロニの定数に関連【1.5.4】
7.2846862171…	ファイゲンバウム-クーレ-トレサー定数の1つ【1.9】
7.3719494907…	c_o, レンジェルの定数に関連【5.7】
7.3719688014…	c_e, レンジェルの定数に関連【5.7】
7.7431319855…	デジタル検索木定数の1つ【5.14】
7.7581602911…	オッターの木数え上げ定数の1つ【5.6】
8.3494991320…	ファイゲンバウム-クーレ-トレサー定数の1つ【1.9】
8.7000366252…	ケプラー-ボウカンプ定数【6.3】
8.7210972…	ファイゲンバウム-クーレ-トレサー定数の1つ【1.9】

$8.83854\cdots$	5D自己回避歩行定数の推定値【5.10】
$9.0803731646\cdots$	ヘンズリー定数【2.18】
$9.27738\cdots$	ファイゲンバウム-クーレ-トレサー定数の1つ【1.9】
$9.2890254919\cdots$	「1/9」定数の逆数【4.5】
$9.2962468327\cdots$	ファイゲンバウム-クーレ-トレサー定数の1つ【1.9】
$9.37\cdots$	3Dグレッチュ環状領域定数【7.8】
$9.576778\cdots$	光探知定数に関連【8.11】
$9.6694754843\cdots$	ベイトマン-グロスワルド c_{04} 定数,ニーヴェン定数に関連【2.6】
$9.7\cdots$	$r_c(4)$ の推定値,k-充足可能定数に関連【5.21】
$10.5101504239\cdots$	ザギエの定数,フレイマンの定数に関連【2.31】
$10.7310157948\cdots$	$\exp(\pi^2/(6\ln(2)))$,レヴィ定数【1.8】
$10.87809\cdots$	6D自己回避歩行定数の推定値【5.10】
$11.0901699437\cdots$	$(11+5\sqrt{5})/2$,強正方形エントロピー定数に関連【5.12】
$12.262874\cdots$	自己回避歩行定数に関連【5.10】
$12.6753318106\cdots$	16・(ラルの定数),ハーディ-リトルウッド定数に関連【2.1】
$14.1347251417\cdots$	ゼータ関数の最初の零点[の虚数部],グレイシャー-キンケリン定数に関連【2.15】
$14.6475663016\cdots$	アーベル群の数え上げ定数の1つ【5.1】
$15.1542622415\cdots$	e^e,反復指数定数の1つ【6.11】
$16.3638968792\cdots$	β,ファイゲンバウム-クーレ-トレサー定数の1つ【1.9】
$16.9787814834\cdots$	ベイトマン-グロスワルド c_{14} 定数,ニーヴェン定数に関連【2.6】
$19.4455760839\cdots$	ベイトマン-グロスワルド c_{05} 定数,ニーヴェン定数に関連【2.6】
$20.9\cdots$	$r_c(5)$ の推定値,k-充足可能定数に関連【5.21】
$21.0220396387\cdots$	ゼータ関数の2番目の零点[の虚数部],グレイシャー-キンケリン定数に関連【2.15】
$22.6\cdots$	4Dグレッチュ環状領域定数【7.8】
$25.0108575801\cdots$	ゼータ関数の3番目の零点[の虚数部],グレイシャー-キンケリン定数に関連【2.15】
$29.576303\cdots$	ファイゲンバウム-クーレ-トレサー定数の1つ【1.9】
$39.1320261423\cdots$	カラビの三角形定数に関連【8.13】
$43.2\cdots$	$r_c(6)$ の推定値,k-充足可能定数に関連【5.21】
$55.247\cdots$	ファイゲンバウム-クーレ-トレサー定数の1つ【1.9】
$118.6924619727\cdots$	アーベル群の数え上げ定数の1つ【5.1】
$137.0359\cdots$	微細構造定数の逆数,ファイゲンバウム-クーレ-トレサー定数に関連【1.9】

定 数 索 引

ア 行

アキェゼル-クレイン-ファヴァール（Achiser-Krein-Favard）定数　257,258
アペリ（Apéry）の定数　39
アーベル（Abel）の定数　239
アポロニウス充填（Apollonian packing）定数　542
アラディ-グリンステッド（Alladi-Grinstead）定数　120
アルキメデス（Archimedes）の定数　17
アルティン（Artin）定数　103,116

一般オイラー（Euler）定数　31
一般化グレイシャー（Glaisher）定数　136

ヴァレー（Vallée）定数　160
ウィットニー-ミクリン（Whitney-Mikhlin）の拡張定数　229
うさぎ（rabbit）定数　443

エルデーシュ（Erdös）の逆数和定数　163
エルデーシュの和違い集合定数　189
エルデーシュ-レーベンソールド（Erdös-Lebensold）定数　186
エルミート（Hermite）の定数　511,512
円による被覆（circular coverage）定数　488

オイラー-ゴンパーツ（Euler-Gopertz）定数　426
オイラー（Euler）定数　14,76
オイラー-マスケローニ（Euler-Mascheroni）の定数　28,116,236,265
オッター（Otter）の木の数え上げ定数　297

カ 行

外半径-内半径（circumradius-inradius）定数　540
ガウス-クズミン-ヴィルジング（Gauss-Kuzmin-Wirsing）定数　151,152
ガウス（Gauss）のレムニスケート定数　423
ガウス双子素数（Gaussian twin prime）定数　89
カーエン（Cahen）の定数　438
掛谷-ベシコヴィッチ（Kakeya-Besicovitch）定数　535,536
過剰数密度（abundant numbers density）定数　125
カタラン（Catalan）の定数　52,100,234
カメロン（Cameron）の無和集合定数　181
カラビ（Calabi）の三角形定数　528
カールソン-レヴィン（Carlson-Levin）の定数　213
カルマール（Kalmár）の合成定数　294

幾何確率（geometric probability）定数　483
ギーゼキング（Gieserking）の定数　234
ギッブス-ウィルブラハム（Gibbs-Wilbraham）定数　250,251
「1/9」予想の定数　261
強正方形（hard square）エントロピー定数　344
強六角形（hard hexagon）エントロピー定数　344
極値（extreme value）定数　365

クネーザー-マーラー多項式（Kneser-Mahler-polynomial）定数　233
クラナー（Klarner）のポリオミノ定数　380
グラハム（Graham）の六角形定数　531
グレイシャー-キンケリン（Glaisher-Kinkelin）定数　135
グレッチュ環状領域（Glötzsch ring）定数　479
グロスマン（Grossman）の定数　433

グロタンディーク（Grothendieck）の定数 237
クロネッカー（Kronecker）定数 93

結合（connective）定数 332
ケプラー-ボウカンプ（Kepler-Bouwkamp）定数 431
ケーベ（Koebe）定数 460

格子（lattice）定数 175
コプソン-ド・ブルイン（Copson-de Bruijn）定数 219
コモルニク-ロレティ（Komornik-Loreti）定数 442
ゴロム-ディックマン（Golomb-Dickman）定数 286
コーン（Korn）定数 227
コンウェイ（Conway）の定数 456
ゴンチャロフ（Goncharov）の定数 465

サ 行

最適停止（optimal stopping）定数 362

シェルピンスキー（Sierpinski）定数 122
自己回避歩行（self-avoiding walk）定数 332
自己数密度（self-numbers density）定数 179
シャピロ-ドリンフェルド（Shapiro-Drinfeld）の定数 210
シャンクス（Shanks）定数 89
シャンパーノウン（Champernowne）定数 446
充填（packing）定数 543
シュタイナー木（Steiner tree）定数 508
シュタイニッツ（Steinitz）定数 241
　　m次元── 243
巡回セールスマン（traveling salesman）定数 502
ジョン（John）定数 469, 470

スティルチェス（Stieltjes）定数 31, 166
ストラルスキー-ハルボルス（Stolarsky-Harborth）定数 145

整係数チェビシェフ（integer Chebyshev）定数 271
正則ホロノーム（regular holonomic）定数 329

双曲的体積（hyperbolic volume）定数 516
素数逆数（prime reciprocal）定数 93

ソファー移動（moving sofa）定数 525
ソボレフ（Sobolev）の等周定数 221
ゾロタリョーフ-シュア（Zolotarev-Schur）定数 231

タ 行

ダイマーに関する定数 234
竹内-プレルバーグ（Takeuchi-Prellberg）定数 322
タット-ベラハ（Tutte-Beraha）定数 419
タムス（Tammes）の定数 513
チェビシェフ（Chebyshev）定数 262
　　整係数── 271
チャイティン（Chaitin）の定数 79, 80
直線的交点（rectilinear crossing）定数 537

出会い（rendezvous）定数 544
ディオファンタス近似（Diophantine approximation）定数 173
デヴィッチ（Devicci）の4次元立方体定数 529
デジタル検索木（digital search free）定数 355
デュボア・レイモン（du Bois Reymond）定数 239, 240

トゥラーン（Turán）の指数和定数 248
特異ホロノーム（singular holonomic）定数 329
ド・ブルイン-ニューマン（de Bruijn-Newman）定数 205, 206
トロット（Trott）の定数 446

ナ 行

ニーヴェン（Niven）定数 111
2分検索木（binary search tree）定数 351

ネイピア（Napier）の定数 14

ハ 行

ハイルブロン（Heilbronn）の三角形定数 532
バクスター（Baxter）の4彩色定数 416
ハーディ-リトルウッド（Hardy-Littlewood）定数 83
パパディミトリアス（Papadimitrious）の定数 505
ハフナー-サルナック-マッカレー（Hafner-Sarnak-McCurlay）定数 109

定 数 索 引

光探知（beam detection）定数　521
ピゾー-ヴィジャヤラハヴァン-サーレム（Pisot-Vijayaraghavan-Salem）定数　193
ピタゴラス3数（Pythagorean triple）定数　278
ピタゴラス（Pythagoras）の定数　1
ヒルベルト（Hilbert）定数　218
ヒンチン（Khintchine）の定数　172
ヒンチン-レヴィ（Khintchine-Lévy）定数　58

ファイゲンバウム-クーレ-トレサー（Feigenbaum-Couller-Tresser）定数　64, 65
ファヴァール（Favard）定数　215, 254, 258
ファン・デル・コルプト（van der Corput）定数　247
フェラー（Feller）の硬貨投げ定数　340
普遍被覆（universal coverage）定数　493
フラジョレット-オドリゾコ（Flajolet-Odlyzko）定数　291
プラスチック（Plastic）定数　9
フランセン-ロビンソン（Fransén-Robinson）定数　264, 265
ブランド（Bland）の定数　504
プルーエ-トゥエ-モース（Prouhet-Thue-Morse）定数　440
プルーフェ（Plouffe）の定数　433
ブルン（Brun）定数　132
フレイマン（Freiman）の定数　201
ブロック（Bloch）の定数　459
ブロック-ランダウ（Bloch-Landau）定数　459
　単葉――　460

ベイカー（Baker）定数　128
ベイトマン（Bateman）定数　89
ベイトマン-グロスワルド（Bateman-Grosswald）定数　113
ヘイマン（Hayman）定数　472
ベキ級数（power series）定数　466
ベリー-エッセン（Berry-Esseen）定数　266
ペル（Pell）定数　118
ペル-スティーヴンハーゲン（Pell-Stevenhagen）定数　118
ベルンシュテイン（Bernstein）定数　259

ホイッタカー（Whittaker）の定数　465
ホイッタカー-ゴンチャロフ（Whittaker-Goncharov）定数　465

ポーター（Porter）定数　156
ポーター-ヘンズリー（Porter-Hensley）定数　156
ポリヤ（Pólya）の酔歩定数　323
ホール-モンゴメリー（Hall-Montgomery）定数　207

マ　行

マイセル-メルテンス（Meissel-Mertens）定数　93
マサー-グラメイン（Masser-Gramain）定数　462
マデラング（Madelung）の定数　74

溝掘り人（trench digger）の定数　523
ミル（Mill）定数　130
ミルベルグ（Myrberg）定数　442
ミンコフスキー-バウアー（Minkowski-Bower）の定数　445

無三重集合（triple-free set）定数　183, 184
村田定数　116

モザー（Moser）のミミズ定数　496
モーデル（Mordell）定数　176
モノマー・ダイマー（monomer-dimer）定数　408

ヤ　行

ヤング-フェイエール-ジャクソン（Young-Fejér-Jackson）定数　244

ラ　行

ラプラス限界（Laplace limit）定数　268
ラル（Lal）定数　89
ランダウ（Landau）の定数　254, 459
ランダウ-コルモゴロフ（Landau-Kolmogorov）の定数　214
ランダウ-ラマヌジャン（Landau-Ramanujan）定数　94, 97, 119

リース-コルモゴロフ（Riesz-Kolmogorov）定数　478
リトルウッド-クルニー-ポメレンケ（Littlewood-Clunie-Pommerenke）定数　475
リーブ（Lieb）の正方氷定数　414
リューヴィル-ロス（Liouville-Roth）定数　171

リンニク (Linnik) 定数　127

ルベーグ (Lebesgue) 定数　252
ルーロー三角形 (Reuleaux triangle) 定数　519

レーニィ (Rényi) の駐車定数　280
レーマー (Lehmer) の定数　436, 437

レムニスケート (lemniscate) 定数　122, 423
レンジェル (Lengyel) 定数　317
レンツ-イジング (Lenz-Ising) 定数　394

ロックス-ポーター (Lochs-Porter) 定数　157
ロバン (Robin) の定数　273

事項索引

A

A 数列　163
abelian square　370
abelian square-free　370
abundant　125
AB 浸透　377
activity　347
acyclic digraph　310
additive composition　294
additive partition　294
adjacent　297
admisible decompositions　120
AGM　423
almost primes　86
alternating sign matrix conjecture　415
antipercolation　377
arcsine law　326
aperiodic sum-free　182
applicants　362
approximate counting　359
asymmetric　301, 302
AVL-木　312

B

B_2 数列　164
B木（B-tree）　312
Bell 数　318
Bernstein の同程度振動予想　255
Bernstein の予想　260
bifurcation ratio　312
binary 無平方語　369
binary Columbian numbers　179

binary incident matrices　409
binary search tree　352
binary self numbers　179
binary splitting　13
binary trees　299
binomial identity　366
blocks　317
Bombieri の最大値ノルム　236
bond percolation model　374
bond percolation theory　377

C

cactus　307
candidate　362
canonical　454
carefree　109
chain　318
Chebyshev 効果　99
Chebyshev 多項式　255
chiral　301
chromatic polynomial　416, 419
chromial　419
circle problem　123
clause　390
Clausen 積分　234
cloud　310
colorable　391
common subsequences　387
common supersequence　387
complete　182
completely additive　209
completely multiplicative　207
complex component　313

computational complexity issues　415
congruent number problem　279
conjugate　194
constitutional isomers　300
continuum percolation　377
convex　382
cover　391
critical determinant　175
critical exponent　333
critical probability　374
critical susceptibility exponent　400
critical temperature　397
cube-free word　369
Curie point　397
cyclically isomorphic　303

D

d-ソーセージ　510
d-単体　510
deficient　125
degree　305
digamma　169
digital search tree　356, 357
digraph　310
dimer　408
dimer arrangement　408
dimer covering　409
Dini-Lipschitz の定理　253
directed graph　310
discriminant　201
distance　305
divisor problem　123
domain of attraction　367
domino　380, 410

double free 183

E

e-数学 19
edge 297
eight-vertex model 415
entropy of folding 416
escape probability 324
Eulerian orientations 414
Euler 数 53
Euler-Maclaurin の総和法 264
Euler-Maclaurin の和公式 29
even polygonal drawing 396
exponential divisor 126
exponent of matrix multiplication 191
extinction probability 312

F

favorite sites 326
FFT 13
first ladder height 327
first-passage percolation 377
first-passage time 325
fixed point 306
forest 297
four-exponentials 予想 197
free tree 297
Friedrichs の不等式 224, 228
friendly 325
full-informationproblem 363

G

gap 203
Gaudin 密度 138
Göbel の数列 450
Goldbach 予想 83
golden root 420
Gorshkov & Wirsing 多項式 272
graph 297
greedily 164
GUE 仮設 138

H

half cube 391
Hall's ray 203
Hardy の不等式 219
hard hexagon 346
hard square 346, 394
hard triangle entropy 345
height 312, 352
height-balanced trees 312
hexagonal grid 410
high-temperature zero-field series 397, 400
Hilbert の不等式 218
holonomic function 329
homeomorphically irreducible 303
house 195
Hurwitz の定理 201
hyperspherical simplices 366

I

ice rule 414
identity tree 302
incomplete 182
increasing subsequence 385
2-increasing subsequence 386
increasing tree 304
indefinite 201
independence number 184
independent 296
intersect 328, 334
interval graph 309
inversion method 268
Ising free energy 397
Ising magnetic susceptibility per site 400
Ising model 394
Ising specific heat 401
iterated exponentials 265

J

jagged 166
Jensen の公式 235

K

k 原始的 (k primitive) 187
k-彩色可能 391
k-充 112
k-充足問題 390
k-文字 390
k-SAT· 390
k-satisfiability 問題 390
Kepler の方程式 268
Khintchine の法則 58
Korn の不等式 227

L

L-当たり数 185
label 304
labeled functional graphs 308
Lagrange スペクトラム 201
Lagrange の逆転法 268
Lagrange の補間多項式 254
Lambert の W 関数 452
largest tree 290
lattice 318
lattice animal 375, 380
lattice gas 345
leaves 300
leftist tree 311
Legendre 級数 254
Lehmer の多項式 194
literal 390
logarithmic divergence 401
longest increasing subsequence problem 367
low-temperature series 395
lozenge tilings 410
Lucas 数列 104
Lyapunov 比 266

M

m-カクタス (cactus) 307
(m, M) 等長写像 469
m^2+1 で表される素数を含む予想 84
Mahler の尺度 235
Mandelbrot 集合 449
mapping patterns 308

事項索引 581

Markov numbers 202
Markov spectrum 201
matching 409
Matsuoka 166
maximal 203, 318
maximal generation size 313
maximal tree 290
meander 335
mean cluster density 373
mean square distance of a monomer from the endpoints 334
mean square end-to-end distance 333
mean square radius of gyration 334
Mian-Chowla 数列 164
Minkowski の？関数 445
Mitrovic 166
mobile 304
mock zeta function 162
monent of extinction 313
monic polynomial 193
monomer 408
monomer-dimer covering 408
Monthly problem 212
multiplicative 209
multiplicative composition 294
multiplicative partition 295
multiplicative spectrum 208

N

n-泉 383
n 弧 521
n 次元同時ディオファンタス近似定数 174
n 単体 516
N 点格子 346
n-fountain 383
N-site lattice with k squares 346
Newton-Cotes 近似 265
nonaveraging 数列 164
non-hypotenuse 数 100
non-reciprocal 195

NP-完全 390

O

odd polygonal drawing 399
optimal stopping 364
ordered trees 303
order parameter 348
order statistics 365
orientation 414
overlap-free 370

P

P モーメント 401
Padé 近似 251
Painlevé III 型微分方程式 403
parallelogram polyominoes 382
parity 定数 440
partially abelian square 370
partially abelian square-free 370
partial ordering 318
partition 317
partition function 345, *346*, 398
Patricia tries 357
Pell 方程式 118
percolation 372, 377
percolation threshold 374
perfect 125
perfect matching 410
periodic 182
permanents 409
physical disorder 378
planar 419
planted homeomorphically irreducible tree 303
polyomino 380
porus solid 372
poset 318
poset of partitions 319
powerfull 112
prefer 195
prime additive partition 295
prime multiplicative compositions 295

prime multiplicative partitions 295
primitive 186, 371
primitive sequences 296
P-V 数 194

Q

q-アナログ（q-analog) 358
q-階乗 318
q-展開（q-tactorial) 442
q-二項係数（q-binomial coefficient) 318
Q モーメント 401
quadtrees 353
quasi-primitive 188

R

r 拡張 229
Ramanujan の連分数 383
random coloring problem 401
random maze 313
register functions 312
2-regular 310
residual entropy for square ice 414
return probability 323
Riemann のゼータ関数 39, 265
Riemann 予想 40
rooted 298
rooted identity tree 302
rooted tree 298
row-convex 382

S

S 許容 175
s-クラスター 373
satisfiable 390
saturation level 352
Schnirelmann 数 87
secretary problem 362
self-avoiding polygons 334
self-avoiding trail 335
self-avoiding walk 332
self-intersecting 328
self-intersection time 328
self-trapping 328

semi-meander 336
series-reduced 303
Sidon 数列 164
silver root 420
simplest 202
site percolation model 373
sixteen-vertex model 415
six-exponentials 定理 197
six-vertex model 415
3-smooth number 197
3-smooth representation 197
Sobolev 不等式 221
spherical triangulation 419
spin-1/2 model 398
square 369
square-diagonal lattice 416
square-free 369
square-full 整数 112
star body 175
stereoisomers 302
Stirling 数 46
Stirling の公式 18, 135
strongly binary tree 299
strongly carefree 109
sum-distinct 189
superadditive 387
Sylvester の数列 439, 448
Szekeres 数列 164

T

Takeuchi 数 322
ternary 無平方語 369
time-homogeneous two-state 340
total progeny 312
travelling salesman problem 502
tree 297
triangular lattice 417
　——with wraparound 410
trimer 282
triple correlation 予想 140
triple-free 183
TSP 502
twenty-vertex model 415

U

Ulam の 1 加法数列 147
unforgeable 371
uniformly distributed 193
union-find algorithm 313
unit interval graph 309
universal 349
unsatisfiable 390

V

valency 305
vicious 325

W

walk 323
Waring 問題 196
weakly binary tree 299
weakly carefree 109
weakly ternary tree 300
Weierstrass のシグマ関数 424
white screen problem 327
word 369
worst 202

X

X に対する n 次のルベーグ定数 255

Y

Young tableaux 386

Z

Z 容認的 176, 541
zero magnetic field case 396

ア　行

アイスルール 414
悪意的 325
アクティビティ 347
アーベルの定理 239
アーベル平方 370
アルファ 475
α-エネルギー 513

イェンセンの公式 235

イジング自由エネルギー 397
　位置ごとの—— 395
イジング比熱 401
イジングモデル 394
位数 288
板の不等式 223
1×2 ドミノ 410
位置ごとのイジング磁化率 400
位置浸透モデル 373
一様分布 193
一列駐車 282
一般ディガンマ関数 169
一般連分数 3
移動被覆 493, 496

ヴィノグラドフ数 87
ウェアリング問題 196
ウェーブレット 251
宇宙論的定理 456
うねり図形 335
ウラムの 1 加法数列 147

枝の浸透モデル 374
枝の浸透理論 377
エノン写像 67
エルゴート理論 58
エルデーシュ-レーニィの進化過程 313
円周率 17
エントロピー 59
円による被覆定数 488
円の問題 123

オイラー数 53
オイラーの公式 21
オイラーの向き 414
オイラー-マクローリンの和公式 29
黄金長方形 6
黄金根 420
黄金比 2, 5, 6
　——円写像 70
応募者 362
覆う 391
お気に入りの場所 326
オーバーラップフリー 370
折り紙数列 443

事項索引

折り曲げエントロピー 416, 417

カ 行

外角 442
回帰確率 323
概周期的無和集合 182
概素数 86
回転数 70
外半径 540
χ 不等式 163
ガウス・ユニタリー集合 138
ガウディン密度 138
可逆的 454
カクタス 307
確率論的数え上げアルゴリズム 441
囲い 195
過剰数 125
価数 305
カタラン数 24
活性度 347
合併発見アルゴリズム 313
加法分解 294
加法分割 294
神の比率 6
カリパス 498
環状領域 480
完 182
完全加法的関数 209
完全グラフ 503
完全乗法的 207
完全情報問題 363
完全数 125
完全楕円積分 23
完全マッチング 410
ガンマ関数 18

木 297
　　――の高さ 352
　　――の飽和レベル 352
幾何確率定数 483
危険関数 428
基数検索トライ 357
偽造不可能 371
奇多角形描画 399
擬等長写像 469
逆正弦法則 326

逆正接積分 56
吸収領域 367
級数既約 303
級数展開 375
「1/9」予想 261
球面三角分割 419
球面導関数 475
キューリー点 397
強 2 分木 299
強三角形エントロピー 345
強磁性の場合 398
強正方形 346, 394
共通部分数列 387
共通優数列 387
行凸 382
強普遍被覆 494, 500
強無関心 109
強無三重 184
共役数 194
行列式 175
行列乗法指数 191
強六角形 346
極大 203, 318
極大木 290
極大世代サイズ 313
極値エネルギー 513
許容される分解 120
距離 305
キラル 301
近似数え上げ 359
近似できない 171

空間充填問題 282
空隙 203
偶多角形描画 396
区塊 317
区間グラフ 309
鎖 318
クビタノビッチ-ファイゲンバウム関数方程式 68
クーポン集めの問題 28
雲 310
クラウゼン積分 234
グラフ 297
グレッチュ環状領域 480
クロミアル 419

計算複雑性問題 415

継続多項式 162
計量エントロピー 59
桁ごとの抽出アルゴリズム 55
ケプラーの方程式 268
ケプラー予想 511
ゲーベルの数列 450
検索失敗 356
検索成功 356
原子 457
原始根 103
原始的 186, 371
原始的数列 296
原始ピタゴラス数 278
原始ヘロン 3 数 279
懸垂線 12

語 369
好意的 325
高温級数展開 396
高温零場級数展開 397, 400
交差 328, 334
格子ガス 345
格子動物 375, 380
格子和 76
合成 457
構造異性体 300
交代ビット集合 148
交代符号行列予想 415
交代ルロス表現 60
恒等木 302
合同数問題 279
候補者 362
氷の規則 414
好む 195
コプソンの不等式 220
コープランド-エルデーシュ数 446
小道 334
ゴルシュコフ&ワーシング多項式 272
ゴールドバッハ予想 83, 86
　　拡張された―― 84
ゴルトン-ワトソン分岐過程 312
ゴンチャロフ多項式 467
コーンの不等式 227

事 項 索 引

サ 行

最悪 202
最近整数による連分数 61
最近接排除モノマー 283
斉時2状態 340
最小架橋木 508
最小全域木 40, 504
最小マッチング 505
最大木 290
最大成分 291
最大平方因数 124
最単純 202
最短の巡回置換の長さ 286
最長増加部分数列問題 367
最長の巡回置換の長さ 286
最長の裾 289
最長の連の長さ 341
最長ローパス 290
最適停止 364
最良一様有理式近似 260
サボテン 307
左翼の木 311
サーレム数 194
3-ソーセージ 509
3円滑数 197
3円滑表現 197
三角形分割 516
三角格子 417
3次元ドミノ 411
3次式で表される素数 88
三重相関予想 140
三重対数関数 46
三重フィボナッチ数列 9
算術幾何平均 18
　　──アルゴリズム 423
3値無平方語 369
残留集合 543

磁化率 399
ジグザグ 497
自己回避小道 335
自己回避多角形 334
自己回避歩行 332
自己交差 328
自己交差時間 328
自己生成連分数 438
自己捕獲 328

指数積分 427
指数的ディオファンタス方程式 80
指数的約数 126
指数塔 452
自然対数の底 12
子孫 311
下から一様有界 245
実対数容量 273
実チェビシェフ多項式 271
実超越直径 273
シドン数列 164
死亡力率 428
弱2分木 299
弱3分木 300
弱型1-1不等式 478
ジャクソンの不等式 260
弱無関心 109
弱無三重 183
写像パターン 308
シャピロの不等式 210
自由木 297
周期的 182
周期倍化作用素 68
充足可能 390
充足不可能 390
充平方整数 112
充ベキ 112
縮小写像 270
シュタイナー極小木 508
シュタイナー点 509
シュタイナー比 509
出力 454
シュニレルマン数 87
巡回セールスマン問題 502
巡回置換 28
巡回同型 303
巡回和 210
循環正則連分数展開 3
準原始的 188
順序統計量 365
順序木 303
上位記録 28
乗法スペクトル 208
乗法的 209
乗法分解 294
乗法分割 295
剰余多項式 467

ジョコヴィッチの予想 212
シルヴェスターの数列 439, 448
浸透確率 375
浸透閾値 374
浸透理論 372

スターリング数 46
スターリングの公式 14, 18, 135, 222
スピゴットアルゴリズム 14
スピン1/2モデル 398
スプライン 251

整係数チェビシェフ多項式 271
正準的 454
整数対数直径 272
整数超越直径 272
生成関数 43
正則連分数展開 2
星体 175
正方対角格子 416
正方氷剰余エントロピー 414
セケレス数列 164
ゼータ関数 21
節 390
絶滅確率 312
絶滅のモーメント 313
0次の変形ベッセル関数 223
漸近的に正規 154
漸近的平均桁数 287
線型トリマー 282
線型ポリマーのモデル 332
全子孫 312
染色多項式 416, 419

増加部分数列 385
増加木 304
相加平均 60
双曲型四面体 234
双曲正弦ゴードン微分方程式 404
相乗平均 58, 60
素数加法分解 295
素数加法分割 295
素数乗法分解 295
素数乗法分割 295

事項索引　　　585

素数定理　14,83
ソーセージ　509,510
ソボレフ不等式　221
ゾロタリョーフの多項式　231

タ 行

第1種完全楕円積分　262
第1種と第2種の完全楕円積分
　　$K(x)$，$E(x)$　238
第2種完全楕円積分　262
第2種のスターリング数　318
対数積分　428
代数的整数　194
対数発散　401
対数容量　273
対数螺線　7
対相関関数　140
ダイマー　282,408
ダイマー配置　408
ダイマー被覆　409
ダイマー問題　53
楕円関数　23
楕円に関するシーウェル問題
　　231
楕円モジュラー関数　424
互いに逆数共役でない　195
高さ　312
高さ平衡木　312
竹内数　322
多孔性固体　372
多重級数　42
多重根号展開　7
多重対数関数　46,47
脱出確率　324
束　318
単位区間グラフ　309
単純　276
単純2次元ずらし変換のエント
　　ロピー　234
単体　510
単体度　516
端点からのモノマーの平均平方
　　距離　334
単葉　460,475

チェビシェフ効果　99
チェビシェフ多項式　231,255
秩序パラメータ　348

駐車問題　281
中心対称　520
中心連分数　61
超安定周期点　65
超安定帯幅　65
超越数　3
超越直径　273
超球面単体　366
超指数数列　452
頂点　297
頂点の次数　305
調和級数　30,39
調和平均　60
直線的交点数　537
直径　460

続きの列　456

ディオファンタス方程式　79
低温級数展開　395
ディガンマ関数　33,245
ディニ-リプシッツの定理　253
定幅　519
ディリクレ（Dirichlet）のベー
　　タ関数　52
デジタル検索木　356,357
デジタル和の指数和　146
デジタル和のベキ和　146
テッセラクト　530
δ-被覆　543
天体力学　268

等角同値　479
等角容量　480
同型　297
同時ディオファンタス近似　4
等周問題　221
同相的既約　303
同相的既約植木　303
ドウソンの積分　518
等ポテンシャル曲線　442
特異関数　445
独立　296
独立数　184
閉じたミミズ　499
凸　382
凸支持線　211
凸閉包　4

ドミノ　380
トムソンの電子問題　514
トライ　357
トリチェリ点　509
トリボナッチ数列　9

ナ 行

内半径　540
長い素数　103
名札　304

2-木（2-tree）　308
2-正則　310
2-増加部分数列　386
2-3 アナログ（2-3 analog）
　　359
2,3-木　311
二項係数　41
二項恒等式　366
二項束　320
二項分割方式　13
2次元ドミノ　409
二重根号　4
20頂点モデル　415
二乗平均平方根　60
2進コロンビア数　179
2進自己数　179
2進接続行列　409
2進抽出アルゴリズム　15
2進展開　19
2進分解方式　29
2進ユークリッドアルゴリズム
　　（互除法）　158
2進隣接行列　409
2値無平方語　369
2分木　299
2分検索木　352
2分検索木定数　351
1/2-アナログ（1/2-analog）
　　358
ニュートン-コーツ近似　265
ニュートン法　1

根　298
根付き木　298
根付き恒等木　302

ハ 行

葉　300
π-数学　19
バイボーン　410
ハウスドルフ次元　543
白銀根　420
白銀比円写像　70
白色画面問題　327
パーコレーション理論　372
端から端までの平均平方　333
はしごの最初の高さ　327
8 項点モデル　415
初通過時間　325
初通過浸透　377
初到達時間　325
ハーディの不等式　219
パデ近似　251
鳩の巣論法　173
パトリシアトライ　357
場の軌跡　442
幅広ミミズ　498
パーマネント　409
林　297,306
パリティー定数　440
半うねり図形　336
半順序　318
半順序集合　318
反浸透　377
バーンズ G-関数　135
半単純　276
反復指数関数　265
反復指数列　452
判別式　201
半立方体　391
パンルヴェ III 型微分方程式
　　403

比較する　160
光探知機　521
菱形　410
菱形充填　410
非斜辺数　100
秘書問題　362
ピゾー-ヴィジャヤラハヴァン
　　数　194
非対称　301,302
ビット複雑度　13

非平均数列　164
ピーベルバッハ予想　476
標準 n 単体　517
標準 n 立方体　517
ヒルベルトの不等式　218

フィボナッチ数列　6
フェケテ点　514
不完全　182
複体成分　313
不斉　301
不足数　125
双子素数予想　83,132
　　拡張された——　84
縁取り数　166
普通交点数　538
普通図　538
物理的無秩序　378
不定　201
不動点　306
部分アーベル平方　370
部分分母　154
普遍ディオファンタス方程式
　　79
普遍的　349
普遍被覆　493,496
普遍被覆定数　493
不偏ミゼールゲーム　454
フリードリックスの不等式
　　224,228
フルヴィッツの定理　201
プルーフェの反復級数　435
ブロック　317
分割　317
分割関数　345,346
分割統治法　147
分割のポセット　319
分岐点　65
分岐幅　65
分岐比　312
分数次元　33
分配関数　398

平均クラスターの大きさ　374
平均クラスター密度　373
平均平方回転半径　334
平行移動被覆　494,500
平行四辺形ポリオミノ　382

平方　369
平面的　419
ベッセル関数級数　269
ペラン数列　8
ベリー-エッセンの不等式　266
ベル数　318
ベルヌーイ数　39,353
ペルの数列　2
ペル方程式　118
ベルンシュテインの同程度振動
　　予想　255
辺　297
変形ゼータ関数　162

棒の不等式　222
放牧山羊の問題　491
補間多項式の近似力　254
歩行　323
ポセット　318
　　分割の——　319
ポリオミノ　380
ポリガンマ関数　33
ボリヤイ-レーニィの表現　60
ホール線　203
ホロノーム関数　329
ボンビエリの最大値ノルム
　　236

マ 行

巻き数　70
巻きつけられた三角格子　410
膜の不等式　223
マッチング　409,505
マッチング問題　13
マーラーの尺度　235
マルコフ数　202
マルコフスペクトル　201
マンデルブロ集合　449

ミアン-チョウラ数列　164
道　335
3 つの素数の組に関する 2 つの
　　予想　84
ミニバー夫人の問題　491
ミンコフスキーの ? 関数　445

無アーベル平方　370
無関心　109

事項索引

向き 414
無限行列積 55
無限積 44
無限余接表現 437
無三重 183
無重複 370
無倍 183
無部分アーベル平方 370
無平方 369
　──の核 110
　──の部分 110
無閉路的 297
無閉路有向グラフ 310
無理数度 171
無立方語 369
無和 163, 181

メアンダー 335
面対面頂点三角形分割 516

モジュラス 480
モックゼータ関数 162
モニック多項式 193
モノマー 283, 408
モノマー・ダイマー被覆 408
モビール 304
森 297
モンゴメリー–オドリズコ法則 138

ヤ 行

約数問題 123
ヤコビの楕円関数 23

優加法性 387
有限半単純結合環 276
有向グラフ 310
有理式近似 261

有理数値 434
弓と矢 522

要素 457
欲張って 164
4つの素数の組に関する2つの予想 84
4指数予想 197
四重対数関数 46
4数ゲーム 9
4分木 353

ラ 行

ラグランジュスペクトル 201
ラグランジュの逆転法 268
ラグランジュの補間多項式 254
ラプラス限界定数 268
ラプラスの偏微分方程式 223
ラベル 304
ラベル付き関数グラフ 308
ラマヌジャンの連分数 383
λ彩色 (λ-coloring) 419
λ状態零温度反強磁性体のポッツ (Potts) モデル 421
ランク付け集合 454
ランダム 80
ランダム鎖 320
ランダムグラフ 40
ランダム彩色問題 401
ランダム写像 289
ランダム漸化式 10
ランダム逐次吸着 282
　──モデル 283
ランダム迷路 313
ランダムヤング図形 386
ランダムリンク TSP 503
ランベルトの W 関数 452

立体異性体 302
リーマン予想 40
リャプノフ比 266
両側リプシッツ写像 469
臨界温度 397, 399
臨界確率 374
臨界行列式 175
臨界磁化率指数 400
臨界指数 333
臨界帯 41
隣接 297

ルカス数列 104
ルジャンドル級数 254
ルドルフの定数 19
ルパート王子の問題 529
ルロス表現 60

零磁場の場合 396
レジスター関数 312
レーマーの多項式 194
連結している 297
連続体浸透 377
連続度 253
連続媒質中の浸透 377
連分数展開 22

6指数定理 197
6頂点モデル 414
ロジスティック写像 64
六角格子 410

ワ 行

ワイエルシュトラスのシグマ関数 424
ワイエルシュトラスのペー関数 425
和違い 189

監訳者略歴

一 松　信
(ひとつまつ　しん)

1926 年　東京都に生まれる
1947 年　東京大学理学部数学科卒業
1969 年　京都大学数理解析研究所教授
1989 年　東京電機大学理工学部教授
現　在　京都大学名誉教授
　　　　理学博士

数学定数事典

2010 年 2 月 20 日　初版第 1 刷
2012 年 3 月 20 日　　　第 2 刷

定価は外函に表示

監訳者　一　松　　　信
発行者　朝　倉　邦　造
発行所　株式会社　朝　倉　書　店
　　　　東京都新宿区新小川町 6-29
　　　　郵便番号　162-8707
　　　　電話　03(3260)0141
　　　　FAX　03(3260)0180
　　　　http://www.asakura.co.jp

〈検印省略〉

© 2010〈無断複写・転載を禁ず〉　　新日本印刷・渡辺製本

ISBN 978-4-254-11126-2　C 3541　　Printed in Japan

〈(社)出版者著作権管理機構 委託出版物〉

本書の無断複写は著作権法上での例外を除き禁じられています．複写される場合は，そのつど事前に，(社)出版者著作権管理機構（電話 03-3513-6969，FAX 03-3513-6979，e-mail: info@jcopy.or.jp）の許諾を得てください．

A.N.コルモゴロフ他編　三宅克哉監訳 **19 世 紀 の 数 学　I** ―数理論理学・代数学・数論・確率論― 11741-7　C3341　　A 5 判 352頁 本体6400円	〔内容〕数理論理学(ライプニッツの記号論理学／ブール代数他)／代数と代数的数論(代数学の進展／代数的数論と可換環論の始まり他)／数論(2次形式の数論／数の幾何学他)／確率論(ラプラスの貢献／ガウスの貢献／数理統計学の起源他)
A.N.コルモゴロフ他編　小林昭七監訳 **19 世 紀 の 数 学　II** ―幾何学・解析関数論― 11742-4　C3341　　A 5 判 368頁 本体6400円	〔内容〕解析幾何と微分幾何／射影幾何学／代数幾何と幾何代数／非ユークリッド幾何／多次元の幾何学／トポロジー／幾何学的変換／解析関数論／複素数／複素積分／コーシーの積分定理，留数／楕円関数／超幾何関数／モジュラー関数／他
A.N.コルモゴロフ他編　藤田　宏監訳 **19 世 紀 の 数 学　III** ―チェビシェフの関数論～差分法― 11743-1　C3341　　A 5 判 432頁 本体7200円	〔内容〕ゼロからのずれが最小の関数／連分数／18世紀の微分方程式／存在と一意性／求積法による方程式の積分／線形微分／微分の解析・定性的理論／19世紀前半・後半の変分法／補間／オイラー-マクローリンの求和公式／差分方程式／他
J.-P.ドゥラエ著　京大 畑　政義訳 **π ― 魅 惑 の 数** 11086-9　C3041　　B 5 判 208頁 本体4600円	「πの探求，それは宇宙の探検だ」古代から現代まで，人々を魅了してきた神秘の数の世界を探る。〔内容〕πとの出会い／πマニア／幾何の時代，解析の時代／手計算からコンピュータへ／πを計算しよう／πは超越的か／πは乱数列か／付録／他
早大 足立恒雄著 **数　　―体系と歴史―** 11088-3　C3041　　A 5 判 224頁 本体3500円	「数」とは何だろうか？一見自明な「数」の体系を，論理から複素数まで歴史を踏まえて考えていく。〔内容〕論理／集合：素朴集合論他／自然数：自然数をめぐるお話他／整数：整数論入門他／有理数／代数系／実数：濃度他／複素数：四元数他／他
I.スチュアート著 聖学院大松原　望監訳　藤野邦夫訳 数学のエッセンス 1 **イアン・スチュアートの 数の世界** 11811-7　C3341　　B 5 判 192頁 本体3800円	多彩な話題で数学の世界を紹介。〔内容〕フィボナッチと植物の生長／彫刻と黄金数／音階の数学／選挙制度と民主的／膨張する宇宙／パスカルのフラクタル／完全数，素数／フェルマーの定理／アルゴリズム／魔方陣／連打される鐘と群論
C.F.ガウス著　九大 高瀬正仁訳 数学史叢書 **ガ ウ ス　整 数 論** 11457-7　C3341　　A 5 判 532頁 本体9800円	数学史上最大の天才であるF.ガウスの主著『整数論』のラテン語原典からの全訳。小学生にも理解可能な冒頭部から書き起こし，一歩一歩進みながら，整数論という領域を構築した記念碑的著作。訳者による豊富な補註を付し読者の理解を助ける
H.ポアンカレ著　元慶大 斎藤利弥訳 数学史叢書 **ポアンカレ　ト ポ ロ ジ ー** 11458-4　C3341　　A 5 判 280頁 本体6200円	「万能の人」ポアンカレが"トポロジー"という分野を構築した原典。図形の定性的な性質を研究する「ゴム風船の幾何学」の端緒。豊富な注・解説付。〔内容〕多様体／同相写像／ホモロジー／ベッチ数／積分の利用／幾何学的表現／基本群／他
N.H.アーベル・E.ガロア著　九大 高瀬正仁訳 数学史叢書 **アーベル／ガロア 楕 円 関 数 論** 11459-1　C3341　　A 5 判 368頁 本体7800円	二人の夭折の天才がその精魂を傾けた楕円関数論の原典。詳細な註記・解説と年譜を付す。〔内容〕〈アーベル〉楕円関数研究／楕円関数の変換／楕円関数論概説／ある種の超越関数の性質／代数的可解方程式／他〈ガロア〉シュヴァリエへの手紙
早大 足立恒雄・元東大 杉浦光夫・ 前放送大 長岡亮介編訳 数学史叢書 **リ ー マ ン 論 文 集** 11460-7　C3341　　A 5 判 388頁 本体7800円	「リーマン幾何」や「リーマン予想」で知られる大数学者の代表論文を編訳し詳細な解説と訳注を付す〔内容〕複素関数論／アーベル関数論／素数の個数／平面波／三角級数論／幾何学の基礎／耳について／心理学／自然哲学／付：リーマンの生涯／他

◈ すうがくの風景 ◈

奥深いテーマを第一線の研究者が平易に開示

慶大 河添 健著
すうがくの風景1
群 上 の 調 和 解 析
11551-2 C3341　　　　A 5 判 200頁 本体3500円

群の表現論とそれを用いたフーリエ変換とウェーブレット変換の,平易で愉快な入門書。元気な高校生なら十分チャレンジできる！〔内容〕調和解析の歩み／位相群の表現論／群上の調和解析／具体的な例／2乗可積分表現とウェーブレット変換

東北大 石田正典著
すうがくの風景2
トーリック多様体入門
——扇の代数幾何——
11552-9 C3341　　　　A 5 判 164頁 本体3200円

本書は,この分野の第一人者が,代数幾何学の予備知識を仮定せずにトーリック多様体の基礎的内容を,何のあいまいさも含めず,丁寧に解説した貴重な書。〔内容〕錐体と双対錐体／扇の代数幾何／2次元の扇／代数的トーラス／扇の多様化

早大 村上 順著
すうがくの風景3
結 び 目 と 量 子 群
11553-6 C3341　　　　A 5 判 200頁 本体3300円

結び目の量子不変量とその背後にある量子群についての入門書。量子不変量がどのように結び目を分類するか,そして量子群のもつ豊かな構造を平明に説く。〔内容〕結び目とその不変量／組紐群と結び目／リー群とリー環／量子群(量子展開環)

神戸大 野海正俊著
すうがくの風景4
パンルヴェ方程式
——対称性からの入門——
11554-3 C3341　　　　A 5 判 216頁 本体3400円

1970年代に復活し,大きく進展しているパンルヴェ方程式の具体的・魅惑的紹介。〔内容〕ベックルント変換とは／対称形式／τ函数／格子上のτ函数／ヤコビ‐トゥルーディ公式／行列式に強くなろう／ガウス分解と双有理変換／ラックス形式

東京女大 大阿久俊則著
すうがくの風景5
D 加群と計算数学
11555-0 C3341　　　　A 5 判 208頁 本体3500円

線形常微分方程式の発展としてのD加群理論の初歩を計算数学の立場から平易に解説〔内容〕微分方程式を線形代数で考える／環と加群の言葉では？／微分作用素環とグレブナー基底／多項式の巾とb関数／D加群の制限と積分／数式処理システム

奈良女大 松澤淳一著
すうがくの風景6
特異点とルート系
11556-7 C3341　　　　A 5 判 224頁 本体3700円

クライン特異点の解説から,正多面体の幾何,正多面体群の構造,特異点解消及び特異点の変形とルート系,リー群・リー環の魅力的世界を活写〔内容〕正多面体／クライン特異点／ルート系／単純リー環とクライン特異点／マッカイ対応

熊本大 原岡喜重著
すうがくの風景7
超 幾 何 関 数
11557-4 C3341　　　　A 5 判 208頁 本体3300円

本書前半ではテイラー展開から大域挙動をつかまえる話をし,後半では三つの顔を手がかりにして最終,微分方程式からの統一理論に進む物語〔内容〕雛形／超幾何関数の三つの顔／超幾何関数の仲間を求めて／積分表示／級数展開／微分方程式

阪大 日比孝之著
すうがくの風景8
グレブナー基底
11558-1 C3341　　　　A 5 判 200頁 本体3300円

組合せ論あるいは可換代数におけるグレブナー基底の理論的な有効性を簡潔に紹介。〔内容〕準備(可換他)／多項式環／グレブナー基底／トーリック環／正規配置と単模被覆／正則三角形分割／単模性と圧搾性／コスツル代数とグレブナー基底

中大 小林道正・東大 小林　研著
LaTeXで数学を
——LaTeX2ε＋AMS-LaTeX入門——
11075-3 C3041　　　　A 5 判 256頁 本体3700円

LaTeX2εを使って数学の文書を作成するための具体例豊富で実用的なわかりやすい入門書。〔内容〕文書の書き方／環境／数式記号／数式の書き方／フォント／AMSの環境／図版の取り入れ方／表の作り方／適用例／英文論文例／マクロ命令

T.H.サイドボサム著　前京大 一松 信訳

はじめからの すうがく 事典

11098-2 C3541　　　　B 5 判 512頁 本体8800円

数学の基礎的な用語を収録した五十音順の辞典。図や例題を豊富に用いて初学者にもわかりやすく工夫した解説がされている。また、ふだん何気なく使用している用語の意味をあらためて確認・学習するのに好適の書である。大学生・研究者から中学・高校の教師、数学愛好者まであらゆるニーズに応える。巻末に索引を付して読者の便宜を図った。〔内容〕1次方程式、因数分解、エラトステネスの篩、円周率、オイラーの公式、折れ線グラフ、括弧の展開、偶関数、他

数学オリンピック財団野口　廣監修
数学オリンピック財団編

数学オリンピック事典
―問題と解法―〔基礎編〕〔演習編〕

11087-6 C3541　　　　B 5 判 864頁 本体18000円

国際数学オリンピックの全問題の他に、日本数学オリンピックの予選・本戦の問題、全米数学オリンピックの本戦・予選の問題を網羅し、さらにロシア(ソ連)・ヨーロッパ諸国の問題を精選して、詳しい解説を加えた。各問題は分野別に分類し、易しい問題を基礎編に、難易度の高い問題を演習編におさめた。基本的な記号、公式、概念など数学の基礎を中学生にもわかるように説明した章を設け、また各分野ごとに体系的な知識が得られるような解説を付けた。世界で初めての集大成

G.ジェームス・R.C.ジェームス編
前京大 一松　信・東海大 伊藤雄二監訳

数　学　辞　典

11057-9 C3541　　　　A 5 判 664頁 本体23000円

数学の全分野にわたる、わかりやすく簡潔で実用的な用語辞典。基礎的な事項から最近のトピックスまで約6000語を収録。学生・研究者から数学にかかわる総ての人に最適。定評あるMathematics Dictionary(VNR社、最新第5版)の翻訳。付録として、多国語索引(英・仏・独・露・西)、記号・公式集などを収載して、読者の便宜をはかった。〔内容〕アインシュタイン／亜群／アフィン空間／アーベルの収束判定法／アラビア数字／アルキメデスの螺線／鞍点／e／移項／位相空間／他

D.ウェルズ著　前京大 宮崎興二・京大 藤井道彦・
京大 日置尋久・京大 山口　哲訳

不思議おもしろ幾何学事典

11089-0 C3541　　　　A 5 判 256頁 本体6500円

世界的に好評を博している幾何学事典の翻訳。円・長方形・3角形から始まりフラクタル・カオスに至るまでの幾何学251項目・428図を50音順に並べ魅力的に解説。高校生でも十分楽しめるようにさまざまな工夫が見られ、従来にない"ふしぎ・おもしろ・びっくり"事典といえよう。〔内容〕アストロイド／アポロニウスのガスケット／アポロニウスの問題／アラベスク／アルキメデスの多面体／アルキメデスのらせん／……／60度で交わる弦／ロバの橋／ローマン曲面／和算の問題

和算研 佐藤健一監修　和算研 山司勝紀・上智大 西田知己編

和　算　の　事　典

11122-4 C3541　　　　A 5 判 544頁 本体14000円

江戸時代に急速に発達した日本固有の数学和算。和算を歴史から紐解き、その生活に根ざした計算法、知的な遊戯としての和算、各地を旅し和算を説いた人々など、さまざまな視点から取り上げる。〔内容〕和算のなりたち／生活数学としての和算／計算法―そろばん・円周率・天元術・整数術・方陣他／和算のひろがり―遊歴算家・流派・免許状／和算と諸科学―暦・測量・土木／和算と近世文化―ままっ子立・さっさ立・目付字他／和算の二大風習―遺題継承・算額奉納／和算書と和算家

上記価格(税別)は 2012 年 2 月現在